A COMPARATIVE
OVERVIEW OF
MAMMALIAN FERTILIZATION

A COMPARATIVE OVERVIEW OF MAMMALIAN FERTILIZATION

Edited by

Bonnie S. Dunbar
Baylor College of Medicine
Houston, Texas

and

Michael G. O'Rand
University of North Carolina at Chapel Hill
Chapel Hill, North Carolina

PLENUM PRESS • NEW YORK AND LONDON

Library of Congress Cataloging-in-Publication Data

A Comparative overview of mammalian fertilization / edited by Bonnie
S. Dunbar and Michael G. O'Rand.
    p.   cm.
  Includes bibliographical references and index.
  ISBN 0-306-43841-0
  1. Fertilization (Biology) 2. Mammals--Physiology.
3. Physiology, Comparative. I. Dunbar, Bonnie S. II. O'Rand,
Michael G.
QP273.C59 1991
599'.03--dc20   91-21138
                                           CIP

ISBN 0-306-43841-0

© 1991 Plenum Press, New York
A Division of Plenum Publishing Corporation
233 Spring Street, New York, N.Y. 10013

Printed in the United States of America

Contributors

Barry D. Bavister   Wisconsin Regional Primate Research Center, University of Wisconsin, Madison, Wisconsin 53715

J. Michael Bedford   Departments of Obstetrics and Gynecology and Cell Biology and Anatomy, Cornell University Medical College, New York, New York 10021

Dorothy E. Boatman   Wisconsin Regional Primate Research Center, University of Wisconsin, Madison, Wisconsin 53715

Gary N. Cherr   Bodega Marine Laboratory, University of California, Davis, Bodega Bay, California 94923

Elizabeth G. Crichton   Department of Anatomy, University of Arizona, Tucson, Arizona 85724

C. J. De Jonge   Department of Obstetrics and Gynecology, Rush University, Rush-Presbyterian-St. Luke's Medical Center, Chicago, Illinois 60612

Erma Z. Drobnis   Departments of Zoology and Obstetrics and Gynecology, University of California, Davis, California 95616

Gil L. Dryden   Biology Department, Slippery Rock University, Slippery Rock, Pennsylvania 16057

Bonnie S. Dunbar   Department of Cell Biology, Baylor College of Medicine, Houston, Texas 77030

N. L. First   Department of Meat and Animal Science, University of Wisconsin, Madison, Wisconsin 53706

George L. Gerton   Division of Reproductive Biology, Department of Obstetrics and Gynecology, University of Pennsylvania School of Medicine, Philadelphia, Pennsylvania 19104-6080

Erwin Goldberg   Department of Biochemistry, Molecular Biology, and Cell Biology, North-western University, Evanston, Illinois 60208

R. H. F. Hunter   Center for Research on Animal Reproduction, Faculty of Veterinary Medicine, University of Montreal, Saint-Hyacinthe, Quebec J2S 7C6, Canada. *Present address*: University of Edinburgh, Edinburgh EH16 5NT, Scotland, United Kingdom

Kristen A. Ivani   Animal Reproduction and Biotechnology Laboratory, Colorado State University, Fort Collins, Colorado 80523

Gregory S. Kopf   Division of Reproductive Biology, Department of Obstetrics and Gynecology, University of Pennsylvania School of Medicine, Philadelphia, Pennsylvania 19104-6080

Philip H. Krutzsch   Department of Anatomy, University of Arizona, Tucson, Arizona 85724

Susan E. Lanzendorf   Division of Reproductive Biology and Behavior, Oregon Regional Primate Research Center, Beaverton, Oregon 97006. *Present address*: The Jones Institute for Reproductive Medicine, Department of Obstetrics/Gynecology, Eastern Virginia Medical School, Norfolk, Virginia 23510

Frank J. Longo   Department of Anatomy, University of Iowa, Iowa City, Iowa 52242

Cherrie A. Mahi-Brown   California Primate Research Center, University of California, Davis, California 95616

Patricia M. Morgan   Wisconsin Regional Primate Research Center, University of Wisconsin, Madison, Wisconsin 53715. *Present address*: Department of Biochemistry, University College, Galway, Ireland

Barbara S. Nikolajczyk   Department of Cell Biology and Anatomy, University of North Carolina at Chapel Hill, Chapel Hill, North Carolina 27599

Gary E. Olson   Department of Cell Biology, Vanderbilt University, Nashville, Tennessee 37232

Michael G. O'Rand   Department of Cell Biology and Anatomy, University of North Carolina at Chapel Hill, Chapel Hill, North Carolina 27599

J. J. Parrish   Department of Meat and Animal Science, University of Wisconsin, Madison, Wisconsin 53706

David M. Phillips   The Population Council, New York, New York 10021

S. V. Prasad   Department of Cell Biology, Baylor College of Medicine, Houston, Texas 77030

John C. Rodger   Department of Biological Sciences, The University of Newcastle, Newcastle 2308, New South Wales, Australia

George E. Seidel, Jr.   Animal Reproduction and Biotechnology Laboratory, Colorado State University, Fort Collins, Colorado 80523

Ruth Shalgi   Department of Embryology and Teratology, Sackler School of Medicine, Tel-Aviv University, Ramat Aviv, Tel-Aviv 69978, Israel

Bayard T. Storey   Division of Reproductive Biology, Department of Obstetrics and Gynecology, University of Pennsylvania School of Medicine, Philadelphia, Pennsylvania 19104-6080

T. M. Timmons   Department of Cell Biology, Baylor College of Medicine, Houston, Texas 77030

Pradeep K. Warikoo   Wisconsin Regional Primate Research Center, University of Wisconsin, Madison, Wisconsin 53715. *Present address*: Department of IVF, Saginaw General Hospital, Saginaw, Michigan 48602

Paul M. Wassarman   Department of Cell and Developmental Biology, Roche Institute of Molecular Biology, Roche Research Center, Nutley, New Jersey 07110

David E. Wildt   National Zoological Park, Smithsonian Institution, Washington, D.C. 20008

Virginia P. Winfrey   Department of Cell Biology, Vanderbilt University, Nashville, Tennessee 37232

Don P. Wolf   Division of Reproductive Biology and Behavior, Oregon Regional Primate Research Center, Beaverton, Oregon 97006, and Department of Obstetrics/Gynecology, Oregon Health Sciences University, Portland, Oregon 97201

Debra J. Wolgemuth   Department of Genetics and Development, and Center for Reproductive Sciences, Columbia University College of Physicians and Surgeons, New York, New York 10032

L. J. D. Zaneveld   Department of Obstetrics and Gynecology, and Department of Biochemistry, Rush University, Rush-Presbyterian-St. Luke's Medical Center, Chicago, Illinois 60612

Preface

In 1964, the Fertilization and Gamete Physiology Research Training Program (FERGAP) was established at the Marine Biological Laboratories, Woods Hole, Massachusetts. Over the course of the next 12 years, under the directorship of Dr. Charles B. Metz, FERGAP brought together, trained, and inspired a generation of students in reproductive biology from all over the world.

As students of C. B. Metz and as FERGAP trainees, we would like to dedicate this collected work on comparative mammalian fertilization to our teacher and mentor, Dr. Charles B. Metz. Like a number of authors contributing to this volume, we have been struck by the significant impact that C. B. Metz and FERGAP had on the development of students of reproductive biology. Applying both the classical and molecular techniques of cell biology and immunology to problems of gamete biology, Dr. Metz emphasized a comparative and analytical approach that was reflected in his own research on fertilization in *Paramecia*, sea urchins, frogs, and mammals.

It is hoped that this volume will serve to stimulate students to discover the myriad of fascinating research problems in gamete and reproductive biology.

Bonnie S. Dunbar
Michael G. O'Rand

Houston, Texas
Chapel Hill, North Carolina

Contents

Part II STUDIES ON MAMMALIAN FERTILIZATION IN SELECTED SPECIES

Part III   NEW APPROACHES FOR STUDYING MAMMALIAN GAMETES

Part I

COMPARATIVE OVERVIEW OF MAMMALIAN GAMETES

1

The Coevolution of Mammalian Gametes

J. Michael Bedford

1. INTRODUCTION

Mammals, grouped as such because of shared features that include hair, homeothermy, and lactation, are composed of *Monotremata* (Prototheria), and the Theria, which include *Marsupialia* (Metatheria), and *Eutheria*. The latter are often referred to as placental mammals, but this is inappropriate in the strict sense since marsupials also briefly develop a functional placenta.

During the evolution of these three distinctive mammalian groups, major differences emerged at several points in the process of conception, and not least in their spermatozoa and eggs. According to recent cladistic analysis, the monotremes appear to be related to the modern therians (Kemp, 1983). Nonetheless, monotreme gametes have retained much of what can be presumed to be the character of their therapsid ancestors' gametes based on those of present-day reptiles and birds. Marsupial and eutherian gametes, by contrast, display several novel features of design that raise questions of concept and interpretation. For example, the process of oocyte maturation in therian mammals can be seen to differ in certain respects from that in other vertebrates, and therian spermatozoa especially undergo a more involved process of maturation after spermiation, partly in the epididymis and partly in the female, the significance of which remains somewhat of a mystery. Finally, eutherian mammals practice what appears to be a unique pattern of interaction during gamete fusion and sperm incorporation by the oocyte.

One major difficulty in attempting to explain and place in perspective the trend to complexity in mammalian gamete organization and function is the sparsity of information about conception in subtherian vertebrates. In the study of fertilization, biologists have focused historically on invertebrate gametes, the material for much of the pioneering work of Lillie and others (Lillie, 1919). However, although increasing attention has been paid in the last 50 years to fertilization and related aspects of gamete function in eutherian mammals, other higher vertebrates have been neglected in this respect. This often means that interpretations or comparisons must rely on observations in only one or two member species of a group. As a result, it is difficult to comprehend fully many of the puzzling features of therian gametes alluded to above.

J. MICHAEL BEDFORD • Departments of Obstetrics and Gynecology and Cell Biology and Anatomy, Cornell University Medical College, New York, New York 10021.

A Comparative Overview of Mammalian Fertilization, edited by Bonnie S. Dunbar and Michael G. O'Rand. Plenum Press, New York, 1991.

To understand the significance of these various developments in therian mammals appears important. It is unsatisfactory in an era of an increasingly molecular approach to sperm–egg interactions to have no reasonable grasp of the implications of what appear to be peculiarly eutherian features of the fertilization process. In addition to academic issues, development of a true understanding of the essential factors required for eutherian spermatozoa to penetrate eggs has practical implications for rational analysis of fertilization failure as well as for efforts to include specific prevention of fertilization among the principles used in new-generation contraceptives.

In approaching these issues, some of which have been touched on by Austin (1976), this chapter emphasizes what has been learned recently of the group-specific features of design and organization that characterize the gametes of monotreme, marsupial, and eutherian mammals, respectively. Then, it considers briefly the functional implications of these anatomic differences and of the novel aspects of gamete maturation and physiology in the scheme that leads to fertilization.

2. GAMETE DESIGN

Evolutionary radiation of the three subclasses of the Mammalia has been accompanied by the development of certain group-specific characteristics in both the male and female gamete (Fig. 1). The general form of both the spermatozoon and egg is not very different in the primitive monotreme mammals from that in many birds and reptiles. On the other hand, both marsupial and eutherian gametes have changed markedly. As the therian (i.e., marsupial and eutherian) egg has become much smaller, so its vestments have become more formidable, especially in Eutheria. In parallel, novel and different group-specific features have appeared in the design of marsupial and eutherian spermatozoa.

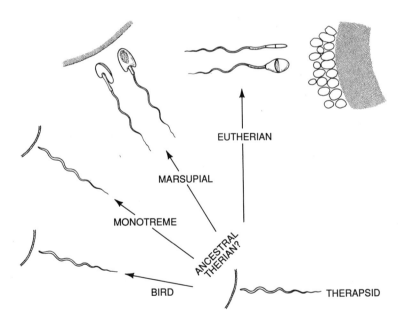

Figure 1.   This figure illustrates in a general way the radiation of gamete form in mammalian evolution.

2.1. Monotremes

This group is now represented by three single-species genera, the echidna or spiny anteater (*Tachyglossus*), the New Guinea spiny anteater (*Zaglossus*), and the duck-billed platypus (*Ornithorhyncus*). Their gamete anatomies are very similar (Griffiths, 1968, 1978).

At fertilization, the monotreme spermatozoon is faced by a very thin zona pellucida (width less than 0.5 μm; Fig. 2), little different from that in reptiles and birds (Hughes, 1977; Hughes and Carrick, 1978). Although the relatively large monotreme egg (diameter 3.5–4 mm) acquires additional coats in the uterus, it presents no other barrier than the thin zona pellucida to the fertilizing spermatozoon. In keeping with this minimal coat, the monotreme sperm head remains wholly primitive (Fig. 3). Monotreme spermatozoa have developed no significant new features over and above those seen in reptiles and nonpasserine bird spermatozoa (Bedford and Rifkin, 1979; Carrick and Hughes, 1982), even lacking a perforatorium—the projection present immediately beneath the apex of the acrosome in both reptilian (Furieri, 1971; Bedford and Calvin, 1974b) and some bird spermatozoa (Nagano, 1962; Asa and Phillips, 1987). Finally, and in marked contrast to that of eutherian spermatozoa, the filamentous monotreme sperm head has a fragility that is readily apparent on treatment with disruptive agents. The nucleus becomes swollen and distorted if spermatozoa are simply left for a period on a glass slide, and, when exposed to 1% SDS, it disperses immediately (Bedford and Rifkin, 1979).

The tail of the monotreme spermatozoon displays no special features of interest either. It has only rudimentary dense fibers and so is in some ways even simpler than that of many reptilian spermatozoa (Bedford and Rifkin, 1979).

2.2. Marsupials

2.2.1. THE OOCYTE

It has been observed in three different marsupial families that ovulated eggs lose all their investing granulosa cells some time before ovulation (Hartman, 1916; Godfrey, 1969; Hughes, 1977; Rodger and Bedford, 1982; Selwood, 1982; Breed and Leigh, 1987). Thus, as in monotremes, a naked zona pellucida forms the sole barrier to the fertilizing spermatozoon in the

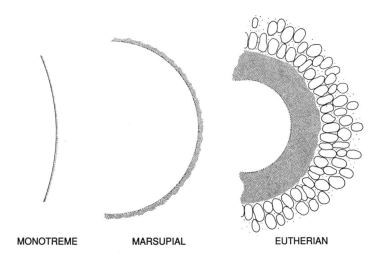

MONOTREME   MARSUPIAL   EUTHERIAN

Figure 2.   The relative size of the egg vestment that faces fertilizing spermatozoa in the various mammalian subclasses.

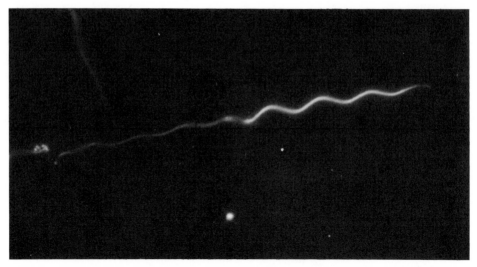

Figure 3.   Spermatozoon of a monotreme, the echidna *Tachyglossus aculeatus*. Monotreme spermatozoa have retained a vermiform shape similar to that of reptile and bird spermatozoa.

marsupials studied so far. The marsupial egg is much smaller (diameter ca. 120–240 μm depending on species), but its zona is generally about 1.0–2.0 μm in thickness and so is distinctly more prominent than that around the monotreme egg (Fig. 2). However, set against the eutherian zona, that of marsupials appears as a trivial structure in at least three respects. When the oocyte is removed, the zona collapses and does not retain its spherical form as a shell (personal observation). In *Didelphis*, the Virginia opossum, the relatively thin zona appears somewhat more diffuse ultrastructurally than that around eutherian eggs (Fig. 8) (Bedford and Calvin, 1974b; Rodger and Bedford, 1982). Furthermore, it is extremely vulnerable to a serine protease such as trypsin, which has a similar action to that of acrosin. Using a phase-contrast microscope I observed the *Didelphis* zona to disappear in about 2 sec when exposed on a slide to 0.1% trypsin at 37°C and pH 7.2 (Rodger and Bedford, 1982).This action of trypsin is illustrated by Talbot and Dicarlantonio (1984), who also point out that hyaluronidase removes matrix filaments of structured extracellular material in the perivitelline space.

2.2.2. THE SPERMATOZOON

The design of the marsupial spermatozoon has moved in a direction that differs in several respects from that taken by the primitive monotreme spermatozoon or the more specialized cell produced by Eutheria.

In regard to the sperm tail, that of marsupials displays several new features, some shared with eutherian spermatozoa. Compared to monotremes, the additional elements include a full set of nine dense fibers and S-S-stabilized mitochondria. The midpiece, in phalangers especially, variously displays very unusual features that divide it into regions (Temple-Smith, 1987). Anteriorly, the midpiece may be occupied according to species by membrane stacks (Harding *et al.*, 1976; Temple-Smith and Bedford, 1976). These are reminiscent of neck membrane complexes seen in some eutherian spermatozoa (Austin, 1976), notably *Galago senegalensis*, the bush baby (Bedford, 1967). In the posterior midpiece, the plasmalemma is underlain by helically arranged bands between which are flask-shaped membrane invaginations, and this banded arrangement is also reflected in the molecular character of the plasmalemmal surface (Olson *et al.*, 1977). As is the case for so many specific features of therian spermatozoa, the functional significance of these unusual characteristics, if any, has yet to be explained.

In the sperm head of marsupials alone, two distinctive novel elements—the disposition of the acrosome and the insertion of the tail—are of special interest in relation to the fertilization process. In almost all marsupial species, the connecting piece of the tail does not abut the posterior limit of the nucleus as is usually the case, but inserts into a fossa in the midventral face of the nucleus (Figs. 4 and 5). In immature epididymal spermatozoa, the angle of its insertion into the nucleus actually forms a T configuration between nucleus and tail, and this is modified during epididymal maturation by an almost 90° pivot of the tail that gives the spermatozoon a more streamlined, spear-like form (Harding *et al.*, 1975; Temple-Smith and Bedford, 1976, 1980). In caenolestids, a New World marsupial group, the tail protrudes at right angles from the nucleus even in mature epididymal spermatozoa (Temple-Smith, 1987). Moreover, a T shape can be assumed again to a variable degree in some marsupial spermatozoa after their release (Austin, 1976) (Fig. 6). Careful observations are needed now to establish the extent to which, and where in the female tract viable spermatozoa adopt such a T shape. Exceptions to this unusual relationship of tail and head are found in the Vombatoidea (wombats and koalas), in which the implantation fossa sits at the posterior border of the nucleus (Harding *et al.*, 1979).

A second distinctive feature involves the acrosome. In other groups this covers the apex of the nucleus symmetrically, but the marsupial acrosome generally sits asymmetrically on the dorsal face of the nucleus and thus opposed to the tail insertion site. It occupies almost the whole

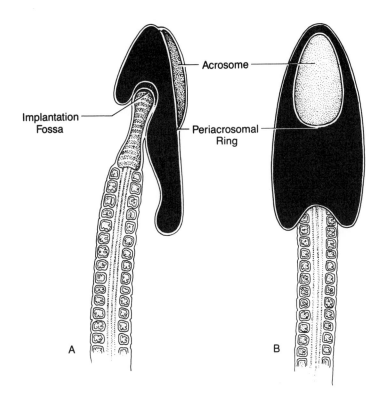

Figure 4. A generalized marsupial spermatozoon in (a) profile and (b) dorsal face views. Unusually, the implantation fossa for the tail is located in the midregion of the nucleus, and the acrosome is positioned on the opposing (dorsal) face. In some species the acrosome occupies most of the area of the dorsal face.

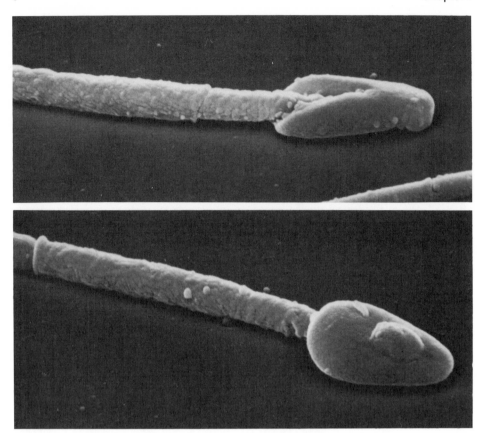

Figure 5. Micrographs of the ventral (upper panel) and dorsal faces of the spermatozoon of an Australian marsupial, the brush-tailed possum (*Trichosurus vulpecula*). As illustrated in Fig. 6, the tail inserts via a groove into the midregion of the head. The swelling on the dorsal face reflects the presence of the underlying acrosome, the limits of which, in reality, extend to the anterior border. SEM courtesy of F. Chretien.

of the dorsal surface in some Dasyuridae and Petauridae (Harding *et al.*, 1979) and in the Didelphidae (Phillips, 1970; Temple-Smith and Bedford, 1980). In the Phalangeridae, Macropodidae, and Peramelidae, the acrosome covers a more limited anterior region of what, nevertheless, is almost entirely dorsal face (Figs. 14 and 15). Significantly, it is that face that establishes contact with the zona pellucida at fertilization. The exceptions again are wombats and koalas, in which the acrosome appears to occupy the inner border of a flexible curved nucleus (Harding *et al.*, 1979; Temple-Smith, 1987).

Little is known of the enzymes in the marsupial acrosome. However, the one study extant (Rodger and Young, 1981) reveals a greater concentration of each of four hydrolases (hyaluronidase, acrosin, N-acetylhexosaminidase, and arylsulfatase B) in the acrosome of the marsupial opossum *Didelphis* than in that of the rabbit (Fig. 7). Finally, it can be noted that as in monotremes, the sperm head is quite fragile in marsupials. It disintegrates readily when exposed to disruptive reagents (Calvin and Bedford, 1971; Temple-Smith and Bedford, 1976) or even when merely kept on a slide (Austin, 1976; Cummins, 1980). Thus, unlike that of the Eutheria (see below), the marsupial sperm head displays no special stability.

To conclude, a modest but distinct enhancement in thickness of the zona pellucida has been

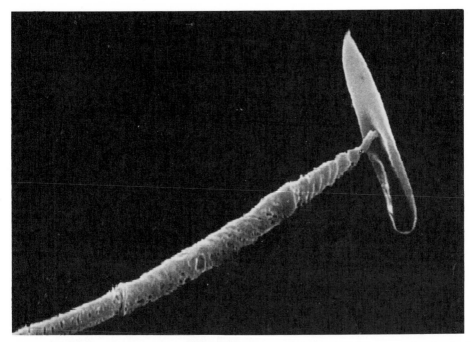

Figure 6. Spermatozoon recovered from epididymis of the Australian rat kangaroo, *Potorous tridactylus*, fixed after swimming in culture for some hours. Note that, compared to those in Figs. 4 and 5, the head has rotated through 90° in relation to the tail. SEM courtesy of D. M. Phillips.

paralleled in almost all marsupial species by an unusual sperm head design. That design appears to favor the sperm's ability to apply the full area of the acrosome to the substance of the zona pellucida.

2.3. Eutherians

2.3.1. THE OOCYTE

From a comparative standpoint the eutherian oocyte is very small (diameter 50–180 μm according to species), and the vestments are unusual in two major respects. Not only is the eutherian zona pellucida more prominent, but it is enveloped at ovulation by the cumulus oophorus, a mass of granulosa cells embedded in a proteoglycan matrix (Fig. 3). This mass and the more closely packed perizonal cells of the corona radiata are reputed to disappear within a very few hours from tubal ova of, for example, sheep, pig, and cow. However, in most species (and probably often in such ungulates as well), the cumulus/corona cell complex is penetrated by the fertilizing spermatozoon before any significant dispersal occurs. At the other extreme, the cumulus in a small insectivore, *Suncus murinus*, is stabilized unusually by intercellular junctions, is not susceptible to hyaluronidase, and does not disintegrate for 15 hr or more after ovulation and fertilization (J. M. Bedford, G. W. Cooper, D. M. Phillips, and G. L. Dryden, unpublished data). Whether a stable cumulus of this type is common in other insectivores remains to be determined.

A second unusual feature of the eutherian egg is the relatively massive zona pellucida (Fig. 8). It varies in thickness from 7 μm in rodent eggs to some 15 μm or so in the larger eggs of rabbit, human, sheep, cow, etc. According to species, it is composed of some three to five different sulfated glycoproteins (Bleil and Wassarman, 1980; Dunbar *et al.*, 1981; Dunbar and Bundman,

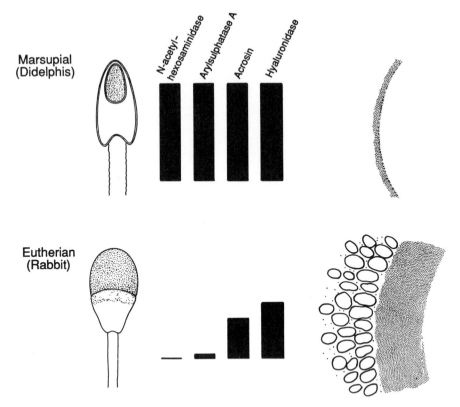

Figure 7. This figure is based on the results of Rodger and Young (1981) and compares the concentration of four acrosomal hydrolases in the *Didelphis* and the rabbit acrosome. The level of each enzyme per spermatozoon is expressed as 100% in the case of *Didelphis*, with the level in the rabbit being depicted as a percentage of that in *Didelphis*. The figure illustrates a disparity between enzyme concentrations and the prominence of the egg coats to be penetrated.

1987; Sacco *et al.*, 1983), some of which appear to be cross-linked to a degree by S-S bonds (Gould *et al.*, 1971; Dunbar *et al.*, 1981; Inoue and Wolf, 1974, 1975). Its physical properties contrast strikingly with that of the marsupial or the monotreme egg. An elastic resilience ensures that the spherical shape of the eutherian zona is often resumed after deformation with a blunt needle, and its shape is maintained as a shell after removal of the oocyte. Moreover, it is much less susceptible to proteases. Solubilization of the zona of small rodent eggs by proteases in the medium requires about 2 min, and that of the larger rabbit, sheep, cow, and pig eggs variably requires at least 5 min before visible changes are induced and at least 20 min for dissolution (see Dunbar, 1983). Thus, the time required for the eutherian zona to dissolve is much longer in all cases than the brief seconds I have observed to be needed for protease dissolution of the zona around the marsupial (*Didelphis*) egg.

2.3.2. THE SPERMATOZOON

The eutherian spermatozoon and its components are as familiar now as those of the egg to biologists interested in mammalian fertilization. However, because the Eutheria alone have been

Figure 8. Transmission electron micrographs taken at the same magnification that allow a comparison of the relative thickness (arrowed) and ultrastructural character of the zona pellucida in *Didelphis* (left panel) and rabbit.

the focus for most investigators, it has not been appreciated how unusual its organization is compared to that of the spermatozoa of other vertebrates.

Although it always displays the same basic organization, the eutherian spermatozoon can differ according to species in its size and its precise form. Thus, the total length can vary from about 260 μm in the Chinese hamster to 50 μm or less in several artiodactyl species (Cummins and Woodall, 1985). Though exceptions occur, there is a general tendency for the largest spermatozoa to be produced by small animals and the smallest by large animals (Cummins, 1983). As can be seen in the discussion of Cummins and Woodall (1985), there is no certain understanding yet of the basis for any of these visible differences among eutherian spermatozoa.

The organization of the sperm tail is almost as complex in the eutherian as in the marsupial spermatozoon. In addition to the basic components of the axoneme, it possesses nine outer dense fibers of variable dimensions according to species, a fibrous mainpiece sheath, and stable mitochondria. Detailed descriptions of the sperm tail and neck have been published by Fawcett and colleagues (Fawcett, 1970, 1975; Fawcett and Phillips, 1969), and their structure has been reviewed recently by Bedford and Hoskins (1990). In drawing comparisons with more primitive spermatozoa, Fawcett (1970) has suggested possible functions for individual organelles on the basis of their disposition and configuration in different-sized tails. He noted that the stiffness of beat relates directly to the cross-sectional size of the dense fibers, and he suggested that greater power may be achieved by acquisition of nine outer dense fibers (see also Rikmenspoel *et al.*, 1973) and an increase in the number of mitochondria, perhaps as an adaptation for function within the female tract. In regard to dense fibers, however, rooster and monotreme spermatozoa form obvious exceptions to that notion. Why the shell of the mitochondrion has been stabilized by S-S bonds in the evolution of the therian mammals (in contrast to subtherian sperm mitochondria; Bedford and Calvin, 1974a) also remains unknown.

Major species variation in the appearance of the sperm head reflects differences in the form of both the nucleus and the acrosome. The nuclear image ranges from the common flat spatulate or spade shape of ungulate and rabbit spermatozoa to the slightly thicker teardrop shape of the human cell or the sickle shape of several murid rodent species (Bishop and Walton, 1960), and nuclear differences may occur even within genera (e.g., Friend, 1936; Breed, 1983) (Fig. 9). Variation in size and shape of the rostral or apical segment of the acrosome is equally common. The development of this segment is most strikingly evident in some hystricomorph or sciurid rodents, but nowhere more so than in a small insectivore, *Suncus murinus*, the musk shrew (Green

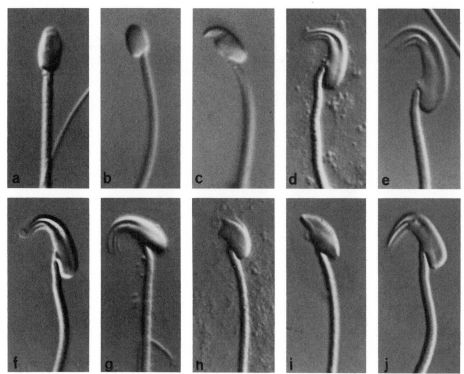

Figure 9. Spermatozoa of two different Australian rodents, *Pseudomys* (a–g) and *Notomys* (h–j). These examples illustrate that a fundamentally different form and size of the sperm head may occur among closely related species. Differential interference contrast microscopy, courtesy of W. G. Breed.

and Dryden, 1976; Cooper and Bedford, 1976; Phillips and Bedford, 1985). Whether this variation in the acrosome relates to its content of specific enzymes is not known.

Notwithstanding these common variations in form, the fundamental organization of the sperm head is essentially the same among all Eutheria. However, when these are set against the spermatozoa of all other vertebrate groups, the novelty of that organization is striking. Fig. 10 shows the distinctive elements of the eutherian sperm head, including these features:

1. Nucleus (head) shape. The nucleus is foreshortened anteroposteriorly and, unusually, is more or less flat in one plane.
2. Disulfide-stabilized perinuclear material is applied over the whole surface of the nucleus (Calvin and Bedford, 1971). This material (not shown in Fig. 10) may act to cement the association between nuclear envelope and the structures external to it.
3. Perinuclear material accumulates at the apex of the nucleus. This is referred to here as the *perforatorium*. Not present in monotremes or marsupials, the perforatorium usually presents an erect tapering profile, and this projection is stabilized by disulfide bonds (Calvin and Bedford, 1971; Olson *et al.*, 1976). Its form ensures that a minimal area of inner acrosomal membrane but a relatively sharp border confronts the substance of the zona pellucida during penetration.
4. Highly resistant chromatin. The stable, rigid quality of this depends on S-S cross-links between the constituent thiol-rich protamine molecules (Borenfreund *et al.*, 1961).

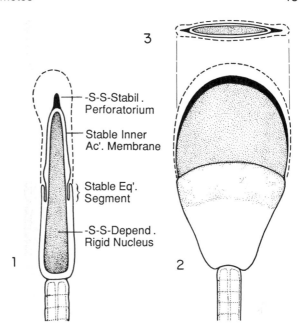

Figure 10. This composite figure illustrates evolutionary features of functional importance that characterize the eutherian sperm head. (1) A section in profile, displaying features that are suggested to relate to the mode of zona penetration or (in the case of the equatorial segment) appear to have evolved as a consequence of that. (2 and 3) The foreshortened flat form of the head, which permits lateral oscillation within the zona. The interrupted outline indicates the limits of the acrosome.

These covalent bonds lend a rigidity to the nucleus (Bedford and Calvin, 1974a; Bedford, 1983a) and are established largely during sperm passage through the epididymis (Calvin and Bedford, 1971).

5. A stable inner acrosomal membrane (IAM). Compared to the fragility of this membrane in chicken or opossum spermatozoa as revealed by mild detergent or ultrasonic treatment (unpublished observations), its stability in all Eutheria studied is especially striking. The IAM persists intact throughout the acrosome reaction, zona penetration, and sperm incorporation by the egg. It also survives conditions and treatments that disrupt or solubilize other sperm head membranes (Wooding, 1973; Srivastava *et al.*, 1974; Bedford, 1983a) as well as fractionation procedures that disperse other head structures (Rahi *et al.*, 1983). The basis of the IAM stability is not yet clear, other than the fact that its integral proteins present as a crystalline array and are underlain in close association by subacrosomal material (Huang and Yanagimachi, 1985). However, a stabilizing role for that material seems questionable, since the membrane's stability does not disappear with a reduction of S-S bonds that leads to dissociation of the material and the nucleus (Bedford, 1983a; Rahi *et al.*, 1983). Thus, its intrinsic organization rather than a stability lent by underlying perinuclear material probably determines the resistant character of the IAM in eutherian spermatozoa.

6. The stable posterior (equatorial) segment of the acrosome. This unique structure fails to take part in the acrosome reaction, and its characteristic stability seems to rest in a core of structural protein that lines and cross-links the inner and outer membranes in this specialized region (Bedford and Cooper, 1978; Russell *et al.*, 1980).

To summarize, with the emergence of eutherian mammals there has occurred a disproportionate growth of the egg vestments. Coincidentally, several reciprocal new features have appeared in the sperm head. Rather than enhancing the enzymic content of the acrosome, these lend a stability and shape appropriate to the deployment of a strategy of physical thrust by the spermatozoon. As will be discussed later, one unique feature, its equatorial segment, may well be a necessary consequence of that strategy.

3. GAMETE MATURATION

As used in the present context, the "maturation" of eggs and spermatozoa occurs at different biological stages. The oocyte maturation discussed here refers to events associated with the resumption of meiosis that parallel the acquisition of fertilizability. Sperm maturation refers to the events that take place in spermatozoa *after* their release from the gonad—events undergone in the epididymis and in the female tract.

3.1. Oocyte Maturation

In considering the maturation of mammalian oocytes, it is unfortunate to find that most functional observations still relate to the Eutheria alone (Masui and Clarke, 1979). In monotremes (reviewed by Griffiths, 1968, 1978), the first polar body is extruded from the egg before ovulation, and, equally conventionally, the second appears after sperm penetration. Both polar bodies then lie together at the margin of a peripheral germinal disk in which the female pronucleus develops. Nothing appears to be known, however, of the penetrability of the immature monotreme oocyte or of the relationship between germinal vesicle (GV) breakdown and the egg's ability to transform the fertilizing monotreme sperm head into a male pronucleus.

Among the respective mammalian groups, the difference in size of the mature oocyte is a reflection of the relative content of yolk, which predominates in the saurian-type eggs of monotremes (Hughes and Carrick, 1978). That ancestral connection is also seen in the fact that there is no antrum formation in the monotreme follicle, and that oocyte and follicular growth are continuous in the monotreme ovary—a pattern that mirrors the situation in reptiles and birds. In the Theria, on the other hand, the growth is biphasic; an initial growth of the small oocyte in parallel with that of the follicle is followed by continuous growth of the follicle alone (Lintern-Moore *et al.*, 1976). A follicular antrum is minimal or absent, however, in many insectivores.

There is considerable species variation in the number of oocytes ovulated in marsupials and eutherians. However, although occasional species shed the first polar body only after ovulation (e.g., *Canis familiaris*: Mahi and Yanagimachi, 1966), the growth patterns of the oocyte and its nucleus are generally similar in both groups. Nonetheless, the events in marsupials may differ in one or two important respects. It is possible that in at least some marsupials complete maturation and oocyte fertilizability are reached after ovulation, following the oocyte's arrival in the oviduct (see Chapter 7). Furthermore, in all of the three marsupial genera examined, a loss of the cumulus oophorus from around the egg begins some hours before and is complete at ovulation. This stands in contrast to the eutherian situation, where a full cumulus invests the oocyte at ovulation and generally at fertilization.

Other than the role proposed by Blandau (1969) in oocyte pickup and transfer to the tubal infundibulum (one clearly not operative in marsupials), it has been difficult to propose a function for the cumulus of the ovulated egg since cumulus-free eggs can be fertilized quite readily. However, in examining the early stages of rat fertilization *in vivo*, in accord with Shalgi and Phillips (1988), we have been impressed (1) that many eggs within cumulus already contain a fertilizing spermatozoon in the absence of other spermatozoa in the cumulus or in the ampullary content, and (2) that some hours later, before dissolution of the cumulus, the great majority of now more numerous ampullary spermatozoa are enmeshed within it. In all, our observations indicate that the cumulus acts as a trap for spermatozoa and favors the meeting of unfertilized eggs with the very few spermatozoa available in the ampulla at the time of ovulation (Bedford and Kim,

1991), as suggested many years ago by Austin (1961). In that regard, it may be pertinent that the *Didelphis* oviduct contains very high numbers of spermatozoa some hours after mating (Bedford *et al.*, 1984), but whether this is true for other marsupials that lack a cumulus oophorus remains to be studied.

Several other observations bear on the fertilizability of oocytes as a function of their maturation status. The resumption of meiosis is accompanied by a change in the physical character of the cumulus oophorus, and this renders the cumulus of the maturing oocyte more easily penetrable by spermatozoa (Trounson *et al.*, 1982; Cuasnicu and Bedford, 1991). By contrast, the zona pellucida appears penetrable in the GV oocyte as well as in subsequent stages of preovulatory maturation (Polge and Dziuk, 1965; Overstreet and Bedford, 1974; Mahi and Yanagimachi, 1976; Lopata and Leung, 1988; Moore and Bedford, 1978; Cuasnicu and Bedford, 1991). However, the resumption of meiosis is paralleled by development in the oocyte of an ability to mount a block to polyspermy, to incorporate (rabbit) spermatozoa normally, and by the appearance of a capacity to transform the fertilizing sperm nucleus into a male pronucleus (Thibault and Gerard, 1977; Usui and Yanagimachi, 1976; Moore and Bedford, 1978; Berrios and Bedford, 1979) which appears to be glutathione-mediated (Perreault *et al.*, 1988). Unfortunately for any consideration of the evolutionary biology of oocytes, such functional aspects of oocyte maturation have been studied almost exclusively in eutherian mammals. Whether the same pattern occurs during oocyte maturation in marsupials, for instance, remains to be determined.

3.2. Sperm Maturation in the Male

An important area for attempts to understand sperm function in mammals is the further maturation undergone in the Wolffian duct or epididymis after spermatozoa leave the testis (see Bedford, 1975, 1979). Unfortunately, the biological significance of that process remains only poorly understood. One problem that precludes the solution of this lies once more in the sparse information available for other vertebrates, against which the events in mammals can be viewed.

As it appears now, spermatozoa of invertebrates and the vertebrate cyclostomes, teleost fish, frogs, and toads for the most part reach a functional state within the testis (Bedford, 1979). By contrast, where fertilization is internal, as in the few reptiles and birds and the mammals examined, a further maturation of variable complexity occurs as spermatozoa traverse the Wolffian duct and epididymis. The complexity of that maturation seems not to have been influenced in any major way by descent of the epididymis as a correlate of scrotal evolution (Jones *et al.*, 1974; Bedford and Millar, 1978).

Among mammals, sperm maturation in the monotreme echidna seems a relatively simple process. It may prove to be most similar to that in some reptiles and birds in which maturation is evidenced only by a change in the character of the sperm surface (Depeiges and Dufaure, 1983; Esponda and Bedford, 1985, 1987; Morris *et al.*, 1987) and in the capacity for motility, as well as in cytoplasmic droplet migration (Bedford, 1979; Depeiges and Dacheux, 1985). In marsupials and eutherians, additional change occurs in the (S-S-dependent) stability of the eutherian sperm head and of several tail organelles, and often in the form of the acrosome. Moreover, the nature of the maturation change manifested in the sperm surface may prove to be somewhat different and perhaps more complex in therian as opposed to subtherian vertebrates (Esponda and Bedford, 1987).

The relative complexity of sperm physiology in the Theria is manifest particularly in an acute dependence of maturing spermatozoa on the epididymis for development of fertilizing ability, and on low (scrotal) temperature and/or androgen-regulated conditions created locally in the terminal region for their storage once maturation is complete. The results of Depeiges and Dacheux (1985) suggest some specific dependence of spermatozoa on the Wolffian duct in the viviparous lizard also. However, this and perhaps other lizards may prove somewhat atypical among reptiles (Fox, 1952). Certainly, the epididymal epithelium in *Lacerta* elaborates distinct secretory granules (Depeiges and Dufaure, 1983) in contrast to the agranular principle cells seen in most higher vertebrates. This dependence in the lizard contrasts with the looser relationship evident in the rooster, at least. Although most rooster spermatozoa mature finally in the Wolffian duct, they can in principle reach a fertile state within the confines of the testis (Munro, 1938b; Howarth, 1983). Moreover, as in certain reptiles (see Bedford, 1979), androgen withdrawal has no effect on the viable life of spermatozoa in the rooster Wolffian duct (Munro, 1938a).

In considering the evolutionary development of posttesticular events in the Wolffian duct, too little attention has been paid to passerine birds (sparrows, finches, etc.), which differ in several important ways from the rooster and related birds. Passerine sperm morphology is quite characteristic ("corkscrew" head, dense fibers, and spiral mitochondria in the tail). Furthermore, passerine (song sparrow) spermatozoa do not acquire the capacity for active motility until they have passed beyond the straight ciliated segment (vas efferens?) that connects the testis and the coiled, nonciliated portion that occupies the cloacal protuberance. Coincident sperm surface change can be demonstrated using a $FeOH_2$ colloid marker at pH 1.8—a possible indication that glycosyl moieties are involved (J. M. Bedford and G. W. Cooper, unpublished observations). Moreover, the scrotum-like arrangement of the paracloacal protuberance (Wolfson, 1954; Bedford, 1979) may well be androgen dependent. Thus, the pattern of events in the male duct of such passerine birds appears curiously similar to that in therian mammals.

The dependence in the relationship between spermatozoa and the monotreme epididymis has not been explored. However, it is probably not correct to view this organ in a similar light to that of testicondid (nonscrotal) therians such as the elephant and hyrax (Jones and Djakiew, 1977; Djakiew, 1982). Because of the primitive sperm characteristics and apparently simpler process of maturation undergone in the monotreme epididymis, the dependence of its spermatozoa on the epididymal environment may finally prove to be less extreme than in therian mammals and even passerine birds, and more akin to that seen in the rooster.

The specific maturation changes in the sperm surface, motility, stability, and morphology that spermatozoa undergo after leaving the testis need to be discussed in some further detail. Wolffian secretory proteins modify the sperm surface in reptiles (Depeiges and Dufaure, 1983; Esponda and Bedford, 1987), in birds (Esponda and Bedford, 1985; Morris *et al.*, 1987), in Eutheria (Eddy, 1989; Bedford and Cooper, 1978), and, by implication from lectin-binding studies, in marsupials (Temple-Smith and Bedford, 1976). The monotreme epididymal epithelium does have prominent secretory characteristics (Temple-Smith, 1974; Carrick and Hughes, 1978; Bedford and Rifkin, 1979), and surface change in echidna spermatozoa has been detected in the acquisition of electron-dense material (Bedford and Rifkin, 1979). However, the absence of lectin-binding change in echidna spermatozoa is consistent with the suggestion that sperm surface modifications in subtherians (with the possible exception of some passerine birds) reflect protein-related rather than the glycosyl-rich elements manifested generally in Eutheria (Esponda and Bedford, 1987).

Interpretation of the role of sperm surface change in the Wolffian duct is made difficult by the fact that among subtherian vertebrates, fertilization performance data exist only for the rooster. The fact that a few fertile spermatozoa exist in the rooster testis indicates that modification of its sperm surface by Wolffian duct proteins is not critical for fertilizing ability *per se*. It is quite possible that this surface change may facilitate sperm function in the female before fertilization, for example, by protecting spermatozoa from phagocytosis or, conversely, protecting the female by masking sperm surface isoantigens.

Although some sperm-coating proteins may fulfill such functions also in therian mammals, others appear to be critical for the ability of therian spermatozoa to fertilize. For reasons that are unclear, sperm surface changes of one type or another have assumed a role in eutherian sperm binding to the zona pellucida (Bedford, 1968; Saling, 1982; Shur and Hall, 1982). They also may mediate the nascent ability of epididymal spermatozoa to swim vigorously for long periods (e.g., Acott and Hoskins, 1978), a process that appears to involve an integrated action of cyclic AMP, calcium, adenosine, and intracellular pH as regulatory factors (Hoskins and Vijayaraghavan, 1990). Finally, a loss of some epididymal sperm-coating proteins from specific regions of the sperm surface during capacitation may set the stage for a switch to the hyperactivated movement and the acrosome reaction seen in fertilizing spermatozoa (see below).

In the rooster, the burgeoning capacity for motility appears in a few sperm before and in most very soon after sperm entry into the Wolffian duct. Thus, motility maturation may be a matter of short-term aging rather than of any specific influence of the duct environment in such birds and perhaps many other subtherians. This is not the case, however, in therian mammals, the spermatozoa of which do not make the transition to a functionally motile state if withheld in the testis. It is not yet clear why therian spermatozoa have assumed such a dependence on the epididymal environment for final maturation of the motility function, although this could relate in part at least to the subsequent events of capacitation (discussed below). However, since spermatozoa of higher vertebrates must survive after ejaculation for relatively long periods in the female, it seems noteworthy that, in addition to the oxidative phosphorylation seen almost universally in spermatozoa, those that fertilize internally and experience a substrate-rich female tract can also generate ATP by glycolysis (Austin, 1976). As evidenced in the discussion by Fraser (1981), the particular functions of these metabolic systems are difficult to distinguish, and it is probable that their relative importance varies according to species (Bedford and Hoskins, 1990).

As for the other changes that spermatozoa undergo in the therian epididymis, disulfide bond formation within the dense fibers, sheath, and mitochondria seems likely to stabilize and to stiffen the tail and so affect the pattern of its beat. The coincident formation of S-S cross-links in the perinuclear material and nucleus of the sperm head in Eutheria alone has sometimes been assumed to further nuclear condensation and perhaps protect the DNA. In fact, the nuclei of caput spermatozoa appear fully condensed, and it is more likely that this cross-linking creates a rigidity of importance to the sperm head in its penetration of the egg. Quite unclear still is the significance of the postspermiation morphological changes undergone variously by the therian acrosome. However, a protective advantage may underlie the associations that develop between the periacrosomal surface of spermatozoa in New World marsupials (Biggers, 1966; Phillips, 1970; Bedford et al., 1984) (Fig. 11) and in a few Eutheria that include the guinea pig and some squirrels (Fawcett and Hollenberg, 1963; Martan and Hruban, 1970).

3.3. Capacitation

The capacitation of mammalian (specifically eutherian) spermatozoa has been reviewed from many standpoints since the first reports of this phenomenon by Austin (1951) and Chang (1951). A general consensus exists now that capacitation probably involves a spectrum of molecular changes in both membrane glycoprotein and lipid components that ultimately modify ion channels in the plasmalemma of mature spermatozoa (Langlais and Roberts, 1985; Yanagimachi, 1989; Eddy, 1989; Bedford and Hoskins, 1990). In turn, such modification of the membrane permits the transmembrane flux of ions that seems to be a key to initiation of the two functional end points of capacitation—hyperactivation and the acrosome reaction (see Bedford, 1983b; Yanagimachi, 1989). Although the time of their onset is similar, these two events are independent of each other and almost certainly involve different region-specific membrane changes in the spermatozoon.

Like those of many invertebrates and lower vertebrates examined, rooster spermatozoa do not require capacitation (Howarth, 1970). This casts reasonable doubt on the capacitation needs of reptile and monotreme spermatozoa, neither of which has been studied in this respect. The

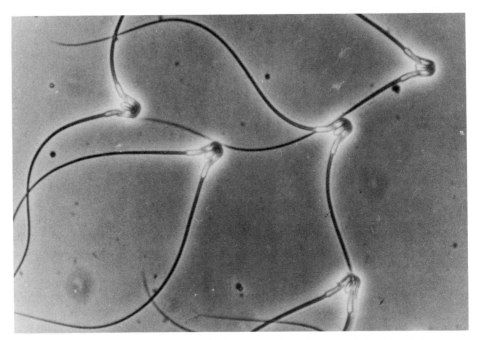

Figure 11.   Sperm pairs released from the cauda epididymidis of the marsupial, *Didelphis virginiana*.

capacitation status of marsupial spermatozoa also has been neglected. However, I have observed that vigorously motile mature single vas deferens spermatozoa of the opossum (*Didelphis virginiana*) do not associate *in vitro* with the naked zona of the opossum egg; they make contact and then swim away. By contrast, opossum spermatozoa flushed from the oviduct after mating adhered to and soon penetrated the zona pellucida *in vitro* (Rodger and Bedford, 1982). Thus, spermatozoa of this marsupial at least develop an ability to associate with and penetrate eggs as a consequence of passing through the female tract. Therefore, capacitation may prove to be a phenomenon that, in one form or another, is confined to eutherian and marsupial mammals.

There is still a major question as to the role that capacitation plays in the conception process: What is its functional significance? Some years ago, I proposed that capacitation may have arisen as a consequence of evolutionary change in the eutherian egg vestments, providing a necessary alternative means of regulating the acrosome reaction and (through hyperactivation) physical thrust of a quality needed for penetration of an unusually resilient zona pellucida (Bedford, 1983b). Now it is necessary to modify those ideas.

In regard to hyperactivation (Yanagimachi, 1970), it is likely that spermatozoa become hyperactivated *in vivo* as they emerge from the isthmus of the oviduct and are exposed to the ampullary environment (Katz and Yanagimachi, 1980; Cummins, 1982). In considering possible functions for hyperactivation, one can note that its onset coincides with the ability of (mouse) spermatozoa to fertilize *in vitro* (Fraser, 1981), and the hyperactivated beat pattern may enhance the thrust a spermatozoon can exert on the zona to a point that it may break covalent bonds (Drobnis *et al.*, 1988). Whether hyperactivation has arisen in response to the allometric development of the eutherian zona pellucida can probably be clarified by comparing the movement of mature spermatozoa from the male tract and that of spermatozoa flushed from the oviduct in marsupials, which have a relatively trivial zona pellucida.

The suggestion that capacitation represents an alternative means of controlling the acrosome reaction (Bedford, 1983b) is no longer appropriate. Good evidence has appeared now that the zona pellucida can stimulate the acrosome reaction in spermatozoa that have been capacitated (e.g., Saling and Storey, 1979; Bleil and Wassarman, 1983; Cherr *et al.*, 1986; Florman and First, 1988). The fact that they will react to cumulus oophorus also is evident in spermatozoa moving within it (Austin, 1960) and in those in cumulus fixed for electron microscopy soon after collection from oviducts (Bedford, 1972; Yanagimachi and Phillips, 1984). In contemplating why there should be a need for capacitation before an acrosome reaction can occur in response to the egg, there are good reasons now to focus on the system of sperm storage in the cauda epididymidis. Its specialized environment maintains the integrity and viability of mature therian spermatozoa, which, in contrast to those of some reptiles, seem relatively fragile and unable to survive elsewhere in the male tract for more than a few days. Such a system, regulated by testicular androgen and in scrotal species also by the low temperature of the scrotum, has been found in all the Theria examined in this respect. By contrast, a regulated system of sperm storage does not seem to operate in most reptiles and nonpasserine birds studied (Bedford, 1979).

The caudal factors on which storage depends have not been specifically characterized. A key element in the present context, however, is the fact that spermatozoa are subject in the cauda (and perhaps even more proximally in some species) to surface modifications not essential for their fertilizing ability. In the rat, for example, a specific caudal protein (HIS) associates with the acrosomal surface as spermatozoa reach the large-diameter segment of the cauda (Rifkin and Olson, 1985). Yet fertile spermatozoa exist more proximally in the lower corpus region, and those in the upper cauda, though devoid of this protein, fertilize at an incidence comparable to that of coated rat spermatozoa from the lower cauda (Dyson and Orgebin-Crist, 1973). In the ram, similarly, many spermatozoa develop the ability to fertilize while in the mid- and the lower corpus epididymidis (Fournier-Delpech *et al.*, 1979) and thus well before any association with a 24-kDa glycoprotein that is secreted by and becomes bound to the sperm surface in the cauda (Dacheux and Voglmayr, 1983). Again, rabbit spermatozoa finally bind a 21-kDa protein secreted in the cauda (Garcia *et al.*, 1988), yet many in the lower corpus epididymidis are already fully fertile. A further glycoprotein secreted by the rabbit corpus epididymidis—acrosome-stabilizing factor (ASF)—will reversibly suppress the fertilizing ability of a capacitated rabbit spermatozoa, probably by blocking the acrosome reaction (Eng and Oliphant, 1978; Thomas *et al.*, 1984). Nonetheless, spermatozoa retained in the midcaput of the epididymis by high ligation above the site of ASF production develop the ability to fertilize and produce normal offspring (Bedford, 1968, 1988). Thus, even ASF makes no essential contribution to rabbit sperm fertilizing ability *per se*.

Since such late sperm surface modifications in the epididymis seem unimportant for fertilizing ability, it is an obvious possibility that they function as an element in sperm storage, perhaps stabilizing the sperm plasmalemma. Therefore, the need for capacitation of epididymal spermatozoa may wholly or in part constitute a necessary escape from a stable state imposed as a facet of sperm storage in the cauda, as suggested by Thomas *et al.* (1986). Support for that idea is suggested by earlier experiments performed in the pig. The fertility profile in the boar epididymis is such that numbers of fertile spermatozoa already exist in the upper corpus (Holtz and Smidt, 1976). When upper corpus spermatozoa were inseminated directly after ovulation into the oviduct, a significant proportion of eggs were penetrated in about 4.5 hr, whereas no potentially fertile cauda spermatozoa could fertilize eggs in the contralateral oviduct in that time (Hunter *et al.*, 1976) (Fig. 12). Thus, upper corpus spermatozoa of the pig become prepared to fertilize sooner than those that have experienced the environment of the cauda. That advantage is annulled, moreover, if corpus spermatozoa are first exposed to caudal secretion (Hunter *et al.*, 1978). A parallel exists in the rabbit, whose cauda fluid reversibly inhibits the capacitated state (Weinmann and Williams, 1964). Preliminary experiments in the hamster provide additional

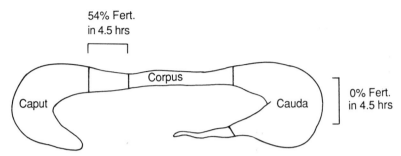

Figure 12. Diagram of the different regions of the boar epididymis from which spermatozoa were removed by Hunter *et al.* (1976). Clearly, by 4.5 hr after tubal insemination, upper corpus spermatozoa had developed the immediate ability to fertilize whereas potentially fertile spermatozoa from the cauda epididymidis, had not.

support for the concept. Knowing that body temperature suppresses some of the proteins in cauda fluid (Esponda and Bedford, 1986), we studied hamster spermatozoa from the cauda reflected to the abdomen. Such spermatozoa undergo hyperactivation and the acrosome reaction sooner *in vitro* and fertilize sooner *in vitro* and *in vivo* (Bedford and Yanagimachi, 1991). This accords with the possibility that temperature-sensitive macromolecules imposed during caudal storage to some degree determine the sperm's subsequent need for capacitation.

To conclude, capacitation is really a collective term for a spectrum of molecular changes in the plasmalemma of the mature spermatozoon normally undergone in the female tract. These allow the independent development of hyperactivated motility and of a responsiveness to egg-related stimuli that induce the acrosome reaction. Whatever the significance of hyperactivation, the need for capacitation before onset of the acrosome reaction in eutherian (and marsupial?) spermatozoa may be linked in part to the evolution of a regulated system of sperm storage in the cauda epididymidis. It must be appreciated that the molecular details of capacitation almost certainly vary among species. Moreover, the determinants of this probably are not confined only to the caudal region. The ability of maturing rat epididymal spermatozoa to bind HIS protein depends on a change that takes place more proximally in the corpus epididymidis (Rifkin and Olson, 1985), the site of ASF acquisition by the rabbit sperm.

4. GAMETE INTERACTION

The cytology of fertilization was established more than 100 years ago by Hertwig, Van Beneden, and then Boveri and others (see Lilli, 1919; Austin and Walton, 1960). In the 1950s, the first ultrastructural observations in several invertebrates confirmed that sperm entry is preceded by an egg-coat-induced exocytosis of the acrosome and that fusion with the oolemma occurs by way of the persisting inner membrane of the reacted acrosome (see Dan, 1967; Colwin and Colwin, 1967). The expectation voiced by Colwin and Colwin (1967) that this mode of fusion would hold for vertebrates has been largely borne out from studies of fertilization in lampreys (Nicander and Sjoden, 1973), anuran and urodele amphibia (Picheral, 1977a,b; Picheral and Charbonneau, 1982), and chickens (Okamura and Nishiyama, 1978). Marsupials also appear to follow this pattern (Rodger and Bedford, 1982; Breed and Leigh, 1987), and because monotreme sperm head organization resembles that in many birds, it seems predictable that monotreme mammals do so as well (Fig. 13).

Given a common pattern in other animals, it was particularly surprising to find in the 1960s that the eutherian spermatozoon fuses with and is incorporated by the oocyte in a strikingly different manner—one that appears to be unique. First sensed in the rat (Piko and Tyler, 1964), full descriptions of the eutherian pattern have since confirmed this for the hamster (Yanagimachi and Noda, 1970a,b), rabbit (Bedford, 1970, 1972), guinea pig (Noda and Yanagimachi, 1976) and most recently man (Sathananthan and Chen, 1986; Sathananthan et al., 1986).

The essence of the visibly different element in eutherian gamete fusion and sperm incorporation is twofold. The spermatozoon fuses with the oocyte by a restricted segment of sperm plasma membrane overlying the equatorial segment in the midregion of the head (Moore and Bedford, 1978; Bedford et al., 1979; Yanagimachi, 1989), not by the inner acrosomal membrane. Second, the oocyte cortex reacts by engulfing the anterior region of the sperm head still encased by inner acrosomal membrane. Consequently, the incorporated sperm nucleus is not naked in Eutheria as it is in other groups; it sits enveloped rostrally by a hybrid vesicle composed of inner acrosomal membrane and externally of egg plasma membrane sequestered during engulfment (Fig. 13).

With insight, the novel mode of gamete interaction in Eutheria could have been seen over the last decades as a red flag—as a clue that in some important respects eutherian fertilization may depend on different mechanisms. Nonetheless, fertilization in eutherian mammals is still thought of very much in terms of the principles that operate in other animals. However, the novel developments in therian gamete design, allied to precise consideration of the way spermatozoa behave at fertilization, now suggest why gamete fusion and sperm incorporation differ uniquely in eutherian mammals. That evidence is consistent with the possibility that the equatorial site of fusion arose as a result of selective stabilization of the inner acrosomal membrane—a stabilization adequate to withstand shear forces involved in penetration of the zona pellucida, but inappropriate for fusion thereafter. If it is true that the forces required for penetration of the eutherian zona pellucida have determined an unusual stability of the inner acrosomal membrane incompatible with a fusion role, this raises a question about the central place of acrosomal lysins in penetration of the zona.

In considering the departure of Eutheria from convention in the matter of gamete fusion and sperm incorporation, it is instructive first to draw a comparison with the gamete design and mode of fertilization in marsupials (Fig. 13), which practice a conventional mode of fusion (by way of the inner acrosomal membrane) and sperm incorporation (Rodger and Bedford, 1982; Breed and Leigh, 1987). The marsupial zona is distinctly thicker than that in subtherian vertebrate eggs (Fig. 2), and the peculiar design of the marsupial sperm head, with its acrosome and the tail insertion on opposed flat surfaces, may favor a direct application of the whole acrosome face to the zona (Figs. 14 and 15). The point at which the marsupial acrosome begins to react needs to be clarified (Rodger and Bedford, 1982; Phillips and Fadem, 1987; Breed and Leigh, 1987). Moreover, the precise approach of the sperm head through the zona may vary according to the limits of the acrosome (cf. *Didelphis* with *Sminthopsis*: Rodger and Bedford, 1982; Breed and Leigh, 1987). Nonetheless, the sperm head organization, the very rapid lysis of the marsupial zona by a protease (Rodger and Bedford, 1982), and the acrosome's relatively high hydrolase levels (Rodger and Young, 1981) together make it likely that the zona gap created by the fertilizing marsupial spermatozoon in large part reflects the action of acrosomal lysins. By implication, a lysin-based mode of penetration would avoid any need for unusual stabilization of the inner acrosomal membrane, and in keeping with this, the inner acrosomal membrane forms the site of fusion with the oolemma in marsupials.

Whereas marsupial sperm design appears to favor the deployment of its acrosome during fertilization, that of the eutherian spermatozoon does not. Although the evidence is very limited, the evolution of a relatively massive zona does not seem to have been matched by any significant enhancement of the inventory of enzymes in the eutherian acrosome (Fig. 7). In fact, the design of the eutherian sperm head, as well as its orientation and behavior in traversing the zona (Fig. 16), seems somewhat inappropriate for any mechanism of penetration dependent to an important

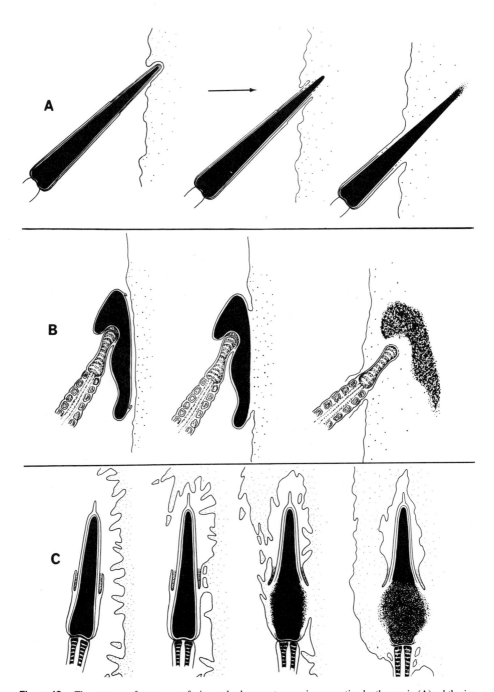

Figure 13. The anatomy of sperm–egg fusion and subsequent sperm incorporation by the egg in (A) subtherian, (B) marsupial, and (C) eutherian gametes, respectively. The inner acrosomal membrane is the site of fusion in all subtherian animals as well as in the marsupials studied. By contrast, the equatorial fusion site with subsequent engulfment of the rostral head region within a membrane-bounded vesicle is a unique departure displayed by all Eutheria.

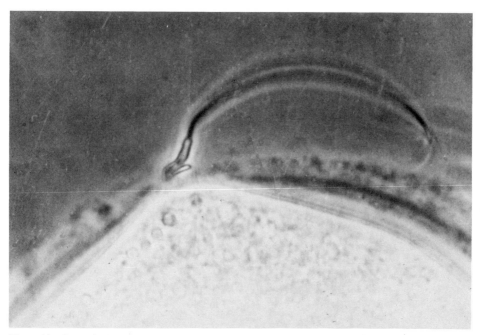

Figure 14.   Interaction between an oocyte and a spermatozoon flushed from the oviduct of a mated opossum soon after ovulation. Note that the whole acrosomal face of the fertilizing spermatozoon bears on the surface of the zona pellucida.

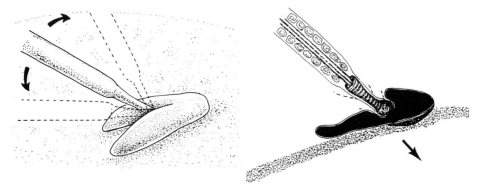

Figure 15.   Diagrammatic illustrations, in perspective (left panel) and in section, of the relationship seen in Fig. 14 between the marsupial sperm head and the zona pellucida immediately prior to and during its penetration. In some species the limits of the acrosome are more extensive than shown here. The disposition of acrosome and tail in relation to sperm head shape allows direct application of the full acrosomal area to the surface of the zona.

degree on lytic enzymes of the acrosome. On the contrary, the novel features that characterize the sperm head point to physical thrust as a main determinant of penetration in eutherian mammals.

The implication that thrust is the primary determinant of zona penetration speaks against what is still the majority view that acrosomal enzymes are central to this in Eutheria, as I suggest they appear to be in marsupials. However, that view for Eutheria was called into question some years ago by the observation that the visible content of the acrosome is lost before spermatozoa make serious inroads into the zona (Bedford, 1968). Such doubts were amplified later by controlled studies with markers for acrosin or acrosomal glycoprotein that were unable to demonstrate their presence on the sheep or rabbit inner acrosome membrane after completion of the acrosome reaction, or in the penetration slit (Harrison *et al.*, 1982; Huneau *et al.*, 1984; Kopecny and Flechon, 1987). It is in fact striking (e.g., in Fig. 2 of Barros *et al.*, 1967) that the dispersing content of the reacting hamster acrosome remains associated with the vesiculating outer membrane, leaving an ultrastructurally "clean" inner membrane of the acrosome. Equally difficult to reconcile are observations that homologous acrosin has no visible effect on the sheep zona (Brown, 1982), that enzyme inhibitors have no influence on penetration once spermatozoa bind to the mouse zona (Saling, 1981), and especially that a significantly increased resistance to acrosin or trypsin in the rabbit zona had no effect on sperm passage through it (Bedford and Cross, 1978). It is also important to bear in mind that at no point does the narrow penetration slit give any hint of erosion or cavitation that might be ascribed to a lytic presence.

It is clear, of course, that acrosomal enzymes can disrupt the matrix of the zona in many Eutheria (Srivastava *et al.*, 1965; Brown, 1982, 1983) or alter it more subtly without visible change (Dunbar *et al.*, 1985). Moreover, several studies report that enzyme inhibitors in the

Figure 16.   Diagram of disposition of the eutherian sperm head tethered by acrosomal carapace in the early stages of zona penetration, as seen in TEM thin sections. Not only is the visible content of the acrosome lost before penetration but only a very limited area of inner acrosomal membrane opposes the intact substance of the zona (arrow) at the leading edge. The sperm head may exert additional shear-thinning force by oscillating in the flat plane of the head (see Fig. 17).

medium prevent or reduce the rate of fertilization (e.g., Beyler and Zaneveld, 1982; Dudkiewicz, 1983). Nevertheless, doubts that enzymes can have a major role in the penetration phase of fertilization in Eutheria are further increased when one examines carefully the way the eutherian sperm head penetrates the zona. The narrowing of the anterior border of the inner acrosomal membrane, exaggerated by virtue of the perforatorium, ensures that only a minimal area of putative enzyme-coated membrane is presented to the opposing zona substance (Fig. 16). Such an arrangement seems to compromise effective deployment of lytic factors there but must focus and so maximize the disruptive effect on the zona of a given thrust. Moreover, the rigidity of the sperm nucleus (an evolutionary development in Eutheria alone: Bedford and Calvin, 1974b) permits the tail's thrust to be translated along the axis of the head to the zona quite directly.

Green (1987) has advanced arguments for the view that the physical thrust that a eutherian (hamster) spermatozoon can mount is insufficient alone to cleave the zona substance. However, his calculation appears to be low by at least an order of magnitude (Drobnis *et al.*, 1988). Furthermore, it does not take into account the important fact that its flat shape allows the penetrating sperm head to oscillate rapidly in that plane (Austin, 1976). In the case of the human sperm head deep within the zona, I have observed subjectively that this oscillates up to eight times per second and that the modest lateral head displacement is generated by movement of a stiff midpiece, a point in the midposterior region of the head being the fulcrum (Fig. 17). The

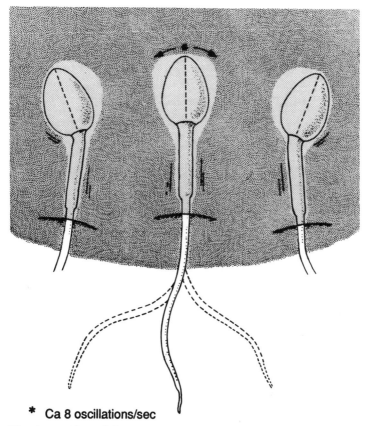

* Ca 8 oscillations/sec

Figure 17.   Three images of a motile human spermatozoon deep within the zona pellucida. These illustrate that the head oscillates laterally in the flat plane about a fulcrum in the posterior region of the head. That movement of the head is created by a vibration of the stiff anterior portion of the tail, generated by the thrashing movement of the posterior segment remaining outside the zona.

oscillation's limited scythe-like movement may create a shear-thinning effect on the zona that maximizes the effectiveness of that thrust.

In accord with the inference that the movement required for zona penetration generates shear forces at the sperm surface, the inner acrosomal membrane has developed an unusual stability which is inappropriate to a fusogenic role. Thus, not only has the fusion site been removed from the leading edge to the midregion of the head, but in several species the disposition of that equatorial site suggests an arrangement that would further minimize its exposure to surface shear forces (Bedford, 1983c). In one group that includes the hamster, rabbit, hare, and bush baby, stable perinuclear material immediately ahead of the equatorial segment forms a discrete shoulder that must divert zona pellucida away from its surface during penetration, and the rostrally flared rat perforatorium may act similarly. Conversely, recession of the equatorial segment within the contour of the nucleus (for example, in musk shrew, marmoset, and perhaps boar spermatozoa) may afford comparable protection.

In summary, it appears that marsupial and eutherian spermatozoa may use different means to penetrate the egg vestment and that the consequences of this are reflected in the events of gamete fusion that follow. The design of the marsupial sperm head lends itself to a mode in which the content of the acrosome can be brought to bear most effectively on the substance of a somewhat more prominent zona pellucida. By contrast, an impressive development of the egg vestment in eutherian evolution does not seem to have been matched by enhancement of the acrosome content or its utilization during penetration. This has been paralleled, however, by the appearance of novel features that would seem to maximize the sperm head's capacity for physical thrust, and it is likely that, together with specific receptors, that is the key to penetration of the eutherian zona. The evidence suggests further that the unique mode of eutherian gamete fusion by way of the equatorial segment is a consequence of the stable state of the inner acrosomal membrane that, in turn, may reflect the physical nature of the zona penetration that precedes the fusion.

5. CONCLUSIONS

The evolution of monotreme, marsupial, and eutherian mammals has brought a striking diversity in the form and design of their respective gametes as well as in the way they interact at fertilization. As yet, however, the implications of this diversity for the mechanisms of fertilization are not certain, one major barrier to such understanding being a lack of relevant comparative information. In part, that is a consequence of the focus on eutherian mammals in this respect, with no more than a handful of studies devoted to reptiles, birds, monotremes, and marsupials. For example, much of the discussion here about bird and marsupial gametes rests on single species, the rooster and the Virginia opossum, respectively. Thus, many of the ideas must be seen as somewhat tentative for the moment. Overall, however, one gains a sense that the conception process has been subject to something of a domino effect, one change being accompanied by adaptive modification of related events. The adoption of internal fertilization, as well as change in the egg vestments, seem likely to have been the major influences in this respect.

Additional facets of sperm maturation in the Wolffian duct beyond those undergone in the testis seem to have emerged with some consistency in most vertebrates that practice internal fertilization, even in elasmobranch fish (Bedford, 1979; Jones et al., 1984). Posttesticular changes in the capacity for motility and in the sperm surface may relate in part to conditions imposed on spermatozoa in the female tract. However, posttesticular maturation appears most complex and most dependent on the Wolffian duct/epididymis in marsupial and in eutherian mammals, where epididymal sperm surface modifications have become an essential element in the ability to bind to and fertilize the egg.

A possible exception to that is presented by men with long-term obstruction of the epididymis, since a few spermatozoa obtained from the proximal caput and even vasa efferentia in selected cases, are able to fertilize *in vitro* (Silber *et al.*, 1990). Whether or not this signals an unusual minimal dependence of human spermatozoa on the epididymis is difficult to say. The point may one day be resolved by an examination for effects of chronic obstruction on secretory patterns and for possible secretory reflux in the proximal regions of the blocked epididymis. Certainly, the process of sperm maturation in the normal human epididymis resembles that in other mammals in every major respect.

A system of regulated storage imposed on spermatozoa in the terminal region of the epididymis has appeared in both marsupials and eutherians, and in order to fertilize they may first need to escape from the stable state that such storage mechanisms bring. Thus, sperm-coating elements linked to caudal storage may be responsible in part for the need that therian spermatozoa have for capacitation in the female tract.

Evolutionary change in sperm tail structure, specifically the appearance of the outer dense fibers and mainpiece sheath and the organization of the mitochondria in the form of a sheath, has been linked tentatively to the adoption of internal fertilization. Yet, although S-S-stabilized dense fibers are represented in reptile and passerine bird spermatozoa, they are not present in some birds, including the rooster, and they are essentially absent in monotreme spermatozoa. Moreover, passerine sperm mitochondria are organized in a spiral fashion along the length of the tail.

As for the sperm head, the major evolutionary trends in its organization and design seem in large part to represent reciprocal developments to adaptive change in the vestment of the egg. In accord with this, monotreme spermatozoa retain the form and organization that are presumed to be those of their therapsid ancestors, whereas unique though very different developments in sperm head organization and design appear in marsupial and eutherian spermatozoa, respectively. The species variations in sperm-head form among the Eutheria especially ømay have little selective advantage or connection with reproductive fitness (Beatty, 1975). However, the idiosyncratic group features of marsupial and eutherian sperm head organization appear to have fundamental implications. In particular, they raise the possibility that somewhat different strategies are used to penetrate the marsupial as opposed to the eutherian egg.

In the face of a moderate further development of the zona pellucida, the novel design of the marsupial sperm head appears to ensure the most effective utilization of the acrosome in penetration of this coat. In Eutheria also, the development of a much more formidable egg vestment has been paralleled by the evolution of novel sperm head features, but these imply a design that favors physical thrust more than lysis. The possibility exists that acrosomal enzymes play some preliminary role in sperm atttachment (Jones *et al.*, 1988) or in the mechanisms of the acrosome reaction itself (Meizel, 1984). Insofar as the process of penetration is concerned, however, the place of these enzymes in eutherian sperm function seems to be in question. If they are capacitated, unreacted spermatozoa can penetrate the cumulus oophorus readily in the hamster (Corselli and Talbot, 1987) and mouse (Storey *et al.*, 1984), and unreacted human spermatozoa penetrate the human cumulus quite rapidly in fertilization conditions (Chen and Sathananthan, 1986; White *et al.*, 1990). Thus, although the acrosomal content, and its hyaluronidase especially, display an impressive ability to dissolve the cumulus matrix (e.g., Rodger and Young, 1981), the release of the acrosomal content is not required for penetration of the cumulus oophorus.

Austin (1976) has raised what may be a specter for some advocates of the zona lysin theory, the possibly vestigial character of certain gamete features. That the essential place of the eutherian acrosome reaction in zona penetration may relate primarily to its modification of the sperm head profile rather than to the enzymes released, is an idea that will not find easy acceptance. Indeed, while it seems questionable now that acrosomal enzymes have an essential function in zona penetration, it would be premature to dismiss such a role for them at this time. Although visible effects have not been demonstrable in the transmission or scanning electron microscope, acrosomal content released at the surface of the zona might possibly have some softening effect on its periphery in certain species. Moreover, not only does the zona response to enzymes vary from species to species, but generalization may be particularly dangerous in the case of fertilization, which, as shown in different chapters here, can vary in other matters of detail. For example, the human spermatozoon seems the least well designed for a penetration strategy based on thrust alone. It has a relatively bulbous coronal profile, it has no perforatorium, and its nuclear protamines, which determine the rigidity of the sperm head, have the lowest cysteine content (and so S-S cross-linking) among the Eutheria examined (Calvin, 1976). Whether these nuances of human sperm head organization reflect a greater role for acrosomal lysins or merely the fact that the human zona pellucida is relatively less resilient is difficult to know yet. However, regardless of such species variation, it seems now that adaptation of the eutherian sperm head (specifically the inner acrosomal membrane) to shear forces created by the sperm's thrust/oscillation movement during zona penetration may have determined the need for the strikingly different mode of gamete fusion via the equatorial region of the sperm head in eutherian mammals alone.

Implicit in such concepts is the question of why, with reduction in size of the therian egg, the zona pellucida has become so prominent in Eutheria. Experiments with zona-free eggs attest to the importance of the zona for survival in the oviduct (Modlinski, 1970), but the disproportionate increase in thickness and resilience of the eutherian zona has not been explained. However, whereas the marsupial zona is covered in the oviduct by albumin and then by a shell as it passes to the uterus after only 1 day, the eutherian egg is retained in the oviduct according to species for 3–4 days to about 2 weeks, and, with the exception of lagomorphs, the zona pellucida is its sole protection. Thus, enhanced zona development and resilience may finally serve in some way to compensate for conditions that eutherian eggs face in their relatively long sojourn in the changing environment of the fallopian tube.

6. REFERENCES

Acott, T. S., and Hoskins, D. D., 1978, Bovine sperm forward motility protein: Partial purification and characterization, *J. Biol. Chem.* **253**:6744–6750.

Asa, C. S., and Phillips, D. M., 1987, Ultrastructure of avian spermatozoa: A short review, in *New Horizons in Sperm Cell Research* (H. Mohri, ed.), Gordon and Breach, New York, pp. 365–374.

Austin, C. R., 1951, Observations on the penetration of the sperm into the mammalian egg, *Aust. J. Sci. Res. Sec. B* **4**:581–596.

Austin, C. R., 1960, Capacitation and the release of hyaluronidase from spermatozoa, *J. Reprod. Fertil.* **1**:310–311.

Austin, C. R., 1961, *The Mammalian Egg*, Blackwell, Oxford.

Austin, C. R., 1976, Specialization of gametes, in: *Reproduction in Mammals*, Vol. 6: *The Evolution of Reproduction* (C. R. Austin and R. V. Short, eds.), Cambridge University Press, Cambridge, pp. 149–182.

Austin, C. R., and Walton, A., 1960, Fertilization, in: *Marshalls Physiology of Reproduction*, Vol. I pt. 2 (A. S. Parkes, ed.), Longmans, Green, London, pp. 310–416.

Barros, C., Bedford, J. M., Franklin, L., and Austin, C. R., 1967, Membrane vesiculation as a feature of the mammalian acrosome reaction. *J. Cell. Biol.* **34**:C1–C5.

Beatty, R. A., 1975, The phenogenetics of spermatozoa, in: *The Biology of the Male Gamete* (J. A. Duckett and P. A. Racey, eds.), Academic Press, New York, pp. 291–299.

Bedford, J. M., 1967, Observations on the fine structure of spermatozoa of the bush baby (*Galago senegalensis*), the African green monkey (*Ceropithecus aethiops*) and man, *Am. J. Anat.* **121**:443–460.

Bedford, J. M., 1968, Ultrastructural changes in the sperm head during fertilization in the rabbit, *Am. J. Anat.* **123**:329–358.

Bedford, J. M., 1970, Sperm capacitation and fertilization in mammals, *Biol. Reprod.* **2**:128–158.

Bedford, J. M., 1972, An electron microscopic study of sperm penetration into the rabbit egg after natural mating, *Am. J. Anat.* **133**:213–254.

Bedford, J. M., 1975, Maturation, transport, and fate of spermatozoa in the epididymis, in: *Handbook of Physiology*, Vol. 5 (D. W. Hamilton and R. O. Greep, eds.), American Physiological Society, Washington, D.C., pp. 303–317.

Bedford, J. M., 1979, Evolution of the sperm maturation and sperm storage functions of the epididymis, in: *The Spermatozoon* (D. W. Fawcett and J. M. Bedford, eds.), Urban and Schwarzenberg, Baltimore, pp. 7–21.

Bedford, J. M., 1983a, Form and function of eutherian spermatozoa in relation to the nature of the egg vestments, in: *Fertilization of the Human Egg in Vitro* (H. M. Beier and H. R. Lindner, eds.), Springer-Verlag, Berlin, pp. 133–146.

Bedford, J. M., 1983b, Significance of the need for sperm capacitation before fertilization in eutherian mammals, *Biol. Reprod.* **28**:108–120.

Bedford, J. M., 1983c, Fertilization, in: *Reproduction in Mammals: 1. Germ Cells and Fertilization* (C. R. Austin and R. V. Short, eds.), Cambridge University Press, Cambridge, pp. 128–163.

Bedford, J. M., 1988, The bearing of epididymal function in strategies for *in vitro* fertilization, *Ann. N.Y. Acad. Sci.* **541**:284–291.

Bedford, J. M., and Calvin, H. I., 1974a, Changes in -S-S- linked structures of the sperm tail during epididymal maturation, with comparative observations in sub-mammalian species, *J. Exp. Zool.*, **187**:181–204.

Bedford, J. M., and Calvin, H. I., 1974b, The occurrence and possible functional significance of -S-S- crosslinks in sperm heads, with particular reference to eutherian mammals, *J. Exp. Zool.* **187**:137–156.

Bedford, J. M., and Cooper, G. W., 1978, Membrane fusion events in the fertilization of vertebrate eggs, in: *Membrane Fusion* (G. Poste and G. L. Nicolson, eds.), Elsevier, Amsterdam, pp. 65–125.

Bedford, J. M., and Cross, N. L., 1978, Normal penetration of rabbit spermatozoa through a trypsin- and acrosin-resistant zona pellucida, *J. Reprod. Fertil.* **54**:385–392.

Bedford, J. M., and Hoskins, D. D., 1990, The mammalian spermatozoon: Morphology, biochemistry and physiology, in: *Marshall's Physiology of Reproduction*, 4th Edition, Vol. 2, (G. E. Lamming, ed.), Churchill Livingston, London, pp. 379–569.

Bedford, J. M., and Kim, H. H., 1991, Sperm distribution and cumulus oophorus function in the ampulla of the oviduct, *J. Exp. Zool.* (submitted).

Bedford, J. M., and Millar, R. P., 1978, The character of sperm maturation in the epididymis of the ascrotal hyrax *Procavia capensis*, and armadillo, *Dasypus novemcinctus*, *Biol. Reprod.* **19**:396–406.

Bedford, J. M., and Rifkin, J. M., 1979, An evolutionary view of the male reproductive tract and sperm maturation in a monotreme mammal—the echidna *Tachyglossus aculeatus*, *Am. J. Anat.* **156**:207–230.

Bedford, J. M., and Yanagimachi, R., 1991, Epididymal storage at abdominal temperature reduces the time required for capacitation of hamster spermatozoa, *J. Reprod. Fertil.* **91**:403–410.

Bedford, J. M., Moore, H. D. M., and Franklin, L. E., 1979, Significance of the equatorial segment of the acrosome of the spermatozoon in eutherian mammals, *Exp. Cell Res.* **119**:119–126.

Bedford, J. M., Rodger, J. C., and Breed, W. G., 1984, Why so many mammalian spermatozoa—a clue from marsupials? *Proc. R. Soc. Lond. [Biol.]* **221**:221–233.

Berrios, M., and Bedford, J. M., 1979, Oocyte maturation: Aberrant post-fusion responses of the rabbit primary oocyte to penetrating spermatozoa, *J. Cell Sci.* **39**:1–12.

Beyler, S. A., and Zaneveld, L. J. D., 1982, Inhibition of *in vitro* fertilization of mouse gametes by proteinase inhibitors, *J. Reprod. Fertil.* **66**:425–431.

Biggers, J. D., 1966, Reproduction in male marsupials, *Symp. Zool. Soc. Lond.* **15**:251–280.

Bishop, M. W. H., and Walton, A., 1960, Spermatogenesis and the structure of mammalian spermatozoa, in: *Marshall's Physiology of Reproduction*, 3rd Edition, Vol. I, part 2 (A. S. Parkes, ed.), Longmans, Green, London, pp. 1–129.

Blandau, R. J., 1969, Gamete transport—comparative aspects, in: *The Mammalian Oviduct* (E. S. E. Hafez and R. J. Blandau, eds.), University of Chicago Press, Chicago, pp. 129–162.

Bleil, J. D., and Wassarman, P. M., 1980, Structure and function of the zona pellucida: Identification and characterization of the proteins of the mouse oocyte's zona pellucida, *Dev. Biol.* **76**:185–202.

Bleil, J. D., and Wassarman, P. M., 1983, Sperm–egg interactions in the mouse: Sequence of events and induction of the acrosome reaction by a zona pellucida glycoprotein, *Dev. Biol.* **95**:317–324.

Borenfreund, E., Fitt, E., and Bendich, A., 1961, Isolation and properties of deoxyribonucleic acid from mammalian sperm, *Nature* **191**:1375–1377.

Breed, W. G., 1983, Variation in sperm morphology in the Australian rodent genus, *Pseudomys* (Muridae), *Cell Tissue Res.* **229**:611–625.

Breed, W. G., and Leigh, C. M., 1987, Studies on *in vivo* fertilization in a dasyurid marsupial, *Sminthopsis crassicaudata*, with special reference to sperm-egg interactions, *Gamete Res.* **19**:131–149.

Brown, C. R., 1982, Effects of ram sperm acrosin on the investments of sheep, pig, mouse and gerbil egg, *J. Reprod. Fertil.* **64**:457–462.

Brown, C. R., 1983, Purification of mouse sperm acrosin, its activation from proacrosin and effect on homologous egg vestments, *J. Reprod. Fertil.* **9**:289–295.

Calvin, H. I., 1976, Comparative analysis of the nuclear basic proteins in rat, human, guinea pig, mouse and rabbit spermatozoa, *Biochim. Biophys. Acta* **434**:377–389.

Calvin, H. I., and Bedford, J. M., 1971, Formation of disulphide bonds in the nucleus and accessory structures of mammalian spermatozoa during epididymal maturation, *J. Reprod. Fertil. [Suppl.]* **13**:65–75.

Carrick, F. N., and Hughes, R. L., 1978, Reproduction in male monotremes, *Aust. Zool.* **20**:211–232.

Carrick, F. N., and Hughes, R. L., 1982, Aspects of the structure and development of monotreme spermatozoa and their relevance to the evolution of mammalian sperm morphology, *Cell. Tissue Res.* **222**:127–141.

Chang, M .C., 1951, Fertilizing capacity of spermatozoa deposited into the fallopian tubes, *Nature* **168**:697–698.

Chen, C., and Sathananthan, A. H., 1986, Early penetration of human sperm through the vestments of human eggs *in vitro*, *Arch. Androl.* **16**:183–197.

Cherr, G. N., Lambert, H., Meizel, S., and Katz, D. F., 1986, *In vitro* studies of the golden hamster sperm acrosome reaction: Completion on the zona pellucida and induction by homologous soluble zonae pellucidae, *Dev. Biol.* **114**:119–131.

Colwin, L. H., and Colwin, A. L., 1967, Membrane fusion in relation to sperm–egg association, in: *Fertilization*, Vol. I (C. B. Metz and A. Monroy, eds.) Academic Press, New York, pp. 295–368.

Cooper, G. W., and Bedford, J. M., 1976, Asymmetry of spermiation and sperm surface charge patterns over the giant acrosome in the musk shrew, *Suncus murinus*, *J. Cell Biol.* **69**:415–428.

Corselli, J., and Talbot, P., 1987, *In vitro* penetration of hamster oocyte–cumulus complexes using physiological numbers of sperm, *Dev. Biol.* **122**:227–242.

Cuasnicu, P., and Bedford, J. M., 1991, Hamster oocyte penetrability during preovulatory maturation, *Molec. Reprod. Dev.* **29**:72–76.

Cummins, J. M., 1980, Decondensation of sperm nuclei of Australian marsupials: Effects of air drying and of calcium and magnesium, *Gamete Res.* **3**:351–367.

Cummins, J. M., 1982, Hyperactivated motility patterns of ram spermatozoa recovered from oviducts of mated ewes, *Gamete Res.* **6**:53–63.

Cummins, J. M., 1983, Sperm size, body mass and reproduction in mammals, in: *The Sperm Cell* (J. Andre, ed.), Martinus Nijhoff, The Hague, pp. 395–398.

Cummins, J. M., and Woodall, P. F., 1985, On mammalian sperm dimensions, *J. Reprod. Fertil.* **75**:153–175.

Dacheux, J. L., and Voglmayr, J. K., 1983, Sequence of sperm cell differentiation and its relationship to exogenous fluid proteins in the ram epididymis, *Biol. Reprod.* **29**:1033–1047.

Dan, J. C., 1967, Acrosome reaction and lysins, in: *Fertilization*, Vol. I (C. B. Metz and A. Monroy, eds.), Academic Press, New York, pp. 163–236.

Depeiges, A., and Dacheux, J. L., 1985, Acquisition of sperm motility and its maintenance during storage in the lizard, *Lacerta vivipara*, *J. Reprod. Fertil.* **74**:23–27.

Depeiges, A., and Dufaure, J. P., 1983, Binding to spermatozoa of a major soluble protein secreted by the epididymis of the lizard, *Lacerta vivipara*, *Gamete Res.* **7**:401–406.

Djakiew, D., 1982, Reproduction in the male echidna (*Tachyglossus aculeatus*) with particular emphasis on the epididymis. Ph.D. Thesis, University of Newcastle, Newcastle, N.S.W. Australia.

Drobnis, E. Z., Yudin, A. I., Cherr, G. N., and Katz, D. F., 1988, Hamster sperm penetration of the zona pellucida: kinematic analysis and mechanical implications, *Dev. Biol.* **130**:311–323.

Dudkiewicz, A. B., 1983, Inhibition of fertilization in the rabbit by anti-acrosin antibodies, *Gamete Res.* **8**:183–197.

Dunbar, B. S., 1983, Morphological, biochemical and immunochemical characterization of the mammalian zona pellucida, in: *Mechanism and Control of Mammalian Fertilization* (J. F. Hartmann, ed.), Academic Press, New York, pp. 140–177.

Dunbar, B. S., and Bundman, D. S., 1987, Evidence for a role of the major glycoprotein in the structural maintenance of the pig zona pellucida, *J. Reprod. Fertil.* **81**:363–376.

Dunbar, B. S., Liu, C., and Sammons, D. W., 1981, Identification of the three major proteins of porcine and rabbit zonae pellucidae by high resolution two-dimensional gel electrophoresis: Comparison with follicular fluid, sera and ovarian cell proteins, *Biol. Reprod.* **24:**1111–1124.

Dunbar, B. S., Dudkiewicz, A. B., and Bundman, D. S., 1985, Proteolysis of specific porcine zona pellucida glycoproteins by boar acrosin, *Biol. Reprod.* **32:**619–630.

Dyson, A. L. M. B., and Orgebin-Crist, M.C., 1973, Effect of hypophysectomy, castration and androgen replacement upon the fertilizing ability of rat epididymal spermatozoa, *Endocrinology* **93:**391–402.

Eddy, E. M., 1989, The spermatozoon, in: *The Physiology of Reproduction*, Vol. I (E. Knobil and J. D. Neill, eds.), Raven Press, New York, pp. 27–68.

Eng, L. A., and Oliphant, G., 1978, Rabbit sperm reversible decapacitation by membrane stabilization with a highly purified glycoprotein from seminal plasma, *Biol. Reprod.* **19:**1083–1094.

Esponda, P., and Bedford, J. M., 1985, The rooster sperm surface changes in passing through the Wolffian duct, *J. Exp. Zool.* **234:**441–449.

Esponda, P., and Bedford, J. M., 1987, Post-testicular changes in the reptile sperm surface with particular reference to the snake, *Natrix fasciata, J. Exp. Zool.* **242:**189–198.

Fawcett, D. W., 1970, A comparative view of sperm ultrastructure, *Biol. Reprod.* [*Suppl.*] **2:**90–127.

Fawcett, D. W., 1975, The mammalian spermatozoon, *Dev. Biol.* **44:**394–436.

Fawcett, D. W., and Hollenberg, R. D., 1963, Changes in the acrosome of guinea pig spermatozoa during passage through the epididymis, *Z. Zellforsch. Mikr. Anat.* **60:**276–292.

Fawcett, D. W., and Phillips, D. M., 1969, The fine structure and development of the neck region of the mammalian spermatozoon, *Anat. Rec.* **165:**153–184.

Florman, H. M., and First, N. L., 1988, The regulation of acrosomal exocytosis. I. Sperm capacitation is required for the induction of acrosome reactions by the bovine zona pellucida *in vitro, Dev. Biol.* **128:**453–463.

Fournier-Delpech, S., Colas, G., Courot, M., Ortavant, R., and Brice, G., 1979, Epididymal sperm maturation in the ram: Motility, fertilizing ability and embryonic survival after uterine artificial insemination in the ewe, *Ann. Biol. Anim. Biochem.* **19:**597–605.

Fox, W., 1952, Seasonal variation in the male reproduction system of Pacific Coast garter snakes, *J. Morphol.* **90:**481–542.

Fraser, L. R., 1981, Dibutyryl cyclic AMP decreases capacitation time *in vitro* in mouse spermatozoa, *J. Reprod. Fertil.* **62:**63–72.

Friend, G. F., 1936, The sperms of British muridae, *Q. J. Microsc. Sci.* **78:**419–443.

Furieri, P., 1971, Sperm morphology in some reptiles: Squamata and Chelonia, in: *Comparative Spermatology* (B. Baccetti, ed.), Academic Press, New York, pp. 115–131.

Garcia, C., Regalado, F., Lopez de Haro, M. S., and Nieto, A., 1988, Ultrastructural localization of epididymal secretory proteins associated with the surface of spermatozoa from rabbit cauda epididymis, *Histochem. J.* **20:**708–714.

Godfrey, K., 1969, Reproduction in a laboratory colony of the marsupial mouse *Sminthopsis larapinta* (Marsupialia: Dasyuridae), *Aust. J. Zool.* **17:**637–654.

Gould, K. G., Zaneveld, G. J. D., Srivastava, P. N., and Williams, W. G., 1971, Biochemical changes in the zona pellucida of rabbit ova induced by fertilization and sperm enzymes, *Proc. Soc. Exp. Biol. Med.* **136:**6–10.

Green, D. P. L., 1987, Mammalian sperm cannot penetrate the zona pellucida solely by force, *Exp. Cell Res.* **169:**31–38.

Green, J. A., and Dryden, G. L., 1976, Ultrastructure of epididymal spermatozoa of the Asiatic musk shrew, *Suncus murinus, Biol. Reprod.* **14:**327–331.

Griffiths, M., 1968, *Echidnas*, Pergamon Press, Oxford.

Griffiths, M., 1978, *The Biology of the Monotremes*, Academic Press, New York.

Harding, H. R., Carrick, F. N., and Shorey, C. D., 1975, Ultrastructural changes in spermatozoa of the brush-tailed possum, *Trichosurus vulpecula* (Marsupialia), during epididymal transit. Part I. The flagellum, *Cell Tissue Res.* **164:**121–132.

Harding, H. R., Carrick, F. N., and Shorey, C. D., 1976, Ultrastructural changes in spermatozoa of the brush-tailed possum, *Trichosurus vulpecula* (Marsupialia), during epididymal transit. Part II. The acrosome, *Cell Tissue Res.* **171:**61–73.

Harding, H. R., Carrick, F. N., and Shorey, C. D., 1979, Special features of sperm structure and function in marsupials, in: *The Spermatozoon* (D. W. Fawcett and J. M. Bedford, eds.) Urban and Schwarzenberg, Baltimore, pp. 289–303.

Harrison, R. A. P., Fléchon, J. E., and Brown, C. R., 1982, The location of acrosin and proacrosin in ram spermatozoa, *J. Reprod. Fertil.* **66:**349–358.

Hartman, C. G., 1916, Studies in the development of the opossum: *Didelphys virginiana* L., I. History of early cleavage. II Formation of the blastocyst, *J. Morphol.* **27:**1–83.

Holtz, W., and Smidt, D., 1976, The fertilizing capacity of epididymal spermatozoa in the pig, *J. Reprod. Fertil.* **46:**227–229.

Hoskins, D. D., and Vijayarhaghavan, S., 1990, A new theory on the acquisition of sperm motility during epididymal transit, in: *Controls of Sperm Motility: Biological and Clinical Aspects* (C. Gagnon, ed.), CRC Press, Boca Raton, pp. 53–62.

Howarth, B., 1970, An examination for sperm capacitation in the fowl, *Biol. Reprod.* **3:**338–341.

Howarth, B., 1983, Fertilizing ability of cock spermatozoa from the testis, epididymis, and vas deferens following intra-magnal insemination, *Biol. Reprod.* **28:**586–590.

Huang, T. F., and Yanagimachi, R., 1985, Inner acrosomal membrane of mammalian spermatozoa; its properties and possible functions in fertilization, *Am. J. Anat.* **174:**249–268.

Hughes, R. L., 1977, Egg membranes and ovarian function during pregnancy in monotremes and marsupials, in: *Reproduction and Evolution* (J. H. Calaby and C. H. Tyndale-Biscoe, eds.), Australian Academy of Sciences, Canberra, pp. 281–291.

Hughes, R. L., and Carrick, F. N., 1978, Reproduction in female monotremes, *Aust. Zool.* **20:**233–253.

Huneau, D., Harrison, R. A. P., and Fléchon, J.-E., 1984, Ultrastructural localization of proacrosin and acrosin in ram spermatozoa, *Gamete Res.* **9:**425–440.

Hunter, R. H. F., Holtz, W., and Henfrey, P. J., 1976, Epididymal function in the boar in relation to the fertilizing ability of spermatozoa, *J. Reprod. Fertil.* **46:**463–466.

Hunter, R. H. F., Holtz, W., and Hermann, H., 1978, Stabilizing role of epididymal plasma in relation to the capacitation time of boar spermatozoa, *Anim. Reprod. Sci.* **1:**161–166.

Inoué, M., and Wolf, D. P., 1974, Comparative solubility properties of the zona pellucida of unfertilized mouse ova, *Biol. Reprod.* **11:**558–565.

Inoué, M., and Wolf, D. P., 1975, Comparative solubility properties of rat and hamster zona pellucida, *Biol. Reprod.* **12:**535–540.

Jones, R., Brown, C. R., and Lancaster, R. T., 1988, Carbohydrate-binding properties of boar sperm acrosin and assessment of its role in sperm–egg recognition and adhesion during fertilization, *Development* **102:**781–792.

Jones, R. C., and Djakiew, D., 1977, The role of the excurrent ducts from the testes of testicondid mammals, *Aust. Zool.* **20:**201–210.

Jones, R. C., Rowlands, I. W., and Skinner, J. D., 1974, Spermatozoa in the genital ducts of the African elephant, *Loxodonta africana*, *J. Reprod. Fertil.* **41:**189–192.

Jones, R. C., Jones, N., and Djakiew, D., 1984, Luminal composition and maturation of spermatozoa in the male genital ducts of the Port Jackson shark, *Heterodontus portusjacksoni*, *J. Exp. Zool.* **230:**417–426.

Katz, D. F., and Yanagimachi, R., 1980, Movement characteristics of hamster spermatozoa within the oviduct, *Biol. Reprod.* **22:**759–764.

Kemp, T. S., 1983, The relationships of mammals, *Zool. J. Linn. Soc.* **77:**353–384.

Kopecny, V., and Fléchon, J.-E., 1987, Ultrastructural localization of labelled acrosomal glycoproteins during *in vivo* fertilization in the rabbit, *Gamete Res.* **17:**35–42.

Langlais, J., and Roberts, K. D., 1985, A molecular model of sperm capacitation and the acrosome reaction of mammalian spermatozoa, *Gamete Res.* **12:**183–224.

Lillie, R. R., 1919, *Problems of Fertilization*, University of Chicago Press, Chicago.

Lintern-Moore, S., Moore, G. P. M., Tyndale-Biscoe, C. H., and Poole, W. E., 1976, The growth of the oocyte and follicle in the ovaries of monotremes and marsupials, *Anat. Rec.* **185:**325–332.

Lopata, A., and Leung, P. C., 1988, The fertilizability of human oocytes at different stages of meiotic maturation, *Ann. N. Y. Acad. Sci.* **541:**324–336.

Mahi, C. A., and Yanagimachi, R., 1976, Maturation and sperm penetration of canine ovarian oocytes *in vitro*, *J. Exp. Zool.* **196:**189–196.

Martan, J., and Hruban, Z., 1970, Unusual spermatozoan formations in the epididymis of the flying squirrel (*Glaucomys volans*), *J. Reprod. Fertil.* **21:**167–170.

Masui, Y., and Clarke, H., 1979, Regulation of oocyte maturation, *Int. Rev. Cytol.* **57:**185–282.

Meizel, S., 1984, The importance of hydrolytic enzymes to an exocytotic event, the mammalian sperm acrosome reaction, *Biol. Rev.* **59:**125–157.

Modlinski, J. A., 1970, The role of the zona pellucida in the development of mouse eggs *in vivo*, *J. Embryol. Exp. Morphol.* **23**:539–547.

Moore, H. D. M., and Bedford, J. M., 1978, Ultrastructure of the equatorial segment of hamster spermatozoa during penetration of oocytes, *J. Ultrastruct. Res.* **62**:110–117.

Morris, S. A., Howarth, B., Crim, J. W., Rodriguez de Cordoba, S., Esponda, P., and Bedford, J. M., 1987, Specificity of sperm-binding Wolffian duct proteins in the rooster and their persistence on spermatozoa in the female host glands, *J. Exp. Zool.* **242**:189–198.

Munro, S. S., 1938a, The effect of testis hormone on the preservation of sperm life in the vas deferens of the fowl, *J. Exp. Zool.* **15**:186–196.

Munro, S. S., 1938b, Functional changes in fowl sperm during their passage through the excurrent ducts of the male, *J. Exp. Zool.* **79**:71–92.

Nagano, T., 1962, Observations on the fine structure of the developing spermatid in the domestic chicken, *J. Cell Biol.* **14**:193–205.

Nicander, L., and Sjoden, I., 1973, An electron microscopical study of the acrosomal complex and its role in fertilization in the river lamprey, *Lampetra fluvialis*, *J. Submicrosc. Cytol.* **3**:309–317.

Noda, Y. D., and Yanagimachi, R., 1976, Electron microscopic observations of guinea pig spermatozoa penetrating eggs *in vitro*, *Dev. Growth Diff.* **18**:15–23.

Okamura, F., and Nishiyama, H., 1978, Penetration of the spermatozoon into the ovum and transformation of the sperm nucleus into the male pronucleus in the domestic fowl, *Gallus gallus*, *Cell Tissue Res.* **190**:89–98.

Olson, G. E., Hamilton, D. W., and Fawcett, D. W., 1976, Isolation and characterization of the perforatorium of rat spermatozoa, *J. Reprod. Fertil.* **47**:293–297.

Olson, G. E., Lifsics, M., Fawcett, D. W., and Hamilton, D. W., 1977, Structural specializations in the flagellar plasma membrane of opossum spermatozoa, *J. Ultrastruct. Res.* **59**:207–221.

Overstreet, J. W., and Bedford, J. M., 1974, Comparison of the penetrability of the egg vestments in follicular oocytes, unfertilized and fertilized ova of the rabbit, *Dev. Biol.* **41**:185–192.

Perreault, S. D., Barbee, R. R., and Slott, V., 1988, Importance of glutathione in the acquisition and maintenance of sperm nuclear decondensing ability in maturing hamster oocytes, *Dev. Biol.* **125**:181–186.

Phillips, D. M., 1970, Ultrastructure of spermatozoa of the woolly opossum, *Caluromys philander*, *J. Ultrastruct. Res.* **33**:381–397.

Phillips, D. M., and Bedford, J. M., 1985, Unusual features of sperm ultrastructure in the musk shrew, *Suncus murinus*, *J. Exp. Zool.* **235**:119–126.

Phillips, D. M., and Fadem, B. H., 1987, The oocytes of a New World marsupial, *Monodelphis domestica*: structure, formation and function of the enveloping mucoid layers, *J. Exp. Zool.* **242**:363–371.

Picheral, B., 1977a, La fécondation chez le Triton Pleurodele I. La traversée des enveloppes de l'oeuf par les spermatozoides, *J. Ultrastruct. Res.* **60**:181–202.

Picheral, B., 1977b, La fécondation chez la Triton Pleurodele II. La pénétration des spermatozoides et la reaction locale de l'oeuf, *J. Ultrastruct. Res.* **60**:106–120.

Picheral, B., and Charbonneau, M., 1982, Anuran fertilization: A morphological reinvestigation of some early events, *J. Ultrastruct. Res.* **81**:306–321.

Piko, L., and Tyler, A., 1964, Fine structural studies of sperm penetration in the rat, in: *Proceedings 5th International Congress of Animal Reproduction Trento*, Italy, Vol. 2, pp. 372–377.

Polge, C., and Dziuk, P., 1965, Recovery of immature eggs penetrated by spermatozoa following induced ovulation in pig, *J. Reprod. Fertil.* **9**:357–358.

Rahi, H., Sheikhnejade, G., and Srivastava, P. N., 1983, Isolation of the inner acrosomal–nuclear membrane complex from rabbit spermatozoa, *Gamete Res.* **7**:215–226.

Rifkin, J. M., and Olson, G. E., 1985, Characterization of maturation-dependent extrinsic proteins of the rat sperm surface, *J. Cell Biol.* **100**:1582–1591.

Rikmenspoel, R., Jacklet, A. C., Orris, S. E., and Lindemann, C., 1973, Control of bull sperm motility. Effects of viscosity, KCN and thiourea, *J. Mechanochem. Cell Motil.* **2**:7–24.

Rodger, J. C., and Bedford, J. M., 1982, Separation of sperm pairs and sperm–egg interaction in the opossum, *Didelphis virginiana*, *J. Reprod. Fertil.* **64**:171–179.

Rodger, J. C., and Young, R. J., 1981, Glycosidase and cumulus dispersal activities of acrosomal extracts from opossum (marsupial) and rabbit (eutherian) spermatozoa, *Gamete Res.* **4**:507–514.

Russell, L. D., Peterson, R. N., and Freund, M., 1980, On the presence of bridges linking the inner and outer acrosomal membranes of boar spermatozoa, *Anat. Rec.* **198**:449–459.

Sacco, A. G., Yurewicz, E. C., and Zhang, S., 1983, Immunoelectrophoretic analysis of the porcine zona pellucida, *J. Reprod. Fertil.* **68**:21–31.

Saling, P. M., 1981, Involvement of trypsin-like activity in binding of mouse spermatozoa to zona pellucida, *Proc. Natl. Acad. Sci. U.S.A.* **78**:6231–6235.

Saling, P. M., 1982, Development of the ability to bind to zona pellucida during epididymal maturation: Reversible immobilization of mouse spermatozoa by lanthanum, *Biol. Reprod.* **26**:429–436.

Saling, P. M., and Storey, B. T., 1979, Mouse gamete interaction during fertilization *in vitro*: Chlortetracycline as a fluorescent probe for the mouse sperm acrosome reactions, *J. Cell Biol.* **83**:544–555.

Sathananthan, A. H., and Chen, C., 1986, Sperm–oocyte membrane fusion in the human during monospermic fertilization, *Gamete Res.* **15**:317–326.

Sathananthan, A. H., Ng, S. C., Edirisinghe, R., Ratnam, S. S., and Wong, P. C., 1986, Human sperm–egg interaction *in vitro*, *Gamete Res.* **15**:177–186.

Selwood, L., 1982, A review of maturation and fertilization in marsupials with special reference to the dasyurid: *Antechinus stuartii*, in: *Carnivorous Marsupials*, Vol. I (M. Archer, ed.), Royal Zoological Society of N.S.W., Mosman, N.S.W., pp. 65–76.

Shalgi, R., and Phillips, D. M., 1988, The motility of rat spermatozoa at the site of fertilization, *Biol. Reprod.* **39**:1207–1213.

Shur, B. D., and Hall, N. G., 1982, Sperm surface galactosyltransferase activities during *in vitro* capacitation, *J. Cell Biol.* **95**:567–573.

Silber, S. J., Ord, T., Balmaceda, J., Patrizio, P., and Asch, R. H., 1990, Congenital absence of the vas deferens—the fertilizing capacity of human epididymal sperm, *New Eng. J. Med.* **323**:1788–1792.

Srivastava, P. N., Adams, C. E., and Hartree, E. F., 1965, Enzymatic action of acrosomal preparations on the rabbit ovum, *J. Reprod. Fertil.* **10**:61–67.

Srivastava, P. N., Munnell, J. F., Yang, C. H., and Foley, C. W., 1974, Sequential release of acrosomal membranes and acrosomal enzymes of ram spermatozoa, *J. Reprod. Fertil.* **36**:363–372.

Storey, B. T., Lee, M. A., Muller, C., Ward, C. R., and Wirtshafter, D. G., 1984, Binding of mouse spermatozoa to the zonae pellucidae of mouse eggs in cumulus: Evidence that the acrosomes remain substantially intact, *Biol. Reprod.* **31**:1119–1128.

Talbot, P., and DiCarlantonio, G., 1984, Ultrastructure of opossum oocyte investing coats and their sensitivity to trypsin and hyaluronidase, *Dev. Biol.* **103**:159–167.

Temple-Smith, P. D., 1974, Seasonal breeding biology of the platypus, *Ornithorhyncus anatinus* with special reference to the male. Ph.d. Thesis, Australian National University, Canberra, Australia.

Temple-Smith, P. D., 1987, Sperm structure and marsupial phylogeny, in: *Possums and Opossum: Studies in Evolution* (M. Archer, ed.), Royal Zoological Society of New South Wales, Mosman, N. S. W., pp. 171–193.

Temple-Smith, P. D. and Bedford, J. M., 1976, The features of sperm maturation in the epididymis of a marsupial, the brush-tailed possum, *Trichosurus vulpecula, Am. J. Anat.* **147**:471–500.

Temple-Smith, P. D., and Bedford, J. M., 1980, Sperm maturation and the formation of sperm pairs in the epididymis of the opossum, *Didelphis virginiana, J. Exp. Zool.* **214**:161–171.

Thibault, C., and Gérard, M., 1977, Facteur cytoplasmique nécessaire à la formation du pronucleus mâle dans l'oocyte de lapine, *C.R. Hebd. Séance Acad. Sci. Paris.* **270**:2025–2026.

Thomas, T. S., Reynolds, A. B., and Oliphant, G., 1984, Evaluation of the site of synthesis of rabbit sperm acrosome stabilizing factor using immunocytochemical and metabolic labelling techniques, *Biol. Reprod.* **30**:693–705.

Thomas, T. S., Wilson, L. S., Reynolds, A. B., and Oliphant, G., 1986, Chemical and physical characteristics of the rabbit sperm acrosome stabilizing factor, *Biol. Reprod.* **35**:691–703.

Trounson, A. O., Mohr, L. R., Wood, C., and Leeton, J. F., 1982, Effects of delayed insemination on *in vitro* fertilization, culture and transfer of human embryos, *J. Reprod. Fertil.* **64**:285–294.

Usui, N., and Yanagimachi, R., 1976, Behaviour of hamster sperm nuclei incorporated into eggs at various stages of maturation, fertilization and early development, *J. Ultrastruct. Res.* **57**:276–288.

Weinmann, D. E., and Williams, W. L., 1964, Mechanism of capacitation of rabbit spermatozoa, *Nature* **203**:423–424.

White, D. R., Phillips, D. M., and Bedford, J. M., 1990, Factors affecting the acrosome reaction in human spermatozoa, *J. Reprod. Fertil.* **90**:71–80.

Wolfson, A., 1954, Sperm storage at lower than body temperature outside the body cavity in some passerine birds, *Science* **161**:176–178.

Wooding, F. B. P., 1973, The effect of Triton X-100 on the ultrastructure of ejaculated bovine sperm, *J. Ultrastruct. Res.* **42**:502–516.

Yanagimachi, R., 1970, The movement of golden hamster spermatozoa before and after capacitation, *J. Reprod. Fertil.* **23:**193–196.

Yanagimachi, R., 1989, Mammalian fertilization, in: *The Physiology of Reproduction* (E. Knobil and J. D. Neill, eds.), Raven Press, New York, pp. 135–185.

Yanagimachi, R., and Noda, Y. D., 1970a, Ultrastructural changes in hamster sperm head during fertilization, *J. Ultrastruct. Res.* **31:**465–485.

Yanagimachi, R., and Noda, Y. D., 1970b, Electron microscope studies of sperm incorporation into the hamster egg, *Am. J. Anat.* **128:**429–462.

Yanagimachi, R., and Phillips, D. M., 1984, The status of acrosomal caps of hamster spermatozoa immediately before fertilization *in vitro*, *Gamete Res.* **9:**1–19.

2

Comparative Morphology of Mammalian Gametes

David M. Phillips and Gil L. Dryden

1. INTRODUCTION

Among mammals there is a wide variation in gamete structure—particularly of male gametes. Interspecies differences exist on the gross level in the lengths and widths of spermatozoa and in the shapes of sperm heads. On the ultrastructural level the morphological characteristics of intracellular organelles vary widely. The degree of morphological variation in spermatozoa between different mammalian groups is so great that a number of workers have used sperm morphology as a phylogenetic trait (Harding *et al.*, 1981, 1982; Vitullo *et al.*, 1988, Breed and Inns, 1985; Friend, 1936; Hirth, 1960; Helm and Bowers, 1973; Linzey and Layne, 1974; Breed and Sarafis, 1979; Feito and Gallardo, 1982). What is the relevance of the wide range of sperm morphology to the fertilization process? Although the morphology of spermatozoa has been characterized in hundreds of mammalian species at the light microscopic level and in dozens of species at the EM level, fertilization has been studied in relatively few species, and even in these species the characterization is largely incomplete. It is easy to see why this is so when one considers the enormous technical problems involved in mating wild mammals and then catching the one fertilizing spermatozoon at the moment of contact with the egg in a thin section. In this chapter we compare the morphology of gametes among mammals. We consider why there is so much variation in sperm morphology and how it could relate to the fertilization process. We do not describe the surface of the mammalian spermatozoon or the sperm tail, because these are discussed in Chapter 3.

DAVID M. PHILLIPS • The Population Council, New York, New York 10021.   GIL L. DRYDEN • Biology Department, Slippery Rock University, Slippery Rock, Pennsylvania 16057.

A Comparative Overview of Mammalian Fertilization, edited by Bonnie S. Dunbar and Michael G. O'Rand. Plenum Press, New York, 1991.

2. LIGHT MICROSCOPY OF SPERMATOZOA

One day, when you're a little bored with your gels and blots, you should go to the library and look through Retzius' beautiful monographs of mammalian spermatozoa (1906, 1909a, b, 1910). Although Retzius was not too concerned about precise dimensions, more recent investigators have been. It is relatively simple to measure the length of a spermatozoon, and the errors in measurements of sperm length tend to be small (van Dujin, 1975; van Dujin and van Voorst, 1971). Recently, Cummins (1983) and Cummins and Woodall (1985) collected data on sperm dimensions of 284 species and correlated these data with the size of these mammals. There is a very wide variation in the length of spermatozoa, ranging from spermatozoa that are 30 to 40 μm to those that are over 300 μm long. Since longer spermatozoa are wider, there are far greater differences in sperm mass. Cummins (1983) found a clear negative correlation between sperm mass and body mass. Larger mammals have smaller spermatozoa.

One can only speculate about the biological significance of this. However, there can be no doubt that a testis of a given size can produce more small spermatozoa than large spermatozoa. It is also known that smaller spermatozoa swim more slowly than large spermatozoa (Katz and Overstreet, 1980). Cummins (1983) suggests that small mammals produce large sperm because large spermatozoa can move faster and more vigorously and there is less need for many spermatozoa because in a small mammal there is less chance of a spermatozoon getting "lost" in the female reproductive tract. Large mammals may have a need to produce more spermatozoa than small mammals, because there is a greater chance of their being lost through dispersion in the large female tract. However, other theoretical arguments regarding sperm size and swimming speeds have also been advanced (Parker, 1970, 1982; Cohen, 1969, 1983; Harcourt et al., 1981).

In a relevant paper Bedford et al. (1984) counted spermatozoa at various locations in the reproductive tract of the American opossum Didelphis virginiana. Although these animals produce very few spermatozoa, a large percentage of these reach the site of fertilization. The probability of an opossum spermatozoon reaching the oviduct is 500 times greater than in the rabbit. Bedford et al. (1984) suggested that differences between numbers of spermatozoa produced by different species could be related to the chances of their remaining viable in the female tract and reaching the site of fertilization. The spermatozoa of New World marsupials, like those of the species Bedford et al. (1984) examined, form pairs, which may serve to protect the acrosome (Fawcett and Phillips, 1970; Olson, 1980; Temple-Smith and Bedford, 1980). This raises the interesting possibility that some interspecies differences in sperm morphology could relate to survival in the female tract.

3. STRUCTURE OF THE SPERM HEAD

As background for the following discussion the reader who is unfamiliar with the general features of mammalian spermatozoa may want to examine some general discussions of the subject (Phillips, 1974, 1975; Fawcett, 1970, 1975; Fawcett and Phillips, 1970). For a description of the morphology and morphogenesis of human spermatozoa, there is a thorough and beautiful atlas by Holstein and Rosen-Runge (1981). Mammalian spermatozoa are characterized by a head containing the nucleus and acrosome, a midpiece where the mitochondria spiral around the 9 + 2 flagellum and associated dense fibers, and a principal piece where the fibrous sheath circumscribes the flagellum.

Characteristically, the mammalian sperm head is flattened (Fig. 1). This is exceptional. Spermatozoa of birds, reptiles, and amphibians have long pencil-shaped heads (Furieri, 1970; Phillips et al., 1987; Asa and Phillips, 1988). Among invertebrates, even those with internal fertilization such as insects (Phillips, 1970; Jamieson, 1987) and certain worms (Sato et al., 1966;

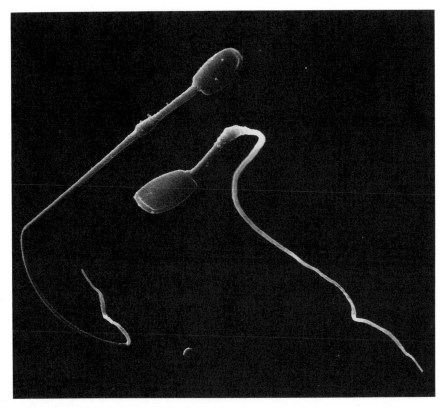

Figure 1.   Scanning electron micrograph of two spermatozoa of an Indian rhinoceros. Mammalian spermatozoa are unique in that the sperm heads are flattened. ×3500.

Anderson *et al.*, 1967; Hendelberg, 1965), sperm heads are also usually pencil-shaped. Thus, the flattened shape of mammalian spermatozoa may relate to a unique feature of fertilization in mammals. Two unique aspects of mammalian fertilization are that the eggs are very small and that eggs are surrounded by a cumulus oophorus which a sperm must traverse. However, there are other unique aspects of fertilization in mammals.

All mammalian spermatozoa have a highly condensed nucleus which is overlain by an acrosome. The acrosome contains two morphologically distinct portions, an acrosome cap and an equatorial segment (Fig. 2). As is described in succeeding chapters of this book, these two regions of the acrosome have very different functions in the fertilization process. Posterior to the acrosome is a region where fibrous material, sometimes termed the postacrosomal sheath (Pedersen, 1972; Olson and Winfrey, 1988; Koehler, 1966; Nicander and Bane, 1966; Olson *et al.*, 1983), lies between the nuclear membrane and the plasma membrane (Fig. 2). In most species the acrosome cap comprises the major portion of the acrosome volume, the equatorial segment being a narrow band between the acrosome cap and the postacrosomal region. However, spermatozoa of the mouse and the rat are exceptional in that the equatorial segments occupy the majority of the sperm head (Phillips, 1977; Yanagamachi and Teichman, 1972). The reader should keep in mind when examining the chapters in this book that discuss fertilization in mouse and rat that the acrosomes of mouse and rat spermatozoa are very unusual in this respect.

Although sperm heads of eutherian mammals are flattened, they have a variety of shapes. One might consider that there are two general types of shapes: symmetrical shapes such as in the

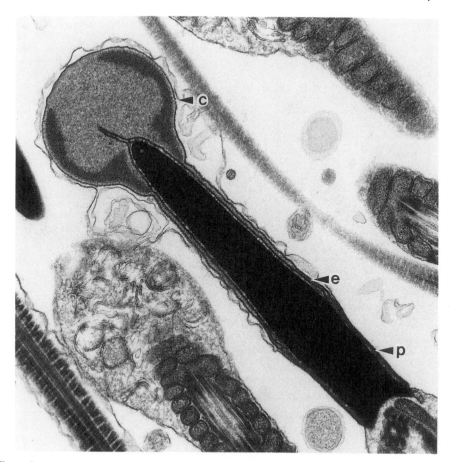

Figure 2. Sperm head of the spiny rat *Proechimys* showing the acrosomal cap (c), equatorial segment (e), and postacrosomal region (p), which characterize the mammalian spermatozoan perforatorium. ×11,000.

rabbit or bull (Bedford and Nicander, 1971; Phillips and Kalay, 1984) and asymmetric shapes such as in the mouse or rat. Although we usually think of asymmetrically shaped sperm heads as characteristic of rodents, asymmetric spermatozoa also occur in mammals that are not related to rodents, including some insectivores (Fig. 3) and bats (Forman and Genoways, 1979). In rodents there are various unusually shaped sperm heads, including paddle-shaped sperm with tail-like structures on one side (Vitullo *et al.*, 1988) and sperm with multiple hoods (Flaherty and Breed, 1983, 1987). Even in some sperm that appear under the light microscope to have symmetrical heads, examination with the SEM reveals that the flagellum is placed slightly asymmetrically. Such is the case in rhinoceros sperm (Fig. 1). Among marsupials there is a wide variety of sperm head shapes, but in both New World and Australian marsupials spermatozoa of virtually all species have asymmetric heads (Harding *et al.*, 1977, 1979, 1982; Rattner, 1972; Krause and Cutts, 1979).

Our own species is an exception to the general rule that in a given species all the male gametes are nearly precisely the same shape. Apparently normal human spermatozoa are irregular in shape, and each one looks slightly different from the next. We are not the only species like this. Recently Suttle *et al.* (1988) observed that spermatozoa from an Australian rodent, *Notomys alexis*, show a wide variety of differently shaped heads. Even in the laboratory

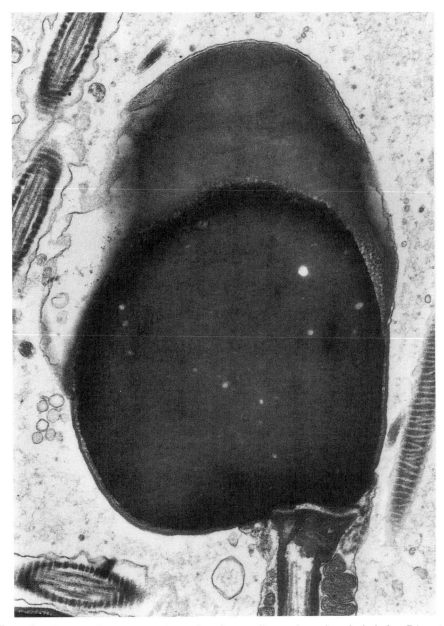

Figure 3.   The head of spermatozoa of a number of mammalian species such as the hedgehog *Erinaceus* is positioned eccentrically. ×7000.

mouse there is variation in sperm morphology among strains (Braden, 1958; Beatty and Sharma, 1959; Illison, 1969; Krzanowska, 1976). A most unusual case of variation in sperm morphology within an African rodent was recently described by Breed *et al.* (1988). One individual produced sperm with a disk-shaped head and complex acrosome while sperm of other individuals of apparently the same species collected from a different region had hook-shaped sperm heads.

Most acrosomes are small, but a few species have giant acrosomes. These species come from different mammalian orders. Giant acrosomes occur in rodents such as guinea pigs (Phillips *et al.*, 1985; Fawcett and Hollenberg, 1963; Fawcett, 1965) and squirrels (Shalgi and Phillips, 1983; Martan and Hruban, 1970), but they also have been observed in shrews, which are insectivores (Green and Dryden, 1976; Phillips and Bedford, 1985; Ploen *et al.*, 1979; Cooper and Bedford, 1976; Koehler, 1977), in the naked-tail armadillo, an edentate (Heath *et al.*, 1987), and in the loris, which is a primate (Phillips and Bedford, 1987). In all of these, except the loris, spermatozoa form rouleaux as they pass through the epididymis. The loris is unusual. In this species we observed that spermatozoa formed rouleaux in the caput epididymis, but the rouleaux disappeared in the cauda epididymis (Phillips and Bedford, 1987). Rouleaux formation is always associated with close apposition of sperm heads, but only in the loris is there a specialized junctional complex. It is possible that rouleaux formation of mammalian spermatozoa serves to protect the acrosome.

There are a number of other morphological components of mammalian spermatozoa that vary among species. At the anterior end of the nucleus between the nucleus and the acrosome there is a structure termed the perforatorium (Baccetti *et al.*, 1980). No one knows its function. In spermatozoa of some species it is minute (Fig. 2), but in others it is large and complicated in shape (Phillips and Bedford, 1985).

4. MAMMALIAN EGGS

At the time of fertilization the mammalian egg is surrounded by an extracellular shell, the zona pellucida. The zona pellucida is in turn surrounded by a cumulus oophorus composed of thousands of cells embedded in a mucus matrix. The egg and its investments are often referred to as the cumulus–oocyte complex (Fig. 4) (Dekel and Phillips, 1979). Although ovulated cumulus–oocyte complexes can be easily obtained from laboratory and some domestic animals, it is not very easy to obtain cumulus–oocyte complexes from wild animals. Thus, compared to spermatozoa, cumulus–oocyte complexes of relatively few species have been studied, and there is less information to compare.

Unlike spermatozoa which vary among mammalian species, mammalian eggs are roughly the same size, 60 to 100 μm, and are covered with numerous microvilli (Phillips and Shalgi, 1980a). In some species such as the laboratory mouse there is a small region that is free of microvilli (Fig. 5). In rodents and rabbits the oocyte cytoplasm contains swirls of lamellar material (Fig. 4) (Koehler *et al.*, 1985; King and Tibbits, 1977). Such inclusions may be unique to the rodent egg. In rodents numerous cortical granules are observed beneath the oolemma (Nicosia *et al.*, 1977; Szollosi, 1967), but the morphology of cortical granules has not been described in detail for enough species to make comparisons possible.

5. THE ZONA PELLUCIDA

All mammalian eggs are enclosed in an extracellular zona pellucida which remains around the embryo until it hatches in the uterus. Viewed in the transmission electron microscope, the zona pellucida is seen as being composed of fibrous material. Some groups of fibers are arranged radially within the zona pellucida (Fig. 6). The arrangement of fibers on the zona surface away from the oocyte is looser than the arrangement adjacent to the oocyte surface (Figs. 5 and 6). Viewed in the scanning electron microscope, the zona surface facing the oocyte and the surface facing the cumulus are very dissimilar (Phillips and Shalgi, 1980b). So far striking differences have not been reported between species, although the zona pellucida of marsupials is much thinner than that in eutherian mammals (Phillips and Fadem, 1987).

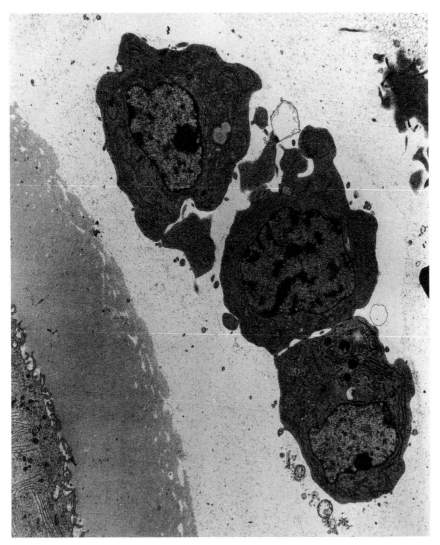

Figure 4.   Oocyte (lower left), zona pellucida, and three cells of the cumulus oophorus of a golden hamster at the time of ovulation. Most of the volume of the cumulus oophorus is composed of mucus. ×2000.

6. THE CUMULUS OOPHORUS

In most eutherian mammals studied the zona pellucida is surrounded by thousands of cells embedded in mucus. The mucus is synthesized by cumulus cells in the follicle during the hours between the LH surge and ovulation (Dekel, 1986). Spermatozoa must pass through the mucus material to reach the zona pellucida. In some mammals, such as humans and rabbits, the cumulus cells immediately adjacent to the zona pellucida are more closely packed together than other cumulus cells. The cells of this layer, which is termed the corona radiata, adhere closely to the zona pellucida such that when the cumulus is dispersed with hyaluronidase, they remain

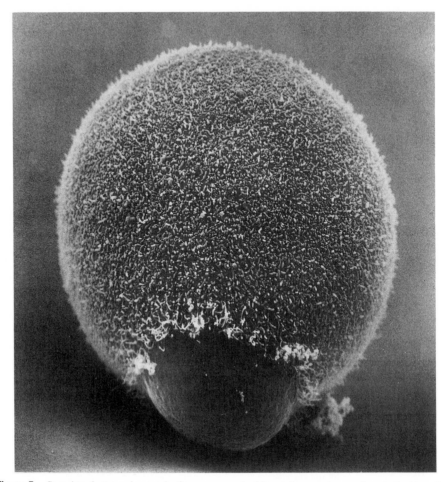

Figure 5. Scanning electron micrograph of a mouse oocyte. Mammalian oocytes are characterized by micro-villi. However, in some species a microvillus-free zona bulges over the site of the meiosis II metaphase spindle. ×800.

associated with the zona. In the rabbit, cells of the corona radiata possess processes that extend into the zona pellucida (Phillips *et al.*, 1985). On the other hand, the cumulus oophorus of most rodents does not have a distinct corona radiata (Fig. 5), and rodent cumuli are completely dispersed with hyaluronidase.

The musk shrew, *Suncus murinus*, has a very unusual cumulus oophorus. The cumulus cells of *Suncus* never produce mucus. At the time of fertilization the cumulus oophorus is composed of cells that are closely opposed together. Each cumulus cell is associated with its neighbors by small gap junctions (Fig. 7). How can the spermatozoon penetrate this cumulus? It is known that in rodents spermatozoa will not penetrate follicular cumuli that have not expanded. Furthermore, the sperm head of *Suncus* is one of the largest mammalian sperm heads (Phillips and Bedford, 1985). The question remains an enigma.

Another interesting situation exists in marsupials. Oocytes are ovulated without a cumulus. However, as they traverse the oviduct they become covered with a thick layer of mucus which is synthesized by the oviductal epithelium (Hill, 1910; Hartman, 1916; Talbot and Dicarlantonio,

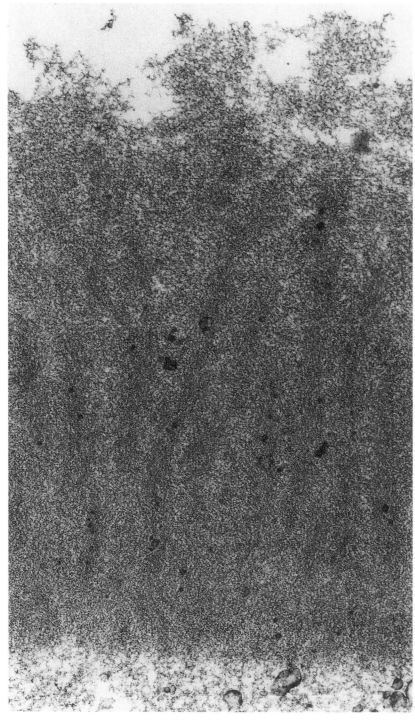

Figure 6. Transmission electron micrograph of the zona pellucida of the shrew *Suncus*. The outer surface of the zona pellucida (above) appears more irregular than the surface facing the oocyte (below). ×7000.

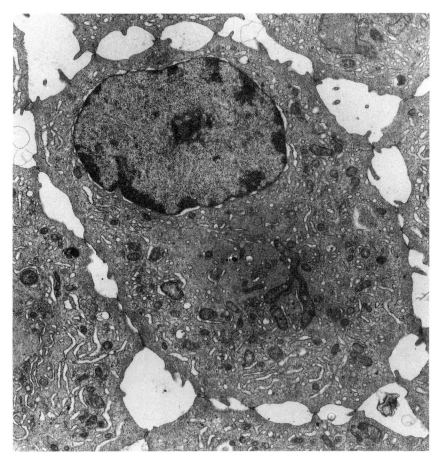

Figure 7. Cumulus cell of the musk shrew *Suncus murinus*. At the time of fertilization cumulus cells are associated with each other by numerous small gap junctions. This is a most unusual exception, since in all other species that have been described, cumulus cells lie in a mucus matrix. ×12,000.

1984; Breed and Leigh, 1988). Viewed in the electron microscope, this mucus is seen to be composed of highly ordered circumferentially arranged material (Fig. 8) (Phillips and Fadem, 1987). This mucus may be functionally analogous to the cumulus oophorus of eutherian mammals.

7. CONCLUSION

It would be nice if the mechanisms of fertilization in mammals were the same in all mammals. However, although fertilization has been studied in very few species, it is already clear that there are interspecific differences in many aspects of the fertilization process. This is perhaps not surprising when we consider the wide interspecies variation in gamete morphology. We hope that as we gain more insights into the fertilization process in different mammalian species, we will be able to develop a clearer understanding of why gametes, particularly male gametes, of different mammals appear so different. What are the advantages and disadvantages of an asymmetrically shaped sperm head? Why do some species produce spermatozoa with giant acrosomes? Why do spermatozoa of some mammals form pairs, and why are spermatozoa of

Figure 8. A mucus matrix secreted by the oviductal epithelium (below) surrounds the zona pellucida (above right) of marsupial, *Monodelphis domestica*, oocytes. ×4000.

some species so much longer than others. Knowledge about fertilization in but a few species limits us primarily to conjecture concerning these and other important questions central to the understanding of mammalian gamete structure.

ACKNOWLEDGMENTS. We would like to thank Vanaja Zacharopoulos and Anne Goldstein for their excellent technical assistance. The research was supported in part by a grant from the Andrew W. Mellon Foundation to D.M.P.

8. REFERENCES

Anderson, W. A., Weissman, A., and Ellis, R. A., 1967, Cytodifferentiation during spermiogenesis in *Lumbricus terrestris*, *J. Cell Biol.* **32**:11–26.

Asa, C. S., and Phillips, D. M., 1988, Nuclear shaping in spermatids of the Thai leaf frog, *Megophrys montana*, *Anat. Rec.* **220**:287–290.

Baccetti, B., Bigliardi, E., and Burrini, A. G., 1980, The morphogenesis of vertebrate perforatorium, *J. Ultrastruct. Res.* **71**:272–287.

Beatty, R. A., and Sharma, K. N., 1959, Genetics of gametes III. Strain differences in spermatozoa from eight inbred strains of mice, *Proc. R. Soc. Edinb. B* **68**:25–51.

Bedford, J. M., and Nicander, L., 1971, Ultrastructural changes in the acrosome and sperm membranes during maturation in the testis and epididymis of the rabbit and monkey, *J. Anat.* **108**:527–543.

Bedford, J. M., Rodger, J. C., and Breed, W. G., 1984, Why so many mammalian spermatozoa—a clue from marsupials, *Proc. R. Soc. Lond. B* **221**:221–233.

Braden, A. W. H., 1958, Strain differences in the morphology of the gamete of the mouse, *Aust. J. Biol. Sci.* **12**:65–71.

Breed, W. G., and Inns, R. W., 1985, Variations in sperm morphology of Australian Vespertilionidae and its possible phylogenetic significance, *Mammalia* **49**:105–108.

Breed, W. G., and Leigh, C. M., 1988, Morphological observations on sperm–egg interactions during *in vivo* fertilization in the Dasyurid marsupial *Sminthopsis crassicaudata*, *Gamete Res.* **19**:131–149.

Breed, W. G., and Sarafis, V., 1979, On the phylogenetic significance of spermatozoal morphology and male reproductive tract anatomy in Australian rodents, *Trans. R. Soc. S. Aust.* **103**:127–135.

Breed, W. G., Cox, G. A., Leigh, C. M., and Hawkins, P., 1988, Sperm head structure of a murid rodent from South Africa: The red veld rat, *Aethomys chrysophilus*, *Gamete Res.* **19**:191–202.

Cohen, J., 1969, Why so many sperms? An essay on the arithmetic of reproduction, *Sci. Prog. Oxf.* **57**:23–41.

Cohen, J., 1983, Selection among spermatozoa, in: *The Sperm Cell* (J. Andre, ed.), Martinus Nijhoff, The Hague, pp. 33–37.

Cooper, G. W., and Bedford, J. M., 1976, Asymmetry of spermiation and sperm surface charge patterns over the giant acrosome in the musk shrew, *Suncus murinus*, *J. Cell Biol.* **69**:415–428.

Cummins, J. M., 1983, Sperm size, body mass and reproduction in mammals, in: *The Sperm Cell* (J. Andre, ed.), Martinus Nijhoff, The Hague, pp. 395–398.

Cummins, J. M., and Woodall, P. F., 1985, On mammalian sperm dimensions, *J. Reprod. Fertil.* **75**:153–175.

Dekel, N., 1986, Hormonal control of ovulation, in: *Biochemical Actions of Hormones*, Academic Press, New York, pp. 57–99.

Dekel, N., and Phillips, D. M., 1979, Maturation of rat cumulus oophorus—a scanning electron microscope study, *Biol. Reprod.* **21**:9–18.

Fawcett, D. W., 1965, The anatomy of the mammalian spermatozoa with particular reference to the guinea pig, *Z. Zellforsch. Mikrosk. Anat.* **67**:279–296.

Fawcett, D. W., 1970, A comparative view of sperm ultrastructure, *Biol. Reprod.* [Suppl.] **2**:90–127.

Fawcett, D. W., 1975, The mammalian spermatozoon, *Dev. Biol.* **44**:394–436.

Fawcett, D. W., and Hollenberg, R., 1963, Changes in the acrosome of the guinea pig spermatozoa during passage through the epididymis, *Z. Zellforsch. Mikrosk. Anat.* **60**:276–292.

Fawcett, D. W., and Phillips, D. M., 1970, Recent observations on the ultrastructure and development of the mammalian spermatozoon, in: *Comparative Spermatology* (B. Baccetti, ed.), Academia de Linceri/Academic Press, Rome, pp. 13–29.

Feito, R., and Gallardo, M., 1982, Sperm morphology of Chilean species of *Ctenomys* (Octodontidae), *J. Mamm.* **63**:658–661.

Flaherty, S. P., and Breed, W. G., 1983, The sperm head of the plains mouse, *Pseudomys australis*: Ultrastructure and effects of chemical treatments, *Gamete Res.* **8**:231–244.

Flaherty, S. P., and Breed, W. G., 1987, Formation of the ventral hooks on the sperm head of the plains mouse, *Pseudomys australis*, *Gamete Res.* **17:**115–129.

Forman, G. L., and Genoways, H. H., 1979, *Biology of Bats of the New World, Phyllostomadtidae Part III Special Publication 16*, The Museum of Texas Tech University Lubbock, pp. 177–204.

Friend, G. F., 1936, The sperms of the British Muridae, *Q. J. Microsc. Sci.* **78:**419–443.

Furieri, P., 1970, Sperm morphology in some reptiles: Squamata and Chelonia, in: *Comparative Spermatology* (B. Bacetti, ed.), Academic Press, New York, pp. 115–132.

Green, J. A., and Dryden, G. L., 1976, Ultrastructure of epididymal spermatozoa of the Asiatic musk shrew, *Suncus murinus*, *Biol. Reprod.* **14:**327–331.

Harcourt, A. H., Harvey, P. H., Larson, S. G., and Short, R. V., 1981, Testis weight, body weight and breeding system in primates, *Nature* **293:**55–57.

Harding, H. R., Carrick, F. N., and Shorey, C. D., 1977, Spermatozoa of Australian marsupials: Ultrastructure and epididymal development, in: *Reproduction and Evolution, Fourth International Symposium on Comparative Biology of Reproduction* (C. H. Tyndale-Biscoe, ed.), Australian Academy of Science, Canberra, pp. 151–152.

Harding, H. R., Carrick, F. N., and Shorey, C. D., 1979, Special features of sperm structure and function in marsupials, in: *The Spermatazoon* (J. M. Bedford, ed.), Urban & Schwarzenberg, Baltimore, pp. 289–303.

Harding, H. R., Carrick, F. N., and Shorey, D. C., 1981, Marsupial phylogeny: New indications from sperm ultrastructure and development in *Tarsipes spenserae*, *Search* **12:**45–47.

Harding, H. R., Woolley, P. A., Shorey, C. D., and Carrick, F. N., 1982, Sperm ultrastructure, spermiogenesis and epididymal sperm maturation in Dasyurid marsupials: Phylogenetic implications, in: *Carnivorous Marsupials* (M. Archer, ed.), Royal Zoological Society of New South Wales, Sydney, pp. 659–673.

Hartman, C. G., 1916, Studies in the development of the opossum: *Didelphis virginiana* I. History of early cleavage II. Formation of the blastocyst, *J. Morphol.* **27:**1–83.

Heath, E., Schaeffer, N., Merit, D. A., and Jeyendran, R. S., 1987, Rouleaux formation by spermatozoa in the naked-tail armadillo, *Cabassous unicinctus*, *J. Reprod. Fertil.* **79:**153–158.

Helm, J. D., and Bowers, J. R., 1973, Spermatozoa of Tylomys and Ototylomys, *J. Mamm.* **54:**769–772.

Hendelberg, J., 1965, On different types of spermatozoa in *Polycladida turbellaria*, *Ark. Zool.* **18:**267–304.

Hill, J. P., 1910, The early development of the marsupials with special reference to the mature cat (*Dasyurus viverrinus*), *G. J. Microsc. Sci.* **56:**1–134.

Hirth, H. F., 1960, The spermatozoa of some North American bats and rodents, *J. Morphol.* **106:**77–83.

Holstein, A. F., and Roosen-Runge, E. C., 1981, *Atlas of Human Spermatozoa*, Grosse Verlag, Berlin.

Illison, L., 1969, Spermatozoal head shape in two inbred strains of mice and their F1 and F2 progenies, *Aust. J. Biol. Sci.* **22:**947–963.

Jamieson, B., 1987, *The Ultrastructure and Phylogeny of Insect Spermatozoa*, Cambridge University Press, Cambridge.

Katz, D. F., and Overstreet, J. W., 1980, Mammalian sperm movement in the secretions of the male and female genital tracts, in: *Testis Development, Structure and Function* (A. Steinberger, ed.), Raven Press, New York, pp. 481–489.

King, B. F., and Tibbits, F. D., 1977, Ultrastructural observations on cytoplasmic lamellar inclusions in oocytes of the rodent *Thomomys*, *Anat. Rec.* **189:**263–272.

Koehler, J. K., 1966, Fine structure observations in frozen-etched bovine spermatozoa, *J. Ultrastruct. Res.* **16:**359–375.

Koehler, J. K., 1977, Fine structure of spermatozoa of the Asiatic musk shrew, *Suncus murinus*, *Am. J. Anat.* **149:**135-151.

Koehler, J. K., Clark, J. M., and Smith, D., 1985, Freeze-fracture observations on mammalian oocytes, *Am. J. Anat.* **174:**317–329.

Krause, W. J., and Cutts, J. H., 1979, Pairing of spermatozoa in the epididymis of the opossum (*Didelphis virginiana*): A scanning electron microscopic study, *Arch Histol. Jpn.* **42:**181–90.

Krzanowska, H., 1976, Types of sperm-head abnormalities in four inbred strains of mice, *Acta. Biol. Cracow (Ser. Zool.)* **19:**79–85.

Linzey, A. V., and Layne, J. N., 1974, Comparative morphology of spermatozoa of the rodent genus *Peromyscus* (Muridae), *Am. Mus. Novitates* **2355:**1–20.

Martan, J., and Hruban, Z., 1970, Unusual spermatozoan formations in the epididymis of the flying squirrel, *Glaucomys volans*, *J. Reprod. Fertil.* **21:**167–170.

Nicander, L., and Bane, A., 1966, Fine structure of the sperm head in some mammals with particular reference to the acrosome and the subacrosomal substance, *Z. Zellforsch.* **72:**496–515.

Nicosia, S. V., Wolf, D. P., and Inoue, M., 1977, Cortical granule distribution and cell surface characteristics in mouse eggs, *Biol. of Reprod.* **57:**56–74.

Olson, G. E., 1980, Changes in intramembranous particle distribution in the plasma membrane of *Didelphis virginiana* spermatozoa during maturation in the epididymis, *Anat. Rec.* **197**:471–486.

Olson, G. E., and Winfrey, V. P., 1988, Characterization of the postacrosomal sheath of bovine spermatozoa, *Gamete Res.* **20**:329–342.

Olson, G. E., Noland, T. D., Winfrey, V. P., and Garbers, D. L., 1983, Substructure of the postacrosomal sheath of bovine spermatozoa, *J. Ultrastruct. Res.* **85**:204–218.

Parker, G. A., 1970, Sperm competition and its evolutionary consequences in the insects, *Biol. Rev.* **45**:525–567.

Parker, G. A., 1982, Why are there so many tiny sperm? Sperm competition and the maintenance of two sexes, *J. Theor. Biol.* **96**:281–294.

Pedersen, H., 1972, The postacrosomal region of the spermatozoa of man and *Macaca arctoides*, *J. Ultrastruct. Res.* **40**:366–377.

Phillips, D. M., 1970, Insect sperm: Their structure and morphogenesis, *J. Cell Biol.* **44**:243–277.

Phillips, D. M., 1974, *Spermiogenesis*, Academic Press, New York.

Phillips, D. M., 1975, Mammalian sperm structure, in: *Handbook of Physiology* (R. Greep and E. Astwood, eds.), Waverly Press, Baltimore, pp. 405–419.

Phillips, D. M., 1977, Surface of the equatorial segment of the mammalian acrosome, *Biol. Reprod.* **16**:128–137.

Phillips, D. M., and Bedford, J. M., 1985, Ultrastructure of spermatozoa of the musk shrew *Suncus murinus*, *J. Exp. Zool.* **235**:119–126.

Phillips, D. M., and Bedford, J. M., 1987, Sperm–sperm associations in the loris, *Gamete Res.* **18**:17–23.

Phillips, D. M., and Fadem, B. H., 1987, The oocytes of a new world marsupial, *Monodelphis domestica*: Structure, formation and function of the enveloping mucoid layers, *J. Exp. Zool.* **242**:363–371.

Phillips, D. M., and Kalay, D., 1984, Observations on mechanisms of flagellar motility deduced from backwards swimming bull sperm, *J. Exp. Zool.* **231**:109–116.

Phillips, D. M., and Shalgi, R., 1980a, Surface properties of the mouse and hamster zona pellucida and oocyte, *J. Ultrastruct. Res.* **72**:172.

Phillips, D. M., and Shalgi, R., 1980b, Surface properties of the zona pellucida, *J. Exp. Zool.* **213**:1–8.

Phillips, D. M., Shalgi, R., and Dekel, N., 1985, Mammalian fertilization as seen with the scanning electron microscope, *Am. J. Anat.* **174**:357–372.

Phillips, D. M., Asa, C., and Stover, J., 1987, Ultrastructure of spermatozoa of the white-naped crane, *J. Submicrosc. Cytol.* **19**:489–494.

Ploen, L., Ekwall, H., and Afzelius, B. A., 1979, Spermiogenesis and the spermatozoa of the European common shrew, *Sorex aruneusl*, *J. Ultrastruct. Res.* **68**:149–159.

Rattner, J. B., 1972, Nuclear shaping in marsupial spermatids, *J. Ultrastruct. Res.* **40**:498–512.

Retzius, G., 1906, Spermatozoa of the Marsupialia, *Biol. Unters. N.F.* **13**:77–86.

Retzius, G., 1909a, Spermatozoa of *Didelphys*, *Biol. Unters. N.F.* **14**:123–126.

Retzius, G., 1909b, Spermatozoa of mammals, *Biol. Unters. N.F.* **14**:163–178.

Retzius, G., 1910, General contribution to the knowledge of spermatozoa with special reference to nuclear material, *Biol. Unters. N.F.* **15**:63–82.

Sato, M., Motow, O., and Sakoda, K., 1966, Electron microscopic study of spermatogenesis in the lung fluke (*Paragonium miyasakii*), *Z. Zellforsch.* **77**:232–243.

Shalgi, R., and Phillips, D. M., 1983, Role of cumulus in zona penetration, in: *Proceedings of the 4th International Symposium on Spermatology-Seillac* (J. Andre, ed.), Martinus Nijhoff, The Hague, pp. 90–93.

Suttle, J. M., Moore, H. D. M., Peirce, E. J., and Breed, W. G., 1988, Quantitative studies on variation in sperm head morphology of the hopping mouse, *Notomys alexis*, *J. Exp. Zool.* **247**:166–171.

Szollosi, D., 1967, Development of cortical granules and the cortical reaction in rat and hamster eggs, *Anat. Rec.* **159**:431–446.

Talbot, P., and Dicarlantonio, G., 1984, Ultrastructure of opossum oocyte investing coats and their sensitivity to trypsin and hyaluronidase, *Dev. Biol.* **103**:159–167.

Temple-Smith, P. D., and Bedford, J. M., 1980, Sperm maturation and the formation of sperm pairs in the epididymis of the opossum, *Didelphis virginia*, *J. Exp. Zool.* **214**:161–171.

van Dujin, C., 1975, Bibliography (with review) on maturation of spermatozoa, *Biol. Reprod.* **25**:241–248.

van Dujin, C., and van Voorst, C., 1971, Precision measurements of dimensions, refractive index and mass of bull spermatozoa in the living state, *Mikroskopie* **27**:142–167.

Vitullo, A. D., Roldan, E. R. S., and Marani, M. S., 1988, On the morphology of spermatozoa of tuco-tucos, *Ctenomys* (Rodentia: Ctenomyidae): New data and its implications for the evolution of the genus, *J. Zool. Lond.* **215**:675–683.

Yanagamachi, R., and Teichman, R. J., 1972, Cytochemical demonstration of acrosomal proteinase in mammalian and avian spermatozoa by a silver proteinase method, *Biol. Reprod.* **6**:87–97.

3

A Comparison of Mammalian Sperm Membranes

Gary E. Olson and Virginia P. Winfrey

1. INTRODUCTION

Mammalian spermatozoa exhibit considerable species differences in their size and shape, yet they all possess the same set of cellular organelles assembled on a common architectural theme. The polarized spermatozoon is partitioned into distinct segments or domains, distinguished by specific subsets of the cellular organelles (Eddy, 1988; Fawcett, 1975). These include the acrosomal and postacrosomal segments of the head, followed by the connecting piece, midpiece, principal piece, and end piece segments of the flagellum. During fertilization, different segments perform specific functions in generating motility, in binding the zona pellucida, in penetrating the egg coats, and in fusing with the egg plasma membrane (Wassarman, 1987; Yanagimachi, 1988). The sperm plasma membrane plays a central role in regulating these functions, and it varies in structure and molecular composition in the different domains. In this chapter we discuss domain-specific properties of the plasma membrane and address mechanisms that may maintain their unique properties.

Membrane proteins are divided into two major classes (Singer and Nicolson, 1972). Integral proteins possess both a hydrophobic domain(s) embedded within the interior of the lipid bilayer and a hydrophilic domain(s) exposed to the aqueous cytoplasmic or extracellular environment. They may be freely diffusible within the plane of the membrane or immobilized by specific protein–protein or protein–lipid interactions. Transmembrane integral proteins play important functions in transport and signal transduction processes. In contrast, peripheral membrane proteins are associated with the membrane surfaces, where they may form a membrane skeleton complex that is linked to the overlying membrane by interactions with specific integral proteins (Marchesi, 1985); this infrastructure functions in maintaining cell shape and in defining the distribution of integral proteins. In the following text we discuss thin-section and freeze-fracture

GARY E. OLSON AND VIRGINIA P. WINFREY • Department of Cell Biology, Vanderbilt University, Nashville, Tennessee 37232.

A Comparative Overview of Mammalian Fertilization, edited by Bonnie S. Dunbar and Michael G. O'Rand. Plenum Press, New York, 1991.

data on mammalian spermatozoa to highlight regional differentiations of the plasma membrane and to provide insights into the distribution of integral and peripheral proteins.

2. PERIACROSOMAL PLASMA MEMBRANE

This domain consists of three subregions corresponding to the apical, principal, and equatorial segments of the acrosome. The plasma membrane over the apical and principal segments contains receptors for the zona pellucida and participates in the membrane fusion events of the acrosome reaction (Meizel, 1985; Wassarman, 1987, 1988), whereas the plasma membrane over the equatorial segment may function in the fusion with the egg plasma membrane (Talbot, 1985; Yanagimachi, 1988). The periacrosomal membrane can be purified (Parks and Hammerstedt, 1985), and several proteins that function in cell–cell recognition, ion transport, and signal transduction processes have been identified. A small-molecular-weight polypeptide secreted by the seminal vesicles, termed caltrin, binds to the periacrosomal plasma membrane of ejaculated bovine and guinea pig spermatozoa; it inhibits calcium uptake and is speculated to prevent premature acrosome reactions (San Agustin et al., 1987). In mouse sperm, galactosyl transferase functions in zona binding and is localized to this domain (Shur and Neely, 1988). A M_r 95,000 polypeptide with receptor activity for ZP-3 has been identified biochemically in the mouse (Leyton and Saling, 1989b) and presumably represents the ZP-3 binding sites demonstrated on the periacrosomal plasma membrane of intact sperm (Bleil and Wassarman, 1986). Other proteins involved in zona pellucida binding have been identified in the acrosomal segment (Benau and Storey, 1987; Lakoski et al., 1988), and epididymal polypeptides implicated in the posttesticular development of zona binding capacity (Fournier-Delpech et al., 1984) presumably interact with this domain. Recent work has implicated specific signal-transducing elements as mediating the response between zona binding and initiation of the acrosome reaction (Endo et al., 1988).

Several structural specializations have been reported in the periacrosomal plasma membrane. In some species, such as the rat and guinea pig, a paracrystalline organization of the glycocalyx is noted exclusively within this domain (Flaherty and Olson, 1988; Friend and Fawcett, 1974; Phillips, 1975; Toyama and Nagano, 1988), and lectin-binding studies have demonstrated its unique composition (Koehler, 1978a; Magargee et al., 1988). Freeze-fracture replicas of the periacrosomal plasma membrane of guinea pig sperm reveal a quilt-like lattice of intramembranous particles that resembles the crystalline organization of the glycocalyx (Friend and Fawcett, 1974). Freeze-fracture also has revealed crystalline protein aggregates within the apical segment of rat, golden hamster, and boar sperm (Reger et al., 1985; Suzuki, 1981; Suzuki and Nagano, 1980a; Toshimori et al., 1987, Toyama and Nagano, 1988). Other species including the bat, bovine, and mouse possess a random distribution of intramembranous particles within the periacrosomal domain; even so it is readily distinguished from adjacent domains on the basis of particle size and numbers (Hoffman et al., 1987; Holt, 1984; Koehler, 1966; Toshimori et al., 1985).

Changes in protein mobility and distribution appear to correlate with the development and expression of functional activity by this domain. Protein redistribution during posttesticular maturation generates the ordered intramembranous particle arrays characteristic of some species (Suzuki, 1981; Suzuki and Nagano, 1980a; Suzuki and Yanagimachi, 1986). In the rat, crystalline particle aggregations appear coincidentally with the binding of a glycocalyx-like material secreted by the epididymal epithelium (Suzuki and Nagano, 1980a), suggesting that externally bound glycoproteins may reorganize the membrane interior. Protein mobility is evident during capacitation and the acrosome reaction, where a redistribution of intramembranous particles precedes the membrane fusion process (Friend et al., 1977; Flechon, 1985; Yanagimachi and Suzuki, 1985). Similarly, at the membrane exterior, a modification of the periacrosomal glycocalyx (Friend, 1984; Koehler, 1978a,b) and changes in the mobility of specific components have been demonstrated during fertilization (O'Rand, 1977; Yanagimachi, 1988). Recent data suggest that ZP-3 induces a redistribution and clustering of sperm membrane receptors that precedes the induction of the acrosome reaction (Leyton and Saling, 1989a).

Cytoskeleton–membrane interactions restrict the lateral mobility of specific transmembrane proteins in somatic cells (Burridge et al., 1988; Marchesi, 1985). Cytoskeletal structures have been identified within the acrosomal segment of several species including the vole and hamster (Koehler, 1978b; Olson and Winfrey, 1985); in the hamster a paracrystalline sheet of 10-nm-diameter filaments is associated with both the plasma membrane and outer acrosomal membrane (Figs. 1–3). In the guinea pig a regular array of filamentous connections between the plasma membrane and outer acrosomal membrane is located exclusively along the ventral surface of the apical segment; these are precisely superimposed over the pillar-like projections of the para-crystalline glycocalyx, and these linked membranes exhibit a reduced tendency to fuse during the acrosome reaction (Flaherty and Olson, 1988). Similar linking elements are noted in the apical segment of bovine spermatozoa (Olson et al., 1990) and human spermatozoa (Escalier, 1984), suggesting that cytoskeletal assemblies that could regulate specific functions of the associated membrane may be a common feature of the acrosomal segment.

The polypeptides comprising the acrosomal cytoskeletal elements remain to be identified. Actin has been reported within the acrosomal segment of boar and human spermatozoa (Camatini et al., 1986; Virtanen et al., 1984), but other studies of these species failed to confirm these results (Clarke et al., 1982; Flaherty et al., 1986); moreover, actin was not found in the acrosomal segment of bovine, guinea pig, and rabbit epididymal spermatozoa (Flaherty et al., 1986; Halenda et al., 1987; Welch and O'Rand, 1985). Similarly, spectrin, a major component of the membrane skeleton of somatic cells, was reported within the acrosomal segment of human spermatozoa (Virtanen et al., 1984), but it was not found in mouse spermatozoa (Damjanov et al., 1986). Finally, vimentin, an intermediate-filament polypeptide, was reported within the equatorial segment of human spermatozoa (Virtanen et al., 1984; Ochs et al., 1986), but it was not found in bovine spermatozoa (Longo et al., 1987). Based on the available data it appears that cytoskeletal elements of the acrosomal segment may be composed of germ-cell-specific polypeptides.

3. POSTACROSOMAL PLASMA MEMBRANE

The postacrosomal plasma membrane extends from the distal margin of the acrosome to the head–tail junction (Figs. 4 and 5). Several studies have implicated this domain in the initial fusion with the egg plasma membrane (Talbot, 1985; Yanagimachi, 1988). The postacrosomal plasma membrane is adherent to a cytoplasmic paracrystalline array of 10 to 12-nm-diameter filaments (Fig. 6) termed the postacrosomal sheath (Olson et al., 1983); it contains a major polypeptide of 58–60 kDa that is not immunologically related to intermediate-filament polypeptides of somatic cells (Longo et al., 1987; Olson and Winfrey, 1988). Specific plasma membrane proteins may be anchored to this filament complex, and their mobility thereby limited.

Freeze-fracture reveals characteristic particle arrays within the postacrosomal plasma membrane (Fig. 4). Anteriorly a random arrangement of intramembranous particles is typical (Friend and Fawcett, 1974; Olson et al., 1983), although in some species paracrystalline particle arrays are noted (Suzuki and Yanagimachi, 1986; Toyama and Nagano, 1983); in the Chinese hamster these crystalline arrays appear in the epididymis. Prominent linear aggregates of intra-membranous particles are located near the head–tail junction; in boar spermatozoa these also develop in the epididymis (Suzuki, 1981). The dynamic properties of this domain are further demonstrated in the guinea pig, where a surface component, termed PH-20, undergoes directed

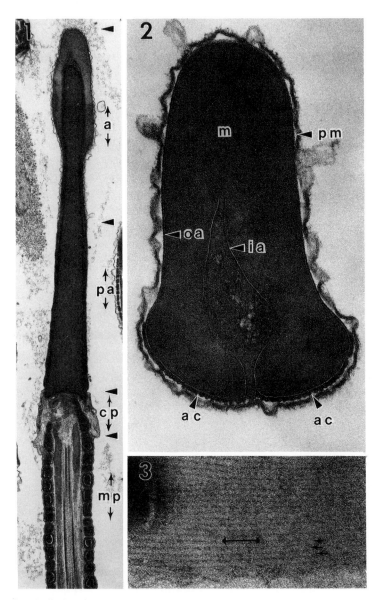

Figure 1. Longitudinal section through hamster spermatozoon showing the acrosomal (a), postacrosomal (pa), connecting piece (cp), and midpiece (mp) segments.

Figure 2. Cross section through acrosomal segment of hamster spermatozoon showing the plasma membrane (pm), outer acrosomal membrane (oa), acrosomal matrix (m), and inner acrosomal membrane (ia). Note that the filaments of the acrosomal cytoskeleton (ac) are located ventrally and associate with specific subdomains of the plasma and acrosomal membranes.

Figure 3. Negatively stained acrosomal cytoskeleton of hamster spermatozoon showing it to be comprised of parallel filaments (short arrows) aligned in the long axis of the sperm (double arrow).

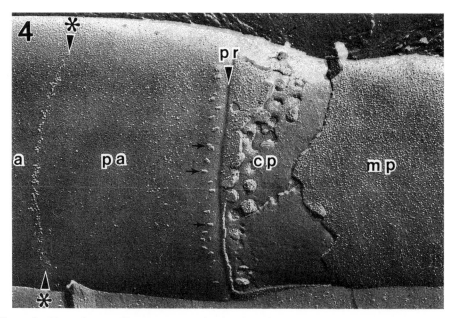

Figure 4. Freeze-fracture of bat spermatozoon showing views of acrosomal (a), postacrosomal (pa), connecting piece (cp), and midpiece segments (mp). Note the distinct band of particles (*) separating the acrosomal and postacrosomal domains. Plaque-like aggregates of particles (arrows) characterize the posterior margin of the postacrosomal domain. The posterior ring (pr) appears as a circumferential groove separating the head and tail domains. The plasma membrane of the midpiece displays a high density of intramembranous particles, some of which are arrayed as obliquely oriented chains.

migration from this domain during the acrosome reaction (Cowan *et al.*, 1987; Phelps *et al.*, 1988).

4. MIDPIECE PLASMA MEMBRANE

The midpiece domain contains an axoneme–outer dense fiber complex that is surrounded by a helically arranged mitochondrial sheath (Figs. 7 and 8). The plasma membrane overlies the mitochondria and is thought to function in regulating energy production and motility. Freeze-fracture replicas of cauda epididymal spermatozoa of several species, including the guinea pig, mouse, bat, and opossum, reveal ordered intramembranous particle arrays within the midpiece segment (Friend and Fawcett, 1974; Koehler and Gaddum-Rosse, 1975; Hoffman *et al.*, 1987; Olson, 1980; Stackpole and Devorkin, 1974). Generally, obliquely oriented linear aggregates of particles are superimposed over the mitochondria. Particle rows appear in the guinea pig during spermiogenesis at the time a close spatial relationship is established between the plasma membrane and mitochondria (Pelletier and Friend, 1983). In marsupial sperm the cytoplasmic face of the midpiece plasma membrane is coated with a prominent layer of electron-dense material that may act as a membrane-skeleton-like stabilizing infrastructure (Fig. 8); this complex forms during sperm maturation in the epididymis concurrently with the appearance of an ordered intramembranous particle distribution (Olson, 1980). It is likely that the interaction of this complex with the integral membrane proteins could function in generating the ordered particle pattern (Fig. 9). In the guinea pig a disruption of the organized particle arrangements is noted during sperm capacitation and may play a role in the alteration of motility pattern (Friend, 1977;

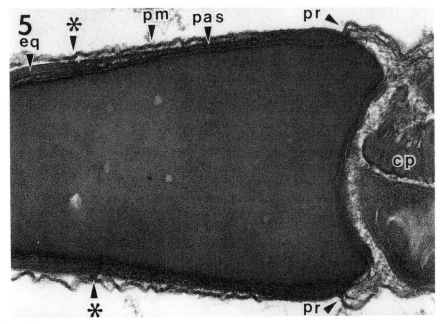

Figure 5. Thin section of the postacrosomal segment of a bat spermatozoon. Note that at the junction of the equatorial segment (eq) and postacrosomal segment (*), a plaque of electron-dense material joins the plasma membrane and outer acrosomal membrane. The postacrosomal sheath (pas) is a layer of electron-dense material associated with the plasma membrane (pm). At the posterior margin of the postacrosomal segment, the posterior ring (pr) is represented as a focal fusion of the plasma membrane with the underlying nuclear membrane.

Koehler and Gaddum-Rosse, 1975). Several surface polypeptides become associated with this domain during epididymal maturation. These may be subsequently modified during capacitation (Eddy *et al.*, 1985; Olson *et al.*, 1987; Vernon *et al.*, 1985), but their role in motility regulation is unresolved.

5. PRINCIPAL PIECE PLASMA MEMBRANE

The principal piece is the longest flagellar segment and consists of a central axoneme–outer dense fiber complex, the fibrous sheath, and a surrounding plasma membrane (Figs. 10–12). In freeze-fracture replicas the principal piece plasma membrane generally displays a random distribution of intramembranous particles, although occasional crystalline aggregates of parti-

Figure 6. Surface replica of the postacrosomal sheath showing it to be comprised of a parallel array of 10- to 12-nm-diameter filaments aligned parallel to the spermatozoons long axis (double arrow).

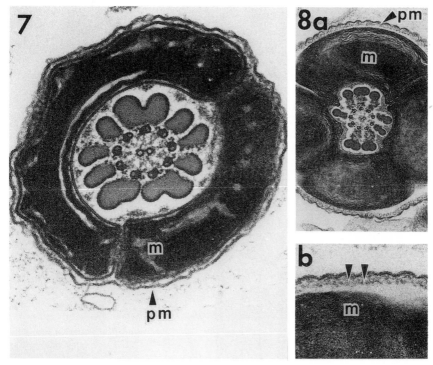

Figure 7.   Cross section of the hamster sperm midpiece showing the close relationship of the plasma membrane (pm) with the underlying mitochondria (m).

Figure 8.   Cross section of midpiece of marsupial sperm from the cauda epididymidis showing the fluted contour of the plasma membrane (pm, in a) and the membrane-skeleton-like complex (arrowheads in b) associated with the cytoplasmic surface of the membrane.

cles are seen (Fawcett, 1975). A characteristic differentiation of the principal piece plasma membrane is a linear aggregate, usually two particles wide, which has been termed the zipper (Friend and Fawcett, 1974). In thin sections the zipper is represented as a focal attachment of the membrane to the underlying fibrous sheath, and it is specifically positioned over doublet microtubule number 1 of the axoneme. The zipper resists detergent extraction and contains glycoproteins that bind the lectins concanavalin A, *Ricinus communis*, and wheat germ agglutinin (Enders *et al.*, 1983). A role for the zipper or other surface components of this segment in the regulation of flagellar motility has yet to be demonstrated.

6. ENDPIECE PLASMA MEMBRANE

The short terminal segment of the flagellum, consisting only of a 9 + 2 axoneme and a surrounding plasma membrane, provides a clear example of membrane–cytoskeleton interactions. Freeze-fracture replicas reveal linear particle aggregates representing sites where the membrane is linked to the underlying microtubules (Suzuki and Nagano, 1980b).

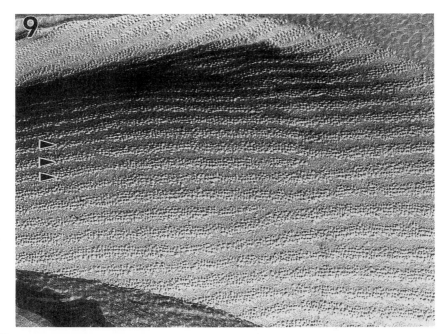

Figure 9. Freeze-fracture of marsupial sperm midpiece showing longitudinally oriented rows of intramembranous particles (arrowheads).

7. STRUCTURES AT DOMAIN INTERFACES

A poorly understood question is how unique domains of the sperm plasma membrane are maintained without substantial molecular intermixing. It has been demonstrated that some molecular components are mobile within a domain but do not diffuse to adjacent domains (Cowan *et al.*, 1987; Myles *et al.*, 1984). Structural studies have identified membrane differentiations at the acrosomal–postacrosomal interface, the postacrosomal–midpiece interface, and at the midpiece–principal piece interface, which could play a barrier function. At the acrosomal–postacrosomal segment freeze-fracture replicas reveal a circumferential band of ridges or particles (Fig. 4), and thin sections show a focal zone of adhesion between the plasma membrane and outer acrosomal membrane at this site (Friend and Fawcett, 1974; Hoffman *et al.*, 1987; Fig. 5).

At the head–tail junction the plasma membrane forms a circumferential focal zone of fusion with the underlying nuclear membrane (Figs. 4 and 5). This structure, termed the posterior ring, has been suggested to create a seal that separates the cytoplasmic compartments of the head and tail. In freeze-fracture it appears as a circumferential particle-free groove with a finely striated substructure (Fawcett, 1975; Koehler, 1966).

Finally, the plasma membrane at the junction of the midpiece and principal piece is adherent to an underlying ring of filaments termed the annulus (Figs. 10 and 11). By freeze-fracture the plasma membrane overlying the annulus possesses a high density of intramembranous particles that may be linked to the annulus. Although it remains to be demonstrated experimentally, it is likely that each of the membrane specializations at the domain boundaries described above could act as a physical barrier to impede movements of proteins between domains.

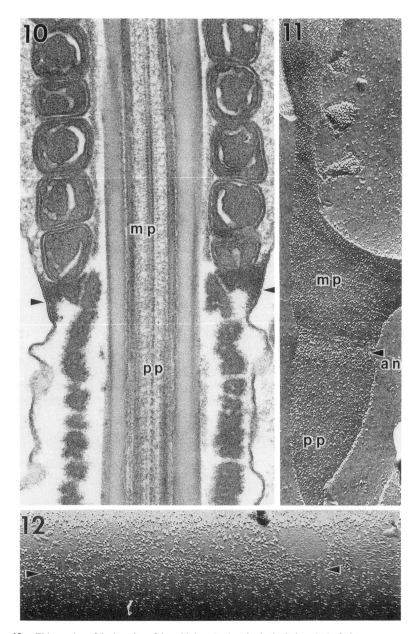

Figure 10.   Thin section of the junction of the midpiece (mp) and principal piece (pp) of a hamster spermatozoon showing the plasma membrane (arrowheads) adherent to the underlying annulus.

Figure 11.   Freeze-fracture replica of the junctional zone of the midpiece (mp) and principal piece (pp) of a bat spermatozoon showing the different particle arrangements in these two domains and the densely packed band of particles at the annulus (an).

Figure 12.   Freeze-fracture replica of the principal piece of a marsupial spermatozoon showing the linear particle aggregate (arrowheads) that characterizes this domain.

8. CONCLUDING COMMENTS

Different surface domains of the mammalian spermatozoon perform clearly defined, biologically distinct roles in fertilization. The geographic extent of the different domains precisely mirrors the underlying segmentation of the spermatozoon, suggesting that membrane interaction with cytoplasmic structures contributes to maintaining distinct domains. Domain-specific cytoskeletal structures and structural specialization at domain interfaces have been identified that may participate in this function. Even though the mosaic construction of the plasma membrane is established during spermiogenesis, it is further modified during posttesticular development. This involves changes in the interaction or assembly of cytoplasmic structures that bind the interior surface of the membrane and thereby modulate fluidity of integral protein components. Alternatively, polypeptides secreted by the excurrent duct system, accessory glands, or lining epithelium of the female tract may bind the exterior surface of selected segments. The end result of this remodeling is a functionally mature sperm whose plasma membrane can participate in the complex set of membrane-mediated interactions with the egg during fertilization.

9. REFERENCES

Benau, D. A., and Storey, B. T., 1987, Zona-binding site sensitive to trypsin inhibitors, *Biol. Reprod.* **32:**282–292.

Bleil, J. D., and Wassarman, P. M., 1986, Autoradiographic visualization of the mouse egg's sperm receptor bound to sperm, *J. Cell. Biol.* **102:**1363–1371.

Burridge, K., Fath, K., Kelly, T., Nuckolls, G., and Turner, C., 1988, Focal adhesions: Transmembrane junctions between the extracellular matrix and the cytoskeleton, *Annu. Rev. Cell Biol.* **4:**487–525.

Camatini, M., Anelli, G., and Casale, A., 1986, Identification of actin in boar spermatids and spermatozoa by immunoelectron microscopy, *Eur. J. Cell Biol.* **42:**311–318.

Clarke, G. N., Clarke, F. M., and Wilson, S., 1982, Actin in human spermatozoa, *Biol. Reprod.* **26:**319–327.

Cowan, A. E., Myles, D. G., and Koppel, D. E., 1987, Lateral diffusion of the PH-20 protein on guinea pig sperm: Evidence that barriers to diffusion maintain plasma membrane domains in mammalian sperm, *J. Cell Biol.* **104:**917–923.

Damjanov, I., Damjanov, A., Lehto, V.-P., and Virtanen, I., 1986, Spectrin in mouse gametogenesis and embryogenesis, *Dev. Biol* **114:**132–140.

Eddy, E. M., 1988, The spermatozoon, in: *The Physiology of Reproduction*, Vol. 1 (E. Knobil and J. D. Neill, eds.), Raven Press, New York, pp. 27–68.

Eddy, E. M., Vernon, R. B., Muller, C. H., Hahnel, A. C., and Fenderson, B. A., 1985, Immunodissection of sperm surface modifications during epididymal maturation, *Am. J. Anat.* **174:**225–238.

Enders, G. C., Werb, Z., and Friend, D. S., 1983, Lectin binding to guinea-pig sperm zipper particles, *J. Cell Sci.* **60:**303–329.

Endo, Y., Lee, M. A., and Kopf, G. S., 1988, Characterization of an islet-activating protein-sensitive site in mouse sperm that is involved in the zona pellucida-induced acrosome reaction, *Dev. Biol.* **129:**12–24.

Escalier, D., 1984, The cytoplasmic matrix of the human spermatozoon: Cross-filaments link the various cell compartments, *Biol. Cell* **51:**347–364.

Fawcett, D. W., 1975, The mammalian spermatozoon, *Dev. Biol.* **44:**394–436.

Flaherty, S. P., and Olson, G. E., 1988, Membrane domains in guinea pig sperm and their role in the membrane fusion events of the acrosome reaction, *Anat. Rec.* **220:**267–280.

Flaherty, S. P., Winfrey, V. P., and Olson, G. E., 1986, Localization of actin in mammalian spermatozoa: A comparison of eight species, *Anat. Rec.* **216:**504–515.

Flechon, J.-E., 1985, Sperm surface changes during the acrosome reaction as observed by freeze-fracture, *Am. J. Anat.* **174:**239–248.

Fournier-Delpech, S., Hamamah, S., Tananis-Anthony,C., Courot, M., and Orgebin-Crist, M.-C., 1984, Hormonal regulation of zona-binding ability and fertilizing ability of rat epididymal spermatozoa, *Gamete Res.* **9:**21–30.

Friend, D. S., 1977, The organization of the spermatozoal membrane, in: *Immunobiology of Gametes* (M. Edidin and M. H. Johnson, eds.), Cambridge University Press, Cambridge, pp. 5–30.

Friend, D. S., 1984, Membrane organization and differentiation in the guinea pig spermatozoon, in: *Ultrastructure of Reproduction* J. Van Blerkom and P. M. Motta, eds.), Martinus Nijhoff, Boston, pp. 75–85.

Friend, D. S., and Fawcett, D. W., 1974, Membrane differentiations in freeze-fractured mammalian sperm, *J. Cell Biol.* **63:**641–664.

Friend, D. S., Orci, L., Perrelet, A., and Yanagimachi, R., 1977, Membrane particle changes attending the acrosome reaction in guinea pig spermatozoa, *J. Cell Biol.* **74:**561–577.

Halenda, R. M., Primakoff, P., and Myles, D. G., 1987, Actin filaments, localized to the region of the developing acrosome during early stages, are lost during later stages of guinea pig spermiogenesis, *Biol. Reprod.* **36:** 491–499.

Hoffman, L. H., Wimsatt, W. A., and Olson, G. E., 1987, Plasma membrane structure of bat spermatozoa: Observations on epididymal and uterine spermatozoa in *Myotis lucifugus*, *Am. J. Anat.* **178:**326–334.

Holt, W. V., 1984, Membrane heterogeneity in the mammalian spermatozoon, *Int. Rev. Cytol.* **87:**159–194.

Koehler, J. K., 1966, Fine structure observations in frozen-etched bovine spermatozoa, *J. Ultrastruct. Res.* **16:** 359–375.

Koehler, J. K., 1978a, The mammalian sperm surface: Studies with specific labeling techniques, *Int. Rev. Cytol.* **54:**73–108.

Koehler, J. K., 1978b, Observations on the fine structure of vole spermatozoa with particular reference to cytoskeletal elements in the mature sperm head, *Gamete Res.* **1:**247–257.

Koehler, J. K., and Gaddum-Rosse, P., 1975, Media induced alterations of the membrane associated particles of the guinea pig sperm tail, *J. Ultrastruct. Res.* **51:**106–118.

Lakoski, K. A., Carron, D. P., Cabot, C. L., and Saling, P. M., 1988, Epididymal maturation and the acrosome reaction in mouse sperm: Response to zona pellucida develops coincident with modification of M42 antigen, *Biol. Reprod.* **38:**221–233.

Leyton, L., and Saling, P., 1989a, Evidence that aggregation of mouse sperm receptors by ZP3 triggers the acrosome reaction, *J. Cell Biol.* **108:**2163–2168.

Leyton, L., and Saling, P., 1989b, 95kd sperm proteins bind ZP3 and serve as tyrosine kinase substrates in response to zona binding, *Cell* **57:**1123–1130.

Longo, F. J., Krohne, G., and Franke, W., 1987, Basic proteins of the perinuclear theca of mammalian spermatozoa and spermatids: A novel class of cytoskeletal elements, *J. Cell Biol.* **105:**1105–1120.

Magargee, S., Kunze, E., and Hammerstedt, R. H., 1988, Changes in lectin-binding features of ram sperm surfaces associated with epididymal maturation and ejaculation, *Biol. Reprod.* **38:**667–685.

Marchesi, V. T., 1985, Stabilizing infrastructure of cell membranes, *Annu. Rev. Cell Biol.* **1:**531–561.

Meizel, S., 1985, Molecules that initiate or help stimulate the acrosome reaction by their interaction with the mammalian sperm surface, *Am. J. Anat.* **174:**285–302.

Myles, D. G., Primakoff, P., and Koppel, D. E., 1984, A localized surface protein of guinea pig sperm exhibits free diffusion in its domain, *J. Cell Biol.* **98:**1905–1909.

Ochs, D., Wolf, D. P., and Ochs, R. L., 1986, Intermediate filament proteins in human sperm heads, *Exp. Cell Res.* **167:**495–504.

Olson, G. E., 1980, Changes in intramembranous particle distribution in the plasma membrane of *Didelphis virginiana* spermatozoa during maturation in the epididymis, *Anat. Rec.* **197:**471–488.

Olson, G. E., and Winfrey, V. P., 1985, Substructure of a cytoskeletal complex associated with the hamster sperm acrosome, *J. Ultrastruct. Res.* **92:**167–179.

Olson, G. E., and Winfrey, V. P., 1988, Characterization of the postacrosomal sheath of bovine spermatozoa, *Gamete Res.* **20:**329–342.

Olson, G. E., Noland, T. D., Winfrey, V. P., and Garbers, D. L., 1983, Substructure of the postacrosomal sheath of bovine spermatozoa, *J. Ultrastruct. Res.* **85:**204–218.

Olson, G. E., Lifsics, M. R., Winfrey, V. P., and Rifkin, J. M., 1987, Modification of the rat sperm flagellar plasma membrane during maturation in the epididymis, *J. Androl.* **8:**129–147.

Olson, G. E., Winfrey, V. P., and Flaherty, S. P., 1990, Membrane–cytoskeleton interactions in the sperm acrosome, in: *Gamete Physiology* (R. Asch, J. Balmaceda, and J. Johnston, eds.), Serono Symposia, Newport Beach, California, pp. 109–118.

O'Rand, M. G., 1977, Restriction of a sperm surface antigen's mobility during capacitation, *Dev. Biol.* **55:** 260–270.

Parks, J. E., and Hammerstedt, R. H., 1985, Developmental changes occurring in the lipids of ram epididymal spermatozoa plasma membranes, *Biol. Reprod.* **32:**653–668.

Pelletier, R.-M., and Friend, D. S., 1983, Development of membrane differentiations in the guinea pig spermatid during spermiogenesis, *Am. J. Anat.* **167:**119–141.

Phelps, B. M., Primakoff, P., Koppel, D. E., Low, M. G., and Myles, D. G., 1988, Restricted lateral diffusion of PH-20, a PI-anchored sperm membrane protein, *Science* **240:**1780–1782.

Phillips, D. M., 1975, Cell surface structure of rodent sperm heads, *J. Exp. Zool.* **191:**1–8.

Reger, J. F., Fain-Maurel, M. A., and Dadoune, J.-P., 1985, A freeze-fracture study on epididymal and ejaculate spermatozoa of the monkey (*Macaca fascicularis*), *J. Submicrosc. Cytol.* **17:**49–56.

San Agustin, J. T., Hughes, P., and Lardy, H. A., 1987, Properties and function of caltrin, the calcium-transport inhibitor of bull seminal plasma, *FASEB J.* **1:**60–66.

Shur, B. D., and Neely, C. A., 1988, Plasma membrane association, and partial characterization of mouse sperm β1,4-galactosyltransferase, *J. Biol. Chem.* **263:**17706–17714.

Singer, S. J., and Nicolson, G. L., 1972, A fluid-mosaic model for the cell membrane, *Science* **175:**720–731.

Stackpole, C. W., and Devorkin, D., 1974, Membrane organization in mouse spermatozoa revealed by freeze-etching, *J. Ultrastruct. Res.* **49:**167–187.

Suzuki, F., 1981, Changes in intramembranous particle distribution in epididymal spermatozoa of the boar, *Anat. Rec.* **199:**361–376.

Suzuki, F., and Nagano, T., 1980a, Epididymal maturation of rat spermatozoa studies by thin sectioning and freeze-fracture, *Biol. Reprod.* **22:**1219–1231.

Suzuki, F., and Nagano, T., 1980b, Morphological relationship between the plasma membrane and the microtubules in the end piece of the boar spermatozoa, *J. Electron Microsc.* **29:**190–192.

Suzuki, F., and Yanagimachi, R., 1986, Membrane changes in Chinese hamster spermatozoa during epididymal maturation, *J. Ultrastruct. Mol. Struct. Res.* **96:**91–104.

Talbot, P., 1985, Sperm penetration through oocyte investments in mammals, *Am. J. Anat.* **174:**331–346.

Toshimori, K., Higashi, R., and Oura, C., 1985, Distribution of intramembranous particles and filipin–sterol complexes in mouse sperm membranes: Polyene antibiotic filipin treatment, *Am. J. Anat.* **174:**455–470.

Toshimori, K., Higashi, R., and Oura, C., 1987, Filipin-sterol complexes in golden hamster sperm membranes with special reference to epididymal maturation, *Cell Tissue Res.* **250:**673–680.

Toyama, Y., and Nagano, T., 1983, Boar spermatozoa observed by rapid-freeze and deep-etch method, *Anat. Rec.* **206:**171–179.

Toyama, Y., and Nagano, T., 1988, Maturation changes of the plasma membrane of rat spermatozoa observed by surface replica, rapid-freeze and deep-etch and freeze-fracture methods, *Anat. Rec.* **220:**43–50.

Vernon, R. B., Hamilton, M. S., and Eddy, E. M., 1985, Effects of *in vivo* and *in vitro* fertilization environments on the expression of a surface antigen of the mouse sperm tail, *Biol. Reprod.* **32:**669–680.

Virtanen, I., Badley, R. A., Paasivuo, R., and Lehto, V.-P., 1984, Distinct cytoskeletal domains revealed in sperm cells, *J. Cell Biol.* **99:**1083–1091.

Wassarman, P. M., 1987, The biology and chemistry of fertilization, *Science* **235:**553–560.

Wassarman, P. M., 1988, Zona pellucida glycoproteins, *Annu. Rev. Biochem.* **57:**415–442.

Welch, J. E., and O'Rand, M. G., 1985, Identification and distribution of actin in spermatogenic cells and spermatozoa of the rabbit, *Dev. Biol.* **109:**411–417.

Yanagimachi, R., 1988, Mammalian fertilization, in: *The Physiology of Reproduction*, Vol. 1 (E. Knobil and J. D. Neill, eds.), Raven Press, New York, pp. 135–186.

Yanagimachi, R., and Suzuki, F., 1985, A further study of lysolecithin-mediated acrosome reaction of guinea pig spermatozoa, *Gamete Res.* **11:**29–40.

4

Mammalian Sperm Acrosomal Enzymes and the Acrosome Reaction

L. J. D. Zaneveld and C. J. De Jonge

1. INTRODUCTION

The sperm acrosome is an intracellular membrane-limited organelle that surrounds the anterior portion of the nucleus. It consists of an inner acrosomal membrane that is closely associated with the nucleus and is continuous with the outer acrosomal membrane. The acrosomal matrix proper is located between the two membranes. The entire spermatozoon is covered by the plasma membrane. The acrosome has several characteristics in common with the lysosome: it contains a number of enzymes, and the internal milieu is normally acidic. Although all mammals possess an acrosome, significant size and shape variations exist.

The ovulated oocyte (egg) is surrounded by several vestments: an outer granulosa (follicle) cell layer and an inner, noncellular layer that consists of glycoproteins, the zona pellucida. The granulosa cell layer from a number of species can be subdivided into an outer cumulus oophorus and an inner corona radiata. Before being able to penetrate the oocyte layers, spermatozoa have to undergo a process called capacitation. Capacitation normally occurs in the female genital tract, but it can be induced *in vitro* using serum, albumin, or other exogenous factors. Capacitation involves biochemical changes in the outer sperm membranes that allow the spermatozoon to undergo a morphological change, the acrosome reaction. This reaction occurs as the spermatozoon migrates through the oocyte's vestments, before or after contacting the zona pellucida. Morphologically, the acrosome reaction occurs in several steps: (1) fusion of the outer acrosomal membrane with the overlying plasma membrane, (2) vesiculation and disappearance of the fused membranes, and (3) release of enzymes and other components contained within the acrosomal matrix. Acrosomal enzymes appear to have an essential role in the induction of the acrosome reaction as well as in sperm binding to and penetration through the layers surrounding the oocyte.

L. J. D. ZANEVELD • Department of Obstetrics and Gynecology, and Department of Biochemistry, Rush University, Rush-Presbyterian-St. Luke's Medical Center, Chicago, Illinois 60612.   C. J. DE JONGE • Department of Obstetrics and Gynecology, Rush University, Rush-Presbyterian-St. Luke's Medical Center, Chicago, Illinois 60612.

A Comparative Overview of Mammalian Fertilization, edited by Bonnie S. Dunbar and Michael G. O'Rand. Plenum Press, New York, 1991.

A large number of publications have appeared on the enzymes of the acrosome and their role in the capacitation/acrosome reaction process and oocyte penetration by spermatozoa, as summarized by a number of recent reviews (Bhattacharyya and Zaneveld, 1982; Rogers and Bentwood, 1982; Clegg, 1983; Monroy and Rosati, 1983; Chang, 1984; Hinrichsen-Kohane *et al.*, 1984; Meizel, 1984, 1985; Austin, 1985; Langlais and Roberts, 1985; Tesarik, 1986; Peterson *et al.*, 1987; Wasserman, 1987; Yanagimachi, 1988; Saling, 1991; Zaneveld *et al.*, 1991) as well as by some older ones (McRorie and Williams, 1974; Zaneveld and Polakoski, 1976; Morton, 1977; Bedford and Cooper, 1978; Stambaugh, 1978; Meizel, 1978; Green, 1978; Parrish and Polakoski, 1979). However, our knowledge of the properties and function of the acrosomal enzymes is still sketchy and controversial. The following is a brief summary of the acrosomal enzymes that have been characterized so far as well as an overview of the biochemical mechanism of the acrosome reaction as it is presently understood. Because of space limitations, only a relatively small number of references have been included, and the reader is referred to the reviews for further information.

2. ACROSOMAL ENZYMES

2.1. Proteinases and Peptidases

2.1.1. ACROSIN

The most widely studied acrosomal enzyme is the endoproteinase acrosin (for reviews, see McRorie and Williams, 1974; Zaneveld *et al.*, 1975; Fritz *et al.*, 1975; Polakoski and Zaneveld, 1976; Morton, 1977; Stambaugh, 1978; Bhattacharya and Zaneveld, 1982; Polakoski and Siegel, 1986). It has certain properties in common with pancreatic trypsin, including a histidine and serine at its active site. Acrosin has been found in all mammalian species studied so far. Although the properties of acrosin from different species are quite similar, quantitative differences are present in regard to the hydrolysis of certain substrates and the interaction with inhibitors. Molecular weight differences and variations in amino acid composition also exist. Depending on the acrosin form that was used to prepare the antibodies, immunologic cross-reactions can occur among species. Acrosin does not cross-react with trypsin.

Acrosin has maximal activity at about pH 8.0 and only cleaves arginine and lysine bonds, preferentially the former. The enzyme hydrolyzes proteins as well as synthetic ester and amide substrates typical for trypsin, although significant differences can exist in the Michaelis constants between acrosin and trypsin. The synthetic substrates most commonly used for the assay of acrosin are α-N-benzoyl-L-arginine ethyl ester (BAEE), α-N-benzoyl-D,L-arginine-β-naphthyl-amide (BANA), α-N-benzoyl-D,L-arginine-*p*-nitroanilide (BAPNA), and *p*-tosyl-L-arginine methyl ester (TAME). Acrosin does not react with substrates for chymotrypsin or exopeptidases.

With a few exceptions (see below), acrosin is inhibited by all the synthetic and natural inhibitors of trypsin, but large kinetic differences can be present between the two enzymes. Inhibitors of acrosin include the synthetic agents diisopropylfluorophosphate (DFP), tosyllysyl-chloromethylketone (TLCK), phenylmethylsulfonylfluoride (PMSF), 4-nitrophenyl-4'-guani-dinobenzoate (NPGB) and other aryl-4-guanidinobenzoates, benzamidine, and aminoben-zamidine; the naturally occurring pancreatic, serum, plant and microbial trypsin inhibitors; certain amidino and diamidino compounds; L-arginine; and cations such as Fe^{2+}, Fe^{3+}, Zn^{2+}, and Hg^{2+}. Calcium stabilizes acrosin and, in some situations, stimulates the activity of the enzyme. The activity of boar acrosin but not that of the human is stimulated by spermine and other polyamines. Monosaccharides, lectins, and sulfated polysaccharides inhibit acrosin but not trypsin because of the glycopeptide nature of acrosin in contrast to trypsin. Approximately 3% (ram), 8% (rabbit), or 10% (boar) of the acrosin molecule consists of carbohydrate.

In ejaculated spermatozoa, the majority of acrosin is present in an inactive form, called proacrosin: man, 93%; boar, 99%; dog, 79%; ram, 95%; guinea pig, 89%; mouse, 73%; and hamster, 92%. Proacrosin appears to be a single-chain polypeptide. Different molecular weight forms have been reported: 70–75 kDa and 45–55 kDa (human); 88 kDa, 64–66 kDa, and 53–55 kDa (boar); 54–56 kDa and 43 kDa (guinea pig); 68–73 kDa and 42.5 kDa (rabbit); and 51–60 kDa (ram). The variations in molecular weight in the same species may result from the method of molecular weight determination, the isolation technique, limited proteolysis, the absence/presence of a binding protein, and/or the presence of different proacrosin forms including a preproacrosin. Much higher molecular weights have also been reported as a result of aggregation. Conversion of isolated proacrosin to acrosin takes place spontaneously at neutral or basic pH within 5–30 min, depending on the species and method of preparation. Conversion appears to involve limited proteolysis through autocatalysis or by the presence of a small amount of acrosin. Activation of proacrosin is prevented at high ionic strength, by benzamidine and some other typical acrosin inhibitors, by polyamines, and by gossypol.

The initial form (m_α) of acrosin that is produced as a result of proacrosin activation can be converted to lower forms, two of which (m_β and m_γ) have been identified in the boar. The various molecular masses of acrosin reported for the boar include 49 kDa, 34–41 kDa and 25–28 kDa; human, 70–75 kDa, 49–55 kDa, and 30–40 kDa; hamster, 44 kDa; guinea pig, 48 kDa and 32–34 kDa; and rabbit, 68 kDa and 34 kDa. The m_β form of boar proacrosin may consist of two chains, a light chain of 4.2 kDa and a heavy chain of 37 kDa, linked by disulfide bonds. The active unit of hamster acrosin has a molecular mass of 8.4 kDa. The various forms of acrosin differ in their kinetic properties only to a relatively small extent, suggesting that the same active site sequence is present. The complete amino acid sequence of human and boar proacrosin has been reported, and the amino acid composition and the partial sequence of acrosin from a number of species have been determined. Acrosin is generally characterized by a high proline content. A majority (77%) of the amino acid sequences that are conserved in other serine proteinases also appear to be conserved in acrosin.

One or more inhibitors of acrosin are present on spermatozoa. After extraction and activation of proacrosin, approximately 90% of the acrosin is complexed to inhibitor. The acrosin inhibitors have molecular masses of 5.6 kDa and 10.1 kDa (human), 8–11.3 kDa (ram), and 6.8 kDa and 13.4 kDa (boar). Much-lower-molecular-weight inhibitors have also been reported. The 10.1-kDa human inhibitor is much more reactive toward trypsin than acrosin, and it is likely that only the 5.6-kDa inhibitor is of physiological importance.

Acrosin/proacrosin is associated with the sperm surface, the acrosomal matrix, and probably with the inner acrosomal membrane. Plasma membrane proacrosin/acrosin appears to be one of the zona receptors on the spermatozoa; i.e., it binds to one or more of the zona glycoproteins (galactosyltransferase is another receptor; see below). Good evidence is also available that acrosin is involved in the acrosome reaction and in sperm passage through the zona pellucida. Such evidence is primarily based on the ability of acrosin inhibitors or antibodies to prevent these events, on acrosin's capacity to lyse the zona and/or the zona glycoproteins, and on the presence of acrosin along the penetration slit through the zona. During the acrosome reaction, a significant portion of proacrosin is converted to acrosin, after which the acrosin is released. It is thought that the proacrosin remaining on the inner acrosomal membrane activates during the penetration of the spermatozoon through the zona pellucida. Conversion of proacrosin may occur autocatalytically or by active acrosin when the pH of the acrosome rises during the initial stages of the acrosome reaction (see below). It has been suggested that a thermolysin-like enzyme ("acrolysin") and/or cathepsin D is involved in proacrosin conversion, but a role for these enzymes remains to be established. The lower-molecular-weight (5–7 kDa) sperm acrosin inhibitor prevents the fertilizing capacity of spermatozoa but is completely or partially removed during the capacitation/acrosome reaction process.

2.1.2. OTHER PROTEINASES AND PEPTIDASES

Spermatozoa have been reported to contain a number of other proteinases besides acrosin, but their association with the acrosome remains to be established in many cases. Care has to be taken with the interpretation of the data until such localization experiments have been performed, because sperm samples may be contaminated with cellular and other debris that can be the source of such enzymes, the enzymes may originate from epididymal or seminal plasma and adhere to spermatozoa, or the enzymes may be associated with the sperm tail. All results obtained to date indicate that acrosin is by far the predominant proteinase associated with the sperm acrosome.

Several enzymes with properties similar to acrosin have been identified. These enzymes may represent one of the acrosin forms or may be unique. Bull spermatozoa possess a proteinase that activates adenylate cyclase, is trypsin-like, and is inhibited by the same agents that inhibit acrosin (Johnson et al., 1985). Human spermatozoa possess a zymogen, called sperminogen, with molecular forms varying from 32 to 36 kDa (Siegel et al., 1987). Sperminogen does not cross-react with a proacrosin antibody. At neutral or basic pH, sperminogen is activated to spermin, which is trypsin-like, hydrolyzes the same substrates as acrosin, and is inhibited by the same agents. Several hours are required for the activation to occur, in contrast to proacrosin.

Bull, rat, boar, and human spermatozoa possess one or more enzymes, called collagenase-like peptidase or Pz-peptidase, with a molecular mass of approximately 110 kDa that hydrolyze 4-phenylazo-Cbz-Pro-Leu-Gly-Pro-D-Arg, producing Pz-Pro-Leu and Pz-Pro. The enzymes are not active against a number of protein substrates such as hemoglobin and serum albumin but do digest gelatin. Two peptidases were identified in bull spermatozoa, Pz-peptidase A and B (Lessley and Garner, 1984). Pz-Peptidase A is inhibited by tosylphenylethylchloromethylketone (TPCK) but not by phosphoramidon and has a pH optimum of approximately 6.0. By contrast, Pz-peptidase B is inhibited by phosphoramidon but not by TPCK and has a pH optimum near 7.0. The activity of Pz-peptidase B is particularly susceptible to detergents. Spermatozoa are the main source of Pz-peptidase B, and the enzyme appears to be associated with the acrosomal region. Bull Pz-peptidase B has many properties in common with a metalloendoproteinase obtained from human spermatozoa (Gottlieb and Meizel, 1987). This metalloendoproteinase was originally thought to be involved in the acrosome reaction, but its role is now in question. Pz-Peptidase B also appears to be the thermolysin-like enzyme ("acrolysin") described by McRorie et al. (1976) that was suggested to stimulate the conversion of proacrosin to acrosin (see Section 2.1.1).

Several sperm proteinases with optimal activity at acid pH against casein, hemoglobin, and other protein substrates have been found in a number of species (for review, see Zaneveld et al., 1975; Morton, 1977). An acid proteinase from boar spermatozoa was shown to have pH optima of 2.8 and 3.5 and a molecular mass of more than 150 kDa. It is inhibited by iodoacetate but not by DFP or ethylenediaminetetraacetic acid (EDTA). Mouse spermatozoa possess an acid proteinase with pH optimum between 3.5 and 5.0 that is inhibited by pepstatin but not by typical trypsin inhibitors. The molecular mass of the enzyme is approximately 40 kDa, and it may be cathepsin D. Cathepsins B and D have also been isolated from rabbit testis. Cathepsin D can catalyze the conversion of proacrosin to acrosin (Srivastava and Ninjoor, 1982). A cysteine proteinase, identified as cathepsin L, with optimal pH of 5.5 has been found in guinea pig spermatozoa (McDonald and Kadhodyan, 1988). The enzyme is present in a latent form that can be unmasked by incubating the spermatozoa at pH 3.5. The proteinase hydrolyzes typical substrates of cathepsin L, is inhibited by thiol reagents, leupepsin, Z-Phe-Phe-CHN$_2$, and gossypol, but not by pepstatin and serine proteinase inhibitors. The acrosome of guinea pig spermatozoa also possesses dipeptidyl peptidase II (Talbot and Dicarlantonio, 1985). This enzyme is an exopeptidase with optimal activity at pH 4.5–5.5 and a molecular mass of 130 kDa. A dipeptidyl carboxypeptidase (angiotensin-converting enzyme, kininase II) of 140 kDa is also associated with the boar sperm acrosome (Yotsumoto et al., 1984).

The neutral proteinase, calpain II, has been localized in the acrosome of boar spermatozoa. The enzyme has a molecular weight of 80,000 and is activated by Ca^{2+} (Schollmeier, 1986). A chymotrypsin-like enzyme was found to be associated with rabbit spermatozoa, and a plasminogen activator with mouse spermatozoa, but these enzymes are probably derived from seminal plasma. Finally, proteolytic activity at a very basic pH (9 or above) was shown to be associated with rabbit, human, bull, ram, and mouse spermatozoa. At least in the rabbit and ram, this enzyme activity is not inhibited by typical acrosin inhibiors, and in the mouse, the activity is associated with a protein of 150 kDa or more. Alkaline proteinases have also been found in the rabbit testis.

2.2. Glycosidases

A number of glycosidases are associated with the sperm acrosome, the best known being hyaluronidase. This enzyme has been found in the spermatozoa from all mammalian species studied. However, large variations exist in hyaluronidase activity. It is highest in buffalo, ram, rabbit, bull, and opossum spermatozoa; less is present in boar, rat, mouse, and human spermatozoa; very little can be found in dog spermatozoa; and the enzyme is almost absent from stallion spermatozoa. The enzyme is primarily associated with the outer acrosomal and plasma membranes of the sperm head and is readily released from viable spermatozoa. The cumulus oophorous cells surrounding the oocyte are held together by a hyaluronic acid matrix, and hyaluronidase is known to have an essential role in sperm penetration through that layer. A putative role (together with acrosin) in sperm passage through the zona pellucida has also been proposed, but good evidence is lacking.

Bull, rabbit, and ram sperm hyaluronidase have been characterized to some extent (Table 1). The properties of hyaluronidase from spermatozoa are identical to those of testicular hyaluronidase but differ from lysosomal hyaluronidase of non-genital-tract tissues. Both bull and rabbit sperm hyaluronidases require salt for activity and are denatured at temperatures above 50°C. Besides the inhibitory agents listed in Table 1, the enzyme is also inhibited by myocrisin (aurothiomalate), fenoprofen, phenylbutazone, and oxyphenbutazone. The enzyme may occur as a dimer, since molecular masses of 110 kDa and 62 kDa have been found under, respectively, nonreducing and reducing conditions. Although many similarities are present among the hyaluronidases from different species, some differences exist. For instance, neither bull nor rabbit sperm hyaluronidase is degraded significantly by cryopreservation or refrigeration, but storage of human spermatozoa can result in enzyme inactivation (Joyce et al., 1985).

Other glycosidases shown to be associated with the sperm acrosome are neuraminidase, β-N-acetylhexosaminidase, α-L-fucosidase, β-glucuronidase, β-glucosidase, and β-galactosidase. The properties of the first two enzymes have been determined, although it should be noted that the source of the enzyme was occasionally seminal plasma rather than spermatozoa (Table 1). Some or all of these glycosidases have been shown to be present in rabbit, ram, rat, bull, and human spermatozoa. β-Glucuronidase and β-N-acetylhexosaminidase can further hydrolyze the hyaluronidase digestion products of hyaluronic acid to monomeric units, and β-N-acetylhexosaminidase inhibitors can prevent cumulus dispersal by sperm extracts. It has been suggested that these enzymes aid hyaluronidase in sperm penetration through the cumulus oophorus, but good evidence is lacking, particularly since it is known that β-glucuronidase and β-N-acetylglucosaminidase inhibitors do not prevent the fertilization of mouse gametes, in contrast to hyaluronidase inhibitors. Similarly, no convincing evidence is available that these enzymes aid acrosin in the penetration of spermatozoa through the zona pellucida, as has been proposed, although they can partially digest the carbohydrate moiety of zona glycoproteins. Treatment of the zona pellucida with sperm neuraminidase causes a hardening of the zona, i.e., makes it less

Table 1. Glycosidases

Enzyme	Species	Properties
Hyaluronidase	Bull[a,b]	pH optimum 3.75; M_r 110,000 (sephadex), 62,000 (SDS electrophoresis); inhibited by Fe^{2+}, Fe^{3+}, heparin, phosphorylated hesperidin; activated by albumin, histones, protamine sulfate, hyamine 2389, spermine, spermidine, cations
	Rabbit[a]	pH optimum 3.75; inhibited by Fe^{2+}, Fe^{3+}, heparin; activated by histones
	Ram[c]	pH optimum 4.3; M_r 62,000 (SDS electrophoresis); inhibited by heparin, chondroitin sulfates A, B, and C
Neuraminidase	Rabbit[d]	Specific for $2 \rightarrow 6'$-linked sialic acid in glycoproteins; pH optimum 5.0 with Cowper's gland mucin and 4.3 with sialyllactose; inhibited by Ca^{2+}, Mg^{2+}, Mn^{2+}, Co^{2+}, and Cu^{2+}; stable to freeze-drying
β-N-Acetylhexosaminidase	Bull[e]	pH optimum 4.5; M_r 190,000; pI 7.96; inhibited by Hg^{2+}, and p-chloromercuribenzoate, and iodoacetamide
	Rabbit[f]	pH optimum 4.0; inhibited by Ag^+, Hg^{2+}, and p-chloromercuribenzoate but not by glutathione; in combination with arylsulfatase, causes swelling of the rabbit zona pellucida

[a]Zaneveld et al. (1973).
[b]Yang and Srivastava (1975).
[c]Yang and Srivastava (1974).
[d]Srivastava and Abou-Issa (1977).
[e]Khar and Anand (1977).
[f]Farooqui and Srivastava (1980).

susceptible to proteinase digestion, and the enzyme may be involved in the zona reaction which constitutes the block to polyspermy.

2.3. Lipases

The sperm acrosome possesses phospholipase A_2 and C. Characteristics of these enzymes are presented in Table 2. Phospholipase A_2 has been found in mouse, hamster, human, and guinea pig spermatozoa. The enzyme hydrolyzes phosphatidylcholine and produces cis-unsaturated fatty acids and lysophospholipids that can act as membrane fusogens. Inhibition by p-bromophenacylbromide indicates that phospholipase A_2 has a histidine at the active site. Phospholipase C has been found in bull, rabbit, and human spermatozoa. Human sperm phospholipase C appears to be specific for phosphoinositides, producing diacylglycerol and inositol phosphates, whereas the bull and rabbit enzyme is phophatidylcholine specific, producing phosphorylcholine (Table 2). Similar to sperm phospholipase A_2, the activity of testicular phospholipase C is stimulated by serine proteinases including acrosin, suggesting the presence of precursor (proenzyme, zymogen) forms of these enzymes. Both phospholipase A_2 and C are activated during the acrosome reaction, and both may have a role in this process (see Section 3).

Phospholipase A_1 appears to be absent from spermatozoa or is present in only very low concentrations. A lysophospholipase may also be present. Based on the ability of cyclooxygenase and lipoxygenase products to induce the acrosome reaction and of cyclooxygenase inhibitors to prevent this reaction (see Section 3), it is likely that the acrosome possesses cyclooxygenase and lipoxygenase, but these enzymes have not as yet been isolated from this organelle and characterized.

2.4. Phosphatases

Several phosphatases are associated with the sperm acrosome, including acid and alkaline phosphatase and ATPases (Table 3). Of these, only the acid phosphatase of rabbit spermatozoa

Table 2. Phospholipases

Enzymes	Species	Properties
Phospholipase A_2	Hamster[a]	Optimal pH range 7.4–7.8; stimulated by Ca^{2+}; inhibited by p-bromophenacylbromide and mepacrine
	Human[b]	pH optimum 7.5; catalyzes the release of fatty acids from the 2 position of phospholipids; Ca^{2+} dependent; inhibited by p-bromophenacylbromide, elevated ionic strength, Zn^{2+}, and Mn^{2+}; biphasic inhibitory effects of trifluoperazine and methoxyverapamil; quinacrine does or does not inhibit; no effect by Cu^{2+}, Cd^{2+}, Mg^{2+}, Sr^{2+}, or mepacrine; heat stable; DMSO has no effect or stimulates; two active enzyme forms; activity stimulated by trypsin and acrosin, suggesting the presence of a proenzyme
	Mouse[c]	pH optimum 8.0, specific for the 2 position of phospholipids, Ca^{2+} dependent; inhibited by Zn^{2+}, indomethacin, meclofenamate, mepacrine, p-bromophenacylbromide, EGTA; no effect by Cu^{2+}, Cd^{2+}, dexamethasone, or aspirin
Phospholipase C	Human[d]	pH optimum 6.0; phosphoinositide specific; Ca^{2+}-dependent; inhibited by EGTA
	Bull[d]	pH optimum 7.2; pI 5.0; hydrolyzes phosphatidylcholine but not phosphoinositide; M_r 69,000 and 55,000 (two subunits); inhibited by EDTA, Cd^{2+}, Pb^{2+}, Ni^{2+}, Fe^{2+}, and Zn^{2+}

[a]Llanos et al. (1982).
[b]Thakkar et al. (1984); Langlais and Roberts (1985); Anderson et al. (1990); Guerette et al. (1988).
[c]Thakkar et al. (1983).
[d]Ribbes et al. (1987).

has been obtained in a highly purified form. This enzyme has also been detected in ram, bull, rat, and human spermatozoa. The acid phosphatase of rabbit spermatozoa exists in at least five multiple forms; however, only two of these forms (S_4 and S_{zn}) are associated with the acrosome. The different forms are distinguished by the effects of NaF, $ZnCl_2$, L-tartaric acid, and creatine phosphate on the hydrolysis of α-naphthyl phosphate by the enzyme at pH 5.0. The function of alkaline and acid phosphatase is unknown.

Sperm membranes can contain Mg^{2+}-ATPase, Ca^{2+}-ATPase, Ca^{2+}, Mg^{2+}-ATPase, and Na^+,K^+-ATPase (Table 3). It remains to be established that the four enzymes are associated with the sperm head membranes of all mammalian species. The enzymes may be located at different sites, e.g., the Ca^{2+}-ATPase may be associated with the outer acrosomal membrane and the Na^+,K^+-ATPase with the plasmalemma. Detailed characterization studies have not been performed with any of the ATPases. The ATPases have generated some interest recently because of their potential role in the acrosome reaction (see Section 3).

2.5. Other Enzymes

A number of other enzymes are associated with the sperm acrosome or the sperm head (Table 4). A total of 11 nonspecific esterases are associated with the bull sperm head based on the hydrolysis of naphthol esters of acetate, butyrate, and caproate (Meizel et al., 1971). One of these enzymes is acetylcholinesterase. The rat caproyl esterase has been partially characterized, although its source was the testis (Table 4). The isolated enzyme disperses the cumulus cells from the mouse oocyte. An esterase (corona-penetrating esterase, CPE) that disperses the corona radiata from rabbit and cat oocytes (Table 4) has been extracted from rabbit and human sperm acrosomes. The enzyme has been purified from the testis and appears to be an arylesterase.

Table 3. Phosphatases

Enzyme	Species[a]	Properties
Mg^{2+}-ATPase	Hamster (1)	Inhibited by Ca^{2+}, N,N'-dicyclohexylcarbodiimide, and NBD; not stimulated by ouabain or cAMP
	Guinea pig (2)	Inhibited by Ca^{2+}, N-ethylmaleimide, and p-chloromer-curibenzoate
	Boar (3,4)	pH optimum 8.5; inhibited by p-cholormercuriphenylsulfo-nate and quercetin but not by NBD or Ca^{2+}
Ca^{2+},Mg^{2+}-ATPase	Boar (4)	Activated by Ca^{2+}; pH optimum 9.5; not inhibited by p-chloromercuriphenylsulfonate and quercitin or by the calmodulin antagonist trifluoperazine
	Human (5)	Inhibited by 2,4-dinitrophenol but not ouabain
	Ram (6)	Activated by Ca^{2+}; inhibited by quercetin
Ca^{2+}-ATPase	Man, rabbit (7)	pH optimum 9.0; inactive at pH 7.0; not inhibited by oubain
	Guinea pig (8)	pH optimum 9.0; not inhibited by p-chloromercuribenzoate or N-ethylmaleimide
	Bull, hamster (9)	pH optimum 7.4; inhibited by 1-anilino-β-naphthalene sulfo-nate, 2-p-toluidinyl-6-naphthalene sulfonate, N,N'-dicyclohexylcarbodiimide, and 5,5'-dithiobis-(2-nitrobenzoic acid); not effected by ouabain; not stimulated by Na+, K+, or Mg^{2+}
Na^+,K^+-ATPase	Rabbit (7)	Active at pH 7.0; requires no divalent ions; inhibited by oua-bain
	Hamster (10)	Inhibited by taurine, hypotaurine, and β-alanine.
	Human (5)	Inhibited by ouabain but not 2,4-dinitrophenol
Acid phosphatase	Human (11,12,13)	pH optimum 3.8–5.6; M_r 50,000–80,000; inhibited by p-chloromercuribenzoate
	Rabbit (11,14)	Two forms (S_4 and S_{zn}); inhibited by NaF and creatine phos-phate
	Ram (15)	pH optimum 4.5
	Bull (11,16)	Inhibited by NaF

[a]References: (1) Working and Meizel (1982); (2) Usui and Yanagimachi (1986); (3) Ashraf *et al*. (1982); (4) Ashraf *et al*. (1984); (5) Abla *et al*. (1974); (6) Breitbart *et al*. (1983); (7) Gordon (1973); (8) Gordon and Barnett (1967); (9) Vijayasarathy *et al*. (1980); (10) Mrsny and Meizel (1985); (11) Bernstein and Teichman (1973); (12) Singer *et al*. (1980); (13) Kipping (1970); (14) Gonzales and Meizel (1973); (15) Allison and Hartree (1970); (16) Teichman and Bernstein (1971).

Corona radiata dispersal by sperm extracts is inhibited by a high-molecular-weight fraction from seminal plasma containing the glycoprotein often referred to as decapacitation factor (DF). Decapacitation factor is present on ejaculated spermatozoa and prevents their fertilizing ability until it is removed during the capacitation process. Evidence that CPE has a role in sperm penetration through the corona radiata was strengthened when it was found that the inhibition of CPE's esterase activity by DF paralleled the ability of DF to prevent corona dispersal by CPE as well as the *in vitro* fertilizing capacity of spermatozoa.

Aspartyl amidase is associated with the human, boar, ram, and squirrel monkey acrosome (Table 4). The acrosome also contains arylsulfatase (Table 4). This sulfatase causes the dispersal of the cumulus oophorus. Both caproyl esterase (see above) and aryl sulfatase have been suggested to aid hyaluronidase in the lysis of the cumulus during sperm penetration, but no convincing evidence is available. A putative role for aryl sulfatase in zona penetration by spermatozoa has also been proposed. Ornithine decarboxylase, the rate-limiting enzyme in the biosynthetic pathway for polyamines, has been found in the acrosomal region of rat spermatozoa (Qian *et al*., 1985).

Galactosyltransferase of the mouse sperm head (Table 4) is an externally oriented compo-

Table 4. Other Enzymes Associated with the Sperm Acrosome

Enzyme	Species[a]	Properties
Caproyl esterase	Rat (1)	pH optimum 5.2; pI 5.1; M_r 57,000–62,000; inhibited by cysteine, dithiothreitol, mercaptoethanol, phenylmethyl sulfonyl fluoride, aurothiomaleate; stimulated by low concn. of Ca^{2+}, Hg, Fe^{2+}, Zn^{2+}, Cu^{2+}, and eserine; disperses the cumulus from mouse oocytes.
Corona-penetrating esterase	Man, rabbit (2,3,4)	Unstable at $-20°C$; stable at 70°C; hydrolyzes p-nitrophenyl acetate at pH 6.5; disperses the corona radiata from rabbit and rat oocytes; inhibited by decapacitation factor (DF)
β-Aspartyl-N-acetylglucosamine amido hydrolase (aspartyl amidase)	Man (5)	pH optimum 7.8; temperature optumum 70°C
Aryl sulfatase	Rabbit (6)	Present as isoenzymes (A and B forms); pH optima 4.8, 5.6, 6.0; temperature optimum 50°C; inhibited by phosphate, sulfite, decapacitation factor (DF); shows higher affinity for p-nitrocatechol sulfate than p-nitrophenyl sulfate; disperses the cumulus from rabbit oocytes.
	Ram (7)	pH optimum 4.9; both A and B forms are present.
	Rat (8)	pH optimum 5.2.
	Boar (9)	The A form was purified; pH optimum 4.2; M_r 65,000; unstable to storage; inhibited by Ag^+; disperses the cumulus from hamster, rabbit, and pig oocytes; no effect on the zona pellucida
β-1,4-Galactosyltransferase	Mouse (10)	Optimal activity requires Mn^{2+}; peak activity at 25°C, activity decreases rapidly above 37°C; inhibited by gastric mucin, N-acetyllactosamine, GM_2 ganglioside, and glucosylceramide; inhibits sperm binding to the zona pellucida

[a]References: (1) Waibel *et al.* (1985); (2) Zaneveld and Williams (1970); (3) Bradford *et al.* (1976a); (4) Bradford *et al.* (1976b); (5) Bhalla *et al.* (1973); (6) Yang *et al.* (1974); (7) Allison and Hartree (1970); (8) Seiguer and Castro (1972); (9) Dudkowicz (1984); (10) Shur and Neely (1988).

nent of the plasma membrane overlying the acrosomal region. The enzyme becomes exposed during capacitation by the removal of high-molecular-weight polylactosaminyl glycosides. These compounds originate from male genital tract fluids and can act as "decapacitation factors" (Shur and Hall, 1982). The enzyme binds noncatalytically to its glycoside substrate in the mouse zona pellucida. Inhibitor studies strongly imply that galactosyltransferase functions as one of the zona receptors on the sperm surface, i.e., in mouse sperm binding to the zona pellucida (proacrosin/ acrosin is another receptor; see above). Some argument is present whether galactosyltransferase functions in this capacity in other species.

Inhibitors and activators of the second messenger system utilizing cAMP respectively inhibit and stimulate the acrosome reaction (see Section 3). These data imply that the enzymes involved in this system, such as adenylate cyclase, phosphodiesterase, and protein kinases, are associated with the acrosome and/or outer sperm head membranes (Hoskins and Casillas, 1975; Garbers and Kopf, 1980). These enzyme systems are also present in the sperm midpiece. The adenylate cyclase from spermatozoa differs from that of somatic cells since it is stimulated by Mn^{2+} rather than Mg^{2+} and since it does not respond to hormones, forskolin, fluoride ion, guanidine nucleotides, and cholera toxin. However, no characterization studies have been performed as yet specifically with the acrosomal enzyme.

3. ACROSOME REACTION

3.1. Surface-Associated Changes

The biochemical events leading up to the acrosome reaction are collectively called capacitation and primarily involve changes in the sperm membrane system. The sperm membrane is a complex, mosaic structure of heterogeneous protein and lipid domains. Capacitation involves the mobilization and/or removal of certain surface components from the sperm plasma membrane such as the following glycoproteins: decapacitation factor (DF), acrosome-stabilizing factor, and acrosin inhibitor. Subsequently, an increase in membrane fluidity and permeability occurs. These events are followed by or are simultaneous with changes in the lipid composition of the sperm membrane, resulting in (1) a decrease in the net negative surface charge, (2) the formation of specialized areas that are devoid of intramembranous proteins and sterols, and (3) increased concentrations of anionic phospholipids (for review, see Langlais and Roberts, 1985). These specialized areas are thought to be the sites where fusion and vesiculation occur during the acrosome reaction (Bearer and Friend, 1982). Thus, capacitation appears to be a phenomenon that involves protein and lipid alterations that modify the sperm plasma membrane to enable the spermatozoa to undergo the acrosome reaction. The following is a brief review of the enzymes known to be involved in the acrosome reaction. For more detail, see Meizel (1984, 1985) and Zaneveld *et al.* (1991).

3.2. Ions and ATPase

Lipid–protein interactions can cause changes in membrane permeability, which, under normal conditions, maintains an electrochemical balance. As a consequence, a host of ion-dependent (e.g., Ca^{2+}, Na^+, K^+) reactions are regulated. Ion transport, specifically Ca^{2+} influx, is essential for exocytosis of most somatic cells. Exocytosis is in many ways analogous to the acrosome reaction, since both processes involve cell membrane events resulting in the release of enzymes and other agents. The movement of calcium from the extracellular to intracellular space is also critical for the acrosome reaction. In rodent species, spermatozoa incubated in calcium-free medium become capacitated but are unable to undergo the acrosome reaction or fertilize eggs until exogenous calcium is added. Capacitated human spermatozoa require treatment with calcium as well as a calcium-transporting agent (e.g., calcium ionophore) before the acrosome reaction takes place. Although a change in membrane permeability to calcium appears to be the primary signal for the start of the acrosome reaction after capacitation, the stimulus for this change remains to be firmly established. Physiologically, one of the factors appears to be a zona glycoprotein.

Alterations in plasma membrane permeability to other ions also occur toward the end of the capacitation process and/or at the initiation of the acrosome reaction. Shapiro *et al.* (1985) have formulated a parallel bidirectional pathway in ion flux for the acrosome reaction in sea urchin spermatozoa. The first pathway is Ca^{2+} independent and involves an increase in intracellular pH via Na^+ influx and H^+ efflux. The second pathway involves a Ca^{2+}-dependent membrane depolarization. These two reactions combined result in the opening of Ca^{2+} channels and a large influx of extracellular calcium. However, with the exception of the Ca^{2+} influx, the sequence of ion exchanges during the fusion reaction in vertebrates remains unclear.

The massive increase in intracellular Ca^{2+} is believed to be ultimately responsible for the fusion between the plasma and outer acrosomal membranes, but the mechanism remains to be established. In somatic cells, calmodulin functions as a Ca^{2+} receptor and mediates most Ca^{2+}-regulated processes, including the activation of Ca^{2+}/calmodulin-dependent protein kinases (Ca-kinases), which phosphorylate proteins. Although calmoldulin has been identified in spermatozoa, its role in the acrosome reaction remains controversial (for review, see Garbers and Kopf, 1980). The requirement for a large Ca^{2+} influx as a stimulant for the acrosome reaction can be bypassed by the use of cAMP analogues, as preliminary data from our laboratory indicate. These results imply that one of the effects of Ca^{2+} influx is the stimulation of the adenylate cyclase system (see Section 3.3). In somatic cells it is known that Ca^{2+}–calmodulin complexes can bind to and regulate enzymes that break down and make cAMP.

Ion movement via ATPases has been demonstrated in spermatozoa from various mammalian species, and these enzymes may have a role in the acrosome reaction (for review, see Meizel, 1984). The properties of sperm ATPases are summarized in Section 2. The Ca^{2+}-ATPase and/or Ca^{2+}, Mg^{2+}-ATPase may function as a calcium pump, keeping the acrosomal calcium levels low until the activity of the pump is either overwhelmed or inhibited at the initiation of the acrosome reaction. There is evidence in hamster spermatozoa that during the later stage of capacitation, just prior to induction of the acrosome reaction, an increase in intraacrosomal pH occurs (the intraacrosomal pH is normally acidic; pH ≥ 5.0). A Mg^{2+}-ATPase may function as a proton pump to maintain elevated H^+ levels in the acrosome during capacitation. Inhibition of enzyme activity at the onset of the acrosome reaction would allow for an increase in acrosomal pH. The increase in intraacrosomal pH may in turn result in an increase in membrane permeability to Ca^{2+} via Ca^{2+}–H^+ exchange. An increase in acrosomal pH, however, does not appear to be an absolute requirement for the acrosome reaction.

Preincubation of rat spermatozoa in medium with a high K^+/Na^+ ratio enhances the fertilization rate. Incubation of hamster spermatozoa under *in vitro* capacitating conditions results in an increase in Na^+, K^+-ATPase activity. These data, in conjunction with evidence showing inhibition of capacitation and the acrosome reaction in hamster spermatozoa by ouabain (a Na^+, K^+-ATPase inhibitor), suggest a role for Na^+, K^+-ATPase activity in events leading up to and including the acrosome reaction (for review, see Meizel, 1984).

3.3. Second Messenger Systems

The transduction of a signal across the cell membrane often results in the receptor-mediated activation of an enzyme system that produces a second messenger. This process generally involves a GTP-binding regulatory protein (G-protein) intermediate. Two common types of second messenger systems have been identified, one depending on the generation of cAMP as a second messenger via the adeylate cyclase system, and the other involving the turnover of inositol phospholipids (phosphoinositides) (Berridge, 1985). Ultimately, the second messengers can induce the activation of protein kinases, the release of internally stored calcium, the formation of arachidonic acid (see below), and/or have some other stimulatory or inhibitory effects. Evidence is accumulating that these two second messenger systems are functional in the acrosome reaction (De Jonge *et al.*, 1991a, 1991b). The role of G-proteins is still argued, although an inhibitory G-protein (G_i) has been identified in spermatozoa (Kopf, 1988) and evidence is accumulating for a role of a stimulatory G-protein (Gs) as well (for a review, see Zaneveld *et al.*, 1991).

Possibly the best description of adenylate cyclase participation in prefertilization events has been elucidated in sea urchin spermatozoa. A substance isolated from components of the egg not only elevates cAMP concentrations and cAMP-dependent protein kinase activity but also causes induction of the acrosome reaction. The effects are entirely Ca^{2+} dependent. In vertebrates, it has been established that an increase in cAMP concentrations occurs toward the end of the capacitation period and/or at the initiation of the acrosome reaction (for review see Hoskins and Casillas, 1975; Garbers and Kopf, 1980). The addition of exogenous cAMP, cAMP analogues, or phosphodiesterase inhibitors (which elevate intracellular cAMP levels) to spermatozoa incubated *in vitro* enhances and/or induces capacitation as well as the acrosome reaction in rodent species. Furthermore, follicular fluid, which stimulates the acrosome reaction in many species, causes an increase in adenylate cyclase activity in porcine spermatozoa. Analogous to the sea urchin, calcium influx as well as phosphodiesterase-inhibitor-stimulated calcium transport enhances adenylate cyclase activity, resulting in an increase in cAMP in spermatozoa from some rodent species (Kopf and Vacquier, 1985; Monks *et al.*, 1986). Stimulation of the acrosome reaction of capacitated human spermatozoa was recently achieved in our laboratory by the addition of forskolin (an adenylate cyclase stimulator), cAMP analogues, or phosphodiesterase inhibitors to capacitated human spermatozoa (De Jonge *et al.*, 1991a). Additionally, the acrosome reaction was prevented by inhibitors of adenylate cyclase. These data provide reasonable evidence that the adenylate cyclase/cAMP second messenger system functions in the acrosome reaction.

The target of cAMP is cAMP-dependent protein kinase (protein kinase A). Recently, it was found in our laboratory that protein kinase A inhibitor prevents the dibutyryl-cAMP-induced acrosome reaction of capacitated human spermatozoa (De Jonge *et al.*, 1991a). If a role for this kinase in the acrosome reaction holds true, a potential sequence of events can be formulated: the initial stimulus for the acrosome reaction is a rapid influx of Ca^{2+} followed by or coincident with the activation of adenylate cyclase and an increase in cAMP levels, which in turn result in the activation of protein kinase A. This kinase may ultimately be responsible for the phosphorylation of proteins that are critical for the acrosome reaction.

The receptor-mediated hydrolysis of membrane-bound inositol phospholipids (phosphoinositides) is known to be another common mechanism for the transduction of extracellular signals across the plasma membrane of somatic cells. This system involves the hydrolysis of membrane-bound phosphatidylinositol 4,5-bisphosphate (PIP_2) by phospholipase C to generate the second messengers inositol 1,4,5-trisphosphate ($InsP_3$) and diacylglycerol (DAG). Phosphatidylinositol, along with other membrane-associated phospholipids, has been isolated from human spermatozoa (Langlais and Roberts, 1985), and spermatozoa are known to possess phospholipase C (see Section 3.4).

The $InsP_3$ molecule is strongly negatively charged. This characteristic of $InsP_3$ taken in conjunction with data demonstrating high concentrations of anionic lipids over fusional zones as spermatozoa become capacitated (Bearer and Friend, 1982) lends credence to the possibility that $InsP_3$ is involved in the acrosome reaction. In sea urchin spermatozoa, egg jelly causes a calcium-dependent increase in $InsP_3$ coincident with the acrosome reaction. This implies that the influx of Ca^{2+} is required for the activation of phospholipase C to produce $InsP_3$. In somatic cells, the primary effect of $InsP_3$ is the release of calcium from intracellular stores. Whether the same occurs in spermatozoa remains to be established.

The alternate branch in the phosphoinositide pathway involves the action of DAG. Diacylglycerol has two potential signaling roles: it can activate protein kinase C, and it can be cleaved to release arachidonic acid. The role of protein kinase C in signal transduction has been shown for secretory and exocytotic processes from a variety of tissues and cells (Nishizuka, 1986). Protein kinase C phosphorylates a number of proteins, which can result in modulating ion conductance of membranes, including that of $Na^+–H^+$ exchange and calcium. Protein kinase C is in an inactive form in most tissues and is present in the soluble fraction. When cells are stimulated, the enzyme is translocated to membranes, a process that is Ca^{2+} dependent. For its activation, protein kinase C requires Ca^{2+} and phospholipids, especially those that are unsaturated and that have a structure similar to phosphatidylinositol. Diacylglycerol, that has the 1-steroyl-2-arachidonyl conformation of phosphatidylinositol in its backbone, increases the affinity of protein kinase C for Ca^{2+}. The DAG disappears rapidly through its conversion back to inositol phospholipids and its further degradation to arachidonic acid. The potential role of arachidonic acid in the acrosome reaction is described in Section 3.4.

The involvement of the DAG/protein kinase C pathway in the acrosome reaction has only recently been investigated. Synthetic diacylglycerols and phorbol diesters, compounds known to activate protein kinase C in somatic cells, appear to effect a transitional state between capacitation and the acrosome reaction of mouse spermatozoa (Lee *et al.*, 1987). Furthermore, phorbol diesters and synthetic diacylglycerols induce the acrosome reaction of capacitated human spermatozoa and a protein kinase C inhibitor prevents this process (De Jonge *et al.*, 1991b).

Other second messenger systems and protein kinases may also be involved in the acrosome reaction such as cGMP-dependent kinase and Ca^{2+}-calmodulin kinase, but this is mostly speculative. More evidence is available for a role of tyrosine kinase, and the zona-induced mouse sperm acrosome reaction may be mediated by this enzyme (Saling, 1991). Tyrosine kinase can activate phospholipase C so that it may act by stimulating the phosphoinositide pathway.

3.4. Phospholipases, Cyclooxygenase, Lipoxygenase, and Their Products

Spermatozoa possess phospholipase A_2 and C (for properties, see Section 2). Both enzymes are stimulated by calcium. Human sperm phospholipase C produces inositol phosphates and DAG from phosphoinositides, whereas bull and rabbit phospholipase C produce phosphoryl-choline from phosphatidylcholine. The potential roles of DAG and inositol phosphates in the acrosome reaction were discussed in the previous section. Phospholipase A_2 converts membrane phospholipids to *cis*-unsaturated fatty acids (including arachidonic acid, see below) and lysophospholipids. Phospholipase A_2 inhibitors such as *p*-bromophenacylbromide and mepacrine prevent the acrosome reaction of rodent spermatozoa, and exogenous phospholipase A_2 enhances the reaction. Lysophospholipids such as lysophosphatidylcholine and lysophosphatidylethanolamine are fusogenic and stimulate the acrosome reaction. Thus, reasonable evidence exists that phospholipase A_2 has a role in the acrosome reaction (for review, see Meizel, 1984; Langlais and Roberts, 1985), although the activity of the enzyme may not be rate limiting (Anderson *et al.*, 1990). The activity of both phospholipase C and A_2 is stimulated by serine proteinases such as acrosin, suggesting a precursor form of the enzyme. Since at least phospholipase A_2 appears to become activated during the acrosome reaction, it is possible that one of the functions of acrosin is to convert the prophospholipases to active forms.

Arachidonic acid is produced by phospholipase A_2 from phosphatidylcholine or can be obtained by the conversion of DAG (see Section 3.3). Arachidonic acid can have several cellular functions. It has fusogenic properties and can directly cause membrane alterations, leading to the acrosome reaction, when added to capacitated spermatozoa (for review, see Meizel, 1984). Additionally, archidonic acid can be metabolized by cyclooxygenase to prostaglandins and other eicosanoids and by lipoxygenase to hydroperoxyeicosatetraenoic acids (HPETE), which are converted to the hydroxy derivatives (HETE) and leukotrienes. Evidence is accumulating that prostaglandins and certain leukotrienes have a role in the acrosome reaction: prostaglandins have been found on the sperm membrane; PGE_2 and $PGF_{2\alpha}$ enhance the acrosome reaction of rodent spermatozoa and stimulate the penetration of human spermatozoa into zona-free hamster oocytes. Cyclooxygenase inhibitors such as indomethacin prevent the guinea pig and hamster acrosome reaction. In regard to the lipoxygenase products, it has been established that HPETE and HETE enhance the hamster acrosome reaction. Recent data from our laboratory show that a cysteine leukotriene antagonist FPL-55712 inhibits mouse fertilization as well as the penetration of human spermatozoa into zona-free hamster oocytes (Basuray *et al.*, 1990). The role of prostaglandins and leukotrienes in the acrosome reaction is not known. Some data indicate that PGE_2 has a calcium-ionophore-like action on human spermatozoa and enhances sperm cAMP levels.

3.5. Acrosin

The properties of acrosin have been presented in Section 2. Acrosin inhibitors such as NPGB and aminobenzamidine prevent the dispersal of the acrosomal matrix from capacitated hamster, guinea pig, and human spermatozoa following addition of calcium and/or calcium ionophore. However, the vesiculation and disappearance of the outer acrosomal membrane and plasma membrane are generally not prevented by these inhibitors. Addition of proteinases to spermatozoa can enhance the capacitation/acrosome reaction process. These data provide reasonable evidence for a role of this enzyme in the acrosome reaction, but its mechanism of action is not clear at this time. Acrosin may digest structural acrosomal proteins and/or cause the activation of zymogen enzymes (see Section 3.4). Proacrosin activation to acrosin occurs during the acrosome reaction, but it remains to be established when this occurs in the cascade of events. Recent evidence from our laboratory indicates that acrosin exerts its activity after adenylate cyclase, since acrosin inhibitors were able to prevent the cAMP-induced acrosome reaction (De Jonge *et al.*, 1989).

ACKNOWLEDGMENTS. The authors appreciate the manuscript preparation by Ms. N. Pabon. Unpublished research was supported by NIH HD 19555.

4. REFERENCES

Abla, A., Mroueh, A., and Durr, I. F., 1974, A divalent cation-dependent ATP-ase in human spermatozoa, *J. Reprod. Fertil* **37**:121–123.

Allison, A. C., and Hartree, E. F., 1970, Lysosomal enzymes in the acrosome and their possible role in fertilization, *J. Reprod. Fert.* **21**:501–515.

Anderson, R. A., Johnson, S. K., Bielfeld, P., Feathergill, K. A., and Zaneveld, L. J. D., 1990, Characterization and inhibitor sensitivity of human sperm phospholipase A_2: evidence against a pivotal involvement of phospholipase A_2 in the acrosome reaction, *Molec. Reprod. Develop.* **27**:305–325.

Ashraf, M., Peterson, R. N., and Russell, L. D., 1982, Activity and location of cation-dependent ATPases on the plasma membrane of boar spermatozoa, *Biochem. Biophys. Res. Commun.* **107**:1273–1278.

Ashraf, M., Peterson, R. N., and Russell, L. D., 1984, Characterization of $(Ca^{2+} + Mg^{2+})$ adenosine triphosphatase activity and calcium transport in boar plasma membrane vesicles and their relation to phosphorylation of plasma membrane proteins, *Biol. Reprod.* **31**:1061–1071.

Austin, C. R., 1985, Sperm maturation in the male and female genital tracts, in: *Biology of Fertilization*, Vol. 2 (C. B. Metz and A. Monroy, eds.), Academic Press, New York, pp. 121–155.

Basuray, R., De Jonge, C. J., Zaneveld, L. J. D., 1990, Evidence for a role of cysteinyl leukotrienes in the fertilization process of mouse and human spermatozoa, *J. Androl.* **11**:47–51.

Bearer, E. L., and Friend, D. S., 1982, Modification of anionic lipid domains preceding membrane fusion in guinea pig sperm, *J. Cell Biol.* **92**:604–615.

Bedford, J. M., and Cooper, G. W., 1978, Membrane fusion events in the fertilization of vertebrate eggs, in: *Membrane Fusion* (G. Poste and G. L. Nicolsen, eds.), North Holland, Amsterdam, pp. 65–125.

Bernstein, M. H. and Teichman, R. J., 1973, A chemical procedure for extraction of the acrosomes of mammalian spermatozoa, *J. Reprod. Fert.* **33**:239–244.

Berridge, M. J., 1985, The molecular basis of communication within the cell, *Sci. Am.* **253**:142–152.

Bhalla, V. K., Tillman, W. L., and Williams, W. L., 1973, Presence of β-aspartyl N-acetyl glucosamine amido hydrolase in mammalian spermatozoa, *J. Reprod. Fertil.* **34**:137–139.

Bhattacharyya, A. K., and Zaneveld, L. J. D., 1982, The sperm head, in: *Biochemistry of Mammalian Fertilization* (L. J. D. Zaneveld and R. T. Chatterton, eds.), John Wiley & Sons, New York, pp. 119–151.

Bradford, M. M., McRorie, R. A., and Williams, W. L., 1976, Involvement of esterases in sperm penetration of the corona radiata of the ovum, *Biol. Reprod.* **15**:102–106.

Bradford, M. M., McRorie, R. A., and Williams, W. L., 1976, A role for esterases in the fertilization process, *J. Exp. Zool.* **197**:297–301.

Breitbart, H., Stern, B., and Rubinstein, S., 1983, Calcium transport and Ca^{2+}-ATPase activity in ram spermatozoa plasma membrane vesicles, *Biochim. Biophys. Acta* **728**:349–355.

Chang, M. C., 1984, Meaning of capacitation: A historical perspective, *J. Androl.* **5**:45–50.

Clegg, E. D., 1983, Mechanisms of mammalian sperm capacitation, in: *Mechanism and Control of Animal Fertilization* (J. F. Hartman, ed.), Academic Press, New York, pp. 177–212.

De Jonge, C. J., Mack, S. R., Zaneveld, L. J. D., 1989, Inhibition of the human sperm acrosome reaction by proteinase inhibitors, *Gamete Res.* **23**:387–397.

De Jonge, C. J., Han, H.-L., Lowrie, H., Mack, S. R., Zaneveld, L. J. D., 1991a, Modulation of the human sperm acrosome reaction by effectors of the adenylate cyclase/cyclic AMP second-messenger pathway, *J. Exp. Zool.* **258**:113–125.

De Jonge, C. J., Han, H.-L., Mack, S. R., and Zaneveld, L. J. D., 1991b, Effect of phorbol esters, synthetic diacylglycerols, and a protein kinase C inhibitor on the human sperm acrosome reaction, *J. Androl.* **12**:62–70.

Dudkiewicz, A. B., 1984, Purification of boar acrosomal arylsulfatase A and possible role in the penetration of cumulus cells, *Biol. Reprod.* **30**:1005–1014.

Farooqui, A. A., and Srivastava, P. N., 1980, Isolation of β-N-acetylhexosamimidase from rabbit semen and its role in fertilization, *Biochem. J.* **191**:827–834.

Fritz, H., Schleuning W.-D., Schiessler, H., Schill, W. B., Wendt, V., and Winkler, G., 1975, Boar, bull and human sperm acrosin–isolation, properties and biological aspects, in: *Proteases and Biological Control* (E. Reich, D. Rifkin, and E. Shaw, eds.), Cold Spring Habor Laboratory, New York, pp. 715–735.

Garbers, D. L., and Kopf, G. S., 1980, The regulation of spermatozoa by calcium and cyclic nucleotides, in: *Advances in Cyclic Nucleotide Research* (P. Greengard and G. A. Robinson, eds.), Raven Press, New York, pp. 251–305.

Gonzales, L. W., and Meizel, S., 1973, Acid phosphatases of rabbit spermatozoa. II. Partial purification and biochemical characterization of the multiple forms of rabbit spermatozoan acid phosphatase, *Biochim. Biophys. Acta* **32**:180–194.

Gordon, M., 1973, Localization of phosphatases activity on the membranes of the mammalian sperm head, *J. Exp. Zool.* **185**:111–119.

Gordon, M., and Barnett, R. H., 1967, Fine structural localization of phosphatases in rat and guinea pig, *Exp. Cell Res.* **48**:395–412.

Gottlieb, W., and Meizel, S., 1987, Biochemical studies of metalloendoproteinase activity in the spermatozoa of three mammalian species, *J. Androl.* **8**:14–24.

Green, D. P. L., 1978, The mechanism of the acrosome reaction, in: *Development in Mammals*, Vol. 3 (M. J. Johnson, ed.), North Holland, New York, pp. 65–80.

Guerette, P., Langlais, J., Antaki, P., Chapdelaine, A., and Roberts, K.D., 1988, Activation of phospholipase A_2 of human spermatozoa by proteases, *Gamete Res.* **19**:203–214.

Hinrichsen-Kohane, A. C., Hinrichsen, M. J., and Schill, W. B., 1984, Molecular events leading to fertilization—a review, *Andrologia* **16**:321–341.

Hoskins, D. D., and Casillas, E. R., 1975, Hormones, second messengers and the mammalian spermatozoon, in: *Advances in Sex Hormone Research* (R. H. Singal and J. A. Thomas, eds.), University Park Press, Baltimore, pp. 283–321.

Johnson, R. A., Jakobs, K. H., and Schultz, G., 1985, Extraction of the adenylate cyclase-activating factor of bovine sperm and its identification as a trypsin-like proteinase, *J. Biol. Chem.* **260**:114–121.

Joyce, C., Jeyendran, R. S., and Zaneveld, L. J. D., 1985, Release, extraction, and stability of hyaluronidase associated with human spermatozoa. Comparisons with the rabbit, *J. Androl.* **6**:152–161.

Khar, A., and Anand, S. R., 1977, Studies on the glycosidases of semen, Purification and properties of β-N-acetylglucosaminidase from bull sperm, *Biochim. Biophys. Acta* **483**:141–151.

Kipping, D., 1970, Zur Kenntnis der sauren Phosphomonoesterase der menschlichen Spermatozoen, *Klinische Wochenschr.* **48**:1127–1128.

Kopf, G. S., 1988, Regulation of sperm function by guanine nucleotide-binding regulatory proteins (G-proteins), in: *Meiotic Inhibition: Molecular Control of Meiosis* (F. P. Hazeltine and F. L. First, eds.), Alan R. Liss, New York, pp. 357–386.

Kopf, G. S., and Vacquier, V. D., 1985, Characteristics of a calcium-modulated adenylate cyclase from abalone spermatozoa. *Biol. Reprod.* **33**:1094–1104.

Langlais, J., and Roberts, K. D., 1985, A molecular membrane model of sperm capacitation and the acrosome reaction of mammalian spermatozoa, *Gamete Res.* **12**:183–224.

Lee, M. A., Kopf, G. S., and Storey, B. T., 1987, Effects of phorbol esters and a diacylglycerol on the mouse sperm acrosome reaction induced by the zona pellucida, *Biol. Reprod.* **36**:617–627.

Lessley, B. A., and Garner, D. L., 1984, Identification and preliminary characterization of two distinct bovine seminal Pz-peptidases, *Biol. Reprod.* **31**:353–369.

Llanos, M. N., Lui, C. W., and Meizel, S., 1982, Studies of phospholipase A_2 related to the hamster sperm acrosome reaction, *J. Exp. Zool.* **221**:107–117.

McDonald, J. K., and Kadhodayen, S., 1988, Cathespin L—a latent proteinase in guinea pig sperm, *Biochem. Biophys. Res. Commun.* **151**:827–835.

McRorie, R. A., and Williams, W. L., 1974, Biochemistry of mammalian fertilization, *Annu. Rev. Biochem.* **43**:777–798.

McRorie, R. A., Turner, R. B., Bradford, M. M., and Williams, W. L., 1976, Acrolysin, the aminoproteinase catalyzing the initial conversion of proacrosin to acrosin in mammalian fertilization, *Biochem. Biophys. Res. Commun.* **71**:492–498.

Meizel, S., 1978, The mammalian sperm acrosome reaction, a biochemical approach, in: *Development in Mammals* (M. H. Johnson, ed.), North Holland, Amsterdam, pp. 1–64.

Meizel, S., 1984, The importance of hydrolytic enzymes to an exocytotic event, the mammalian sperm acrosome reaction, *Biol. Rev.* **59**:125–157.

Meizel, S., 1985, Molecules that initiate or help stimulate the acrosome reaction by their interaction with the mammalian sperm surface, *Am. J. Anat.* **174**: 285–302.

Meizel, S., Boggs, D., and Cotham, J., 1971, Electrophoretic studies of esterases of bull spermatozoa, cytoplasmic droplets and seminal plasma, *J. Histochem. Cytochem.* **19**:226–231.

Monks, N. J., Stein, D. M., and Fraser, L. R., 1986, Adenylate cyclase activity of mouse sperm during capacitation *in vitro*; effect of calcium and a GTP analogue. *Int. J. Androl.* **9**:67–76.

Monroy, A., and Rosati, F., 1983, A comparative analysis of sperm–egg interaction, *Gamete Res.* **7**:85–102.

Morton, D. B., 1977, The occurrence and function of proteolytic enzymes in the reproductive tract of mammals, in: *Proteinases in Mammalian Cells and Tissues* (A. J. Barret, ed.), North Holland, Amsterdam, pp. 445–500.

Mrsny, R. J., and Meizel, S., 1985, Inhibition of hamster sperm Na^+, K^+-ATPase activity by taurine and hypotaurine, *Life Sci.* **36**:271–275.

Nishizuka, Y., 1986, Studies and perspectives of protein kinase C. *Science* **233**:305–312.

Parrish, R. F., and Polakoski, K. L., 1979, Mammalian sperm proacrosin–acrosin system, *Int. J. Biochem.* **10**: 391–395.

Peterson, R. N., Hunt, W. P., and Saxena, N., 1987, Role of spermatozoa membranes in sperm capacitation and oocyte recognition, *CRC Crit. Rev. Anat. Sci.* **1**:1–14.

Polakoski, K. L., and Siegel, M. S., 1986, The proacrosin–acrosin system, in: *Andrology, Male Fertility and Sterility* (J. D. Paulson, A. Nigro-Vilar, E. Lucena, and L. Martin, eds.), Academic Press, New York, pp. 359–375.

Polakoski, K. L., and Zaneveld, L. J. D., 1976, Proacrosin, in: *Methods in Enzymology,* Vol. LV, *Proteolytic Enzymes,* Part B (L. Lorand, ed.), Academic Press, New York. pp. 325–331.

Qian, Z.-U., Tsai, Y.-H., Steinberger, A., Lu, M., Greenfield, A. R. L., and Haddox, M. K., 1985, Localization of ornithine decarboxylase in rat testicular cells and epididymal spermatozoa, *Biol. Reprod.* **33**:1189–1195.

Ribbes, H., Plantavid, M., Bennet, P. J., Chap, H., and Douste-Blazy, L., 1987, Phospholipase C from human sperm specific for phosphoinositides, *Biochim. Biophys. Acta* **919**:245–254.

Rogers, B. J., and Bentwood, B. J., 1982, Capacitation, acrosome reaction and fertilization, in: *Biochemistry of Mammalian Fertilization* (L. J. D. Zaneveld and R. T. Chatterton, eds.), John Wiley & Sons, New York, pp. 203–230.

Saling, P. M., 1991, How the egg regulates sperm function during gamete interaction: facts and fantasies, *Biol. Reprod.* **44**:246–251.

Schollmeier, J. E., 1986, Identification of calpain II in porcine sperm, *Biol. Reprod.* **34**:721–731.

Seiguer, A. C., and Castro, A. E., 1972, Electron microscope demonstration of arylsulfatase activity during acrosome formation in the rat, *Biol. Reprod.* **7**:31–42.

Shapiro, B. M., Schackmann, R. W., Tomber, R. M., and Kazazoglau, T., 1985, Coupled ionic and enzymatic regulation of sperm behavior, *Curr. Top. Cell Regul.* **26**:97–113.

Shur, B. D., and Hall, N. G., 1982, Sperm surface galactosyltransferase activities during *in vitro* capacitation, *J. Cell Biol.* **95**:567–673

Shur, B. D., and Neely, C. A., 1988, Plasma membrane association, purification and partial characterization of mouse sperm $\beta 1,4$-galactosyltransferase, *J. Biol. Chem.* **263**:17706–17714.

Siegel, M. S., Bechtold, D. S., Willard, J. L., and Polakoski, K. L., 1987, Partial purification and characterization of human sperminogen, *Biol. Reprod.* **36**:1063–1068.

Singer, R., Barnet, M., Allalouf, D., Schwartzman, S., Sagiv, M., Landau, B., Segenreich, E., and Servadio, C., 1980, Some properties of acid and alkaline phosphatase in seminal fluid and isolated sperm, *Arch. Androl.* **5**:195–199.

Srivastava, P. N., and Abou-Issa, H., 1977, Purification and properties of rabbit spermatozoal acrosomal neuraminidase, *Biochem. J.* **161**:193–200.

Srivastava, P. N., and Ninjoor, V., 1982, Isolation of rabbit testicular cathepsin D and its role in the activation of proacrosin, *Biochem. Biophys. Res. Commun.* **109**:63–69.

Stambaugh, R., 1978, Enzymatic and morphological events in mammalian fertilization, *Gamete Res.* **1**:65–85.

Talbot, P., and Dicarlantonio, G., 1985, Cytochemical localization of dipeptidyl peptidase II (DPP-II) in mature guinea pig sperm, *J. Histochem. Cytochem.* **33**:1169–1172.

Teichman, R. J., and Bernstein, M. H., 1971, Fine structure localization of acid phosphatase in rabbit and bull sperm heads, *J. Reprod. Fert.* **27**:243–248.

Tesarik, J., 1986, From the cellular to the molecular dimension: The actual challenge for human fertilization research, *Gamete Res.* **13**:47–89.

Thakkar, J. K., East, J., Seyler, D., and Francson, R. C., 1983, Surface-active phospholipase A_2 in mouse spermatozoa, *Biochim. Biophys. Acta* **754**:44–50.

Thakkar, J. K., East, J., and Francson, R. C., 1984, Modulation of phospholipase A_2 activity associated with human sperm membranes by divalent cations and calcium antagonists, *Biol. Reprod.* **30**:679–686.

Usui, N., and Yanagimachi, R., 1986, Cytochemical localization of membrane-bound Mg^{2+}-dependent ATPase activity in guinea pig sperm head before and during the acrosome reaction, *Gamete Res.* **13**:271–280.

Vijayasarathy, S., Shivaji, S., and Balaram, P., 1980, Plasma membrane bound Ca^{2+}-ATPase activity in bull sperm, *FEBS Lett.* **114**:45–49.

Waibel, R., Granet, R., Ficsor, G., and Ginsberg, L., 1985, Caproyl esterase from rat testis: purification and action on cumulus cells, *Gamete Res.* **12**:75–84.

Wasserman, P. M., 1987, Early events in mammalian fertilization, *Annu. Rev. Cell Biol.* **3**:109–142.

Working, P. K., and Meizel, S., 1982, Preliminary characterization of a MG^{2+}-ATPase in hamster sperm head membranes, *Biochem. Biophys. Res. Commun.* **104**:1060–1065.

Yang, C.-H., and Srivastava, P. N., 1974, Separation and properties of hyaluronidase from ram sperm acrosomes, *J. Reprod. Fertil.* **37**:17–25.

Yang, C.-H., and Srivastava, P. N., 1975, Purification and properties of hyaluronidase from bull sperm, *J. Biol. Chem.* **250**:79–83.

Yanigmachi, R., 1988, Mammalian fertilization, in: *The Physiology of Reproduction* (E. Knobil and J. Neill, eds.), Raven Press, New York, pp. 135–185.

Yotsumoto, H., Sato, S., and Shibuya, M., 1984, Localization of angiotension converting enzyme (dipeptidyl carboxypeptidase) in swine sperm by immunofluorescence, *Life Sci.* **35**:1257–1261.

Zaneveld, L. J. D., and Polakoski, K. L., 1976, Biochemistry of human spermatozoa, in: *Human Semen and Fertility Regulation in Men* (E. S. E. Haffez, ed.), C. V. Mosby, St. Louis, pp. 167–175.

Zaneveld, L. J. D., and Williams, W. L., 1970, A sperm enzyme that disperses the corona radiata and its inhibition by decapacitation factor, *Biol. Reprod.* **2**:363–368.

Zaneveld, L. J. D., Polakoski, K. L., and Schumacher, G. F. B., 1973, Properties of acrosomal hyaluronidase from bull spermatozoa, *J. Biol. Chem.* **248**:564–570.

Zaneveld, L. J. D., Polakoski, K. L., and Shumacher, G. F. B., 1975, The proteolytic enzyme systems of mammalian genital tract secretions and spermatozoa, in: *Proteases and Biological Control* (E. Reich, D. Rifkin, and E. Shaws, eds.), Cold Spring Harbor Laboratory, New York, pp. 683–706.

Zaneveld, L. J. D., De Jonge, C. J., Anderson, R. A., and Mack, S. R., 1991, Human sperm capacitation and the acrosome reaction, *Hum. Reprod.* (In press).

5

Morphogenesis of the Mammalian Egg Cortex

Frank J. Longo

1. INTRODUCTION

The final maturation processes of the mammalian oocyte leading to ovulation include events of germinal vesicle breakdown, i.e., the disappearance of nucleoli, nuclear envelope breakdown, and chromosome condensation. The chromosomes are assembled on the first meiotic spindle, which then migrates to the oocyte cortex; this is followed by the formation of the first polar body and development of the second meiotic spindle, at which time meiosis is arrested until the egg is fertilized. Recent studies have demonstrated that these nuclear changes are accompanied by, and in some instances causal to, specific rearrangements of organelles and cytoskeletal elements comprising the cell cortex (Longo and Chen, 1984, 1985; Maro et al., 1984, 1986c; Van Blerkom and Bell, 1986; Longo, 1987).

Operationally, the term "egg cortex" often refers to the layer of cytoplasm lying directly subjacent to the plasma membrane (Vacquier, 1981). However, its demarcation from the underlying cytoplasm has never been satisfactorily identified, thus making it fundamentally misleading (Schroeder, 1981). Classically, this region has been shown to be associated with determinants that specify the future of embryonic cells and limit their potential long before the cells in question manifest their fates (Davidson, 1980). Because the cytoskeleton has been shown to be an integral feature of somatic cell structure, possibly involved in the localization of specific macromolecules, it has been suggested that cytoskeletal elements may be essential to the regulation of the totipotent egg (Showman et al., 1982; Moon et al., 1983; Jeffrey et al., 1983). Changes in cortical morphology of maturing eggs are comparable to those observed in somatic cells, where, in response to external and/or internal stimuli, the cytoskeleton undergoes regulated changes in architecture conferring on the cell its shape, adhesivity, and ability to affect the level of specific macromolecules (Penman et al., 1983).

FRANK J. LONGO • Department of Anatomy, University of Iowa, Iowa City, Iowa 52242.

A Comparative Overview of Mammalian Fertilization, edited by Bonnie S. Dunbar and Michael G. O'Rand. Plenum Press, New York, 1991.

Considerable inroads have been made in regard to our understanding of the morphological, biochemical, and physiological aspects of mammalian oocyte maturation (Liebfried-Rutledge *et al.*, 1989), and investigations examining dynamics of the egg cortex and its relationship to physiological changes of oocyte maturation are yielding new insights into this important part of the egg cytoplasm. Because of the relative scarcity of material that can feasibly be employed for study, isolation of mammal egg and oocyte cortical preparations for biochemical studies are fraught with difficulties and are, in many instances, Herculean if not impossible tasks. Consequently, analyses relying basically on the examination of single specimens prepared with immunochemical or pharmacological probes, vital fluorescence stains, and video image intensification, in combination with conventional techniques of light, transmission, and scanning electron microscopy, have been utilized to circumvent this problem (Albertini, 1984; Longo and Chen, 1984, 1985; Koehler *et al.*, 1985; G. Schatten *et al.*, 1985, 19886; Capco and McGaughey, 1986; Maro *et al.*, 1986a,b,c; Longo, 1987; Ducibella *et al.*, 1988a). The results of such efforts, when integrated, provide a high-resolution, global view of dynamic changes the oocyte undergoes as it prepares for ovulation and fertilization. In the present review three major components of the maturing oocyte and their relationship to one another are discussed: the plasma membrane, the underlying first few micrometers of cytoplasm that constitutes the "cortex," and the meiotic spindle.

2. PLASMA MEMBRANE AND MICROVILLI

The mammalian oocyte at the germinal vesicle stage of meiosis is covered entirely with microvilli (Fig. 1; Calarco and Epstein, 1973; Eager *et al.*, 1976; Phillips and Shalgi, 1980). In the mouse, oocyte microvilli are approximately 1 μm in length by 0.1 μm in diameter, possess a core of actin microfilaments, and, because of their density (number/unit surface area; 7.2/μm^2), effectively doubled the surface area of the oocyte (Longo and Chen, 1985; Longo, 1987). In the hamster, microvilli measure 0.6 to 1.0 μm in length by 0.16 μm in diameter and have a density of 7.3/μm^2 (Ebensperger and Barros, 1984). The surfaces of ovulated mammalian eggs studied thus far (e.g., mouse, rat, rabbit, pig, and hamster) are also projected into microvilli 0.3 to 1.0μm in length, which are distributed in a polarized fashion; i.e., they are uniformly distributed along the surface of the egg except for an area overlying the meiotic spindle (Fig. 2; Odor and Renninger, 1960; Yanagimachi and Chang, 1961; Stefanini *et al.*, 1969; Zamboni, 1970, 1971; Norberg, 1973; Thompson *et al.*, 1974; Longo, 1974; Gulyas, 1976; Calarco and Epstein, 1973; Gould, 1973; Szollosi, 1976; Nicosia *et al.*, 1977, 1978; Wablik-Sliz and Kujat, 1979; Phillips and Shalgi, 1980; Schmell *et al.*, 1983; Longo and Chen, 1985). The region devoid of microvilli is referred to as the microvillus-free area (Fig. 2; Longo, 1985) or "nipple" (Allworth and Ziomek, 1988). Depending on the species in question, the microvillus-free area is of variable size and shows an irregular or corrugated appearance but is, nevertheless, relatively smoother than the remainder of the egg surface. In ovulated mouse eggs this area is projected into a truncated cone approximately 15 μm (r_1) $\times$ 5.5 μm (r_2) $\times$ 8 μm (l) and encompasses about 1300 μm^2 of the ovum's surface (Longo and Chen, 1984, 1985). The microvillus-free area of metaphase I hamster eggs is larger and measures about 19 mm in diameter and has a microvillar density of 1.25/μm^2 (Ebensperger and Barros, 1984). At metaphase II, this area measures 19 to 21 μm in diameter and possesses long cortical projections with lengths between 5.1 and 9.1 μm and diameters of 2.38 and 0.33 μm at their bases and tips, respectively. Interestingly, the conical projections are present in ovulated eggs recovered 15 hr after hCG injection but are gone in eggs incubated *in vitro* (Ebensperger and Barros, 1984).

The appearance of the microvillus-free area is directly correlated with meiotic maturation and the position of the maternal chromosomes. Longo and Chen (1984, 1985) have demonstrated

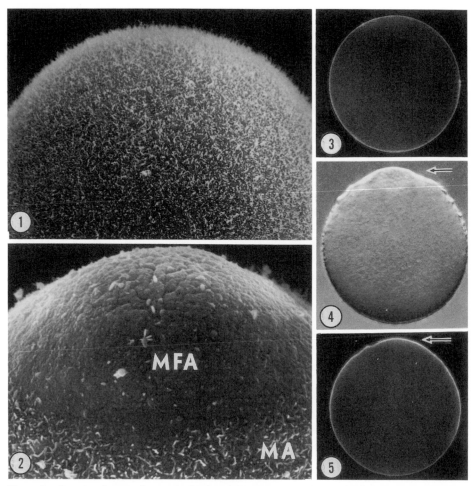

Figure 1. Scanning electron micrograph of an immature mouse oocyte characterized by numerous microvilli; a microvillus-free area is not present. From Longo and Chen (1985).

Figure 2. Scanning electron micrograph of a mature, oviductal mouse egg showing the microvillus-free area (MFA) and a portion of the microvillus region (MA). From Longo and Chen (1985).

Figure 3. Mouse oocyte treated with rabbit antiactin serum and FITC-labeled goat antirabbit antibody demonstrating a uniform cortical layer of actin. From Longo and Chen (1985).

Figure 4. Ovulated mouse egg photographed with differential contrast optics. The arrow points to the microvillus-free area and underlying meiotic spindle. From Longo and Chen (1985).

Figure 5. Same specimen shown in Fig. 4 treated with rabbit antiactin serum and FITC-labeled goat antirabbit antibody demonstrating cortically located actin. The fluorescent layer is enhanced in the region of the microvillus-free area (arrow). From Longo and Chen (1985).

that cortices of mouse oocytes matured *in vitro* undergo changes leading to a polarized morphology. When germinal vesicle oocytes are examined 17 hr after culture, the ovum cortex possesses a well-defined microvillus-free area comparable in size and structure to that of a mature oviductal egg. Microvilli along the remainder of the egg surface (microvillous area) are structurally similar to and have a density ($4.2/\mu m^2$) equal to that of mature oviductal eggs (Longo and Chen, 1985). When immature oocytes are incubated with cytochalasin B, germinal vesicle breakdown occurs;

however, the meiotic spindle fails to move to the egg cortex, and a microvillus-free area is not formed. Inhibition of these processes is reversed when cytochalasin-B-treated eggs are washed and cultured in fresh medium (Longo and Chen, 1984, 1985). In some eggs a nuclear envelope forms along chromosomes localized within the cortex. In these instances a microvillus-free area does not develop, suggesting that whatever affect the chromosomes have on the egg surface, it is blocked or disappears when the chromosomes become enclosed within membrane.

That oocytes are capable of forming microvillus-free areas only when meiosis is initiated and the chromosomes become localized to the egg cortex (Longo and Chen, 1984, 1985; Longo, 1987) indicates that cortical changes normally associated with oocyte maturation are not autonomous cytoplasmic events independent of nuclear-derived structures and events. Further observations, injecting chromosomes into eggs, or disrupting the chromosome mass of the meiotic spindle of metaphase I or II mouse eggs with nocodazole or colchicine demonstrates that when small groups of chromosomes become confined to the egg periphery, the overlying cortex takes on the same morphological characteristics associated with the cortex superficial to the meiotic spindle of untreated/uninjected ova (Longo and Chen, 1985; Van Blerkom and Bell, 1986). These experiments demonstrate that chromosomes are sufficient to establish localized intracellular changes requisite to the development of a microvillus-free area at virtually any region of the oocyte surface. As discussed below, this modification in the egg surface involves the reorganization of cortical components, notably actin.

The appearance of a microvillus-free area results in a loss of approximately 5000 microvilli (Longo, 1985; Longo and Chen, 1985) and represents a significant decrease in surface area, which is relevant to metabolic studies analyzing the incorporation of precursors (Wassarman *et al.*, 1981). The mechanism by which microvilli are reduced is unknown, but it may involve the retraction of the microvillar cytoskeleton and the diffusion of membrane components. Hence, the membrane area of the microvillus is spread into a "smooth" plasma membrane. Asymmetric expression of microvilli has been reported for other cells, such as lymphocytes and embryonic blastomeres (Ducibella *et al.*, 1977; Loor, 1981). Loor (1981) described the development of microvillar asymmetry in lymphocytes as a flow of microvilli toward one pole of the cell. Such a mechanism seems unlikely to account fully for the change in microvilli distribution at oocyte maturation, as the density of microvilli is much higher in oocytes than in mature eggs (Longo and Chen, 1985). The surface areas of the mouse oocyte and mature egg have been calculated to be 58,000 and 36,900 μm^2, respectively (Longo and Chen, 1985; see also Ducibella *et al.*, 1988a,b). Hence, when the oocyte matures to the second metaphase of meiosis, there is a "loss" or accommodation of approximately 21,000 μm^2 of surface area. Formation of the first polar body and the small vesicles that are produced in conjunction with this structure (Zamboni, 1971) would include some of this "extra" surface area, but not all, as the amount of membrane to be accommodated is in excess of that delimiting the first polar body. What role processes such as endocytosis might have in the regulation of surface area changes of maturing oocytes has not been determined.

Investigations utilizing individual specimens have characterized different properties of the egg/oocyte plasma membrane. The plasma membranes of rabbit and hamster eggs are abundant in acidic anionic residues as determined by colloidal iron hydroxide staining (Cooper and Bedford, 1971; Yanagimachi *et al.*, 1973). Plasma membranes delimiting the microvillous and microvillus-free areas of mouse ova show differences in concanavalin A binding that are related to the presence of microvilli (Johnson *et al.*, 1975; Eager *et al.*, 1976). Investigations measuring the diffusion of proteins labeled with concanavalin A or TNBS by the FRAP technique show that each label has the same diffusion coefficient on microvillus-free and microvillous areas (Wolf and Ziomek, 1983). That is, there is a significant absolute increase in the diffusing fraction in the two regions, but the molecular protein fraction is increased in the microvillus-free compared to the microvillous area. Wolf and Ziomek (1983) conclude that this dichotomy reflects true differences in membrane diffusibility in the two regions of the egg. Factors other than lipid viscosity may be involved here, since the protein diffusion rates are two orders of magnitude slower (Peters, 1981) than predicted by fluid dynamics (Saffman and Delbruck, 1975). Differences in cytoskeletal components present in the two regions might contribute to this difference in protein diffusibility.

Fluidity differences in the microvillous and microvillus-free areas of mouse oocyte plasma membranes have been demonstrated by employing the lipid probe diI-C18 (Wolf et al., 1981). The mean value of the diffusion coefficient for the unfertilized mouse egg is 1.9×10^{-8} cm^2/sec, i.e., of the same magnitude as that reported for other cells. No consistent variation is seen when diI-C18 diffusion is compared at different regions of the unfertilized egg surface. However, small changes in the diffusion rates and functional dependency of the rates on acyl chain length are observed, which may reflect the organization of the membrane into lipid domains (Wolf et al., 1981).

Freeze-fracture replicas of the mature hamster egg through the microvillous region demonstrate that the P-face of the plasma membrane has a large number of intramembranous particles, whereas the E-face possesses fewer particles (Koehler et al., 1985). Interestingly, the number of particles in plasmalemma of the microvillous area is greater than that along the microvillus-free area, indicating structural differences in the plasma membrane of these two areas. Specialized regions within the plane of the membrane that might mark the presence of underlying cortical structures, e.g., cortical granules, are not apparent (Longo, 1981). When eggs are incubated in filipin, the P-face of plasma membrane delimiting microvilli demonstrates multiple protrusions (filipin-sterol complexes), indicating the presence of 3-β-hydroxysterols. Regions between microvilli possess fewer complexes (Koehler et al., 1985; see also Pratt, 1985).

The localization of 3-β-hydroxysterols has been examined in mouse eggs by use of the fluorescence of filipin–sterol complexes (Allworth and Ziomek, 1988). A punctate staining pattern is noted along the surfaces of oocytes, probably indicating projections of the cumulus cells. Unfertilized eggs have a lower concentration of filipin staining in the microvillus-free area in comparison to membrane associated with the microvillous area, as assessed by the intensity of the staining and the rapidity with which it bleaches. When examined with scanning electron microscopy, filipin incorporation has a pronounced effect on the appearance of the unfertilized egg (Allworth and Ziomek, 1988); microvillar number is reduced, and those that are present are shorter than normal. A peripheral ring of membrane is observed at the base of the microvillus-free area that may represent a framework or barrier between cytoplasms of the microvillus and microvillus-free areas, excluding organelles from the cortical region associated with the meiotic spindle (Eager et al., 1976; Nicosia et al., 1977; Longo and Chen, 1985).

3. CORTICAL CYTOSKELETON

Ultrastructural studies of mammalian eggs have demonstrated a polarity of the cytoplasm in which the cytoskeleton maintains, directly or indirectly, most organelles from the cytoplasmic region associated with the microvillus-free area via an anchorage to the cell surface (Odor and Renninger, 1960; Eager et al., 1976). Mammalian oocytes and eggs reacted with antiactin antibodies or rhodamine-phalloidin demonstrate a rim of cortical actin staining (Fig. 3; Longo and Chen, 1984, 1985; Maro et al., 1984; Longo, 1985; Le Guen et al., 1989). When extracted and incubated with heavy meromyosin, a rich meshwork of actin microfilaments, which originate from dense, punctate aggregations on the cytoplasmic leaflet of the plasma membrane, fill the cortices of germinal vesicle oocytes (Longo, 1987; Le Geun et al., 1989). Localization of actin to other parts of the oocyte is not observed (Longo and Chen, 1985; Longo, 1987).

With meiotic maturation in hamster and mouse ova there is a change in the distribution of cortical actin. Concomitant with the movement of the meiotic spindle is an increase in actin along the pole to which the spindle moves and a concomitant loss of actin at the opposite pole (Maro *et al.*, 1984; Karasiewicz and Soltynska, 1985; Longo, 1985; Van Blerkom and Bell, 1986). As a result of this cortical reorganization, a dense meshwork of filaments up to ~800 nm in thickness occupies the space between the meiotic spindle and the plasma membrane (Figs. 4 and 5; Szollosi, 1967; Zamboni, 1971; Gulyas, 1976). Microfilaments are present along other portions of egg cortex; however, they do not form as prominent a layer as seen in the region of the meiotic spindle. The change in actin distribution during oocyte maturation is consistent with a transloca-tion of macromolecules from one part of the cell to the other (Longo, 1987). A similar transforma-tion in cortical actin distribution occurs during cap formation, where actin, which was formerly distributed along the cell periphery, becomes concentrated to one locus of the cell surface (Condeelis, 1979). This rearrangement is believed to be a result of a "sliding" of actin filaments that are bound to membrane receptors (Geiger, 1983), thereby reducing the percentage of plasma membrane surface that is associated with actin.

By use of various agents that affect cytoskeletal components, it has been shown that the appearance of the microvillus-free area is correlated with the deposition of actin in the vicinity of the meiotic spindle, thus indicating an intimate relationship between the cytoskeleton and the conformation of the egg surface (Longo and Chen, 1984, 1985). The shape of the mouse egg, as well as its elasticity, may depend on cell-surface proteins whose interactions reinforce the egg plasma membrane with a deformable meshwork (Longo and Chen, 1984, 1985).

The accumulation of actin toward the pole occupied by the meiotic spindle has been shown to be dependent on the presence of chromosomes (Longo and Chen, 1984, 1985; Maro *et al.*, 1986c; Van Blerkom and Bell, 1986). When mouse eggs are treated with agents to disrupt the meiotic spindle (e.g., colchicine or nocodazole) or injected with chromosomes, the chromosomes become localized to different regions along the egg cortex and initiate the development of a microvillus-free area associated with an accumulation of actin filaments. It is interesting to note that latrunculin, an inhibitor of microfilament-mediated processes, prevents chromosome disper-sion induced by colcemid, suggesting that chromosome dispersal is mediated by microfilaments (G. Schatten *et al.*, 1986). These observations also demonstrate that there is no special or predetermined site to which the meiotic spindle must migrate in order to form a microvillus-free area and associated actin meshwork. Hence, the entire oocyte cortex appears to be competent for the elaboration of these surface/cortical specializations. This effect of chromosomes on the cell cortex may be unique to mammalian eggs, as such changes do not appear to have been reported for other cell types. It has been suggested that the thickened band of microfilaments in association with the meiotic spindle of mammalian eggs serves to maintain the spindle in a cortical position (Webb *et al.*, 1986; Le Guen *et al.*, 1989). Support for this suggestion comes from experiments in which cytochalasins appear to disrupt the spatial relationship of the meiotic spindle and its associated layer and microfilaments as well as preventing spindle rotation and polar body formation (Maro *et al.*, 1984; Longo and Chen, 1985; Le Guen *et al.*, 1989).

The polarized deposition of actin observed in mouse and hamster eggs may not be a feature common to all mammalian ova. Rat, sheep, human, and porcine eggs reportedly possess a continuous band of actin (Battaglia and Gaddum-Rosse, 1986, 1987; Abertini *et al.*, 1987; Pickering *et al.*, 1988; Le Guen *et al.*, 1989). Observations of human ova have not verified a cortical polarity similar to that of mouse eggs (Zamboni, 1972; Zamboni *et al.*, 1972; Sathan-anthan and Lopata, 1980; Sathananthan and Trounson, 1982a,b).

Spectrin-like protein has been identified with fluorescent indirect antibody techniques in mouse ova (Reima and Lehtonen, 1985; H. Schatten *et al.*, 1986; Damjanov *et al.*, 1986). These observations stand in contrast to those of Sobel and Alliegro (1985), who did not find a positive staining for this cytoskeletal protein in mouse eggs. This difference may be caused by the antibodies employed as well as the techniques for handling and processing of specimens. Staining with antibodies to spectrin reveals a positive layer in association with the plasma membrane that is most prominent at the region where the polar body and egg are in contact (H. Schatten *et al.*, 1986; Reima and Lehtonen, 1985) as well as a diffuse cytoplasmic staining. Damjanov *et al.*, (1986) claim that maturation of the oocyte is accompanied by a loss of cortical spectrin and its redistribution throughout the cytoplasm. Based on its localization within the mouse egg, H. Schatten *et al.* (1986) suggest that spectrin may be involved in maintaining the meiotic spindle in its correct position, polar body formation, and/or determining the axis of cleavage.

Ten-nanometer filaments have been detected in unfertilized mouse eggs (Lehtonen *et al.*, 1983; Lehtonen and Virtanen, 1985; Lehtonen, 1985) and in association with crystalline arrays of detergent-extracted eggs observed with high-voltage electron microscopy (Capco and McGaughey, 1986). Because of similar structural characteristics and their resistance to detergent extraction, it is believed that the filaments are composed of intermediate-filament protein (Lehtonen, 1987). Immunofluorescence studies employing antibodies to intermediate-filament proteins (Lehtonen *et al.*, 1983) reveal a diffuse or punctate staining throughout the cytoplasm depending on the antibody employed. Lehtonen (1987) suggests that the paracrystalline arrays present in the egg cytoplasm contain cytoskeleton-related proteins. Experiments by other groups, however, indicate that mammalian oocytes and eggs do not contain intermediate filaments (Franke *et al.*, 1982; Osborn and Weber, 1983).

The distribution of clathrin has been examined in mouse oocytes and shown to be localized to the area of the meiotic spindle (Maro *et al.*, 1985b). The function of clathrin in this particular region of the egg has not been determined. An association between microtubules or tubulin and coated vesicles or clathrin has also been observed in mitotic and interphase cells (Louvard and Reggio, 1981; Imhof *et al.*, 1983; Kelly *et al.*, 1983; Pfeffer *et al.*, 1983; Maro *et al.*, 1985b).

4. CORTICAL GRANULES

The cortical granules of mammalian eggs are 100- to 500-nm-diameter secretory granules composed of specialized enzymes and glycoproteins that originate from the Golgi apparatus. Following their formation, cortical granules migrate to the cortex during oocyte growth (Baca and Zamboni, 1967; Szollosi, 1967; Zamboni, 1970; Suzuki *et al.*, 1981; Sathananthan and Trouson, 1982a,b; Cran and Cheng, 1985). The structure, chemistry, and function of these organelles in mammalian eggs have been reviewed (Gulyas, 1980; Guraya, 1982; Schuel, 1985); the following deals with changes in their distribution during oocyte maturation.

Two types of cortical granules have been described in mouse eggs based on their electron density in electron microscopic preparations, light and dark (Nicosia *et al.*, 1977; Ducibella *et al.*, 1988a,b). It is not known whether these cortical granules are functionally different or whether one is the precursor of the other. The mean distance of light granules from the plasma membrane is slightly less in germinal vesicle oocytes than in mature eggs, whereas the mean distance of dark granules is slightly greater in germinal vesicle oocytes (Ducibella *et al.*, 1988b). Cortical granules are rarely observed within 0.1 μm or in direct contact with the egg plasma membrane (Nicosia *et al.*, 1977; Ducibella, 1988a,b). They are almost evenly distributed within the cortex except for the area immediately overlying the meiotic spindle, which is referred to as the cortical granule-free domain (Szollosi, 1967; Yanagimachi and Noda, 1970; Nicosia *et al.*, 1977; Okada *et al.*, 1986; Ducibella *et al.*, 1988a).

In maturing hamster oocytes the cortical granule-free domain appears twice during the course of meiotic maturation: at metaphase I and then at metaphase II (Okada *et al.*, 1986). The cortical granule-free domain at metaphase I is about 250 μm² in area and is believed to form as a result of peripheral migration and exocytosis of cortical granules. During first polar body formation both the cortical granule-free domain and the adjacent cortex rich in cortical granules are incorporated into the first polar body (Szollosi, 1967; Zamboni *et al.*, 1972; Sathananthan and Lopata, 1980; Lopata *et al.*, 1980; Cran and Cheng, 1985; Okada *et al.*, 1986). Okada *et al.* (1986) suggest that the cortical granule-free domain at metaphase II (960–1500 μm^2) forms, in part, by the exocytosis of cortical granules. They show that this domain continues to increase in size following ovulation. The extrusion of cortical granules that reportedly occurs with the formation of the cortical granule-free domain in hamster eggs is reminiscent of that described by Nicosia *et al.* (1977) for mouse ova, which undergo a "premature" release of about 25% of their cortical granules by 30 min after *in vitro* insemination. In such cases a specific population of cortical granules is released, i.e., the light cortical granules. Premature cortical granule exocytosis reportedly continues throughout oocyte maturation (Zamboni, 1974).

Recent studies, taking advantage of the presence of glycoproteins in the cortical granules of mammalian eggs that react with lentil lectin (Ducibella *et al.*, 1988a; Cherr *et al.*, 1988), have localized and quantitated the distribution of cortical granules in mouse eggs and oocytes. In the mouse cortical granules occupy the entire cortex of the oocyte. With meiotic maturation these organelles disappear from the cortical region occupied by the meiotic spindle such that a cortical granule-free domain comprising 39% to 41% of the entire egg cortex or about 6800 μm^2 is formed. Estimates of the size of the cortical granule-free domain range from 17% to 24% of the cortex (Szollosi, 1967; Zamboni, 1970). The reason for this variability is not known, but it may be related to differences in gamete ages, strains of mice, etc.

The density of cortical granules, i.e., the number per unit area, is lower in mouse oocytes than in mature eggs. This is consistent with reports of cortical granule migration and production near the time of ovulation in mouse as well as other mammals (Baca and Zamboni, 1967; Zamboni, 1970; Cran and Cheng, 1985; Sathananthan and Trounson, 1982a,b). It is possible that the cortical granule-free domain is formed, in part, by the redistribution of cortical granules within the cortex, i.e., a movement away from the cortical region in which the meiotic spindle is located. This, however, does not account for the decrease in the total number of cortical granules that occurs during oocyte maturation (Ducibella *et al.*, 1988a). The lower number of cortical granules in mature eggs could be related to formation of the first polar body, although this structure does not possess enough granules to account for the decrease (Ducibella *et al.*, 1988a). Other possibilities that have been proposed include premature cortical granule release, turnover, and/or the inability of the probe to stain all cortical granules. If the loss of cortical granule number involves their exocytosis, a release of this magnitude (~45% of the oocyte's complement) is paradoxial, as one might expect the zona block to polyspermy to be initiated (Gulyas, 1980). Ducibella *et al.* (1988a) suggest that in this instance there might be the release of a subpopulation of granules that do not affect sperm binding but prevent sperm penetration at the cortical granule-free domain of the egg.

5. ORGANIZATION OF THE MEIOTIC SPINDLE

Structural events concerning the breakdown of the germinal vesicle and formation of the first meiotic spindle in mammalian oocytes have been reviewed (Zamboni, 1971; Szollosi, 1976). Once formed, the meiotic spindle of the mouse egg moves to the cortex, where it becomes closely associated with the thickened actin filament layer and microvillus-free area. The meiotic spindle is a barrel-shaped structure composed of microtubules but lacks centrioles and asters; at its poles are bands of pericentriolar material (Szollosi *et al.*, 1972; Wassarman and Fugiwara, 1978; Calarco-Gilliam *et al.*, 1983; Maro *et al.*, 1985a, 1986b; Schatten *et al.*, 1985). The meiotic spindles of mouse, rat, and hamster eggs lie parallel to, whereas in sheep, cow, pig, rabbit, and human eggs the spindle is oriented perpendicular to the egg surface (Zamboni, 1970; Thibault *et al.*, 1987; Pickering *et al.*, 1988; Le Guen *et al.*, 1989). Maro *et al.* (1988) have successfully isolated meiotic spindles from mouse eggs, and further investigations employing such preparations may provide new insights into the structure and functions of this organelle.

Foci of pericentriolar material are also located in the cytoplasm (Maro *et al.*, 1985a), but in unfertilized eggs these sites are inactive as microtubule-organizing centers. However, they can organize microtubule arrays in the presence of taxol and give rise to the formation of multiple asters (Maro *et al.*, 1985a; Schatten *et al.*, 1985; see also Pickering *et al.*, 1988). Clusters of dispersed chromosomes have also been shown to induce the polymerization of tubulin (Maro *et al.*, 1986b; Johnson and Pickering, 1987). The effect of meiotic chromosomes on microtubules is similar to that observed in other cells that have been shown to be capable of lowering the critical concentration of tubulin to promote microtubule polymerization (Karsenti *et al.*, 1984). The chromosome–microtubule complex recruits adjacent pericentriolar material, which appear to be involved in the spatial organization of the spindle and the alignment of microtubules (Maro *et al.*, 1986b; Pickering and Johnson, 1987).

Acetylated α-tubulins have been localized at the poles of the meiotic spindle of unfertilized mouse eggs; microtubules in other spindle regions are unlabeled (Schatten *et al.*, 1988). At anaphase, spindle microtubules are more heavily decorated with antibody to acetylated α-tubulin, but at the completion of the second meiotic division only the midbody microtubules label. These observations indicate that meiotic spindle microtubules are differentially acetylated along their lengths and that this localization changes with meiosis. When eggs are treated with low temperature or griseofulvin, the meiotic spindle disappears except for a few microtubules that remain, which are acetylated. When reversed, two or more meiotic spindles may form in which acetylated α-tubulin is localized to their poles. Taxol-stabilized microtubules in the unfertilized egg are not heavily acetylated. Acetylation of α-tubulin, a posttranslational modification, has been observed in numerous cell types (L'Hernault and Rosenbaum, 1985; Piperno and Fuller, 1985; Bulinski *et al.*, 1988; Wolf *et al.*, 1988). A correlation between acetylation and microtubule stability has been noted (Piperno *et al.*, 1987; Schulze *et al.*, 1987); however, the connection between posttranslational modification and stability is not well understood particularly with respect to the dynamics of the mammalian egg meiotic spindle.

In invertebrates (Longo, 1972; Shimizu, 1981) polar body formation has been shown to consist of two processes: (1) formation of a cytoplasmic projection that contains chromosomes to be emitted from the egg and (2) the subsequent development of a cleavage furrow, which eventually severs the projection (Longo, 1972; Shimizu, 1981). Both processes appear to be microfilament dependent (Longo, 1972; Peaucellier *et al.*, 1974; Shimizu, 1981). Although polar body formation in mammalian eggs is susceptible to cytochalasins (Battaglia and Gaddum-Rosse, 1986), morphogenesis of this structure appears to differ in detail from that of nonmammals eggs (Maro *et al.*, 1984). Observations by Okada *et al.* (1986) and Longo (1987) indicate that the structure and dynamics of processes carried out at first meiosis are similar to those of the second meiotic division, although additional observations are needed to substantiate this.

Maro *et al.* (1984, 1986c) have examined changes in cortical actin during the formation of the second polar body in mouse eggs and have shown that concomitant with spindle rotation, there is a furrowing of the surface overlying the spindle equator (Fig. 6; Sato and Blandau, 1979). As a result, the thickened layer of actin associated with the meiotic spindle is bisected so that two shoulders of actin are formed, and one decreases in size while the second enlarges. This differential in cortical actin is believed to help bring about the rotation of the spindle such that the latter becomes situated with its equator at a plane tangential to the egg surface. Distal to this plane lies the forming polar body containing chromosomes and a half-spindle. Proximally lies the remaining half-spindle and the maternal chromosomes that will develop into a female pronucleus.

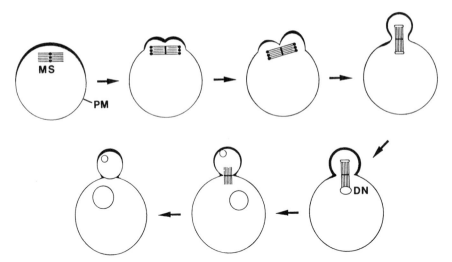

Figure 6. Diagrammatic representation of the second meiotic division and the changing pattern of actin associated with second polar body formation in fertilized mouse eggs. Modified from Maro *et al.* (1984). Events involving the incorporation of the sperm nucleus and its transformations are not depicted. The thickening associated with the plasma membrane (PM) represents the relative distribution of actin. Refer to text for details of polar body formation. MS, meiotic spindle; DN, developing nuclei.

The plasma membrane region that forms the leading edge of the cleavage furrow ultimately constricts to yield an actin-lined polar body containing the distal set of chromosomes.

6. ORGANIZATION OF ENDOPLASMIC RETICULUM AND ENDOSOMES

In mouse eggs, the cortical endoplasmic reticulum has been shown to be closely associated with both the plasma membrane and cortical granules (Luttmer and Longo, 1985). Similar cisternae have been observed in the eggs of other species, where it has been shown that these elements share common features with the sarcoplasmic reticulum of muscle cells. In *Xenopus* and sea urchin eggs there is evidence that the cortical endoplasmic reticulum represents sites of calcium flux that are involved in the cortical granule reaction (Gardiner and Grey, 1983). Recent investigations with hamster eggs have demonstrated that there is a propagated release of calcium from internal stores at fertilization (Miyazaki *et al.*, 1986; Miyazaki, 1988). On the basis of this evidence and investigations of muscle cells and nonmammalian eggs (Gardiner and Grey, 1983; Luttmer and Longo, 1985), such internal stores may include the cortical endoplasmic reticulum.

Endocytotic organelles, vesicular clusters, are localized almost exclusively to the cortex of rat and mouse eggs and oocytes (Pratt, 1985; Fleming and Pickering, 1985). Following incubation in horseradish peroxidase these organelles are virtually the only intracellular structures to accumulate tracer. Based on their morphology and ability to accumulate exogenous substances, it has been suggested that these organelles are functionally similar to endosomes (Helenius *et al.*, 1983). Although the role these structures play during oocyte maturation has not been determined, their structural similarities to vesicular elements formed as a part of endocytosis in somatic cells suggests that they may function in the metabolism of exogenous materials incorporated into eggs and oocytes.

7. CONCLUSIONS

The investigations discussed here characterize cellular components involved with alterations of the egg cortex prior to and following meiotic maturation and define the sequence of events attending this process. The nature of the interrelationships of cortical components, the precise character of the signals, and the actual mechanisms that are involved in producing the cytoarchitectural changes characteristic of the egg at its different stages of maturation remain to be determined. Van Blerkom (1985) suggests that protein derived from the germinal vesicle may activate cytoskeletal-associated enzymes that in turn function in the stage-related posttranslation modifications of oocyte proteins (Van Blerkom and Bell, 1986). The activation of enzymes such as kinases and phosphatases (Schultz *et al.*, 1983) could lead to alterations in cytoarchitecture to allow for the dynamic cortical changes, movements of the meiotic spindle, and redistribution of organelles that characterize the maturing mammalian oocyte. Future investigations along such lines will no doubt further our understanding of cortical changes taking place during oocyte maturation.

ACKNOWLEDGMENTS. The assistance of Ms. Susan Cook during the writing of this review is gratefully acknowledged. Work carried out in the author's laboratory was supported by research funds from the Rockefeller Foundation and the NIH.

8. REFERENCES

Albertini, D., 1984, Novel morphological approaches for the study of oocyte-maturation, *Biol. Reprod.* **30**:13–28.

Albertini, D. F., Overstrom, E. W., and Ebert, K. M., 1987, Changes in the organization of the actin cytoskeleton during perimplantation development of the pig embryo, *Biol. Reprod.* **37**:441–451.

Allworth, A., and Ziomek, C. A., 1988, Filipin-labeled complexes are polarized in their distribution in the cytoplasm of meiotically mature mouse eggs, *Gamete Res.* **20**:574–489.

Baca, M., and Zamboni, L., 1967, The fine structure of human follicular oocytes, *J. Ultrastruct. Res.* **19**:354–381.

Battaglia, D. E., and Gaddum-Rosse, P., 1986, The distribution of polymerized actin in the rat egg and its sensitivity to cytochalasin B during fertilization, *J. Exp. Zool.* **237**:97–105.

Battaglia, D. E., and Gaddum-Rosse, P., 1987, Influence of the calcium ionophore A23187 on rat egg behavior and cortical F-actin, *Gamete Res.* **18**:141–152.

Bulinski, J. C., Richards, J. E., and Piperno, G., 1988, Post-translational modifications of α-tubulin: Detyrosination and acetylation differentiate populations of interphase microtubules in cultured cells, *J. Cell Biol.* **106**:1213–1220.

Calarco, P. G., and Epstein, C. J., 1973, Cell surface changes during preimplantation development in the mouse, *Dev. Biol.* **32**:208–213.

Calarco-Gillam, P. G., Siebert, M. C., Hubble, R., Mitchison, T., and Kirschner, M., 1983, Centrosome development in early mouse embryos as defined by an auto-antibody against pericentriolar material, *Cell* **35**:621–629.

Capco, D. G., and McGaughey, R. W., 1986, Cytoskeletal reorganization during early mammalian development: Analysis using embedment-free sections, *Dev. Biol.* **115**:446–458.

Cherr, G. N., Drobnis, E. Z., and Katz, D. F., 1988, Localization of cortical granule constituents before and after exocytosis in the hamster egg, *J. Exp. Zool.* **246**:81–93.

Condeelis, J., 1979, Isolation of concanavalin A caps during various stages of formation and their association with actin and myosin, *J. Cell Biol.* **80**:751–758.

Cooper, C. W., and Bedford, J. M., 1971, Charge density change in the vitelline surface following fertilization of the rabbit egg, *J. Reprod. Fertil.* **25**:431–436.

Cran, D., and Cheng, W., 1985, Changes in cortical granules during porcine oocyte maturation, *Gamete Res.* **11**:311–319.

Damjanov, I., Damjanov, A., Lehto, V.-P., and Vertanen, I., 1986, Spectrin in mouse gametogenesis and embryogenesis, *Dev. Biol.* **114**:132–140.

Davidson, E. H., 1980, *Gene Activity in Early Development*, Academic Press, New York.

Ducibella, T., Ukena, T., Karnovsky, M., and Anderson, E., 1977, Changes in cell surface and cytoplasmic organization during early embryogenesis in the preimplantation mouse embryo, *J. Cell Biol.* **74:**153–167.

Ducibella, T., Anderson, E., Albertini, D. F., Aalberg, J., and Rangarajan, S., 1988a, Quantitative studies of changes in cortical granule number and distribution in the mouse oocyte during meiotic maturation, *Dev. Biol.* **130:**184–197.

Ducibella, T., Rangarajan, S., and Anderson, A., 1988b, The development of mouse oocyte cortical reaction competence is accompanied by major changes in cortical vesicles and not cortical granule depth, *Dev. Biol.* **130:**789–792.

Eager, D. D., Johnson, M. H., and Thurley, K. W., 1976, Ultrastructural studies on the surface membrane of the mouse egg, *J. Cell Sci.* **22:**345–353.

Ebensperger, C., and Barros, C., 1984, Changes at the hamster oocyte surface from the germinal vesicle stage to ovulation, *Gamete Res.* **9:**387–397.

Fleming, T. P., and Pickering, S. J., 1985, Maturation and polarization of the endocytotic system in outside blastomeres during mouse preimplantation development, *J. Embryol. Exp. Morphol.* **89:**175–208.

Franke, W. W., Schmid, E., Schiller, D. L., Winter, W., Jarasch, E. D., Moll, R., Denk, H., Jackson, B. W., and Illmensee, K., 1982, Differentiation-related patterns of expression of proteins of intermediate-size filaments in tissues and cultured cells, *Cold Spring Harbor Symp. Quant. Biol.* **46:**431–453.

Gardiner, D. M., and Grey, R. D., 1983, Membrane junctions in *Xenopus* eggs: their distribution suggests a role in calcium regulation, *J. Cell Biol.* **96:**1159–1163.

Geiger, B., 1983, Membrane–cytoskeleton interaction, *Biochim. Biophys. Acta* **737:**305–341.

Gould, K. G., 1973, Preparation of mammalian gametes and reproductive tract tissues for scanning electron microscopy, *Fertil. Steril.* **24:**448–456.

Gulyas, B. J., 1976, Ultrastructural observations on rabbit, hamster and mouse eggs following electrical stimulation *in vitro*, *Am. J. Anat.* **147:**203–217.

Gulyas, B. J., 1980, Cortical granules of mammalian eggs, *Int. Rev. Cytol.* **63:**357–392.

Guraya, S. S., 1982, Recent progress in the structure, origin, composition and function of cortical granules in animal eggs, *Int. Rev. Cytol.* **78:**257–359.

Helenius, A., Mellman, I., Wall, D., and Hubbard, A., 1983, Endosomes, *Trends Biochem. Sci.* **8:**245–250.

Imhof, B. A., Martin, U., Boller, K., Frank, H., and Birchmeier, W., 1983, Association between coated vesicles and microtubules, *Exp. Cell Res.* **145:**199–207.

Jeffrey, W. R., Tomlinson, D. R., and Brodeur, R. D., 1983, Localization of actin messenger RNA during early ascidian development, *Dev. Biol.* **99:**408–417.

Johnson, M. H., and Pickering, S. J., 1987, The effect of dimethylsulphoxide on the microtubular system of the mouse oocyte, *Development* **100:**313–324.

Johnson, M. H., Eager, D., and Muggleton-Harris, A., 1975, Mosiacism in organization of concanavalin A receptor on surface membrane of mouse eggs, *Nature* **257:**321–322.

Karasiewicz, J., and Soltynska, M. S., 1985, Ultrastructural evidence for the presence of actin filaments in mouse eggs at fertilization, *Wilhelm Rouxs Arch. Dev. Biol.* **194:**369–372.

Karsenti, E., Newport, J., Hubble, R., and Kirschner, M., 1984, Interconversion of metaphase and interphase microtubule arrays as studied by the injection of centrosomes and nuclei into *Xenopus* eggs, *J. Cell Biol.* **98:**1730–1745.

Kelly, W. G., Passaniti, A., Woods, J. W., Baiss, J. L., and Roth, T. F., 1983, Tubulin as a molecular component of coated vesicles, *J. Cell Biol.* **97:**1191–1199.

Koehler, J. K., Clark, J. M., and Smith, D., 1985, Freeze-fracture observations on mammalian oocytes, *Am. J. Anat.* **174:**317–329.

Le Guen, P., Crozet, N., Huneau, D., and Gall, L., 1989, Distribution and role of microfilaments during events of sheep fertilization, *Gamete Res.* **22:**411–425.

Lehtonen, E., 1985, A monoclonal antibody against mouse oocyte cytoskeleton recognizing cytokeratin-type filaments, *J. Embryol. Exp. Morphol.* **90:**197–209.

Lehtonen, E., 1987, Cytokeratins in oocytes and preimplantation embryos of the mouse, in: *Current Topics in Developmental Biology*, Vol. 22. *The Molecular and Developmental Biology of Keratins* (A. A. Moscona and A. Monroy, eds.), Academic Press, New York, pp. 153–173.

Lehtonen, E., and Vertanen, I., 1985, Evidence for the presence of cytokeratin-like protein in preimplantation mouse embryos, *Ann. N.Y. Acad. Sci.* **455:**744–747.

Lehtonen, E., Lehto, V.-P., Vartio, T., Badley, R. A., and Virtanen, I., 1983, Expression of cytokertin polypeptides in mouse oocytes and preimplantation embryos, *Dev. Biol.* **100:**158–165.

L'Hernault, S. W., and Rosenbaum, J. L., 1985, Reversal of the post-translational modification on *Chlamydomonas* flagellar α-tubulin occurs during flagellar resorption, *J. Cell Biol.* **100:**457–462.

Liebfried-Rutledge, M. L., Florman, H. M., and First, N. L., 1989, The molecular biology of mammalian oocyte maturation, in: *The Molecular Biology of Fertilization* (H. Schatten and G. Schatten, eds.), Academic Press, New York, pp. 259–301.

Longo, F. J., 1972, The effects of cyotchalasin-B on the events of fertilization in the surf clam *Spisula solidissima*, I. Polar body formation, *J. Exp. Zool.* **182:**321–344.

Longo, F. J., 1974, An ultrastructural analysis of spontaneous activation of hamster eggs aged *in vivo*, *Anat. Rec.* **179:**27–56.

Longo, F. J., 1981, Morphological features of the surface of the sea urchin (*Arbacia punctulata*) egg: Oolemma–cortical granule association, *Dev. Biol.* **83:**173–181.

Longo, F. J., 1985, Fine structure of the mammalian egg cortex. *Am. J. Anat.* **174:**303–315.

Longo, F. J., 1987, Actin–plasma membrane associations in mouse eggs and oocytes, *J. Exp. Zool.* **243:** 299–309.

Longo, F. J., and Chen, D.Y., 1984, Development of surface polarity in mouse eggs, *Scanning Electron Microsc.* **2:**703–716.

Longo, F. J., and Chen, D. Y., 1985, Development of cortical polarity in mouse eggs: Involvement in the meiotic apparatus, *Dev. Biol.* **107:**382–294.

Loor, F., 1981, Cell surface–cell cortex transmembranous interactions with special reference to lymphocyte functions, in: *Cytoskeletal Elements and Plasma Membrane Organization.* (G. Poste and G. L. Nicolson, eds.), Elsevier/North Holland, Amsterdam, pp. 255–335.

Lopata, A., Sathananthan, A. H., McBain, J. C., Johnston, W. I. H., and Speirs, A. L., 1980, The ultrastructure of preovulatory human egg fertilized *in vitro*, *Fertil. Steril.* **33:**12–20.

Louvard, D., and Reggio, H., 1981, Role des microtubules dans l'organization due complexe de Golgi, *Ann. Endocrinol.* **42:**349–362.

Luttmer, S., and Longo, F. J., 1985, Ultrastructural and morphometric observations of cortical endoplasmic reticulum in *Arbacia, Spisula* and mouse eggs, *Dev. Growth Differ.* **27:**349–359.

Maro, B., Johnson, M. H., Pickering, S. J., and Flach, G., 1984, Changes in actin distribution during fertilization of mouse egg, *J. Embryol. Exp. Morphol.* **81:**211–237.

Maro, B., Howlett, S. K., and Webb, M., 1985a, Non-specific microtubule organizing centers in metaphase II-arrested mouse oocytes, *J. Cell Biol.* **101:**1665–1672.

Maro, B., Johnson, M. H., Pickering, S. J., and Louvard, D., 1985b, Changes in the distribution of membranous organelles during mouse early development, *J. Embryol. Exp. Morphol.* **90:**287–309.

Maro, B., Howlett, S. K., and Johnson, M. H., 1986a, Cellular and molecular interpretation of mouse early development: The first cell cycle, in: *Gametogenesis and the Early Embryo* (J. G. Gall, ed.), Alan R. Liss, New York, pp. 389–407.

Maro, B., Howlett, S. K., and Houliston, E., 1986b, Cytoskeletal dynamics in the mouse egg. *J. Cell Sci. [Suppl.]* **5:**343-359.

Maro, B., Johnson, M. H., Webb, M., and Flach, G., 1986c, Mechanism of polar body formation in the mouse oocyte: An interaction between the chromosomes, the cytoskeleton and the plasma membrane, *J. Embryol. Exp. Morphol.* **92:**11–32.

Maro, B., Houliston, E. H., and Paintrand, M., 1988, Purification of meiotic spindles and cytoplasmic asters from mouse oocytes, *Dev. Biol.* **129:**275–282.

Miyazaki, S., 1988, Inositol 1,4,5 triphosphate-induced calcium release and guanine nucleotide-binding protein-mediated periodic calcium rises in golden hamster eggs, *J. Cell Biol.* **106:**345–353.

Miyazaki, S., Hashimoto, N., Yoshimoto, Y., Kishimoto, T., Igusa, Y., and Hiramoto, Y., 1986, Temporal and spatial dynamics of the periodic increase in intracellular free calcium at fertilization of golden hamster eggs, *Dev. Biol.* **118:**259–267.

Moon, R. T., Nicosia, R. F., Olsen, C., Hille, M. B., and Jeffrey, W. R., 1983, The cytoskeletal framework of sea urchin eggs and embryos: Developmental changes in the association of messenger RNA, *Dev. Biol.* **95:**447–458.

Nicosia, S. V., Wolf, D. P., and Inoue, M., 1977, Cortical granule distribution and cell surface characteristics in mouse eggs, *Dev. Biol.* **57:**56–74.

Nicosia, S. V., Wolf, D. P., and Mastroianni, L., 1978, Surface topography of mouse eggs before and after insemination, *Gamete Res.* **1:**145–155.

Norberg, H. S., 1973, Ultrastructure of pig tubal ova, *Z. Zellforsch* **141:**103–122.

Odor, D. L., and Renninger, D. F., 1960, Polar body formation in the rat oocyte as observed with the electron microscope, *Anat. Rec.* **147:**13–23.

Okada, A., Yangimachi, R., and Yangimachi, H., 1986, Development of a cortical granule-free area of cortex and the perivitelline space in the hamster oocyte during maturation and following ovulation, *J. Submicrosc. Cytol.* **18:**233–247.

Osborn, M., and Weber, K., 1983, Tumor diagnosis by intermediate filament typing: A novel tool for surgical pathology, *Lab. Invest.* **48:**372–394.

Peaucellier, G., Guerrier, P., and Bergerard, J., 1974, Effects of cytochalasin B on meiosis and development of fertilized and activated eggs of *Sabellaria alveolata* (Polychaete Annelid), *J. Embryol. Exp. Morphol.* **31:**61–74.

Penman, S., Capco, D. G., Fey, E. G., Chatterjee, P., Reiter, T., Ermish, S., and Wang, K., 1983, The three-dimensional structural networks of cytoplasm and nucleus: Function in cells and tissue, in: *Modern Cell Biology*, Vol. 2 (J. R. MacIntosh, ed.). Alan R. Liss, New York, pp. 385–415.

Peters, R., 1981, Translation diffusion in the plasma membrane of single cells as studied by fluorescence microphotolysis, *Cell Biol. Int. Rep.* **5:**733–760.

Pfeffer, S. R., Drubin, D. G., and Kelly, R. B., 1983, Identification of three coated vesicle components as alpha and beta tubulin linked to a phosphorylated 50,000 dalton polypeptide, *J. Cell Biol.* **97:**40–47.

Phillips, D. M., and Shalgi, R., 1980, Surface architecture of the mouse and hamster zona pellucida and oocyte, *J. Ultrastruct. Res.* **72:**1–12.

Pickering, S. J., and Johnson, M. H., 1987, The influence of cooling on the organization of the meiotic spindle of mouse oocytes, *Hum. Reprod.* **2:**207–216.

Pickering, S. J., Johnson, M. H., Broude, P. R., and Houliston, E., 1988, Cytoskeletal organization in fresh, aged and spontaneously activated human oocytes, *Hum. Reprod.* **3:**978–989.

Piperno, G., and Fuller, M. T., 1985, Monoclonal antibodies specific for an acetylated form of α-tubulin recognize the antigen in cilia and flagella from a variety of organisms, *J. Cell Biol.* **101:**1665–1672.

Piperno, G., LeDizet, M., and Chang, X., 1987, Microtubules containing acetylated α-tubulin in mammalian cells in culture, *J. Cell Biol.* **104:**289–302.

Pratt, H. P. M., 1985, Membrane organization in the preimplantation mouse embryo, *J. Embryol. Exp. Morphol.* **90:**101–121.

Reima, I., and Lehtonen, E., 1985, Localization of monerythroid spectrin and actin in mouse oocytes and preimplantation embryos, *Differentiation* **30:**68–75.

Saffman, P. G., and Delbruck, M., 1975, Brownian motion in biological membranes, *Proc. Natl. Acad. Sci. U.S.A.* **72:**3111–3113.

Sathananthan, A. H., and Lopata, L., 1980, Ultrastructure of human eggs: Aspirated preovulatory mature ova prior to fertilization, *Micron* **11:**469–470.

Sathananthan, A. H., and Trounson, A. O., 1982a, Ultrastructural observations on cortical granules in human follicular oocytes cultured *in vitro*, *Gamete Res.* **5:**191–198.

Sathananthan, A. H., and Trounson, A. O., 1982b, Ultrastructure of cortical granule release and zona interaction in monospermic and polyspermic human ova fertilized *in vitro*, *Gamete Res.* **6:**225–234.

Sato, K., and Blandau, R. J., 1979, Second meiotic division and polar body formation in mouse eggs fertilized *in vitro*, *Gamete Res.* **2:**283–293.

Schatten, G., Simerly, C., and Schatten, H., 1985, Microtubule configurations during fertilization, mitosis and early development in the mouse and the requirement for egg microtubule-mediated motility during mammalian fertilization, *Proc. Natl. Acad. Sci. U.S.A.* **82:**4152–4256.

Schatten, G., Schatten, H., Spector, I., Cline, C., Paweletz, N., Simerly, C., and Petzelt, C., 1986, Latrunculin inhibits the microfilament-mediated processes during fertilization, cleavage and early development in sea urchins and mice, *Exp. Cell Res.* **166:**191–208.

Schatten, G., Simerly, C., Asai, D. J., Szoke, E., Cooke, P., and Schatten, H., 1988, Acetylated α-tubulin in microtubules during mouse fertilization and early development, *Dev. Biol.* **130:**74–86.

Schatten, H., Cheney, R., Balczon, R., Willard, M., Cline, C., Simerly, C., and Schatten, G., 1986, Localization of fodrin during fertilization and early development of sea urchins and mice, *Dev. Biol.* **118:** 457–466.

Schmell, E. O., Gulyas, B. J., and Hedrick, J. L., 1983, Egg surface changes during fertilization and the molecular mechanisms of the block to polyspermy, in: *Mechanism and Control of Animal Fertilization* (J. F. Hartmann, ed.), Academic Press, New York, pp. 365–413.

Schroeder, R. E., 1981, Interrelations between the cell surface and the cytoskeleton in cleaving sea urchin eggs, in: *Cytoskeletal Elements and Plasma Membrane Organization* (G. Poste and G. L. Nicolson, eds.), Elsevier/North-Holland Biomedical Press, New York, pp. 169–216.

Schuel, H., 1985, Functions of egg cortical granules, in: *Biology of Fertilization*, Vol. 3 (C. B. Metz and A. Monroy, eds.), Academic Press, New York, pp. 1–44.

Schultz, R. M., Montgomery, R., and Belanoff, J. R., 1983, Regulation of mouse oocyte meiotic maturation: Implication of a decrease in oocyte cAMP and protein dephosphorylation in commitment to resume meiosis, *Dev. Biol.* **97:**264–273.

Schulze, E., Asai, D. J., Bulinski, J. C., and Kirschner, M., 1987, Posttranslational modification and microtubule stability, *J. Cell Biol.* **105**:2167–2177.

Shimizu, T., 1981, Cortical differentiation of the animal pole during maturation division in fertilized eggs of *Tubifex* (Annelida, Oligochaeta), *Dev. Biol.* **85**:7–88.

Showman, R. M., Wells, D. E., Anstrom, J., Hursch, D. A., and Raff, R. A., 1982, Message specific sequestration of maternal histone in RNA in the sea urchin egg, *Proc. Natl. Acad. Sci. U.S.A.* **79**:5944–5947.

Sobel, J. S., and Alliegro, M. A., 1985, Changes in the distribution of a spectrin-like protein during development of the preimplantation mouse embryo, *J. Cell Biol.* **100**:333–336.

Stefanini, M., Oura, C., and Zamboni, L., 1969, Ultrastructural of fertilization in the mouse. 2. Penetration of the sperm into the ovum, *J. Submicrosc. Cytol.* **1**:1–23.

Suzuki, S., Kitai, H., Tojo, R., Seki, K., Oba, M., Fujiwara, T, and Iizuka, R., 1981, Ultrastructural and some biologic properties of human oocytes and granulosa cells cultured *in vitro*, *Fertil. Steril.* **35**:142–148.

Szollosi, D., 1967, Development of cortical granules and the cortical reaction in rat and hamster eggs, *Anat. Rec.* **159**:431–446.

Szollosi, D., 1976, Oocyte maturation and paternal contribution to the embryo in mammals, in: *Current Topics in Pathology* (E. Grundmann and W. H. Kirsten, eds.), Springer-Verlag, New York, pp. 9–27.

Szollosi, D., Calarco, P. G., and Donahue, R. P., 1972, Absence of centrioles in the first and second meiotic spindles of mouse oocytes, *J. Cell Sci.* **11**:521–541.

Thibault, C., Szollosi, D., and Gerard, M., 1987, Mammalian oocyte maturation, *Reprod. Natr. Dev.* **27**:865–896.

Thompson, R. S., Moore-Smith, D., and Zamboni, L., 1974, Fertilization of mouse ova *in vitro*: an electron microscopy study, *Fertil. Steril.* **25**:222–249.

Vacquier, V. D., 1981, Dynamic changes of the egg cortex, *Dev. Biol.* **84**:1–26.

Van Blerkom, J., 1985, Extragenomic regulation and autonomous expression of a developmental program in the early mouse embryo, *Ann. N.Y. Acad. Sci.* **442**:58–72.

Van Blerkom, J., and Bell, H., 1986, Regulation of development in the fully grown mouse oocyte: Chromosome-mediated temporal and spatial differentiation of the cytoplasm and plasma membrane, *J. Embryol. Exp. Morphol.* **93**:213–238.

Wablik-Sliz, B., and Kujat, R., 1979, The surface of mouse oocytes from two inbred strains differing in efficiency of fertilization, as revealed by scanning electron microscopy, *Biol. Reprod.* **20**:405–408.

Wassarman, P. M., and Fugiwara, K., 1978, Immunofluorescent anti-tubulin staining of spindles during meiotic maturation of mouse oocytes *in vitro*, *J. Cell Sci.* **29**:171–188.

Wassarman, P. M., Bleil, J. D., Caseio, S. M., La Marca, M. J., Letourneau, G. E., Mrozak, S. C., and Schultz, R. M., 1981, Programming of gene expression during mammalian oogenesis. in: *Bioregulators of Reproduction* (G. Jagiello and H. J. Vogel, eds.), Academic Press, New York, pp. 119–150.

Webb, M., Howlett, S. K., and Maro, B., 1986, Parthenogenesis and cytoskeletal organization in aging mouse eggs, *J. Embryol. Exp. Morphol.* **95**:131–145.

Wolf, D. E., and Ziomek, C. A., 1983, Regionalizaion and lateral diffusion of membrane proteins in unfertilized and fertilized mouse eggs, *J. Cell Biol.* **96**:1786–1790.

Wolf, D. E., Edidin, M., and Handyside, A. H., 1981, Changes in the organization of the mouse egg plasma membrane upon fertilization and first cleavage: Indication from the lateral diffusion rates of fluorescent lipid analogs, *Dev. Biol.* **85**:195–198.

Wolf, N., Regan, C. L., and Fuller, M. T., 1988, Temporal and spatial pattern of differences in microtubule behavior during *Drosophila* embryogenesis revealed by distribution of a tubulin isoform, *Development* **102**:311–324.

Yanagimachi, R., and Chang, M. C., 1961, Fertilizable life of golden hamster ova and their morphological changes at the time of losing fertility, *J. Exp. Zool.* **148**:185–203.

Yanagimachi, R., and Noda, Y. D., 1970, Electron microscopic studies of sperm incorporation into the golden hamster egg, *Am. J. Anat.* **128**:429–462.

Yanagimachi, R., Nicolson, G. L., Noda, Y. D., and Fujimoto, M., 1973, Electron microscopic observations of the distribution of acidic anionic residues on hamster spermatozoa and eggs before and during fertilization, *J. Ultrastruct. Res.* **43**:344–353.

Zamboni, L., 1970, Ultrastructural of mammalian oocytes and ova, *Biol. Reprod. [Suppl.]* **2**:44–63.

Zamboni, L., 1971, *Fine Morphology of Mammalian Fertilization*, Harper & Row, New York.

Zamboni, L., 1972, Comparative studies on the ultrastructure of mammalian oocytes, in: *Oogenesis* (J. D. Biggers and A. W. Schuetz, eds.), University Park Press, Baltimore, pp. 5–45.

Zamboni, L., 1974, Fine morphology of the follicle wall and follicle cell–oocyte association, *Biol. Reprod.* **10**:125–149.

Zamboni, L., Thompson, R. S., and Smith, D. M.,1972, Fine morphology of human oocyte maturation *in vitro*, *Biol. Reprod.* **7**:425–457.

6

Comparative Structure and Function of Mammalian Zonae Pellucidae

Bonnie S. Dunbar, S. V. Prasad, and T. M. Timmons

1. INTRODUCTION

The process of fertilization, like other aspects of mammalian reproduction, varies markedly among different species. Of the many complex events that accompany this process, the interaction between the sperm and the zona pellucida (ZP) is one of the most significant. The mammalian ZP is an extracellular glycoprotein structure that is formed around the oocyte during the early stages of ovarian follicular development. This matrix is important in the initial stages of fertilization because the sperm must bind to and penetrate it before fusing with the oocyte plasma membrane (Austin and Braden, 1956; Austin, 1975). The ZP is also involved in the block to polyspermy in many species and further serves to protect the embryo after fertilization until implantation in the uterine wall. Species differences have been reported for several steps in the fertilization process, including sperm capacitation (the changes spermatozoa undergo in the female reproductive tract in order to develop the capacity to fertilize the egg) and sperm binding to and penetration of the ZP. Many of these differences can be attributed to morphological and biochemical variations of the gametes themselves (Yanagimachi, 1977; Bedford, 1974; other chapters in this text).

2. MORPHOLOGICAL PROPERTIES OF ZONAE PELLUCIDAE

The morphological properties of mammalian ZP have been studied by light microscopy and by scanning and transmission electron microscopy. These studies, in addition to immunocytochemical localization studies, suggest that the ZP of nonrodent species such as the human (Stegner and Wartenberg, 1961), pig (Dickmann and Dziuk, 1964; Hedrick and Frye, 1980), rabbit (Sacco, 1981; Skinner and Dunbar, 1986), and cat (Konecny, 1959) can be resolved more distinctly into multiple layers than the ZP of the mouse (Cholewa-Stewart and Massaro, 1972) or

BONNIE S. DUNBAR, S. V. PRASAD, AND T. M. TIMMONS • Department of Cell Biology, Baylor College of Medicine, Houston, Texas 77030.

A Comparative Overview of Mammalian Fertilization, edited by Bonnie S. Dunbar and Michael G. O'Rand. Plenum Press, New York, 1991.

the rat (Sacco, 1981). If ruthenium red is used in histochemical electron micrographic studies, however, the ZP of both the mouse (Baranska *et al.*, 1974) and rat (Dunbar and Wolgemuth, 1984) can be resolved into two distinct layers. Figure 1 illustrates the layers visualized within the ZP of different species using various microscopic techniques.

The inner (oocyte-proximal) and outer (oocyte-distal) boundaries of all ZP examined to date can easily be distinguished by scanning electron microscopy (Dunbar, 1983; Phillips and Shalgi, 1980) (Fig. 2). Morphological observations have shown that the ZP of different species also vary

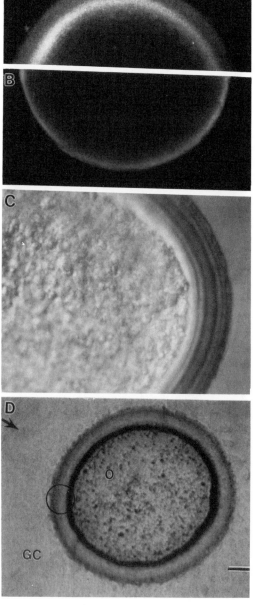

Figure 1. Visualization of the "layered" structure of the mammalian ZP matrix. (A) Pig ZP, incubated with fluorescein-labeled wheat-germ agglutinin (WGA). (B) Pig ZP, incubated with fluorescein-labeled *Ricinus communis* agglutinin (RCA). (C) Pig ZP, incubated with monoclonal antibody PS1 that recognizes a carbohydrate antigen (see text), viewed under modulation-contrast optics. (D) Section (5 μm thick) of rabbit ovary embedded in Epon and incubated with monoclonal antibody (PS1) that recognizes a carbohydrate antigen, followed by a peroxidase-labeled second antibody and DAB. Circle encloses ZP.

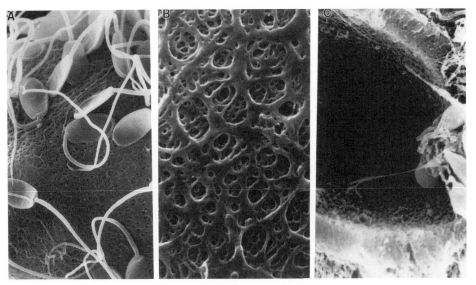

Figure 2. Morphology of matrix by scanning electron microscopy. (A) Rabbit sperm bound to the ZP (from Swenson and Dunbar, 1982). At higher magnification (B and C), the porcine ZP matrix appears to be a complex but porous meshwork as observed by scanning electron microscopy. (B) Outer surface. (C) Inner surface (D. M. Phillips and B. S. Dunbar, unpublished observations).

considerably in width. For example, ZP of mouse oocytes is relatively thin compared to the human, pig, guinea pig, or rabbit ZP (Fig. 3; Table 1). Such simple structural differences reflect the quantities and complexities of constituent glycoproteins, resulting in observed differences in physicochemical parameters (ease of solubilization by chemicals or enzymes, for example).

Further histochemical studies have been carried out to examine the lectin-binding properties of different mammalian ZP. These have demonstrated the uniform binding of some lectins throughout the ZP, whereas other lectins localize within discrete domains of the matrix (see

Table 1. Summary of the Relative Sizes of "Mature" Zonae Pellucidae from a Variety of Mammalian Species[a]

Species	Zona width (μm)[b]	References
Opossum	1–2	J. M. Bedford (personal communication)
Mouse	5	Lowenstein and Cohen (1964)
Hamster	8	Austin (1961)
Human	13	Austin (1961)
Dog	13	B. S. Dunbar (unpublished)
Baboon	13	B. S. Dunbar (unpublished)
Sheep	14.5	Wright et al. (1977)
Cat	15	B. S. Dunbar (unpublished)
Rabbit	15	Austin (1965)
Pig	16	B. S. Dunbar (unpublished)
Cow	27	Wright et al. (1977)

[a]Modified from Dunbar (1983).
[b]Values estimated from photographs.

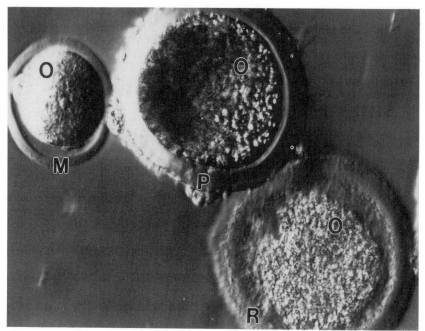

Figure 3. Comparison of the size of mammalian ZP. M, mouse ZP of mature mouse ovum (O); P, pig ZP of follicular ovum (O); R, rabbit ZP of follicular ovum (O); 15 μm.

reviews by Dunbar, 1980; Dunbar and Wolgemuth, 1984). The variety of lectins and methods of fixation and histological preparations used in these studies suggests that the asymmetric binding of certain lectins reflects true biological heterogeneity of the ZP. However, the identification of distinct glycoproteins within the matrix is greatly hampered by the lack of lectin-binding specificity and their limited ability to permeate the ZP matrix.

3. PHYSICOCHEMICAL PROPERTIES OF ZONAE PELLUCIDAE

Studies have been carried out to examine the efficacy of a variety of conditions to cause dissolution of zonae pellucidae, as determined by light microscopy. Although visual evaluation of ZP matrix dissolution does not necessarily correlate with biochemical criteria (Dunbar *et al.*, 1980, 1985), such information is useful to compare the ZP of different mammalian species and the susceptibility of the intact structures to the action of different enzymes. The effects of proteolytic enzymes on different species of ZP have been described in detail (Dunbar *et al.*, 1980). It is a general rule that "thinner" ZP structures are more sensitive to a variety of physical and enzymatic treatments. For example, the macromolecular structure of rodent ZP appears to be more sensitive to pH extremes and to proteolytic enzymes than those of other species, as evaluated by microscopic criteria.

4. GLYCOPROTEINS OF MAMMALIAN ZONAE PELLUCIDAE

The individual glycoprotein components of the ZP number between three and five, depending on the species examined and on the resolution and sensitivity of the protein separation and detection methods used (Bleil and Wassarman, 1980b; Dunbar *et al.*, 1981, 1985; Sacco *et al.*, 1981). In addition, in the case of hamster, although three species of glycoproteins have been demonstrated by various investigators (Ahuja and Botwell, 1983; Oikawa *et al.*, 1988; Moller *et al.*, 1990), there is a wide discrepancy in the reported molecular weights of these glycoproteins, suggesting the need for resolution of these proteins by two-dimensional electrophoresis. Porcine and rabbit ZP glycoproteins were the first analyzed by two-dimensional polyacrylamide gel electrophoresis (2D-PAGE) and were resolved as protein "families" with a common polypeptide backbone, each of which exhibits extensive charge and molecular weight heterogeneity (Dunbar *et al.*, 1981; Sacco *et al.*, 1981). This microheterogeneity is caused by the presence of highly charged oligosaccharide side chains, and this complex structure has given rise to a confusing and inconsistent nomenclature (Timmons and Dunbar, 1988). In a system based on the molecular weight of the least-glycosylated members of a particular family (Yurewicz *et al.*, 1987), the porcine ZP glycoproteins can be identified as ZP1 (82K), ZP2 (61K), ZP3-α (55K-α), and ZP3-β (55K-β). An alternate method that has been used more consistently to identify ZP glycoproteins is based on the estimation of molecular weight of the glycoprotein after partial deglycosylation by endo-β-galactosidase. The estimated molecular weights of ZP components of different species by this method are summarized in Table 2.

It should be noted that unlike the ZP components of nonrodent mammalian species, the glycosylated mouse ZP glycoproteins can be resolved into three major bands by one-dimensional nonreducing PAGE, presumably because the specific posttranslational modifications present do not introduce the same degree of molecular weight heterogeneity seen in other mammalian species (Sacco *et al.*, 1981).

5. CARBOHYDRATE COMPOSITION OF ZONAE PELLUCIDAE

Information on the ZP oligosaccharides from several species is now available (Yurewicz *et al.*, 1987; Hedrick and Wardrip, 1987; Salzmann *et al.*, 1983). Carbohydrate analysis of the mouse ZP glycoproteins indicates that all three glycoproteins contain asparagine-linked complex-type oligosaccharides, and at least two also contain serine/threonine-linked carbohydrates (Wassarman, 1988). A specific class of O-linked oligosaccharides on the ZP3 molecule (83K) containing an α-linked galactose residue at the nonreducing terminus functions as the primary initial attachment site for sperm (Bleil and Wassarman, 1988), possibly through the action of galactosyltransferase activity residing on the sperm surface (Shur and Hall, 1982).

Precise oligosaccharide structures are not yet known for all species, but most of the ZP glycoproteins carry both N- and O-linked sugars (Yurewicz *et al.*, 1987; Florman and Wassarman, 1985; Timmons and Dunbar, 1990a). Trifluoromethane sulfonic acid completely removes both types of carbohydrate chains, leaving only the galactose residue still linked to the serine or threonine residue in the protein backbone (Karp *et al.*, 1982). A variety of glycosidases are effective in removing carbohydrate moieties from the ZP glycoproteins (Timmons and Dunbar, 1990a; Wassarman, 1988), including both exo- and endoglycosidases. Endo-β-galactosidase (EBGD) has proved especially useful, since the ZP components of all species examined contain lactosaminoglycan (LAG) structures (T. M. Timmons and B. S. Dunbar, unpublished observations). If the reaction is allowed to go to completion, the partially deglycosylated glycoproteins that still contain the high-mannose core carbohydrates and other non-LAG structures can be easily and reproducibly resolved by 1D-PAGE (Table 2).

A monoclonal antibody (PS1) that recognizes a carbohydrate epitope in multiple ZP glycoproteins of several mammalian species has been generated through the use of a porcine ZP glycoprotein immunogen excised from a silver-stained polyacrylamide gel (Drell and Dunbar, 1984; Timmons and Dunbar, 1990a). Although the immunogen was a single protein charge

Table 2. Estimated Molecular Weights of ZP Proteins
from Different Mammalian Species[a]

Species	Molecular weight[b] ($\times 10^3$ kda)	Molecular weight[c] ($\times 10^3$ kda)	References[e]
Pig	40–110	45,47	1,2,3
	70–110	62	
	95–118	80	
Rabbit	68–125	55	1,4
	81–100	75	
	100–132	85	
Baboon	---	50	4
	---	70	
	---	95	
Cat	50–110	50	4,5
	90–100	65	
		90	
Dog	50–100	48	4,5
	70–95	70	
	90–100	73	
		88	
Mouse[d]	200	--	3,6
	120	--	
	83	--	
Human	57–73	--	7
	64–78	--	
	90–110	--	

[a]Modified from Timmons and Dunbar (1988).
[b]Values are determined by 2D-SDS-PAGE under reducing conditions.
[c]Values represent ZP glycoproteins following partial deglycosylation by endo-β-galactosidase.
[d]Values represent nonreducing conditions on 1D-PAGE.
[e](1) Dunbar *et al*. (1981); (2) Dunbar *et al*. (1985); (3) Sacco *et al*. (1981); (4) T. M. Timmons and B. S. Dunbar, unpublished; (5) Maresh and Dunbar (1987); (6) Bleil and Wassarman (1980a); (7) Shabanowitz and O'Rand (1988).

species of the most abundant porcine ZP protein from a two-dimensional polyacrylamide gel, ,the antibody reacts with the most acidic (the most extensively glycosylated and/or sulfated) members of all three porcine ZP glycoprotein families resolved by two-dimensional polyacrylamide gel electrophoresis (2D-PAGE) in the presence of reducing agents. In the nomenclature described above, these are ZP1 (82K), ZP2 (61K), and ZP3-α (55K). PS1 also recognizes at least one of the ZP glycoproteins of several other mammalian species, including rabbit, cat, dog, and baboon (Maresh and Dunbar, 1987; Timmons and Dunbar, 1990b), but not the rodent ZP. Preliminary investigations into the biochemical structure of the epitope have demonstrated that it is associated with an N-linked rather than an O-linked carbohydrate chain in the pig, since PS1 recognition of ZP glycoproteins in immunoblots is destroyed following treatment with N-glycosidase F. PS1 antibody binding is also sensitive to treatment of the glycoproteins with endo-β-galactosidase (EBGD), an enzyme specific for the degradation of LAG structures (Fukuda and Matsumura, 1976). The carbohydrate content of porcine ZP glycoproteins has been reported, and each contains approximately equimolar amounts of galactose and N-acetyl glucosamine (Dunbar *et al*., 1980; Hedrick and Wardrip, 1987). These repeating Gal(β)1→4GlcNAc(β)1→3 carbohydrate structures are commonly found as part of glycolipids and glycoproteins on cell surfaces and are known to function in cell adhesion and cell–cell interaction in several systems (Macek and Shur, 1988; Dutt *et al*., 1987; Fukuda, 1985; Hakomori, 1981).

These studies have also demonstrated that the PS1 antigen is present on an N-linked oligosaccharide associated with the surface of the porcine ZP. They further suggest that it is also involved in the initial stages of sperm–ZP interaction, since antibodies to this antigen will inhibit boar sperm from binding to the pig ZP (T. M. Timmons and B. S. Dunbar, unpublished observations). Detailed biochemical analysis of its structure will help elucidate the mechanism by which homologous sperm–ZP attachment, binding, and penetration occur during the process of mammalian fertilization.

To date, all glycoproteins reported to contain polylactosaminoglycans are membrane anchored with the exception of secretory glycoproteins such as erythropoietin (Sasaki *et al.*, 1987) and α-(1)-acid glycoprotein (Yoshima *et al.*, 1981), which contain only one or two N-acetyllactosamine repeat units. Based on carbohydrate analysis and on the magnitude of the decrease in apparent molecular weight as measured by 1D-SDS-PAGE after EBGD treatment (Table 2), ZP glycoproteins contain numerous repeat units per molecule. Therefore, they represent another class of molecules carrying LAG structures: non-membrane-bound extracellular matrix constituents containing large LAG chains. Polylactosaminoglycans also carry important cell surface determinants such as the ABO blood group and specific embryonal antigens, and these carbohydrate structures can exhibit striking changes during development and cellular differentiation (Fukuda, 1985). By analogy, dramatic changes in the polylactosaminoglycan structures of the ZP that are involved in extracellular matrix interactions with the oocyte and granulosa cells might be expected during follicular development.

6. MOLECULAR ANALYSIS OF ZP PROTEINS

Genes encoding ZP proteins are uniquely expressed during ovarian follicular development. By use of cDNA and oligomer probes, the gene for ZP3 of mouse has been isolated (Kinloch *et al.*, 1988; Chamberlain and Dean, 1989). This gene has been shown to contain eight exons spanning approximately 8.6 kilobases as revealed by S1 analysis, and it appears to be present in the mouse genome as a single-copy gene located on chromosome number 5 (Lunsford *et al.*, 1990). Comparison of restriction digests of ovarian and somatic DNA do not reveal any rearrangements. Sequence analysis of the genomic clones has revealed six imperfect 54-bp tandem repeats in the 5' flanking region and five in the seventh exon. At the 3' end, 43 base pairs downstream of polyadenylation site, a 12-nucleotide sequence is tandemly repeated 11 times (Chamberlain and Dean, 1989). Kinloch *et al.* (1988) observed similar repeats 50 base pairs downstream of the 3' end that were reiterated 21 times. The reason for this discrepancy is not clear.

Southern blots of genomic DNA from various species have been probed with mouse ZP3 cDNA to investigate the degree of DNA homology between various species. Restriction digests of genomic DNA from human, mouse, rat, rabbit, pig, dog, cow, and chicken were probed with mouse ZP3 cDNA under conditions that detect sequences of >78% homology. Mouse ZP3 cDNA hybridized strongly to human, rat, and mouse, weakly to chicken and pig, and not to rabbit (Ringuette *et al.*, 1986, 1988). However, when the blots were probed under lower stringency conditions, in which >58% homologies would be detected, mouse ZP3 cDNA hybridized to both pig and rabbit. It has been inferred from hybridization analyses that the binding of mouse ZP3 to chicken genomic DNA is not significant, whereas that to rabbit is, although ZP3 bound to rabbit DNA at a lower stringency (Ringuette *et al.*, 1988). These observations need to be confirmed by probing mouse genomic blots with cDNAs encoding different ZP proteins of other species and by comparing cDNA sequences directly. This would also reveal if cDNAs from the mouse and other species have conserved sequences at the DNA level, even though antigens are not shared between mouse and other mammalian species tested to date.

More recently, cDNA clones for the mouse ZP2 protein (Liang *et al.*, 1990) and for the rabbit 55K (Schwoebel *et al.*, 1991; Timmons and Dunbar, 1990b) and 75 ZP proteins (Dunbar, *et al.*, unpublished observations) have been obtained. Although the deduced amino acid sequences of the proteins encoded by these cDNAs show little similarity to the mouse ZP3 cDNA, there are considerable similarities between the mouse ZP2 and the rabbit 75K proteins. Although these similarities may be sufficient to demonstrate protein relatedness, these proteins are sufficiently different to be immunogenically distinct. The rabbit cDNA (rc55) that codes for the 55K (EBGD) ZP protein has been isolated and has been used to probe genomic Southern blots of human, pig, cat, dog, and rabbit (B. S. Dunbar, unpublished observations). The rc55 cDNA clone hybridized to the DNA of all species with the same intensity.

7. SYNTHESIS OF ZP GLYCOPROTEINS AND FORMATION OF THE EXTRACELLULAR MATRIX

The formation of the ZP matrix is developmentally regulated, since it occurs during specific stages of oogenesis that take place during differentiation. To date, the signals responsible for initiation of ZP synthesis are unknown. These signals could be hormonal or may result from cell–cell interaction and the establishment of communication between the oocyte and follicular cells. Whatever the signal, the regulation of ZP formation could occur at different levels, including posttranslational modification, translation, transcription, and/or genome organization. The molecular basis of regulation remains unknown for ZP formation in any species.

Even the cellular origin of the various ZP proteins in different species has been the subject of some controversy: the oocyte, granulosa cells, or both have been implicated by various investigators. Since intraspecies biochemical, immunologic, and structural differences are present in the ZP, it seems likely that both the site of synthesis and the posttranslational modifications may vary among species. This question is of particular interest, since a dual cellular origin of the ZP proteins or their posttranslational modifications would represent a coordinated production of a complicated extracellular matrix by very different cells, i.e., germ cells and differentiated somatic cells.

The Golgi complexes and the cortical granules of the oocyte may have a central role in the formation of the rat, mouse, and rabbit ZP (Odor, 1960; Zamboni and Mastroianni, 1966; Martinek and Krausova, 1972; Kang, 1974; Baranska et al., 1975), since they migrate from a juxtanuclear to a subcortical location simultaneously with the deposition of the ZP. Bousquet and colleagues (1981) and Leveille et al. (1987) have observed ZP antibody immunofluorescent labeling of oocyte but not granulosa cell cytoplasm in sections of human and hamster ovaries. Other studies on the mouse have shown that the oocyte is responsible for the synthesis and secretion of mouse ZP proteins.

Most of the recent biochemical studies suggesting the oocyte as the sole origin of the ZP have been performed using rodents. In the mouse, in vitro biosynthetic studies have demonstrated the synthesis and secretion of ZP proteins by growing oocytes (Haddad and Nagai, 1977; Flechon et al., 1974; Bleil and Wassarman, 1980b; Philpott et al., 1987; Greve et al., 1982; Shimizu et al., 1983; Salzmann et al., 1983). The observation that the oocyte can synthesize ZP proteins has been confirmed by detection of mRNA in the oocyte using a molecular probe for ZP3 (Ringuette et al., 1986; Philpott et al., 1987).

In contrast, granulosa cells have been identified as the origin of ZP material in several species (rabbit, mouse, rat, guinea pig, hamster, cat) (Trujillo-Cenoz and Sotelo, 1959; Chiquoine, 1960). The granulosa cells of primary and secondary follicles (the stages during which the ZP is first synthesized) are, like the growing oocyte, actively synthesizing proteins, and the rough endoplasmic reticulum is well developed in comparison to the oocyte (Merker, 1961). Wartenberg and Stegner (1960) and Tesoriero (1984) have found glycoprotein-staining components in dog, rabbit, and mouse granulosa cell cytoplasm that resemble the staining of ZP. Examination of the granulosa cell processes has revealed the presence of flocculent material in the cytoplasm of these processes in human ovaries that is similar to that found in the ZP surrounding them (Guraya, 1974).

Finally, a number of studies have suggested roles for both the oocyte and the granulosa cells in the production of the ZP. These include reports in which staining of the oocyte cytoplasm can be detected early in follicular development in the rabbit ovary using antibodies against ZP (Wolgemuth *et al.*, 1984; Skinner *et al.*, 1990). Zona pellucida proteins in the cytoplasm of inner layers of follicular cells can be detected at the two- to three-cell-layer stage (early secondary) but not in mature follicies. These observations suggest that both cell types could contribute to the formation of the rabbit ZP. Further studies confirming the role of granulosa cells in ZP protein synthesis have been carried out using cell culture techniques (see discussion below).

It is well established that the constituent molecules of extracellular matrices are responsible for many aspects of cell differentiation, structure, and function in a number of different cell types (see review by Reid and Jefferson, 1984). During the development of the growing ovarian follicle, the differentiating granulosa cells form two compartments: one that is in close opposition to the extracellular matrix ("basement" membrane) and the inner layers, which are in contact with the oocyte and its unique extracellular matrix, the ZP.

8. DEVELOPMENTAL EXPRESSION OF ZP PROTEINS

As stated earlier, the ZP is formed during specific stages of oogenesis and follicular development. In the mouse, *in situ* hybridization and Northern analysis using the mouse ZP3 as a probe have revealed the presence of ZP3 transcripts in growing oocytes and not in granulosa cells (Philpott *et al.*, 1987). In addition, ZP3 transcripts were not seen in resting oocytes but were found to accumulate in growing oocytes to 0.1–0.2% of total polyA$^+$ RNA. The transcript level decreases in later stages and during meiotic maturation and is very low in ovulated eggs (<0.04%) (Philpott *et al.*, 1987). In the rabbit, Northern analysis has revealed a 600-fold decrease in rc55 transcripts in the adult ovary compared to the immature 6-week-old ovary (B. S. Dunbar, unpublished observations). This reflects the abundance of growing follicles that are undergoing atresia at this stage of ovarian development in the rabbit.

Using antibody to rabbit ZP proteins, Wolgemuth *et al.* (1984) have shown the localization of ZP proteins in the cytoplasm at the periphery of oocytes surrounded by a thin squamous follicular cell layer. As the oocytes grow and the follicular cells form a multilayer, ZP protein is localized within the cytoplasm of the inner layers of these follicular cells and then diminishes in cells of preantral follicles. Gaining a more complete understanding of the site of synthesis of ZP proteins and their developmental regulation in the rabbit should be possible using the molecular probes now available for all three rabbit ZP proteins (Timmons *et al.*, 1990) in Northern analysis and *in situ* hybridization.

9. OVARIAN FOLLICULAR DEVELOPMENT AND FORMATION OF THE ZONA PELLUCIDA

Ovarian follicular development consists of a complex, highly regulated process of cell differentiation during which the oocyte and follicular cells undergo vast structural and functional changes. To date, most studies have been limited to the regulation of follicular cell structure and function just prior to and during antral formation, ovulation, and corpora lutea development (Austin and Short, 1982; Centola, 1983; Jones, 1978; Peters and McNatty, 1980; Rolland *et al.*, 1982). Although the early stages of follicular development involve the proliferation and differen-tiation of multiple cell types, comparatively little is known about the mechanisms that initiate and regulate this process. One of the most significant events of the early stages of ovarian follicular development is the synthesis and secretion of ZP proteins and the extracellular assembly of this matrix.

In order to study these events, we have established a unique ovarian follicle cell culture system to analyze the effects of the extracellular matrix on early stages of granulosa cell differentiation and ZP synthesis. Primary and early secondary follicles isolated from ovaries of sexually immature rabbits (6–7 weeks) were grown on poly-D-lysine or Englebreth–Holm–Swarm (EHS) basement membrane biomatrix substrata in a serum-free, hormonally defined medium (Maresh *et al.*, 1990). The secretion of the ZP proteins, which occurs during early stages of follicular development, was analyzed using ELISA assays and immunoblots of 1D- and 2D-PAGE separations of secreted proteins. Monoclonal and epitope-selected polyclonal antibodies demonstrated the expression of ZP proteins by granulosa cells *in vitro*. The expression of two specific ZP proteins was altered by the substrate used for cell culture: $55K_{EBGD}$ was secreted by cells grown on either EHS biomatrix or poly-D-lysine, but a greater amount of $75K_{EBGD}$ was secreted by cells grown on poly-D-lysine. These studies are the first to show that granulosa cells isolated from early-stage ovarian follicles express ZP proteins *in vitro* in the absence of oocytes, although it appears that proper posttranslational modification has not occurred. These studies should provide valuable information on the roles of the ZP molecules in follicle differentiation and in the fertilization process.

10. SPERM–ZP INTERACTION DURING FERTILIZATION

Because many of these studies are outlined in detail in other chapters of this text, only an overview is provided here. Initial studies using rodent species suggested that sperm attachment to the zona is species specific (Austin and Braden, 1956; Yanagimachi, 1977). These studies prompted the search for specific "sperm receptors" on the surface of the zona. Such receptors have been described in analogous structures of invertebrate species, such as sea urchin vitelline envelopes (Aketa, 1973; Kinsey and Lennarz, 1981; Ruiz-Bravo and Lennarz, 1989). Although it is apparent that there is a critical need for specificity of species recognition between gametes in environments in which external fertilization occurs, the need may be less critical in mammalian species in which physiological and behavioral barriers exist to prevent gametes from contacting each other. Although limited evidence has been presented for such species-specific receptors in mammals (Gwatkin and Williams, 1977; Peterson *et al.*, 1981), the universality of this phenomenon is yet to be proved. For instance, the species specificity of sperm–zona interaction has been described for some mammals, including mouse, hamster, guinea pig, and rat (Austin and Braden, 1956; Yanagimachi, 1977; Schmell and Gulyas, 1980), but less specificity is exhibited by sperm of other mammalian species such as the rabbit (Bedford, 1977; Swenson and Dunbar, 1982).

Other factors affecting sperm–zona interaction, including sperm capacitation, also vary among species (see other chapters in this text). Sperm capacitation is required for binding *in vitro* to zonae of mice (Saling *et al.*, 1978) and for physiologically significant binding in hamsters (Hartmann and Hutchison, 1974, 1977) but not in pig or rabbit (Peterson *et al.*, 1980; Swenson and Dunbar, 1982). To date, the most detailed information on sperm–ZP interaction is available for the mouse (see Chapters 9 and 10 in this text) because this has been the most cost-effective system and because the mouse *in vitro* fertilization system is well established. These results need to be carefully interpreted, however, because the studies have been limited to the use of epididymal sperm. These studies do not take into account other factors in seminal plasma that may be involved, either negatively or positively, in the fertilization process.

Recently, there have been reports that the sperm acrosomal enzyme acrosin has both zona-pellucida- and fucose-binding properties (Topfer-Petersen and Henschen, 1987, 1988). The studies are exciting because they imply that this acrosomal enzyme may facilitate sperm binding to and penetration of the ZP during fertilization. They may further explain many of the species-specific events that are associated with the fertilization process, since the acrosin molecules also have species-specific properties (see Chapter 5 of this text).

11. SPERM PENETRATION OF THE ZONA PELLUCIDA DURING FERTILIZATION

Sperm penetration of the zona pellucida is another critical step in the fertilization process of mammals and may also play an important role in the species specificity of fertilization. Although a variety of mechanisms for sperm–egg interaction have been proposed (see other chapters in this text), these have been difficult to evaluate because of the limited numbers of oocytes available for investigation and because of the extensive morphological and biochemical variations among gametes of different species (see reviews by Dunbar, 1983; Moore and Bedford, 1983; Dunbar and Wolgemuth, 1984). The mechanism of sperm penetration of the ZP is believed to be related to limited proteolysis of the ZP glycoprotein matrix (Stambaugh, 1978; see review by McRorie and Williams, 1974). Whereas numerous studies have provided evidence for the role of a trypsin-like acrosomal proteinase (acrosin; EC 3.4.2.10) in mediating penetration of the ZP during fertilization (Srivastiva et al., 1965; Stambaugh and Buckley, 1969; Zaneveld et al., 1971; Schleuning et al., 1973; Polakoski and McRorie, 1973; Brown, 1974), other investigators have questioned the role of this enzyme (Bedford and Cross, 1978; Brown, 1982). These controversies can be attributed in part to the use of microscopic criteria to evaluate dissolution of the ZP matrix in response to sperm enzymes and to considerable species variation in the morphological, physicochemical, and biochemical properties of the ZP. Other investigators have suggested that acrosin may also be involved in the initiation of the acrosome reaction in the hamster sperm (Meizel and Liu, 1976).

More recently, proteolysis of specific glycoproteins of porcine ZP was monitored by high-resolution two-dimensional polyacrylamide gel electrophoresis (2D-PAGE) (Dunbar et al., 1985). Although these enzymes do not alter the macroscopic properties of the ZP matrix as observed by light microscopy, the 2D-PAGE ZP protein patterns were markedly altered (Fig. 4). The two high-molecular-weight glycoprotein families were sensitive to proteolytic digestion, whereas the major glycoprotein family of the porcine ZP (ZP3) was only partially proteolyzed by acrosin and trypsin. Furthermore, it was demonstrated that acrosin had unique substrate specificity compared to that of trypsin, since the ZP peptides generated by these enzymes were distinct. These studies have demonstrated that certain integral glycoproteins of the native ZP matrix are specifically proteolyzed by acrosin from the homologous species, and that this proteolysis occurs without the dissolution of the native porcine matrix. It is also likely that the composition of different enzymes of spermatozoa of different species (see Chapter 5 of this text) plays a complex role in the penetration of the ZP matrix.

12. CHANGES IN THE ZONA PELLUCIDA FOLLOWING FERTILIZATION

Cellular and molecular mechanisms have evolved in most species to ensure that the egg will be fertilized by only one sperm. This process is commonly referred to as the "block to polyspermy" (see reviews by Wolf, 1981; Schmell et al., 1983). Detailed studies on this process

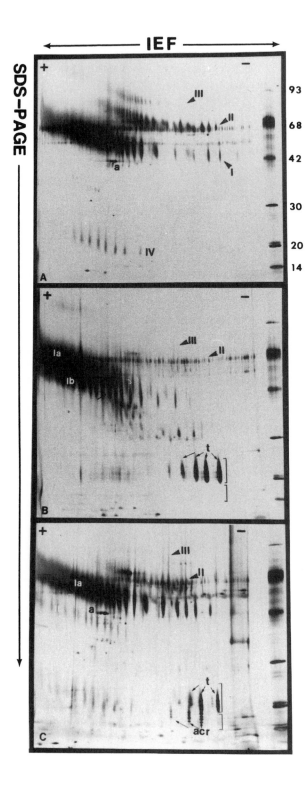

Figure 4. Analysis of the proteolysis of porcine ZP glycoproteins by trypsin and acrosin by high-resolution two-dimensional PAGE. The glycoproteins of the intact porcine ZP are sensitive to digestion by proteolytic enzymes. Protein patterns of control porcine ZP (A) were compared with those treated with either trypsin (B) or porcine acrosin (C). The ZP were solubilized and analyzed by 2D-PAGE and silver staining. Although all three preparations of ZP appeared morphologically intact, the lowest-molecular-weight porcine ZP glycoprotein had been specifically proteolyzed by these two enzymes. The right-hand column of each gel contains molecular weight standards in the second dimension (modified from Dunbar *et al.*, 1985).

have been carried out using invertebrate or amphibian species, since large numbers of gametes are available and fertilization can easily be carried out *in vitro*. Shapiro and co-workers have shown that peroxidase activity is localized in the cortical granules of the unfertilized egg. At fertilization, this enzyme is released with other proteases and causes cross-linking between proteins of the vitelline envelope, resulting in envelope hardening. However, such a peroxidase activity has not been found associated with the cortical granules of the *Xenopus laevis* egg. In this species, a cortical granule "lectin" is released from the granules and interacts with the jelly coat that interfaces the vitelline envelope (Greve and Hedrick, 1978). This reaction makes the envelope impenetrable to sperm (Grey *et al.*, 1976). Although the fundamental mechanisms of the block to polyspermy may be similar, the cellular level of the block and the molecules involved in such a reaction appear to vary among species.

Since the morphology and the molecular properties of the mammalian ZP are remarkably different from those of lower vertebrates and invertebrates, it is difficult to propose a universal mechanism for the block to polyspermy. Braden and Austin (1954) and Braden *et al.* (1954) first observed that the hamster ZP became less permeable or totally impermeable to supernumerary sperm, a phenomenon they referred to as the "zona reaction." This reaction has been attributed to the release of contents from the cortical granules at fertilization. As is true with other aspects of fertilization, the observed zona reaction depends not only on the species but on the methods employed by investigators to study this phenomenon. The time necessary for this reaction to occur and its effectiveness in preventing polyspermy also appear to vary among species. For example, the zona reaction in the hamster occurs within 8–35 min (Barros and Yanigamachi, 1971; Gwatkin *et al.*, 1973), whereas it requires 60–120 min in the mouse (Inoue and Wolf, 1975). The zona reaction in the rabbit appears to be different than in the rodents, since more than one sperm will penetrate the rabbit zona (Bedford and Cross, 1978), and multiple sperm bind to the ZP of fertilized eggs. Recent studies of McCulloh *et al.* (1987) have demonstrated, however, that the incidence of polyspermy is increased following removal of the ZP; therefore, these investigators conclude that the ZP is important in this process. In other species, such as the pig or human, the zona reaction is only effective after sperm have contacted and begun to penetrate the zona (Dickman and Dziuk, 1964). The zonae from the fertilized ova of some species also have been observed to develop an increased resistance to chemical dissolution, a phenomenon referred to as "zona hardening."

The precise mechanism for the zona reaction in mammals is still not known, but its physiological function may be related to the block to polyspermy. Studies on the mouse reveal changes in ZP molecules following fertilization that result in the inhibition of sperm binding to the ZP surface (see elsewhere in this text). However, the rabbit ZP undergoes changes at fertilization that lead to its increased resistance to enzymatic digestion but still allow ZP penetration by spermatozoa.

Recently, Shabanowitz and O'Rand (1989) have shown that fertilization-associated modifications also occur in the human ZP. These studies have used electrophoretic procedures to demonstrate that there is a fertilization-associated modification of the human ZP1 molecule (mol. wt. 90,000–111,000), and it was suggested that this modification may be effected by cortical granule dehiscence after fertilization. Advanced techniques for analysis of ZP proteins by 2D-PAGE and for the detection of small amounts of protein will greatly aid in understanding the molecular events that accompany the fertilization process in different mammalian species.

13. ANTIGENIC COMPOSITION OF MAMMALIAN ZONAE PELLUCIDAE

It is clear from the numerous reports characterizing antibodies against the ZP that the major ZP antigenic determinants consist of amino acid sequences, conformational or structural deter minants, and carbohydrate structures (Drell *et al.*, 1984; Maresh and Dunbar, 1987; Timmons *et al.*, 1987; Sacco *et al.*, 1981) and that these determinants vary among species (Maresh and Dunbar, 1987). Because there has been a great deal of interest in the use of ZP antigens as contraceptive vaccines, much research has focused on the generation of anti-ZP antibodies that inhibit fertilization without altering ovarian function and on the identification and characterization of these specific epitopes (see reviews by Timmons and Dunbar, 1988; Dunbar, 1989; Skinner *et al.*, 1990; Timmons and Dunbar, 1990b). A number of these antigens are species-specific and likely reflect differences in the molecular composition of ZP carbohydrates and proteins, and in the macromolecular "suprastructure" of the matrix itself. There are other determinants associated with ZP proteins that are shared among many species (Maresh and Dunbar, 1987; Timmons *et al.*, 1987). Standard immunological techniques have not detected any rodent ZP glycoprotein antigens that are shared with other mammalian ZP (Sacco *et al.*, 1981; Maresh and Dunbar, 1987; Millar *et al.*, 1989). However, mouse ZP proteins isolated from large numbers of superovulated eggs have recently been analyzed using improved minigel immunoblotting methods and sensitive visualization techniques (Timmons and Dunbar, unpublished observations). The results indicate the presence of mouse ZP antigens recognized by rabbit, sheep, and guinea pig antisera against pig and rabbit ZP. Identification of these shared antigenic determinants will provide insight into the roles of individual ZP proteins in mammals.

14. REFERENCES

Ahuja, K. K., and Botwell, G. P., 1983, Probable asymmetry in the organization of components of the hamster zona pellucida, *J. Reprod. Fertil.* **69:**49–55.

Aketa, K., 1973, Physiological studies on the egg surfaces component responsible for sperm–egg binding in sea urchin fertilization. I. Effect of sperm-binding protein on the fertilizing capacity of sperm, *Exp. Cell. Res.* **70:**439–441.

Austin, C. R., 1961, *The Mammalian Egg*, Springfield, IL: Charles C. Thomas.

Austin, C. R., 1965, *Fertilization*, Prentice-Hall, Englewood Cliffs, NJ.

Austin, C. R., 1975, Membrane fusion events in fertilization, *J. Reprod. Fertil.* **44:**155–166.

Austin, C. R., and Braden, A. W. H., 1956, Early reactions of the rodent egg to spermatozoa penetration, *J. Exp. Biol.* **33:**358–365.

Austin, C. R., and Short, R. V., eds., 1982, *Reproduction in Mammals: Germ Cells and Fertilization*, Cambridge University Press, Cambridge.

Baranska, W., Konwinski, M., and Kujawa, M., 1974, Fine structure of the zona pellucida of unfertilized egg cells and embryos, *J. Exp. Zool.* **192:**193–202.

Baranska, W., Konwinski, M., and Kujawa, M., 1975, Fine structure of the zona pellucida of unfertilized egg cells and embryos, *J. Exp. Zool.* **192:**193–202.

Barros, C., and Yanagimachi, R., 1971, Induction of zona reaction in golden hamster eggs by cortical granule material, *Nature* **233:**268–269.

Bedford, J. M., 1974, Mechanisms involved in penetration of spermatozoa through the vestments of the mammalian egg, in: *Physiology and Genetics of Reproduction*, Part B (E. M. Coutino and F. Fuch, eds.), Plenum Press, New York, pp. 55–68.

Bedford, J. M., 1977, Sperm/egg interaction: The specificity of human spermatozoa, *Anat. Rec.* **188:**477–488.

Bedford, J. M., and Cross, N. L., 1978, Normal penetration of rabbit spermatozoa through a trypsin- and acrosin-resistant zona pellucida, *J. Reprod. Fertil.* **54:**385–392.

Bleil, J. D., and Wassarman, P., 1980a, Mammalian sperm-egg interaction: Identification of a glycoprotein in mouse egg zonae pellucidae possessing receptor activity for sperm, *Cell* **20:**873–882.

Bleil, J. D., and Wassarman, P. M., 1980b, Synthesis of zona pellucida proteins by denuded and follicle-enclosed mouse oocytes during culture *in vitro*, *Proc. Natl. Acad. Sci. U.S.A.* **77:**1029–1033.

Bleil, J. D., and Wassarman, P. M., 1988, Galactose at the nonreducing terminus of O-linked oligosaccharides of mouse egg zona pellucida glycoprotein ZP3 is essential for the glycoprotein's sperm receptor activity, *Proc. Natl. Acad. Sci. U.S.A.* **85:**6778.

Bousquet, D., Leveille, M. C., Roberts, K. D., Chapdelaine, A., and Bleau, G., 1981, The cellular origin of the zona pellucida antigen in the human and hamster, *J. Exp. Zool.* **215:**215–218.

Braden, A. W. H., and Austin, C. R., 1954, The number of sperm about the eggs in mammals and its significance for normal fertilization, *Aust. J. Biol. Sci.* **7:**543–551.

Braden, A. W. H., Austin, C. R., and David, H. A., 1954, The reaction of zona pellucida to sperm penetration, *Aust. J. Biol. Sci.* **7:**391–409.

Brown, C. R., 1982, Effects of ram sperm acrosin on the investments of sheep, pig, mouse, and gerbil eggs, *J. Reprod. Fertil.* **64:**457–462.

Brown, C. R., 1974, Distribution of trypsin-like proteinase in the ram spermatozoa, *J. Reprod. Fertil.* **36:**195–198.

Centola, G. M., 1983, Structural changes: Follicular development and hormonal requirements, in *Comprehensive Endocrinology: The Ovary* (G. B. Serra, (ed.), Raven Press, New York, p. 95.

Chamberlain, M. E., and Dean, J. 1989, Genomic organization of a sex specific gene: The primary sperm receptor of the mouse zona pellucida, *Dev. Biol.* **131:**207–214.

Chiquoine, A. D., 1960, The development of the zona pellucida of the mammalian ovum, *Am. J. Anat.* **106:**149–169.

Cholewa-Stewart, J., and Massaro, E. J., 1972, Thermally induced dissolution of the murine zona pellucida, *Biol. Reprod.* **7:**166–169.

Dickmann, Z., and Dziuk, P. J., 1964, Sperm penetration of the zona pellucida of the pig egg, *J. Exp. Biol.* **41:**603–608.

Drell, D. M., and Dunbar, B. S., 1984, Monoclonal antibodies to rabbit and pig zonae pellucidae distinguish species-specific and shared antigenic determinants, *Biol. Reprod.* **30:**445.

Drell, D. W., Wood, D. M., Bundman, D. S., and Dunbar, B. S., 1984, Immunological comparison of antibodies to porcine zonae pellucidae in rats and rabbits, *Biol. Reprod.* **30:**435–444.

Dunbar, B. S., 1980, Model systems to study the relationship between antibodies to zonae and infertility: Comparison to rabbit and porcine zonae pellucidae. *Ninth International Congress on Animal Reproduction and Artificial Insemination II*, Spanish Ministry of Agriculture, Madrid, Spain, pp. 191–199.

Dunbar, B. S., 1983, Morphological, biochemical, and immunochemical characterization of the mammalian zonae pellucidae, in: *Mechanism and Control of Animal Fertilization* (J. Hartmann, ed.), Academic Press, New York, pp. 139–175.

Dunbar, B. S., 1989, Ovarian antigens and infertility, *Am. J. Reprod. Immunol.* **21:**28–31.

Dunbar, B. S., and Wolgemuth, D. J., 1984, Structure and function of the mammalian zona pellucida, a unique extracellular matrix, in: *Modern Cell Biology* (B. Satir, ed.), Alan R. Liss, New York, pp. 77–111.

Dunbar, B. S., Wardrip, N. J., and Hedrick, J. L., 1980, Isolation, physicochemical properties and macromolecular composition of zona pellucida from porcine oocytes, *Biochemistry* **19:**356.

Dunbar, B. S., Liu, C., and Sammons, D. W., 1981, Identification of the three major proteins of porcine and rabbit zonae pellucidae by high resolution two-dimensional gel electrophoresis: Comparison with follicular fluid, sera, and ovarian cell proteins, *Biol. Reprod.* **24:**1111–1124.

Dunbar, B. S., Dudkiewicz, A., and Bundman, D. S., 1985, Proteolysis of specific porcine zona pellucida glycoproteins by boar acrosin, *Biol. Reprod.* **32:**619–630.

Dutt, A., Tang, J.-P., and Carson, D. D., 1987, Lactosaminoglycans are involved in uterine epithelial cell adhesion in vitro, *Dev. Biol.* **119:**27–37.

Flechon, J. E., Pavlok, A., and Kopecny, V., 1974, Dynamics of zona pellucida formation by the mouse oocyte. An autoradiographic study, *Biol. Cell* **51:**403–406.

Florman, H. M., and Wassarman, P. M., 1985, O-Linked oligosaccharides of mouse egg ZP3 account for its sperm receptor activity, *Cell* **41:**313.

Fukuda, M., 1985, Cell surface glycoconjugates as onco-differentiation markers in hematopoietic cells, *Biochim. Biophys. Acta* **780:**119–150.

Fukuda, M. N., and Matsumura, G., 1976, Purification and endoglycosidic action on keratan sulfates, oligosaccharides, and blood group active glycoprotein, *J. Biol. Chem.* **251:**6218–6225.

Greve, L. C., and Hedrick, J. L., 1978, An immunocytochemical localization of the cortical granule lectin in fertilized and unfertilized eggs of *Xenopus laevis*, *Gamete Res.* **1:**13–18.

Greve, J. J., Salzman, G. S., Roller, R. J., and Wassarman, P. M., 1982, Biosynthesis of the major zona pellucida glycoprotein secreted by oocytes during mammalian oogenesis, *Cell* **31:**749–759.

Grey, R. D., Working, P. K., and Hedrick, J. L., 1976, Evidence that the fertilization envelope blocks sperm entry in eggs of Zenopus laevis: Interaction of sperm with isolated envelopes, *Dev. Biol.* **54:**52–60.

Guraya, S. S., 1974, Morphology, histochemistry, and biochemistry of human oogenesis and ovulation, in: *International Review of Cytology* (G. H. Bourne and J. F. Danielli, eds.), Academic Press, New York, pp. 121–151.

Gwatkin, R. B. L., and Williams, D. T., 1977, Receptor activity of the hamster and mouse solubilized zona pellucida before and after the zona reaction, *J. Reprod. Fertil.* **49:**55–59.

Gwatkin, R. B. L., Williams, D. T., Hartmann, J. F., and Kniazuk, M., 1973, The zona reaction of hamster and mouse eggs: Production *in vitro* by a trypsin-like protease from cortical granules, *J. Reprod. Fertil.* **32:**259–265.

Haddad, A., and Nagai, M. E. T., 1977, Radioautographic study of glycoprotein biosynthesis and renewal in the ovarian follicles of mice and the origin of the zona pellucida, *Cell Tissue Res.* **177**:347–369.

Hakomori, S., 1981, Glycosphingolipids in cellular interaction, differentiation, and oncogenesis, *Annu. Rev. Biochem.* **50**:733–764.

Hartmann, J. F., and Hutchison, C. F., 1974, Nature of the pre-penetration contact interaction between hamster gametes *in vitro*, *J. Reprod. Fertil.* **36**:49–57.

Hartmann, J. F., and Hutchison, C. F., 1977, Involvement of two carbohydrate-containing components in the binding of uncapacitated spermatozoa to eggs of the golden hamster *in vitro*, *J. Exp. Zool.* **201**:383–390.

Hedrick, J. L., and Frye, G. N., 1980, Immunocytochemical studies on the porcine zona pellucida, *J. Cell Biol.* **87**:136a.

Hedrick, J. L., and Wardrip, N. J., 1987, On the macromolecular composition of the zona pellucida from porcine oocytes, *Dev. Biol.* **121**:478–488.

Inoue, M., and Wolf, D. P., 1975, Comparative solubility properties of rat and hamster zonae pellucidae, *Biol. Reprod.* **12**:535–540.

Jones, R. E., 1978, *The Vertebrate Ovary*, Plenum Press, New York.

Kang, Y. H., 1974, Development of the zona pellucida in the rat oocyte, *Am. J. Anat.* **139**:535–566.

Karp, D. R., Atkinson, J. P., and Shreffler, D. C., 1982, Genetic variation in glycosylation of the fourth component of murine complement, *J. Biol. Chem.* **257**:7330–7335.

Kinloch, R. A., Roller, R. J., Fimiani, C. M., Wassarman, D. A., and Wassarman, P. M., 1988, Primary structure of the mouse sperm receptor polypeptide determined by genomic cloning, *Proc. Natl. Acad. Sci. U.S.A.* **85**:6409–6413.

Kinsey, W. H., and Lennarz, W. J., 1981, Isolation of a glycopeptide fraction from surface of the sea urchin egg that inhibits sperm-egg binding and fertilization, *J. Cell Biol.* **91**:325–331.

Konecny, M., 1959, Etude histochimique de la zone pellucide des ovules de chatte, *C.R. Soc. Biol.* **153**:893–894.

Leveille, M. C., Roberts, K. D., Chevalier, S., Chapdelaine, A., and Bleau, G., 1987, Formation of the hamster zona pellucida in relation to ovarian differentiation and follicular growth, *J. Reprod. Fertil.* **79**:173–183.

Liang, L.-F., Chamow, S. M., and Dean, J., 1990, Oocyte-specific expression of mouse ZP2: Developmental regulation of the zona pellucida genes, *Mol. Cell Biol.* **10**:1507–1515.

Lunsford, R. D., Jenkins, N. A., Kozak, C. A., Liang, L.-F., Silan, C. M., Copeland, N. G., and Dean, J., 1990, Genomic mapping of murine ZP-2 and ZP-3, two oocyte-specific loci encoding zona pellucida proteins, *Genomics* **6**:184–187.

Macek, M. B., and Shur, B. D., 1988, Protein–carbohydrate complementarity in mammalian gamete recognition, *Gamete Res.* **20**:93–102.

Maresh, G. A., and Dunbar, B. S., 1987, Antigenic comparison of five species of zonae pellucidae, *J. Exp. Zool.* **244**:299–307.

Maresh, G. A., Timmons, T. M., and Dunbar, B. S., 1990, Effects of extracellular matrix on the expression of specific ovarian proteins, *Biol. Reprod.* **43**:965–976.

Martinek, J., and Krausova, H., 1972, Development of the zona pellucida in the rat, *Fol. Morphol.* **20**:73–75.

McCulloh, D., Wall, R. J., and Levitan, H., 1987, Fertilization of rabbit ova and the role of ovum investments in the block to polyspermy, *Dev. Biol.* **120**:385–391.

McRorie, R. A., and Williams, W. L., 1974, Biochemistry of mammalian fertilization, *Annu. Rev. Biochem.* **43**: 777–801.

Meizel, S., and Liu, C. W., 1976, Evidence for a role of a trypsin-like enzyme in the hamster acrosome reaction, *J. Exp. Zool.* **195**:137–144.

Merker, H. J., 1961, Electronmikroskopishce untersuchungen uber die bildung der zona pellucida in den follikeln des kaninchenovars, *Z. Zellforsch.* **65**:677–688.

Millar, S. E., Chamon, S. M., Baur, A. W., Oliver, C., Robey, F., and Dean, J., 1989, Vaccination with a synthetic zona pellucida peptide produces long-term contraception in female mice, *Science* **246**:935–938.

Moller, C. C., Bleil, J. D., Kinloch, R. A., and Wassarman, P. M., 1990, Structural and functional relationships between mouse and hamster zona pellucida glycoproteins, *Dev. Biol.* **137**:276–286.

Moore, H. D. M., and Bedford, J., 1983, The interaction of mammalian gametes in the female, in: *Mechanism and Control of Animal Fertilization* (J. Hartmann, ed.), Academic Press, New York, pp. 453–497.

Odor, D. L., 1960, Electron microscope studies on the ovarian oocytes and unfertilized tubal ova in the rat, *J. Biophys. Biochem. Cytol.* **7**:567–574.

Oikawa, T., Sendai, Y., Kurata, S., and Yanagimachi, R., 1988, A glycoprotein of oviductal origin alters biochemical properties of the zona pellucida of hamster egg, *Gamete Res.* **19**:113–122.

Peters, H., and McNatty, K. P., 1980, *The Ovary: A Correlation of Structure and Function in Mammals*, University of California Press, Berkeley.

Peterson, R. N., Russell, L. D., Bundman, D., and Freund, M., 1980, Sperm–egg interaction: Evidence for boar sperm plasma membrane receptors for porcine zona pellucida, *Science* **207**:73–74.

Peterson, R. N., Russell, L. D., Bundman, D., Conway, M., and Freund, M., 1981, The interaction of living boar sperm and sperm plasma membrane vesicles with the porcine zona pellucida, *Dev. Biol.* **84**:144–156.

Phillips, D. M., and Shalgi, R. M., 1980, Surface properties of the zona pellucida, *J. Exp. Zool.* **213**:1–8.

Philpott, C. C., Ringuette, M. J., and Dean, J., 1987, Oocyte-specific expression and developmental regulation of ZP3, the sperm receptor of the mouse zona pellucida, *Dev. Biol.* **121**:568–575.

Polakoski, K. L., and McRorie, R. A., 1973, Boar acrosin. II. Classification, inhibition, and specificity studies of a proteinase from sperm acrosomes, *J. Biol. Chem.* **248**:8183–8188.

Reid, L. M., and Jefferson, D. M., 1984, Cell culture studies using extracts of extracellular matrix to study growth and differentiation in mammalian cells, in: *Mammalian Cell Culture* (J. Mathur, ed.), Plenum Press, New York, p. 239.

Ringuette, M. J., and Sobieski, D. A., Chamow, S. M., and Dean, J., 1986, Oocyte-specific gene expression: molecular characterization of a cDNA coding for ZP-3, the sperm receptor of the mouse zona pellucida, *Proc. Natl. Acad. Sci. U.S.A.* **83**:4341–4345.

Ringuette, M. J., Chamberlain, M. E., Baur, A. W., Sobieski, D. A., and Dean, J., 1988, Molecular analysis of cDNA coding for ZP3, a sperm binding protein of the mouse zona pellucida, *Dev. Biol.* **127**:287–295.

Rolland, R., Van Hall, E. V., Hillier, S. G., McNatty, K. P., and Schoemaker, J., eds., 1982, *Follicular Maturation and Ovulation*, Excerpta Medica, Amsterdam.

Ruiz-Bravo, N., and Lennarz, W. J., 1989, Receptors and membrane interactions in fertilization, in: *The Molecular Biology of Fertilization* (H. Schatter and G. Schatter, eds.), Academic Press, San Diego, pp. 21–36.

Sacco, A. G., 1981, Immunocontraception: Consideration of the zona pellucida as a target antigen, *Obstet. Gynecol.* **10**:1–26.

Sacco, A. G., Yurewicz, E. C., Subramanian, M. G., and DeMayo, F. J., 1981, Zona pellucida composition: Species crossreactivity and contraceptive potential of antiserum to a purified peg zona antigen (PPZA), *Biol. Reprod.* **25**:997–1008.

Saling, P. M., Storey, B. T., and Wolf, D. P., 1978, Calcium-dependent binding of mouse epididymal spermatozoa to the zona pellucida, *Dev. Biol.* **65**:515–525.

Salzmann, G. S., Greve, J. M., Roller, R. J., and Wassarman, P. M., 1983, Biosynthesis of the sperm receptor during oogenesis in the mouse, *EMBO J.* **2**:1451–1456.

Sasaki, H., Bothner, B., Dell, A., and Fukuda, M., 1987, Carbohydrate structure of erythropoietin expressed in Chinese hamster ovary cells by a human erythropoietin cDNA, *J. Biol. Chem.* **262**:12059–12076.

Schleuning, M. D., Schiessler, H., and Fritz, H., 1973, Highly purified acrosomal proteinase (boar acrosin): Isolation by affinity chromatography using benzamidine-cellulose and stabilization, *Hoppe-Seylers Z. Physiol. Chem.* **354**:550–554.

Schmell, E. D., Gulyas, B. J., and Hedrick, J. L., 1983, Egg surface changes during fertilization and the molecular mechanism of the block to polyspermy, in: *Mechanism and Control of Animal Fertilization* (J. Hartmann, ed.), Academic Press, New York, pp. 365–503.

Schwoebel, E., Prasad, S., Timmons, T., Cook, R., Kimura, H., Niu, E., Cheung, P., Skinner, S., Avery, S., Wilkins, B., and Dunbar, B., 1991, Isolation and characterization of a full length cDNA encoding the 55K rabbit ZP protein, *J. Biol. Chem.* **266**:7214–7219.

Shabanowitz, R. B., and O'Rand, M. G., 1988, Characterization of the human zona pellucida from fertilized and unfertilized eggs, *J. Reprod. Fertil.* **82**:151–161.

Shabanowitz, R. B., and O'Rand, M. G., 1989, Molecular changes in the human zona pellucida associated with fertilization and human sperm-zona interactions, *Ann. N. Y. Acad. Sci.* **541**:621–623.

Shimizu, S., Tsuji, M., and Dean, J., 1983, *In vitro* biosynthesis of three sulfated glycoproteins of murine zonae pellucidae by oocytes grown in follicle culture, *J. Biol. Chem.* **258**:5858–5863.

Shur, B. D., and Hall, N. G., 1982, A role for mouse sperm galactosyltransferase in sperm binding to the egg zona pellucida, *J. Cell Biol.* **95**:574–579.

Skinner, S. M., and Dunbar, B. S., 1986, Species variation in the zona pellucida, in: *Immunological Approaches to Contraception and Promotion of Fertility* (G. P. Talwar, ed.), Plenum Press, New York, pp. 251–268.

Skinner, S. M., Timmons, T. M., Schwoebel, E. D., and Dunbar, B. S., 1990, The role of zona pellucida antigens in fertility and infertility, *Immunol. Allergy Clin. North Am.* **10**:185–197.

Schmell, E. D., and Gulyas, B. J., 1980, Mammalian sperm–egg recognition and binding *in vitro*. Specificity of sperm interactions with live and fixed eggs in homologous and heterologous inseminations of hamster, mouse, and guinea pig oocytes, *Biol. Reprod.* **23**:1075–1085.

Srivastava, P. N., Adams, C. E., and Hartree, E. F., 1965, Enzymic action of acrosomal preparation on the rabbit ovum *in vitro*, *J. Reprod. Fertil.* **10**:61–67.

Stambaugh, R., 1978, Enzymatic and morphological events in mammalian fertilization, *Gamete Res.* **1**:65–85.

Stambaugh, R., and Buckley, J., 1969, Identification and subcellular localization of the enzymes effecting penetration of the zona pellucida by rabbit spermatozoa, *J. Reprod. Fertil.* **19**:423–432.

Stegner, H. E., and Wartenberg, H., 1961, Elektronenmikroskopische und histochemische Untersuchen uber Struktur und Bildung der Zona pellucida menshlicher lin Zellen, *Z. Zellforsch. Mikrosk. Anat.* **53**:702–713.

Swenson, C. E., and Dunbar, B. S., 1982, Specificity of sperm–zona interaction, *J. Exp. Zool.* **219**:97–104.

Tesoriero, J. V., 1984, Comparative cytochemistry of the developing ovarian follicles of the dog, rabbit and mouse: Origin of the zona pellucida, *Gamete Res.* **10**:301–318.

Timmons, T. M., and Dunbar, B. S., 1988, Antigens of mammalian zona pellucida, in: *Perspectives in Immuno-reproduction: Conception and Contraception* (S. Mathur and C. M. Fredericks, eds.), Hemisphere, New York, pp. 242–260.

Timmons, T. M., and Dunbar, B. S., 1990a, Analysis of the structure and function of a carbohydrate antigen of the mammalian zonae pellucidae, *J. Cell Biol.* **111 (5, Pt. 2)**:365a.

Timmons, T. M., and Dunbar, B. S., 1990b, Glycosylation and maturation of the mammalian zona pellucida, in: *Gamete Interaction: Prospects for Immunocontraception* (N. Alexander, D. Griffin, G. Waites, and J. Spieler, eds.), John Wiley & Sons, Inc., New York, pp. 277–292.

Timmons, T. M., Maresh, G. A., Bundman, D. S., and Dunbar, B. S., 1987, Use of specific monoclonal and polyclonal antibodies to define distinct antigens of the porcine zonae pellucidae, *Biol. Reprod.* **36**:1275–1287.

Topfer-Petersen, E., and Henschen, A., 1987, Acrosin shows zona and fucose binding, novel properties for a serine protease, *FEBS Lett.* **226**:38–42.

Topfer-Petersen, E., and Henschen, A., 1988, Zona pellucida-binding and fucose-binding of boar sperm acrosin is not correlated with proteolytic activity, *Hoppe-Seylers Z. Physiol Chem.* **369**:69–76.

Trujillo-Cenoz, O., and Sotelo, J. R., 1959, Relationships of the ovular surface with follicle cells and origin of the zona pellucida in rabbit oocytes, *J. Biophys. Biochem. Cytol.* **5**:347–348.

Wartenberg, H., and Stegner, H. E., 1960, Uber die elektronenmikroshopische Feinstruktur des menschlichen Ovarialeies, *Z. Zellforsch.* **52**:450–474.

Wassarman, P. M., 1988, Zona pellucida glycoproteins, *Annu. Rev. Biochem.* **57**:415–522.

Wolf, D. P., 1981, The mammalian eggs' block to polyspermy, in: *Fertilization and Embryonic Development in Vitro* (L. Mastroianni, Jr., and J. D. Biggers, eds.), Plenum Press, New York, pp. 183–197.

Wolgemuth, D. J., Celenza, J., Bundman, D. S., and Dunbar, B. S., 1984, Formation of the rabbit zona pellucida and its relationship to ovarian follicular development, *Dev. Biol.* **106**:1–14.

Wright, R. W., Jr., Cupps, P. T., Goskins, C. T., and Hillers, J. K., 1977, Comparative solubility properties of the zona pellucida of unfertilized murine, ovine, and bovine ova, *J. Anim. Sci.* **44**:850–853.

Yanagimachi, R., 1977, Specificity of sperm-egg interaction, in: *Immunobiology of Gametes* (M. Edidin and M. H. Johnson, eds.), Cambridge University Press, London, pp. 187–207.

Yoshima, H., Matsumoto, A., Mizuochi, T., Kawasaki, T., and Kobata, A., 1981, Comparative study of the carbohydrate moieties of rat and human plasma alpha(1)-acid glycoproteins, *J. Biol. Chem.* **256**:8476–8482.

Yurewicz, E. C., Sacco, A. G., and Subramanian, M. G., 1987, Structural characterization of the $M_r = 55,000$ antigen (ZP3) of porcine oocyte zona pellucida, *J. Biol. Chem.* **262**:564–571.

Zamboni, L., and Mastroianni, L., 1966, Electron microscopic studies on rabbit ova. I. The follicular oocyte, *J. Ultrastruct. Res.* **14**:95–117.

Zaneveld, L. J. D., Robertson, R. T., Kessler, M., and Williams, W. L., 1971, Inhibition of fertilization *in vivo* by pancreatic and seminal plasma trypsin inhibitors, *J. Reprod. Fertil.* **25**:387–392.

Part II

STUDIES ON MAMMALIAN FERTILIZATION IN SELECTED SPECIES

7

Fertilization of Marsupials

John C. Rodger

1. BACKGROUND

1.1. Marsupials

There are three distinct mammalian lineages; the monotremes, which lay eggs (e.g., platypus); marsupials, which give birth to embryo-like young (e.g., koala and kangaroos); and the eutherians, which include all the familiar "placental" mammals. For many years a hierarchical view of their relationships held scientific and popular sway that saw mammalian evolution as an ascending climb to placental status in the eutherians. Contemporary thought recognizes that both eutherians and marsupials are placental, but with different emphasis on the roles of the uterus and lactation in development (for review, see Tyndale-Biscoe and Renfree, 1987). Put very simply, the fetal stage of marsupial development is completed after leaving the uterus, while the young is suckled, usually in a pouch (see reviews in Tyndale-Biscoe and Janssens, 1988). It was the presence of this "second uterus" or delphis that gave the first marsupial described by Western science its generic name *Didelphis*.

Marsupials account for only around 6% of extant mammalian species. Although perhaps the best known are Australian, of the approximately 250 living marsupials about one-third are South/Central American and generally known as opossums. The Virginia opossum, a recent immigrant from Central America, is the only marsupial found in the Northern Hemisphere. Australian species are rather more diverse ranging from large ruminant-like herbivores (kangaroos; up to 60 kg) to dog- and hyena-like carnivores (dasyurids) and tiny shrew-like insectivores. Table 1 outlines the major groups of living marsupials and lists common and scientific names of species discussed here.

1.2. Marsupial Fertilization

Living marsupials are as modern as their eutherian counterparts and should not be seen as museum-like repositories of ancestral characteristics. However, they do represent an alternative

JOHN C. RODGER • Department of Biological Sciences, The University of Newcastle, Newcastle 2308, New South Wales, Australia.

A Comparative Overview of Mammalian Fertilization, edited by Bonnie S. Dunbar and Michael G. O'Rand. Plenum Press, New York, 1991.

Table 1. Major Groups of American and Australian Marsupials and
Animals Referred to Specifically in this Chapter

Family	Species		
	Common name	Scientific name	Size[a]
American Marsupials			
Didelphidae	Virginia opossum	*Didelphis virginiana*	4 kg
	Gray short-tailed opossum	*Monodelphis domestica*	100 g
	Woolly opossum	*Caluromys philander*	300 g
Microbiotheriidae		*Dromiciops australis*	30 g
Caenolestidae	Shrew opossums		30 g
Australian Marsupials			
Macropodidae (kangaroos)	Tammer wallaby	*Macropus eugenii*	5 kg
Potoroidae (rat kangaroos)	Brush-tailed bettong	*Bettongia penicillata*	1.5 kg
Phalangeridae (possums)	Brush-tailed possum	*Trichosurus vulpecula*	2.5 kg
Phascolarctidae	Koala	*Phascolarctos cinereus*	6 kg
Dasyuridae	Dunnart	*Sminthopsis crassicaudata*	15 g
	Antechinus	*Antechinus stuartii*	30 g
	Quoll	*Dasyurus viverrinus*	1 kg
Tarsipedidae	Honey possum	*Tarsipes rostratus*	10 g
Other Australian Families			
Peramelidae (bandicoots), Petauridae (gliders), Pseudocheridae (ringtail possums), Vombatidae (wombats)			

[a]Approximate weight of a mature female. Modified from Tyndale-Biscoe and Renfree (1987) and P. Temple-Smith (unpublished data).

path from these ancestors and thus are an important part of our understanding of the evolution of mammalian fertilization mechanisms. Study of fertilization in marsupials is still in its early days, and only recently have marsupial gametes and their interactions begun to be examined experimentally. The first papers to describe fertilization and related events in a marsupial were only published in 1982 (*Didelphis*: Rodger and Bedford, 1982a,b), and since then gamete interaction has only been reported for one other species (*Sminthopsis*; Breed and Leigh, 1988). However, it is appropriate that marsupials are included in this volume on the comparative aspects of mammalian fertilization because despite the general "mammalian" character of their gametes, they differ in several features that play critical roles in eutherian fertilization.

The present chapter provides an overview of marsupial gametes and their form and function in contrast with that described for eutherians in the rest of the volume. Perhaps in some subsequent edition the final group of mammals, the egg-laying monotremes, may also be considered, but at present essentially nothing is known of gamete interaction in this group.

2. MARSUPIAL SPERMATOZOA

2.1. General Features

Marsupial sperm are all unmistakably mammalian in character, with a compact nucleus, a midpiece (with spiral mitochondrial helix), and a well-formed tail with accessory fibers peripheral to the axoneme (reviews: Harding *et al.*, 1979, 1983; Harding, 1987; Temple-Smith, 1987). As in the case for eutherians, major groups of marsupials have their own distinct sperm head morphology, which has proven a useful phylogenetic indicator (Hughes, 1965; Temple-Smith, 1987). Marsupial sperm heads tend to be relatively small and streamlined so that one is first impressed by the tail structures rather than the head. In bandicoots, *Tarsipes*, and dasyurids, this is also because of the great length of the tail (around 200–300 μm; Cummins and Woodall, 1985; Harding *et al.*, 1990). Exceptions to this general pattern are the koala and wombat, and the American didelphids. Both groups have sperm with conspicuous heads with processes that project away from the streamlined head–tail orientation (see below).

2.1.1. HEAD–TAIL RELATIONSHIP

A characteristic feature of all marsupial sperm (an apparent exception to this is discussed later) is that the neck of the tail is inserted into one side of the sperm head rather than at its trailing edge. How this occurs in spermiogenesis is discussed in Section 2.1.4. As a result the spermatozoa of most marsupials have a distinct dorsal–ventral orientation with the tail inserted into the implantation fossa on the ventral surface of the head (Cleland and Rothschild, 1959). The underside or ventral side of the head is deeply grooved, and in didelphids the neck is bent so that the head and tail can achieve a streamlined profile.

2.1.2. SPERM HEAD CHARACTER

Although the sperm head is compact, it is not stabilized by disulfide linkages either in the testis or during epididymal transit. This was first described by studies using detergents and disulfide-cleaving reagents (Bedford and Calvin, 1974; Temple-Smith and Bedford, 1976, 1980), but it has more recently been shown that marsupial protamines lack cysteine entirely (Balhorn *et al.*, 1988, Fifis *et al.*, 1990). As a result the marsupial sperm head is a far more fragile structure than the eutherian sperm head and can be disrupted by detergents, air drying, or even high concentrations of divalent cations (>0.25 M Ca or Mg: Cummins, 1980). The acrosome remains intact after such treatment as assessed by both light and electron microscopy (Cummins, 1980; Mate and Rodger, 1991).

2.1.3. THE ACROSOME

The marsupial acrosome lies on the dorsal surface of the head opposite the tail insertion (Fig. 1). It never forms a cap over the leading edge of the head and appears not to be divided into regions (Harding *et al.*, 1979). In particular, there is no morphologically distinct equatorial segment, the site of sperm–oocyte membrane fusion in the eutherians (Yanagimachi, 1988; and relevant chapters of this volume). The acrosome ranges in shape from a small button-like structure positioned toward the leading end of the sperm head and restricted to approximately 25% of the length of the head to a rather thin layer covering almost all the dorsal surface (Harding, 1987; Temple-Smith, 1987). The nearest approximation to an acrosomal cap, found in Australian species, is a small lip of acrosome that projects from the dorsal side of the head over its leading surface (*Tarsipes*: Harding *et al.*, 1984) or where anterior displacement of the acrosome leaves it projecting forward from the rostral end of the head (bandicoots: Harding *et al.*, 1990).

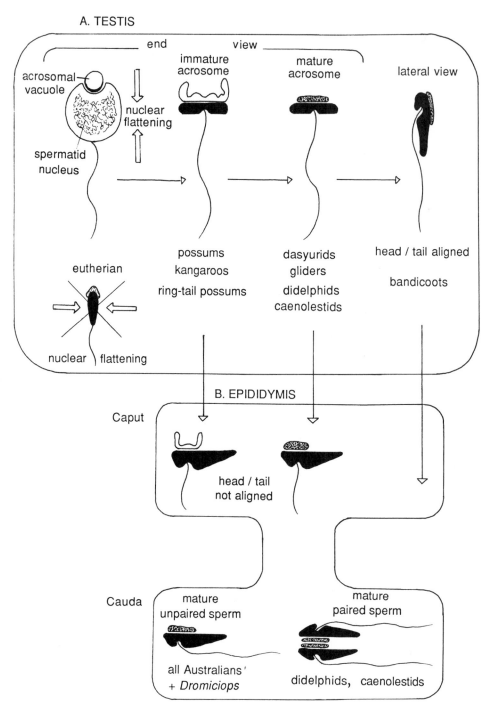

Figure 1. A diagrammatic representation of (A) spermiogenesis in the testis and (B) sperm maturation in the epididymis of marsupials.

2.1.4. NUCLEAR FLATTENING IN SPERMIOGENESIS

The unique arrangement of the tail insertion and the location and form of the acrosome arise because of the manner of nuclear flattening during spermiogenesis (Sapsford *et al.*, 1967, 1969; Phillips, 1970a; Harding *et al.*, 1976a). In eutherians the spermatid nucleus flattens laterally, that is, in the same plane as the forming sperm tail, and the acrosome attached to its anterior surface is then wrapped around the forming head as the familiar acrosomal cap (Fig. 1A). In marsupials the head flattens perpendicular to the tail to produce a flattened carpet-tack-like sperm head with a flat acrosome (Fig. 1A) with tail attached at right angles to the long axis of the head. At this stage the head is not at all streamlined, and to achieve this it must rotate through approximately 90° so that its long axis lies parallel to the tail. This rotation is initiated in the testis and can be essentially complete at spermiation (e.g., bandicoots, Fig. 1A), but in most species the process is only completed during epididymal transit (Fig. 1B). In some species this situation is further complicated in that the head, after partial rotation, returns to the right-angle configuration at spermiation to be restreamlined during epididymal transit (possum, Fig. 1B).

Despite all these apparent specific complexities of sperm head structure, there is an underlying general pattern. The manner of head flattening leads to the characteristic ventral tail insertion and the dorsal position of the acrosome as a vesicle or layer rather than a cap. Subsequently the head rotates relative to the tail to produce the streamlined mature spermatozoon. The only marsupials to differ apparently significantly from this are the koala and its closest relative, the wombat. These animals have hooked spermatozoa superficially rodent-like in appearance (Harding *et al.*, 1987; Temple-Smith, 1987). However, even in this very confusing and poorly understood situation, flattening is still at right angles to the forming tail, and the acrosome is limited to the opposite "dorsal" surface of the head (Harding *et al.*, 1987). The situation in the koala is further complicated by a marked amount of heterogeneity of sperm head form and by the apparent occurrence of posttesticular sperm head shaping (Temple-Smith and Taggart, 1990).

2.2. Epididymal Maturation

Marsupial sperm undergo very similar maturational events to eutherian sperm during epididymal transit. Full progressive motility is achieved in the epididymis, and sperm undergo significant changes in their surface characteristics (Cummins, 1976; Temple-Smith and Bedford, 1976, 1980). Although the manner of cytoplasmic droplet loss is different in marsupials, this final cytoplasmic reduction also occurs during passage through the epididymis (Temple-Smith, 1984). In contrast to the eutherians, however, there is no significant increase in the disulfide stabilization of sperm head or membranes during epididymal transit, and thus, as mentioned earlier, the marsupial sperm head is a far more fragile structure than the eutherian sperm head (Temple-Smith and Bedford, 1976, 1980; Cummins, 1980).

2.2.1. MORPHOLOGICAL CHANGES

In contrast to the mainly functional maturation events described for eutherian sperm, marsupial sperm undergo quite obvious morphological changes in the epididymis. The alignment of the head-and-tail axis that produces a fully streamlined sperm has already been mentioned, but, in addition, in many Australian possums and kangaroos there is also a major reorganization of the acrosome. In these species the acrosome projects as a rather exotic veil-like structure from the sperm head at spermiation and undergoes contraction and membrane loss during transit through the epididymis to produce the mature button-like acrosome (Fig. 1B) (Harding *et al.*, 1976b; Temple-Smith and Bedford, 1976; Cummins, 1976). Where this occurs, mature cauda sperm usually have quite small compact acrosomes restricted to less than 50% of the dorsal nuclear surface (Harding, 1987). In most marsupials this process is completed in the testis (Fig. 1B), but in either case the final form and location of the acrosome are distinctly marsupial.

2.2.2. SPERM PAIRING

A further and quite unique epididymal maturation phenomenon is the pairing of sperm, which until recently was thought to occur in all American marsupials (Biggers and DeLamater, 1965; Rodger, 1982). A striking exception to this rule has recently been reported for *Dromiciops*, possibly the only surviving member of the South American microbiotherid family. This species has single sperm that are quite startlingly similar in head shape to the sperm of Australian possums and kangaroos (Temple-Smith, 1987). Pairing of sperm has never been reported for an Australian marsupial (review: Temple-Smith, 1987).

Sperm pairing brings together, in a precisely organized manner, two presumably quite unrelated and randomly selected cells in a relationship akin to a very large gap junction (Phillips, 1970b; Olson and Hamilton, 1976; Olson, 1980; Temple-Smith and Bedford, 1980). It is particularly intriguing that pairing is via the dorsal surfaces of the sperm heads such that their acrosomes are locked away within the paired complex (Fig. 1B). Once paired, the sperm swim quite normally as a single unit with the tails beating in an Irish-harp-like pattern (Temple-Smith and Bedford, 1980; Rodger and Bedford, 1982a). Because of the occurrence of the head–tail realignment, the tails are facing in the same direction, not in opposition (Fig. 1B). However, in one American marsupial group, the caenolestids, there does appear to be an opposition of the tails of paired cauda sperm at 180° (Temple-Smith, 1987). Live sperm of any caenolestid have yet to be observed, so it is not known whether they swim in a coordinated manner despite this arrangement or if their tails become aligned on release from the epididymis or in the female tract.

3. MARSUPIAL OOCYTES

3.1. Oocyte Maturation

The marsupial oocyte, like that of eutherians, is held in meiotic arrest at the germinal vesicle stage until the immediately preovulatory period, when one oocyte, or several if the species is polyovulatory, enters the final phase of follicle growth and the completion of meiosis. Prior to this the arrested oocyte and its follicle go through the typical mammalian two-phase growth pattern to become an antral Graafian follicle (Lintern-Moore *et al.*, 1976; Lintern-Moore and Moore, 1977; Panyaniti *et al.*, 1985).

In all species so far examined the oocyte is ovulated after completing the first meiotic division at the first-polar-body stage (Rodger and Bedford, 1982b, Breed and Leigh, 1988; Baggott *et al.*, 1987, Tyndale-Biscoe and Renfree, 1987). Marsupials appear to be spontaneous ovulators except perhaps one rat kangaroo (*Bettongia*: M. Smith and L. Hinds, unpublished data). As is the normal mammalian pattern, fertilization triggers the completion of meiosis and the release of the second polar body (Rodger and Bedford, 1982b; Tyndale-Biscoe and Renfree, 1987).

3.2. Oocyte Vestments

At the time of fertilization it appears that, as a general rule, the marsupial oocyte has lost all cellular vestments and is surrounded by only a relatively thin zona pellucida of the order of 1–4 μm thick (Hughes, 1974; Rodger and Bedford, 1982b; Breed and Leigh, 1988; Tyndale-Biscoe and Renfree, 1987). Not only is there no cellular shroud around the ovulated marsupial oocyte but even the inner layer of granulosa cells equivalent to the corona radiata is lost during the preovulatory maturation phase in the follicle. The marsupial zona has not been studied bio-chemically or immunologically, but it is susceptible to crude acrosomal enzyme extracts (Rodger and Bedford, 1982b; Rodger and Young, 1981) and to trypsin (Talbot and DiCarlantonio, 1984). In addition to the morphologically distinct zona pellucida, the perivitelline space of opossum and *Sminthopsis* oocytes contains ruthenium-red-staining extracellular matrix material that may act as a further vestment for the fertilizing spermatozoon to penetrate (Talbot and DiCarlantonio, 1984; Breed and Leigh, 1988). The matrix material appears to be formed of protein granules and hyaluronic acid filaments (Talbot and DiCarlantonio, 1984).

4. COLLECTION OF GAMETES

4.1. Induction of Ovulation

The marsupial ovary responds to exogenous gonadotrophins and an endogenous luteinizing hormone (LH) release. Oocyte maturation and ovulation can be induced in gonadotrophin-primed animals by synthetic gonadotrophin-releasing hormone (GnRH) (Rodger and Mate, 1988; Rodger, 1990; J. Rodger and L. Hinds, unpublished data). In the best-understood species, the brush-tailed possum (*Trichosurus*), prophase I through metaphase I chromosomes are found a mean of 16.2 hr after GnRH, polar body 1 formation around 20 hr, and ovulation usually by 24 hr (Hayman and Rodger, 1990; K. Mate, S. Cousins, and J. Rodger, unpublished data). Some polyovulatory species will ovulate spontaneously after only gonadotrophin treatment (single injection of PMSG), but all monovular species examined and at least one polyovulatory species appear to absolutely require a specific LH release stimulus (Rodger, 1990). The encouraging aspect of these superovulation studies is that relatively large numbers of oocytes are now available, and experimental study of marsupial gametes and their maturation and interaction *in vitro* is now feasible.

4.2. Semen Collection

Semen can be collected from marsupials by electroejaculation (Rodger and White, 1978; Rodger and Pollitt, 1981), and the sperm removed from the abundant accessory gland secretions by conventional techniques such as "swim up" (Mate and Rodger, 1991). Marsupial semen contains N-acetylglucosamine or glycogen/glucose rather than fructose as the major sugar (Rodger, 1976; Rodger and White, 1974, 1980). However, as in eutherians, the sperm readily use exogenously supplied glucose as an energy source when incubated *in vitro* (Rodger and Suter, 1978).

5. GAMETES IN THE FEMALE TRACT

5.1. The Anatomy of the Female Tract

Before discussing gamete function in the female, it is probably of use to describe the quite unfamiliar form of the marsupial female reproductive tract (review: Tyndale-Biscoe and Renfree, 1987). In marsupials the developing ureters lie central to the Mullerian ducts, and the female tract does not fuse in the characteristic eutherian manner. This not only means that the uteri and cervices are completely separate; it also produces a complex three-section Grecian-urn-like vagina (Fig. 2). The lateral arms carry semen from the site of semen deposition at the upper part of the urogenital sinus to the twin cervices. The lateral vaginae and upper parts of the vaginal sac undergo hypertrophy at estrus and become highly secretory, producing 3–4 ml of mucus in the

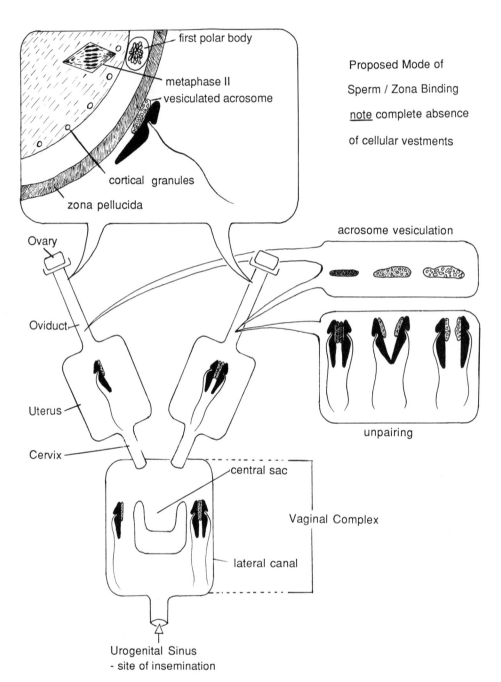

first polar body

metaphase II

vesiculated acrosome

cortical granules

zona pellucida

Proposed Mode of

Sperm / Zona Binding

note complete absence

of cellular vestments

acrosome vesiculation

unpairing

Ovary

Oviduct

Uterus

Cervix

central sac

Vaginal Complex

lateral canal

Urogenital Sinus
- site of insemination

Figure 2. A diagrammatic representation of sperm transport, function, and fertilization events in the marsupial female tract.

possum *Trichosurus*, which is presumably mixed with semen at mating (Hughes and Rodger, 1971). The central sac forms a usually temporary patent connection to the urogenital sinus at parturition and serves as the birth canal. In some Australian species such as kangaroos this remains as a permanent patent birth canal and central third vagina. In at least one monovulatory species, the tammar wallaby, the complete separation of left and right cervix–uterus–oviduct has significant implications for fertility. In this animal, and possibly many monovulatory marsupials, ovulation is alternate in successive estrous cycles, and the animals conceive at a postpartum estrus (Tyndale-Biscoe, 1984). However, in this case only the ovulatory (not postpartum) side is able to effectively transport sperm (Tyndale-Biscoe and Rodger, 1978). The recently pregnant and nonovulatory side is infertile until the next cycle. This is a pregnancy effect because in cycling but nonpregnant animals both the ovulatory and nonovulatory sides of the female tract transport sperm.

5.2. Timing of Mating and Ovulation

There is information on the interval from mating to ovulation and fertilization for several species. In most species mating precedes ovulation and fertilization by around 12–30 hr (tammar; Tyndale-Biscoe and Rodger, 1978; *Didelphis*, Rodger and Bedford, 1982a; *Sminthopsis*, Breed and Leigh, 1988). However, in certain dasyurids sperm are stored in the female tract for many days awaiting ovulation (*Dasyurus*; Hill and O'Donoghue, 1913; *Antechinus*, Selwood and McCallum, 1987). In *Antechinus* the interval from mating to ovulation varies from 1 to 20 days, but peak fertility is at 9.5 days (Selwood and McCallum, 1987). Marsupial sperm thus reside in the female tract for moderate to extended periods in much the same way as seen in eutherians. In both opossums and dasyurids this sperm storage appears to occur in the lower oviduct in subluminal crypts lined by low, non-mucoid-secreting epithelium (Rodger and Bedford, 1982a; Breed and Leigh, 1988).

The above observations *in vivo* suggest that marsupial sperm probably require a period of capacitation, as appears to be the case for all eutherian sperm so far studied (Bedford, 1983). However, in this respect no direct experiments have been carried out to determine the minimum period sperm must reside in the oviduct to achieve fertilization except for the *Antechinus* study, which found low fertility if sperm had not resided in the oviduct for more than 5 days prior to ovulation (Selwood and McCallum, 1987). There is evidence for capacitation from *in vitro* incubation studies. Washed ejaculated possum (*Trichosurus*) sperm completely fail to bind to the zona pellucida when sperm and follicular oocytes are incubated together at 33°C (possum body temperature) (K. E. Mate and J. C. Rodger, unpublished data). Thus, it appears that a sperm maturation event (capacitation?) is required before ejaculated sperm can interact with the egg surface. Capacitation is discussed further in Sections 6 and 7.

5.3. Unpairing of Opossum Sperm

In *Didelphis* and *Monodelphis*, unpairing of sperm occurs in the lower oviduct in crypts lying below the oviduct lumen proper and lined by non-mucoid-secreting and nonciliated cells (Rodger and Bedford, 1982a; Bedford *et al.*, 1984; J. C. Rodger and J. VandeBerg, unpublished data). This process can be observed in sperm flushed from the oviduct in simple defined culture medium (Rodger and Bedford, 1982b). Unpairing is initiated by loss of the peripheral membrane junction while the sperm heads remain adherent over the broad area of plasma membrane overlying the acrosomes. The leading anterior surfaces of the two sperm heads then begins to split open in a scissors-like manner until eventually the two heads are held together only at their most dorsal trailing extremities (Fig. 2). Separated sperm are fully capable of progressive motility.

There is no accepted explanation of the role of sperm pairing, but it has been suggested that it protects the opossum sperm acrosome from the environment of the female tract and prevents its premature activation. This view was based on observations that found that species with sperm pairing ejaculated low numbers of sperm, which were then very efficiently transported so that around 5% of the ejaculate reached the oviducts (Bedford *et al.*, 1984). The majority of Australian marsupials ejaculate sperm in similar numbers to eutherian mammals and a have similarly "inefficient" sperm transport. The exception to this pattern are the Australian dasyurids, which have very large single sperm and have reduced sperm numbers to an even greater extreme than the opossums (Woolley, 1975). However, in the isthmus of the oviduct of both dasyurids and opossums there are specialised sperm storage crypts lined by non-mucoid-secreting epithelium, and these may also have a role in protecting the acrosome, particularly in those species that store sperm for several days awaiting ovulation (Bedford *et al.*, 1984).

6. POSTEJACULATORY SPERM FUNCTION

Whether marsupial sperm require a period of prefertilization incubation equivalent to capacitation remains to be established. However, there is intriguing evidence from work with brush-tailed possum (*Trichosurus*) sperm incubated in "capacitating" media and with agents that induce acrosome reactions in eutherian sperm to suggest that major changes are occurring in marsupial sperm that precede sperm–egg interaction and presumably are involved in preparation of the sperm for fertilization (Mate and Rodger, 1991).

6.1. Attempts to Induce Marsupial Sperm to Acrosome React

Initially experiments directed at characterizing the marsupial acrosome reaction began with the simple premise that agents such as calcium ionophores would induce an exocytotic event similar at least in principle to that seen in all eutherian sperm so far examined, or indeed in most vertebrates and invertebrates studied, but this is not the case. The possum acrosome is remarkably resistant to all acrosome-inducing reagents tested and remains intact even at very high concentrations that disrupt sperm membrane integrity and sperm head stability (Table 2). The marsupial acrosome is not unaffected by incubation *in vitro*, but the response occurs within the acrosomal matrix and is evident only at the electron microscope level (Table 2). The acrosomal contents undergo a vesiculation process that ultimately replaces the amorphous granular matrix with membrane-like material (Fig. 3A). During this process the acrosome swells slightly, but none of these changes are evident at the light microscope level as assessed by phase-contrast microscopy or acrosomal strains.

This transformation of the acrosomal contents is seen to a reasonable degree in around 30% of sperm after 1 hr of incubation in control media at 33°C and is not apparently influenced by acrosome-reaction-inducing agents such as a calcium ionophore. Marked vesiculation of around 60% of sperm occurred when sperm were incubated in a "capacitating" protocol based on experience in a variety of eutherian species (e.g., Oliphant *et al.*, 1985). This consisted of 0.5 hr in high-ionic-strength (380 mosm) adjusted Hank's balanced salt solution (HBSS) followed by 0.5 hr in HBSS with 3% BSA (HIS/BSA).

6.2. Acrosome Changes *in Vivo*

Apparently identical changes in acrosomal shape and contents have been described as a correlate of *in vivo* sperm unpairing in *Didelphis* (Fig. 3B) (Rodger and Bedford, 1982b). This agreement between *in vitro* and *in vivo* studies argues strongly that the acrosomal transformation

Table 2. The Effects on the Brush-Tailed Possum Spermatozoon Acrosome of Incubation with Agents that Capacitate or Acrosome React Eutherian Spermatoza

Treatment	Light microscope: % acrosome		Electron microscope: % acrosome vesiculation		
	Present	Absent	Nil	Slight	Marked
HBSS, time 0	99	1	77	12	11
HBSS, 1hr	99	1	33	36	31
HIS/BSA	97	3	20	22	58
Control	99	1	38	26	36
1.0 μM A23187	100	0	36	36	29
10 μM A23187	98	2	18	20	36[a]
100 μM A23187	98	2		nd	

Other Agents Tested: Dibutyryl cAMP and cGMP (0.01–10 μM); dioctanoyl glycerol (0.1–50 μM); inositol triphosphate (0.01–50 μM); phorbol 12-myristate 13-acetate (1–1000 nM); lysophosphatidylcholine (1.4–85 μMg/ml); trypsin (0.5–5.0%)

[a]The remaining 26% acrosomes were damaged by ionophore treatment and were not classified.

or metamorphosis described here is an integral part of sperm function prior to gamete interaction. Whether this is a morphological equivalent of the functionally described eutherian phenomenon, capacitation, is an intriguing possibility. If this is so, it would suggest that unpairing of sperm may be part of capacitation, since acrosomal vesiculation is only seen at the time sperm begin to unpair, and the process is complete when sperm are single. In species where sperm do not pair, presumably there is a capacitation system based on modification of membrane proteins or lipids, as is generally thought to be the case for eutherian sperm (reviews: Bedford, 1983; Meizel, 1985; Oliphant *et al.*, 1985).

6.3. The Acrosome and Sperm–Egg Interaction

Presumably the acrosome is eventually lost or reacted in some way after reaching the egg surface, but there has been no direct observation of these events in any marsupial to date. It is known, however, that primary contact between the sperm surface and zona is on the acrosome-covered surface of the *Didelphis* and *Monodelphis* sperm head (Rodger and Bedford, 1982b; L. Baggott and H. Moore, unpublished data), and in *Sminthopsis* the heads of sperm penetrating the zona pellucida lack an acrosome (Breed and Leigh, 1988). The marsupial acrosome appears to contain the normal spectrum of proteolytic and glycosidase enzyme activities seen in eutherian spermatozoa (Rodger and Young, 1981).

6.4. Do Marsupial Sperm Undergo Prefertilization Motility Changes?

There is no definitive evidence to date as to whether marsupial sperm undergo any change in their pattern of motility as fertilization is approached comparable with the hyperactivation described for eutherian spermatozoa (Yanagimachi, 1988), although incubated possum sperm occasionally are seen to swim in a nonprogressive figure-eight pattern (S. Cousins and J. Rodger, unpublished data). There is, however, a fairly clear change in the relationship between tail and head when sperm attach to the surface of *Didelphis* oocytes (Rodger and Bedford, 1982b). The head attaches via its dorsal surface, and the tail reorientates to its more immature right-angle arrangement, apparently exerting force directly down on the acrosomal surface apposed to the zona pellucida (Fig. 2). As noted by Breed and Leigh (1988), space restrictions within the oviduct

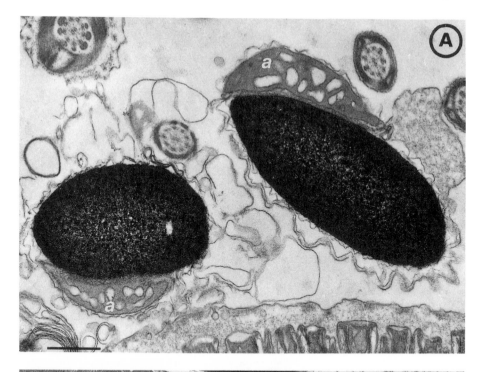

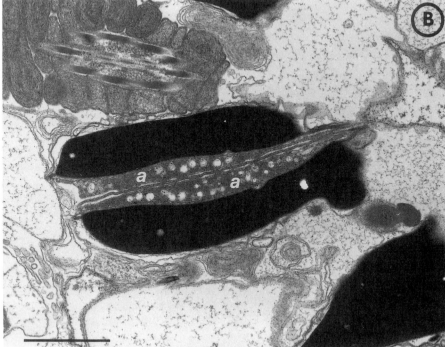

Figure 3. Electron micrographs of vesiculating acrosomal matrix (a) of (A) brush-tailed possum spermatozoa incubated under the HIS/BSA regimen for a total of 1 hr at 33°C and (B) unpairing Virginia opossum spermatozoa in an oviducal crypt (bars = 0.5 μm).

may limit the possible orientation of sperm and oocyte in species like *Sminthopsis* with very large gametes and a narrow oviduct.

7. GAMETE INTERACTION

7.1. Sperm–Egg Binding

In all marsupials it is likely, based on observations in *Didelphis* and *Monodelphis* and on the structure of the gametes, that the marsupial sperm, probably in its acrosome-altered state, meets and binds to the surface of the naked zona pellucida by the outer acrosomal surface (Fig. 2). This has been visualized at the phase-contrast light microscope level in *Didelphis virginiana*, where the acrosomal surface of the head has a particularly distinct morphology (Rodger and Bedford, 1982b).

Trichosurus sperm washed free of seminal plasma by "swim up" do not bind to the zona pellucida when incubated under a variety of conditions with mature oocytes collected from preovulatory follicles (K. Mate and J. Rodger, unpublished data). The same result has been obtained with *Sminthopsis* gametes using epididymal sperm (H. Moore and W. Breed, unpublished data). In contrast, *Didelphis* sperm flushed from the oviducts around 12 hr after mating bind rapidly *in vitro* to the surface of such oocytes and penetrate the zona pellucida. In this case the sperm presumably had quite adequate opportunity to undergo capacitation *in vivo* prior to their collection from the oviducts. Fertilization of *Sminthopsis* oocytes *in vitro* has also been achieved using sperm collected from the oviducts several hours after mating, but unfortunately this has not been able to be repeated (W. Breed, unpublished data). These observations are further evidence that marsupial sperm do require a period of capacitation prior to achieving the ability to bind to the zona.

Alternatively, or in addition, it is possible that the oocyte and zona pellucida are undergoing subtle maturational changes that are required before sperm can bind. Evidence that the oocyte is undergoing postovulatory maturation comes from observations on cortical granules in the brush-tailed possum and tammar wallaby. When oocytes from incubation experiments were examined in the electron microscope, no cortical granules were evident, and it was assumed that handling or incubation had induced premature cortical granule release, blocking fertilization. However, subsequent examination of follicular and ovulated oocytes indicated that morphologically recognizable cortical granules only began to be found around ovulation, and it currently appears that their major formation is a postovulatory event (K. E. Mate, I. Giles, and J. C. Rodger, unpublished data).

7.2. Zona Pellucida Penetration

Studies in *Didelphis* and *Sminthopsis* suggest two quite different modes of zona penetration.

7.2.1. IN OPOSSUMS

In *Didelphis* (Rodger and Bedford, 1982b) the sperm attaches to the zona by its relatively broad acrosomal surface, applying an essentially flat acrosome-covered head to the zona. Presumably an exocytotic acrosome reaction occurs at this site, and the released enzymes digest a path through the very thin zona pellucida. These events have not been observed directly but in eggs in the process of sperm penetration into the cytoplasm, with the major part of the tail still projecting through the zona, it is clear that a large hole has been produced by zona penetration. The size of this hole may in part be a fixation artifact, but in living eggs observed under phase contrast it is clear that a significant gap exists in the zona, and a cytoplasmic bleb is easily forced through it, usually containing the head and midpiece of the penetrating sperm. How the species then manages to prevent polyspermy is dealt with in Section 8.

7.2.2. IN SMINTHOPSIS

In *Sminthopsis* (Breed and Leigh, 1988) a narrow penetration slit similar to that found in eutherian fertilization has been described. To what extent these differences reflect fundamental features of marsupial fertilization or differences in individual gamete size and morphology remains to be examined. *Sminthopsis* has extremely large sperm operating in a very confined oviducal space, and this may account for the differences observed.

7.3. Sperm–Oocyte Membrane Fusion

Sperm–oocyte membrane fusion has yet to be examined at the ultrastructural level, although it has been observed under phase contrast for *Didelphis*. This is a critical issue because it will settle the question of the presence or absence of the equatorial segment and definitively confirm that fusion is by the inner acrosomal membrane as suggested by Rodger and Bedford (1982b). This is an important question because of the hypothesis (Bedford, 1982) that the equatorial segment has arisen in eutherian mammals as part of the general disulfide stabilization of the sperm head correlated with the robust character of the eutherian zona pellucida. Marsupial gametes lack both the disulfide-stabilized sperm head and a thick zona pellucida.

Since definitive descriptions of acrosome-reacted marsupial sperm and sperm–egg membrane fusion remain to be achieved, it is probably wise to be cautious in completely accepting the absence of a specialized fusion region in marsupials equivalent to the equatorial segment. Although the morphologically distinct equatorial segment of eutherian spermatozoa may seem obvious in hindsight, it was in the first instance described in acrosome-reacted sperm (Barros *et al.*, 1967). Further, in many marsupial species there is a degree of regionality of the acrosome [e.g., see figures in Harding (1987) and Temple-Smith (1987)]. Usually this takes the form a main thicker body and thinner peripheral regions. The koala acrosome has the most equatorial-segment-like form with thin peripheral regions composed of little more than parallel inner and outer acrosomal membrane (illustrated in Harding *et al.*, 1986).

7.4. Sperm Incorporation by the Egg

Sperm incorporation in marsupials seems to follow the conventional mammalian pattern, and the whole sperm is incorporated into the oocyte cytoplasm. The head begins very rapidly to decondense or at least lose its shape and swell on entering the egg cytoplasm. Midpiece mitochondria and tail elements persist for some hours in the cytoplasm and may still be visible in eggs nearing syngamy. In *Sminthopsis* the sperm tail is so large (approx. 100 μm) that it is clearly visible inside living fertilized oocytes under phase contrast (Fig. 4A). Membrane-bound pronuclei are formed after the sperm head and egg chromosomes decondense, but the manner of syngamy has not been observed for any marsupial (Selwood, 1982; J. C. Rodger and J. M. Bedford, unpublished data).

8. THE BLOCK TO POLYSPERMY

The block to polyspermy appears very efficient in both species examined, with all supernumerary sperm restricted to the surface of the zona pellucida or possibly more importantly the mucoid layer deposited by the oviduct on the zona surface. In *Didelphis* and *Monodelphis* mucoid deposition on the zona is so closely correlated temporally with fertilization that it may play a part

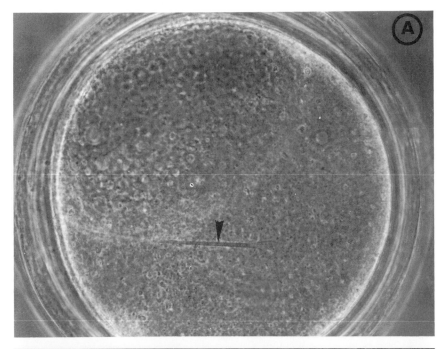

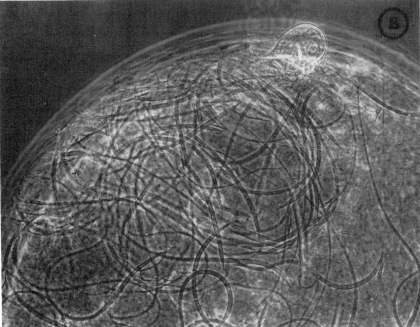

Figure 4. Phase-contrast light micrographs of living fertilized eggs recovered from the oviduct of the dasyurid *Sminthopsis crassicaudata* showing (A) the tail of the fertilizing spermatozoon (arrowed) within an oocyte essentially devoid of mucoid and (B) supernumerary spermatozoa trapped in the mucoid covering the surface of a fertilized egg.

in the block to polyspermy (Rodger and Bedford, 1982b; Baggott *et al.*, 1987). This was suggested because in the opossum sperm penetration leaves such a large hole in the zona pellucida and because all eggs examined in the early stages of gamete interaction and penetration already had some mucoid on their surface, albeit on the surface away from the site of zona–sperm interaction. In some *Didelphis* oocytes fertilized *in vitro* by oviductal sperm, multiple zona penetration sites were found, suggesting that at least under these artificial conditions supernumerary sperm could bind to and penetrate the zona. No evidence of polyspermy was ever seen in eggs fertilized *in vivo* (Rodger and Bedford, 1982b).

Whatever the role of the mucoid in the block to polyspermy, it invariably traps a large number of sperm on the egg surface. These sperm are clearly visible even under phase contrast and are a very good indication of fertilization. Eggs lacking sperm in the mucoid are invariably unfertilized (Selwood, 1982). In *Sminthopsis*, because of the large size of the sperm, fertilized eggs often have a basketwork-like appearance because of the many supernumerary sperm wrapped around their surface, trapped in the oviduct-secreted mucoid (Fig. 4B).

9. EGG TRANSPORT

In all marsupials studied, the fertilized egg travels very rapidly down the oviduct and, as a rule, begins cleavage in the uterus. During oviduct transit, which takes at most 24 hr and possibly as little as 12 hr, the mucoid coat is completed, and perhaps the first part of the outer shell coat is deposited, although the latter is essentially a uterine secretion (Hughes, 1974; Rodger and Bedford, 1982b; Tyndale-Biscoe and Renfree, 1987).

10. CONCLUSION

Fertilization in marsupials thus parallels events described for eutherian mammals, but with important differences the significance of which we at present are only beginning to appreciate. The current state of knowledge suggests that work on marsupial gametes and their maturation and interaction will contribute significantly to our understanding of mammalian fertilization and its evolution in the following important areas: (1) the origin and significance of capacitation; (2) the zona pellucida and its role in gamete recognition; (3) mechanism(s) of zona penetration; (4) the equatorial segment as a specialized site of sperm–egg membrane fusion; and (4) the zona block to polyspermy.

ACKNOWLEDGMENTS. I am indebted to my colleagues who are cited in the text for their contributions to the collaborative work described and for allowing me to discuss their as yet unpublished research. I am particularly thankful to Karen Mate and Peter Temple-Smith for their efforts in reviewing the manuscript and for constructive suggestions. The drawings in Figs. 1 and 2 are the excellent work of Maureen Conroy of the University of Newcastle. Figure 4 was obtained with the assistance of Bill Breed as part of a collaborative study at the University of Adelaide. The author's work reported here was carried out with the financial assistance of the Australian Research Council and the Research Committee of the University of Newcastle.

11. REFERENCES

Baggott, L. M., Davis-Butler, S., and Moore, H. D. M., 1987, Characterization of oestrus and timed collection of oocytes in the grey short-tailed opossum, *Monodelphis domestica, J. Reprod. Fertil.* **79**:105–114.

Balhorn, R., Mazrimas, J. A., Corzett, M., Cummins, J., and Fadem, B., 1988, Analysis of protamines isolated from two marsupials, the ring-tailed wallaby and gray short-tailed opossum, *J. Cell. Biol.* **107:**167A.

Barros, C., Bedford, J. M., Franklin, L. E., and Austin, C. R., 1967, Membrane vesiculation as a feature of the mammalian acrosome reaction, *J. Cell Biol.* **34:**C1–C5.

Bedford, J. M., 1982, Fertilization, in: *Reproduction in Mammals*, Vol. 1, *Germ Cells and Fertilization*, 2nd edn. (C. R. Austin and R. V. Short, eds.), Cambridge, University Press, Cambridge, pp. 128–163.

Bedford, J. M., 1983, Significance of the need for sperm capacitation before fertilization in eutherian mammals, *Biol. Reprod.* **28:**108–120.

Bedford, J. M., and Calvin, H. I., 1974, The occurrence and possible functional significance of -S-S- crosslinks in sperm heads, with particular reference to eutherian mammals, *J. Exp. Zool.* **188:**137–156.

Bedford, J. M., Rodger, J. C., and Breed, W. G., 1984, Why so many mammalian spermatozoa—a clue from marsupials? *Proc. R. Soc. Lond. [Biol.]* **221:**221–233.

Biggers, J. D., and DeLamater, E. D., 1965, Marsupial spermatozoa pairing in the epididymis of American forms. *Nature* **208:**402–404.

Breed, W. G., and Leigh, C. M., 1988, Morphological observations on sperm–egg interactions during *in vivo* fertilization in the dasyurid marsupial *Sminthopsis crassicaudata, Gamete Res.* **19:**131–149.

Cleland, K. W., and Rothschild, Lord, 1959, The bandicoot spermatozoan: An electron microscope study of the tail, *Proc. R. Soc. London. [Biol.]* **150:**24–42.

Cummins, J. M., 1976, Epididymal maturation of spermatozoa in the marsupial *Trichosurus vulpecula*: Changes in motility and gross morphology, *Aust. J. Zool.* **24:**499–511.

Cummins, J. M., 1980, Decondensation of sperm nuclei of Australian marsupials: Effects of air drying and of calcium and magnesium, *Gamete Res.* **3:**351–367.

Cummins, J. M., and Woodall, P. F., 1985, On mammalian sperm dimensions, *J. Reprod. Fertil.* **75:**153–175.

Fifis, T., Cooper, D. W., and Hill, R. J., 1990, Characterization of the protamines of the tammar wallaby (*Macropus eugenii*), *Comp. Biochem. Physiol.* **95B:**571–575.

Harding, H. R., 1987, Interrelationships of the families of the diprotodonta—a view based on spermatozoan ultrastructure, in: *Possums and Opossums: Studies in Evolution* (M. Archer, ed.), Surrey Beatty and Sons and the Royal Zoological Society NSW, Sydney, pp. 195–216.

Harding, H. R., Carrick, F. N., and Shorey, C. D., 1976a, Spermiogenesis in the brush-tailed possum, *Trichosurus vulpecula* (Marsupialia): The development of the acrosome, *Cell Tissue Res.* **171:**75–90.

Harding, H. R., Carrick, F. N., and Shorey, C. D., 1976b, Ultrastructural changes in spermatozoa of the brush-tailed possum, *Trichosurus vulpecula* (Marsupialia), during epididymal transit, Part II: The acrosome, *Cell Tissue Res.* **171:**61–73.

Harding, H. R., Carrick, F. N., and Shorey, C. D., 1979, Special features of sperm structure and function in marsupials, in: *The Spermatozoon* (D. W. Fawcett and J. M. Bedford, eds.), Urban & Schwarzenberg, Baltimore, pp. 289–303.

Harding, H. R., Carrick, F. N., and Shorey, C. D., 1983, Acrosome development during spermiogenesis and epididymal sperm maturation in Australian marsupials, in: *The Sperm Cell* (J. Andre, ed.), Martinus Nijhoff, The Hague, pp. 411–414.

Harding, H. R., Carrick, F. N., and Shorey, C. D., 1984, Sperm ultrastructure and development in the honey possum, *Tarsipes rostratus*, in: *Possums and Gliders* (A. P. Smith and I. D. Hume, eds.), Surrey Beatty and Sons and the Australian Mammal Society, Sydney, pp. 451–461.

Harding, H. R., Carrick, F. N., and Shorey, C. D., 1986, The affinities of the koala *Phascolarctos cinereus* (Marsupialia: Phascolarctidae) on the basis of sperm ultrastructure and development, in: *Possums and Opossums: Studies in Evolution* (M. Archer, ed.), Surrey Beatty and Sons and the Royal Zoological Society, NSW, Sydney, pp. 353–364.

Harding, H. R., Shorey, C. D., and Cleland, K. W., 1990, Ultrastructure of spermatozoa and epididymal sperm maturation in some perameloids, in: *Bandicoots and Bilbies* (J. H. Seebeck, P. R. Brown, R. L. Wallis, and C. M. Kemper, eds.), Surrey Beatty and Sons, Sydney, pp. 235–250.

Hayman, D. L., and Rodger, J. C., 1990, Meiosis in male and female *Trichosurus vulpecula* (Marsupialia), *Heredity* **65:**251–254.

Hill, J. P., and O'Donoghue, C. H., 1913, The reproductive cycle in the marsupial *Dasyurus viverrinus*, *Q. J. Microsc. Sci.* **59:**133–174.

Hughes, R. L., 1965, Comparative morphology of spermatozoa from five marsupial families, *Aust. J. Zool.* **13:**533–543.

Hughes, R. L., 1974, Morphological studies on implantation in marsupials, *J. Reprod. Fertil.* **39:**173–186.

Hughes, R. L., and Rodger, J. C., 1971, Studies on the vaginal mucus of the marsupial *Trichosurus vulpecula*, *Aust. J. Zool.* **19**:19–33.

Lintern-Moore, S., and Moore, G. P. M., 1977, Comparative aspects of oocyte growth in mammals, in: *Reproduction and Evolution* (J. H. Calaby and C. H. Tyndale-Biscoe, eds.), Australian Academy of Sciences, Canberra, pp. 215–219.

Lintern-Moore, S., Moore, G. P. M., Tyndale-Biscoe, C. H., and Poole, W. E., 1976, The growth of oocyte and follicle in the ovaries of monotremes and marsupials, *Anat. Rec.* **185**:325–332.

Mate, K. E., and Rodger, J. C. 1991, Stability of the acrosome of the brush-tailed possum (*Trichosurus vulpecula*) and tammar wallaby (*Macropus eugenii*) *in vitro* and after exposure to conditions and agents known to cause capacitation or acrosome reaction of eutherian spermatozoa, *J. Reprod. Fertil.* **91**:41–48.

Meizel, S., 1985, Molecules that initiate or help stimulate the acrosome reaction by their interaction with the mammalian sperm surface, *Am. J. Anat.* **174**:285–302.

Oliphant, G., Reynolds, A. B., and Thomas, T. S., 1985, Sperm surface components involved in the control of the acrosome reaction, *Am. J. Anat.* **174**:269–283.

Olson, G. E., 1980, Changes in intramembranous particle distribution in the plasma membrane of *Didelphis virginiana* spermatozoa during maturation in the epididymis, *Anat. Rec.* **197**:471–488.

Olson, G. E., and Hamilton, D. W., 1976, Morphological changes in the midpiece of wooly opossum spermatozoa during epididymal transit, *Anat. Rec.* **186**:387–404.

Panyaniti, W., Carpenter, S. M., and Tyndale-Biscoe, C. H., 1985, Effects of hypophysectomy on folliculogenesis in the tammar *Macropus eugenii* (Marsupialia: Macropodidae), *Aust. J. Zool.* **33**:303–311.

Phillips, D. M., 1970a, Development of spermatozoa in the wooly opossum with special reference to the shaping of the sperm head, *J. Ultrastruct. Res.* **33**:369–380.

Phillips, D. M., 1970b, Ultrastructure of spermatozoa of the wooly opossum *Caluromys philander*, *J. Ultrastruct. Res.* **33**:381–397.

Rodger, J. C., 1976, Comparative aspects of the accessory sex glands and seminal biochemistry of mammals, *Comp. Biochem. Physiol.* **55B**:1–8.

Rodger, J. C., 1982, The testis and its excurrent ducts in American caenolestid and didelphid marsupials, *Am. J. Anat.* **163**:269–282.

Rodger, J. C., 1990, Prospects for the artificial manipulation of marsupial reproduction and its application in research and conservation, *Aust. J. Zool.* 37:249–258.

Rodger, J. C., and Bedford, J. M., 1982a, Induction of oestrus, recovery of gametes, and the timing of fertilization events in the opossum, *Didelphis virginiana*, *J. Reprod. Fertil.* **64**:159–169.

Rodger, J. C., and Bedford, J. M., 1982b, Separation of sperm pairs and sperm–egg interaction in the opossum, *Didelphis virginiana*, *J. Reprod. Fertil.* **64**:171–179.

Rodger, J. C., and Mate, K. E., 1988, A PMSG/GnRH method for the superovulation of the monovulatory brush-tailed possum (*Trichosurus vulpecula*), *J. Reprod. Fertil.* **83**:885–891.

Rodger, J. C., and Pollitt, C. C., 1981, Radiographic examination of electroejaculation in marsupials, *Biol. Reprod.* **24**:1125–1134.

Rodger, J. C., and Suter, D. A. I., 1978, Respiration rates and sugar utilization by marsupial spermatozoa, *Gamete Res.* **1**:111–116.

Rodger, J. C., and White, I. G., 1974, Free N-acetylglucosamine in marsupial semen, *J. Reprod. Fertil.* **39**:383–386.

Rodger, J. C., and White, I. G., 1978, The collection, handling and some properties of marsupial semen, in: *The Artificial Breeding of Non-domestic Animals* (P. F. Watson, ed.), Academic Press, London, pp. 289–301.

Rodger, J. C., and White, I. G., 1980, Glycogen not N-acetylglucosamine the prostatic carbohydrate of three Australian and American marsupials, and patterns of these sugars in Marsupialia, *Comp. Biochem. Physiol.* **67B**:109–113.

Rodger, J. C., and Young, R. J., 1981, Glycosidase and cumulus dispersal activities of acrosomal extracts from opossum (marsupial) and rabbit (eutherian) spermatozoa, *Gamete Res.* **4**:507–514.

Sapsford, C. S., Rae, C. A., and Cleland, K. W., 1967, Ultrastructural studies on spermatids and Sertoli cells during early spermiogenesis in the bandicoot *Perameles nasuta* Geoffroy (Marsupialia), *Aust. J. Zool.* **15**:881–909.

Sapsford, C. S., Rae, C. A., and Cleland, K. W., 1969, Ultrastructural studies, on maturing spermatids and on Sertoli cells in the bandicoot *Perameles nasuta* Geoffroy (Marsupialia), *Aust. J. Zool.* **17**:195–292.

Selwood, L., 1982, A review of maturation and fertilization in marsupials with special reference to the dasyurid: *Antechinus stuartii*, in: *Carnivorous Marsupials*, Vol. 1 (M. Archer, ed.), Royal Zoological Society of NSW, Sydney, pp. 65–76.

Selwood, L., and McCallum, F., 1987, Relationship between longevity of spermatozoa after insemination and the percentage of normal embryos in brown marsupial mice (*Antechinus stuartii*), *J. Reprod. Fertil.* **79**:495–503.

Talbot, P., and DiCarlantonio, G., 1984, Ultrastructure of opossum oocyte investing coats and their sensitivity to trypsin and hyaluronidase. *Dev. Biol.* **103:**159–167.

Temple-Smith, P. D., 1984, Phagocytosis of sperm cytoplasmic droplets by a specialized region in the epididymis of the brushtailed possum, *Trichosurus vulpecula, Biol. Reprod.* **30:**707–720.

Temple-Smith, P. D., 1987, Sperm structure and marsupial phylogeny, in: *Possums and Opossums: Studies in Evolution* (M. Archer, ed.), Surrey Beatty and Sons and the Royal Zoological Society of NSW, Sydney, pp. 171–193.

Temple-Smith, P. D., and Bedford, J. M., 1976, The features of sperm maturation in the epididymis of a marsupial, the brushtailed possum *Trichosurus vulpecula, Am. J. Anat.* **147:**471–500.

Temple-Smith, P. D., and Bedford, J. M., 1980, Sperm maturation and the formation of sperm pairs in the epididymis of the opossum, *Didelphis virginiana, J. Exp. Zool.* **214:**161–171.

Temple-Smith, P. D., and Taggart, D. A., 1990, On the male generative organs of the koala (*Phascolarctos cinereus*): An update, in: *Biology of the Koala* (A. K. Lee, K. A. Handasyde, and G. D. Sanson, eds.), Surrey Beatty and Sons, Sydney, pp. 33–54.

Tyndale-Biscoe, C. H., 1984, Mammals: Marsupials, in: *Marshall's Physiology of Reproduction*, Vol. 1, 4th ed. (G. E. Lamming, ed.), Churchill Livingstone, Edinburgh, pp. 386–454.

Tyndale-Biscoe, C. H., and Janssens, P. A., 1988, *The Developing Marsupial, Models for Biomedical Research*, Springer-Verlag, Berlin.

Tyndale-Biscoe, C. H., and Renfree, M. B., 1987, *Reproductive Physiology of Marsupials*, Cambridge University Press, Cambridge.

Tyndale-Biscoe, C. H., and Rodger, J. C., 1978, Differential transport of spermatozoa into the two sides of the genital tract of a monovular marsupial, the tammar wallaby (*Macropus eugenii*), *J. Reprod. Fertil.* **52:**37–43.

Woolley, P., 1975, The seminiferous tubules in dasyurid marsupials, *J. Reprod. Fertil.* **45:**255–261.

Yanagimachi, R., 1988, Mammalian fertilization, in: *The Physiology of Reproduction*, Vol. 1 (E. Knobil and J. D. Neill, eds.), Raven Press, New York, pp. 135–185.

8

Fertilization in Bats

Philip H. Krutzsch and Elizabeth G. Crichton

1. INTRODUCTION

The process of reproduction among the exceedingly diverse and specialized group of flying mammals that comprise the order Chiroptera is remarkably variable in both anatomic and physiological characteristics. The ubiquitous distribution, almost worldwide, and varied habitat and niche selection of the some 847 known species (Koopman and Jones, 1970) have resulted in the expression of many different reproductive patterns, of which monestry, polyestry, asymmetry of female reproductive tract function, dyssynchrony between male primary and secondary sex organ function, long gestations, and delays in ovulation, implantation, or development are interesting aspects that have been described (see reviews in the *Journal of Reproduction and Fertility*, Symposium 14, and the *American Journal of Anatomy*, volume 178).

2. PATTERNS OF REPRODUCTION IN CHIROPTERA

2.1. Asynchrony of Gamete Production and Sperm Storage

It is relevant to this review to recognize that insemination and ovulation do not always coincide in bats (Pagenstecher, 1859; Hartman, 1933; Wimsatt, 1960, 1969; Krutzsch, 1979; Oxberry, 1979). Thus, in many species, sperm production and testicular regression precede ovulation by a period of up to many months. Viable spermatozoa are stored in the caudae epididymides and in the female reproductive tracts throughout this prolonged period (Hiraiwa and Uchida, 1956a,b; Racey, 1973, 1979; Uchida *et al.*, 1988). Although further inseminations can and do occur, experiments isolating inseminated females from males (Wimsatt, 1942, 1944) have demonstrated that sperm introduced at the beginning of the sperm storage season can fertilize ova released at the end of the season, an interval of 6 months or more. An interesting juxtapositional relationship between sperm stored in the oviduct and uterotubal junction and the adjacent epithelium has been described; it could provide a route for metabolic-supportive,

PHILIP H. KRUTZSCH AND ELIZABETH G. CRICHTON • Department of Anatomy, University of Arizona, Tucson, Arizona 85724.

A Comparative Overview of Mammalian Fertilization, edited by Bonnie S. Dunbar and Michael G. O'Rand. Plenum Press, New York, 1991.

inhibitory, or capacitating substances to this portion of the stored sperm (Racey, 1979; Krutzsch *et al.*, 1982; Racey *et al.*, 1987). Some physiological parameters of the storage environments have also been measured (Racey, 1975; Crichton *et al.*, 1981, 1982; Krutzsch *et al.*, 1984). To date, however, nothing is known of the mechanisms of the sperm storage phenomenon.

Most of the species showing sperm longevity belong to the families Vespertilionidae and Rhinolophidae. These bats live in geographic regions in which the winter environment elicits either extended torpor or sustained hibernation. Sperm storage throughout the winter would thus appear to represent a physiological adaptation to prolonged periods of reduced metabolism. Nevertheless, limited periods of arousal and foraging occur during the sperm storage interval. Furthermore, some heterothermic vespertilionid species living in tropical environments also store sperm (Gopalakrishna and Madhavan, 1971; Medway, 1972; Racey *et al.*, 1975). Thus, "cold storage" cannot be totally responsible for prolonged gamete survival. It is important to keep in mind that the spermatozoa for which we have the most information (*Pipistrellus, Myotis, Plecotus, Antrozous*, and *Eptesicus* species) all derive from sperm-storing species. Thus, they may be specialized for storage and differ biologically from the spermatozoa of non-sperm-storing chiropteran and other mammalian species.

2.2. Synchrony of Gamete Production

An interesting exception to the asynchronous production of gametes occurs in one genus of vespertilionid bats. In temperate regions of its range, *Miniopterus schreibersii* ovulates and conceives at the time of fall copulation; implantation of the blastocyst is delayed during winter torpor (Courrier, 1927; Richardson, 1977). Even so, sperm remain in the caudae epididymides for some 7 months after conception (at 37° S lat.: E. G. Crichton, unpublished observations) and for a more limited time in the female tracts before (Mori and Uchida, 1980) and after fall conception (E. G. Crichton, unpublished observations). These stored sperm remain motile, although (based on observations at 37° S lat.: P. H. Krutzsch and E. G. Crichton, unpublished data) their later insemination is unlikely, since the accessory glands are involuted and lack secretion. As far as we know, assemblages of bats that do not hibernate (eg., *Macrotus*: Phyllostomatidae; *Tadarida*: Molossidae) display synchronous timing of the male and female reproductive cycles and gamete production and thus follow the typical mammalian reproductive pattern.

3. MORPHOLOGY OF THE MALE GAMETE

The morphology of chiropteran spermatozoa has been described, at various levels of detail, for only a few species. Depending on the genus, the acrosome varies in size and shape from a small organelle closely applied to the nucleus (*Myotis, Eptesicus, Antrozous, Macrotus*) to a large structure extending some distance anteriorly (*Miniopterus, Pteropus, Syconycteris, Rhino-lophus*) (Fawcett and Ito, 1965; Wimsatt *et al.*, 1966; Uchida and Mori, 1972; Racey, 1979; Mori and Uchida, 1980, 1981, 1982; Mori *et al.*, 1982; Breed and Inns, 1985; Cummins and Woodall, 1985; Rouse and Robson, 1986). There seem to be no differences between the sperm of storing and nonstoring species except in the size, number, and arrangement of mitochondria. Interestingly, there are many more mitochondria in the midpiece of sperm-storing species (Fawcett and Ito, 1965; Uchida and Mori, 1972; Racey, 1979). Furthermore, pairs of these organelles are arranged end to end in successive turns, apparently aligned along the dorsal and ventral aspects of the midpiece; this contrasts with their random orientation in most mammals (Fawcett and Ito, 1965; Wimsatt *et al.*, 1966; Fawcett, 1970). The functional significance of this unusual arrangement has not been established. The presence of four larger dense fibers instead of the usual three is the only fine-structural difference from other mammalian spermatozoa that has been noted (Fawcett and Ito, 1965; Wimsatt *et al.*, 1966; Uchida and Mori, 1972). Interestingly, stored uterine sperm of *Myotis lucifugus* and *Eptesicus fuscus* lack the conspicuous membranous neck scroll characteristic of epididymal sperm. Freeze-fracture studies have revealed unusual aggregates of particles between the acrosomal and postacrosomal regions of the head of a sperm-storing species, *Myotis lucifugus* (Hoffman *et al.*, 1987). These various differences may be related to the unique longevity of these gametes, but in ways that can only be speculated on at present.

4. MORPHOLOGY OF THE FEMALE GAMETE

The female gamete may also be modified in some species. In bats that store sperm, the female remains in proestrus throughout the storage period. The cumulus oophorus surrounding the oocyte of the preovulatory follicle contains large stores of glycogen (Wimsatt, 1944; Wimsatt and Kallen, 1957; Wimsatt and Parks, 1966; Oxberry, 1979). Ultrastructural observations in our laboratory have disclosed lipid stores and abundant mitochondria in the oocytes of hibernating bat species (Fig. 1). The conspicuous microtubules that are characteristic of hamster oocytes are not visible. As in other mammals, it appears that the ovum is in metaphase or telophase of the second meiotic division at fertilization (Hiraiwa and Uchida, 1956a; Mori and Uchida, 1981).

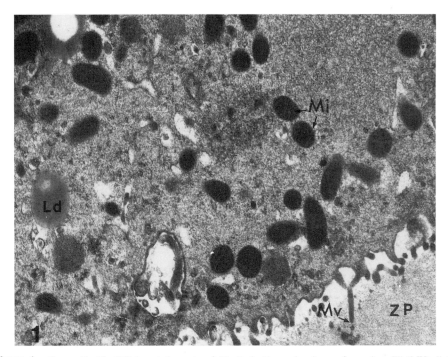

Figure 1. Zona pellucida (ZP)-invested oocyte of *Myotis lucifugus* showing surface microvilli (MV), lipid droplets (Ld), and abundant mitochondria (Mi) in the ooplasm.

5. THE ACROSOME REACTION

Some information is available on prefertilization changes (acrosome reaction) in bat sperm *in vivo*. Most of the data derive from *Miniopterus schreibersii*, which may be unique among Chiroptera in that the acrosome reaction appears to involve the equatorial segment in addition to the anterior acrosome (Uchida and Mori, 1974; Mori and Uchida, 1981). According to these authors, this finding contrasts with "partially" acrosome-reacting sperm in the oviduct, wherein the acrosome reaction does not usually encompass this segment. However, we have obtained views of epididymal sperm of this species in which the segment is not visible following what appears to have been the orderly vesiculation and dissolution processes characteristic of the true acrosome reaction (Fig. 2).

In vitro observations suggest that the acrosome reaction in bat sperm is similar to that in other mammals, involving a fusion between the outer acrosomal and overlying plasma membranes, followed by the formation of hybrid vesicles and the gradual dissolution of the acrosome (*Myotis velifer*, *Tadarida braziliensis*, *Eptesicus fuscus*, personal observations, Figs. 3–5; *Pteropus poliocephalus*, Cummins *et al.*, 1986). In these species, the equatorial segment remains intact.

6. FERTILIZATION *IN VIVO*

Information on fertilization in bats is limited and encompasses only a few of the many species. In general, fertilization occurs either in the ampulla (*Miniopterus schreibersii*, Uchida,

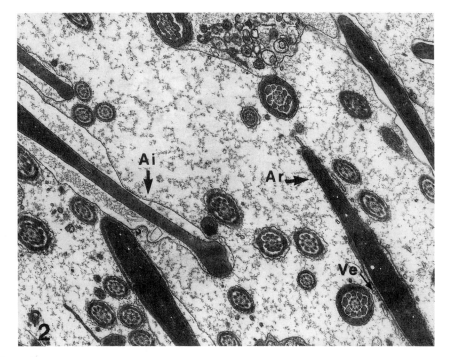

Figure 2. An acrosome-intact (Ai) and acrosome-reacted (Ar) sperm from *Miniopterus schreibersii*. Note presence of vesicles (Ve) adjacent to the inner acrosomal membrane.

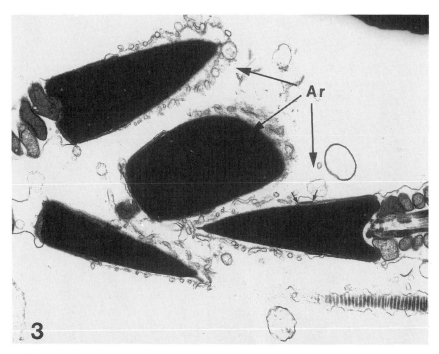

Figure 3. Four acrosome-reacted sperm of *Myotis velifer* in BWW culture medium. Note shroud of vesicles anterior and lateral to sperm heads; these represent fusions between the plasma and outer acrosomal membranes. Small arrows mark the equatorial segment.

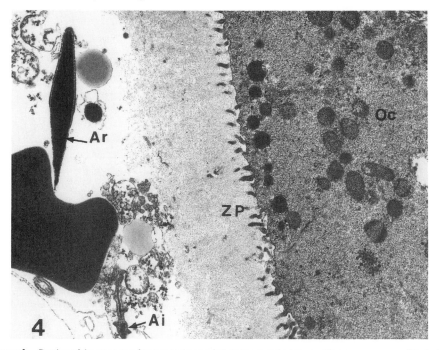

Figure 4. Portion of the acrosome from an acrosome-intact sperm and a sperm without an acrosome of *Tadarida braziliensis* adjacent to an oocyte from the same species.

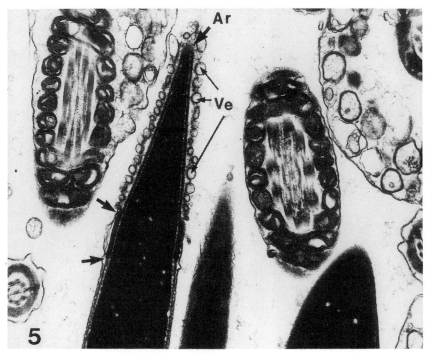

Figure 5.   Sperm of *Eptesicus fuscus* showing hybrid vesicles of true acrosome reaction. Note that the equatorial segment (arrowed) remains intact.

1957; Uchida and Mori, 1974; Mori and Uchida, 1981; *Myotis lucifugus*, Buchanan, 1987; *Carollia sp.*, Bonilla and Rasweiler, 1974; *Desmodus rotundus*, Quintero and Rasweiler, 1974; *Lyroderma lyra*, Ramakrishna, 1951; *Corynorhinus [= Plecotus] rafinesquei*, Pearson *et al.*, 1952; *Noctilio albiventris*, Rasweiler, 1977, 1978; *Peropteryx kappleri*, Rasweiler, 1982; *Glossophaga soricina*, Rasweiler, 1972, 1979) or perhaps even in the bursa or oviductal fimbrae (*Chalinolobus morio*, Kitchener and Coster, 1981; *Myotis lucifugus*, Buchanan, 1987; *Pipistrellus tralatitius*, Uchida, 1953). However, fertilization is reported to occur toward the uterine end of the oviduct in *Scotophilus heathi* (Gopalakrishna and Madhavan, 1978), *Pipistrellus mimus mimus* (Karim, 1975), and *Corynhorinus [= Plecotus] rafinesquei* (Pearson *et al.*, 1952).

Little is known of the events following sperm penetration of the zona pellucida in bats. Mori and Uchida (1981) noted a curved penetration slit left by a spermatozoon in the zona of an oocyte of *Miniopterus, in vivo*. Uchida and Mori (1974) described the sperm head in *M. schreibersii* as invested only by the inner acrosomal membrane as it reached the zona pellucida. Mori and Uchida (1981) described sperm parts within the ooplasm in *M. schreibersii in vivo*. They also observed a fertilization cone with an incorporated male pronucleus. It appears that in *M. schreibersii* the block to polyspermy occurs at the level of the vitelline membrane, since the loss of the cortical granules did not preclude the entry of acrosome-reacted sperm into the perivitelline space. In *M. schreibersii*, excess sperm are removed from the perivitelline space by phagocytosis by the developing blastocyst (Mori and Uchida, 1981), and in *M. velifer*, our *in vitro* studies show sperm being phagocytosed from the surrounding milieu by cumulus cells (unpublished observations).

7. FERTILIZATION *IN VITRO*

Experiments in our laboratory to define some of the parameters of stored bat sperm in culture and to view fertilization events have met with varied success. *In vitro* work has suggested that bat sperm (both hibernating and nonhibernating species) are remarkably resistant to capacitation and the acrosome reaction. These processes, which occur in the female reproductive tract in most mammals soon after insemination, can normally be readily induced *in vitro* in the sperm of most mammals, using culture media. However, despite the assertions of Lambert (1981) that bat sperm (*Myotis lucifugus*) can decondense within hamster ooplasm in a temperature-dependent manner, we find that the gametes of bat species (*M. velifer, Antrozous pallidus, Macrotus californicus, Tadarida braziliensis*) are particularly difficult to acrosome-react in culture, at least by the conventional means employed. Thus, despite the application of several culture media (BWW; TALP II and IV; canine culture medium of Mahi and Yanagimachi, 1978; TEST yolk buffer, Irvine Scientific), we have been unable to demonstrate decondensing sperm within hamster ooplasm or even to produce a significant acrosome reaction response.

Furthermore, the success rate has not been improved by the variation within these media of the concentrations of such components as calcium, albumin (BSA up to 5.0% as well as the usage of heat-inactivated fetal calf and bat sera), glucose (a sugar known to be present in the female reproductive tracts of *M. velifer, Plecotus townsendii, and Antrozous pallidus* in significantly higher levels during the sperm-storing season: unpublished observations), or fructose (also known to be present in higher levels in sperm-storing female tracts of *M. velifer* and *M. lucifugus*, Crichton *et al.*, 1981). Neither has the addition of caffeine (5.0 mmol), acetylcarnitine (20.0 μmol, present in high levels in sperm-filled caudae epididymides and in sperm-containing female tracts of *Myotis* sp., Krutzsch *et al.*, 1984), heparin (10.0 μg/ml), cAMP (5.0 mmol), or EDTA (1.0 mg/ml)—all known to play roles in sperm biology—produced fertilizations.

The time and temperature of preincubation and postincubation of sperm have also been varied within a wide range without result. Likewise, no sperm head decondensations within hamster ooplasm have been observed following treatment of sperm with various proteases (pronase, trypsin, collagenase + neuraminidase + glycosidase, trypsin + collagenase + neuraminidase + glycosidase, 1.0 mg/ml for 20 min each treatment) or following overnight preincubation (washing) of sperm in BWW or saline prior to IVF experiments. Interestingly, under all experimental conditions, a marked trophic response of sperm to the oocytes in culture, and sperm remain motile within all media for hours, sometimes days. Ultrastructural analyses of sperm after these experiments reveal that although occasional acrosome-reacted sperm are seen in the medium (Fig. 3), all sperm adjacent to the oocyte maintain normal acrosome morphology despite the treatment (Fig. 6). Mouse sperm controls run simultaneously decondensed within hamster ooplasm.

We have seen two exceptions to the consistent observation that bat sperm do not undergo the acrosome reaction in response to our various experimental procedures. Following treatment with Ca ionophore (2.0 μmol: Fig. 7) or incubation of sperm with bat oocytes of the homologous species (from PMS/HCG-treated bats and cultured to maturity overnight in M199 + 30% inactivated fetal calf serum), sperm in various stages of acrosomal vesiculation have been observed adjacent to ova (bat and hamster) or among bat cumulus cells (Fig. 8); in one instance, a spermatozoan head freed of its investments was seen within the zona pellucida of a bat oocyte (Fig. 9), and in another, phase microscopy revealed a sperm in the perivitelline space. These observations demonstrate that the acrosome reaction can be produced in bat sperm if appropriate inducing or natural conditions are employed. In addition, bat sperm appear to be incompatible with hamster oocytes, since ionophore-induced acrosome-reacted sperm were seen adjacent to, but never within, ooplasm; if further observations substantiate these preliminary data, then the sperm of the bat and Chinese hamster (Yanagimachi, personal communication) will be the only

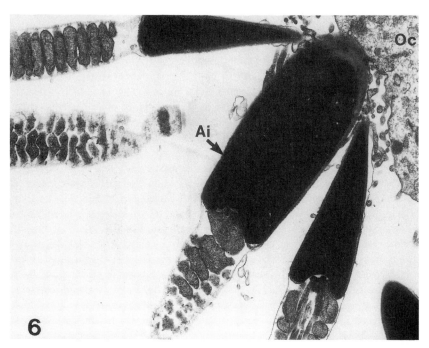

Figure 6.   Three acrosome-intact sperm of *Myotis velifer* at the surface of a zona-pellucida-free hamster oocyte (Oc).

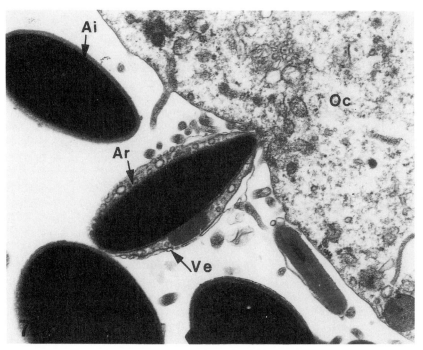

Figure 7.   Sperm of *Myotis lucifugus* that has been treated with 2.0 μmol Ca ionophore. Note vesiculation points of acrosome reaction (Ar) in one gamete. The sperm are lying adjacent to a hamster oocyte.

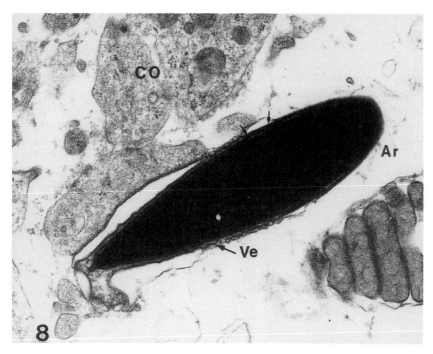

Figure 8. Sperm of *Myotis lucifugus* undergoing an acrosome reaction in the vicinity of cumulus cells (CO) surrounding an oocyte of the same species. The persistent equatorial segment is arrowed.

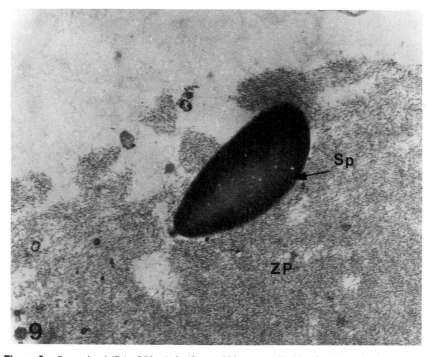

Figure 9. Sperm head (Sp) of *Myotis lucifugus* within zona pellucida of a *Myotis lucifugus* oocyte.

species that resist the acrosome reaction in standard culture media used in IVF experiments. Furthermore, the bat and dog (Yanagimachi, 1988) will be the only mammalian species known to date to be unable to penetrate the zona-free hamster oocyte; therefore, this heterologous system will be unsuitable for monitoring their biology.

At this time, therefore, nothing is known of the specific molecules required to promote capacitation and the acrosome reaction in bat sperm. It is clear, however, that in order to express the prefertilization changes necessary for egg penetration, these sperm require a heretofore-untried component added to the culture medium or other variation in methods. Future experimentation may reveal a specific requirement, likely peculiar to bat sperm, that must be included within the capacitating medium to produce positive results for these species. Only when such components are identified will we have some idea of the peculiar biology of these gametes and the mechanisms of their prolonged longevity and capacitation. In this regard, the recent findings of Uchida and Mori (1987) and Uchida et al. (1988) are provocative and suggest that the female reproductive tract plays an important role in the preparation of sperm for fertilization. Supporting the speculations of Krutzsch et al. (1982) that sperm storage in bats reflects a delay in the capacitation process, these experiments indicate that bat sperm require long residency in the female tract (at least 50 days) before they are capable of fertilization, and that their ability to fertilize (capacitate) increases with the time of storage. Opposing this idea, however, are the experiments of Racey (1973) in which conceptions resulted from the artificial insemination of bats in March: such sperm could have resided in the female tract for only a few days prior to fertilization. No indication has been gained from our laboratory that sperm obtained late in the storage season are any more vulnerable to capacitation than those obtained earlier in the season.

8. CONCLUSIONS AND FUTURE DIRECTIONS

Extrapolating from all the data at hand, we speculate that an inherent structural integrity conferring resistance to capacitating change (and thereby longevity) may be part of the make-up of bat sperm. It seems likely that the induction of prefertilization changes (at least in the gametes of sperm-storing species) is a function of an environmentally produced cascade of events initiated (following arousal from hibernation) by estrus and ovulation. Specific substances (eg., raised albumin levels) appearing in the female tracts at this time could act on stored (or artificially inseminated) sperm to release them from suppression by eliciting conformational changes in the plasma membrane preparatory for capacitation and the acrosome reaction. Equally, the provision of certain substances may release sperm from inhibition, allowing them to participate in prefertilization events. Two such likely aspects of sperm biology, the content of zinc and plasma membrane lipids, are currently being investigated in our laboratory as logical candidates that, once removed from sperm, could allow Ca influx and capacitation to occur. Plasma membrane lipids have been selected for consideration based on reports that time to capacitation in mammals is related to cholesterol : phospholipid levels in the sperm plasma membrane (Davis, 1981) and that such lipids also afford cold shock protection to sperm. Zinc has been selected for further study since we have previously established high levels of this metallic ion in caudae epididymides and uteri of bats during the sperm storage season (Crichton et al., 1982).

Research along these lines offers hope that a system will be devised whereby fertilization in these interesting and unique species can be observed in culture. In addition, further information will probably be gained from the use of oocytes and sperm from homologous bat species. At the same time, however, gathering sufficient numbers of oocytes from these mostly monotocous, monestrous species and their successful culturing to the right maturation stage to support fertilization present a challenging area of research in itself.

9. REFERENCES

Bonilla, H. de, and Rasweiler, J. J. IV, 1974, Breeding activity, preimplantation development, and oviduct histology of the short-tailed fruit bat, *Carollia*, in captivity, *Anat. Rec.* **179**:385–404.

Breed, W. G., and Inns, R. W., 1985, Variation in sperm morphology of Australian Vespertilionidae and its possible phylogenetic significance, *Mammalia* **49**:105–108.

Buchanan, G. D., 1987, Timing of ovulation and early embryonic development in *Myotis lucifugus* (Chiroptera: Vespertilionidae) from northern central Ontario, *Am. J. Anat.* **178**:335–340.

Courrier, R., 1927, Etude sur la determinisme des caracteres sexuel secondaries chez quelques mammiferes a activite testiculaire periodique, *Arch. Biol. Paris* **37**:173–334.

Crichton, E. G., Krutzsch, P. H., and Wimsatt, W. A., 1981, Studies on prolonged spermatozoa survival in Chiroptera—I. The role of uterine free fructose in the spermatozoa storage phenomenon, *Comp. Biochem. Physiol.* **70A**:387–395.

Crichton, E. G., Krutzsch, P. H., and Chvapil, M., 1982, Studies on prolonged spermatozoa survival in Chiroptera—II. The role of zinc in the spermatozoa storage phenomenon, *Comp. Biochem. Physiol.* **71A**:71–77.

Cummins, J. M., and Woodall, P. F., 1985, On mammalian sperm dimensions, *J. Reprod. Fertil.* **75**:153–175.

Cummins, J. M., Robson, S. K., Rouse, G. W., and Graydon, M., 1986, Ultrastructure of the ionophore-induced acrosome reaction in spermatozoa of the grey flying fox, *Pteropus poliocephalus, Aust. Soc. Reprod. Biol.* **18**:73.

Davis, B. K., 1981, Timing of fertilization in mammals: Sperm cholesterol/phospholipid ratio as a determinant of the capacitation interval, *Proc. Natl. Acad. Sci. U.S.A.* **78**:7560–7564.

Fawcett, D. W., 1970, A comparative review of sperm ultrastructure, *Biol. Reprod.* **2**:90–127.

Fawcett, D. W., and Ito, S., 1965, The fine structure of bat spermatozoa, *Am. J. Anat.* **116**:567–610.

Gopalakrishna, A., and Madhavan, A., 1971, Survival of spermatozoa in the female genital tract of the Indian vespertilionid bat, *Pipistrellus ceylonicus chrysothrix* (Wroughton), *Proc. Indian Acad. Sci.* **73**:43–49.

Gopalakrishna, A., and Madhavan, A., 1978, Viability of inseminated spermatozoa in the Indian vespertilionid bat, *Scotophilus heathi* (Horsefield), *Indian J. Exp. Biol.* **16**:852–854.

Hartman, C. Z., 1933, On the survival of spermatozoa in the female genital tract of the bat, *Q. Rev. Biol.* **8**:185–193.

Hiraiwa, Y. K., and Uchida, T. A., 1956a, Fertilization in the bat, *Pipistrellus abramus abramus* (Temminck) III. Fertilizing capacity of spermatozoa stored in the uterus after copulation in the fall, *Sci. Bull. Fac. Agr. Kyushu Univ.* **15**:565–574.

Hiraiwa, Y. K., and Uchida, T. A., 1956b, Fertilization in the bat, *Pipistrellus abramus*. A successful example of artificial insemination with epididymal spermatozoa in autumn, *Science Tokyo* **26**:535.

Hoffman, L. H., Wimsatt, W. A., and Olson, G. E., 1987, Plasma membrane structure of bat spermatozoa: Observations on epididymal and uterine spermatozoa in *Myotis lucifugus, Am. J. Anat.* **178**:326–334.

Karim, K. B., 1975, Early development of the embryo and implantation in the Indian vespertilionid bat, *Pipistrellus mimus mimus* (Wroughton), *Z. Zool. Soc. India* **27**:119–136.

Kitchener, D. J., and Coster, P., 1981, Reproduction in female *Chalinolobus morio* (Gray) (Vepertilionidae) in southwestern Australia, *Aust. J. Zool.* **29**:305–320.

Koopman, K. F., and Jones, J. K., Jr., 1970, Classification of bats, in: *About Bats* (B. H. Slaughter and D. W. Walton, eds.), Southern Methodist University Press, Dallas, pp. 22–28.

Krutzsch, P. H., 1979, Male reproductive patterns in nonhibernating bats, *J. Reprod. Fertil.* **56**:333–344.

Krutzsch, P. H., Crichton, E. G., and Nagle, R. B., 1982, Studies on prolonged spermatozoa survival in Chiroptera: A morphological examination of storage and clearance of intrauterine and cauda epididymal spermatozoa in the bats *Myotis lucifugus* and *M. velifer, Am. J. Anat.* **165**:421–434.

Krutzsch, P. H., Crichton, E. G., Lennon, D. L. F., Stratman, F. W., and Carter, A. L., 1984, Studies on prolonged spermatozoa survival in Chiroptera—III. Preliminary data on carnitine, *Andrologia* **16**:34–37.

Lambert, H., 1981, Temperature dependence on capacitation in bat sperm monitored by zona-free hamster ova, *Gamete Res.* **4**:525–533.

Mahi, C. A., and Yanagimachi, R., 1978, Capacitation, acrosome reaction, and egg penetration by canine spermatozoa in a single defined medium, *Gamete Res.* **1**:101–109.

Medway, Lord, 1972, Reproductive cycles of the flat-headed bats *Tylonycteris pachypus* and *T. robustula* (Chiroptera, Vespertilionidae) in a humid equatorial environment, *Zool. J. Linn. Soc.* **51**:33–61.

Mori, T., and Uchida, T. A., 1980, Sperm storage in the reproductive tract of the female Japanese long-fingered bat, *Miniopterus schreibersii fuliginosus, J. Reprod. Fertil.* **58**:429–433.

Mori, T., and Uchida, T. A., 1981, Ultrastructural observations of fertilization in the Japanese long-fingered bat *Miniopterus schreibersii fuliginosus, J. Reprod. Fertil.* **63**:231–235.

Mori, T., and Uchida, T. A., 1982, Changes in the morphology and behaviour of spermatozoa between copulation and fertilization in the Japanese long-fingered bat, *Miniopterus schreibersii fuliginosus*, *J. Reprod. Fertil.* **65**:23–28.

Mori, T., Oh, Y. K., and Uchida, T. A., 1982, Sperm storage in the oviduct of the Japanese greater horseshoe bat *Rhinolophus ferrumequinum nippon*, *Kyushu Univ. Fac. Agr. J.* **27**:47–53.

Oxberry, B. A., 1979, Female reproductive patterns in hibernating bats, *J. Reprod. Fertil.* **56**:359–367.

Pagenstecher, H. A., 1859, Uber die Begattung von *Vesperugo pipistrellus*, *Ver. Naturh.-Med. Ver. Heidelb.* **1**:194–195.

Pearson, O. P., Koford, M. R., and Pearson, A. K., 1952, Reproduction of the lump nosed bat (*Corynorhinus rafinesquei*) in California, *J. Mamm.* **33**:273–320.

Quintero, H. F., and Rasweiler, J. J. IV, 1974, Ovulation and early embryonic development in the captive vampire bat, *Desmodus rotundus*, *J. Reprod. Fertil.* **41**:265–273.

Racey, P. A., 1973, The viability of spermatozoa after prolonged storage by male and female European bats, *Period. Biol.* **75**:201–205.

Racey, P. A., 1975, The prolonged survival of spermatozoa in bats, in: *The Biology of the Male Gamete* (J. G. Duckett and P. A. Racey, eds.) Academic Press, London, pp. 385–416.

Racey, P. A., 1979, The prolonged storage and survival of spermatozoa in Chiroptera, *J. Reprod. Fertil.* **56**:391–402.

Racey, P. A., Suzuki, F., and Medway, Lord, 1975, The relationship between stored spermatozoa and the oviductal epithelium in bats of the genus *Tylonycteris*, in: *The Biology of Spermatozoa: Transport, Survival and Fertilizing Capacity* (E. S. E. Hafez and C. G. Thibault, eds.), S. Karger, Basel, pp. 123–133.

Racey, P. A., Uchida, T. A., Mori, T., Avery, M. I., and Fenton, M. B., 1987, Sperm–epithelial relationships in relation to the time of insemination in little brown bats (*Myotis lucifugus*), *J. Reprod. Fertil.* **80**:445–454.

Ramakrishna, P. A., 1951, Studies on reproduction in bats. 1. Some aspects of reproduction in the oriental vampires *Lyroderma lyra lyra* Geoff and *Megaderma spasma* (Linn.), *J. Mysore Univ. B Science* **2**:107–118.

Rasweiler, J. J. IV, 1972, Reproduction in the long-tongued bat, *Glossophaga soricina*. I Preimplantation development and histology of the oviduct, *J. Reprod. Fertil.* **31**:249–262.

Rasweiler, J. J. IV, 1977, Preimplantation development fate of the zona pellucida, and observations on the glycogen-rich oviduct of the little bulldog bat, *Noctilio albiventris*, *Am. J. Anat.* **150**:269–300.

Rasweiler, J. J. IV, 1978, Unilateral oviductal and uterine reactions in the little bulldog bat, *Noctilio albiventris*, *Biol. Reprod.* **19**:467–492.

Rasweiler, J. J. IV, 1979, Differential transport of embryos and degenerating ova by the oviducts of the long-tongued bat, *Glossophaga soricina*, *J. Reprod. Fertil.* **55**:329–334.

Rasweiler, J. J. IV, 1982, The contribution of observations on early pregnancy in the little sac-winged bat, *Peropteryx kappleri*, to an understanding of the evolution of reproductive mechanisms in monovular bats, *Biol. Reprod.* **27**:681–702.

Richardson, E. G., 1977, The biology and evolution of the reproductive cycle of *Miniopterus schreibersii* and *M. australis* (Chiroptera: Vespertilionidae), *J. Zool. Lond.* **183**:353–375.

Rouse, G. W., and Robson, S. K., 1986, An ultrastructural study of megachiropteran (Mammalia: Chiroptera) spermatozoa: Implications for chiropteran phylogeny, *J. Submicrosc. Cytol.* **18**:137–152.

Uchida, T. A., 1953, Studies on the embryology of the Japanese house bat, *Pipistrellus tralatitius abramus* (Temminck). II From the maturation of the ova to the fertilization, especially on the behaviour of the follicle cells at the period of fertilization, *Sci. Bull. Fac. Agr. Kyushu Univ.* **14**:153–168.

Uchida, T. A., 1957, Fertilization and hibernation in bats, *Heredity (Tokyo)* **11**:14–17.

Uchida, T. A., and Mori, T., 1972, Electron microscope studies on the fine structure of germ cells in Chiroptera. 1. Spermiogenesis in some bats and notes on its phylogenetic significance, *Sci. Bull. Fac. Agr. Kyushu Univ.* **26**:399–418.

Uchida, T. A., and Mori, T., 1974, Electron microscopic analysis of the mechanisms of fertilization in Chiroptera. 1. Acrosomal reaction and consequence to death of the sperm in the Japanese long-fingered bat, *Miniopterus schreibersi fuliginosus*, *Sci. Bull. Fac. Agr. Kyushu Univ.* **28**:177–184.

Uchida, T. A., and Mori, T., 1987, Prolonged storage of spermatozoa in hibernating bats, in: *Recent Advances in the Study of Bats* (M. B. Fenton, P. A. Racey, and J. M. V. Rayner, eds.), Cambridge University Press, Cambridge, pp. 351–365.

Uchida, T. A., Mori, T., and Son, S. W., 1988, Delayed capacitation of sperm in the Japanese house bat, *Pipistrellus abramus*, *J. Mamm. Soc. Jpn.* **13**:1–10.

Wimsatt, W. A., 1942, Survival of spermatozoa in the female reproductive tract of the bat, *Anat. Rec.* **83**:299–307.

Wimsatt, W. A., 1944, Further studies on the survival of spermatozoa in the female reproductive tract of the bat, *Anat. Rec.* **88**:193–204.

Wimsatt, W. A., 1960, Some problems of reproduction in relation to hibernation in bats, *Bull. Mus. Comp. Zool. Harvard Univ.* **124**:249–267.

Wimsatt, W. A., 1969, Some interrelations of reproduction and hibernation in mammals, *Symp. Exp. Biol.* **23**:511–549.

Wimsatt, W. A., and Kallen, F. C., 1957, The unique maturation response of the Graafian follicles of hibernating vespertilionid bats and the question of its significance, *Anat. Rec.* **129**:115–132.

Wimsatt, W. A., and Parks, H. F., 1966, Ultrastructure of the surviving follicle of hibernation and of the ovum–follicle cell relationship in the vespertilinoid bat *Myotis lucifugus*, *Symp. Zool. Soc. Lond.* **15**:419–454.

Wimsatt, W. A., Krutzsch, P. H., and Napolitano, L., 1966, Studies on sperm survival mechanisms in the female reproductive tract of hibernating bats. 1. Cytology and ultra-structure of intra-uterine spermatozoa in *Myotis lucifugus*, *Am. J. Anat.* **119**:25–60.

Yanagimachi, R., 1988, Mammalian fertilization, in: *Physiology of Reproduction*, Vol. I (E. Knobil and J. D. Neill, eds.), Raven Press Ltd., New York, pp. 135–185.

9

Fertilization in the Mouse

I. The Egg

Paul M. Wassarman

1. INTRODUCTORY REMARKS

The mouse has been used extensively in studies of mammalian fertilization, especially during the last 20 years or so. Among the many reasons for choosing mice for such studies are (1) the tremendous wealth of knowledge available about mouse developmental and reproductive biology, (2) the relatively low cost and ease with which mice can be obtained, housed, and handled, (3) the well-established protocols available for reliably obtaining and culturing relatively large numbers of mouse gametes, (4) the well-established protocols available for carrying out fertilization with mice, both *in vivo* and *in vitro*, and (5) the firm belief that fertilization in mice is an appropriate model for understanding many aspects of human fertilization. These and other factors have made the mouse a principal player in mammalian fertilization research and, consequently, a major contributor to our understanding of the mammalian fertilization process. Several manuals are available that describe mouse husbandry and experimental manipulation of mouse gametes and embryos (Rafferty, 1970; Daniel, 1971, 1978; Hogan *et al.*, 1986; Burki, 1986; Monk, 1987).

Here, I review briefly some aspects of mouse egg development (oogenesis) and participation of the egg in the fertilization process. Emphasis is placed on *in vitro* rather than *in vivo* studies of fertilization in mice and on involvement of the egg zona pellucida in the fertilization process. I have not attempted to reference all of the relevant literature; consequently, the reader is referred to several other monographs and reviews for more extensive coverage of the subject (Biggers and Schuetz, 1972; Gwatkin, 1977; Zuckerman and Weir, 1977; Jones, 1978; Mastroianni and Biggers, 1981; Austin and Short, 1982; Hartmann, 1983; Metz and Monroy, 1985; Yanagimachi, 1988; Knobil and Neill, 1988; Wassarman, 1988a, 1990a). In addition, whenever possible, the subject of mouse sperm and its involvement in the fertilization process has been left by and large to another chapter in this volume (Chapter 10).

PAUL M. WASSARMAN • Department of Cell and Developmental Biology, Roche Institute of Molecular Biology, Roche Research Center, Nutley, New Jersey 07110.

A Comparative Overview of Mammalian Fertilization, edited by Bonnie S. Dunbar and Michael G. O'Rand. Plenum Press, New York, 1991.

2. DEVELOPMENT OF UNFERTILIZED MOUSE EGGS

The unfertilized mouse egg is the end product of oogenesis, a process that begins during fetal development and ends months to years later in the sexually mature adult (Fig. 1). Oogenesis begins with the appearance of primordial germ cells (day 7–9 fetus), which become the oogonia that populate fetal ovaries (day 11–12 fetus) and, in turn, become nongrowing oocytes (day 12–14 fetus) that populate ovaries of neonatal mice. The transition from oogonia to oocytes involves a change from mitotic to meiotic cells. Progression through the first meiotic prophase (leptotene, zygotene, pachytene, diplotene) with pairing of homologous chromosomes, crossing over, and recombination takes 4–5 days. Shortly after birth (day 21 post-coitus), nearly all oocytes are arrested in late diplotene (dictyate stage), where they remain until stimulated to resume meiotic progression at the time of ovulation. This pool of small, nongrowing oocytes is the sole source of unfertilized eggs in the sexually mature mouse (about 6 weeks of age). It should be noted that as much as 50% of the oocyte population present in the ovary at birth is lost during the first week following birth.

In sexually mature mice, each ovary contains about 8000 nongrowing oocytes. Each nongrowing oocyte (12–15 μm in diameter) is contained within a cellular follicle that grows concomitantly with the oocyte, from a single layer of a few epithelial-like cells to three layers of cuboidal granulosa cells ($\sim$900 cells; $\sim$125 μm in diameter) by the time the oocyte has completed its growth (80–85 μm in diameter). During this growth phase ($\sim$2 weeks), oocytes are continually arrested at the dictyate stage of the first meiotic prophase. The dictyate stage is characterized by very diffuse chromosomes, and oocyte growth is characterized by high rates of transcription and translation. The theca is first distinguishable outside of and separated by a basement membrane from the granulosa cells when the granulosa region is two cell layers thick ($\sim$400 cells; $\sim$100 μm in diameter). Over several days, while the oocyte remains a constant size, the follicular cells undergo rapid division, increasing to more than 50,000 cells and resulting in a Graafian follicle more than 600 μm in diameter. The follicle exhibits an incipient antrum when it is several layers thick ($\sim$6000 cells; $\sim$250 μm in diameter), and, as the antrum expands, the oocyte takes up an acentric position surrounded by two or more layers of granulosa cells (cumulus cells). The innermost layer becomes columnar in shape and constitutes the corona radiata. These innermost follicle cells communicate with both the oocyte and other follicle cells through an extensive network of gap junctions. Apparently, many metabolic precursors and other small molecules ($\leq$1000 M_r) required by the growing and fully grown oocyte pass through these gap junctions from follicle cells into the oocyte.

Fully grown oocytes in Graafian follicles resume meiosis and complete the first meiotic reductive division (meiotic maturation) just prior to ovulation (Fig. 2). Resumption of meiosis can be mediated by a hormonal stimulus *in vivo* (surge in the level of lutenizing hormone, LH) or simply by release of oocytes from their ovarian follicles into a suitable culture medium *in vitro*. Meiotic maturation takes 12–14 hr and involves nuclear progression from the dictyate stage of the first meiotic prophase to metaphase II (second meiotic division) (Fig. 3). Unfertilized eggs display 20 chromosomes, each composed of two chromatids, aligned on the metaphase II spindle and a small polar body containing 20 homologous chromosomes. The ovulated eggs complete meiosis, with separation of chromatids and emission of a second polar body containing one-half the chromosomal complement, on fertilization or artificial activation.

Under normal laboratory housing conditions, a sexually mature mouse ovulates once every 4 days. In a natural ovulation, a mouse releases 8–12 eggs, whereas a superovulated mouse (injected with pregnant mare's serum, PMS, followed by human chorionic gonadotropin, hCG) releases 20–60 eggs (these numbers are very dependent on the mouse strain). Eggs are released from the ovarian follicle, enter the opening (ostium) of the oviduct (fallopian tube), and move to the lower ampulla region of the oviduct, where fertilization probably takes place. It has been

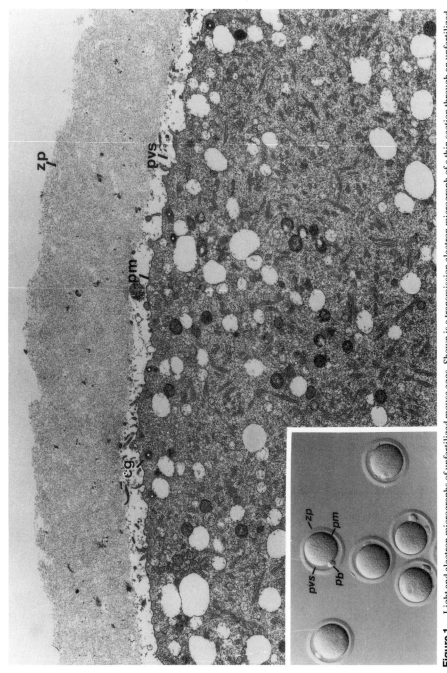

Figure 1. Light and electron micrographs of unfertilized mouse eggs. Shown is a transmission electron micrograph of a thin section through an unfertilized mouse egg (ovulated *in vivo*). Inset is a light micrograph (Nomarski differential interference contrast) of a group of unfertilized mouse eggs (ovulated *in vivo*). cg, cortical granule (several cg near the surface of the egg are noted with white dots); pb, polar body; pm, plasma membrane; pvs, perivitelline space; zp, zona pellucida.

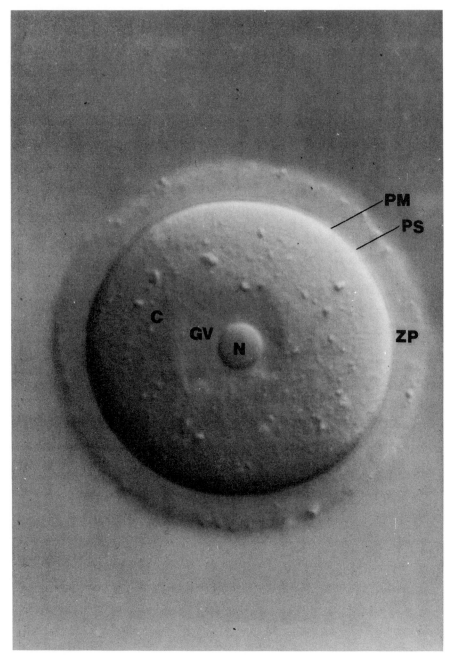

Figure 2. Light micrograph of a fully grown mouse oocyte. The micrograph was taken using Nomarski differential interference contrast microscopy. The oocyte is approximately 80 μm in diameter. C, cytoplasm; GV, germinal vesicle; N, nucleolus; PM, plasma membrane; PS, perivitelline space; ZP, zona pellucida.

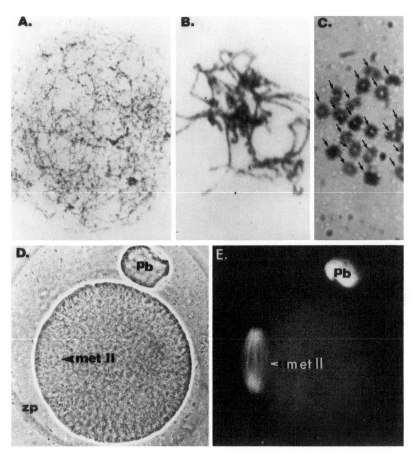

Figure 3. Aspects of meiotic maturation of mouse oocytes. (A) Chromosome spread of diffuse dictyate chromatin in a fully grown oocyte. (B) Chromosome spread of condensing chromatin during meiotic maturation of an oocyte *in vitro*. (C) Chromosome spread of compact bivalents during meiotic maturation of an oocyte *in vitro* (stage just prior to alignment of chromosomes on metaphase I spindle). (D) Light micrograph of an oocyte having completed meiotic maturation *in vitro* (unfertilized egg). The position of the metaphase II spindle is indicated. (E) Immunofluorescent image of the oocyte shown in D following staining with an antibody directed against tubulin. The position of the metaphase II spindle is indicated. pb, polar body; zp, zona pellucida.

estimated that mouse eggs remain capable of being fertilized and giving rise to normal offspring for about 8–12 hr following ovulation.

3. FERTILIZATION OF MOUSE EGGS

The pathway to fertilization of mouse eggs consists of several steps that take place in a compulsory order (Wassarman, 1987a,b, 1988b, 1990b). Initial contact between gametes occurs when sperm attach to the unfertilized egg extracellular coat or zona pellucida (ZP). Attachment is a relatively loose association between gametes and is not species specific. However, association of gametes can progress from attachment to a state in which sperm are bound to the egg ZP (Fig. 4). Binding is a very tight association between gametes and is relatively species specific. Sperm bind to the ZP via plasma membrane overlying the anterior region of the sperm head. Binding is

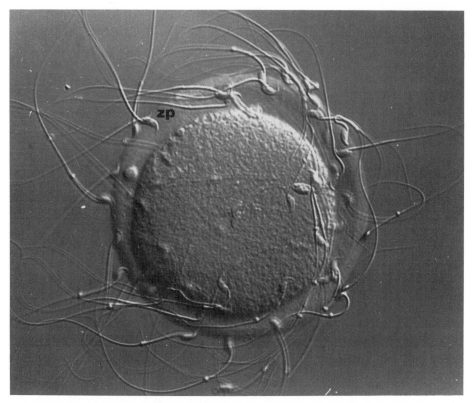

Figure 4.   Light micrograph of mouse sperm bound to an unfertilized mouse egg *in vitro*. The micrograph was taken using Nomarski differential interference contrast microscopy. zp, zona pellucida.

attributable to sperm receptors present in the ZP and complementary egg-binding proteins present in sperm plasma membrane. At very high sperm concentrations, as many as 1500 sperm can bind to a mouse egg *in vitro*. Bound sperm then undergo the acrosome reaction (AR), a form of signal-transduced exocytosis. The acrosome is a lysosome-like organelle located in the sperm head, overlying the nucleus. The AR involves fusion of outer acrosomal membrane and sperm plasma membrane at many sites. The small, hybrid membrane vesicles that form are released from the sperm head, and inner acrosomal membrane, with its associated enzymes, is exposed. Acrosome-reacted sperm penetrate the ZP, probably using a proteinase, reach the perivitelline space between the ZP and egg plasma membrane, and then fuse with egg plasma membrane to form a zygote. Plasma membrane above the equatorial segment of the sperm fuses first with egg plasma membrane to initiate formation of a zygote.

Fusion of sperm and egg results in activation of the egg, and development ensues. In addition, fertilization by a single sperm induces eggs to undergo the cortical reaction (CR), which in turn induces the zona reaction (ZR). Like the acrosome, cortical granules (CG) are membrane-bound lysosome-like organelles that occupy a region of egg cytoplasm just beneath the plasma membrane. The CR involves fusion of CG membrane with egg plasma membrane and, consequently, release of CG contents, including various enzymes, into the perivitelline space (CG exocytosis). The contents enter the ZP, which is a very porous extracellular matrix, and induce the ZR some minutes after fertilization. The ZR includes the so-called "hardening" of the ZP as well as loss of its ability to bind sperm. These changes in properties of the ZP are thought to constitute a "slow" or secondary block to polyspermy. The "fast" or primary block to polyspermy probably occurs at the level of egg plasma membrane as a change in membrane potential.

In vitro studies have provided a rough timetable for the fertilization process in mice (Sato and Blandau, 1979; Gaddum-Rosse *et al.*, 1982; Gaddum-Rosse, 1985; Florman and Storey, 1982; Bleil and Wassarman, 1983). Binding of some sperm to eggs occurs as early as 1–2 min after combining gametes *in vitro* (maximum binding is observed within 10–20 min). Once bound to the ZP, some sperm undergo the AR within another 10–20 min. It takes at least 15 min for a bound, acrosome-reacted sperm to penetrate the ZP (mean time, 20 min) and reach the perivitelline space. The sperm head is incorporated into the egg cytoplasm within another 10 min or so. In mice, the sperm head, midpiece, and large portion of the tail enter the fertilized egg cytoplasm. Thus, sperm enter egg cytoplasm as early as 1–2 hr after mouse gametes are combined *in vitro*. It should be noted that mouse eggs fertilized *in vitro* give rise to viable offspring following transfer to oviducts of foster mothers.

4. MOUSE EGG COMPONENTS DIRECTLY INVOLVED IN FERTILIZATION

At least three mouse egg components are directly involved in the fertilization process: (1) the ZP, which is the site of species-specific primary sperm receptors, the AR-inducer, secondary sperm receptors, and the secondary block to polyspermy; (2) the egg plasma membrane, which is the site of both fusion of eggs with sperm and the primary block to polyspermy; and (3) the CG, which fuse with egg plasma membrane in response to fertilization; their contents modify the ZP, making it refractory to sperm binding and penetration. By and large, all three components are products of the growing oocyte itself as it prepares for ovulation, fertilization, and early development.

4.1. Zona Pellucida

The mouse egg ZP is about 7 μm thick and contains 3–4 ng of protein (Bleil and Wassarman, 1980a; Wassarman, 1988c; Wassarman *et al.*, 1985a, 1989; Dietl, 1989) (Fig. 1). It is composed of three relatively acidic glycoproteins, called ZP1 (200,000 M_r), ZP2 (120,000 M_r), and ZP3 (83,000 M_r), that exhibit considerable heterogeneity on SDS-PAGE. All three glycoproteins are synthesized and secreted by oocytes during their 2 to 3-week growth phase, the period when the ZP first appears. The glycoproteins are products of the oocyte Golgi, which undergoes tremendous expansion during growth of the oocyte (Wassarman and Josefowicz, 1978). The ZP increases in thickness as the oocyte increases in diameter. The ZP consists of long filaments that are composed of ZP2 and ZP3 (located every 15 nm or so) and are cross-linked by ZP1 to form a very porous, three-dimensional matrix (Fig. 5). The ZP glycoproteins interact with one another via noncovalent bonds.

4.2. Plasma Membrane

The mouse egg plasma membrane is extensive since it is associated with thousands of long (0.3–1 μm) microvilli (7.22 microvilli per square micrometer: Longo, 1985). Only the surface of the egg where emission of the first polar body occurred is free of microvilli. During oocyte growth the number and length of microvilli increase markedly as the oocyte increases in diameter (Fig. 6). Since the oocyte increases about 300-fold in volume during its growth phase, it is essential that the growing oocyte maintain an acceptable surface-to-volume ratio by projecting increasing amounts of microvillar membrane during this 2 to 3-week period. Furthermore, fusion between

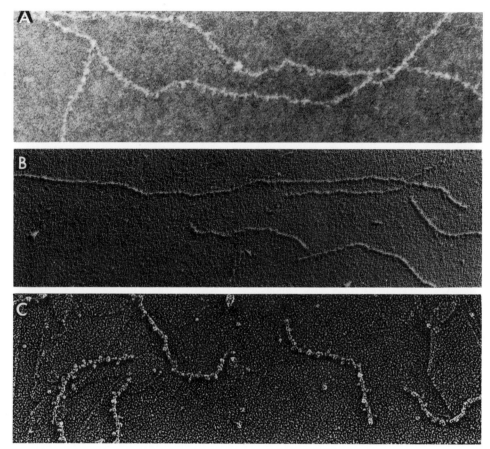

Figure 5. Transmission electron micrographs of mouse zona pellucida filaments prepared from enzyme-solubilized oocyte zonae pellucidae. (A) Filaments adsorbed to grids and negatively stained. (B) Filaments sprayed onto grids and unidirectionally shadowed. (C) Filaments freeze-dried and shadowed. The micrograph shown in C was taken by Dr. John E. Heuser, Washington University School of Medicine.

gametes nearly always involves egg microvillar membranes, probably because they have a low radius of curvature (as compared with amicrovillar plasma membrane) that permits maximum apposition of sperm and egg. In general, the egg plasma membrane does not appear to confer species specificity on gamete fusion, and there does not appear to be a specific site on the egg surface where gamete fusion must occur. However, these last two points are certainly worthy of further investigation.

4.3. Cortical Granules

The mouse egg contains about 4500 CG ($\sim$30 CG per 100 μm plasma membrane) ranging from 200 nm to 600 nm in diameter and located within about 2 μm of the plasma membrane (Nicosia *et al.*, 1977) (Fig. 1). The CG are heterogeneous in appearance and, in some cases, appear to be attached to the cytoplasmic face of the egg plasma membrane. Like the ZP, CG first

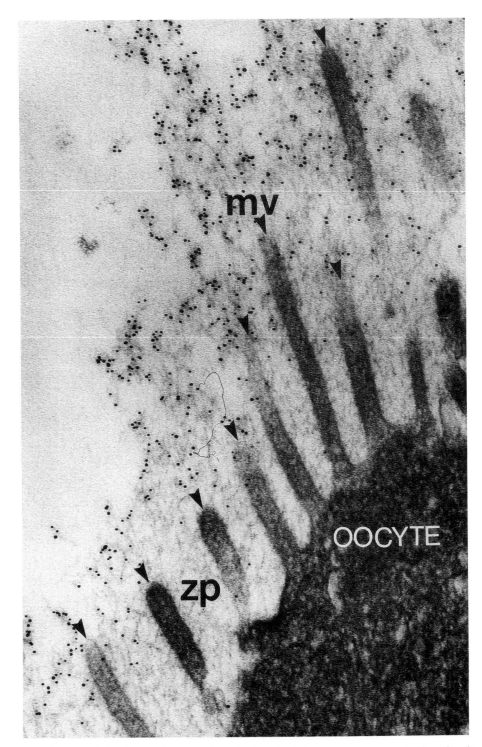

Figure 6. Transmission electron micrograph of a section through a fully grown mouse oocyte. In this section, the zona pellucida is labeled with immunogold (black dots). mv, microvilli (indicated with arrowheads); zp, zona pellucida.

appear in the oocyte during its growth phase, as a product of the Golgi, and increase in number as the oocyte increases in diameter (Gulyas, 1980; Guraya, 1982). A variety of experimental evidence suggests that CG exocytosis (CR) following fertilization involves a signal transduction system that is similar to that of somatic cells (Ducibella, 1990). It is likely that the egg acquires the ability to undergo the CR relatively late in oogenesis and maintains this ability for a relatively short time after ovulation (Ducibella, 1990).

5. MOUSE EGG PRIMARY SPERM RECEPTOR

In order to initiate the fertilization process, acrosome-intact sperm must bind to a component of the egg ZP in a relatively species-specific manner. The mouse egg primary sperm receptor is ZP3, an 83,000 M_r glycoprotein present in more than a billion copies throughout the ZP (Bleil and Wassarman, 1980b, 1988; Florman and Wassarman, 1985; Wassarman, 1987a,b, 1988b,c, 1989, 1990b; Wassarman *et al.*, 1985a,b, 1989) (Fig. 7). There are millions of copies of ZP3 at the ZP surface where sperm bind to the egg. ZP3 possesses a unique 44,000 M_r polypeptide chain (402 amino acids) as well as both asparagine-linked (N-linked), complex-type oligosaccharides and serine/threonine-linked (O-linked) oligosaccharides. ZP3 oligosaccharides are responsible for the glycoprotein's heterogeneous appearance on SDS-PAGE and its relatively low isoelectric point. A specific size class of ZP3 O-linked oligosaccharides (3900 M_r), not polypeptide chain, accounts for the ability of ZP3 to function as a sperm receptor. A galactose residue located in α-linkage with the penultimate sugar at the nonreducing terminus of the O-linked oligosaccharides is essential for ZP3 sperm receptor function. There are tens of thousands of complementary binding sites for ZP3 on the head (plasma membrane) of acrosome-intact sperm. These binding sites are apparently lost following the AR. Thus, gamete recognition and adhesion in mice are mediated, at least on the part of the egg, by carbohydrate. This is consistent with findings that ZP3 sperm receptor activity is resistant to exposure to denaturants, detergents, and high temperature.

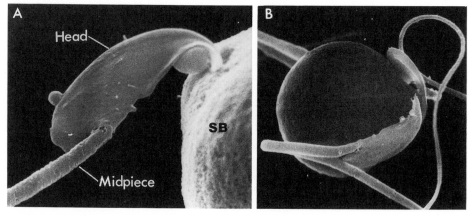

Figure 7.   Scanning electron micrographs of mouse sperm bound to silica beads to which ZP3 is covalently linked. The silica beads are 10 μm in diameter, on average. sb, silica bead.

6. MOUSE EGG ACROSOME REACTION INDUCER

Sperm bound to the egg ZP must undergo the AR in order to penetrate the ZP and fuse with egg plasma membrane. The sperm receptor, ZP3, also serves as an AR inducer following the binding of acrosome-intact sperm to the egg ZP (Bleil and Wassarman, 1983; Florman *et al.*, 1984; Wassarman, 1987a,b, 1988b, 1989, 1989b; Wassarman *et al.*, 1985b, 1989). Purified ZP3 is as effective as ionophore A23187 (Fig. 8) in inducing sperm to undergo the AR *in vitro* (in terms of percentage of sperm acrosome-reacted). This aspect of ZP3 function depends on the glycoprotein's O-linked oligosaccharides and polypeptide chain. Certain evidence suggests that induction of the AR depends on multivalent interactions between ZP3 and a sperm plasma membrane component and may involve aggregation ("capping" or "patching") of the sperm component (Wassarman *et al.*, 1985a,b; Leyton and Saling, 1989). In any case, ZP3 apparently transmits its effect through sperm G proteins and protein kinase C (Kopf and Gerton, 1990; Wassarman, 1990b), regulators of receptor-mediated exocytosis in somatic cells (Freissmuth *et al.*, 1989).

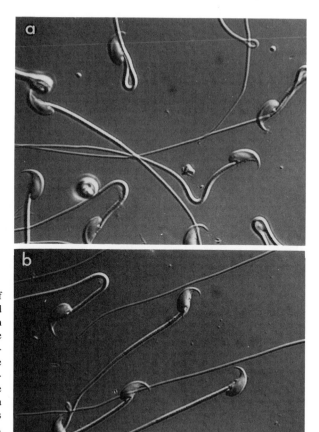

Figure 8. Light micrographs of acrosome-intact and acrosome-reacted mouse sperm. Micrographs were taken using Nomarski differential interference contrast microscopy. (a) Sperm incubated in EGTA to prevent the acrosome reaction. (b) Sperm treated with ionophore A23187 to induce the acrosome reaction. Acrosome-intact sperm have a prominent ridge (1.2 μm thick at its widest point) associated with their head, whereas acrosome-reacted sperm do not have as prominent a ridge (0.5 μm thick at its widest point).

7. MOUSE EGG SECONDARY SPERM RECEPTOR

Mouse sperm that undergo the AR following binding to the egg ZP must remain bound in order to penetrate the extracellular coat. In fact, many sperm are released from the ZP after undergoing the AR on the ZP *in vitro*. Maintenance of sperm binding is achieved by interaction of acrosome-reacted sperm, via their inner acrosomal membrane, with ZP2 (Bleil and Wassarman, 1986; Bleil *et al.*, 1988). Thus, ZP2 serves as a secondary receptor for sperm. Like ZP3, ZP2 is present every 15 nm or so along ZP filaments, placing millions of copies at the ZP surface. Certain evidence suggests that binding of acrosome-reacted sperm to ZP2 is supported by a trypsin-like proteinase associated with sperm inner acrosomal membrane (Bleil *et al.*, 1988). In contrast to primary binding of sperm to ZP3, apparently secondary binding to ZP2 is stabilized by relatively weak interactions. Since bound, acrosome-reacted sperm must penetrate the ZP, relatively weak interactions could be advantageous for a sperm's progress through the extracellular coat. Once sperm begin to penetrate the ZP, their forward swimming motion and location within the extracellular matrix provide additional constraints on the cell.

8. MOUSE EGG SECONDARY BLOCK TO POLYSPERMY

Following fusion of sperm and egg, the ZR takes place, causing the ZP to become refractory to both binding of free-swimming sperm and penetration by sperm that had partially penetrated the extracellular coat prior to fertilization (Fig. 9). The former is caused by inactivation of ZP3 as a sperm receptor (and AR inducer), and the latter apparently results from modification of ZP2. In both cases, CG enzymes released into the ZP as a result of the CR are responsible for changes in ZP glycoproteins that constitute a secondary block to polyspermy.

Recent results suggest that, following fertilization, ZP3 is inactivated by a CG glycosidase(s) that modifies the glycoprotein's O-linked oligosaccharides that are responsible for gamete recognition and adhesion (J. D. Bleil and P. M. Wassarman, unpublished results). At the same time, ZP2 undergoes limited proteolysis, which is catalyzed by a CG proteinase (21,000–34,000 M_r) that is insensitive to a wide range of inhibitors (Bleil *et al.*, 1981; Moller and Wassarman, 1989). Proteolysis of ZP2 probably causes a structural rearrangement of the ZP that promotes filament–filament interactions and, thereby, makes the ZP more insoluble ("harder"). In addition, proteolysis of ZP2 may preclude maintenance of binding of acrosome-reacted sperm to the ZP by inactivating ZP2 as a secondary sperm receptor.

9. CONCLUDING REMARKS

In this brief review, I have attempted to highlight certain aspects of egg involvement during fertilization in mice. In particular, the importance of oocyte growth is emphasized, since it is during this relatively short period of oogenesis that several egg components (e.g., ZP, plasma membrane, CG) are laid down that subsequently participate in the fertilization process. It is also noteworthy that mouse egg components participate in three membrane fusion events during the fertilization process: sperm exocytosis (AR), fusion of sperm and egg, and CG exocytosis (CR). Egg ZP glycoproteins play principal roles at several steps in the fertilization pathway, serving as sperm receptors, mediators of signal transduction, and substrates for egg and sperm enzymes. Thus, mouse ZP glycoproteins continue to be very attractive candidates for future studies of gamete recognition and adhesion and of receptor-mediated signal transduction during mammalian fertilization. The availability of a variety of molecular probes directed against these glycoproteins raises the likelihood that such studies will be fruitful.

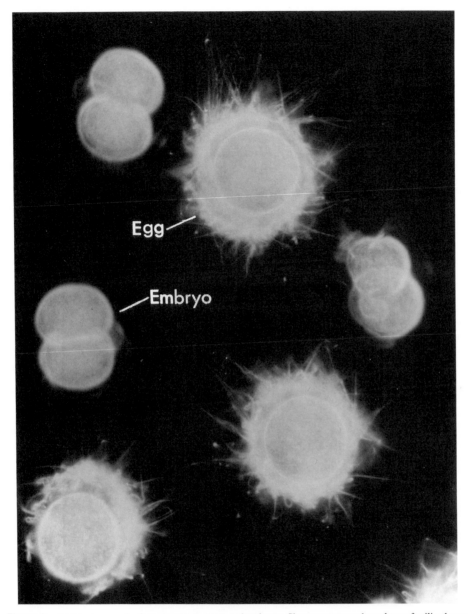

Figure 9. Light micrograph of mouse sperm, eggs, and embryos. Shown are sperm bound to unfertilized eggs but not to two-cell embryos because of the zona reaction. It should be noted that although a zona pellucida is present around the eggs and embryos, it is not visible in dark-field microscopy.

ACKNOWLEDGMENTS. It is a great pleasure for me to acknowledge past and present members of my laboratory who made valuable contributions to both experimental and conceptual aspects of much of the research summarized here. This research was supported in part by the NICHD, NSF, Rockefeller Foundation, and Hoffmann–La Roche Inc.

10. REFERENCES

Austin, C. R., and Short, R. V., eds., 1983, *Reproduction in Mammals: 1. Germ Cells and Fertilization*, Vol. 1, Cambridge University Press, Cambridge.

Biggers, J. D., and Schuetz, A. W., eds., 1972, *Oogenesis*, University Park Press, Baltimore.

Bleil, J. D., and Wassarman, P. M., 1980a, Structure and function of the zona pellucida: Identification and characterization of the proteins of the mouse oocyte's zona pellucida, *Dev. Biol.* **76:**185–203.

Bleil, J. D., and Wassarman, P. M., 1980b, Mammalian sperm–egg interaction: Identification of a glycoprotein in mouse egg zonae pellucidae possessing receptor activity for sperm, *Cell* **20:**873–882.

Bleil, J. D., and Wassarman, P. M., 1983, Sperm–egg interactions in the mouse: Sequence of events and induction of the acrosome reaction by a zona pellucida glycoprotein, *Dev. Biol.* **95:**317–324.

Bleil, J. D., and Wassarman, P. M., 1986, Autoradiographic visualization of the mouse egg's sperm receptor bound to sperm, *J. Cell Biol.* **102:**1363–1371.

Bleil, J. D., and Wassarman, P. M., 1988, Galactose at the nonreducing terminus of O-linked oligosaccharides of mouse egg zona pellucida glycoprotein ZP3 is essential for the glycoprotein's sperm receptor activity, *Proc. Natl. Acad. Sci. U.S.A.* **85:**6778–6782.

Bleil, J. D., Beall, C. F., and Wassarman, P. M., 1981, Mammalian sperm–egg interaction: Fertilization of mouse eggs triggers modification of the major zona pellucida glycoprotein, *Dev. Biol.* **86:**189–197.

Bleil, J. D., Greve, J. M., and Wassarman, P. M., 1988, Identification of a secondary sperm receptor in the mouse egg zona pellucida: Role in maintenance of binding of acrosome-reacted sperm, *Dev. Biol.* **128:**376–385.

Burki, K., 1986, *Experimental Embryology of the Mouse*, S. Karger, Basel.

Daniel, J. C., ed., 1971, *Methods in Mammalian Embryology*, Vol. 1, Academic Press, New York.

Daniel, J. C., ed., 1978, *Methods in Mammalian Embryology*, Vol. 2, Academic Press, New York.

Dietl, J., ed., 1989, *The Mammalian Egg Coat: Structure and Function*, Springer-Verlag, Berlin.

Ducibella, T., 1990, Mammalian egg cortical granules and the cortical reaction, in: *Elements of Mammalian Fertilization* Vol. 1, (P. M. Wassarman, ed.), CRC Press, Boca Raton, pp. 205–232.

Florman, H. M., and Storey, B. T., 1982, Mouse gamete interactions: The zona pellucida is the site of the acrosome reaction leading to fertilization *in vitro*, *Dev. Biol.* **91:**121–130.

Florman, H. M., and Wassarman, P. M., 1985, O-linked oligosaccharides of mouse egg ZP3 account for its sperm receptor activity, *Cell* **41:**313–324.

Florman, H. M., Bechtol, K. B., and Wassarman, P. M., 1984, Enzymatic dissection of the functions of the mouse egg's receptor for sperm, *Dev. Biol.* **106:**243–255.

Freissmuth, M., Casey, P. J., and Gilman, A.G., 1989, G proteins control diverse pathways of transmembrane signaling, *FASEB J.* **3:**2125–2131.

Gaddum-Rosse, P., 1985, Mammalian gamete interactions: What can be gained from observations on living eggs? *Am. J. Anat.* **174:**347–356.

Gaddum-Rosse, P., Blandau, R. J., Langley, L. B., and Sato, K., 1982, Sperm tail entry into the mouse egg *in vitro*, *Gamete Res.* **6:**215–223.

Gulyas, B., 1980, Cortical granules of mammalian eggs, *Int. Rev. Cytol.* **63:**357–392.

Guraya, S. S., 1982, Recent progress in the structure, origin, composition, and function of cortical granules in animal eggs, *Int. Rev. Cytol.* **78:**257–282.

Gwatkin, R. B. L., 1977, *Fertilization Mechanisms in Man and Mammals*, Plenum Press, New York.

Hartmann, J. F., ed., 1983, *Mechanism and Control of Animal Fertilization*, Academic Press, New York.

Hogan, B., Costantini, F., and Lacy, E., 1986, *Manipulating the Mouse Embryo: A Laboratory Manual*, Cold Spring Harbor Laboratory, New York.

Jones, R., ed., 1978, *The Vertebrate Ovary*, Plenum Press, New York.

Knobil, E., and Neill, J. D., eds., 1988, *The Physiology of Reproduction*, Vols. 1 and 2, Raven Press, New York.

Kopf, G. S., and Gerton, G. L., 1990, The mammalian sperm acrosome and the acrosome reaction, in: *Elements of Mammalian Fertilization* Vol. 1 (P. M. Wassarman, ed.), CRC Press, Boca Raton, pp. 153–204.

Leyton, L. and Saling, P., 1989, Evidence that aggregation of mouse sperm receptors by ZP3 triggers the acrosome reaction, *J. Cell Biol.* **108:**2163–2168.

Longo, F. J., 1985, Fine structure of the mammalian egg cortex, *Am. J. Anat.* **174:**303–316.

Mastroianni, L., and Biggers, J. D., eds., 1981, *Fertilization and Embryonic Development*, Plenum Press, New York.

Metz, C. B., and Monroy, A., eds., 1985, *Biology of Fertilization*, Vols. 1–3, Academic Press, New York.

Moller, C. C., and Wassarman, P. M., 1989, Characterization of a proteinase that cleaves zona pellucida glycoprotein ZP2 following activation of mouse eggs, *Dev. Biol.* **103:**103–112.

Monk, M., ed., 1987, *Mammalian Development: A Practical Approach*, IRL Press, Oxford.

Nicosia, S., Wolf, D., and Inoue, M., 1977, Cortical granule distribution and cell surface characteristics in mouse eggs, *Dev. Biol.* **57:**56–74.

Rafferty, K., 1970, *Methods in Experimental Embryology in the Mouse*, Johns Hopkins Press, Baltimore.

Sato, K., and Blandau, R. J., 1979, Time and process of sperm penetration in cumulus-free mouse eggs fertilized *in vitro*, *Gamete Res.* **2:**283–293.

Wassarman, P. M., 1987a, The biology and chemistry of fertilization, *Science* **235:**553–560.

Wassarman, P. M., 1987b, Early events in mammalian fertilization, *Annu. Rev. Cell Biol.* **3:**109–142.

Wassarman, P. M., 1988a, The mammalian ovum, in: *The Physiology of Reproduction*, Vol. 1 (E. Knobil and J. D. Neill, eds.), Raven Press, New York, pp. 69–102.

Wassarman, P. M., 1988b, Fertilization in mammals, *Sci. Am.* **256**(Dec.):78–84.

Wassarman, P. M., 1988c, Zona pellucida glycoproteins, *Annu. Rev. Biochem.* **57:**415–442.

Wassarman, P. M., ed., 1990a, *Elements of Mammalian Fertilization*, Vols. 1 and 2, CRC Press, Boca Raton.

Wassarman, P. M., 1990b, Profile of a mammalian sperm receptor, *Development* **108:**1–17.

Wassarman, P. M., 1989, Role of carbohydrates in receptor-mediated fertilization in mammals, in: *Carbohydrate Recognition in Cellular Function* (G. Bock and S. Harnett, eds.), Ciba Foundation Symposium 145, John Wiley & Sons, Chichester, pp. 135–155.

Wassarman, P. M., and Josefowicz, W. J., 1978, Oocyte development in the mouse: An ultrastructural comparison of oocytes isolated at various stages of growth and meiotic competence, *J. Morphol.* **156:**209–236.

Wassarman, P. M., Bleil, J. D., Florman, H. M., Greve, J. M., Roller, R. J., Salzmann, G. S., and Samuels, F. G., 1985a, The mouse egg's receptor for sperm. What is it and how does it work? *Cold Spring Harbor Symp. Quant. Biol.* **50:**11–19.

Wassarman, P. M., Florman, H. M., and Greve, J. M., 1985b, Receptor-mediated sperm–egg interactions in mammals, in: *Biology of Fertilization*, Vol. 2 (C. B. Metz and A. Monroy, eds.), Academic Press, New York, pp. 341–360.

Wassarman, P., Bleil, J., Fimiani, C., Florman, H., Greve, J., Kinloch, R., Moller, C., Mortillo, S., Roller, R., Salzmann, G., and Vazquez, M., 1989, The mouse egg receptor for sperm: A multifunctional zona pellucida glycoprotein, in: *The Mammalian Egg Coat: Structure and Function* (J. Dietl, ed.), Springer-Verlag, Berlin, pp. 18–37.

Yanagimachi, R., 1988, Mammalian fertilization, in: *The Physiology of Reproduction*, Vol. 1 (E. Knobil and J. D. Neill, eds.), Raven Press, New York, pp. 135–185.

Zuckerman, S., and Weir, B. J., eds., 1977, *The Ovary*, Vol. 1, Academic Press, New York.

10

Fertilization in the Mouse

II. Spermatozoa

Bayard T. Storey and Gregory S. Kopf

1. INTRODUCTION

To the murine spermatozoon, its conspecific egg *in vivo* must seem as remote and screened by obstacles as was the Holy Grail to the Knights of Chivalry. In order that the egg be fertilized, the sperm must overcome obstacles found in the female reproductive tract and surrounding the egg itself as well as prepare itself for the membrane fusion events that lead to its entry into the egg's cytoplasm. In the laboratory, the technique of *in vitro* fertilization removes most of the naturally occurring impediments posed by the female reproductive tract and provides almost unobstructed access for the sperm to the egg and an unobstructed view for the investigator. The process by which the sperm and egg interact from time of initial contact to formation of male and female pronuclei may, as a result, be studied in detail. In the mouse, successful *in vitro* fertilization of isolated eggs was accomplished over 20 years ago (Whittingham, 1968). In the intervening period, clarification of the sequence of the reactions involved in mouse sperm–egg interaction leading to fertilization has begun, and the number and complexity of those reactions are now more fully appreciated. However, too much emphasis on *in vitro* experiments may serve to obscure the biology of fertilization. Section 2 of this chapter is, therefore, concerned with the natural route taken by the sperm to reach the egg and the ensuing events as fertilization occurs *in vivo*. Section 3 then describes *in vitro* studies aimed at understanding the reactions involved in direct sperm–egg interaction that culminates in fertilization of the egg. It is hoped that information gained from both *in vivo* and *in vitro* studies will permit the construction of a unifying description of the process of fertilization in the mammal.

We wish to emphasize at the outset that the focus of this chapter is on the interaction of mouse gametes from the sperm's viewpoint. Many observations in other mammalian species on this topic have been omitted in the interest of maintaining this focus as well as of keeping the presentation within bounds. Comprehensive and critical reviews relative to gamete interaction in

BAYARD T. STOREY AND GREGORY S. KOPF • Division of Reproductive Biology, Department of Obstetrics and Gynecology, University of Pennsylvania School of Medicine, Philadelphia, Pennsylvania 19104-6080.

A Comparative Overview of Mammalian Fertilization, edited by Bonnie S. Dunbar and Michael G. O'Rand. Plenum Press, New York, 1991.

other species are those of Yanagimachi (1981, 1988), Moore and Bedford (1983), Overstreet (1983), Fraser (1984b), Katz *et al.* (1989), and Saling (1989). This list is far from inclusive, but all topics with regard to mouse gametes in this chapter are covered with regard to gametes of other mammalian species in those reviews.

A schematic view of the mouse sperm cell, with linear dimensions, is shown in Fig. 1A. A diagram of the sperm head is shown in Fig. 1B. Murine spermatozoa, in common with other rodent spermatozoa, are approximately twice the length of sperm of the large mammals, such as bovine, equine, ovine, and human (Cummins and Woodall, 1985). The sperm head has the characteristic hooked shape shown as seen in side view. This general shape is also seen in rat and hamster sperm.

Spermatozoa from the cauda epididymis are normally used for studies of mouse egg fertilization *in vitro*, since they are readily isolated as a highly pure population of cells. Mouse ejaculates are exceedingly difficult to obtain, as might be expected. These caudal epididymal sperm are capable of fertilizing eggs, but before they can do so, they must undergo the process of capacitation (Austin, 1951, 1952; Chang, 1951). This term derives from the original concept of Austin (1951) and Chang (1951) that sperm must reside in the female reproductive tract for a

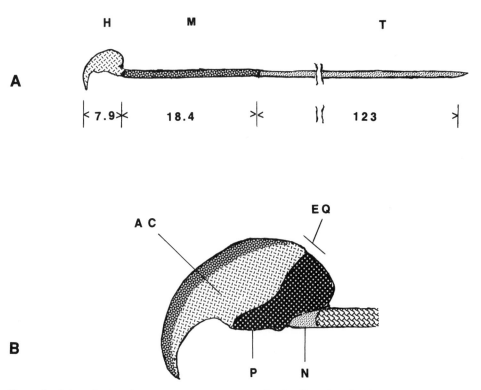

Figure 1.   Schematic view of a mature mouse spermatozoon showing dimensions (in micrometers) of the sperm head (H), midpiece (M), and tail (T) in the side view (A). The dimensions are taken from the compilation of Cummins and Woodall (1985). Details of the regions of the mouse sperm head in side view are depicted in B. Only a short portion of the midpiece is shown. The neckpiece and basal portion of the nucleus covered by the plasma membrane is designated N. The postacrosomal region, the equatorial segment, and the acrosomal cap are designagted, P, EQ, and AC, respectively. The darker shading along the dorsal portion of the acrosomal cap in B is the intact acrosome covered by the plasma membrane. The drawing in B is adapted from Figs. 1 and 2 in Stefanini *et al.* (1969).

characteristic time in order to acquire the "capacity" to fertilize the egg. Mouse sperm may be capacitated by incubation *in vitro* in media containing bovine serum albumin (BSA) (see review by Go and Wolf, 1983). This process and its possible mechanisms are discussed in Section 3, dealing with *in vitro* fertilization. The reader is also directed to the excellent review by Florman and Babcock (1990) that addresses this subject in greater detail.

A reaction crucial to fertilization is the exocytotic release of the contents of the acrosome in a reaction unique to spermatozoa, the acrosome reaction. In mammalian spermatozoa, the plasma membrane overlying the acrosome and the outer acrosomal membrane join, fuse, and vesiculate. This is shown schematically in Fig. 2. The stages of the acrosome reaction as shown are taken to correspond to the stages of exocytosis observed in mast cells (Zimmerberg, 1987); this is a convenient formalism and is not at present supported by direct experimental evidence. The relationship between capacitation and the acrosome reaction in mammalian sperm has been the focus of ardent investigation over the last three decades (for reviews, see Yanagimachi, 1981, 1988). This relationship is discussed primarily in Section 3 on *in vitro* investigations of mouse fertilization. The functional need for the acrosome reaction is generally agreed on: this reaction is

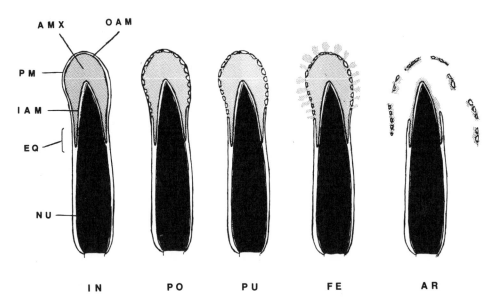

Figure 2.   Different states of the mouse sperm acrosome as seen in a sagittal section through the head, which is 3.2 μm wide in this section (Cummins and Woodall, 1985). The solid black body in all views is the nucleus, designated NU. Five states, corresponding to five putative stages of the acrosome reaction, are depicted. The first is the intact acrosome (IN). The second shows the outer acrosomal membrane (OAM) apposed at different points to the cytosolic face of the plasma membrane (PM). This stage is designated "poised" (PO). The third is a schematic rendering of pore formation at the points of apposition of the OAM and the PM. This stage is designated "punctate" (PU). At the punctate stage, the pores are presumed to be small enough that only ions and small molecules can pass through. The fourth shows enlargement of the pores to the extent that large molecules, including parts of the acrosomal matrix (AMX), shown by stippling, can leave the acrosomal compartment. This stage is designated "fenestrate" (FE). The last depicts loss of the vesicles formed by fusion of the OAM and PM from the sperm cell, leaving the inner acrosomal membrane (IAM) as the limiting membrane over the anterior part of the head. Some acrosomal matrix material is shown left behind at this stage, but this is conjectural. This stage is designated "acrosome-reacted" (AR). The equatorial segment EQ is still covered by plasma membrane and is the point at which the sperm PM fuses with that of the egg. The drawing of the intact stage was made from a direct tracing of an electron micrograph (P. M. Saling, J. Sowinski, and B. T. Storey, unpublished). The subsequent stages were depicted by modifications of this drawing with adaptions from both Zimmerberg (1987) and Yanagimachi (1988).

necessary for sperm penetration of the zona pellucida, entry of the sperm into the perivitelline space, and eventual gamete membrane fusion.

Certain processes peculiar to mouse sperm, such as the mode of capacitation, the zona-pellucida-mediated acrosome reaction, and questions regarding the block to polyspermy, have received primary emphasis in this review. Treatment of these topics has been deliberately selective in order to maintain the overall fertilization process in perspective. Recent reviews highly recommended for further probing of these topics are the following: Wassarman (1987, 1988), Saling (1989), Florman and Babcock (1990), and Kopf and Gerton (1990). Wassarman in this volume has been the advocate for the mouse egg (see Chapter 9). In addition, reviews by Yanagimachi (1988) and Kopf *et al.* (1989) describe the events of fertilization once the sperm has entered the egg. These events are at that point under the control of the egg, and it is there that we must abandon the mouse sperm to its fate.

2. FERTILIZATION *IN VIVO*

2.1. Sperm Distribution in the Female Reproductive Tract

The major difficulty in studying fertilization in eutherian mammals *in vivo* is, of course, that the process is so well hidden. The experimental challenge is to study the events from insemination to fertilization in such a manner as to render them visible with minimal artifactual perturbation. Removal of gametes and zygotes for examination by flushing the uterine and tubal lumen is not satisfactory in this regard (Zamboni, 1972). A method for rapid fixation to "freeze" all cellular processes for subsequent examination is required. An appropriate technique was developed by Zamboni and Stefanini (1968), originally for study of sperm *in situ* in the epididymis. It utilizes vascular perfusion with 2.5% glutaraldehyde in 0.1 M cacodylate buffer for 5 min to "freeze" all cellular processes by preliminary fixation, followed by complete fixation for 3 to 8 hr. Stefanini *et al.* (1969) and Zamboni (1972) applied this fixation by perfusion to female mice at different intervals after mating to obtain the time sequence of sperm progression through the uterus and oviduct and contact with and penetration of the egg. The small size of the mouse made this species particularly adaptable to this method of whole-animal fixation. After pairing with proven breeder males, the females were inspected every 30 min for the presence of vaginal plugs. Zero time for insemination could thereby by resolved to within this period. With hormonal induction, the time of ovulation could also be set. The females were rapidly anesthetized and perfused with fixative at the desired time after insemination, and the various parts of the reproductive tract were examined by both light and electron microscopy.

The results revealed a drastic diminution in the number of sperm proceeding from the uterus to the distal oviduct. The sperm are rapidly transported from the uterus, their point of deposition during mating, to the oviduct. Sperm were found at the uterotubal junction and in the caudal lumen of the oviduct within 20 min after formation of the vaginal plug. The uterotubal junction remains widely patent for at least 1 hr after coitus. Sections taken through the colliculus tubaricus showed that sperm movement through the junction is not at all impeded. Bacteria and exfoliated cells from the vaginal epithelium were found with the sperm; the presence of those nonmotile cells indicates that sperm transport to the oviduct is mediated primarily through the contractions of the musculature of the uterus and uterotubal junction. Sperm motility plays little if any role in this process. The uterotubal junction remains open for only a limited period; 4 hr after coitus it is tightly closed. The epithelial cells lining the lumen develop microvilli, which interdigitate and form a tight seal through which sperm cannot pass. It is this closing of the junction that limits the number of sperm that enter the oviduct. Those sperm left behind in the uterus, which form a large percentage of the ejaculate, degenerate and are eventually phagocytosed by the leukocytes entering the uterine lumen in large numbers by 8 hr after coitus.

Zamboni (1972) also determined the distribution of the fraction of the sperm that had passed through the uterotubal junction by the rapid whole-animal fixation method. Most of the sperm were found in the caudal coils of the isthmus and were most numerous in the intramural and extramural segments. The same result was obtained by Tessler and Olds-Clarke (1981), who used rapid fixation by superfusion of the exposed genital tract. Suarez (1987) excised intact oviducts containing live sperm at different times post-coitus and recorded the numbers and motions of the sperm in the lumen on videotape. The mouse oviduct is sufficiently transparent that it can be transilluminated to show the sperm inside. Manipulation of the excised oviduct was assiduously avoided to prevent contractions induced by handling, which would redistribute the sperm within the lumen. The results were very similar to those obtained by fixation: at 1–2 hr post-coitus, sperm were observed in the extramural segment of the uterotubal junction. Most of these sperm appeared to stick to the walls of the epithelium, but a number of motile sperm were also observed, as were previously motionless sperm suddenly detaching from the epithelial wall and swimming into the lumen. The intramural segment was harder to observe and appeared to contain fewer sperm. The lower isthmus contained the majority of the sperm at this time. At the two later time points examined, 4 and 6 hr post-coitus, corresponding to 1–2 hr before and 1–2 hr after ovulation, the lower isthmus still contained more than half the sperm observed. A most surprising observation was made during this examination of the lower isthmus: a defined zone of immotile sperm was present in the lumen. The zone was 100 to 300 μm long and contained immotile and very slowly moving sperm. Motile sperm were seen on either side of the zone. The width of the lumen in the lower isthmus is less than the 120-μm length of a mouse spermatozoon (Cummins and Woodall, 1985), and so the flagellar motion is necessarily restricted. The motion that was observed in the motile sperm population was consistent with the onset of hyperactivated motility, indicating that this form of flagellar motion is essentially fully developed in mouse sperm as they enter the oviduct (Suarez, 1987; Suarez and Osman, 1987).

Suarez (1987) also found that sperm in the upper isthmus were present in much lower numbers than in the lower isthmus at the two later time points examined, and none were observed 1 to 2 hr post-coitus. The few sperm observed appeared to be hyperactivated. No sperm were observed in the ampulla 1 to 2 hr post-coitus, and at the later two time points corresponding to the pre- and postovulatory period, only three or fewer hyperactivated sperm were observed. One sperm was observed in the perivitelline space of the ovulated egg. The results with transilluminated oviducts provide striking corroboration of the results obtained with rapid whole-animal fixation by Stefanini et al. (1969) and Zamboni (1972). The earlier studies had shown that the ampullae were essentially devoid of sperm and that the number of sperm found equaled the number of eggs. Only after the eggs had been penetrated were any free sperm found in the oviduct, and these were few in number. The conclusion drawn was that, in mouse egg fertilization in vivo, the sperm/egg ratio in the oviduct at the site of fertilization was essentially unity. This is apparently a result of rapid transit of the sperm to the ampulla. If a sperm does not encounter an egg, it passes rapidly through this section of the oviduct. It a sperm does encounter an egg, it cannot avoid this obstacle, and so it binds. Zamboni (1972) noted that sperm were found within the fimbria of the oviduct, supporting the concept of rapid ampullary transit.

2.2. Sperm Capacitation *in Vivo*

Penetration of the egg by the sperm, which leads to fertilization, requires capacitation of the sperm, acrosome reaction of the sperm, penetration of the zona pellucida by the sperm, entry of the sperm into the perivitelline space, and fusion of the sperm plasma membrane overlying the equatorial region (Fig. 1B) with the egg plasma membrane. Except for the last of these, study of these reactions has been carried out mostly *in vitro* and is discussed in more detail in Section 3 below. In examining ampullary eggs fertilized *in vitro*, Stefanini *et al*. (1969) and Zamboni (1972) reported that the earliest observable stage of sperm–egg interaction was incipient fusion of the two gamete membranes. Despite the large number of ampullary cumulus-intact eggs that were serially sectioned and examined for sperm, no progression of a fertilizing sperm through the cumulus oophorus or zona pellucida was captured. This attests to the low number of sperm present and the rapidity of the transit through the ampulla. Questions of capacitation and of the point of occurrence of the acrosome reaction relative to the penetration of the cumulus oophorus and the zona pellucida, therefore, could not be studied as a result of the failure of this search.

The timing of sperm entry into the oviduct relative to the time of penetration as deduced from the experiments *in vivo* (Stefanini *et al*., 1969; Zamboni, 1972) is consistent with either uterine or tubal capacitation. Sperm arrive at the uterotubal junction within an hour of coitus, and sufficient numbers appear in the caudal coils of the isthmus to provide a reservoir for the small number of sperm that actually do get to the ampulla. As pointed out above, the uterotubal junction is tightly shut by 4 hr post-coitus. Sperm penetration of mouse eggs occurs between 5.5 and 7 hr post-coitus. Ovulation occurs about 3 hr post-coitus, so the eggs remain in the ampulla from a minimum of 2.5 hr to a maximum of 4 hr before interacting with spermatozoa. Even the lower figure of 2.5 hr should be ample time for mouse sperm capacitation in the oviduct, based on *in vitro* studies of the process (Toyoda *et al*., 1971a,b; Miyamoto and Chang 1973; Inoue and Wolf, 1975b; Wolf and Hamada, 1979; Go and Wolf, 1983, 1985; Ward and Storey, 1984).

Although sperm capacitation in the oviduct is a plausible scenario for mouse fertilization, there is no experimental evidence to sustain this plausibility from *in vivo* studies. Suarez and Osman (1987) determined flagellar curvature ratios for mouse sperm in freshly and carefully excised oviducts that were transilluminated to make possible videotape imaging. The flagellar curvature ratios of sperm observed in the region extending from colliculus tubarius through the intra- and extramural junctions to the lower isthmus area were similar and characteristic of those of mouse sperm incubated *in vitro* under capacitating conditions. The ratios indicated that hyperactivation of flagellar motion had occurred. If one could equate onset of hyperactivation of mouse sperm with capacitation, these observations would imply that the sperm had become capacitated in the uterus. But analysis of the effects of bicarbonate and BSA on capacitation and hyperactivation by Neill and Olds-Clarke (1987) indicates that hyperactivation can be induced by bicarbonate in the absence of capacitation. The two processes do not appear to be tightly coupled. At present, the details of *in vivo* capacitation of mouse sperm are still in question because of the experimental difficulties involved in assessing such a process. A large number of *in vitro* experiments showing the requirement for capacitation of mouse sperm if they are to fertilize eggs imply that the process does occur *in vivo*. No experimental evidence has been provided to date showing that it does not occur *in vivo*.

2.3. Sperm Acrosome Reaction *in Vivo*

The inability to capture the progression of sperm through the cumulus oophorus of ampullary eggs or to find them on or near the zona pellucida makes it impossible to determine the precise point of occurrence of the acrosome reaction in the fertilizing sperm in the ampulla (Stefanini *et al*., 1969; Zamboni, 1972). From evidence obtained primarily with rabbit sperm (reviewed by Bedford, 1970), it had been assumed that the acrosome reaction occurred at or in the cumulus oophorus in most if not all mammalian species. This concept was rendered plausible by the indications that the acrosome reaction would liberate hyaluronidase, presumed necessary for depolymerization of the hyaluronic acid cementing together the cells of the cumulus oophorus (Austin, 1960; Bedford, 1970; Yanagimachi and Noda, 1970). Further biochemical studies summarized in the review of McRorie and Williams (1974), again based primarily on work with rabbit gametes, reinforced this view, which subsequently became generalized to all mammalian sperm. Zamboni (1972), however, pointed out that vesiculated remnants of the plasma and outer acrosomal membranes are found in the acrosome-reacted mouse sperm as fusion of sperm and egg begins *in vivo*. The presence of these membranes would be difficult to reconcile with the occurrence of the acrosome reaction in a location as remote as the cumulus oophorus.

The need for the mouse sperm acrosome reaction as a means for giving the sperm a passage through the cumulus of the mouse egg was also obviated by the observations of Zamboni (1970) on the state of the eggs in the ampulla. In the 2.5 to 4 hr following ovulation available to the mouse egg in the ampulla before it encounters its fertilizing sperm, a final maturation process occurs that results in loosening of the cumulus cells. Sperm with intact membranes have access to the zona pellucida without the need for hyaluronidase action. The ultrastructural evidence provided by Zamboni (1970, 1972) and Stefanini *et al*. (1969) strongly indicated that mouse sperm bind to the mouse egg zona pellucida with intact plasma membranes *in vivo*, and then undergo the acrosome reaction prior to entering the perivitelline space. This was fully confirmed by the *in vivo* studies of Bryan (1974) utilizing a specific histochemical stain for intact acrosomes and by the later *in vitro* studies discussed below in Section 3. The earlier evidence from ultrastructural work with rabbit sperm *in vivo* indicating the occurrence of the acrosome reaction at an earlier point in sperm progression (Bedford, 1968, 1970) has also been confirmed by later *in vitro* studies (Kuzan *et al*., 1984). This comparison emphasizes the need for experimental evidence in each system before one can generalize on the basis of plausibility (Zamboni, 1971).

2.4. Sperm–Egg Interaction *in Vivo*

The first aspect of gamete interaction captured *in vivo* by the whole-animal fixation method was contact between the sperm head and the surface of the egg plasma membrane (Stefanini *et al*., 1969; Zamboni, 1972). The sperm head lies flat over the egg surface, and the flagellum is wholly within the perivitelline space, wrapped around the egg. Membrane fusion occurs in the region immediately posterior to the area from which the plasma and outer acrosomal membranes have been lost in the course of the acrosome reaction. The flat angle of contact with the egg of the sperm head indicates that sperm penetration through the zona pellucida occurs at an oblique angle, as has been observed in other species (Yamagimachi, 1981, 1988). *In vitro* studies by Thompson *et al*. (1974) of zona pellucida penetration indicate an oblique orientation even in the absence of the constraints imposed by the presence of the cumulus cells. Fusion of the membranes is observed at the equatorial segment very shortly after gamete contact. A single thick lumina forms and then disappears, giving the sperm nucleus direct access to the ooplasm (Zamboni, 1972). As membrane fusion continues, the anterior portion of the head, whose limiting membrane is now the inner acrosome membrane, becomes engulfed by the egg cytoplasm with formation of a prominent incorporation cone (Stefanini *et al*., 1969). No fusion takes place between egg plasma membrane and inner acrosomal membrane, so that this part of the head resides in an ooplasmic vacuole. Filamentous dispersion of the sperm chromatin occurs during this process. By the time this chromatin mass is dispersed, the membraneous elements of the anterior head, along with the wall of the vacuole containing them, remain as a vestigial sac in the ooplasm close to the dispersed chromatin as the male pronucleus forms. This serves to mark the male pronucleus, as fully formed male and female pronuclei of the newly fertilized mouse egg are otherwise difficult to distinguish morphologically (Zamboni, 1972).

As gamete fusion proceeds, the flagellum is drawn into the ooplasm and there degenerates. Stefanini *et al*. (1969) noted that this degeneration occurred earlier in mouse sperm at this stage than with sperm of other species. The mitochondria show degenerative changes in the form of swollen cristae and apparent loss of matrix material. Not all the mitochondria in the midpiece seem to be so affected, but it is clear that any paternal inheritance of the mitochondrial genome must be vanishingly small in the mouse. The disorganization and fragmentation of the fibrous and microtubular elements of the flagellum, which still remains near the forming male pronucleus, continue throughout formation of the zygote. By the time the two-cell embryo has formed, the flagellar structural elements are no longer recognizable. The mouse sperm has performed its fertilizing function and has ceased to exist as a cell.

2.5. Summary of *in Vivo* Observations

The observation of the events of mouse fertilization *in vivo* is of necessity an exercise in morphology. The objectives of such investigations have been to determine the time course of the changes of sperm distribution in the oviduct and the description of the changes in the gametes as they finally meet and fuse. For the observations to be physiologically valid, perturbation of the system allowing for observations to be made must be kept to a minimum. The whole-animal fixation method (Zamboni and Stefanini, 1968; Stefanini *et al.*, 1969) fulfills the dual criteria of minimal perturbation and rapid arrest of all cellular action at the chosen time points (Zamboni, 1972). Transillumination of the excised oviduct for videotaping also fulfills these criteria and provides a means of examining live sperm within the oviduct (Suarez, 1987). Five salient points emerged from these investigations with regard to mouse spermatozoa after coitus. The first point is that the access of sperm to the oviduct is not regulated by sperm motility and uterotubal tortuosity but by uterine contractile processes and the tight closing on fixed schedule of the previously widely patent uterotubal junction. The second point is that a very limited number of sperm enter the oviduct, and most of these remain in the caudal segment. Of particular interest is the zone in the lower isthmus in which the motility of a portion of this group of sperm is suppressed. The third point is that the sperm traverse the ampulla rapidly so that few are found in this region of the oviduct. The fourth point, which appears to be a consequence of the third, is that the sperm : egg ratio in the ampulla at fertilization is unity. A spermatozoon either traverses the ampulla rapidly without encountering a mature egg, or it collides with the mature egg in the ampulla because it cannot avoid it. The fifth point is that the morphological evidence supports the occurrence of the sperm acrosome reaction at the egg zona pellucida rather than outside or within the cumulus oophorus.

In order to resolve the individual biochemical reactions involved in fertilization of the mouse egg, sperm–egg interaction must be studied *in vitro*. Successful *in vitro* fertilization of mouse eggs has been established for over two decades (Whittingham, 1968). An advantage to this early success is that the system is well established, but a disadvantage lies in the fact that many different procedures have become available whose variables must be sorted out. One important difference between the two approaches remains that the sperm : egg ratio *in vitro* is always orders of magnitude greater than the one : one relation found *in vivo*. Another important difference is that caudal epididymal, rather than ejaculated, sperm are used *in vitro*. One should keep these differences in mind when considering the investigations of the reactions of mouse egg fertilization by *in vitro* techniques. These investigations and their results are described in the following section.

3. FERTILIZATION *IN VITRO*

3.1. Early Studies: Establishment of an *in Vitro* System

The first unequivocal demonstration of *in vitro* fertilization in mammals was provided by Chang (1959) using rabbits. The biological endpoint was the birth of live young after transplantation of the *in vitro* fertilized eggs into hormonally prepared surrogate does. *In vitro* fertilization in the mouse was first carried out in explanted fallopian tubes by Brinster and Biggers (1965). Development to morulae and blastocysts occurred in 10% of the eggs so fertilized. Pavlok (1967) refined this method to obtain percentages of fertilized mouse eggs between 53% and 73%. Fertilization of mouse eggs completely removed from the oviduct and inseminated in a defined medium under oil was reported the following year by Whittingam (1968). The overall percentage of two-cell embryos obtained by this procedure was 25% (seven experiments), with a range of values from 10% to 41% in each of the experiments. In one set of experiments, eggs from Fl hybrids (C57B1 × Balb/c) were fertilized with sperm from randomly bred Swiss white males and then implanted in pseudopregnant Swiss white females. Examination of the fetuses showed nearly a 1 : 1 ratio of black-eyed (hybrid) to pink-eyed (Swiss) progeny, as expected if the *in vitro* fertilized eggs from the Fl hybrids were the ones developing into fetuses.

In these early studies, spermatozoa used for insemination were retrieved from the uterus shortly after mating. In the experiments with mouse gametes, the sperm recovery time was 1 to 2 hr after mating. This procedure was adopted as a means of capacitating the sperm *in vivo*. Capacitation of sperm was first identified by Austin (1951) and Chang (1951) as a process required for fertilization of mammalian eggs from the observations that rabbit sperm would not fertilize rabbit eggs *in vivo* if their residence time in the female reproductive tract prior to ovulation was less than 6 hr. Similar results were obtained with rat sperm; the minimum residence time in this case was 4 hr (Austin, 1985). The term capacitation was introduced the next year by Austin (1952) and was taken to encompass the physiological changes undergone by the spermatozoon before it could penetrate the egg. The question of capacitation *in vivo* has been previously considered (Section 2.2), and the functional correlates of mouse sperm capacitation as inferred from *in vitro* studies are discussed in Section 3.2. For the purpose of this account of early studies of mouse *in vitro* fertilization, it suffices to point out that *in vitro* capacitation of epididymal mouse sperm was achieved by Iwamatsu and Chang (1969, 1970), using heated bovine follicular fluid. This result was of crucial importance to subsequent studies in the mouse, since ejaculated sperm are exceedingly difficult to obtain from mice other than by recovery from the uterus after mating; the supply of such sperm, therefore, is severely limited. Sperm from the cauda epididymis are obtainable as a population of motile cells virtually free of contamination by other cells in much greater quantities than uterine sperm. Iwamatsu and Chang (1970) also established that preincubation of mouse sperm in medium containing 33% bovine follicular fluid for 1 hr was sufficient to obtain a largely capacitated population in that the time course of egg penetration after insemination was considerably shortened.

A completely defined medium for fertilization of mouse eggs *in vitro* was provided by Toyoda *et al.* (1971a,b). The key components in this medium were BSA and the energy substrates glucose and pyruvate; the remaining components were those used to make Krebs-Ringer bicarbonate medium. A systematic examination of the optimal concentrations of BSA and metabolizable substrates was undertaken by Miyamoto and Chang (1973), who showed that lactate was also a useful metabolizable substrate in the system. The medium of Toyoda *et al.* (1971a) was shown to be able to capacitate sperm as well as to provide a medium supportive of the early processes of egg fertilization. This medium, along with the refinements suggested by Miyamoto and Chang (1973), has proven to be the "workhorse" of mouse *in vitro* fertilization studies. The advent of this medium and the methods described by Toyoda *et al.* (1971a) mark the beginning of the systematic studies of the reactions leading to fertilization of mouse eggs *in vitro*. The ultrastructural study of Thompson *et al.* (1974) verified that there were no morphological aberrations occurring in this *in vitro* fertilization system for mouse eggs, thus providing confidence that these investigations would retain physiological relevance.

In the following sections, we follow the progress *in vitro* of the mouse sperm from its preparation to meet the mouse egg through to the point of fusion of the gamete membranes and penetration of the sperm into the egg cytoplasm. The processes to be considered are capacitation, sperm binding to the zona pellucida, the sperm acrosome reaction induced by the zona pellucida, sperm penetration of the zona pellucida and entry into the perivitelline space, and binding of the sperm to the egg plasma membrane with subsequent fusion of both gamete plasma membranes to produce the fertilized egg.

3.2. *IN VITRO* CAPACITATION OF MOUSE SPERMATOZOA

3.2.1. CAPACITATION: PROBLEMS IN DEFINING THE PROCESS

The aforementioned studies, which contributed to the development of an experimental system capable of supporting mouse egg fertilization *in vitro*, of necessity had to deal with the problem of *in vitro* capacitation. This was done by a circular route, in effect, since the endpoint of sperm capacitation was the successful fertilization of eggs in the shortest time by capacitated sperm. Using this criterion, Iwamatsu and Chang (1970) demonstrated that the process was rapid in mouse sperm, requiring only about 1 hr. These studies, carried out with epididymal sperm, showed clearly that capacitation is the first process that mouse sperm must undergo in order to penetrate the egg.

The process in the sequence leading to egg penetration affected by capacitation was first identified as zona pellucida penetration by Pavlok and McLaren (1972). They compared the fertilization rate of eggs freed of cumulus cells, both with intact zonae pellucidae and with zonae pellucidae removed with chymotrypsin. Epididymal sperm, which were supposedly not capacitated, yielded a low fertilization rate with zona-pellucida-intact eggs but a normal rate with zona-pellucida-denuded eggs. They concluded that capacitation was required for sperm penetration of the zona pellucida but not for penetration of the egg itself. This provided one definition of sperm capacitation.

At the time this conclusion was reached (Pavlok and McLaren, 1972), the role of the acrosome reaction in zona pellucida penetration, as opposed to penetration of the cumulus, was still far from clarified. As work on *in vitro* fertilization in mouse and other laboratory mammals progressed, so of necessity did work on *in vitro* capacitation, with the result that the definition of capacitation became both complicated and controversial. Chang (1957) demonstrated that seminal plasma could reverse the effects of capacitation and called this process decapacitation. There was considerable debate as to whether the definition of capacitation should include the acrosome reaction or should refer only to those reactions leading up to the acrosome reaction. Only the latter definition allows reversibility by the process of decapacitation. Austin (1975) advocated in an important review article the strict separation of the two processes. Chang and Hunter (1975), in an equally important review chapter published the same year, grouped both processes as part of capacitation. The controversy has been summarized in a recent article by Chang (1984), and the reader is referred to this clear and elegant presentation of the topic. In this chapter, we take the view that capacitation is the process that readies the mouse sperm to undergo the zona-pellucida-induced acrosome reaction and so does not include the acrosome reaction in the process.

3.2.2. CAPACITATION AND SPERM BINDING TO THE ZONA PELLUCIDA

As part of their extensive investigation of mouse sperm–egg interactions, Inoue and Wolf (1975b) examined the binding of fresh, uncapacitated sperm to cumulus-free, zona-pellucida-intact unfertilized eggs. They observed that the eggs, inseminated for 5 min in culture medium with such sperm, bound very few sperm. The number bound was equivalent to that bound by two-cell embryos fertilized *in vivo*. Insemination for 30 min yielded maximal binding. The culture medium for insemination was also demonstrated to support sperm capacitation. When fresh sperm were preincubated for 5 and 30 min in this medium, added to fresh eggs with a residence time of 5 min, and then scored for binding ability, the 5-min sperm bound poorly, but the 30-min sperm bound at full capacity. The acquired ability of sperm to bind to the zona pellucida was thus considered to be an early aspect of the capacitation process (Inoue and Wolf, 1975b). Further studies (Wolf and Inoue, 1976) on the effect of sperm concentration on sperm binding to the zona pellucida and egg penetration showed that binding increased as the 0.75 power of the sperm concentration. On the other hand, the ratio of sperm bound to sperm penetrated decreased with decreasing sperm concentration, approaching 1.0 at low sperm concentrations. In these experiments, the sperm were capacitated *in vitro* for 1–3 hr prior to use. From a comparison of the available sperm in the suspension and the number of sperm bound, Wolf and Inoue (1976) calculated that only a maximum of 15% of the sperm could be counted as capacitated, if only capacitated sperm could bind.

These results emphasize an important point to be considered in studies of *in vitro* fertilization: the heterogeneity of the sperm population inseminating the eggs. A figure of 15% capacitated sperm would indicate rather considerable heterogeneity, but this figure depends on the assumption that capacitation is required for binding to the zona pellucida. This point was examined in greater detail by Saling *et al.* (1978), who determined the time course and effect of medium composition on the binding process. The one essential component in the medium required for zona pellucida binding was found to be Ca^{2+}. Both the time course and extent of sperm binding to the zonae pellucidae of cumulus-free, zona-pellucida-intact unfertilized eggs was the same in buffer consisting solely of Tris-HCl, NaCl, and $CaCl_2$ at pH 7.4 (medium TNC) as in the fully constituted culture medium (medium CM; Toyoda *et al.*, 1971a,b; Inoue and Wolf, 1974a,b) used for *in vitro* fertilization. If, however, $CaCl_2$ was omitted from medium TNC (medium TN), little or no binding was observed. The sperm were isolated from two caudae epididymides in medium TN using a 10-min interval for sperm to swim up from the punctured tissue. The suspension was then diluted fivefold in either medium TN, TNC, or CM and incubated for 1 hr prior to the determination of zona pellucida binding.

Only medium CM contained albumin and so was capable of capacitation. The sperm in TNC remained uncapacitated, showing that this process in its entirety was not necessary for sperm binding. The binding was rapid, with a half-time of 6 min, and maximal binding was achieved by 20 min. This indicated that the fresh epididymal sperm had acquired the ability to bind during preincubation in both medium TNC and medium CM. The time required for fresh epididymal sperm to acquire the ability to bind to the zona pellucida was determined in medium CM by Heffner and Storey (1982). The half-time is about 5 min, and acquisition of binding ability is complete within 15 min. These results are similar to those of Inoue and Wolf (1975b). The time is presumably similar in medium TNC. It is of interest that Shur and Hall (1982a) found a rapid dissociation of poly-N-acetyllactosamine glycosidases from the mouse sperm surface, which occurred within that 15-min time frame. We postulate that it is this dissociation that confers the ability of mouse sperm to bind to mouse zonae pellucidae. Although this reaction is clearly a necessary prerequisite for sperm/egg interaction, its short time course and its lack of a requirement for what is generally regarded as a capacitating medium better qualifies it as a precursor reaction to the capacitation process rather than as a part of the process itself.

3.2.3. CAPACITATION AND HYPERACTIVATION OF MOTILITY

The flagellar beat pattern in sperm from hamster and guinea pig changes, as the sperm become capacitated, from a low amplitude favoring progressive motility to a high amplitude favoring frenzied activity with little progression (Yanagimachi, 1970, 1972). This was originally designated "activation" of sperm motility (Yanagimachi, 1970). The designation of "hyperactivation" was later suggested by Yanagimachi (1981) to avoid confusion with other uses of the term "activation." The effect has been reported in sperm from a number of mammalian species (reviewed by Yanagimachi, 1981, 1988), but there are evidently considerable differences between species with regard to the magnitude of the change in motility pattern. Hyperactivation of motility in mouse sperm during capacitation was first noted by Fraser (1977), who reported that sperm from the TO strain of mice acquired a "figure-of-eight" pattern of motion on incubation in a medium known to induce capacitation. Calcium was essential to this transformation, since the omission of this component alone from the medium prevented it. Aounuma et al. (1980) also reported the requirement of Ca^{2+} in the capacitation medium for induction of motility pattern change, but the transformed pattern was described as sigmoidal. There appear to be considerable differences between mouse strains with regard to patterns of hyperactivation, to the extent that onset of hyperactivation in some strains may result in motility changes so subtle as to pass unnoticed. This seems to be the case with outbred Swiss mice, since the extensive investigations using this strain by Wolf and co-workers (Inoue and Wolf, 1974a,b, 1975b; Wolf and Inoue, 1976; Wolf et al., 1976, 1977; Wolf, 1977; Saling et al., 1978; Wolf and Armstrong, 1978; Wolf and Hamada, 1979) on the interaction of capacitated sperm with eggs ranging in state from cumulus intact to zona pellucida denuded carry no mention of hyperactivated motility.

There are two issues with regard to hyperactivation of motility in mouse sperm. The first is an objective description of the motility pattern; the second is the coupling between onset of hyperactivation and capacitation of the sperm. The most useful description appears to be nonprogressive motility. This was quantitated for sperm from several mouse strains by Tessler et al. (1981), using both the area traverse method and the time exposure method of Katz and Dott (1975). One pattern of erratic and convoluted trajectories in which the sperm swam back and forth with vigor in a confined area was readily observed in one particular strain, $t^{W32}/+$ males. The authors proposed that this activity be called "dancing"; unfortunately, this highly descriptive term has been subsumed in the more ponderous but more precise term "nonprogressive motility." The onset of this nonprogressive motility is calcium dependent (Olds-Clarke, 1983). The same dependence on calcium for development of hyperactivation of sperm from outbred C3H and LACG strains was clearly shown by Cooper (1984) using time exposure under stroboscopic illumination. The divalent cation specificity was such that only Ca^{2+} and Sr^{2+} would support the transformation in motility pattern. Strontium had previously been shown to be specific for maintenance of mouse sperm motility but unable to mediate binding, whereas calcium is optimal for both activities (Heffner et al., 1980). Hyperactivation in these strains was readily recognized as movement in tight spirals; again, as in the observations of Tessler et al. (1981), much activity was confined to a small area of the observation field. A numerical index system for classifying mouse sperm motility patterns has been provided by Tessler and Olds-Clarke (1985), using a computer-assisted velocity measurement system, in the form of ratios of the following velocity parameters: V_n, the net displacement velocity; V_c, the curvilinear velocity of the sperm head; and V_a, a five-point moving average of the sperm head track. The progressiveness ratio (PR) is defined as V_n/V_c, the curvilinear progressiveness ratio (PR_c) as V_a/V_c, and the linear index (LI) as V_n/V_a. Further refinement of the computer-assisted system enabled Neill and Olds-Clarke (1987) to choose PR_c as the parameter best suited for identifying hyperactivated mouse sperm: $PR_c < 0.56$ and $V_c > 169$ μm/sec selects hyperactivated motility. This provides a satisfactory means of resolving the issue of describing hyperactivation in mouse sperm.

The second issue concerns the link between onset of hyperactivation and capacitation. Are the two processes obligatorily linked? Or are the two processes independent but normally occurring in the same time frame prior to sperm–egg interaction? If the two processes are independent, then it should be possible to uncouple them under appropriate experimental conditions. Neill and Olds-Clarke (1987) were able to accomplish this by manipulation of the incubation medium and utilization of separate methods for assaying hyperactivation and capacitation. The criteria for hyperactivated motility were $PR_c < 0.56$ and $V_c > 169$ μm/sec, as indicated above. The criterion for sperm capacitation was the acquisition of a particular chlortetracycline

(CTC) fluorescence pattern in which the anterior head and midpiece show bright fluorescence separated by a characteristic dark band over the posterior head. This pattern, designated "B," originally observed by Saling and Storey (1979), was shown by Ward and Storey (1984) to be diagnostic of capacitated, acrosome-intact mouse sperm. Omission of bicarbonate from the incubation medium effectively blocked acquisition of hyperactivation and the characteristic B fluorescence pattern. Omission of BSA, which would be expected to prevent capacitation of mouse sperm (Go and Wolf, 1983, 1985), in contrast, did block acquisition of the B pattern but did not affect onset of hyperactivation. This observation leads to the conclusion that although both processes normally proceed in parallel and require similar conditions, hyperactivation can proceed in the absence of capacitation.

Olds-Clarke (1989) utilized a genetic approach to this problem with sperm from two strains of mice, BTG and F1, each carrying the t complex. She showed that sperm from males carrying the t complex, $t^{W32}/+$ sperm, showed rapid acquisition of this type of motility in 50–60% of the population, while the balance remained nonhyperactivated and had lower velocities than their $+/+$ counterparts from the same strain. Yet capacitation as monitored by the CTC fluorescence pattern B proceeded normally and at the same rate in both strains, with no evident discrimination in the $t^{W32}/+$ between hyperactivated and nonhyperactivated sperm. This observation leads to the conclusion that capacitation can occur with a different time course than hyperactivation or can occur in the complete absence of hyperactivation. Lee and Storey (1986) showed that sperm from outbred Swiss males showed partial capacitation in the absence of bicarbonate by the CTC fluorescence method; in this strain hyperactivation is not normally observed. Taken together, these results strongly favor the concept that capacitation and hyperactivation occur as independent processes in mouse sperm and that it is capacitation that is the necessary precursor to fertilization.

3.2.4. CAPACITATION AND THE ACROSOME REACTION

As pointed out in Section 3.2.1 above, we have adopted Austin's (1975) point of view that capacitation and the acrosome reaction should be considered sequential but separate processes. Differentiation between the two processes is possible in mouse sperm because of an independent means to assess capacitation with the CTC fluorescence assay, which was alluded to in the previous discussion of hyperactivation. Conditions for capacitation of mouse sperm had been established originally by Iwamatsu and Chang (1970) and Miyamoto and Chang (1973). Extensive studies by Wolf and co-workers (Inoue and Wolf, 1974b, 1975b; Wolf and Inoue, 1976; Wolf et al., 1976) established the times and conditions using the incubation medium of Toyoda et al. (1971a,b). The basic time frame and the requirement for BSA (Wolf, 1979; Go and Wolf, 1983) were 60 to 90 min and 20 mg/ml of albumin, respectively; lower albumin concentrations required longer incubation times. Capacitation and acrosomal status was monitored by a CTC fluorescence assay (Ward and Storey, 1984), which was adapted from an earlier CTC fluorescence probe technique originally developed to track the acrosome reaction in mouse sperm (Saling and Storey, 1979). The fluorescence patterns observed using this assay are illustrated in Fig. 3. The B pattern characteristic of capacitated sperm increases from less than 20% at the first observable time point of 10 min to over 80% at 60 min. This percentage remains quite constant for the next 30 min and then declines as the acrosome reaction occurs spontaneously at a slow rate. No spontaneous acrosome reaction is observed in uncapacitated sperm. Capacitation of mouse sperm is, therefore, required for the spontaneous acrosome reaction to occur.

A much clearer differentiation between capacitation and the acrosome reaction is obtained in the case of the induction of acrosomal exocytosis in mouse sperm by the zona pellucida of the mouse egg. Mouse sperm bind to zonae pellucidae of eggs with plasma membrane intact (Saling et al., 1979) and, if capacitated, exhibit the B fluorescence pattern (Fig. 3) in the presence of CTC

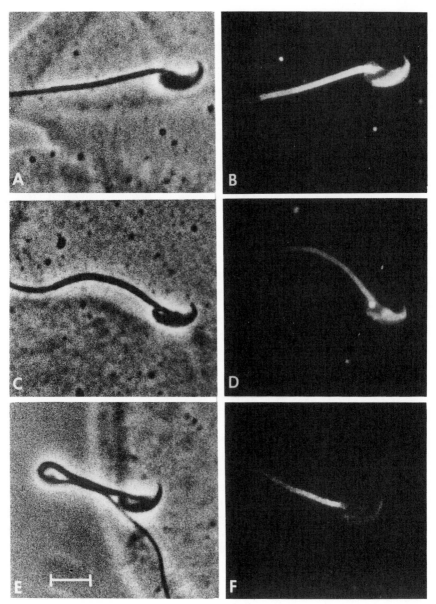

Figure 3. Paired phase-contrast and epifluorescence photomicrographs of capacitated mouse sperm bound to isolated zonae pellucidae from mouse eggs and treated with CTC according to Ward and Storey (1984). The scale bar represents 10 mm. (A) Phase-contrast photograph of a spermatozoon bound to a zona pellucida 15 min post-binding. (B) Epifluorescence photograph of the sperm in A, with fluorescence excitation at 405 nm and emission at wavelengths greater than 470 nm. Note the bright fluorescence over the anterior portion of the head and midpiece and the dark band over the postacrosomal region. This is designated pattern B. (C) Phase-contrast photograph of a spermatozoon 120 min post-binding. (D) Epifluorescence photograph of the sperm in C illuminated as in B. Note the fluorescence on the midpiece and the irregular group of bright flourescent spots on the head. This is designated pattern S. (E) Phase-contrast photograph of a spermatozoon 210 min post-binding. (F) Epifluorescence photograph of the sperm in E illuminated as in B. Note the fluorescence on the midpiece in contrast to the very faint fluorescence seen on the head. This is designated patter AR. Photographs reproduced from Lee and Storey (1985) by permission of *Biology of Reproduction*.

(Saling and Storey, 1979). Once bound to the zona pellucida, the sperm undergo the acrosome reaction as shown by loss of the B pattern and eventual acquisition of the AR pattern (Fig. 3). This reaction is inhibited in the presence of the compound, 3-quinuclidinyl benzilate (QNB), an analogue of acetylcholine that is routinely used as an antagonist of the muscarinic class of cholinergic neurorecptors (Florman and Storey, 1982a,b). In the presence of QNB, the sperm remain attached to the zona pellucida; a high percentage of the cells exhibit the B fluorescence pattern with only a slow decline of this percentage over time, corresponding to the rate of the spontaneous acrosomal reaction. The compound QNB does not inhibit the spontaneous reaction. The acrosome reaction induced in sperm bound to the zona pellucida is presumed to be the physiological form of the acrosome reaction, since it occurs in the sperm best located to penetrate the zona pellucida as a result of this exocytotic process. This reaction occurs only in capacitated sperm displaying the B fluorescence pattern with CTC; uncapacitated sperm do not undergo the zona-pellucida-induced acrosome reaction (Florman and Storey, 1982a; Ward and Storey, 1984). Inhibition of the zona-pellucida-induced acrosome reaction in capacitated sperm by QNB allows a clear separation between the process of capacitation and the acrosome reaction. The requirement that the sperm be capacitated for the zona-pellucida-induced reaction to occur provides a useful operational definition of capacitation. Mouse spermatozoa may be termed capacitated if they undergo the relatively rapid zona-pellucida-induced acrosome reaction, which is sensitive to inhibition by QNB.

Possible mechanisms of the zona-pellucida-induced acrosome reaction and details of sperm–zona pellucida interactions involving the zona pellucida glycoprotein ZP3 specific to the induction reaction (Bleil and Wassarman, 1983) are considered in Section 3.4 of this chapter. At this point, it is useful to emphasize that induction of the acrosome reaction in sperm by the zona pellucida of the egg in a given mammalian species has been fully documented only in mouse. There are indications that this scenario holds for the majority of mammalian species (Saling, 1989; Kopf and Gerton, 1990), but three exceptions are noted. The first is the guinea pig, in which strong evidence for binding of both acrosome-reacted and acrosome-intact sperm to the zona pellucida has been presented (Huang et al., 1981; Myles et al., 1987). The second is in the human, in which binding to the zona pellucida can be initiated by both acrosome-intact and acrosome-reacted sperm (Morales et al., 1989). The third is the rabbit, in which sperm recovered from the perivitelline space, and so presumed to be acrosome reacted, were shown to be able to penetrate the zona pellucida on reinsemination (Kuzan et al., 1984). For an overview of the differences among species with regard to capacitation and the acrosome reaction, the reader is referred to the recent reviews by Yanagimachi (1988), Saling (1989), Kopf and Gerton (1990), and Florman and Babcock (1990).

3.2.5. CAPACITATION: REQUIREMENTS AND REACTIONS

As described earlier, media used for capacitation of mouse sperm and in vitro fertilization of mouse eggs are usually minor variants of the medium described by Toyoda et al. (1971a). This is a Krebs–Ringer bicarbonate medium in which bicarbonate is the primary buffer; phosphate may be absent and, if present, does not exceed 1 mM. Metabolizable substrates such as glucose, lactate, and pyruvate are normally present. Albumin, usually BSA because of its ready availability, is always present. Metal cations present are those normally in Krebs–Ringer bicarbonate medium: Na^+, K^+, Mg^{2+}, and Ca^{2+}. Of these components, the two that most investigators agree are necessary and specific for capacitation are Ca^{2+} and albumin. The evidence supporting this consensus and the reactions that these two components may mediate are discussed in turn below.

The obligatory requirement for the presence of Ca^{2+} in the mouse system was first established by Iwamatsu and Chang (1971). The optimal concentration was found to be 1.8 mM; at 0.9 mM Ca^{2+} or less, fertilization was severely inhibited. At 4.1 mM Ca^{2+}, inhibition of fertilization was also noted. The Ca^{2+} is 1.71 mM in the medium of Toyoda *et al.* (1971a,b) and so is optimal. Miyamoto and Ishibashi (1975) examined the effects of Ca^{2+} on *in vitro* fertilization of both mouse and rat eggs and showed that, in both species, Ca^{2+} concentrations less than 0.3 mM and greater than 5 mM resulted in no fertilization. At 0.9 mM, the percentage of mouse egg fertilization was little more than half that observed at 1.7 mM.

Fraser (1987a) dissected the Ca^{2+} requirements for the mouse system into the different processes of capacitation, hyperactivated motility, the acrosome reaction, zona pellucida penetration, and gamete fusion. By varying concentrations in the preincubation of sperm needed for capacitation and then adding the sperm suspension to eggs in medium with 1.8 mM Ca^{2+}, she deduced the following minimum requirements. Capacitation requires at least 0.09 mM. The acrosome reaction and gamete fusion need 0.9 mM, whereas hyperactivated motility and zona pellucida penetration require 1.8 mM. High Ca^{2+} concentrations were again found to be inhibitory. The concentrations of Ca^{2+} quoted in these studies are the total in the medium. It is the free Ca^{2+} that is important, however. Neill and Olds-Clarke (1988) were able to demonstrate that part of the apparent inhibitory effect of lactate on mouse sperm capacitation was caused by lactate chelation of medium Ca^{2+}. The divalent cation specificity for Ca^{2+} of these various processes appears to extend in part to Sr^{2+}. Fraser (1987b) reported that Sr^{2+} could substitute for all these processes, from capacitation to gamete fusion, and so could sustain fertilization of mouse eggs. This result is not consistent with that of Heffner *et al.* (1980), who showed that Sr^{2+} could maintain the motility of mouse sperm but could not substitute for Ca^{2+} in mediating sperm binding to the zona pellucida. Clearly zona penetration could not occur in the absence of such binding. Heffner *et al.* (1980) used a simple NaCl–Tris buffer for the binding and motility measurements, whereas Fraser (1987a) used a complete culture medium capable of supporting *in vitro* fertilization. A more extensive study of the ionic specificity of mouse sperm capacitation is needed to settle these discrepancies. To date, this question has, with the exception of the studies just mentioned, been virtually ignored.

The time requirements for the presence of Ca^{2+} in the medium have been worked out by Fraser (1982). This was done by adding Ca^{2+} at defined times during the 120-min incubation period used for capacitation and checking the ability of the sperm to fertilize eggs and undergo the acrosome reaction induced by the divalent cation ionophore A23187. Neither 5 min nor 30 min of Ca^{2+} exposure proved sufficient, but 60 min or longer was fully effective. These results are fully consistent with the time frames for capacitation established by Iwamatsu and Chang (1970), Inoue and Wolf (1974b), Aonuma *et al.* (1980), and Ward and Storey (1984).

The need for albumin to obtain successful *in vitro* fertilization of mouse eggs was inferred from the early experiments of Iwamatsu and Chang (1970) and of Toyoda *et al.* (1971a,b). A systematic examination of serum albumins from different sources by Miyamoto and Chang (1973) revealed a lack of specificity in this regard. Albumins from bovine, equine, rabbit, and human sera were all equally effective in promoting *in vitro* fertilization of mouse eggs. Presently, BSA is the one mostly used for mouse sperm capacitation work because of its availability. The time required for capacitation depends on the concentration of albumin over the range 2–20 mg/ ml (Wolf, 1979; Ward and Storey, 1984).

Before discussing the possible modes of action by which albumin is proposed to mediate capacitation of mouse sperm, we should point out that opinion is not unanimous concerning the requirement for this protein in the process. Fraser (1985) has reported that mouse sperm may become capacitated in albumin-free medium but that the acrosome reaction required for fertilization does require albumin. Sperm preincubated in the absence of albumin and used for insemination of both cumulus-intact and cumulus-free zona-pellucida-intact eggs gave normal egg penetration if albumin was present in the fertilizing medium. If albumin was omitted from the fertilizing medium, penetration of cumulus-intact eggs was reduced, and cumulus-free eggs exhibited very low rates of penetration. The presence of albumin also increased the rate of spontaneous acrosome reactions. A similar conclusion was reached with regard to hamster sperm by Andrews and Bavister (1989b): albumin was required for induction of the acrosome reaction by the zona pellucida in sperm capacitated in the absence of albumin. Andrews and Bavister (1989a) found that 0.5 mM D-penicillamine can apparently substitute for albumin in the 4-hr preincubation normally utilized for capacitation. Egg penetration after a further 1.5-hr insemination time in the presence of albumin was as successful as with albumin-capacitated sperm. Omission of albumin in the insemination step resulted in no egg penetration (Andrews and Bavister, 1989b). In both sets of experiments using mouse (Fraser, 1985) and hamster (Andrews and Bavister, 1989a,b) gametes, the sperm were exposed for a finite period to albumin in those cases where egg penetration occurred.

This raises the question of whether different rates of capacitation induced by albumin among the cells in the sperm population might not account for the results. Another question is also raised concerning the definition of capacitation. The operational definition proposed in this chapter (Section 3.2.4) states that mouse sperm may be termed capacitated if they undergo the acrosome reaction induced by the zona pellucida, which is sensitive to inhibition by QNB. If albumin is required for the zona-pellucida-induced acrosome reaction, then it becomes functionally difficult to assay capacitation in the absence of albumin. It is our opinion that albumin is required for mouse sperm capacitation *in vitro* as well as for capacitation of other mammalian sperm, from considerations of the mechanism of the process in the following discussion.

The mechanism by which albumin promotes capacitation in mammalian sperm has received considerable attention. The evidence presented to date supports the concept put forth by Davis (1980) that a decrease in the cholesterol content of the sperm plasma membrane mediated by albumin, as shown with rat sperm by Davis *et al*. (1980), is a key component of the capacitation process. Davis (1981) extended this observation of albumin-mediated cholesterol removal to sperm from ram, boar, bull, rabbit, and man. An excellent correlation was found between the capacitation interval, defined as the half-maximum time for egg penetration by sperm, and the ratio of cholesterol to phospholipid. Capacitation by this interpretation corresponds to lowering of the cholesterol/phospholipid ratio from initial values between 0.4 and 1.0, depending on species, to an apparent critical value of 0.28. In mouse sperm, loss of cholesterol caused by BSA was documented by Go and Wolf (1983, 1985); the rate of loss was dependent on both the concentration and the type of albumin preparation used.

The question of whether the cholesterol/phospholipid ratio or the actual cholesterol content of the sperm plasma membrane is the key factor in mouse sperm capacitation remains unresolved, however. Go and Wolf (1985) found that the phospholipid content of mouse sperm remained unchanged during capacitation. Since cholesterol was removed during the process, the cholesterol/phospholipid ratio inevitably decreased. Yet it is not clear whether it is the ratio of cholesterol to phospholipid, which is directly proportional to the mole fraction of cholesterol in total lipid, or the total membrane content of cholesterol that is the key factor in determining if the sperm plasma membrane has reached a state of capacitation. A derivative of cholesterol that has been put forward as another key factor in capacitation is cholesterol sulfate. Langlais *et al*. (1981) reported that this cholesterol ester is a normal constituent of human sperm and hypothesized that this compound, a known inhibitor of acrosin, acts as both membrane stabilizer and protease inhibitor. Cleavage of the sulfate groups in the female tract by the appropriate sulfatase activity would then allow lowering of the cholesterol mediated by albumin in the tract lumen. This hypothesis has been given full formulation by Langlais and Roberts (1985). As a refinement of the mechanism for removal of cholesterol by albumin, Langlais *et al*. (1988) have proposed that lysophosphatidylcholine bound to albumin is the cholesterol-binding agent, a concept consistent with the finding of Go and Wolf (1985) that albumin-containing phospholipid was more effective in cholesterol extraction than delipidated albumin.

In addition to the loss of cholesterol, there have been reports of the loss of other surface components from the mouse sperm surface during capacitation. Loss of antigenic proteins from the cell surface of rabbit sperm was an important feature of the hyperosmolal capacitation medium developed by Brackett and Oliphant (1975) for these sperm (as reviewed by Saling, 1989). The nature of the components lost from mouse sperm has never been fully defined. One candidate is a protease inhibitor from murine seminal plasma described by Poirier and co-workers (Aarons *et al.*, 1984; Poirier *et al.*, 1986; Robinson *et al.*, 1987). This component, however, appears to mask a site on the sperm cell required for zona pellucida binding (Poirier *et al.*, 1986), and so its loss from the sperm is probably involved with the acquisition of binding ability. Fraser (1984a) reported that a factor inhibitory to fertilization could be washed off mouse epididymal sperm, but further characterization of this factor has apparently not been undertaken. At present, it seems reasonable to postulate the existence of such factors associated with mouse sperm prior to capacitation, but the nature of these factors and their role in the capacitation process both remain unknown.

Changes in the properties of proteins integral to the mouse sperm plasma membrane during capacitation of mouse sperm had not been documented until very recently. Leyton and Saling (1989a) recently demonstrated that phosphorylation of a tyrosine residue of a plasma membrane protein of M_r 95,000 in mouse sperm is enhanced under incubation conditions conducive to capacitation. This protein was also shown to bind the zona pellucida glycoprotein ZP3 but not ZP2. Since ZP3 is the zona pellucida glycoprotein that induces the acrosome reaction (Bleil and Wassarman, 1983), it seems plausible to postulate that this phosphorylation coincides with the acquired ability of the sperm to undergo the acrosome reaction. Two other proteins containing phosphotyrosine residues also were found in capacitated but not in uncapacitated sperm (Leyton and Saling, 1989a). These have molecular masses of M_r 75,000 and M_r 52,000. Since these proteins do not interact with zona pellucida glycoproteins on SDS gels as does the M_r 95,000 protein, it is not yet possible to tell if these are also plasma membrane proteins. The M_r 95,000 protein seems to be a major player in the signal transduction pathway in the sperm cell leading to acrosomal exocytosis (Kopf, 1989; Kopf and Gerton, 1990). The change in its phosphorylation state with capacitation provides the first possibility that this process may be monitored by a specific biochemical reaction.

3.3. *In Vitro* Binding of Mouse Spermatozoa to the Zona Pellucida

3.3.1. DEFINITIONS AND ASSAYS OF SPERM BINDING

The transition between a freely swimming spermatozoon in the immediate vicinity of the zona pellucida and the same spermatozoon tightly bound to the zona pellucida does not represent a simple collision and sudden adhesion process. In the hamster, at least two stages of attachment and binding with an intricate set of factors affecting the process were recognized by Hartmann and Hutchison (1974a,b,c, 1980; reviewed by Gwatkin, 1977; Hartmann, 1983). In the mouse, two stages of sperm binding are experimentally distinguishable. In order to make this distinction, the assays used to determine sperm binding to the zona pellucida *in vitro* must first be considered.

Inoue and Wolf (1975b) allowed sperm to interact with cumulus-free, zonae-pellucidae-intact eggs for chosen time periods and then transferred the eggs through three volumes of medium using micropipettes with diameters about 1.5 that of the eggs. The shear forces generated by the pipetting procedure served to dislodge loosely bound sperm from the zona pellucida. Using this method, they demonstrated the need for a short incubation of sperm and eggs in capacitation medium before binding took place. The control for sperm binding in these experiments was the two-cell embryo, in which sperm binding to the zona pellucida did not occur. The pipetting procedure was shown to remove all sperm from the embryos. A similar method was used by Bleil and Wassarman (1980c) to determine the sperm-binding activity associated with each of the three zona pellucida glycoproteins ZP1, ZP2, and ZP3. This particular method has the advantage of simplicity and internal self-consistency. Its disadvantage is that the shear forces generated depend on the geometry of the micropipette as well as the suction pressure used to transfer the eggs into the pipettes and to expel them into the wash volumes. These parameters are highly dependent on the experimenter, and so comparisons between results from different individuals and different laboratories have proven difficult.

The second method utilized for the determination of sperm–zona pellucida binding is the step-gradient centrifugation method developed by Saling *et al.* (1978). After incubation, the suspension of sperm and eggs is centrifuged through a step gradient of medium containing dextran such that the unbound sperm stay in the upper layer of low dextran concentration while the eggs with sperm bound are sedimented into the lower layer of high dextran concentration. The lower layer also contained glutaraldehyde in the early experiments to fix the zona-pellucida-bound sperm, and so the method was termed "stop-fix." In later work, the glutaraldehyde was shown to be unnecessary (Saling and Storey, 1979; Florman and Storey, 1982a). This method has the advantage that uniform shear forces are exerted on all samples and thus provides a more "objective" set of conditions for comparison of results among investigators. The disadvantage of the method is that near-quantitative recovery of eggs is required to obtain accurate results. In the original study (Saling *et al.*, 1978), this was accomplished by a lengthy procedure to clean and condition the surface of the glass centrifuge tubes to prevent sticking of the eggs to the side of the tubes. The recent availability of heparin-coated microcentrifuge tubes has made this unnecessary, and a recently reported modification of the "stop-fix" method utilizing these tubes provides excellent recovery with no washing of tubes (Benau and Storey, 1987, 1988).

Both assay procedures depend on shearing forces of unknown magnitude and orientation to remove from the zonae pellucidae sperm that are "loosely" bound. Sperm bound to zonae pellucidae are, in effect, defined as those that survive either procedure. The terms "loose" or "tight" carry little quantitative significance in terms of either assay. We prefer to use the term "mechanical binding" to define sperm adhesion to the zona pellucida after either micropipette or stop-fix centrifugation processing. This binding may be subsequently dissected into two stages, sequential in time. In the first stage, sperm binding may be reversed or prevented by addition of certain inhibitors (see below). In the second and final stage, sensitivity to these inhibitors is lost.

The first stage, in which binding is rapid with no delay in capacitated sperm and rapid with a delay of but a few minutes in uncapacitated sperm, requires Ca^{2+} and is reversible by the calcium chelator EGTA (Saling *et al.*, 1978; Heffner and Storey, 1982; Florman *et al.*, 1982). The time to half-maximal binding is 5–6 min at 37°C. The requirement for Ca^{2+} appears to be both necessary and sufficient. This stage of binding occurs at the same rate and with the same sperm/egg binding ratio in NaCl–Tris medium containing 1.7 mM Ca^{2+} as it does in the complete culture medium used for *in vitro* fertilization (Saling *et al.*, 1978). No binding is observed if Ca^{2+} is omitted from either of the two media. Addition of EGTA at early times of sperm–egg interaction results in a decrease of sperm binding, and sperm bound to the eggs fall off. As the process proceeds, the effect of EGTA diminishes until, at 20 min, the sperm bound are completely insensitive to removal by EGTA (Florman *et al.*, 1982). In addition to EGTA, trypsin inhibitors (Saling, 1981) and UDP-galactose (Lopez *et al.*, 1985) effectively block binding but lose their effectiveness after 20 min. Soldani and Rosati (1987) also demonstrated EGTA reversibility in this system at early times but also noted a diminution in sperm binding just at the transition between the first (inhibitor-sensitive) and second (inhibitor-insensitive) stages.

For the remainder of this chapter, we use the term binding to denote mechanical attachment of sperm to the zona pellucida; no distinctions between loose and tight or between attachment and binding are made. Only the distinction between the first and second stages, as defined above, is maintained.

3.3.2. SPERM-ASSOCIATED SITES FOR ZONA PELLUCIDA LIGANDS INVOLVED IN BINDING

With the demonstration that mouse sperm bind to the mouse zona pellucida with their plasma membrane intact (Saling and Storey, 1979; Saling *et al.*, 1979), it became reasonable to search for sperm cell surface components that act as sites for interaction with the zona pellucida glycoproteins to mediate mechanical binding. The zona pellucida glycoprotein involved in the process was shown to be ZP3 as defined by Bleil and Wassarman (1980a,b). The interaction occurs between functional groups associated with both the carbohydrate side chains and the polypeptide chain of ZP3 and receptor-like molecules on or in the sperm plasma membrane. In this discussion, we designate the zona pellucida groups as ligands and the sperm plasma-membrane-associated binding moieties for ZP2 and ZP3 as receptors.

Four types of mouse sperm-binding site may be differentiated experimentally, based on the ability of agents to inhibit sperm binding to the zona pellucida. An oddity of this differentiation is that it is done by assays of enzymatic activities, which themselves do not appear to be involved in the binding process. As a result, each site has acquired a designation cast in terms of a particular enzyme. The four sites are the trypsin-inhibitor-sensitive (TI) site (Saling, 1981; Benau and Storey, 1987), the galactosyl transferase (GT) site (Shur and Hall, 1982a,b; Lopez *et al.*, 1985; McLaughlin and Shur, 1987; Shur and Neely, 1988), the fucosyl transferase (FT) site (Apter *et al.*, 1988), and the mannosidase (Ma) site (Cornwall *et al.*, 1989). The first two sites have received fuller characterization and are considered in some detail. The last two sites, for which little published documentation is as of yet available, are described only briefly.

The TI site is perhaps the most aptly named. It derives from the observation by Saling (1981) that both large- and small-molecular-weight inhibitors of the protease trypsin blocked sperm binding to the zona pellucida. Soybean and lima bean trypsin inhibitors are examples of large-molecular-weight protein inhibitors, and *p*-nitrophenylguanidino benzoate (NPGB) and N-tosyl lysylchloromethyl ketone (TLCK) are examples of small-molecular-weight inhibitors. Both types are effective only on the first stage of binding. No enzymatic activity corresponding to that of trypsin or of acrosin (whose substrate specificity is very similar to that of trypsin) is associated with the site. Hydrolysis of trypsin or acrosin model substrates was not observed. Instead, an esterase activity with substrate specificity confined to guanidinobenzoate esters was found (Benau and Storey, 1987). These esters, including the *p*-nitrophenyl ester (NPGB) used by Saling (1981) and the methylumbelliferyl ester (MUGB) used by Benau and Storey (1987), were developed as active site titrants for the active sites of trypsin or acrosin by effecting the single-turnover release of the chromogenic or fluorogenic phenol followed by enzyme inactivation (Chase and Shaw, 1970; Jameson *et al.*, 1973). The TI site, in contrast, catalyzes the steady hydrolysis of both NPGB and MUGB. The latter substrate, because of the high sensitivity of the fluorometric method, is routinely used to assay the site. Hydrolysis of MUGB mediated by the TI site may be inhibited by NPGB, soybean trypsin inhibitor, and acid-solubilized mouse zonae pellucidae protein, with a concentration dependence matching that of inhibition of sperm binding to zonae pellucidae by these compounds (Benau and Storey, 1987). This implies that a ligand present in ZP3 in the acid-solubilized zona pellucida preparation competes with MUGB for the same site. This is presumably one of the ligands mediating binding in the structurally intact zona pellucida. What the nature of this ligand might be and how it might resemble the guanidino-benzoate molecule remain both unknown and unclear.

The GT site is the best characterized of the four. Shur and Hall (1982a,b) demonstrated the expression of galactosyltransferase activity in the presence of Mn^{2+} and the donor/acceptor pair UDP-galactose/N-acetylglucosamine in mouse sperm. They also demonstrated that the loss of its enzymatic activity by the GT inhibitors UDP-aldehyde and α-lactalbumin occurred in parallel to the loss of sperm binding to zonae pellucidae. Inhibition of both binding and enzymatic activity by UDP-aldehyde displayed the same concentration dependence on each inhibitor. Inhibition of milk galactosyltransferase by α-lactalbumin is actually an induced change in enzyme acceptor specificity: glucose becomes the acceptor in place of N-acetylgucosamine. Shur and Hall (1982b) demonstrated that the mouse sperm surface GT enzyme activity was similarly affected by α-lactalbumin. As the activity toward N-acetylglucosamine as acceptor diminished with increasing concentrations of α-lactalbumin, the activity toward glucose as acceptor increased accordingly. The decline in sperm binding followed the decline in activity towards N-acetylglucosamine.

Further evidence for participation of a GT-like protein in sperm binding to zonae pellucidae was provided by Lopez *et al.* (1985) using the soluble, affinity-purified enzyme from bovine milk. The milk enzyme was a potent inhibitor of sperm binding, as was a monospecific polyclonal antibody to the enzyme. Indirect immunofluorescence using this antibody localized the sperm surface GT site to the anterior portion of the sperm head; this is the location expected for a plasma membrane protein mediating sperm binding to the zona pellucida. The donor for the GT-catalyzed transfer reaction is UDP-galactose, regardless of whether N-acetylglucosamine or glucose acts as an acceptor. UDP-galactose was shown to inhibit binding of sperm to zonae pellucidae if added early in the binding reaction (Lopez *et al.*, 1985), indicating that it may act at the first stage like EGTA. Comparison of the concentration dependence on UDP-galactose of increasing enzyme activity and decreasing sperm binding showed that both processes were saturable with similar K_{Mapp} values (Benau and Storey, 1988). The sperm plasma membrane GT-like molecule has recently been isolated by selective extraction from the intact cells and purified to homogeneity by Shur and Neely (1988). It is a M_r 60,000 protein that could be recovered after SDS polyacrylamide gel purification, renatured, and demonstrated to express the enzymatic activity expected for GT with appropriate substrate specificities.

These results argue strongly for a sperm binding site for appropriate zona pellucida ligands that has all the properties of a galactosyltransferase enzyme. The question of whether expression of sperm GT enzymatic activity is required for the zona pellucida binding reaction has not been definitively resolved (Saling, 1989), but the experimental evidence reported to date speaks against such a requirement. For expression of GT enzymatic activity, Mn^{2+} is obligatory (Shur and Bennett, 1979; Shur and Hall, 1982a; Shur and Neely, 1988). Yet mouse sperm bind to mouse zonae pellucidae in the complete absence of Mn^{2+} as long as either Ca^{2+} or La^{3+} is present (Saling *et al.*, 1978; Heffner *et al.*, 1980; Saling, 1982). Binding does not occur in the presence of Mn^{2+} as sole divalent cation, and if Ca^{2+} is present to mediate binding, Mn^{2+} inhibits the binding with a concentration dependence similar to that for its activation of enzymatic activity (Benau *et al.*, 1990). This indicates that expression of GT enzymatic activity and sperm binding to the zona pellucida may be antagonistic.

Shur and Hall (1982b) reported that solubilized and intact zonae pellucidae acted as acceptors in the GT enzymatic reaction using UDP-galactose as donor, again in the presence of Mn^{2+}. Since neither donor nor metal cation activator is present in the usual sperm binding assay, this aspect of GT activity does not seem relevant to the binding reaction. Further evidence that GT enzymatic activity and GT-mediated sperm binding are antagonistic is provided by the disparity in the inhibition of binding and the inhibition of the GT-catalyzed galactosylation of N-acetylglucosamine by acid-solubilized zonae pellucidae. As pointed out above in the description of the TI site, acid-solubilized zonae pellucidae inhibit sperm binding directly in parallel with inhibition of sperm-catalyzed MUGB hydrolysis, implying effective competition of a zona pellucida ligand for the site that binds MUGB. Under conditions optimal for expression of GT enzymatic activity, however, acid-solubilized zonae pellucidae have but a weak inhibitory effect on the enzymatic reaction (Benau and Storey, 1988). The extent of inhibition at a particular solubilized zonae pellucidae concentration is far less than that expected from inhibition of binding and is very similar to that exerted on the GT enzymatic reaction by soybean trypsin inhibitor. Because soybean trypsin inhibitor appears to interact with the sperm at the TI site, its inhibition of GT enzymatic activity is most reasonably attributed to mild steric hindrance. The TI and GT sites appear to be in close enough proximity that such hindrance can occur at one site by large molecules bound to the other site; at the level of small-molecule inhibitors the sites are mutually independent (Benau and Storey, 1988). Solubilized zonae pellucidae glycoproteins, the active component being ZP3 (Bleil and Wassarman, 1980a,b), also bind to the TI site but evidently interact with the GT site only by mild steric hindrance. The implication is that, under conditions allowing expression of GT enzymatic activity, the zona pellucida ligand normally interacting at the GT site in the binding process can no longer do so effectively. From these observations we conclude that the GT site acts in a "substrate-binding" mode for sperm binding; induction of turnover in the enzymatic mode is antagonistic to binding (Benau and Storey, 1988; Benau *et al.*, 1990). Further details concerning this aspect of the GT site are discussed below in connection with the zona pellucida ligands involved in sperm binding.

By contrast with the TI and GT sites, experimental identification of the FT and Ma sites is very recent, and so detailed reports are not yet available. The FT site (Apter *et al.*, 1988) appears to be very similar to the GT site in that GDP-fucose, the donor in the enzymatic reaction, inhibits sperm binding and that the expression of enzymatic activity requires Mn^{2+}. Asialiofetuin, the glycoprotein acceptor used in the enzymatic assay with GDP-fucose as donor, also inhibits sperm binding (Apter *et al.*, 1988). A similar result was obtained by Benau *et al.* (1990) using asialo/ agalacto-α_1-acid glycoprotein, which is a potent inhibitor of sperm binding to zonae pellucidae and an acceptor in the GT enzymatic reaction.

The Ma site was characterized originally in rat sperm by Tulsiani *et al.* (1989) and shown to be present in mouse sperm by Cornwall *et al.* (1991). Enzymatic activity, as expressed with a mannose oligomer substrate, was inhibited by various sugars in parallel with inhibition of sperm binding to zonae pellucidae. Conditions for expression of enzymatic activity and for sperm binding to the zona pellucida are apparently the same in this case.

One point regarding these four sperm sites deserves particular emphasis. The inhibition of each one of the sites individually by an inhibitor specific to that site in the assay for sperm binding to zonae pellucidae results in inhibition of binding. For instance, if full inhibition of either GT or TI enzymatic activity is observed at a given concentration of UDP-aldehyde or NPGB, respectively, then that concentration of UDP-aldehyde alone, or NPGB alone, will result in full inhibition of sperm binding (Benau and Storey, 1987, 1988). This result is not the one expected if the sites were functioning independently. Possible interpretations first require discussion of the zona pellucida ligands putatively involved in sperm binding.

3.3.3. ZONA-PELLUCIDA-ASSOCIATED LIGANDS MEDIATING SPERM BINDING

The zona pellucida of the mouse egg is composed of three sulfated glycoproteins designated ZP1, ZP2, and ZP3 and is truly an egg-associated product since it is synthesized and secreted throughout the period of oocyte growth (Bleil and Wassarman, 1980a,b; Shimizu *et al.*, 1983). These individual glycoproteins have specific functions. ZP1 (M_r 200,000) exists as a dimer connected by intermolecular disulfide bonds and appears to function to maintain the three-dimensional structure of the zona pellucida by cross-linking filaments composed of repeating ZP2/ZP3 heterodimers. ZP2 (M_r 120,000 under nonreducing and reducing conditions) may mediate the binding of acrosome-reacted sperm to the zona pellucida (Bleil *et al.*, 1988; Wassarman, 1988). ZP2 is converted to a form called $ZP2_f$ when the egg is fertilized (Wassarman,

1987, 1988). This modification is brought about by the action of a protease that is most likely secreted from the egg as a consequence of cortical granule exocytosis (Moller and Wassarman, 1989). $ZP2_f$ has a $M_r = 120,000$ under nonreducing conditions, which shifts under reducing conditions to $M_r = 90,000$, suggesting that the proteolysis of the ZP2 molecule results in the generation of fragments that are held together by disulfide bonds. The biological consequence of the conversion of ZP2 to $ZP2_f$ is that $ZP2_f$ no longer binds to acrosome-reacted sperm (Bleil et al., 1988). ZP3 (M_r 83,000) accounts for both the sperm-binding and the acrosome-reaction-inducing activities of the zona pellucida of unfertilized eggs (Bleil and Wassarman, 1980a,b, 1983, 1986). The amino acid sequence of ZP3 has been established through the nucleotide sequence coding for it (Ringuette et al., 1988; Kinloch et al., 1988). The polypeptide backbone accounts for only 44,000 daltons of the 83,000-dalton molecular mass of the intact glycoprotein; N- and O-linked oligosaccharides account for the rest. Fertilization is associated with a loss of both the sperm-binding and acrosome-reaction-inducing activities of ZP3 (Wassarman, 1987, 1988). The loss of these two important biological activities is associated with a minor biochemical modification of the ZP3 molecule, since the electrophoretic mobility of ZP3 from fertilized eggs is similar to that of ZP3 from unfertilized eggs.

ZP3 possesses biological properties that make it ideally suited as a ligand to mediate the initial steps of sperm binding and the subsequent steps of sperm activation that culminate in acrosomal exocytosis. First, ZP3 is synthesized only by the growing oocyte (Bleil and Wassarman, 1980b), consistent with the fact that its site and mode of action are restricted to events associated with fertilization and preimplantation development. Second, there is little apparent amino acid sequence homology between ZP3 and any other known proteins or glycoproteins thus far examined (Ringuette et al., 1988), suggesting that it serves a unique function. Third, ZP3 is an immobilized ligand that acts at a localized and restricted distance because of its presence in the zona pellucida as a ZP2/ZP3 heterodimer cross-linked in an orderly fashion by ZP1 (Greve and Wassarman, 1985; Wassarman, 1988). Fourth, ZP3 is biologically potent since both the sperm-binding and acrosome-reaction-inducing activities of this molecule are observed in the nanomolar range (Florman and Wassarman, 1985; Bleil and Wassarman, 1986). Fifth, acrosome-intact mouse sperm appear to possess complementary binding sites for ZP3 that are localized to the acrosomal cap region and are present in numbers (10,000–50,000 binding sites/cell) similar to those observed for receptor numbers present in many hormone-responsive cells (Bleil and Wassarman, 1986). Sixth, acrosome-intact mouse sperm bind to ZP3 immobilized onto silica beads via the sperm head (Vazquez et al., 1989), further supporting the idea that the sperm-associated ZP3 binding sites are associated with the plasma membrane overlying the acrosomal region, that region of initial contact where sperm–pellucida interaction first occurs.

The nature of the zona pellucida ligands interacting with the sites on the sperm plasma membrane in the process of mechanical binding to the zona pellucida is not known at present. The experimental results presented below enable one to make some educated guesses, which we present more as gentle provocation than as hard information.

A key role for ZP3 oligosaccharides in the binding of sperm to the zona pellucida could be demonstrated by cleavage of the glycoprotein into small glycopeptides that retain full binding activity (Florman et al., 1984; Florman and Wassarman, 1985). Binding activity of the glycopeptides was assessed by measuring the inhibitory effects of the glycopeptides on the ability of sperm to bind to cumulus-free, zona-pellucida-intact eggs. Florman and Wassarman (1985) isolated an O-linked oligosaccharide fraction from ZP3 with a molecular mass in the range 3400–4500 daltons that retained sperm-binding activity. In addition, they showed that the N-linked oligosaccharides of ZP3 did not possess sperm-binding activity. Involvement of a terminal α-1-linked galactose residue in the binding process was subsequently inferred by Bleil and Wassarman (1988) from experiments with galactose oxidase. Treatment of either ZP3 or the isolated O-linked oligosaccharide possessing sperm-binding activity with this enzyme abolished binding activity. Restoration of activity was obtained on subsequent treatment with sodium borohydride to reduce the C-6 aldehyde produced by the galactose oxidase back to the hydroxyl group.

Bleil and Wassarman (1989) have shown further that a protein of M_r 56,000 extracted from mouse sperm binds to both ZP3 and galactose affinity columns but not to ZP2 affinity columns. The identity of this protein is not yet known, and there is as yet no evidence either for or against the concept that is might be the GT site protein isolated by Shur and Neely (1988), which has a molecular mass of 60,000 daltons. Involvement of a galactose residue, particularly a terminal galactose with the α-1-linkage to the penultimate group of the oligosaccharide, is fully consistent with the observed inhibition of binding by UDP-galactose, in which the sugar has the same linkage to the nucleotide. It would also be consistent with the GT site operating in the substrate-binding mode, as postulated earlier. This mode of operation would also predict that terminal N-acetylglucosamine residues on one or more of zone pellucida oligosaccharides should function in binding. The potent inhibitory effect of glycoproteins containing such residues, in particular asialoagalacto-α-1-acid glycoprotein, on zona pellucida binding (Benau et al., 1990) provides indirect evidence for this prediction. No terminal N-acetylglucosamine on ZP3 has yet been reported, however. Since asialoagalacto-α-1-acid glycoprotein is also a good acceptor in the GT enzymatic reaction, we present here the educated guess that both α-1-linked terminal galactose and β-1-linked terminal N-acetylglucosamine (the linkage in the treated acid glycoprotein) provide the ligands required for interaction at the GT site. In accord with the results of Bleil and Wassarman (1988) favoring the α-1-linkage for the galactose, asialo-α-1-acid glycoprotein with β-1-linked terminal galactose residues does not inhibit sperm binding to zonae pellucidae (Benau et al., 1990). The steric configuration of the terminal saccharide must, therefore, be important. In this regard, it is worthy of note that the individual monosaccharides galactose, N-acetyl-glucosamine, and fucose have little or no inhibitory effect on sperm binding, even at concentrations in the 10–50 mM range (Lambert and Le, 1984; Benau and Storey, 1988; Cornwall et al., 1991).

Bleil and Wassarman (1988) also reported that treatment of ZP3 with fucosidase diminished binding activity, although release of fucose could not be detected. This is at least consistent with functioning of the FT site. The question of whether mannose residues are found on ZP3 and, if so, how they interact with the Ma site is at present totally obscure. Lambert and Le (1984) found that a horseshoe crab lectin specific for sialic acid agglutinated mouse sperm and that treatment of these sperm with neuraminidase inhibited their ability to bind to zonae pellucidae. Sialic acid residues on ZP3 have not at present been implicated in binding, however, nor has a binding site specific for this residue and assayable by an enzymatic activity been reported. The zona pellucida ligand interacting with the TI site poses the greatest puzzle of all. We present another educated guess that the ligand is not an oligosaccharide but a segment of the polypeptide chain of ZP3. The C-terminal end of the chain has a strongly hydrophobic region at the end of which is found arginine at position 410 followed by lysine at position 411; a relatively hydrophilic set of 13 amino acids then complete the C terminus (Kinloch et al., 1988; Ringuette et al., 1988). The properties of the polypeptide chain around the arginine at position 410 might well be mimicked by arylesters of guanidinobenzoic acid such as MUGB and NPGB. They could also be mimicked by the arginine-containing tetrapeptides shown by Bunch and Saling (1988) to inhibit sperm binding to the zona pellucida. This postulate, although plausible, at present requires experimental support.

A further point concerning the zona pellucida ligands mediating mechanical binding of sperm cells to the zona pellucida is that of cooperativity. The experimental observation that a specific inhibitor of any one zona pellucida ligand–sperm site interaction effectively blocks sperm binding indicates that some form of cooperativity may be operative. A possible mechanism for such cooperativity would be the induction of rapid binding of zona-pellucida-associated ligands by initial binding of one ligand. No binding studies of solubilized zona pellucida proteins or purified ZP3 leading to Scatchard or Hill plots have yet been reported, so that experimental differentiation between various forms of independent and cooperative ligand binding does not exist. The question of ordered, as compared to random, ligand interaction with the sperm sites also remains unresolved. Calculations made by Baltz and Cardullo (1989) based on the properties of the mouse sperm GT and TI sites indicate that relatively few binding reactions are needed to anchor the sperm to the zona pellucida. The total number of bonds is of the order of 100 in one set of calculations and between one and ten in another set. If one takes 100 as the total number of interactive bonds required and four as the number of different kinds of sites acting as a single unit, then 25 such units would be needed. This relatively small number would make possible the anchoring of a patch of sperm plasma membrane to a group of adjacent ZP3 molecules, which are arranged like beads on a string in an alternating arrangement with ZP2 on the cross-linked filaments comprising the intact mouse zona pellucida (Greve and Wassarman, 1985). The potential role of multivalent interactions in mediating sperm–zona pellucida (or ZP3) interactions is addressed in greater detail in Section 3.4.3b.

3.4. The Acrosome Reaction Induced by the Zona Pellucida *in Vitro*

3.4.1. EARLY STUDIES: DEFINING THE COMPONENTS OF THE ACROSOME REACTION

The concept of zona-pellucida-induced acrosome reaction is quite recent. In the mid-1970s, there was little support for the idea that mouse zona pellucida played any role at all in the mouse sperm acrosome reaction (Yanagimachi, 1981). As pointed out in Section 2.3 above, the scenario generally agreed on at that time for the mammalian sperm acrosome reaction was that it occurred near or at the cumulus oophorus, thus enabling the sperm to utilize its acrosomal hyaluronidase to penetrate the cumulus matrix (McRorie and Williams, 1974). According to this scheme, acrosomal exocytosis would be completed by the time the sperm reached the zona pellucida; binding to and penetration of the zona pellucida would be mediated by the sperm inner acrosomal membrane. The *in vivo* studies of Zamboni (1970, 1972) strongly indicated that the physiologically relevant acrosome reaction in mouse sperm occurred at the zona pellucida, a position defended by Zamboni (1971) with some vigor but later ignored. Yet, the interaction of a cell with an extracellular matrix by means of an internal membrane would be highly unusual; one would expect such an interaction to occur at the surface of the plasma membrane. The *in vitro* studies of Saling *et al.* (1979) showed, by electron microscopy, that those sperm bound to the outside of the zona pellucida were all acrosome intact, whereas those that had penetrated the zona pellucida and reached the perivitelline space were all acrosome reacted. This result demonstrated that, for the mouse gamete system at least, the acrosome reaction could occur in sperm bound to the zona pellucida and that its occurrence was a prerequisite to penetration of the zona pellucida. If an intact sperm plasma membrane was a prerequisite not only for binding but also for the physiological acrosome reaction mediating sperm penetration of the zona pellucida, one would then predict that acrosome-reacted sperm should not bind to the zona pellucida. This prediction was borne out by the observation of Saling and Storey (1979) that zona-pellucida-intact eggs selectively bound only acrosome-intact sperm from a suspension containing 50–80% acrosome-reacted sperm.

The need for an intact plasma membrane in the sequence of zona pellucida binding and penetration by the sperm implied the possible active participation of the zona pellucida in the acrosome reaction. This participation was first established by the finding that QNB acted as a potent and rather specific inhibitor of the acrosome reaction in sperm bond to zonae pellucidae (Florman and Storey, 1981, 1982a,b; see Section 3.2.4. above). When QNB was added at increasingly later times after the sperm had bound to zonae pellucidae, its inhibitory potency was lost. Conversely, when QNB was added at the beginning of the binding process, and the eggs with bound sperm then transferred to QNB-free medium, the inhibitory effect of QNB was retained (Florman *et al.*, 1982). These results were interpreted as showing that the sperm plasma membrane contained a receptor site for a ligand provided by the zona pellucida and that a ligand–receptor initiation of the acrosome reaction occurred. The block to this initiation by both QNB and atropine, both potent antagonists of the muscarinic class of cholinergic receptors, indicated that the receptor might be of the muscarinic type, although significant differences in receptor pharmacology were found with regard to muscarinic antagonists and agonists (Florman and Storey, 1982b). Occlusion of the site by the zona pellucida over time could either exclude QNB from the site if it were added after a delay or prevent it from dissociating and so allow it to retain its inhibitory potency in the absence of QNB in the medium.

The concept that the zona pellucida could act as an agonist to induce the acrosome reaction through a sperm plasma membrane receptor received further credence from the demonstration by Bleil and Wassarman (1983) that the zona pellucida glycoprotein ZP3 alone could act as inducer. The ZP2 glycoprotein, in contrast, could not. The mouse sperm acrosome reaction could now be formulated in terms of an agonist–receptor process, with ZP3 acting as an immobilized ligand (Bleil and Wassarman, 1983) and the sperm QNB site (Florman and Storey, 1982a,b) acting as the receptor. This formulation raised the important questions of how a three-dimensional, insoluble matrix, the form of the zona pellucida presented to the sperm, could provide ZP3 ligands to the sperm receptors. The answer could be found in the earlier determination by Bleil and Wassarman (1980b) that ZP3 was also the zona pellucida glycoprotein that mediated the mechanical binding of the sperm to the zona pellucida. Once the sperm was secured to the filamentous matrix of the zona pellucida by interaction with ZP3, the stage would be set for the integration of an agonist ligand associated with the same ZP3 molecule with the sperm's muscarinic-like receptor to initiate the intracellular events leading to the acrosome reaction.

The formulation of the zona-pellucida-induced acrosome reaction as an agonist–receptor-mediated process immediately implied the existence in mouse sperm of a transmembrane signaling and intracellular signal transduction system similar to that operating through such receptors in somatic cells (Berridge, 1986; Kopf, 1988, 1989; Saling, 1989). Intracellular signal transduction pathways operating in this manner proceed through a series of sequential reactions involving enzymes and second messengers that usually form a complex net of interactions (Berridge, 1986). Thus, determination of the time courses of the different stages of the processes triggered by initial ligand–receptor interactions and of the reaction sequences defining these processes is necessary. Reactions need reactants in order to proceed, and so the identification of the reactants participating in the signal transduction pathways is equally necessary. The experimental determination of the time courses of the stages of the zona-pellucida-induced acrosome reaction is briefly reviewed in the section immediately following. The components that appear to be involved in the signal transduction system are then examined.

3.4.2. TIME COURSE AND STAGES OF THE ZONA-PELLUCIDA-INDUCED ACROSOME REACTION *IN VITRO*

Three different experimental systems utilizing whole zonae pellucidae have been used to study the time course of the zona-pellucida-induced acrosome reaction in mouse sperm *in vitro*. The first utilizes zona-pellucida-intact eggs in which the cumulus oophorus cells have been removed by hyaluronidase (Inoue and Wolf, 1974a,b; Saling *et al.*, 1978). The second system utilizes structurally intact zonae pellucidae removed from the egg by shearing it in a micropipette (Wolf *et al.*, 1976; Saling and Storey, 1979). The shearing tears the zona pellucida such that the egg is mechanically extruded; with care, zona-pellucida-free eggs can also be isolated by this method. The third preparation utilizes solubilized zonae pellucidae prepared either directly from cumulus-free, zonae-pellucidae-intact eggs incubated at pH 4.0 (Inoue and Wolf, 1974b; Florman and Storey, 1982a; Endo *et al.*, 1988) or from isolated zonae pellucidae. The latter approach allows a greater range of solubilization conditions. Solubilization rate is increased as the pH is decreased (Inoue and Wolf, 1974a). Bleil and Wassarman (1980b, 1983) used phosphate buffer at pH 2.5 to obtain rapid solubilization. Alternatively, the zonae pellucidae dissolution rate also increases rapidly with temperature at a given pH (Inoue and Wolf, 1974a), so that solubilization may be effected at pH 7.4 by heating for 5 min at 60°C (Leyton and Saling, 1989a,b). All of the solubilized zonae pellucidae preparations retain acrosome reaction inducing activity.

The first set of time course determinations of this exocytotic reaction utilized all three systems for comparison, with the solubilized zonae pellucidae being obtained by direct dissolution from eggs at pH 4.0 (Florman and Storey, 1982a). A striking feature of the time course was that no acrosome reactions occurred in sperm bound to structurally intact zonae pellucidae, beyond a slow spontaneous rate, until a lag time of about 25 min had passed. No difference was observed in this regard between zonae pellucidae on the eggs and isolated, structurally intact zonae pellucidae. In contrast, no lag time was observed when solubilized zonae pellucidae were added to a suspension of sperm. In all three systems, the zona-pellucida-induced acrosome reaction could be expressed as an exponential function of time once the reaction had begun in a population of cells. The spontaneous acrosome reaction was, in contrast, far slower and linear with time. The zona-pellucida-induced acrosome reaction had essentially the same half-time of about 40 min after the lag period in sperm bound to structurally intact zonae pellucidae, either isolated or still eggs. This half-time was decreased to 16 min in experiments utilizing solubilized zonae pellucidae.

The actual reaction being followed in the experiments described above was the loss of CTC pattern (Fig. 3), characteristic of capacitated sperm with intact plasma membranes. In those experiments, 10 μM CTC was used, a concentration low enough to avoid perturbation of the system. *In vitro* fertilization was found to proceed normally at this dye concentration (Saling and Storey, 1979). Use of CTC at 250 μM in a fluorescent assay in which the sperm were fixed with 0.1% glutaraldehyde (Ward and Storey, 1984) revealed an intermediate pattern, designated S (Fig. 3), whose appearance in time during the zona-pellucida-induced acrosome reaction followed pattern B but preceded pattern AR (Fig. 3), diagnostic of the completed acrosome reaction (Lee and Storey, 1985). This observation allowed the zona-pellucida-induced acrosome reaction to be resolved into two sequential stages, based on this progression of CTC fluorescence patterns. The first stage is marked by a transition of B fluorescence pattern to the S pattern. For convenience, this has been termed the B-to-S transition (Endo *et al.*, 1987b; Lee *et al.*, 1987). The second stage is marked by a transition from the S fluorescence pattern to the AR pattern, termed the S-to-AR transition. The time courses of the two sequential transitions can be modified, further supporting the concept of two stages following in sequence. Treatment of sperm bound to structurally intact zonae pellucidae with 65 nM tetradecanoyl phorbol acetate (TPA) accelerates the B-to-S transition but slows the S-to-AR transition (Lee *et al.*, 1987). For TPA to exert this effect, the sperm must be bound to zonae pellucidae; sperm in suspension in the absence of zonae pellucidae are unaffected by TPA concentrations up to 10 μM. Pertussis toxin and QNB (described earlier) are two agents that effectively block the B-to-S transition. The B-to-S transition may proceed in the absence of the S-to-AR transition if the eggs from which the zonae pellucidae are obtained are first treated with TPA (Endo *et al.*, 1987a,c; Schultz, *et al.*, 1988). This treatment causes an egg-induced modification of ZP3 such that it binds sperm but can no longer induce a complete acrosome reaction. The mechanistic implications of this observation are discussed in Section 3.4.3b following. For the purpose of this discussion, it suffices to point out that this particular protocol provides an experimental means of isolating the first stage of the zona-pellucida-induced acrosome reaction for further characterization.

The use of CTC fluorescent patterns to distinguish the two sequential stages of the zona-pellucida-induced acrosome reaction provides the desired time course data but offers little insight into the changes being monitored empirically by this fluorescent dye. The patterns arise from the property of the CTC–Ca^{2+} complex; enhancement of its fluorescence occurs if it is bound to membranes (Caswell and Hutchinson, 1971). Changes in CTC fluorescence patterns indicate that sites for binding the CTC–Ca^{2+} complex on the sperm plasma membrane are changing. The nature of these changes is as yet unknown. In order to correlate the CTC fluorescence patterns with physiological events, two fluorescent probes for H^+ and one for Ca^{2+} were examined in this system.

The H^+ fluorescent probes were both 9-amino acridine derivatives: 9-amino-3-chloro-7-methoxyacridine (ACMA) and N-(n-dodecyl)-9-amino acridine (NDAA) (Lee and Storey, 1985, 1989). 9-Amino acridines are weak bases that accumulate in intracellular compartments that are acidic relative to the suspending medium, with resulting quenching or enhancement of fluorescence, depending on probe structure and polyanion content of the compartment (Kraayenhof *et al.*, 1976). In the case of the mouse sperm head, ACMA or NDAA fluorescence enhancement is observed in the following patterns. With ACMA, bright fluorescence is seen over the entire sperm head, but with NDAA bright fluorescence is seen over only the anterior portion of the sperm head corresponding to the acrosomal compartment (Lee and Storey, 1985, 1989). This fluorescence enhancement is similar to what has been previously observed with hamster sperm acrosomes (Meizel and Deamer, 1978) and chromaffin granules (Johnson *et al.*, 1980). The loss of fluorescence of ACMA and NDAA in sperm bound to zonae pellucidae occurs with a time course identical for both dyes and is similar to that observed for the loss of CTC pattern B in the B-to-S transition. The time course is not quite the same, however; the loss of pattern B occurs sooner than loss of ACMA and NDAA fluorescence. The two processes are linked, since TPA also accelerates the loss of ACMA fluorescence as well as the B-to-S transition, and pertussis toxin inhibits both the B-to-S transition and the loss of ACMA fluorescence (Endo *et al.*, 1987b). Loss of ACMA and NDAA fluorescence under these conditions implies the loss of membrane permeability barriers, which would allow the equilibration of H^+ between the sperm cell interior compartments and the suspending medium. In the case of NDAA, the compartment appears to be the acrosome.

The question of whether equilibration of H^+ between compartments is peculiar to H^+ or represents a generalized loss of membrane permeability barriers toward cations was examined with a fluorescent probe for Ca^{2+}, fura-2 (Grynkiewicz *et al.*, 1985). Under appropriate experimental conditions, this dye may be loaded as a membrane-permeant ester into the acrosome of the mouse sperm, where it is hydrolyzed to the membrane-impermeant indicator (Lee and Storey, 1989). When excited with light of appropriate wavelength for the Ca^{2+}-free dye, fluorescence is seen only in the acrosomal compartment. This fluorescence is lost in sperm bound to zonae pellucidae with a time course identical to that observed for loss of NDAA fluorescence (Lee and Storey, 1989). This observation strongly supports the concept that the loss of fluorescence results from a generalized loss of membrane permeability barriers. Furthermore, it appears that the barrier loss marks the end of the B-to-S transition and the start of the S-to-AR transition. Support for this latter concept comes from the observation that the B-to-S transition seen in sperm bound to zonae pellucidae from TPA-treated eggs does not result in loss of NDAA fluorescence (Kligman *et al.*, 1991). Under these conditions, the B-to-S transition is completed with regard to the CTC fluorescence pattern but not with regard to the associated loss of the membrane permeability barriers.

Resolution of the zona-pellucida-induced acrosome reaction in mouse sperm by means of these time course studies allows one to formulate a scheme for five successive stages of the reaction, as shown in Fig. 2. This scheme is essentially identical to that proposed by Zimmerberg (1987) for the five successive steps of mast cell granule exocytosis. In our view, the acrosome reaction is an exocytotic process very similar to that observed in mast cell degranulation. The transition from the poised condition to the punctate condition in Fig. 2 corresponds to the transition between granule–membrane adhesion and pore formation in mast cell degranulation

(Zimmerberg, 1987). This transition in the sperm cell results in the loss of permeability barriers and equilibration of H+ and Ca²⁺ between the acrosome and the suspending medium. This equilibration is the irreversible step in the acrosome reaction. Once it occurs, membrane vesiculation and loss of the fused plasma membrane outer acrosomal membrane vesicles inevitably occurs. This process corresponds to completion of the S-to-AR transition. The scheme in Fig. 2 provides an overall framework for consideration of the zona-pellucida-induced acrosome reaction as a receptor-mediated stimulus–secretion coupling process. The components that react and interact within that framework are considered in the following section.

3.4.3. COMPONENTS OF THE ACROSOME REACTION: RECEPTORS, G PROTEINS, AND SECOND MESSENGERS

3.4.3a. Sperm-Associated Receptors for Zona Pellucida Glycoproteins. Although sea urchin sperm remain the sole experimental system in which receptor-mediated signal transduction has been unequivocally demonstrated (Garbers, 1989), studies in the mouse are beginning to yield information regarding the nature of sperm-associated receptors for components of the zona pellucida. Two different approaches have been utilized to describe putative sperm-associated receptors for the zona pellucida.

The first and more direct approach involves the elucidation of components of the sperm surface that interact directly with purified zona pellucida glycoproteins. Bleil and Wassarman (1986) have used this approach to examine the specific binding of [¹²⁵I]ZP3 to mouse sperm by whole-mount autoradiography. They demonstrated that binding is associated solely with the plasma membrane overlying the acrosomal region of the sperm and is not observed on acrosome-reacted sperm (Bleil and Wassarman, 1986). Binding of this radiolabeled ligand is competed by unlabeled ZP3 but not by ZP2. ZP3 binding does not occur on somatic cells, thus demonstrating target cell specificity. Similar techniques have been used to localize binding sites for ZP2 to the inner acrosomal membranes of acrosome-reacted, but not acrosome-intact, mouse sperm (Bleil *et al.*, 1988). Although the methodology utilized in these studies has limitations with regard to the characterization of the sperm-associated entities involved in ZP3 and ZP2 binding, these aforementioned observations suggest that specific sperm-associated binding sites for these ligands exist in discrete domains on the plasma membrane of this highly differentiated cell. Recently, these investigators demonstrated that either purified ZP3 or glycopeptides of ZP3 possessing sperm-binding activity can be specifically cross-linked to a M_r 56,000 protein of acrosome-intact mouse sperm (Bleil and Wassarman, 1989). This protein interacts specifically with ZP3, but not ZP2, affinity columns. Whole-mount autoradiography utilizing radiolabeled, cross-linked ZP3 glycopeptides that possess sperm-binding activity demonstrates a localization to the head region of acrosome-intact sperm. This experimental approach may provide a great deal of promise with regard to establishing the molecular identity of such sperm-associated receptors for zonae pellucidae glycoproteins.

Recently, Leyton and Saling (1989a) demonstrated that antiphosphotyrosine antibodies react with mouse sperm plasma membrane proteins of M_r 52,000, 75,000, and 95,000. Indirect immunofluorescence using this antibody demonstrated that positive immunoreactivity is localized to the acrosomal region of the sperm head (Bunch *et al.*, 1990). [¹²⁵I]ZP3 binds, on nitrocellulose blots, to a M_r 95,000 protein, and these workers have inferred that the M_r 95,000 protein is the same protein that reacts with the antiphosphotyrosine antibodies. Although it is suggested that this M_r 95,000 protein may be a receptor for ZP3 that either acts as a substrate for tyrosine phosphorylation or possesses tyrosine kinase activity itself, additional studies will be required to demonstrate clearly the specificity of the ZP3-binding activity of this protein. These results are provocative in light of the fact that somatic cell receptors for a variety of ligands possess intrinsic protein kinase activity (Yarden and Ullrich, 1988).

The second, more indirect approach involves the delineation of sperm-associated binding sites for the zona pellucida glycoproteins by examining the ability of specific agents to interfere with the sperm–zona pellucida interaction. Sperm-surface-associated protease-inhibitor-sensitive sites (Saling, 1981; Aarons *et al.*, 1984; Benau and Storey, 1987), galactosyltransferase activity (Shur and Hall, 1982a,b; Lopez *et al.*, 1985; McLaughlin and Shur, 1987; Shur and Neely, 1988), fucosyltransferase activity (Apter *et al.*, 1988), and mannosidase (Cornwall *et al.*, 1989) have all been implicated in the binding of sperm to the zona pellucida, presumably through ZP3. The involvement of these sites and enzymes in sperm-zona pellucida binding has been reviewed in detail in Section 3.3.2 of this chapter and so is not discussed at this juncture. The relationship of these sites to one another, and to those sites defined by the ZP3-binding experiments described above, is not clear at this time.

The role of a sperm-associated receptor of the cholinergic type in mediating acrosomal exocytosis induced by the zona pellucida has also been implied by previous studies using QNB (see Sections 3.2.4 and 3.4.1). Since QNB inhibits the acrosome reaction by blocking its first stage, the B-to-S transition, a reasonable postulate is that QNB binds as an antagonist to that site. Atropine and scopolamine, both muscarinic receptor antagonists like QNB, also inhibit the zona-pellucida-induced acrosome reaction (Florman and Storey, 1982b), so that one could stretch this postulate to include the idea that the receptor resembles the muscarinic class of cholinergic receptors. The pharmacology of the mouse sperm QNB-binding site indicates that the resemblance is only partial, since propylbenzilyl and benzilyl choline mustards do not bind to mouse sperm, nor do they displace [³H]QNB from its site on sperm (Florman and Storey, 1982b). These two compounds, as well as QNB, are potent antagonists of all classes of muscarinic receptors (Young *et al.*, 1972; Birdsall and Hulme, 1976). The mustard compounds form covalent bonds with the receptor sites and so are useful in monitoring the receptors during isolation. The lack of reactivity of the sperm site with the mustard antagonists has been a serious impediment to the isolation and characterization of this site. Experiments aimed at its isolation have shown that there is but one QNB site in mouse sperm, which yields a linear Scatchard plot with QNB whether intact cells, cell homogenates, plasma membrane fractions, or soluble digitonin extracts are assayed (C. R. Ward and B. T. Storey, unpublished data). The Scatchard plots yield a $K_d = 5$ nM with 6000 sites per cell (Florman and Story, 1982b; C. R. Ward and B. T. Storey, unpublished data). This site has not yet been further characterized.

3.4.3b. Nature of the Interaction between the Zona Pellucida and the Sperm Surface to Induce Acrosomal Exocytosis. As noted in the previous sections and paragraphs of this chapter, little is known about the molecular nature of the ligands associated with ZP2 and ZP3 as well as their complementary binding sites on the sperm surface, which mediate sperm binding (of both acrosome-intact and acrosome-reacted sperm) and acrosomal exocytosis. A discussion of the nature of the interactions between the sperm surface and the zona pellucida to mediate these important biological events, therefore, might seem inappropriate at this time. However, data have emerged recently from a few laboratories that may delineate the type of interactions that occur between the mouse sperm surface and ZP3 to mediate these important prefertilization sperm–egg interactions. These studies have evolved from previous observations that ZP3 has dual functions in initial sperm–egg interaction; e.g., it is responsible for both the binding of acrosome-intact sperm and for initiating acrosomal exocytosis.

Previous investigations have demonstrated that the sperm-binding and acrosome-reaction-inducing activities of ZP3 can be either partially or completely dissociated from one another by agents that act either at the level of the sperm or at the level of ZP3. In cases where acrosomal exocytosis is blocked, the inhibition by these agents appears to occur at a specific stage of the zona pellucida (or ZP3)-induced acrosome reaction. In these particular experiments the acrosomal status of mouse sperm was monitored using the fluorescent probe CTC (Saling and Storey,

1979; Ward and Storey, 1984; Lee and Storey, 1985). The mechanism by which this fluorescent probe is postulated to bind to mouse sperm, as well as the characteristic dye-binding patterns (B, S, and AR patterns) and their correlation with capacitation and acrosomal status, have been outlined in detail in Sections 3.2.4. and 3.4.2 of this chapter. The zona-pellucida-induced acrosome reaction in mouse sperm, as monitored by this fluorescence assay, consists of B-to-S and S-to-AR transitions.

QNB represents the first agent that has been demonstrated to dissociate the two biological activities of the ZP3 molecule, and this dissociation occurs at the level of the sperm. This classical muscarinic-cholinergic antagonist binds to a single class of high-affinity sites on mouse sperm and inhibits the zona-pellucida-induced acrosome reaction without affecting the ability of the sperm to bind to the zona pellucida (Florman and Storey, 1981, 1982a,b). Sperm incubated with QNB are inhibited from undergoing a B-to-S transition in response to either solubilized zonae pellucidae (Lee and Storey, 1985) or purified ZP3 (Endo et al., 1988). A second reagent that has been demonstrated to inhibit the zona pellucida (and ZP3)-induced acrosome reaction without affecting sperm binding to this extracellular matrix is M-42, a monoclonal antibody directed against a M_r 200,^{000}/$_{220}$,000 mouse sperm protein (Saling and Lakoski, 1985; Saling, 1986; Leyton et al., 1989). The inhibitory effect of this antibody is at the B-to-S transition, does not block the more distal events leading to the S-to-AR transition, and is effective using both solubilized zonae pellucidae and purified ZP3. Pertussis toxin (PT) represents a third agent that, when added to capacitated mouse sperm, inhibits specifically the B-to-S transition induced by structurally intact zonae pellucidae, solubilized zonae pellucidae, or purified ZP3 without affecting sperm binding (Endo et al., 1987b, 1988; see also Section 3.4.2). This bacterial toxin functionally inactivates a number of guanine nucleotide-binding regulatory proteins (G proteins) by its ability to catalyze the ADP-ribosylation of the α subunit of these proteins (Ross, 1989; Freissmuth et al., 1989) and has been shown to inactivate the G_i-like protein present in mouse sperm (Kopf et al., 1986; Endo et al., 1987b; see Section 3.4.3c). All three of these agents, when added to sperm, inhibit a specific stage(s) of the acrosome reaction induced solely by the zona pellucida (or ZP3) without affecting reactions involved in sperm–zona pellucida binding mediated by ZP3.

In addition, incubation of mouse sperm with either biologically active phorbol diesters or diacylglycerols, which are activators of the Ca^{2+}- and phospholipid-dependent protein kinase (protein kinase C; PK-C), has dramatic effects on the kinetics of the different stages comprising the zona-pellucida-induced acrosome reaction (Lee et al., 1987) but do not affect sperm binding to the zona pellucida (Endo et al., 1987a) (see Section 3.4.2). Both agents, when incubated with sperm in the presence of zonae pellucidae, accelerate the B-to-S transition but inhibit the S-to-AR transition, suggesting that these two transitions can be independently regulated (Lee et al., 1987).

Biologically active phorbol diesters and diacylglycerols also bring about egg-induced dissociations of the two biological activities of ZP3, and these observations have likewise provided insight to the nature of sperm–ZP3 interaction. These agents, when added to zona-pellucida-intact mouse eggs, cause egg-induced modifications of ZP2 to ZP2$_f$ and a modification of ZP3 such that the sperm-binding activity of ZP3 is completely retained and the acrosome-reaction-inducing component of ZP3 is partially modified (Endo et al., 1987a,c). It is likely that these agents are acting to stimulate cortical granule exocytosis, since preliminary experiments indicate that treatment of eggs with these PK-C activators results in a partial reduction of cortical granules (T. Ducibella, G. S. Kopf, and R. M. Schultz, unpublished observations) and the secretion of a protease that modifies ZP2 to ZP2$_f$ (Y. Endo, R. M., Schultz, and G. S. Kopf, unpublished observations). Sperm incubated with either solubilized zonae pellucidae or purified ZP3 from eggs treated with these PK-C activators bind normally and undergo the B-to-S transition with normal kinetics but are inhibited from completing the acrosome reaction as a consequence of their inability to undergo an S-to-AR transition. Sperm suspended in the intermediate S pattern are not irreversibly stuck in this stage, since solubilized zonae pellucidae from untreated eggs (but not treated or fertilized eggs) can bring about an S-to-AR transition of those S-pattern sperm (Kligman *et al.*, 1991; Leyton *et al.*, 1989).

Taken together, these observations demonstrate that (1) sperm binding to the zona pellucida and the zona-pellucida-induced acrosome reaction are two independent processes and (2) the zona-pellucida-induced acrosome reaction in the mouse may consist of discrete, independently regulated events, as monitored by the CTC assay. The reactions driving these different events can be inhibited by a number of agents in a very specific manner. The concept that this exocytotic process proceeds in discrete steps is consistent with known regulatory events comprising exocytotic reactions in other cells (Zimmerberg, 1987; Gomperts and Tatham, 1988).

Further insight into the nature of the interaction of ZP3 with the sperm surface to modulate both sperm binding and the interaction of a complete acrosome reaction can be obtained from attempts to explain the effects of PK-C activators on the aforementioned egg-induced modifications of ZP3. A model has been presented to explain the results of these PK-C activators on the apparent partial modifications of ZP3 described above (Kopf *et al.*, 1989) and takes into account the properties of sperm binding and the induction of the acrosome reaction observed with varying concentrations of solubilized zonae pellucidae or purified ZP3 (Bleil and Wassarman, 1983). This model is based on the fact that multiple interactions between ZP3 molecules and sperm surface sites may be required *in toto* for sperm binding and the induction of acrosomal exocytosis. The model also predicts that there are differences in the concentrations of ZP3 required for the expression of sperm-binding activity and acrosome-reaction-inducing activity. Specifically, the concentration–response curve for ZP3 acrosome-reaction-inducing activity is shifted to the right of the concentration–response curve for ZP3 sperm-binding activity. Finally, this model predicts that ZP3 is composed of multiple "functional ligands" and that the interaction of these ligands with the sperm surface is responsible for both the sperm-binding activity and the ability to induce a complete acrosome reaction. Successful sperm binding to ZP3 requires, at the minimum, interaction with a single ligand of the ZP3 molecule. Experimental indications are that multiple ligands are involved. The questions remain which ZP3 ligand binds first, and which ligand is rate limiting. Once the interactions to mediate sperm binding have been completed, interactions of the sperm with multiple ligands of the ZP3 molecule may then occur to induce a complete acrosome reaction. This is consistent with the observation that the acrosome reaction may be stalled in an intermediate stage (Endo *et al.*, 1987a,b; Kligman *et al.*, 1991; Leyton *et al.*, 1989).

Fertilization normally results in a modification or inactivation of a majority, or all, of the ligands associated with the ZP3 molecule and hence the loss of both biological activities. In contrast, the partial release of cortical granules that is stimulated by PK-C activators results in a submaximal modification of these ligands. Consequently, ZP3 still possesses sperm-binding activity, but the suboptimal number of active ligands remaining is not sufficient to permit the multiple interactions with the sperm surface necessary to induce a complete acrosome reaction. As a result, sperm initiate (B-to-S transition) but do not complete the acrosome reaction because they remain arrested at the S pattern. This partial inhibition can be relieved when zonae pellucidae (or ZP3) that contain a sufficient number of active ligands (e.g., from unfertilized eggs) are now added to the sperm arrested at the S pattern (Kligman *et al.*, 1988; Leyton *et al.*, 1989). Induction of the complete acrosome reaction, based on this model, involves multiple interactions between sperm-associated ZP3-binding components and the ZP3 molecule.

Recently, similar conclusions have been reached using a totally different experimental approach (Leyton and Saling, 1989b). On the basis of studies using monospecific polyclonal antibodies to ZP2 and ZP3, it was hypothesized that the aggregation of sperm-associated receptors for ZP3 may play a role in the ZP3-induced acrosome reaction. ZP3 glycopeptides, which possess sperm-binding activity but do not possess the ability to induce an acrosome reaction when added to sperm alone, bring about an acrosome reaction when incubated with sperm in the presence of anti-ZP3 IgG. This effect is not observed with anti-ZP2 IgG or with monovalent anti-ZP3 Fab fragments. If sperm are incubated with ZP3 glycoproteins and anti-ZP3 Fab fragments in the presence of an anti-IgG, the ability of the sperm to undergo acrosome reactions is restored. Although the addition of anti-ZP3 IgG to capacitated B-pattern sperm does not induce an acrosome reaction, the addition of this antibody to sperm that are suspended in the S-pattern (obtained by incubation of sperm with zonae pellucidae from phorbol-diester-treated eggs) brings about this exocytotic event. These investigators concluded that the aggregation of sperm-associated receptors for ZP3 that were recognized by the ZP3 glycopeptides or by zonae pellucidae from phorbol-diester-treated eggs and then cross-linked by the addition of the anti-ZP3 IgG fraction is required for the induction of the complete acrosome reaction.

The idea that ligand-dependent signal transduction is mediated through receptor aggregation is not unique (Yarden and Schlessinger, 1987). Extracellular matrices from a variety of tissues have been demonstrated to contain domains that aggregate acetylcholine receptors (Godfrey *et al.*, 1988). It is possible, therefore, that the zona pellucida, itself an extracellular matrix, also possesses this property. These issues are discussed in greater detail elsewhere (Kopf and Gerton, 1990).

3.4.3c. The Role of G Proteins in Zona-Pellucida-Mediated Acrosomal Exocytosis.

The guanine-nucleotide-binding regulatory proteins (G proteins), which include G_s, G_i, G_o, G_z, and transducin, occupy critical roles as signal-transducing elements in coupling many ligand–receptor interactions with the generation of intracellular second messengers (Casey and Gilman, 1988; Ross, 1989). These heterotrimeric plasma-membrane-associated proteins are composed of distinct α subunits, which contain a GTP-binding domain, and more highly conserved β and γ subunits. The α subunits can be covalently modified by ADP-ribosylation in the presence NAD^+ by the action of a variety of bacterial toxins, including cholera toxin, pertussis toxin, and botulinum toxin. The ability to undergo toxin-catalyzed ADP-ribosylation has been used by many investigators to identify general classes of G proteins. Activation of G proteins occurs upon GTP binding to the α subunit, causing dissociation of α-GTP from the $\beta\gamma$ dimer. These dissociated subunits then exert their regulatory actions on respective targets, ultimately resulting in ligand coupling to intracellular signaling systems. G proteins regulate the hormonally responsive adenylate cyclase of somatic cells, the cGMP phosphodiesterase of retinal rod outer segments, atrial K^+ channels, Ca^{2+} channels, and have been implicated in regulating phospholipase C, phospholipase A_2, and phospholipase D (Casey and Gilman, 1988; Ross, 1989; Freissmuth *et al.*, 1989).

Until recently, sperm were regarded as "unique" cells in that they apparently lacked G proteins. These conclusions were based largely on the properties of the sperm adenylate cyclase, which in somatic cells is regulated by guanine nucleotides, a stimulatory G protein (G_s), and an inhibitory G protein (G_i) (Casey and Gilman, 1988). The sperm adenylate cyclase is particulate, and the activity of the enzyme does not appear to be affected by fluoride ions, cholera toxin plus NAD^+, or forskolin, agents that are known to stimulate G-protein-regulated adenylate cyclases in other cells (Garbers and Kopf, 1980; Ross and Gilman, 1980). Catalytic activity, for the most part, is poorly supported by Mg^{2+}, a property similar to the adenylate cyclase activity from the S-49 lymphoma cell line, which is devoid of G_s regulated enzymatic activity. These properties are consistent with an enzyme either devoid of, or uncoupled from, G proteins (Garbers and Kopf, 1980; Ross and Gilman, 1980). Hildebrandt *et al.* (1985) obtained confirmatory results with bovine sperm and concluded that mammalian sperm lack both G_s and G_i and that the sperm adenylate cyclase does not interact with G_s from exogenous sources. Although guanine nucleotides have been reported to stimulate sperm adenylate cyclases, the nature of these stimulatory effects remains unclear, since there is no apparent effect of cholera toxin (Cheng and Boettcher, 1979; Casillas *et al.*, 1980; Hyne and Lopata, 1982; Stein *et al.*, 1986). The activity of the sperm enzyme has, however, been demonstrated to be regulated by extracellular effectors of sperm function, including egg-associated components, Ca^{2+}, and calmodulin (Garbers and Kopf, 1980; Kopf, 1988). Sperm adenylate cyclase activity, therefore, can be regulated by biologically relevant factors, but the mode of this regulation is presently unclear.

The question of whether sperm contain G proteins was addressed independently by Kopf *et al.* (1986) and Bentley *et al.* (1986). Extracts from both invertebrate and mammalian sperm were assessed for the presence of the α subunits of G proteins by both pertussis toxin (PT)- and cholera-toxin (CT)-catalyzed ADP ribosylation. Pertussis toxin catalyzes the ADP-ribosylation of the α subunits of G_i, G_o, and transducin and will functionally inactivate these proteins. Both invertebrate (abalone) and mammalian (human, mouse, guinea pig, bovine) sperm contain a single PT substrate of M_r 41,000. Peptide mapping of the mouse and human sperm M_r 41,000 PT substrates by limited proteolytic digestion demonstrate a similarity to α_i of mouse S-49 lymphoma cells (Kopf *et al.*, 1986). These data suggest that the mouse and human sperm M_r 41,000 PT substrates are α_i-like. Subsequent experiments by Jones *et al.* (1989) demonstrated that mouse sperm contain two closely migrating PT substrates that can only be resolved when sodium dodecyl sulfate containing a substantial portion of C_{14} and C_{16} alkyl sulfates is used for polyacrylamide gel electrophoresis. The nature of these multiple forms is not presently known. The presence of G proteins in mouse sperm was further substantiated using an antiserum directed against the β subunit common to the different heterotrimeric G proteins (Kopf *et al.*, 1986). Analysis of one-dimensional polyacrylamide gels of mouse sperm membrane extracts by immunoblotting techniques revealed the presence of a single immunoreactive band of $M_r = 35,000/36,000$, which comigrates with the β subunit of G proteins from somatic cell membranes. These data indicate that mammalian sperm contain G proteins with properties similar to that of somatic cell G_i. Additional studies will be required to determine the nature of the G_i subtype(s) (i.e., $G_{i\alpha1}$, $G_{i\alpha2}$, $G_{i\alpha3}$) present in sperm, since antisera specific to the different forms are now becoming available.

Sperm have also been analyzed for the presence of specific substrates for cholera-toxin-catalyzed ADP-ribosylation, indicative of the presence of G_S (Kopf *et al.*, 1986). No specific substrates are detected in any of the fractions assayed under a variety of conditions. The inability to detect G_S in sperm is consistent with the data of others (Hildebrandt *et al.*, 1985) and is also supported by the work of investigators who have examined the guanine nucleotide regulation of the sperm adenylate cyclases (Garbers and Kopf, 1980). It is possible that G_S is present in sperm but that the current methods used to probe for its existence are inadequate at this time. However, the inability to reconstitute sperm adenylate cyclase activity with purified G_S from an exogenous source implies that this enzyme may have unique regulatory properties and that G_S may not play a regulatory role in sperm function (Hildebrandt *et al.*, 1985).

Experiments utilizing polyclonal antisera generated against conserved peptide sequences common to the α subunits of various G proteins were designed to characterize further the nature of mammalian sperm G proteins and to determine whether they display distinct regionalization in this highly differentiated cell (Glassner *et al.*, 1989). Both mouse and guinea pig sperm display positive immunofluorescence in the acrosomal region using an antiserum directed against specific peptide regions common to all G_α proteins. This fluorescence disappears when sperm undergo the acrosome reaction, either spontaneously or in response to zonae pellucidae glycoproteins, suggesting that the immunoreactive material is associated with the plasma/outer acrosomal membrane overlying the acrosome. Negligible midpiece and tail staining is observed with this antiserum. This antiserum also immunoprecipitates a M_r 41,000 PT-catalyzed [^{32}P]ADP-ribosylated protein(s) from mouse sperm. The association of G proteins with the sperm acrosome was further confirmed in guinea pig sperm by demonstrating that purified plasma/outer acrosomal membrane fractions isolated from cells acrosome-reacted with the ionophore A23187 contain a M_r 41,000 substrate for PT-catalyzed ADP-ribosylation. These data demonstrate that G protein α subunits are present in the acrosomal region of mammalian sperm, consistent with their postulated role in regulating zona-pellucida-mediated acrosomal exocytosis (see below).

The physiological role of the sperm G_i-like protein in the zona-pellucida-induced acrosome reaction of the mouse has also been examined (Endo *et al.*, 1987b, 1988). In these experiments, sperm were treated with PT, and the ability of the sperm to bind to structurally intact zonae pellucidae and undergo the acrosome reaction was then examined. Pertussis toxin has been demonstrated to cross the plasma membrane and functionally inactivate G_i by ADP-ribosylating its α subunit (Casey and Gilman, 1988). Intact mouse sperm were incubated with PT under conditions conducive to capacitation, and it was demonstrated that the toxin could enter the cell and ADP-ribosylate the G_i-like protein. The PT does not affect the ability of the mouse sperm to become capacitated, and these sperm bind to structurally intact zonae pellucidae to the same extent as control sperm incubated in the absence of the toxin. Likewise, sperm motility and viability are not affected at any PT concentration tested. However, the zona-pellucida-induced acrosome reaction, as monitored by the CTC assay, is inhibited in a concentration-dependent manner by PT, with half-maximal effects at 0.1–1.0 ng/ml PT. The effect of PT to inhibit the acrosome reaction results from an effect on the ability of the sperm to progress from the B pattern to the intermediate S pattern; thus, the sperm remain in B. Inactivated PT does not inhibit the zona-pellucida-induced acrosome reaction, and cholera toxin is also without effect. These inhibitory effects of PT on the zona-pellucida-induced acrosome reaction are also confirmed at the electron microscopic level (Endo *et al.*, 1988). These data suggest that functional inactivation of the mouse sperm G_i-like protein by PT prevents the zona-pellucida-induced acrosome reaction and that the inhibition appears, at least, to be at the level of the B-to-S transition.

The specificity of the PT effect on the zona-pellucida-induced sperm acrosome reaction was further defined by examining the effect of GTP and GDP analogues on PT-mediated inhibition of this exocytotic reaction. The β-subunit of the G_i $\alpha\beta\gamma$ heterotrimer appears to be required for the PT-catalyzed ADP-ribosylation of the α_i subunit (Neer *et al.*, 1984). The slowly hydrolyzable analogue of GTP guanosine-5'-O-3-thiotriphosphate (GTPγS) binds to the α_i subunit with high affinity and dissociates the heterotrimer into GTPγS-α_i and $\beta\gamma$ subunits (Casey and Gilman, 1988). It would be expected that if GTPγS penetrated intact sperm and bound to the α subunit of the sperm G_i-like protein, the resultant subunit dissociation would prevent subsequent PT-catalyzed ADP-ribosylation of the α-subunit and, as a result, prevent functional inactivation of the protein. There is precedent for cellular permeability of GTPγS (Minke and Stephenson, 1985). In contrast, incubation of sperm with the nonhydrolyzable GDP analogue guanosine-5'-O-2-thiodiphosphate (GDPβS) would not be expected to inhibit the PT-catalyzed ADP-ribosylation, since this analogue does not cause subunit dissociation. When intact sperm are capacitated in the presence of 100 μM GTPγS prior to the addition of 10 ng/ml PT, sperm binding to the zona pellucida and the subsequent acrosome reaction occur to the same extent as control sperm incubated in the absence of both of these agents; this contrasts sharply with the inhibitory effects on the acrosome reaction observed when sperm are incubated with PT alone. Capacitation of sperm in the presence of 100 μM GDPβS prior to the addition of PT, on the other hand, does not abolish the PT-mediated inhibition of the zona-pellucida-induced acrosome reaction. These data strengthen support for the concept that the PT-sensitive component of mouse sperm is a G_i-like protein that, on activation by GTPγS (but not GDPβS), loses its ability to become functionally inactivated by PT-catalyzed ADP-ribosylation. The only PT substrate found in mouse sperm is the M_r 41,000 α subunit of the G_i-like protein, implying that this protein is the PT-sensitive component involved in mediating the zona-pellucida-induced acrosome reaction.

Similar experiments were carried out using purified ZP3 (Endo *et al.*, 1988). When compared to sperm capacitated in the absence of PT, sperm capacitated in the presence of PT are inhibited from undergoing the acrosome reaction in response to ZP3. As with structurally intact zonae pellucidae, this inhibition occurs at the B-to-S transition. In contrast to the inhibitory effects of PT on the ZP3-induced acrosome reaction, the ability of sperm to undergo either a spontaneous acrosome reaction (i.e., in the absence of solubilized ZP3) or a nonphysiologically induced acrosome reaction (A23187 induced) is insensitive to PT treatment. These data demonstrate that only acrosome reactions induced by the physiologically relevant ZP3 molecule are inhibited by PT. These results are consistent with the idea that the PT-sensitive site (G_i-like protein) in mouse sperm plays an important intermediary role in the acrosome reaction induced specifically by ZP3, the biologically relevant molecule present in the zona pellucida. Similar observations in the mouse using ZP3 have subsequently been reported by other investigators (Vazquez et al., 1989).

If one considers the zona-pellucida-induced acrosome reaction as an example of stimulus–secretion coupling that occurs in a receptor-mediated fashion, one would propose that receptor–G-protein interaction subsequently leads to the generation of intracellular second messengers and/or the modulation of ionic changes within the sperm. Two criteria would have to be met to establish the signal-transducing function of this GTP-binding protein in the mouse sperm acrosome reaction. First, the occupation of a putative sperm receptor for the acrosome-reaction-inducing component of ZP3 should result in G-protein activation in a manner described for other ligand–receptor–G-protein interactions. Second, the receptor–G-protein interaction should then modulate second messenger/ionic changes distal to receptor occupancy (see Section 3.4.3d). There is presently evidence to support both of these criteria.

The first criterion has been addressed by measuring the activity of a high-affinity GTPase activity in mouse sperm homogenates incubated in both the absence and presence of acid-solubilized zonae pellucidae glycoproteins in order to determine whether this extracellular matrix can activate the sperm-associated G_i-like protein (Wilde and Kopf, 1989). This particular experimental approach is based on the fact that, in other systems, receptor activation of G proteins is accompanied by the dissociation of the α subunit from the $\beta\gamma$ subunits of the heterotrimeric G protein, which results in the expression of a latent high-affinity GTPase activity associated with the α subunit. An increase in GTP hydrolysis of approximately 50% over basal activity is observed when sperm homogenates are incubated in the presence of acid-solubilized zonae pellucidae. The zona-pellucida-induced activation of this enzyme occurs as a consequence of an increase in the V_{max} of the enzyme, with little effect on the K_{Mapp} of the enzyme for GTP, indicating that increased GTP turnover occurs in response to this extracellular matrix. Accompanying this increase in enzyme activity is a reduction in the ability of PT to catalyze in vitro [^{32}P]ADP-ribosylation of the M_r 41,000 G_i-like protein, suggesting that the increase in GTPase activity is associated with the activation of a PT-sensitive sperm G protein(s), since an intact $\alpha\beta\gamma$ heterotrimer is required for PT-catalyzed ADP-ribosylation (Neer et al., 1984). Concentrations of 0.2 zonae pellucidae/μl are required for half-maximal activation of the enzyme, which is similar to the concentration required for sperm–zona pellucida binding but is an order of magnitude below the concentration required for zona-pellucida-induced acrosomal exocytosis. These data support the reaction scheme in which a ZP3 ligand binds to a sperm surface receptor and activates a G protein, which is in turn coupled to intracellular second messenger cascades required for induction of acrosomal exocytosis. The nature of these second messenger systems is not known at present. However, it is presumed that one or more of these systems are involved in the steps resulting in apposition of the outer acrosomal and plasma membranes, followed by pore formation leading to loss of permeability to Ca^{2+} and H^+ (see Section 3.4.2 and Fig. 2).

3.4.3d. Intracellular Effector Systems Modulating Zona-Pellucida-Mediated Acrosomal Exocytosis.

Since the acrosome reaction has many elements of a classical receptor-mediated exocytotic event, the nature of the intracellular signals that are generated in sperm in response to the zona pellucida and result in membrane fusion and vesiculation will likely be similar to those signals that have been postulated to regulate exocytotic events in other cells. Such signals include changes in ionic conductance, changes in cyclic nucleotide metabolism, and changes in phospholipid metabolism. Although there have been innumerable reports describing the ions and/or second messenger systems proposed to play a role in the mammalian sperm acrosome reaction (as reviewed by Yanagimachi, 1988; Kopf and Gerton, 1990), there is little known regarding what system(s) may be operating physiologically, since there has been little attention paid to the effects of biologically relevant acrosome-reaction-inducing ligands on sperm signaling. Such studies, therefore, are still in their infancy but are worthy of note at the present time. It should also be noted that although the mouse sperm-associated G_i-like protein appears to play an important intermediary role in the ZP3-induced acrosome reaction, the intracellular signaling systems that this particular signal-transducing protein is coupled to are not presently known.

Using the lipophilic fluorescent probe 9-amino-3-chloro-7-methoxyacridine (ACMA) to monitor transmembrane pH gradients in mouse sperm, Lee and Storey (1985) demonstrated that the zona-pellucida-induced B-to-S transition, as monitored using CTC fluorescence, is accompanied by a parallel abolition of a transmembrane pH gradient as evidenced by the loss of ACMA fluorescence associated with the intact sperm. The loss of this transmembrane pH gradient is not a consequence of major membrane changes such as the loss of the plasma and outer acrosomal membranes. It can be concluded, however, that the zona-pellucida-induced B-to-S transition is associated with a modification of ion permeability in these cells. Since PT specifically inhibits the B-to-S transition, the effects of this toxin on the loss of ACMA fluorescence of sperm bound to structurally intact zonae pellucidae was examined (Endo et al., 1988). It was demonstrated that the PT-induced inhibition of the B-to-S transition is accompanied by a parallel retention of ACMA fluorescence at all of the time points tested. These data demonstrate that the PT-sensitive site is upstream from those events associated with the loss of the transmembrane pH gradient normally accompanying the zona-pellucida-induced B-to-S transition and suggest that the G_i-like protein might modulate such changes in ion permeability either directly or indirectly. Additional studies are needed to confirm whether these changes in ionic permeability, as well as other ionic changes, are coupled to this signal-transducing element. This is important in light of two recent observations. Lee and Storey (1989) have demonstrated that the zona-pellucida-induced acrosome reaction in the mouse is accompanied by changes in Ca^{2+} permeability. In addition, Florman et al. (1989) have recently demonstrated that the pH and Ca^{2+} changes observed in bovine sperm incubated in the presence of solubilized zonae pellucidae can be partially inhibited by PT, suggesting that the G_i-like protein in bovine sperm may regulate a host of ionic changes in response to zonae pellucidae.

Studies from other laboratories have suggested that alterations in phospholipid metabolism and/or cyclic nucleotide metabolism may play important intermediary roles in the sperm acrosome reaction (Kopf and Gerton, 1990). Lee et al. (1987) demonstrated that biologically active phorbol diesters and diacylglycerols can alter the kinetics of the different stages of the zona-pellucida-mediated acrosome reaction in mouse sperm, thus providing evidence for the potential role of protein kinase C in regulating this exocytotic event. However, the products of poly-phosphoinositide turnover (e.g., inositol 1,4,5,-trisphosphate and 1,2,-diacylglycerol) have not yet been examined in sperm challenged with zonae pellucidae. These PK-C activators could completely overcome the inhibitory effects of PT on the B-to-S transition (and thus the inhibitory effect of PT on the zona-pellucida-mediated acrosome reaction), suggesting that the site of action of these PK-C activators might either be downstream from or independent of the PT-sensitive site (i.e., the sperm G_i-like protein) (Endo et al., 1988). The interrelationship between the sperm G_i-like protein and this intracellular signaling system will, therefore, be of great interest for future studies.

Noland et al. (1988) have reported that solubilized zonae pellucidae from mouse eggs induce transient elevations in mouse sperm cAMP concentrations, which are dependent on the presence of extracellular Ca^{2+}. These cAMP elevations of four- to sixfold over buffer controls occur prior to, and appear correlated with, the induction of the acrosome reaction and suggest that cAMP may be a potential participant in the signaling pathway(s) leading to acrosomal exocytosis. Additional work will be required to determine the interrelationship between this intracellular second messenger and the induction of the acrosome reaction induced by the zona pellucida. For instance, it is not known whether zona-pellucida-induced cAMP elevations occur in noncapacitated sperm. It is also not known which zona pellucida glycoprotein (ZP1, ZP2, or ZP3) is responsible for the cAMP elevations; it is presumed that ZP3 is the active moiety, but this must be tested. Furthermore, it remains to be demonstrated whether cAMP-dependent changes in the phosphorylation state of specific proteins mediated by the activation of a cAMP-dependent protein kinase occur in response to zona pellucida binding. The relationship between extracellular Ca^{2+} and sperm cAMP metabolism has also yet to be established in this system. Since sperm adenylate cyclases and cyclic nucleotide phosphodiesterases have been demonstrated to be regulated by Ca^{2+} and/or Ca^{2+}–calmodulin interactions, it is possible that calmodulin, which is present in sperm, plays an important regulatory role in modulating the effects of Ca^{2+} on the cyclic nucleotide metabolism (Garbers and Kopf, 1980; Tash and Means, 1983; Kopf and Gerton, 1990). Finally, since this intracellular signaling system is coupled to G proteins in a receptor-mediated fashion in other cell types, it will be of interest to determine whether this second messenger system is modulated by the sperm G_i-like protein.

3.5. Zona Pellucida Penetration *in Vitro*

Zamboni (1972) was unable to catch mouse sperm in transit through the zona pellucida during his extensive *in vivo* studies of the fertilization process (see Section 2.2.2). He concluded that sperm penetration of the zona pellucida was rapid. The actual transit of the mouse egg zona pellucida by sperm *in vitro* has not been, to date, observed or tracked in time, having proved to be both a rare and rapid event under conditions designed to mimic those *in vivo* (Wolf *et al.*, 1977; Sato and Blandau, 1979). The rate of zona pellucida penetration of cumulus-free, zona-pellucida-intact eggs by capacitated sperm was markedly increased by the addition of 1–5 μM ionophore A23187, as first observed by Wolf (observation cited in Saling *et al.*, 1979). This observation was utilized by Saling *et al.* (1979) to demonstrate by electron microscopy that sperm within the perivitelline space had undergone the acrosome reaction, whereas sperm bound to the outside of the zona pellucida were acrosome intact. The ionophore accelerated the entry of sufficient numbers of sperm through the zona pellucida so that sampling of the perivitelline space for the acrosomal status of those sperm by electron microscopy became experimentally feasible.

The results from this study provided experimental support for the acrosome reaction as the process prerequisite to sperm penetration of the zona pellucida in the mouse system, in accord with results reported in other mammalian systems (Yanagimachi, 1981). Further experimental support for the acrosome reaction as the rate-determining step in zona pellucida penetration comes from observations made with isolated, structurally intact zonae pellucidae. Lee and Storey (1985) followed the progress of the acrosome reaction in sperm bound to these isolated extracellular matrices using CTC as the fluorescent probe. They found that as the percentage of acrosome-reacted sperm increased (as monitored by a concomitant percentage increase in the AR pattern), the number of sperm bound per zona pellucida decreased. They interpreted this result as a loss of sperm by penetration through the isolated zona pellucida structure followed by escape of the still motile sperm. Addition of 5 μM A23187 to this system resulted in a dramatic increase in the rate of acrosome reactions occurring in the sperm bound to isolated zonae pellucidae. This increase in the rate of acrosome reactions provides the plausible explanation for the equally dramatic increase in the rate of zona pellucida penetration of zona-pellucida-intact eggs by bound sperm caused by addition of the ionophore (Saling *et al.*, 1979), as described above.

An additional piece of evidence implicating the partial or completed acrosome reaction in the zona pellucida penetration process by mouse sperm is provided by the study of Bleil and

Wassarman (1986) on the binding of the two zona pellucida glycoproteins, ZP2 and ZP3, to the sperm. Using ^{125}I-labeled ZP2 and ZP3, each free of contamination by the other, they demonstrated by autoradiography that ZP3 bound to acrosome-intact but not to acrosome-reacted sperm. The reverse was found with ZP2. This latter glycoprotein bound to acrosome-reacted but not to acrosome-intact sperm. The ultrastructural characterization of the intact zona pellucida by Greve and Wassarman (1985) showed that this insoluble matrix consists of filaments comprised of ZP2/ZP3 heterodimers cross-linked by ZP1. A reasonable scenario for zona pellucida penetration may be constructed from these observations as follows. Sperm bind to the zona pellucida filaments by multiple attachments to ZP3. The appropriate ligands on ZP3 trigger the acrosome reaction in the bound sperm (Florman and Storey, 1982a; Bleil and Wassarman, 1983). The ZP2 molecules provide binding sites for the acrosome-reacted sperm (Bleil et al., 1988), whose lytic enzymes are presumed to cause highly specific cleavages of ZP2 and possibly ZP1. Highly specific cleavages of the zona pellucida glycoproteins by sperm have not yet been demonstrated in the mouse but have been demonstrated to occur by the action of porcine sperm acrosin on the glycoproteins of the porcine zona pellucida (Brown and Cheng, 1985; Dunbar et al., 1985; Urch et al., 1985; Hedrick et al., 1989). This specificity would result in the clean, quasisurgical path taken by mammalian sperm through the zona pellucida (Yanagimachi, 1988). Species specificity of this process may be inferred from the report of Brown (1986), who demonstrated that ram acrosin produces specific cleavages in the structure of the sheep zona pellucida, whereas this enzyme causes complete dissolution of the mouse zona pellucida.

It is evident that the process of zona pellucida penetration by sperm during the first major encounter of the two mouse gametes has not been completely characterized. The nature of the binding sites on the acrosome-reacted sperm for ZP2 are unknown, as are the binding ligands associated with ZP2. It has not yet been demonstrated whether mouse sperm acrosin is the actual zona pellucida lysin involved in clearing the path through the zona pellucida for the sperm. Mouse acrosin has proved difficult to purify in quantities sufficient for detailed examination of its properties. Preliminary results using the small amounts available indicate that purified mouse acrosin could dissolve the entire mouse zona pellucida (Brown, 1983), thus differing from the action of ram and boar acrosin on their respective zonae pellucidae. Complete dissolution of the mouse zona pellucida by mouse acrosin requires 6 hr, however (Brown, 1983), implying that highly specific cuts of the zona pellucida may normally occur at the much shorter times characteristic of sperm penetration of the zona pellucida. This question remains to be examined.

3.6. Sperm–Egg Fusion

The documentation of the zona-pellucida-induced acrosome reaction in mouse sperm (Florman and Storey, 1982a) and the subsequent characterization of this induced exocytotic reaction as a ligand–receptor-mediated signal transduction process (Kopf and Gerton, 1990; Kopf, 1988, 1989) appear to have left the sperm stranded on or in the zona pellucida matrix, in terms of experimental investigation. The final stages of sperm–egg interaction have, by way of contrast, received far less attention. These stages include the binding of the acrosome-reacted sperm to the egg plasma membrane after sperm entry into the perivitelline space, fusion of the sperm and egg plasma membranes, and entry of the sperm into the egg. Recent reviews on this topic have been provided by Fraser (1984b) and Yanagimachi (1988). It should be noted from those reviews that relevant studies with mouse gametes have been quite sparse.

3.6.1. SPERM–EGG BINDING

The possible role of egg plasma membrane surface proteins in the binding of sperm to zona pellucida-free mouse eggs was first observed by Wolf et al. (1976). Comparison of sperm penetration rates of eggs in which the zona pellucida was removed by protease or mechanical treatment demonstrated that the latter procedure yielded eggs readily susceptible to penetration with a high percentage of polyspermy. Pronase treatment yielded eggs that were poorly penetrated. Chymotrypsin treatment had a less severe effect on the denuded eggs as compared to pronase, but both percentage penetration and polyspermy were reduced as compared to eggs in which the zona pellucida was removed mechanically with micropipettes. The time course of penetration in these experiments was long enough that both capacitation and spontaneous acrosome reactions would have been expected to occur, but this point was not explicitly addressed (Wolf *et al.*, 1976, 1977).

Although mechanical removal of zonae pellucidae does not seem to perturb the plasma membrane of the egg, the micropipetting procedure is demanding of the investigator as well as the cells, resulting inevitably in low yields of intact, zona-pellucida-free eggs. A much improved isolation procedure has been reported by Boldt and Wolf (1986). It utilizes a combined chymotrypsin/mechanical treatment, in which the protease at low concentrations serves to weaken and distend the zona pellucida structure such that gentle mechanical pipetting with large-bore micropipettes may be used. Yields of zona-pellucida-free eggs are high, and no perturbation of the egg surface by the enzyme could be detected. This procedure has made possible a more detailed examination of those egg plasma membrane surface molecules that may mediate sperm binding. Boldt *et al.* (1988) treated zona-pellucida-free eggs with purified enzymes of different specificities in both concentration- and time-dependent protocols and then evaluated the effects of the treatments on sperm binding. In these experiments, sperm were capacitated for 60 min and incubated with eggs for an additional 20 min. Under these conditions, a low percentage of spontaneously acrosome-reacted sperm would be expected to be present in the sperm suspension (Ward and Storey, 1984), and the question of whether the bound sperm were acrosome-intact or acrosome-reacted was not directly addressed. The results showed that both trypsin and chymotrypsin decreased sperm binding in a concentration- and time-dependent manner and that phospholipases A_2 and C, when added to the suspension of gametes, also inhibited binding at certain concentrations. Only one polypeptide of $M_r = 94,000$ was released by both chymotrypsin and trypsin, implicating it as a potential player in sperm binding to the zona-pellucida-free egg (Boldt *et al.*, 1989a).

Hyaluronidase also inhibited binding, but only at high concentrations, and neuraminidase and the glycosidases had no detectable effect. The egg plasma membrane contains available mannose groups, since concanavalin A (ConA), a lectin specific for the α-mannosides, was shown to bind readily to the plasma membrane. [^{125}I]ConA was found to be a useful marker for isolation of plasma membrane fragments (Boldt and Wolf, 1987). This mannoside moiety does not appear to be involved in sperm binding, however. ConA binding was inhibited by 0.1 M 2-α-methyl mannoside to the extent of 98%, but 0.05 M 2-α-methyl mannoside inhibited zona-pellucida-free egg penetration by only 30%. Of the other saccharides examined for involvement in sperm binding, only fucose, galactose, and N-acetylglucosamine appear to be effective (Boldt *et al.*, 1989b). Treatment of the sperm with fucosidase also blocked egg penetration, implying that fucose residues on the mouse sperm membrane, presumed to be the plasma membrane of the equatorial segment or the inner acrosomal membrane, are involved in initial sperm–egg interaction. It is of interest that fucose has been implicated as the carbohydrate moiety on the oligosaccharide linked to boar acrosin that may play a role in binding porcine sperm to porcine eggs (Jones *et al.*, 1988; Topfer-Peterson and Henschen, 1988). One could propose that the fucose effect on mouse sperm binding reported by Boldt *et al.* (1989b) might result from the presence of this saccharide residue on mouse acrosin bound to the inner acrosomal membrane, as suggested by the work of Topfer-Peterson. However, this proposal remains without experimental support until mouse acrosin itself and its possible presence on the inner acrosomal membrane receive further investigative attention.

3.6.2. SPERM PENETRATION

Although attempts to catch the mouse sperm in the process of penetrating the zona pellucida have not been successful, the larger target presented by the mouse egg has made it possible to observe its penetration by the sperm. Wolf and Armstrong (1978) filmed this event and found that about 30 min was required from the time that the sperm attached to the plasma membrane of the egg until the sperm head had disappeared into the egg. Flagellar rotation around the entrance point continued as the head moved deeper into the egg cytoplasm and head decondensation began. Prior to sperm penetration, the egg plasma membrane of the zona-pellucida-free egg was observed to be smooth or to have a mixed smooth and ruffled appearance, whereas the egg prior to zona pellucida removal exhibited numerous microvilli on its surface. After sperm penetration, zona-pellucida-free eggs again exhibited microvilli on their surface and displayed ruffled and smooth regions with sharp delineations between them (Nicosia *et al.*, 1978). Sperm binding to the egg surface occurs on the lateral face of the head, with the firm point of attachment between the sperm and egg plasma membranes occurring at the equatorial segment (Wolf and Armstrong, 1978). It is difficult to tell how much binding, if any, occurs at the inner acrosomal membrane. In other mammalian species, the inner acrosomal membrane does not appear to interact with the egg plasma membrane (Bedford and Cooper, 1978; Yanagimachi, 1988). Further visual investigations of the fusion process utilizing either film or video image processing have not been carried out, so the fine points of the kinetics and possible actions of the egg cytoskeleton in the fusion process remain as untouched topics for further study.

Proteins are undoubtedly involved in mediating the mouse gamete membrane fusion reaction leading to sperm penetration, but little is known about them. Wolf (1977) reported that trypsin inhibitors reduced sperm penetration of both zona-pellucida-intact and zona-pellucida-free mouse eggs. Saling (1981) showed that the inhibition seen with zona-pellucida-intact eggs could be attributed to the effect of the trypsin inhibitors on blocking sperm binding to zona pellucida (see Section 3.2.2). These inhibitors did not block sperm binding to zona-pellucida-free eggs, however (Saling, 1981), so that the reduction in penetration observed by Wolf (1977) may be attributed to a cessation of the membrane fusion process. No protein with trypsin inhibitor binding capacity has been reported in connection with this process. Saling *et al.* (1983, 1985) have isolated a monoclonal antibody raised to sperm antigens, designated M29, that blocks the membrane fusion process but does not block sperm binding to the zona pellucida. This antibody recognized a protein of M_r 40,000 on immunoblots obtained by detergent extraction of caudal epididymal mouse sperm. Since this antigen could be localized by indirect immunofluorescence to the equatorial segment, it is at least properly positioned to play a role in the membrane fusion process. This protein has not been further characterized. The egg protein affected by trypsin inhibitors (Wolf, 1977) and the M29 sperm antigen are, at present, the only two proteins characterized that may play a role in sperm penetration of the egg. Sperm penetration may also have specific ion requirements. Boldt *et al.* (1989c) have reported that K^+ promotes membrane fusion in a manner not requiring K^+ channel activation. Whether K^+ is a cofactor for an enzymatic reaction remains to be clarified.

3.6.3. BLOCK TO POLYSPERMY

Fertilization of the mammalian egg by one spermatozoan sets in motion a series of reactions to deny entry to more sperm, thus avoiding the lethal consequences of polyspermy. The block to polyspermy may occur at the level of the egg plasma membrane, the zona pellucida, or both (Yanagimachi, 1988). In the mouse egg, the zona pellucida, the egg plasma membrane, and the egg itself all appear to participate in the prevention of polyspermy. Zonae pellucidae from fertilized eggs become "hardened" relative to those of unfertilized eggs, as shown by a reduced solubilization with proteases, reducing agents, or acid treatment (Chang and Hunt, 1956; Inoue and Wolf, 1974a,b). A more spectacular effect is the loss of the ability of the zona pellucida to bind sperm after fertilization (Inoue and Wolf, 1975b), an effect shown to be caused by modifications of the zona pellucida glycoproteins ZP3 and ZP2 (Bleil and Wassarman, 1980b).

Sperm binding to zona-pellucida-free eggs decreases markedly after sperm penetration, and part of this decrease may be attributed to detachment of previously bound sperm that did not penetrate (Wolf and Hamada, 1979). It was estimated that 40 min was required to establish this plasma membrane block to sperm binding and penetration (Wolf, 1978). In many invertebrate eggs a plasma membrane block to polyspermy is electrically mediated (Jaffe, 1976). However, Jaffe *et al.* (1983) examined whether mouse eggs displayed such an electrically mediated polyspermy block and found no compelling evidence for its existence.

The mouse egg appears to have one final line of defense against polyspermy. This is the remarkable process of blebbing of the egg cytoplasm enclosing the sperm head with consequent removal of the sperm head (Yu and Wolf, 1981). These blebs form in polyspermic eggs at about the same time that the second polar body is released. Only the one report of Yu and Wolf (1981) has appeared describing this process. It is not clear whether this phenomenon has been observed in eggs of other mammalian species.

The process that mediates the block to polyspermy at the level of the zona pellucida appears to be mediated by the egg cortical granules. Wolf and Hamada (1977) were able to show a significant decrease in both sperm penetration and fertilization of zona-pellucida-intact eggs with cortical granule exudates obtained from inseminated zona-pellucida-free eggs. This effect of the cortical granule exudate could be blocked by the addition of trypsin inhibitors to the exudate prior to incubation with the zona-pellucida-intact eggs. The yield of mouse egg cortical granule exudate obtained by insemination was far too low to allow further characterization of its enzymatic activities. The recent work of Moller and Wassarman (1989) remedied this problem, in part, by isolating the exudate from mouse eggs activated by ionophore A23187. This exudate contained a protease that cleaves the zona pellucida glycoprotein ZP2 to $ZP2_f$, the form that is found in fertilized eggs. The identity of this and other cortical granule components that modify the biochemical and biological properties of the zona pellucida after fertilization remain to be investigated. The mechanism by which sperm–egg fusion initiates those reactions leading to cortical granule exocytosis is beyond the scope of this chapter but is outlined in the review by Kopf *et al.* (1989). The reader is also encouraged to examine the review in the present volume by Wassarman (Chapter 9).

3.7. Envoi

If one follows the mouse sperm through its various interactions with the mouse egg, one may be impressed by the considerable amount of information that has been uncovered recently with regard to the characterization of the process. However, one is even more struck by the large informational voids that still remain to be filled. It has been our purpose to present this information to the best of our ability for the edification of our readers, and it remains our hope that this exposition will stimulate them to further thought and experimentation.

ACKNOWLEDGMENTS. We gratefully acknowledge the members of our laboratories, whose hard work and dedication are greatly appreciated. We also wish to thank our scientific colleagues at the University of Pennsylvania for their support. The authors are supported by grants from the National Institutes of Health.

4. REFERENCES

Aarons, D., Speake, J. L.,and Poirier, G. R., 1984, Evidence for a proteinase inhibitor binding component associated with murine spermatozoa, *Biol. Reprod.* **31**:811–817.

Andrews, J. C., and Bavister, B. D., 1989a, Capacitation of hamster spermatozoa with the divalent cation chelators, D-penicillamine, L-histidine, and L-cysteine in a protein-free culture medium, *Gamete Res.* **23**:159–170.

Andrews, J. C., and Bavister, B. D., 1989b, Hamster zonae pellucidae cannot induce physiological acrosome reactions in chemically capacitated hamster spermatozoa in the absence of albumin, *Biol. Reprod.* **40**:117–122.

Aounuma, S., Okabe, M., Kawaguchi, M., and Kishi, Y., 1980, Studies on sperm capacitation. IX. Movement characteristics of spermatozoa in relation to capacitation, *Chem Pharm. Bull.* **28**:1497–1502.

Apter, F. M., Baltz, J. M., and Millette, C. F., 1988, A possible role for cell surface fucosyltransferase (FT) activity during sperm–zona pellucida binding in the mouse, *J. Cell. Biol.* **107**:175a.

Austin, C. R., 1951, Observations on the penetration of the sperm into the mammalian egg, *Aust. J. Sci. Res. B.* **4**:581–596.

Austin, C. R., 1952, The capacitation of mammalian sperm, *Nature* **170**:326.

Austin, C. R., 1960, Capacitation and the release of hyaluronidase from spermatozoa, *J. Reprod. Fertil.* **1**:310–311.

Austin, C. R., 1975, Membrane fusion events in fertilization. *J. Reprod. Fertil.* **44**:155–166.

Austin, C. R., 1985, Sperm maturation in the male and female genital tracts, in: *Biology of Fertilization*, Vol. 2 (C. B. Metz and A. Monroy eds.), Academic Press, New York, pp. 121–155.

Baltz, J. M.,and Cardullo, R. A., 1989, On the number and rate of formation of sperm–zona bonds in the mouse, *Gamete Res.* **24**:1–8.

Bedford, J. M., 1968, Ultrastructural changes in the sperm head during fertilization in the rabbit, *Am. J. Anat.* **123**:329–358.

Bedford, J. M., 1970, Sperm capacitation and fertilization in mammals, *Biol. Reprod. [Suppl.]* **2**:128–158.

Bedford, J. M., and Cooper, G. W., 1978, Membrane fusion events in the fertilization of vertebrate eggs, in: *Cell Surface Reviews*, Vol. 5 (G. Poste and G. L. Nicholson eds.), Elsevier North Holland, Amsterdam, pp. 65–125.

Benau, D. A., and Storey, B. T., 1987, Characterization of the mouse sperm plasma membrane zona-binding site sensitive to trypsin inhibitors, *Biol. Reprod.* **36**:282–292.

Benau, D. A., and Storey, B. T., 1988, Relationship between two types of mouse sperm surface sites that mediate binding of sperm to the zona pellucida, *Biol. Reprod.* **39**:235–244.

Benau, D. A., McGuire, E. J., and Storey, B. T., 1990, Further characterization of the mouse sperm surface zona-binding site with galactosyltransferase activity, *Mol. Reprod. Dev.* **25**:393–399.

Bentley, J. K., Garbers, D. L., Domino, S. E., Noland, T. D., and VanDop, C., 1986, Spermatozoa contain a guanine nucleotide binding protein ADP-ribosylated by pertussis toxin, *Biochem. Biophys. Res. Commun.* **138**:728–734.

Berridge, M. J., 1986, Cell signalling through phospholipid metabolism, *J. Cell Sci. [Suppl.]* **4**:137–153.

Birdsall, N. J. M., and Hulme, E. C., 1976, Biochemical studies on muscarinic acetylcholine receptors, *J. Neurochem.* **27**:7–16.

Bleil, J. D., and Wassarman, P. M., 1980a, Structure and function of the zona pellucida: Identification and characterization of the proteins of the mouse oocyte's zona pellucida, *Dev. Biol.* **76**:185–202.

Bleil, J. D., and Wassarman, P. M., 1980b, Synthesis of zona pellucida proteins by denuded and follicle-enclosed mouse oocytes during culture *in vitro, Proc. Natl. Acad. Sci. U.S.A.* **77**:1029–1033.

Bleil, J. D., and Wassarman, P. M., 1980c, Mammalian sperm–egg interaction: Identification of a glycoprotein in mouse egg zonae pellucidae possessing receptor activity for sperm, *Cell* **20**:873–882.

Bleil, J. D., and Wassarman, P. M., 1983, Sperm–egg interactions in the mouse: Sequence of events and induction of the acrosome reaction by a zona pellucida glycoprotein, *Dev. Biol.* **95**:317–324.

Bleil, J. D., and Wassarman, P. M., 1986, Autoradiographic visualization of the mouse egg's sperm receptor bound to sperm, *J. Cell Biol.* **102**:1363–1371.

Bleil, J. D., and Wassarman, P. M., 1988, Galactose at the non-reducing terminus of O-linked oligosaccharides of mouse egg zona pellucida glycoprotein ZP3 is essential for the glycoprotein's sperm receptor activity. *Proc. Natl. Acad. Sci. U.S.A.* **85**:6778–6782.

Bleil, J. D., and Wassarman, P. M., 1989, Identification of a mouse sperm protein that recognizes ZP3, *J. Cell Biol.* **109**:125a.

Bleil, J. D., Greve, J. M., and Wassarman, P. M., 1988, Identification of a secondary sperm receptor in the mouse egg zona pellucida: Role in maintenance of binding of acrosome-reacted sperm to eggs, *Dev. Biol.* **128**:376–385.

Boldt, J., and Wolf, D. P., 1986, An improved method for isolation of fertile zona-free mouse eggs, *Gamete Res.* **13**:213–222.

Boldt, J., and Wolf, D. P., 1987, Isolation of ^{125}I-concanavalin A-labeled plasma membrane from unfertilized mouse eggs, *Gamete Res.* **16:**303–310.

Boldt, J., Howe, A. M., and Preble, J., 1988, Enzymatic alteration of the ability of mouse egg plasma membrane to interact with sperm, *Biol. Reprod.* **39:**19–27.

Boldt, J., Gunter, L. E., and Howe, A. M., 1989a, Characterization of cell surface polypeptides of unfertilized, fertilized, and protease-treated zona-free mouse eggs, *Gamete Res.* **23:**91–101.

Boldt, J., Howe, A. M., Parkerson, J. B., Gunter, L. E., and Kuehn, E., 1989b, Carbohydrate involvement in sperm–egg fusion in mice, *Biol. Reprod.* **40:**887–896.

Boldt, J., Casas, A., Whaley, E., and Lewis, J. B., 1989c, Sperm–egg fusion in mice is potassium dependent, *J. Cell Biol.* **109:**125a.

Brackett, B. G., and Oliphant, G., 1975, Capacitation of rabbit spermatozoa *in vitro*, *Biol. Reprod.* **12:**260–274.

Brinster, R. L., and Biggers, J. D., 1965, *In vitro* fertilization of mouse ova within the explanted fallopian tube. *J. Reprod. Fertil.* **10:**277–279.

Brown, C. R., 1983, Purification of mouse sperm acrosin, its activation from proacrosin and effect on homologous egg investments, *J. Reprod. Fertil.* **69:**289–295.

Brown, C. R., 1986, The morphological and molecular susceptibility of sheep and mouse zona pellucida to acrosin, *J. Reprod. Fertil.* **77:**411–417.

Brown, C. R., and Cheng, W. T. K., 1985, Limited proteolysis of the porcine zona pellucida by homologous sperm acrosin, *J. Reprod. Fertil.* **74:**257–260.

Bryan, J. H. D., 1974, Capacitation in the mouse: The response of murine acrosomes to the environment of the female reproductive tract, *Biol. Reprod.* **10:**414–421.

Bunch, D. O., and Saling, P. M., 1988, The recognition sequence, RGDS, disrupts mouse sperm–zona pellucida interaction. *J. Cell Biol.* **107:**175a.

Bunch, D. O., Le Guen, P., and Saling, P. M., 1990, P95, a ZP3 receptor with tyrosine kinase activity, fractionates with mouse sperm membranes that inhibit sperm-zona binding, *J. Androl.* **11:**24-P.

Casey, P. J., and Gilman, A. G., 1988, G protein involvement in receptor–effector coupling, *J. Biol. Chem.* **263:**2577–2580.

Casillas, E. R., Elder, C. M., and Hoskins, D. D., 1980, Adenylate cyclase activity of bovine spermatozoa during maturation in the epididymis and the activation of sperm particulate adenylate cyclase by GTP and polyamines, *J. Reprod. Fertil.* **59:**297–302.

Caswell, A. H., and Hutchinson, J. D., 1971, Selectivity of cation chelation to tetracyclines: Evidence for special conformation of calcium chelate, *Biochem. Biophys. Res. Commun.* **43:**525–530.

Chang, M. C., 1951, Fertilizing capacity of spermatozoa deposited in the fallopian tubes, *Nature* **168:**697–698.

Chang, M. C., 1957, A detrimental effect of seminal plasma on the fertilizing capacity of sperm, *Nature* **184:**466–467.

Chang, M. C., 1959, Fertilization of rabbit ova *in vitro*, *Nature* **184:**466–467.

Chang, M. C., 1984, The meaning of sperm capacitation. A historical perspective, *J. Androl.* **5:**45–50.

Chang, M. C., and Hunt, D. M., 1956, Effects of proteolytic enzymes on the zona pellucida of fertilized and unfertilized mammalian eggs, *Exp. Cell Res.* **11:**497–499.

Chang, M. C., and Hunter, R. F. H., 1975, Capacitation of mammalian sperm: Biological and experimental aspects, in: *Handbook of Physiology*, Section 7: *Endocrinology*. Vol. V, *Male Reproductive System* (R. O. Greep and E. B. Astwood senior eds., D. W. Hamilton and R. O. Greep volume eds.), American Physiological Society, Washington, DC, pp. 339–351.

Chase, T., Jr., and Shaw, E., 1970, Titration of trypsin, plasmin, and thrombin with *p*-nitrophenyl *p*'-guanidinobenzoate HC1, *Methods Enzymol.* **19:**20–27.

Cheng, C. Y., and Boettcher, B., 1979, Effects of cholera toxin and 5'-guanylylimidodiphosphate on human spermatozoal adenylate cyclase activity, *Biochem. Biophys. Res. Commun.* **91:**1–9.

Cooper, T. G., 1984, The onset and maintenance of hyperactivated motility of spermatozoa from the mouse, *Gamete Res.* **9:**55–74.

Cornwall, G. A., Tulsiani, D. R. P., and Orgebin-Crist, M.-C., 1991, Inhibition of the mouse sperm surface α-D-mannosidase inhibits sperm-egg binding *in vitro*, *Biol. Reprod.* **44:**913–921.

Cummins, J. M., and Woodall, P. F., 1985, On mammalian sperm dimensions, *J. Reprod. Fertil.* **75:**153–175.

Davis, B. K., 1980, Interaction of lipids with the plasma membrane of sperm cells. I. The antifertilization action of cholesterol, *Arch. Androl.* **5:**249–254.

Davis, B. K., 1981, Timing of fertilization in mammals: Sperm cholesterol/phospholipid ratio as a determinant of the capacitation interval, *Proc. Natl. Acad. Sci. U.S.A.* **78:**7560–7564.

Davis, B. K., Byrne, R., and Bedigan, K., 1980, Studies on the mechanism of capacitation: Albumin-mediated changes in plasma membrane lipids during *in vitro* incubation of rat sperm cells, *Proc. Natl. Acad. Sci. U.S.A.* **77**:1546–1550.

Dunbar, B. S., Dudkiewicz, A. B., and Bundman, D. S., 1985, Proteolysis of specific porcine zona pellucida glycoproteins by boar acrosin, *Biol. Reprod.* **332**:619–630.

Endo, Y., Schultz, R. M., and Kopf, G. S., 1987a, Effects of phorbol esters and a diacylglycerol on mouse eggs: Inhibition of fertilization and modification of the zona pellucida, *Dev. Biol.* **119**:199–209.

Endo, Y., Lee, M.A., and Kopf, G. S., 1987b, Evidence for the role of a guanine nucleotide-binding regulatory protein in the zona pellucida-induced mouse sperm acrosome reaction, *Dev. Biol.* **119**:210–216.

Endo, Y., Mattei, P., Kopf, G. S., and Schultz, R. M., 1987c, Effects of a phorbol ester on mouse eggs: Dissociation of sperm receptor activity from acrosome reaction-inducing activity of the mouse zona pellucida protein, ZP3, *Dev. Biol.* **123**:574–577.

Endo, Y., Lee, M. A., and Kopf, G. S., 1988, Characterization of an islet activating protein-sensitive site in mouse sperm that is involved in the zona pellucida-induced acrosome reaction, *Dev. Biol.* **129**:12–24.

Florman, H. M., and Babcock, D. F., 1990, Progress towards understanding the molecular basis of capacitation, in: *The Biology and Chemistry of Mammalian Fertilization* (P. M. Wassarman, ed.), CRC Press, Boca Raton, FL, pp. 105–132.

Florman, H. M., and Storey, B. T., 1981, Inhibition of *in vitro* fertilization of mouse eggs: 3-Quinuclidinyl benzilate specifically blocks penetration of zonae pellucidae by mouse spermatozoa, *J. Exp. Zool.* **216**:159–167.

Florman, H. M., and Storey, B. T., 1982a, Mouse gamete interactions: The zona pellucida is the site of the acrosome reaction leading to fertilization *in vitro*, *Dev. Biol.* **91**:121–130.

Florman, H. M., and Storey, B. T., 1982b, Characterization of cholinomimetic agents that inhibit *in vitro* fertilization in the mouse, *J. Androl.* **3**:157–164.

Florman, H. M., and Wassarman, P. M., 1985, O-Linked oligosaccharides of mouse egg ZP3 account for its sperm receptor activity, *Cell* **41**:313–324.

Florman, H. M., Saling, P. M., and Storey, B. T., 1982, Fertilization of mouse eggs *in vitro*. Time resolution of the reactions preceding penetration of the zona pellucida, *J. Androl.* **3**:373–381.

Florman, H. M., Bechtol, K. B., and Wassarman, P. M., 1984, Enzymatic digestion of the functions of the mouse egg's receptor for sperm, *Dev. Biol.* **106**:243–255.

Florman, H. M., Tombes, R. M., First, N. L., and Babcock, D. F., 1989, An adhesion-associated agonist from the zona pellucida activates G protein-promoted elevations of internal Ca^{2+} and pH that mediate mammalian sperm acrosomal exocytosis, *Dev. Biol.* **135**:133–146.

Fraser, L. R., 1977, Motility patterns in mouse spermatozoa before and after capacitation, *J. Exp. Zool.* **202**:439–444.

Fraser, L. R., 1982, Ca^{2+} is required for mouse sperm capacitation and fertilization *in vitro*, *J. Androl.* **3**:412–419.

Fraser, L. R., 1984a, Mouse sperm capacitation involves loss of a surface associated component, *J. Reprod. Fertil.* **72**:373–384.

Fraser, L. R., 1984b, Mechanisms controlling mammalian fertilization, *Oxford Rev. Reprod. Biol.* **6**:174–225.

Fraser, L. R., 1985, Albumin is required to support the acrosome reaction but not capacitation in mouse spermatozoa *in vitro*, *J. Reprod. Fertil.* **74**:185–196.

Fraser, L. R., 1987a, Minimum and maximum extracellular Ca^{2+} requirements during mouse sperm capacitation and fertilization *in vitro*, *J. Reprod. Fertil.* **81**:77–89.

Fraser, L. R., 1987b, Strontium supports capacitation and the acrosome reaction in mouse sperm and rapidly activates mouse eggs, *Gamete Res.* **18**:363–374.

Freissmuth, M., Casey, P. J., and Gilman, A. G., 1989, G proteins control diverse pathways of transmembrane signaling, *FASEB J.* **3**:2125–2131.

Garbers, D. L., 1989, Molecular basis of fertilization, *Annu. Rev. Biochem.* **58**:719–742.

Garbers, D. L., and Kopf, G. S., 1980, The regulation of spermatozoa by calcium and cyclic nucleotides, *Adv. Cyclic Nucleotide Res.* **13**:251–306.

Glassner, M., Abisogun, A. O., Kligman, I., Woolkalis, M. J., Gerton, G. L., and Kopf, G. S., 1989, Immuno-cytochemical and biochemical analysis of guanine nucleotide-binding regulatory proteins (G proteins) in mammalian spermatozoa, *J. Cell Biol.* **109**:250a.

Go, K. J., and Wolf, D. P., 1983, The role of sterols in sperm capacitation, *Adv. Lipid Res.* **30**:317–330.

Go, K. J., and Wolf, D. P., 1985, Albumin-mediated changes in sperm sterol content during capacitation, *Biol. Reprod.* **32**:145–153.

Godfrey, E. W., Dietz, M. E., Morstad, A. L., Wallskog, P. A., and Yorde, D. E., 1988, Acetylcholine receptor-aggregating proteins are associated with the extracellular matrix of many tissues in *Torpedo*, *J. Cell Biol.* **106**:1263–1272.

Gomperts, B. D., and Tatham, P. E. R., 1988, GTP-binding proteins in the control of exocytosis, *Cold Spring Harbor Symp. Quant. Biol.* **53**:983–992.

Greve, J. M., and Wassarman, P. M., 1985, Mouse egg extracellular coat is a matrix of interconnected filaments possessing a structural repeat, *J. Mol. Biol.* **181**:253–264.

Grynkiewicz, G., Poenie, M., and Tsien, R. Y., 1985, A new generation of Ca^{2+} indicators with greatly improved fluorescence properties, *J. Biol. Chem.* **260**:3440–3450.

Gwatkin, R. B. L., 1977, *Fertilization Mechanisms in Man and Mammals*, Plenum Press, New York.

Hartmann, J. R., 1983, Mammalian fertilization: Gamete surface interaction *in vitro*, in: *Mechanisms and Control of Animal Fertilization* (J. F. Hartmann, ed.), Academic Press, New York, pp. 325–364.

Hartmann, J. F., and Hutchison, C. F., 1974a, Nature of the pre-penetration contact interactions between hamster gametes *in vitro*, *J. Reprod. Fertil.* **36**:49–57.

Hartmann, J. F., and Hutchison, C. F., 1974b, Mammalian fertilization *in vitro*: Sperm induced preparation of the zona pellucida of hamster ova for final binding, *J. Reprod. Fertil.* **37**:44–46.

Hartmann, J. F., and Hutchison, C. F., 1974c, Contact between hamster spermatozoa and the zona pellucida releases a factor which influences early binding stages, *J. Reprod. Fertil.* **37**:61–66.

Hartmann, J. F., and Hutchison, C. F., 1980, Nature and fate of the factors released during early contact interactions between hamster sperm and egg prior to fertilization *in vitro*, *Dev. Biol.* **78**:380–393.

Hedrick, J. L., Urch, U. A., and Hardy, D. M., 1989, Structure–function properties of the sperm enzyme acrosin, in: *Biocatalysis in Agricultural Biotechnology* (J. R. Whitaker and P. E. Sonnet, eds.), American Chemical Society, Washington, DC, pp. 215–229.

Heffner, L. J., and Storey, B. T., 1982, Cold lability of sperm binding to zona pellucida, *J. Exp. Zool.* **219**:155–161.

Heffner, L. J., Saling, P. M., and Storey, B. T., 1980, Separation of calcium effects on motility and zona binding ability in mouse spermatozoa, *J. Exp. Zool.* **212**:53–59.

Hildebrandt, J. D., Codina, J., Tash, J. S., Kirchick, H. J., Lipschultz, L., Sekura, R. D., and Birnbaumer, L., 1985, The membrane-bound spermatozoal adenylyl cyclase system does not share coupling characteristics with somatic cell adenylyl cyclases, *Endocrinology* **116**:1357–1366.

Huang, T. T. F., Fleming, A. D., and Yanagimachi, R., 1981, Only acrosome-reacted spermatozoa can bind to and penetrate zona pellucida: a study using the guinea pig, *J. Exp. Zool.* **217**:287–290.

Hyne, R. V., and Lopata, A., 1982, Calcium and adenosine affect human sperm adenylate cyclase activity, *Gamete Res.* **6**:81–89.

Inoue, M., and Wolf, D. P., 1974a, Solubility properties of the murine zona pellucida, *Biol. Reprod.* **10**:512–518.

Inoue, M., and Wolf, D. P., 1974b, Comparative solubility properties of the zonae pellucidae of unfertilized and fertilized mouse ova, *Biol. Reprod.* **11**:558–565.

Inoue, M., and Wolf, D. P., 1975a, Comparative solubility properties of rat and hamster zonae pellucidae, *Biol. Reprod.* **12**:535–540.

Inoue, M., and Wolf, D. P., 1975b, Sperm binding characteristics of the murine zona pellucida, *Biol. Reprod.* **13**:340–348.

Iwamatsu, T., and Chang, M. C., 1969, *In vitro* fertilization of mouse eggs in the presence of bovine follicular fluid, *Nature* **224**:919–920.

Iwamatsu, H., and Chang, M. C., 1970, Further investigation of capacitation of sperm and fertilization of mouse eggs *in vitro*, *J. Exp. Zool.* **175**:271–282.

Iwamatsu, H., and Chang, M. C., 1971, Factors involved in the fertilization of mouse eggs *in vitro*, *J. Reprod. Fertil.* **26**:197–208.

Jaffe, L. A., 1976, Fast block to polyspermy in sea urchin eggs is electrically mediated, *Nature* **261**:68–71.

Jaffe, L. A., Sharp, A. P., and Wolf, D. P., 1983, Absence of an electrical polyspermy block in the mouse, *Dev. Biol.* **96**:317–323.

Jameson, G. W., Roberts, D. V., Adams, R. W., Kyle, W. S. A., and Elmore, D. T., 1973, Determination of the operational molariy of solutions of bovine alpha-chymoptrypsin, trypsin, thrombin, and factor Xa by spectrofluorimetric titration, *Biochem. J.* **131**:107–117.

Johnson, R. G., Carty, S. E., Fingerhood, B. J., and Scarpa, A., 1980, The internal pH of mast cell granules, *FEBS Lett.* **120**:75–79.

Jones, J., Kopf, G. S., and Schultz, R. M., 1989, Variability in electrophoretic mobility of G_i-like proteins; effect of SDS, *FEBS. Lett.* **243**:409–412.

Jones, R., Brown, C. R., and Lancaster, R. T., 1988, Carbohydrate-binding properties of boar sperm proacrosin and assessment of its role in sperm–egg recognition and adhesion during fertilization, *Development* **102**:781–792.

Katz, D. F., and Dott, H. M., 1975, Methods of measuring swimming speed of spermatozoa, *J. Reprod. Fertil.* **45**:263–272.

Katz, D. F., Drobnis, E. Z., and Overstreet, J. W., 1989, Factors regulating mammalian sperm migration through the female reproductive tract and oocyte vestments, *Gamete Res.* **22**:443–469.

Kinloch, R. A., Roller, R. J., Fimiani, C. M., Wassarman, D. A., and Wassarman, P. M., 1988, Primary structure of the mouse sperm receptor polypeptide determined by gene cloning, *Proc. Natl. Acad. Sci. U.S.A.* **85**:6409–6413.

Kligman, I., Glassner, M., Storey, B. T., and Kopf, G. S., 1988, Zona pellucida-mediated acrosomal exocytosis in mouse spermatozoa: Characterization of an intermediate stage prior to the completion of the acrosome reaction, *Dev. Biol.* **145**:344–355.

Kopf, G. S., 1988, Regulation of sperm function by guanine nucleotide-binding regulatory proteins (G-proteins), in: *Meiotic Inhibition: Molecular Control of Meiosis* (F. Haseltine and N. First, eds.), Alan R. Liss, New York, pp. 357–386.

Kopf, G. S., 1989, Mechanisms of signal transduction in mouse spermatozoa, *Ann. N. Y. Acad. Sci.* **564**:289–302.

Kopf, G. S., and Gerton, G. L., 1990, The mammalian sperm acrosome and the acrosome reaction, in: *The Biology and Chemistry of Mammalian Fertilization* (P. M. Wassarman, ed.), CRC Press, Boca Ratan, FL, pp.153–203.

Kopf, G. S., Woolkalis, M. J., and Gerton, G. L., 1986, Evidence for a guanine nucleotide-binding regulatory protein in invertebrate and mammalian sperm: Identification by islet-activating protein-catalyzed ADP-ribosylation and immunochemical methods, *J. Biol. Chem.* **261**:7327–7331.

Kopf, G. S., Endo, Y., Mattei, P., Kurasawa, S., and Schultz, R. M., 1989, Egg-induced modifications of the murine zona pellucida, in: *Mechanisms of Egg Activation* (R. L. Nuccitelli, W. H. Clark, and G. N. Cherr, eds.), Plenum Press, New York, pp. 249–272.

Kraayenhof, R., Brocklehurst, J. R., and Lee, C. P., 1976, Fluorescent probes for the energized state in biological membranes, in: *Concepts in Biochemical Fluorescence* (R. F. Chen and E. Edelhoch eds.), Marcel Dekker, New York, pp. 767–807.

Kuzan, F., Fleming, A. D., and Seidel, G., 1984, Successful fertilization *in vitro* of fresh intact oocytes by perivitelline (acrosome-reacted) spermatozoa in the rabbit, *Fertil. Steril.* **41**:766–770.

Lambert, H., and Le, A. V., 1984, Possible involvement of a sialylated component of the mouse sperm plasma membrane in sperm–zona interaction in the mouse, *Gamete Res.* **10**:153–163.

Langlais, J., and Roberts, K. D., 1985, A molecular membrane model of sperm capacitation and the acrosome reaction of mammalian spermatozoa, *Gamete Res.* **12**:183–224.

Langlais, J., Zollinger, M., Plante, L., Chapdelaine, A., Bleau, G., and Roberts, K. D., 1981, Localization of cholesteryl sulfate in human spermatozoa in support of a hypothesis for the mechanism of capacitation, *Proc. Natl. Acad. Sci. U.S.A.* **78**:7266–7270.

Langlais, J., Kan, F. W. K., Granger, L., Raymond, L., Bleau, G., and Roberts, K. D., 1988, Identification of sterol acceptors that stimulate cholesterol efflux from human soermatozoa during *in vitro* capacitation, *Gamete Res.* **20**:185–201.

Lee, M. A., and Storey, B. T., 1985, Evidence for plasma membrane impermeability to small ions in acrosome-intact mouse spermatozoa bound to mouse zonae pellucidae, using an aminoacridine fluorescent probe: Time course of the zona-induced acrosome reaction monitored by both chlortetracycline and pH probe fluorescence, *Biol. Reprod.* **33**:235–246.

Lee, M. A., and Storey, B. T., 1986, Bicarbonate is essential for fertilization of mouse eggs. Mouse sperm require it to undergo the acrosome reaction, *Biol. Reprod.* **34**:349–356.

Lee, M. A., and Storey, B. T., 1989, Endpoint of first stage of zona pellucida-induced acrosome reaction in mouse spermatozoa characterized by acrosomal H^+ and Ca^{2+} permeability: Population and single cell kinetics, *Gamete Res.* **24**:303–326.

Lee, M. A., Kopf, G. S., and Storey, B. T., 1987, Effects of phorbol esters and a diacylglycerol on the mouse sperm acrosome reaction induced by the zona pellucida, *Biol. Reprod.* **36**:617–627.

Leyton, L., and Saling, P., 1989a, 95 kD sperm proteins bind ZP3 and serve as tyrosine kinase substrates in response to zona binding, *Cell* **57**:123–130.

Leyton, L., and Saling, P., 1989b, Evidence that aggregation of mouse sperm receptors by ZP3 triggers the acrosome reaction, *J. Cell Biol.* **108**:2163–2168.

Leyton, L., Robinson, A., and Saling, P. M., 1989, Relationship between the M42 antigen of mouse sperm and the acrosome reaction induced by ZP3, *Dev. Biol.* **132**:174–178.

Lopez, L. C., Bayna, E. M., Litoff, D., Shaper, N. L., Shaper, J. H., and Shur, B. D., 1985, Receptor function of mouse sperm surface galactosyltransferase during fertilization, *J. Cell Biol.* **101**:1501–1510.

McLaughlin, J. D., and Shur, B. D., 1987, Binding of caput epididymal mouse sperm to the zona pellucida, *Dev. Biol.* **124**:557–561.

McRorie, R. A., and Williams, W. L., 1974, Biochemistry of fertilization, *Annu. Rev. Biochem.* **43**:777–803.

Meizel, S., and Deamer, D. W., 1978, The pH of the hamster acrosome, *J. Histochem. Cytochem.* **26**:98–105.

Minke, B., and Stephenson, R. S., 1985, The characteristics of chemically induced noise in *Musca* photoreceptors, *J. Comp. Physiol. A* **156**:339–356.

Miyamoto, H., and Chang, M. C., 1973, The importance of serum albumin and metabolic intermediates for capacitation of spermatozoa and fertilization of mouse eggs *in vitro*, *J. Reprod. Fertil.* **32**:193–205.

Miyamoto, H., and Ishibashi, T., 1975, The role of calcium ions in fertilization of mouse and rat eggs *in vitro*, *J. Reprod. Fertil.* **45**:523–526.

Moller, C. C., and Wassarman, P. M., 1989, Characterization of a proteinase that cleaves zona pellucida glycoprotein ZP2 following activation of mouse eggs, *Dev. Biol.* **132**:103–112.

Moore, H. D. M., and Bedford, J. M., 1983, The interaction of mammalian gametes in the female, in: *Mechanism and Control of Animal Fertilization* (J. F. Hartmann, ed.), Academic Press, New York, pp. 453–497.

Morales, P., Cross, N. L., Overstreet, J. W., and Hanson, F. W., 1989, Acrosome-intact and acrosome-reacted human sperm can initiate binding to the zona pellucida, *Dev. Biol.* **133**:385–392.

Myles, D. G., Hyatt, H., and Primakoff, P., 1987, Binding of both acrosome-intact and acrosome-reacted guinea pig sperm to the zona pellucida during *in vitro* fertilization, *Dev. Biol.* **121**:559–567.

Neer, E. J., Lok, J. M., and Wolf, L. G., 1984, Purification and properties of the inhibitory guanine nucleotide regulatory unit of brain adenylate cyclase, *J. Biol. Chem.* **259**:14222–14229.

Neill, J. M., and Olds-Clarke, P., 1987, A computer-assisted assay for mouse sperm hyperactivation demonstrates that bicarbonate but not bovine serum albumin is required, *Gamete Res.* **18**:121–140.

Neill, J. M., and Olds-Clarke, P., 1988, Incubation of mouse sperm with lactate delays capacitation and hyperactivation and lowers fertilization levels *in vitro*, *Gamete Res.* **20**:459–473.

Nicosia, S. V., Wolf, D. P., and Mastroianni, L., Jr., 1978, Surface topography of mouse eggs before and after insemination, *Gamete Res.* **1**:145–155.

Noland, T. D., Garbers, D. L., and Kopf, G. S., 1988, An elevation in cyclic AMP concentration precedes the zona pellucida-induced acrosome reaction of mouse spermatozoa, *Biol. Reprod.* **38**(Suppl.):94.

Olds-Clarke, P., 1983, The nonprogressive motility of sperm populations from mice with a t^{w32} haplotype, *J. Androl.* **4**:136–143.

Olds-Clarke, P., 1989, Sperm from $t^{w32}/+$ mice: Capacitation is normal, but hyperactivation is premature and nonhyperactivated sperm are slow, *Dev. Biol.* **131**:475–482.

Overstreet, J. W., 1983, Transport of gametes in the reproductive tract of the female mammal, in: *Mechanism and Control of Animal Fertilization* (J. F. Hartmann, ed.), Academic Press, New York, pp. 499–543.

Pavlok, A., 1967, Development of mouse ova in explanted oviducts: Fertilization, cultivation, and transplantation, *Science* **157**:1457–1458.

Pavlok, A., and McLaren, A., 1972, The role of cumulus cells and the zona pellucida in fertilization of mouse eggs *in vitro*, *J. Reprod. Fertil.* **29**:91–97.

Poirier, G. R., Robinson, R., Richardson, R., Hinds, K., and Clayton, D., 1986, Evidence for a binding site on the sperm plasma membrane which recognizes the zona pellucida: A binding site on the sperm plasma membrane, *Gamete Res.* **14**:235–243.

Ringuette, M. J., Chamberlin, M. E., Baur, A. W., Sbieski, D. A., and Dean, J., 1988, Molecular anlysis of cDNA coding for ZP3, a sperm binding protein of the mouse zona pellucida, *Dev. Biol.* **127**:287–295.

Robinson, R., Richardson, R., Hinds, K., Clayton, D., and Poirier, G. R., 1987, Features of a seminal proteinase inhibitor-zona pellucida-binding component on murine spermatozoa, *Gamete Res.* **16**:217–228.

Ross, E. M., 1989, Signal sorting and amplification through G protein-coupled receptors, *Neuron* **3**:141–152.

Ross, E. M., and Gilman, A. G., 1980, Biochemical properties of hormone-sensitive adenylate cyclase, *Annu. Rev. Biochem.* **49**:533–564.

Saling, P. M., 1981, Involvement of trypsin-like activity in binding of mouse spermatozoa to zonae pellucidae, *Proc. Natl. Acad. Sci. U.S.A.* **78**:6231–6235.

Saling, P. M., 1982, Development of the ability to bind to zonae pellucidae during epididymal maturation: Reversible immobilization of mouse spermatozoa by lanthanum, *Biol. Reprod.* **26**:429–436.

Saling, P. M., 1986, Mouse sperm antigens that participate in fertilization. IV. A monoclonal antibody prevents zona penetration by inhibition of the acrosome reaction, *Dev. Biol.* **177**:511–519.

Saling, P. M., 1989, Mammalian sperm interaction with extracellular matrices of the egg, *Oxford Rev. Reprod. Biol.* **11:**339–388.

Saling, P. M., and Lakoski, K. A., 1985, Mouse sperm antigens that participate in fertilization. II. Inhibition of sperm penetration through the zona pellucida using monoclonal antibodies, *Biol. Reprod.* **33:**527–536.

Saling, P. M., and Storey, B. T., 1979, Mouse gamete interactions during fertilization *in vitro*: Chlortetracycline as fluorescent probe for the mouse sperm acrosome reaction, *J. Cell Biol.* **83:**544–555.

Saling, P. M., Storey, B.T., and Wolf, D. P., 1978, Calcium-dependent binding of mouse epididymal spermatozoa to the zona pellucida, *Dev. Biol.* **65:**515–525.

Saling, P. M., Sowinski, J., and Storey, B. T., 1979, An ultrastructural study of epididymal mouse spermatozoa binding to zonae pellucidae *in vitro*: Sequential relationship to the acrosome reaction, *J. Exp. Zool.* **209:**229–238.

Saling, P. M., Raines, L. M., and O'Rand, M. G., 1983, Monoclonal antibody against mouse sperm blocks a specific event in the fertilization process, *J. Exp. Zool.* **227:**481–486.

Saling, P. M., Irons, G., and Waibel, R., 1985, Mouse sperm antigens that participate in fertilization. I. Inhibition of sperm fusion with the egg plasma membrane using monoclonal antibodies, *Biol. Reprod.* **33:**515–526.

Sato, K., and Blandau, R. J., 1979, Time and process of sperm penetration into cumulus-free eggs fertilized *in vitro*, *Gamete Res.* **2:**295–304.

Schultz, R. M., Endo, Y., Mattei, P., Kurasawa, S., and Kopf, G. S., 1988, Egg-induced modifications of the mouse zona pellucida, in: *Cellular Factors in Development and Differentiation—Embryos, Teratocarcinomas, and Differentiated Tissue* (S. Harris and G. Sato, eds.), Alan R. Liss, New York, pp. 77–92.

Shimizu, S., Tsuji, M., and Dean, J., 1983, *In vitro* biosynthesis of three sulfated glycoproteins of murine zonae pellucidae by oocytes grown in follicle culture, *J. Biol. Chem.* **258:**5858–5863.

Shur, B. D., and Bennett, D., 1979, A specific defect in galactosyltransferase regulation on sperm bearing mutant alleles of the *T/t* locus, *Dev. Biol.* **71:**243–259.

Shur, B. D., and Hall, N. G., 1982a, Sperm surface galactosyltransferase activities during *in vitro* capacitation, *J. Cell Biol.* **95:**567–573.

Shur, B. D., and Hall, N. G., 1982b, A role for mouse sperm surface galactosyltransferase in sperm binding to the egg zona pellucida, *J. Cell Biol.* **95:**574–579.

Shur, B. D., and Neely, C. A., 1988, Plasma membrane association, purification, and partial characterization of mouse sperm β-1,4-galactosyltransferase, *J. Biol. Chem.* **263:**17706–17714.

Soldani, P., and Rosati, F., 1987, Sperm–egg interaction in the mouse using live and glutaraldehyde-fixed eggs, *Gamete Res.* **18:**225–235.

Stefanini, M., Oura, C., and Zamboni, L., 1969, Ultrastructure of fertilization in the mouse. Penetration of the sperm into the ovum, *J. Submicrosc. Cytol.* **1:**1–23.

Stein, D. M., Fraser, L. R., and Monks, N. J., 1986, Adenosine and Gpp(NH)p modulate mouse sperm adenylate cyclase, *Gamete Res.* **13:**151–158.

Suarez, S. S., 1987, Sperm transport and motility in the mouse oviduct: Observations *in situ*, *Biol. Reprod.* **36:**203–210.

Suarez, S. S., and Osman, R. A., 1987, Initiation of hyperactivated flagellar bending in mouse sperm within the female reproductive tract, *Biol. Reprod.* **36:**1191–1198.

Tash, J. S., and Means, A. R., 1983, Cyclic adenosine 3′,5′-monophosphate, calcium and protein phosphorylation in flagellar motility, *Biol. Reprod.* **28:**75–104.

Tessler, S., and Olds-Clarke, P., 1981, Male genotype influences sperm transport in female mice, *Biol. Reprod.* **24:**806–813.

Tessler, S., and Olds-Clarke, P., 1985, Linear and non linear mouse sperm motility patterns, a quantitative classification, *J. Androl.* **6:**35–44.

Tessler, S., Carey, J. E., and Olds-Clarke, P., 1981, Mouse sperm motility affected by factors in the *T/t* complex, *J. Exp. Zool.* **217:**277–285.

Thompson, R. S., Smith, D. M., and Zamboni, L., 1974, Fertilization of mouse ova *in vitro*: An electron microscopic study, *Fertil. Steril.* **25:**222–249.

Topfer-Peterson, E., and Henschen, A., 1988, Zona pellucida-binding and fucose-binding of boar sperm acrosin is not correlated with proteolytic activity, *Hoppe Seylers Z. Physiol. Chem.* **369:**69–76.

Toyoda, Y., Yokoyama, M., and Hosi, T., 1971a, Studies on the fertilization of mouse eggs *in vitro*. I. *In vitro* fertilization of mouse eggs by fresh epididymal semen, *Jpn. J. Anim. Reprod.* **16:**147–151.

Toyoda, Y., Yokoyama, M., and Hosi, T., 1971b, Studies on the fertilization of mouse eggs *in vitro*. II. Effects of *in vitro* incubation of spermatozoa on time of sperm penetration of mouse eggs *in vitro*, *Jpn. J. Anim. Reprod.* **16:**152–157.

Tulsiani, D. R. P., Skudlarek, M. D., and Orgebin-Crist, M.-C., 1989, Novel α-D-mannosidase of rat sperm plasma membranes: Characterization and potential role in sperm–egg interactions, *J. Cell Biol.* **109:**1257–1267.

Urch, U. A., Wardrip, N. J., and Hedrick, J. L., 1985, Limited and specific proteolysis of the zona pellucida by acrosin, *J. Exp. Zool.* **233**:479–483.

Vazquez, M. H., Phillips, D. M., and Wassarman, P. M., 1989, Interaction of mouse sperm with purified sperm receptors covalently linked to silica beads, *J. Cell Sci.* **92**:713–722.

Ward, C. R., and Storey, B. T., 1984, Determination of the time course of capacitation in mouse spermatozoa using a chlortetracycline fluorescence assay, *Dev. Biol.* **104**:287–296.

Wassarman, P. M., 1987, Early events in mammalian fertilization, *Annu. Rev. Cell Biol.* **3**:109–142.

Wassarman, P. M., 1988, Zona pellucida glycoproteins, *Annu. Rev. Biochem.* **57**:415–442.

Whittingham, D. G., 1968, Fertilization of mouse eggs *in vitro*, *Nature* **220**:592–593.

Wilde, M. W., and Kopf, G. S., 1989, Activation of a G-protein in mammalian sperm by an egg-associated extracellular matrix, the zona pellucida, *J. Cell Biol.* **109**:251a.

Wolf, D. P., 1977, Involvement of a trypsin-like activity in sperm penetration of zona-free mouse ova, *J. Exp. Zool.* **199**:149–156.

Wolf, D. P., 1978, The block to sperm penetration in zona-free mouse eggs, *Dev. Biol.* **64**:1–10.

Wolf, D. P., 1979, Mammalian fertilization, in: *The Biology of the Fluids of the Female Genital Tract* (F. K. Beller and G. F. B. Schumacher, eds.), Elsevier/North-Holland, Amsterdam, pp. 407–414.

Wolf, D. P., and Armstrong, P. B., 1978, Penetration of the zona-free mouse egg by capacitated epididymal sperm: Cinemicrographic observations, *Gamete Res.* **1**:39–46.

Wolf, D. P., and Hamada, M., 1977, Induction of zonal and egg plasma membrane blocks to sperm penetration in mouse eggs with cortical granule exudate, *Biol. Reprod.* **17**:350–354.

Wolf, D. P., and Hamada, M., 1979, Sperm binding to the mouse egg plasmalemma, *Biol. Reprod.* **21**:205–211.

Wolf, D. P., and Inoue, M., 1976, Sperm concentration dependency in the fertilization and zona sperm binding properties of mouse eggs inseminated *in vitro*, *J. Exp. Zool.* **196**:27–38.

Wolf, D. P., Inoue, M., and Stark, R. A., 1976, Penetration of zona-free mouse ova, *Biol. Reprod.* **15**:213–221.

Wolf, D. P., Hamada, M., and Inoue, M., 1977, Kinetics of sperm penetration into and the zona reaction of mouse ova inseminated *in vitro*, *J. Exp. Zool.* **201**:29–36.

Yanagimachi, R., 1970, *In vitro* capacitation of golden hamster spermatozoa by homologous and heterologous blood sera, *Biol. Reprod.* **3**:147–153.

Yanagimachi, R., 1972, Fertilization of guinea pig eggs *in vitro*, *Anat. Rec.* **174**:9–20.

Yanagimachi, R., 1981, Mechanisms of fertilization in mammals, in: *Fertilization and Embryonic Development in Vitro* (L. Mastroianni and J. D. Biggers, eds.), Plenum Press, New York, pp. 81–182.

Yanagimachi, R., 1988, Mammalian fertilization, in: *The Physiology of Reproduction* (E. Knobil and J. D. Neill, eds.), Raven Press, New York, pp. 135–185.

Yanagimachi, R., and Noda, Y. D., 1970, Ultrastructural changes in the hamster sperm head during fertilization, *J. Ultrastruct. Res.* **31**:465–484.

Yarden, Y., and Schlessinger, J., 1987, Epidermal growth factor induces rapid, reversible aggregation of the purified epidermal growth factor receptor, *Biochemistry* **26**:1443–1451.

Yarden, Y., and Ullrich, A., 1988, Growth factor receptor tyrosine kinases, *Annu. Rev. Biochem.* **57**:443–448.

Young, J. M., Hiley, J. R., and Burger, A. S. V., 1972, Homologues of benzilylcholine mustard, *J. Pharm. Pharmacol.* **24**:950–954.

Yu, S.-F., and Wolf, D. P., 1981, Polyspermic eggs can dispose of supernumerary sperm, *Dev. Biol.* **82**:203–210.

Zamboni, L., 1970, Ultrastructure of mammalian oocytes and ova, *Biol. Reprod. [Suppl.]* **2**:44–63.

Zamboni, L., 1971, Acrosome loss in fertilizing mammalian spermatozoa; A clarification, *J. Ultrastruct. Res.* **34**:401–405.

Zamboni, L., 1972, Fertilization in the mouse, in: *Biology of Mammalian Fertilization and Implantation* (K. S. Moghissi and E. S. E. Hafez, eds.), Charles C. Thomas, Springfield, IL, pp. 213–262.

Zamboni, L., and Stefanini, M., 1968, On the configuration of the plasma membane of the mature spermatozoon, *Fertil. Steril.* **19**:570–579.

Zimmerberg, J., 1987, Molecular mechanisms of membrane fusion: Steps during phospholipid and exocytotic membrane fusion, *Biosci. Rep.* **7**:251–268.

11

Fertilization in the Golden Hamster

Gary N. Cherr and Erma Z. Drobnis

1. INTRODUCTION

Early studies of fertilization in mammals involved observations of gametes that had been flushed from the reproductive tract of mated females. The development in the early 1960s of methods for the *in vitro* capacitation of sperm and the *in vitro* fertilization for the golden hamster (*Mesocricetus auratus*) eggs allowed investigators better control of the conditions of gamete interaction. It became possible to test hypotheses using well-controlled experiments in this species. Subsequently, hamsters were used extensively in studies of mammalian fertilization.

In this review we mention some of the advantages and disadvantages of the hamster as a model species for studies of mammalian fertilization. We then discuss the major events of fertilization through fusion of the gametes and establishment of the block to polyspermy. For each topic, we focus on what is known in the hamster system (i.e., how well characterized it is), how much it has been used relative to other species, and what its potential use is in future experiments. Unless mentioned specifically, the term hamster refers to the golden (or Syrian) hamster, *Mesocricetus auratus*.

2. ADVANTAGES OF THE HAMSTER MODEL

There are many advantages of using the hamster as an experimental model for studies of mammalian fertilization. As outlined in this section, some advantages are shared with the murine rodents, laboratory rat and mouse, whereas others are specific to the hamster.

GARY N. CHERR • Bodega Marine Laboratory, University of California, Davis, Bodega Bay, California 94923.   ERMA Z. DROBNIS • Departments of Zoology and Obstetrics and Gynecology, University of California, Davis, California 95616.

A Comparative Overview of Mammalian Fertilization, edited by Bonnie S. Dunbar and Michael G. O'Rand. Plenum Press, New York, 1991.

2.1. Advantages in Common with Murine Rodents

Similar to rats and mice, hamsters are inexpensive and are readily available from multiple commercial sources. For studies in reproductive biology, these species have the advantage of strict 4-day estrous cycles. The cycle, and therefore the timing of the onset of estrus and of ovulation, can be controlled by adjusting the photoperiod. Most important, the time of ovulation relative to the onset of the dark period and relative to the onset of estrus is constant within an hour or two (Ward, 1946; Boyer, 1953; Orsini, 1961).

Rodent spermatozoa are relatively large (e.g., the hamster sperm head has approximately four times the surface area of the human sperm head and approximately three and a half times the total length), which is one reason for their popularity for early observations of mammalian fertilization. Sperm of hamsters, as well as those of rats and mice, can be capacitated *in vitro*, and *in vitro* fertilization is possible both under culture conditions (Yanagimachi and Chang, 1963, 1964; Bavister, 1989) and during videomicroscopic observation (Drobnis and Katz, 1990). The distinct motility changes that occur during epididymal maturation and capacitation have been characterized, particularly for hamsters (Yanagimachi, 1981; Drobnis and Katz, 1990) but also for mice. The oviduct in rodents is quite transparent, and in mice and hamsters, this has permitted visualization of spermatozoa within the oviduct during the periovulatory period (Section 4).

Until recently, it was not possible to culture hamster embryos *in vitro* past the two-cell stage of development, and this has limited the measurement of reproductive outcome in this species. Recently, the so-called "two-cell block" to development has been overcome, with reasonable success, for eggs fertilized *in vitro* (Schini and Bavister, 1988a,b), and further work in this area is ongoing (Bavister, 1989).

2.2. Advantages Specific to the Hamster

The estrous cycle of hamsters can be timed accurately by simple external observation of postestrus discharge (Orsini, 1961); vaginal smears and behavior with the male are not necessary. Alternatively, estrus can be detected easily by placing each female with a male and observing for lordosis behavior. Indeed, mating can be observed without disturbing the process, and mating involving a single ejaculation can be timed exactly (Yamanaka and Soderwall, 1960). In addition to having regular estrous cycles, hamsters, in contrast to some species, can be superovulated; that is, the number of oocytes produced can be more than doubled by stimulation with follicle-stimulating hormone or pregnant mare serum gonadotropin. Of the rodent species, the hamster responds particularly well to superovulation, regularly producing 30–100 oocytes per female. Superovulation is usually combined with induction of ovulation using human chorionic gonado-tropin, and this technique can be used without follicle stimulation to improve the synchronization of ovulation.

The hamster is unique among inexpensive model species in having a large acrosome, which can be visualized easily at the level of the light microscope. This large acrosome is one of the primary reasons for the extensive use of hamsters in studies of mammalian fertilization. Of the species commonly used in this field, only in the guinea pig and, to a lesser degree, in the rabbit can the acrosome reaction (AR) be observed in living spermatozoa as they interact with oocytes.

The hamster is among those species having an extremely rapid and efficient block to polyspermic fertilization at the level of the zona pellucida (Section 14), an attribute it shares with dogs and ruminants. This makes the hamster a good candidate for use in studies of this phenomenon. Although the zona block prevents multiple sperm from penetrating the zona, multiple penetrations of the zona can be studied by using salt-stored zonae (Boatman *et al.*, 1988). In contrast to its strong block to polyspermy at the zona, the hamster egg seems to be unique in exhibiting little or no block to polyspermy at the level of the plasma membrane. In consequence, multiple observations of sperm fusion with the oolemma, incorporation of the sperm, and sperm decondensation can be made on a single oocyte.

Another feature of gamete fusion in hamsters is a reduced level of a species-specificity block to sperm fusion at the oolemma. Sperm from many species can fuse with the zona-free hamster oocyte (Yanagimachi, 1981), and this attribute has been exploited as a test of sperm function for many species. The successful penetration assay (SPA) is useful for determining if sperm have completed capacitation (and probably the AR) under given conditions. The SPA has also been used extensively in a clinical setting (human and veterinary) to determine if sperm from a given male are capable of capacitation, the AR, and fusion with the oolemma *in vitro*.

2.3. Disadvantages of the Hamster Model

Although the hamster has been used extensively to study fertilization in mammals, there are a number of research areas for which it may not be the best species. For studies at the molecular level, the hamster is not as well characterized as the mouse (and, for some studies, the guinea pig). However, because the hamster is relatively well characterized with respect to fertilization events *in vivo*, it is a reasonable second species in which to repeat the work completed under *in vitro* conditions in other species. Another disadvantage that the hamster shares with all mammals except the mouse is that no inbred strains are available. This limits its value for genetic/molecular studies in which differences among strains can be used to test mechanistic hypotheses, and differences can be reduced eventually to the molecular level. This disadvantage may be eliminated by the application of transgenic techniques to the hamster. Finally, the hamster may be dissimilar to the human in the sequence of sperm binding to the zona pellucida and the AR (Section 8). The hamster will be an important control during the continuing studies to resolve this question. Moreover, if differences do exist, the differences themselves can be exploited when testing hypotheses regarding these events.

3. OVULATION

The hamster has been used as a model to study the cellular processes involved in ovulation *in vivo* and *in vitro*. Talbot and co-workers have developed a valuable *in vitro* ovulation system to investigate the sequence of events leading to expulsion of the oocyte–cumulus complex from the ovary and the mechanisms involved in this process (Martin and Talbot, 1981a; Talbot and Chacon, 1982; Schroeder and Talbot, 1982; Talbot, 1983). Prior to rupture, the follicle transforms from a low to a tall profile. The size of the antrum decreases, probably as a result of the movement of follicular fluid through the developing rupture site at the apex. There is a thinning of the apical wall of the antrum, and the wall is weakened by the action of hydrolytic enzymes such as collagenase and plasmin (Murdoch and Cavender, 1987) and an evacuation of the cumulus-surrounded oocyte (Talbot, 1983).

It had been suggested that smooth muscle contractions are responsible for expulsion of the oocyte–cumulus complex from the follicle in mammals; however, the exact role of follicular contractions in ovulation is still unclear (Murdoch and Cavender, 1987). Martin and Talbot (1981a) investigated the distribution of smooth muscle cells in the walls of preovulatory follicles of hamster ovaries and correlated the contraction of these cells with the changes in shape of follicles during ovulation. In hamsters, each follicle is surrounded by one to three layers of spindle-shaped cells that are both smooth muscle and fibroblasts. Ultrastructurally, these two cell types can be distinguished from each other; Martin and Talbot demonstrated that thick filaments (myosin) were present in smooth muscle. As predicted based on function, the smooth muscle cells were more abundant in the theca externa in the basal hemisphere of the preovulatory follicle; this distribution was inversely related to the distribution of fibroblasts.

In correlating the morphological changes in the smooth muscle cells with changes in shape of the antrum and ultimately its rupture, Martin and Talbot (1981a) found that prior to the antrum becoming taller, the smooth muscle cells at the base of the follicle appear relaxed (stretched). Smooth muscle at the base of follicles undergoing various degrees of constriction just prior to rupture typically showed ultrastructural evidence of contraction. These data indicate that follicular smooth muscle contraction leads to a morphological change in the follicle, and this alteration in morphology may result in rupture at ovulation (Martin and Talbot, 1981a). Further support for the role of smooth muscle in the ovulation process comes from studies that used drugs to block smooth muscle contraction *in vivo* (Martin and Talbot, 1981b). In addition to blocking smooth muscle contraction, these drugs prevented constriction of the follicle base and ovulation.

Thus, perhaps the best evidence for a direct role for smooth muscle in ovulation comes from the hamster model. Smooth muscle contraction appears to squeeze the cumulus mass toward the apex of the follicle; the follicle wall, which is already weakened by hydrolytic enzymes, ruptures at the apex. Continued contraction of the smooth muscle cells following rupture may pull the follicle wall down, collapsing the follicle and insuring that the cumulus is evacuated (Martin and Talbot, 1981a).

The hamster system has been utilized for studying the timing of the events of ovulation *in vitro* (Talbot, 1983). By culturing ovaries from hormonally stimulated animals and videotaping the ovulating follicles, Talbot was able to quantitatively investigate the entire ovulation process. Many of the details of the ovulation process can only be captured in such a system. For example, the details of the evacuation of the cumulus mass were described, and it was found that the oocyte is positioned apically and is extruded with the first portions of cumulus material. This type of *in vitro* system can be extremely useful for investigations of the mechanisms of ovulation.

4. SPERM TRANSPORT

In mammals, fertilization occurs in a complex microenvironment within the female reproductive tract. Although it is difficult to observe sperm and oocytes at the site of fertilization, experiments *in vivo* are important to our understanding of fertilization in mammals (Moore and Bedford, 1983). Studies of sperm transport in the female reproductive tract are particularly important, since spermatozoa are morphologically and physiologically heterogeneous in mammals (Moore and Bedford, 1983; Bedford, 1983). Millions of sperm are deposited at coitus, and only one or a few participate in fertilization. This makes it difficult, when studying sperm collected from the male, to determine which subpopulations of sperm have attributes similar to those of the fertilizing sperm. The gradual establishment of sperm reservoirs within the female reproductive tract, which may continue over several days in some species, and the release or escape of fertilizing sperm from the oviductal isthmus into the ampulla once ovulation has occurred are complex processes controlled by both the sperm and the female reproductive tract (see reviews: Bishop, 1969; Blandau, 1969, 1973; Bedford, 1970b, 1983; Hamner, 1972; Freund, 1973; Hunter, 1975, 1980, 1987; Overstreet and Katz, 1977; Harper, 1982; Overstreet, 1983; Mortimer, 1983; Hawk, 1987; Katz *et al.*, 1989). It is not known which characteristics of a spermatozoon (e.g., surface properties, motility, energy charge) will result in its residence for given amounts of time in given reservoirs or its eventual participation in fertilization. Moreover, the mechanisms in the female reproductive tract that control the establishment and release of sperm reservoirs are largely uncharacterized.

Although considerable information is available for the golden hamster, mammalian sperm transport has been best characterized in rabbits (Overstreet, 1983). This animal model is convenient for studying sperm transport because it is a reflex ovulator, so that the time interval between mating and ovulation is constant. Another important factor in choice of an experimental animal is the site of semen deposition. In most mammals, semen is deposited during coitus either directly into the uterus (e.g., dogs, horses, pigs) or into the vagina, where it is rapidly drawn into the uterus (rats and perhaps other rodents). In contrast, semen is deposited in the vagina of primates, ruminants, and rabbits; the fertilizing sperm must migrate into the cervical mucus, which may serve as a sperm reservoir from which sperm continue to migrate to the upper reproductive tract. Rabbits have vaginal semen deposition and at least rudimentary mucus-laden cervices (Overstreet *et al.*, 1978) and may therefore provide a better model for sperm transport in ruminants and humans than do rodents. In the rabbit, it is known when and where reservoirs of sperm are established and how long they persist. Also, the behavior of sperm collected from different sites has been studied.

Hamsters have three advantages as experimental models for sperm transport, the first two of which are shared with other rodents: first, the timing of estrus and ovulation are determined by the photoperiod and are regular without hormonal induction (which may alter sperm transport), and second, the tissues of the female reproductive tract are quite transparent, permitting direct visualization of sperm within the excised tract. In addition to these advantages, hyperactivation of sperm motility (see Section 6), which may have an important role in the final stages of sperm transport, has been studied extensively in the golden hamster relative to other mammalian species.

Sperm transport in hamsters has been well characterized with respect to sperm numbers in the reproductive tract over time after mating and to a lesser extent with respect to the behavior of sperm during interaction with the female reproductive tract. The events are quite similar to those in other mammals studied. A few sperm are rapidly transported to the oviductal ampulla shortly after mating (Yamanaka and Soderwall, 1960; Smith *et al.*, 1987), but these sperm do not participate in fertilization (Smith *et al.*, 1987). One hour after mating, sperm begin to colonize the oviductal isthmus (Yanagimachi and Chang, 1963; Smith *et al.*, 1987). In mammals, this site is believed to be an important reservoir in which sperm are maintained in a quiescent state and released into the ampulla by some unknown mechanism. Transluminal observation of sperm within the oviducts excised from mated mice (Suarez, 1987) and hamsters (Smith *et al.*, 1987) provided the first evidence that sperm are quiescent within the isthmus and that they are closely associated, perhaps stuck, to the isthmic epithelium. Sperm continue to enter the isthmus from the uterus, reaching maximum sperm numbers in the oviduct 7 hr after mating (Smith *et al.*, 1987). Even when mating occurs well before ovulation, very few sperm are present in the cranial isthmus or ampulla until after fertilization is complete, and none are present until 3 hr after mating (Strauss, 1956; Smith *et al.*, 1987). Artificial insemination is possible in hamsters, so that sperm could be manipulated prior to insemination (e.g., labeled subpopulations). Sperm transport following artificial insemination is very similar to that following natural mating (Cummins and Yanagimachi, 1982; Smith *et al.*, 1987). Gamete fusion begins 4–7 hr after mating and is complete by 7–10 hr (Cummins and Yanagimachi, 1982); the interval between mating and fertilization is shorter if mating is late in estrus. For the interval during which fertilization occurs, the ratio of motile sperm to oocytes in the ampulla is less that 1 : 1 (Cummins and Yanagimachi, 1986). It has been proposed that oviductal factors, for example abovarian peristaltic waves (Battalia and Yanagimachi, 1979), may participate in restricting sperm to the isthmus until the products of ovulation in the ampulla stimulate a change resulting in the release of sperm. However, in a recent study, hamsters were mated at the end of estrus after ovulation, and sperm still required 2–3 hr residence within the isthmus before they entered the ampulla to effect fertilization. This suggests that residence in the isthmus is required by sperm and that changes in the sperm during this interval, for example, changes in surface characteristics and/or motility, may allow them to escape to the ampulla. If this is true, then the isthmus of the oviduct is the site where the fertilizing sperm complete capacitation *in vivo*.

Hyperactivated motility (Section 6) was first described for hamster sperm collected from the site of fertilization and for sperm undergoing capacitation *in vitro* (Yanagimachi, 1969; Gwatkin and Anderson, 1969). These original investigators and others have proposed that hyperactivated motility is important for penetration of the oocyte-investing layers and for escape of sperm from the isthmic reservoir. In order for this motility pattern to function in these roles, it must be present in the small number of sperm in the ampulla at the time of fertilization. Transluminal illumination of the excised hamster oviduct was used to observe sperm in the ampulla (Katz and Yanagimachi, 1980), and, indeed, they displayed hyperactivated motility (Fig. 1).

Clearly, control of sperm transport is an important goal both for treating subfertility and for developing contraceptive methods. The control mechanisms for the establishment of sperm reservoirs in the female reproductive tract are poorly understood. The hamster would be an ideal model species for establishment of the isthmic reservoir and release of sperm into the ampulla for fertilization.

5. SPERM CAPACITATION

To date, all mammalian sperm investigated have been found to require a period of residence in the female reproductive tract before acquiring the ability to undergo the AR and penetrate the egg. The hamster has proven to be an outstanding model system for understanding the process of capacitation in mammals. Both *in vitro* and *in vivo* studies with the hamster, dating back to the early 1960s, have provided the basis for much of our recent success with *in vitro* capacitation of sperm from a number of other species, including primates.

Yanagimachi and Chang (1963) obtained fertilization of hamster eggs *in vitro* by recovering sperm from the female tract after mating and, most importantly, by preincubating epididymal sperm with oocyte–cumulus masses and oviduct secretions. These observations demonstrated that sperm capacitation in mammals could be accomplished *in vitro*. Later, Yanagimachi (1969) and Gwatkin and Anderson (1969) investigated bovine follicular fluid and found that it supported hamster sperm capacitation *in vitro*. Yanagimachi (1969) further determined that bovine follicular fluid could be fractionated into a heat-stable dialyzable component that enhanced sperm motility and a heat-labile nondialyzable component that induced capacitation when present with the first component. Subsequently, components with the same properties were identified in blood serum (see Austin, 1985; Yanagimachi, 1981; Bavister, 1986). A great deal of research on the molecules and mechanisms involved in capacitation of hamster sperm has occurred as a result of these early observations.

5.1. Factors Regulating Capacitation *in Vitro*

In many instances it is not appropriate to assume that factors that induce capacitation *in vitro* function (or are even present) *in vivo* (Moore and Bedford, 1983; Bedford, 1983). Nevertheless, a fundamental understanding of the mechanisms of capacitation can only be achieved in defined media under experimentally controllable *in vitro* conditions. Sperm of many mammals can survive and capacitate in media similar to a modified Krebs–Ringers solution (Yanagimachi, 1988b); however, golden hamster sperm will become senescent and die in such a medium. Hamster sperm appear to be unusually "leaky" and require the addition of "sperm motility factors" to the media in order to achieve capacitation *in vitro* (Meizel, 1985; Bavister, 1986; Yanagimachi, 1988b).

Following up on the ability of serum and tissue components to support capacitation, Meizel and co-workers found that the motility factors were biogenic amines (see Meizel, 1985) and β-amino acids (taurine and hypotaurine) (Meizel *et al.*, 1980). The mechanisms by which biogenic amines and β-amino acids act on hamster sperm are not yet known. It has been suggested that taurine and hypotaurine "protect" sperm *in vitro* by inhibiting lipid peroxidation, but this has yet to be demonstrated in the hamster (see Bavister, 1986; Yanagimachi, 1988b). The most effective of the biogenic amines on capacitation in hamster sperm is epinephrine. α-Adrenergic agonists are also effective, whereas β agonists are not; this suggests that an α-adrenergic receptor is involved in hamster sperm capacitation (Meizel, 1985). This stimulation may, in turn, also stimulate motility (see Bavister, 1986; Meizel, 1985). Interestingly, hamster sperm capacitated in the absence of biogenic amines are still quite capable of undergoing the AR in the presence of a natural inducer, the zona pellucida (Cherr *et al.*, 1986). The mechanisms of α-adrenergic stimulation in hamster sperm is unclear. Effects on cytosolic Ca^{2+} as well as Na^+, K^+-ATPase have been suggested, but additional studies are needed.

Albumin is the major protein constituent in the female tract, is probably a significant component in capacitation *in vivo*, and is required *in vitro* (Dow and Bavister, 1989). However, the role of serum albumin in hamster sperm capacitation is not clear. Albumin is included in nearly all capacitation recipes and may facilitate capacitation by removing fatty acids and/or cholesterol from sperm membranes, by binding biogenic amines and prostaglandins, and by its ability to bind Zn^{2+} (which inhibits hamster sperm capacitation) (see Meizel, 1978, 1984).

Bavister and co-workers have found that hamster serum albumin is more effective in supporting hamster fertilization than bovine serum albumin (which is typically used in virtually all hamster studies); however, the reason for this increased activity is unknown (Bavister, 1986).

5.2. Cellular Changes during Capacitation

As capacitation proceeds in hamster sperm, an evolution of motility can be observed, which eventually results in a very vigorous asymmetric flagellar motion called "hyperactivation" (Katz *et al.*, 1986; see Yanagimachi, 1981) (Section 6). Dibutyryl cAMP promotes hyperactivation in hamster sperm (Mrsny and Meizel, 1980). Sperm adenylate cyclase has also been shown to increase its activity during capacitation (Morton and Albagli, 1973; Mrsny *et al.*, 1984). Such increases could stimulate cAMP-dependent protein kinase and in turn alter the physiological properties of the membranes (see Yanagimachi, 1988a; Fraser and Ahuja, 1988). Meizel and co-workers have suggested that an increase in Na^+, K^+-ATPase activity is required for capacitation and may be responsible for the K^+-ion influx that occurs during capacitation (Meizel, 1984, 1985). Furthermore, cGMP may mediate this activity (Mrsny *et al.*, 1984). However, how these physiological changes translate into changes in sperm function is unclear.

It has been demonstrated in a number of species that a modification and/or removal of sperm surface components occurs during capacitation (see Fraser and Ahuja, 1988; Yanagimachi, 1988a). Plant lectins have been employed in studies of hamster sperm during capacitation in order to monitor cell surface changes throughout the capacitation process (Ahuja, 1985). By assessing the number and regional specificity of lectin-coated bead binding to sperm, Ahuja (1984) showed that changes in surface glycoconjugates occurred in a time-dependent manner. More meaningful was the observation that at the time hamster sperm exhibit hyperactivated motility, the tail bound large numbers of ConA-coated beads. This increase in tail binding, as well as hyperactivated motility, could also be induced precociously by β-amino acids and could be inhibited by metabolic inhibitors and local anesthetics (see Fraser and Ahuja, 1988). Thus, the correlation between the "appearance" (unmasking?) of specific sperm surface glycoconjugates and hyperactivated motility suggests, at the very least, that metabolic changes during capacitation are linked in some way. Whether the ConA receptors on the hamster sperm tail are ion channels or are just useful, nonfunctional capacitational markers remains to be seen.

The hamster has been commonly used in studies of sperm–zona binding. Many of these have examined the acquisition of zona binding by sperm at various stages of capacitation; this is discussed in Section 8. The majority of detailed studies on dynamic alterations in membrane components during capacitation have focused on other species (such as the guinea pig) (see Yanagimachi, 1988a). However, the hamster could provide useful material for future molecular studies on the evolution of motility or cell surface components involved in the interaction with the oocyte vestments, primarily the cumulus extracellular matrix and the zona pellucida.

6. HYPERACTIVATION

Hyperactivated motility, first described for hamster sperm (Yanagimachi, 1969; Gwatkin and Anderson, 1969), is an extremely vigorous, nonprogressive form of motility displayed primarily by capacitated mammalian spermatozoa (Fig. 1). This motility has been described for sperm collected from the female reproductive tract or capacitated *in vitro* in at least 15 species (Yanagimachi, 1981, 1988b; Drobnis and Katz, 1990); thus, it is likely a universal feature of mammalian fertilization. During hyperactivation, sperm produce extremely high-curvature flagellar bends, which increase greatly the forces they generate (Katz *et al.*, 1987, 1989). These increased forces are probably required by sperm to penetrate the zona pellucida (see Section 10) and may also be involved in the escape of sperm from the isthmic reservoir (see Section 4). Although this distinctive form of motility is considered critical for the completion of fertilization, *in vitro* as well as *in vivo*, the molecular basis of hyperactivation has not been characterized. This is a research area of considerable current interest.

Hamster sperm continue to be a prevalent model species in studies of motility hyperactivation, particularly in relation to other events of capacitation (Section 5) and sperm transport (Section 4). Hyperactivated motility has been observed within the oviductal ampulla of the hamster, demonstrating that it is present in the very small numbers of sperm present at the site of fertilization.

The head hook in hamster spermatozoa, as well as those of murine rodents, indicates the direction of bending with respect to the axonemal elements (Woolley, 1977; Mohri and Yano, 1980; Ishijima, and Mohri, 1985). The hamster is the only species for which flagellar kinematics have been well characterized for epididymal sperm (Suarez, 1988), sperm capacitating in low-viscosity medium (Katz *et al.*, 1978, 1986; Suarez *et al.*, 1984; Corselli and Talbot, 1986; Drobnis *et al.*, 1988c; Suarez, 1988), sperm collected from or within the female reproductive tract (Katz and Yanagimachi, 1980; Cummins and Yanagimachi, 1982), and sperm penetrating the cumulus cell matrix (Section 7) and the zona pellucida (Section 10).

One important method used to study the control of sperm motility is removing the membranes with detergent, then reactivating motility by addition of ATP and appropriate modulators of axonemal function (Gibbons, 1981). Such work has revealed the importance of cAMP, Ca^{2+}, pH, and other factors in the control of the pattern and vigor of flagellar flagellation. For mammalian sperm, these studies have used primarily bovine, dog, and rat sperm. However, because the events that require hyperactivated motility, particularly penetration of the zona pellucida, are best characterized in the hamster, this should be a species of choice in continued studies of the molecular basis of motility control.

7. SPERM PENETRATION OF THE CUMULUS CELL MATRIX

With the exception of ruminants, mammalian sperm that leave the oviductal isthmus and enter the ampulla are confronted with the cumulus cell complex (the cumulus cells and their associated extracellular matrix, hereafter referred to as the cumulus), which must be traversed in order to gain access to the zona pellucida. Because this complex is optically opaque, bio-

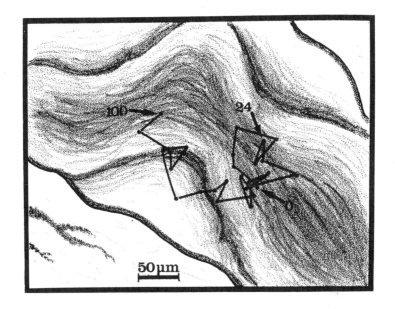

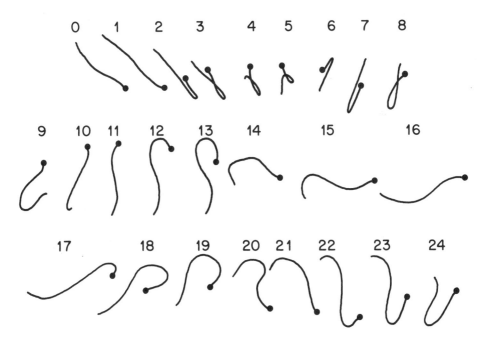

Figure 1. Sketch made from the video monitor of the path taken by a hamster spermatozoon within the oviductal ampulla from a mated female. The oviduct was removed and mounted in a slide preparation for observation. Tracings of the sperm flagellum at each video frame are shown below. From Katz and Yanagimachi (1980).

chemically complex, and poorly characterized, it is usually removed in order to simplify studies of mammalian fertilization. In consequence, there is a paucity of information regarding penetration of the cumulus by sperm. Two related areas are of primary interest: (1) the physiological attributes required by capacitated sperm to enable penetration of the cumulus, and (2) the mechanism of cumulus penetration.

What is required for penetration of the cumulus is a very important question. For example, it is possible that events that have been studied on oocytes or zonae from which the cumulus has been removed are not relevant to conditions *in vivo* because the sperm interacting with the zona may not have been able to gain access to the zona in the presence of the cumulus. The mechanism of cumulus penetration is also important. It is widely accepted that the enzyme hyaluronidase, which is associated with mammalian sperm and is capable of dispersing the cumulus, enables the sperm to penetrate the cumulus matrix. However, this hypothesis is controversial (Yanagimachi, 1981, 1988b; Talbot, 1985) and requires further investigation. The controversy centers around the site of the AR (Section 9). Since most of the hyaluronidase activity is associated with the acrosome, and since it is a predominant hypothesis that the acrosome reaction occurs at the surface of the zona pellucida, how can the acrosomal hyaluronidase participate in dissolution of the cumulus? Recent studies suggest that both acrosin and hyaluronidase in the sperm acrosome can alter the hyaluronate-acid-rich cumulus matrix (Yudin *et al.*, 1988; Cherr *et al.*, 1990).

Penetration of the cumulus by sperm has been studied principally in the hamster, since this species produces large numbers of oocytes and has an acrosome visible by light microscopy. Studies involving the site of the acrosome reaction and evidence against a requirement for enzymatic degradation of the matrix for sperm penetration are discussed below (Section 9.1). The sperm attributes required for penetration of the cumulus have also been studied in hamsters. There is a discrete time interval during which hamster sperm being incubated under capacitating conditions are capable of cumulus penetration: freshly collected sperm and sperm that have completed the AR and lost their acrosomal caps are not capable of initiating cumulus penetration; rather, they adhere to cumulus cells and/or the matrix material. As discussed above (Section 5), surface changes accompany capacitation, and these may enable penetration of the cumulus. Alternatively, several investigators have proposed that the increased forces produced by hyperactivated sperm are important in penetration not only of the zona pellucida but also of the cumulus (Yanagimachi, 1981; Talbot, 1985). However, direct observations of cumulus penetration by hamster sperm have demonstrated that sperm penetrate the surface of the cumulus mass by means of low-amplitude, high-frequency sinusoidal beats; they appear to burrow their way into the matrix. Once within the matrix, the high-curvature bends typical of hyperactivation resume (Fig. 2), and these may be involved in deflecting or tearing the highly elastic matrix, since these high-curvature bends are usually followed by low-amplitude flagellation, during which sperm progress. Electron microscopy of this process shows no evidence of dissolution of the matrix; instead, the acrosome-intact sperm appear to stretch the matrix (Fig. 3) (Yudin *et al.*, 1988). To elucidate the molecular mechanisms of cumulus penetration, more research is required. This area is central, involving capacitation and induction of the AR. Work should continue using the hamster model.

8. SPERM-ZONA PELLUCIDA INTERACTION

8.1. Morphology of Sperm–Zona Interaction

Sperm–egg interactions have been investigated extensively in the hamster. The hamster has been a valuable system for these studies, particularly those that have taken advantage of the differing affinities for the zona of sperm in different capacitational states; in addition, the ease of

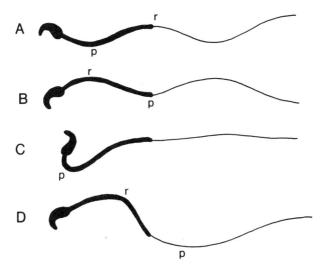

Figure 2. Sketch of hamster spermatozoa showing the direction of flagellar bending relative to the head hook and beat shapes typical of sperm during penetration of the oocyte investments. A and B show one symmetrical, sinusoidal beat cycle. Note that the principal bend initiated in A has propagated along the flagellum and is followed by a reverse bend of similar curvature formed in B. C shows an extreme principal bend such as those involved in "hatchet strokes" in low-viscosity medium and in the cumulus matrix and in "lever strokes" during penetration of the zona pellucida. These distinctive bends form slowly, are released very rapidly, and do not seem to be propagated along with flagellum. D shows an extreme reverse bend. These bends, which are initiated more distally than are extreme principal bends, are never displayed during zona penetration. These reverse bends become quite extreme in some sperm incubating under capacitating conditions. Since they are followed by an extremely low-curvature principal bend, the resulting asymmetric beat produces a star-shaped, circular, or figure-eight trajectory. From Katz *et al.* (1989).

observing the acrosome during zona interaction and the quantitative aspects of flagellar movement are other aspects that make the hamster useful for sperm–zona investigations. The hamster, in comparison with the mouse, has only recently been utilized for molecular studies on sperm–zona interaction (Moller *et al.*, 1990).

The strength of sperm–zona interaction has been a key endpoint for most studies conducted in mammalian gamete interaction, and studies using the hamster are no exception. In the hamster the strength of sperm–zona binding has been qualitatively examined by repeatedly pipetting eggs with adhering sperm through a large-bore pipet and counting the number of sperm still attached; this is compared to the number prior to pipetting or to those in various treatments (Hartmann *et al.*, 1972; Yanagimachi, 1981). It has been demonstrated that two types of sperm adhesion to the zona occur: a loose *attachment* in which sperm can be removed by pipetting, and a tight *binding* in which sperm cannot; presumably both of these types of adhesion occur in sperm prior to their undergoing the AR, but this has not been well documented in the hamster (Hartmann, 1983; Wassarman, 1987).

During gamete interaction, hamster sperm first "loosely" attach to the zona, and the lateral portion of the sperm head contacts the zona material; these sperm swivel back and forth against the zona and may detach and subsequently reattach (Hartmann, 1983; Suarez *et al.*, 1984; Katz *et al.*, 1986; Drobnis *et al.*, 1988b). Following a period of time, which may vary from a few minutes to 30 min (depending on the sperm preparation and the laboratory reporting the results), sperm

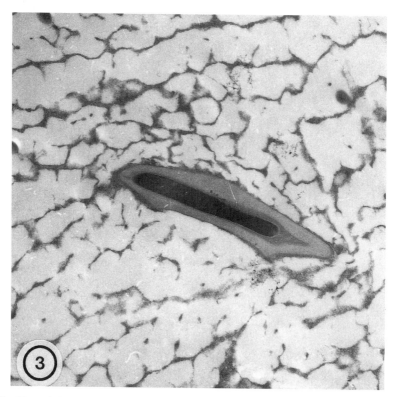

Figure 3. Transmission electron micrograph of rapidly frozen hamster sperm within the cumulus extracellular matrix. The acrosome of this sperm is intact, and the compression of the matrix material at the leading edge of the sperm can be observed. From Yudin *et al.* (1988).

bind tightly to the zona and do not detach. Although it is unclear if in the different studies "bound" sperm have initiated or completed the AR, it has been documented that the movement of the sperm head against the zona is less pronounced in AR sperm, suggesting greater head restriction because of a higher-affinity interaction (Suarez *et al.*, 1984; Katz *et al.*, 1986; Drobnis *et al.*, 1988b) (Section 10). Head movement increases again once the sperm begins to penetrate the zona but is restricted to a small angle of rotation (Drobnis *et al.*, 1988b). It is likely that tight binding to the zona closely precedes and/or is associated with the AR; however, this relationship has not been thoroughly investigated in the hamster.

Once hamster sperm complete the AR on the zona, the fenestrated acrosomal cap (referred to as a ghost or shroud) remains tightly attached to the zona and provides an anchoring mechanism for the sperm as it penetrates the zona (Fig. 4) (Yanagimachi and Phillips, 1984; Cherr *et al.*, 1986). It is possible that the remnants of the cap region are what are responsible for some binding ability of AR sperm observed *in vitro*. The analysis of the cap region in sperm and the molecules involved in bending should be feasible since the caps can be isolated (Zao and Talbot, 1986).

8.2. Zona Receptor Activity in Sperm

Sperm–zona binding, but not attachment, is species specific, reflecting the ligand-receptor interaction that probably occurs in binding and not in attachment (Gulyas and Schmell, 1981; Wassarman, 1987). Gwatkin and Williams (1977) first demonstrated that soluble hamster zonae in

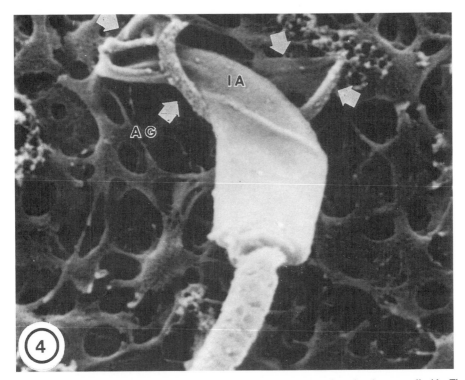

Figure 4. Scanning electron micrograph of acrosome-reacted hamster sperm bound to the zona pellucida. The remnants of the acrosomal cap (acrosomal "ghost"; AG) are adhering to the zona. The inner acrosomal membrane (IA) is exposed and is the leading edge of the sperm as it begins to penetrate the zona. From Cherr *et al.* (1986).

the fertilization medium inhibited capacitated sperm from binding to homologous zonae. Furthermore, solubilized zonae from fertilized eggs do not possess this inhibition of sperm binding. Sperm can attach but do not bind to intact cumulus-free fertilized eggs. These data indicate that, as in the mouse zona, the binding capability of the hamster zona is altered following egg activation (Gwatkin, 1989) (Section 11).

Hamster sperm's ability to bind to zonae is dependent on their degree of capacitation (Hartmann and Gwatkin, 1971; Hartmann, 1983). Noncapacitated sperm attach to the zona, and this attachment may involve an epididymal component and terminal glucuronic acid residues (Hartmann, 1983). Capacitated and noncapacitated hamster sperm exhibit differing affinities for zona solutions, since these solutions only inhibit capacitated sperm from binding to eggs (Gwatkin and Williams, 1977). The sperm surface components involved in zona binding are probably "exposed" in some way as a result of capacitation; this certainly must be associated with the ability of sperm to undergo the zona-induced AR. The hamster sperm component that appears to exhibit affinity for the hamster zona is a 26-kDa glycoprotein; it is believed that this is a plasma membrane component (Sullivan and Bleau, 1985). As the molecular details of sperm surface dynamics during capacitation are uncovered, this will most likely lead to an understanding of sperm–zona recognition and the induction of the AR.

Investigations of the molecular details of sperm–zona binding in the hamster are limited compared to the murine system. Gwatkin and co-workers have demonstrated that protease treatment of hamster zonae inhibits sperm binding (but not attachment) (Gwatkin, 1977). Furthermore, various lectins that bind to specific sugar residues on the zona also block binding and fertilization (Oikawa *et al.*, 1973, 1974). However, since the zona is a glycoprotein coat, these results do not specifically differentiate among the various zona components. More recently, Ahuja (1985) used specific haptens to inhibit sperm–zona recognition that involves sugar moieties. Complex polysaccharides containing fucose and galactose inhibit sperm–egg binding. In addition, treatment of sperm (but not eggs) with glycosidases reduces sperm binding, suggesting that the sperm surface component(s) involved in zona recognition in the hamster is a carbohydrate. However, nonspecific effects of the enzyme treatments have not been completely ruled out. Most recently, Moller and co-workers (1990) have demonstrated that a 56-kDa glycoprotein from the hamster zona, ZP3, possesses all of the sperm receptor activity for the zona. This hamster zona glycoprotein is homologous to the mouse ZP3, which also possesses murine receptor and AR-inducing activities. Although soluble hamster zonae in the Moller *et al.* study induced acrosome reactions in sperm, purified ZP3 did not. At present it is not clear if AR-inducing activity, in addition to sperm receptor activity, is associated with hamster ZP3.

9. THE ACROSOME REACTION

The hamster has been used a great deal to study the mechanisms involved in the exocytotic event, the AR. Some of the reasons for the hamster's popularity for these studies are: (1) the *in vitro* capacitation system for hamster sperm is well defined; (2) one can obtain cells that are synchronously capacitated but have yet to be induced to undergo the AR; (3) there is typically a lower level of background reactions in the cell suspensions than in other mammals; (4) inducers (both natural and artificial) of the reaction can be added to capacitated unreacted cells, and ARs can occur rapidly (minutes); and (5) the acrosome of hamster sperm can be easily observed in the light microscope, and its alteration as a result of the membrane fusion can be assessed in motile sperm (Fig. 5). The hamster sperm AR has been proposed as a potential model system for understanding the molecular mechanisms involved in exocytosis in many cell types (Meizel, 1984, 1985). Much of the literature has focused on the regulation of the AR by endogenous molecules through the use of exogenous and/or synthetic inhibitors and inducers. Only more recently have some of the natural inducers, derived from oocyte–cumulus complexes, been utilized. Since the sperm AR is believed to be an absolute requirement for zona penetration (Section 10), sperm–egg fusion (Section 12), and in some mammals to play a role in cumulus penetration, the location of the *in vivo* AR and its regulation by the natural inducer(s) are important to our basic understanding and clinical manipulation of the fertilization process.

As discussed by Meizel (1978, 1984), it should be realized that it may be difficult to separate the molecular events of capacitation and the AR since it is difficult to determine when capacitation ends and the AR begins. A useful definition may be that capacitation prepares the sperm for undergoing the AR, as proposed by Bedford (1970a).

9.1. Induction of the Acrosome Reaction during Interaction with the Oocyte–Cumulus Complex

The hamster has been perhaps the most widely utilized mammalian system for investigating the occurrence of the sperm AR during interaction with oocyte–cumulus investments (Meizel, 1985; Bavister, 1986). Although a number of "natural" and synthetic inducers of the hamster sperm AR have been documented, the site of the AR of the fertilizing sperm has been very difficult to address. Since a great deal of inference arises from studies of the AR that utilize inducers *in vitro*, attempts to conduct studies that can elucidate the site of the AR *in vivo* need to continue in order to achieve a better understanding of the functional role of the AR.

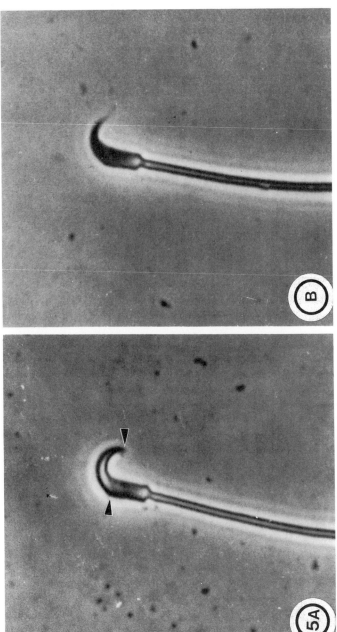

Figure 5. Phase-contrast (40× objective lens) micrographs of acrosome-intact (A) and acrosome-reacted (B) hamster sperm. The hook of the head is greater in the unreacted sperm, and the acrosomal cap (arrows) is highly refractile. This difference is easy to detect even in highly motile sperm. From Cherr *et al.* (1986).

It has been demonstrated that hamster sperm that have completed the AR cannot initiate entry of the cumulus matrix (Suarez *et al.*, 1984; Cherr *et al.*, 1986; Cummins and Yanagimachi, 1986; Corselli and Talbot, 1986). From experiments in which sperm in oocyte cumulus complexes were recovered from the oviduct, it has been suggested that the AR was initiated, completed, or, at the very least, the acrosomes were "modified" (as reviewed by light microscopy) when sperm were in the cumulus (Cummins and Yanagimachi, 1982; Yanagimachi and Phillips, 1984; see Bavister, 1986; Yanagimachi, 1988b). Unfortunately, in studies of this type, it is almost impossible to determine the residence time of the sperm in the cumulus prior to recovery and AR assessment (i.e., age of sperm) or if these sperm are supernumerary. Since capacitated hamster sperm spontaneously undergo the AR as they age, and excess sperm may not be capable of zona penetration, the assumption that the inducer was a component of the cumulus may not be appropriate. In more recent *in vitro* studies that controlled sperm residence time in the cumulus, it was found that sperm did not initiate and/or complete the AR in the cumulus (Fig. 3) (Cherr *et al.*, 1986; Cummins and Yanagimachi, 1986; Yudin *et al.*, 1988).

Although there is still some uncertainty as to the AR-initiating (rather than induction and completion) capability of the cumulus, it is now established in the hamster that the zona is a potent inducer of the AR (Cherr *et al.*, 1986; Yoshimatsu and Yanagimachi, 1988; Uto *et al.*, 1988; Moller *et al.*, 1990). Nevertheless, the presence of the cumulus extracellular matrix is important in modulating (increasing) the competency of sperm to complete the AR on the zona during sperm interaction with the oocyte–cumulus complex (Cherr *et al.*, 1986). Although the cumulus could act synergistically with the zona to induce the AR, it is most likely a "barrier" or "filter" that restricts sperm that possess surface properties characteristic of either a lesser degree of capacitation or are less capable of initiating the AR from reaching the zona (Suarez *et al.*, 1984; Yudin *et al.*, 1988; Cherr *et al.*, 1986, 1990; see Meizel, 1985, Bavister, 1986).

9.2. Stimulation and Regulation of the Acrosome Reaction *in Vitro*

A number of reviews of the exogenous molecules that stimulate the hamster AR *in vitro* suggest that a number of potentially relevant biological compounds are capable of initiating a response (Meizel, 1978, 1984, 1985; Bavister, 1986). Serum albumin appears to be an important molecule in the hamster AR (Dow and Bavister, 1989). Albumin certainly is critical to capacitation; therefore, its role in the AR may be secondary. However, a number of active stimulators of the AR have been found to be bound by albumin, suggesting that albumin may act as a vehicle for these inducers. In addition, albumin may perturb the lipid bilayer of the sperm membranes, thereby inducing membrane destabilization; however, more research on the specific role of the albumin molecule in the AR is required (see Meizel, 1985). Serotonin has been shown to stimulate the hamster AR *in vitro* (Meizel, 1985). This stimulation is in a more direct manner (with respect to the AR) than that by catecholamines, which increase capacitation. Whether this stimulation involves an increase in membrane Ca^{2+} permeability, as in other systems, remains to be seen.

For both the zona-induced and spontaneous hamster AR, specific extracellular ions in the medium are required (Yanagimachi, 1981, 1982; Yoshimatsu and Yanagimachi, 1988). For example, Ca^{2+} influx is necessary in order for the AR to occur (Yanagimachi 1982). There is evidence that Na^+, K^+-ATPase activity is important and that K^+ influx occurs during the AR (Meizel, 1984). Meizel and co-workers have also demonstrated that an alkalinization of the hamster acrosomal compartment must occur in order for the AR to proceed (see Meizel, 1984). These ionic changes may modulate hydrolytic enzymes that are activated during the hamster AR. Meizel and co-workers have also demonstrated that trypsin-like enzyme activity is required for the membrane fusion events of the AR (Dravland *et al.*, 1984; Meizel, 1984); this suggests that the dispersion of the acrosomal matrix by trypsin-like activity may be only a secondary action of these enzymes. The activation of a trypsin-like protease in the acrosome corresponds to the conversion of proacrosin to acrosin, perhaps as a result of increased pH. Phospholipase A_2 activity may also be important for the hamster AR. Fusogenic lysophospholipids and *cis*-unsaturated fatty acids that can result from phospholipase A_2 activity are capable of inducing the hamster AR (Meizel, 1984). There also is evidence that arachidonic acid and its metabolism via cyclooxygenase and lipoxygenase pathways may be involved in the AR (Meizel, 1984). Such oxidative products are hydroxyeicosatetraenoic acids and prostaglandins stimulate the AR, and arachidonic acid stimulation of the AR can be inhibited by cyclooxygenase and lipoxygenase inhibitors.

The hamster has been an extremely useful model for investigating the AR from a molecular as well as a reproductive biological perspective. There are numerous advantages to use of the hamster for AR studies, and it should continue to be one of the key systems for future research efforts.

10. SPERM PENETRATION OF THE ZONA PELLUCIDA

Among the most active areas of research on mammalian fertilization are those involving the interaction of sperm with the zona pellucida. As discussed above, this includes binding of sperm to the zona (Section 8) and induction of the AR (Section 9). The penetration of sperm into the zona matrix has received less attention, and the mechanism by which the sperm penetrates this mechanically tough material is unknown; most investigators believe that both enzymatic cleavage and mechanical forces are involved.

There is controversy, analogous to that for penetration of the cumulus cell matrix (Section 7), regarding the mechanism of penetration of mammalian sperm into the material of the zona pellucida (Yanagimachi, 1981, 1988b; Moore and Bedford, 1983; Talbot, 1985; Green, 1988; Katz *et al.*, 1987, 1989). Associated with mammalian spermatozoa is a high activity of acrosin, a trypsin-like protease capable of dissolving the zona material under some conditions. It is not clear that acrosin activity is localized on the surface of AR (Yanagimachi, 1981; 1988b; Moore and Bedford, 1983; Talbot, 1985), which would be requisite for participation in penetration of the zona.

The role of sperm forces in zona penetration is also controversial (Green, 1988; Katz *et al.*, 1989; Katz and Drobnis, 1990). The zona pellucida is a structure unique to mammals. Moore and Bedford (1983) have proposed that unique morphological features of mammalian sperm—the extremely flattened head and the sharp leading edge of the sperm head, with its specialized hardened structure, the perforatorium—allow the sperm to cut, knife-like, into the zona material. Early estimations of the forces exerted by spermatozoa assumed symmetrical, sinusoidal flagellar waveforms or were estimated experimentally using sperm having this motility (Katz and Drobnis, 1990; Drobnis and Katz, 1990).

Direct observations of zona penetration by hamster sperm have shed some light on the mechanical forces involved in penetration of the zona. During zona penetration, hamster sperm alternate between symmetrical, sinusoidal beats and asymmetric "lever" beats in which the flagellum remains straight (Fig. 6). The latter beats involve formation and release of extreme principal bends, and they enable the sperm to exert peak forces against the zona that are more than an order of magnitude greater than the forces produced during sinusoidal flagellation (Drobnis *et al.*, 1988b; Katz and Drobnis, 1990). Although similar bends are seen in sperm penetrating the cumulus ("hatchet" beats) and in some hyperactivated sperm swimming in low-viscosity medium, they have the opposite curvature to the extreme reverse bends produced by sperm swimming in circles and star-shaped trajectories in low-viscosity medium. Thus, the direction of the flagellar bends are important in studies of sperm motility, and this is an advantage to making such observations using hamster, mouse, or rat spermatozoa.

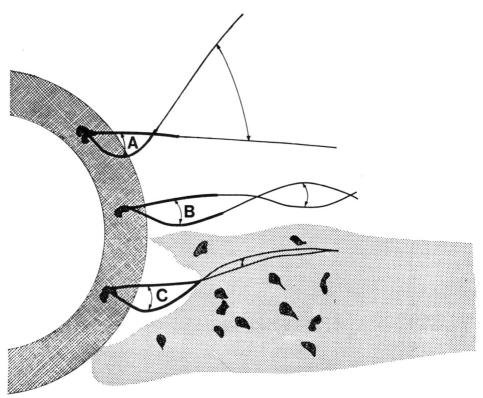

Figure 6. Sketch showing the kinematics of hamster spermatozoa during penetration of the hamster zona pellucida. The "lever beat" shown in A, during which the flagellum remains relatively straight, enables the sperm to exert large peak forces against the zona material. These asymmetric lever beats alternate with symmetrical, sinusoidal beats shown in B. C shows the expression of lever motility for a spermatozoon enclosed in the cumulus cell matrix. Under these conditions, the flagellum is constrained, with minimal deflection in the distal flagellum. Drawn by A. I. Yudin. From Drobnis *et al.* (1988b).

Future studies will continue to focus on the requirements for sperm enzyme activity and strong sperm forces during zona penetration. The hamster is an excellent model species for such studies, although the results obtained in the hamster should be applied to other species as well.

11. ZONA PELLUCIDA STRUCTURE AND COMPOSITION

The hamster zona pellucida is a an extracellular coat that appears to be composed of filamentous material and is very similar in morphology to the mouse zona (Phillips and Shalgi, 1980; Talbot, 1984; Yudin *et al.*, 1988). When viewed using rapid freezing techniques to avoid artifactual compaction, the hamster zona appears to possess large crevices or pores that are greatest in number in the outer third; in addition, the zona material exhibits an overall porous structure (Fig. 7) (Yudin *et al.*, 1988).

Various lectins have been used to probe the molecular nature of the zona. Generally, these studies have indicated that there may be an asymmetry of zona carbohydrates, since increased labeling is typically observed in the outer regions of the zona (Nicolson *et al.*, 1975; Ahuja and

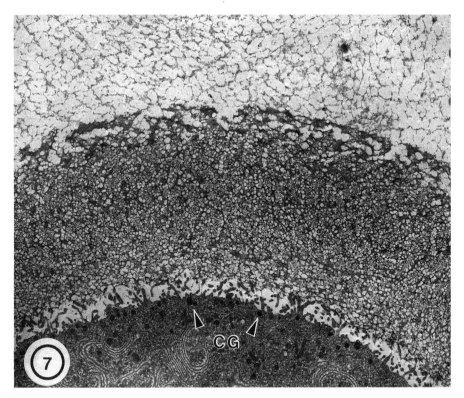

Figure 7. Transmission electron micrograph of a rapidly frozen hamster egg showing the porous nature of the zona pellucida and the extension of the cumulus extracellular matrix into the crevices. This is a different view of the zona ultrastructure from the one observed using conventional fixation methods. Cortical granules (CG) in the egg cytoplasm and the numerous microvilli on the egg surface can also be observed. From Yudin *et al.* (1988).

Bolwell, 1983). Since many of these studies labeled intact zonae rather than histological sections, it is possible that the label was only able to penetrate the outer regions of the zona, which possess the greatest pore diameter. In addition, as label begins to increase in the outer zona region, it may block additional label from penetrating. Based on lectin labeling, the hamster zona possesses N-acetylglucosamine, mannose, galactose, and N-acetylgalactosamine residues (Nicolson *et al.*, 1975, Ahuja and Bolwell, 1983; Oikawa *et al.*, 1988).

Analyses of the protein composition of the hamster zona suggest that three major components are present (Ahuja and Bolwell, 1983; Oikawa *et al.*, 1988; Moller *et al.*, 1990). Ahuja and Bolwell (1983) reported that the three components they found under reducing conditions were glycoproteins with molecular masses of 240 kDa, 150 kDa, and 80 kDa. Oikawa and co-workers (1988) found zona components ranging from 55 to 110 kDa in molecular mass. More recently, Moller *et al.* (1990) found that under reducing conditions, radioiodinated hamster zonae had components with molecular masses of 208 kDa, 103 kDa, and 56 kDa. When the hamster zona components were probed with a radiolabeled mouse ZP3 cDNA it was also found that the probe hybridized specifically to a 1.5-kb RNA in hamster poly(A)$^+$ RNA preparations; this is consistent with the size of the murine ZP3 mRNA. Overall, Moller and co-workers believe that the hamster zona glycoproteins are structurally and functionally related to the murine zona components, which have been so thoroughly characterized (Wassarman, 1987).

In addition to the three major structural zona components, oviductal components appear to be associated with zona from eggs that have been in oviductal fluid (Oikawa *et al.*, 1988; Kan *et al.*, 1988; Bleau and St. Jacques, 1989). Oikawa and co-workers (1988) have described a glycoprotein, ZP0, that has a molecular mass of 200–240 kDa and is added to the zona in the oviduct. This glycoprotein appears to possess galactose and/or N-acetylgalactosamine residues. Bleau and co-workers have also identified a hamster oviductal glycoprotein that associates with the zona, which they call oviductin (Bleau and St. Jacques, 1989). Under reducing conditions, oviductin has a molecular weight of 200,000; this glycoprotein is probably the same as the one described by Oikawa *et al.* (1988). The functional significance of oviductal components in the zona is unclear; there is no evidence to suggest they play a role in fertilization *per se*. It may be that their function is in facilitating transport of the fertilized egg through the oviduct or in embryonic development.

12. SPERM–EGG FUSION

A significant number of studies on sperm interaction with the vitellus of living hamsters eggs have been conducted. Following sperm penetration of the zona pellucida, the sperm head enters the perivitelline space and immediately contacts the vitellus. Within 5–15 sec after contacting the vitellus, hamster sperm motility declines (Yanagimachi, 1978). Complete sperm head incorporation occurs within 15 min in a phagocytic manner (Yanagimachi, 1988a). The incorporation of the flagellum occurs shortly following incorporation of the sperm head (Yanagimachi, 1981).

In the hamster, as in other eutherian mammals, the plasma membrane, not the inner acrosomal membrane, fuses with the vitellus. Specifically, the plasma membrane above the sperm's equatorial segment, presumably as a result of the AR, possesses the ability to fuse with the egg's microvilli; acrosome intact sperm cannot fuse (Shalgi and Phillips, 1980; Yanagimachi, 1981, 1988a,b). Sperm generally do not fuse in the region of the egg over the metaphase spindle, since this region lacks microvilli and probably possesses different membrane and cortical properties.

The zona-free hamster egg has been used extensively in studies of sperm–egg fusion at both a fundamental level and, most commonly, at the clinical level (Yanagimachi, 1981; Section 14.2). Extracellular Ca^{2+} is required for hamster sperm–egg fusion, as in other mammals (Yanagimachi, 1982). In Ca^{2+} deficient medium, sperm attach to the vitellus but do not fuse; Mg^{2+}, Ba^{2+}, and Sr^{2+} can substitute for Ca^{2+} but are not as effective with respect to supporting fusion (Yanagimachi, 1988b). Proteolytic treatment of zona-free hamster eggs does not affect the ability of sperm to fuse with the egg plasma membrane; in fact, numerous enzyme treatments do not affect fusion with the exception of phospholipase C (Hirao and Yanagimachi, 1978). Following fusion, the hamster sperm plasma membrane intermixes with egg plasma membrane components (Yanagimachi, 1981).

The plasma membrane of mammalian eggs is far less species specific than the zona. The hamster plasma membrane appears to possess the lowest degree of specificity among mammals (Section 14.2). Yanagimachi (1981) has shown that although the hamster egg is permissive to heterologous sperm, it will demonstrate a preference for homologous sperm when challenged with a mixed-species solution of AR sperm. It has become common practice to use the zona-free hamster egg to assess sperm-fusing parameters in heterologous sperm, since it may be difficult to obtain homologous eggs and the hamster egg is widely permissive to sperm from many species. However, caution must be taken, since mechanisms involved in regulating homologous sperm–egg fusion may not apply to a heterologous systems. Little is known about mechanisms involved in the regulation of sperm-egg fusion, but there are indications that heterologous systems may differ from homologous ones. For example, Fukuda and Chang (1978) showed that when heterologous sperm (mouse) fused with zona-free hamster eggs, an incomplete, delayed exocytosis of cortical granules occurred. This suggests that a homologous receptor-mediated event other than just membrane fusion is required to achieve normal egg activation (Section 13); such a discrepancy could dramatically alter the interpretations of numerous heterologous "mechanistic" studies.

13. EGG ACTIVATION

In the hamster, as in other animals, the fertilizing sperm triggers the egg to undergo a series of complex morphological and physiological changes. The most obvious indication of egg activation is the "disappearance" (exocytosis) of cortical granules (Barros and Yanagimachi, 1972; Cherr et al., 1988). The exocytosis of cortical granule material is necessary for alterations in the zona pellucida that are critical for polyspermy prevention and embryonic development. Much of this has recently been reviewed (Miyazaki, 1989; Cherr and Ducibella, 1990). The first morphological evidence of activation of the hamster egg are alterations in the egg surface; these include an increase in space between microvilli and the appearance of cortical granule fusion sites (Cherr and Ducibella, 1990). The cortical granules, which lie beneath the plasma membrane, secrete their contents into the perivitelline space within a few minutes after sperm–egg fusion or on artificial stimulation (e.g., electrical stimulation, ionophores, membrane-active agents).

The hamster has recently become the mammalian system of choice for investigating the transduction of the activation signal in the egg from the sperm, and as a result, this research has increased our understanding of signal transduction in cells. Miyazaki and co-workers have published a series of articles describing their investigations on the events and mechanisms of egg activation and signal transduction in the hamster. The fertilization response in the hamster egg is a series of recurring transient negative-going hyperpolarizations (Miyazaki and Igusa, 1981). This is the opposite of the sea urchin egg, which exhibits a positive-going depolarization at activation. The resting potential of the hamster egg is between -20 and -25 mV prior to insemination and gradually shifts to -40 mV during a series of hyperpolarizing responses; these are a series of Ca^{2+} transients occurring at 40 to 120 sec intervals, and the entire response may take up to 2 hr (Miyazaki, 1989). The hyperpolarizing response has been shown to be related to a K^+ current activated by an increase in Ca^{2+} (Miyazaki and Igusa, 1982).

In general, the biochemical events and the endogenous regulators of egg activation in mammals appear similar to what has been described in the more thoroughly studied invertebrate systems (Cherr and Ducibella, 1990). Sperm in some way activate a phosphatidylinositol bisphosphate (PIP_2) cascade with the production of inositol trisphosphate (IP_3) followed by a Ca^{2+} activation of cortical granule exocytosis. Since ionophores can activate hamster eggs in the absence of extracellular Ca^{2+} (Steinhardt et al., 1974) and IP_3 can induce intracellular Ca^{2+} release (Miyazaki, 1988), it is believed that the mobilization of the egg's own Ca^{2+} stores is adequate for cortical granule release. Finally, there is evidence that GTP-binding protein activation is involved in Ca^{2+} release and cortical granule exocytosis (Miyazaki, 1988; Cran et al., 1988). Although intracellular Ca^{2+} stores appear sufficient to initiate activation of the hamster egg, it appears that Ca^{2+} influx is necessary for a normal series of hyperpolarization responses (Miyazaki, 1989).

Recent research on the hamster (as well as the mouse) has suggested that the mammalian egg is excellent for investigating mechanisms of activation. Future molecular studies on signal transduction and signal-induced Ca^{2+} events at fertilization will probably utilize the hamster system in some capacity.

14. THE ZONA REACTION AND VITELLINE BLOCK TO POLYSPERMIC FERTILIZATION

In mammalian oocytes, the block to polyspermic fertilization initiated by oocyte activation (Section 13) occurs at two levels: (1) the level of oolemma, which we call *the vitelline block*, and (2) the level of the zona pellucida, which we call the *zona block*. Since there are differences among mammalian species in the relative efficacy of these blocks as well as the time required for their formation, this aspect of fertilization must be studied separately for each species (Gwatkin, 1977; Yanagimachi, 1981, 1988b; Wolf, 1981; Schmell *et al.*, 1983).

14.1. The Zona Reaction in Hamster Eggs

Among mammals, the golden hamster has a particularly rapid and efficacious zona block (Wolf, 1981). Within 10 min of fertilization, penetration of the zona is no longer initiated by sperm. In consequence, it is very rare to find sperm in the perivitelline space of fertilized hamster eggs collected from mated females. In contrast, the rabbit zona is regularly penetrated by numerous sperm, which persist in the perivitelline space of early embryos. The mechanism of the zona block is unknown but probably involves both loss of sperm receptors from the surface of the zona and a decrease in penetrability of the zona material.

Loss of sperm receptors from the zona surface, similar to the loss of receptors from the egg envelope in invertebrates, may be involved in the zona reaction (Schmell *et al.*, 1983). As in the sea urchin, cortical granules of hamster oocytes contain proteases that are released during egg activation (Section 13), and these may be involved in modification of the zona surface (Cherr *et al.*, 1988). However, the zona reaction involves more than the loss of receptors (or of inducers of the AR) from the surface of the zona, since in hamsters, and in other species, sperm that have initiated zona penetration are immobilized in the zona material, unable to complete penetration. These sperm, which are unable to complete penetration of the zona, are found in fertilized oocytes flushed from the oviducts. Zona hardening, first described in the mouse, describes the increase in resistance to dissolution (or dispersion) of the zona pellucida following fertilization. Initially, some investigators thought that an increased resistance to dissolution by sperm proteases and/or an increased mechanical resistance to sperm forces might be involved in the zona block. This hypothesis was later rejected because, in spite of the uniquely strong zona block in hamsters, there was apparently no zona hardening; that is, the zona of embryos was as easily removed chemically as was the zona of oocytes. Conversely, the rabbit zona does harden following fertilization but does not become resistant to penetration by sperm. More recently, using resistance to mechanical deformation, one of us showed that although the hamster zona does not increase its resistance to dissolution following fertilization, it does become mechanically stiffer (Drobnis *et al.*, 1988a). This reopens the possibility that the zona block involves an increase in the resistance of the zona to mechanical cleavage by sperm.

In contrast to the hamster, most species have a slower, less complete zona block and rely instead on a rapid and complete vitelline block. In these species (i.e., mice, rats, rabbits), more than one sperm rarely fuse with an oocyte, even under conditions of *in vitro* fertilization, where large numbers of sperm are present; however, multiple sperm are occasionally found in the perivitelline space, having crossed the zona but being unable to fuse with the oolemma. The most extreme case is the rabbit, which does not have a zona block. There can be hundreds of perivitelline sperm surrounding a fertilized rabbit oocyte.

14.2. The Vitelline Block to Polyspermy in the Hamster

In contrast to its efficacious zona block, the golden hamster has little or no vitelline block or, at the very least, a transient one (Cherr *et al.*, 1988). In fact, not only is it possible for multiple hamster sperm to fuse with the zona-free vitellus, but sperm from a large number of species are able to fuse with the hamster vitellus. This fact is the basis of the successful penetration assay (SPA) or zona-free hamster egg assay for sperm function. In humans, cattle, and a variety of other species, capacitated sperm are incubated with hamster vitelli, and their ability to fuse and form male pronuclei is assessed. In studies of the mechanism of sperm fusion with the oolemma, the hamster is clearly the species to use as a negative control when testing putative mechanisms.

ACKNOWLEDGMENTS. The authors would like to thank Elisabeth Clark for her help with the typing of this manuscript. We also thank Murali Pillai and Eleanor Uhlinger for their technical assistance.

15. REFERENCES

Ahuja, K. K., 1984, Lectin-coated agarose beads in the investigation of sperm capacitation in the hamster, *Dev. Biol.* **104**:131–142.

Ahuja, K. K., 1985, Carbohydrate determinants involved in mammalian fertilization, *Am. J. Anat.* **174**:207–224.

Ahuja, K. K., and Bolwell, G. P., 1983, Probable asymmetry in the organization of components of the hamster zona pellucida, *J. Reprod. Fertil.* **69**:49–55.

Austin, C. R., 1985, Sperm maturation in the male and female genital tracts, in: *Biology of Fertilization*, Vol. 2 (C. B. Metz and A. Monroy, eds.), Academic Press, New York, pp. 121–155.

Barros, C., and Yanagimachi, R., 1972, Polyspermy preventing mechanisms in the golden hamster egg, *J. Exp. Zool.* **180**:251–266.

Battalia, D. B., and Yanagimachi, R., 1979, Enhanced and co-ordinated movement of the hamster oviduct during the periovulatory period, *J. Reprod. Fertil.* **56**:515–520.

Bavister, B., 1986, Animal *in vitro* fertilization and embryo development, in: *Manipulation of Mammalian Development* (R. B. L. Gwatkin, ed.), Plenum Press, New York, pp. 81–148.

Bavister, B. D., 1989, A consistently successful procedure for *in vitro* fertilization of golden hamster eggs, *Gamete Res.* **23**:139–158.

Bedford, J. M., 1970a, Sperm capacitation and fertilization in mammals, *Biol. Reprod.* **2**:128–158.

Bedford, J. M., 1970b, The saga of mammalian sperm from ejaculation to syngamy, in: *Mammalian Reproduction* (H. Gibian and E. J. Plotz, eds.), Springer Verlag, New York, pp. 124–182.

Bedford, J. M., 1983, Significance of the need for sperm capacitation before fertilization in eutherian mammals, *Biol. Reprod.* **28**:108–120.

Bishop, D. W., 1969, Sperm physiology in relation to the oviduct, in: *The Mammalian Oviduct* (E. S. E. Hafez and R. J. Blandau, eds.), University of Chicago Press, Chicago, pp. 231–250.

Blandau, R. J., 1969, Gamete transport—comparative aspects, in: *Mammalian Oviduct* (E. S. E. Hafez and R. J. Blandau, eds.), University of Chicago Press, Chicago, pp. 129–162.

Blandau, R J., 1973, Gamete transport in the female mammal, in: *Handbook of Physiology, Section 7, Endocrinology II* (R. O. Greep and E. B. Astwood, eds.), American Physiological Society, Washington, DC, pp. 153–163.

Bleau, G., and St. Jacques, S., 1989, Transfer of oviductal proteins to the zona pellucida, in: *The Mammalian Egg Coat Structure and Function* (J. Dietl, eds.), Springer-Verlag, Berlin, pp. 99–110.

Boatman, D. E., Andrews, J. C., and Bavister, B. D., 1988, A quantitative assay for capacitation: Evaluation of multiple sperm penetration through the zona pellucida of salt-stored hamster eggs, *Gamete Res.* **19**:19–29.

Boyer, C. C., 1953, Chronology of development for the golden hamster, *J. Morphol.* **92**:1–38.

Cherr, G. N., and Ducibella, T., 1990, Activation of the mammalian egg: Cortical granule distribution, exocytosis, and the block to polyspermy, in: *Fertilization in Mammals* (B. D. Bavister, J. Cummins, and E. R. S. Roldan, eds.), Serono Symposia, Norwell, Massachusetts, pp. 309–330.

Cherr, G. N., Lambert, H., Meizel, S., and Katz, D. F., 1986, *In vitro* studies of the golden hamster sperm acrosome reaction: Completion on the zona pellucida and induction by homologous solubilized zonae pellucidae, *Dev. Biol.* **114**:119–131.

Cherr, G. N., Drobnis, E. Z., and Katz, D. F., 1988, Localization of cortical granule constituents before and after exocytosis in the hamster egg, *J. Exp. Zool.* **246**:81–93.

Cherr, G. N., Yudin, A. I., and Katz, D. F., 1990, Organization of the hamster cumulus extracellular matrix: A hyaluronate–glycoprotein gel which modulates sperm access to the oocyte, *Dev. Growth Differ.* **32**: 353–365.

Corselli, J., and Talbot, P., 1986, An *in vitro* technique to study penetration of hamster oocyte–cumulus complexes by using physiological numbers of sperm, *Gamete Res.* **13**:293–308.

Cran, D. G., Moor, R. M., and Irvine, R. F., 1988, Initiation of the cortical reaction in hamster and sheep oocytes in response to inositol trisphosphate, *J. Cell Sci.* **91**:139–144.

Cummins, J. M., and Yanagimachi, R., 1982, Sperm–egg ratios and the site of the acrosome reaction during *in vivo* fertilization in the hamster, *Gamete Res.* **5**:239–256.

Cummins, J. M., and Yanagimachi, R., 1986, Development of ability to penetrate the cumulus oophorus by hamster spermatozoa capacitated *in vitro*, in relation to the timing of the acrosome reaction, *Gamete Res.* **15**:187–212.

Dow, M. P. D., and Bavister, B. D., 1989, Direct contact is required between serum albumin and hamster spermatozoa for capacitation *in vitro*, *Gamete Res.* **23**:171–180.

Dravland, J. E., Llanus, M. N., Munn, R. J., and Meizel, S., 1984, Evidence for the involvement of a sperm trypsin-like enzyme in the membrane events of the hamster acrosome reaction, *J. Exp. Zool.* **232**:117–128.

Drobnis, E. Z., and Katz, D. F., 1990, Videomicroscopy of mammalian fertilization, in: *The Biology and Chemistry of Mammalian Fertilization* (P. M. Wassarman, ed.), CRC Press, New York, pp. 269–300.

Drobnis, E. Z., Andrew, J. B., and Katz, D. F., 1988a, Biophysical properties of the zona pellucida measured by capillary suction: Is zona hardening a mechanical phenomenon?, *J. Exp. Zool.* **245**:206–219.

Drobnis, E. Z., Yudin, A. I., Cherr, G. N., and Katz, D. F., 1988b, Hamster sperm penetration of the zona pellucida: Kinematic analysis and mechanical implications, *Dev. Biol.* **130**:311–323.

Drobnis, E. Z., Yudin, A. I., Cherr, G. N., and Katz, D. F., 1988c, Kinematics of hamster sperm during penetration of the cumulus cell matrix, *Gamete Res.* **21**:367–383.

Fraser, L. R., and Ahuja, K. K., 1988, Metabolic surface events in fertilization, *Gamete Res.* **20**:491–519.

Freund, M., 1973, Mechanisms and problems of sperm transport, in: *The Regulation of Mammalian Reproduction* (S. J. Segal, R. Crozier, P. A. Corfman, and P. G. Condliff, eds.), Charles C. Thomas, Springfield, IL, pp. 352–361.

Fukuda, Y., and Chang, M. C., 1978, The time of cortical granule breakdown and sperm penetration in mouse and hamster eggs inseminated *in vitro*, *Biol. Reprod.* **19**:261–266.

Gibbons, I. R., 1981, Cilia and flagella of eukaryotes, *J. Cell Biol.* **91**:107–124.

Green, D. P. L., 1988, Sperm thrusts and the problem of penetration, *Biol. Rev.* **63**:79–105.

Gulyas, B. J., and Schmell, E. D., 1981, Sperm–egg recognition and binding in mammals, in: *Bioregulators of Reproduction* (G. Jagiello, and H. Vogel, eds.), Academic Press, New York, pp. 499–519.

Gwatkin, R. B. L., 1977, *Fertilization Mechanisms in Man and Mammals*, Plenum Press, New York.

Gwatkin, R. B. L., 1989, Zona binding sites of the spermatozoon, in: *The Mammalian Egg Coat, Structure and Function* (J. Dietl, ed.), Springer-Verlag, Berlin, pp. 61–74.

Gwatkin, R. B. L., and Anderson, O. F., 1969, Capacitation of hamster spermatozoa by bovine follicular fluid, *Nature* **224**:1111–1112.

Gwatkin, R. B. L., and Williams, D. T., 1977, Receptor activity of the hamster and mouse solubilized zona pellucida before and after the zona reaction, *J. Reprod. Fertil.* **49**:55–59.

Hamner, C. E., 1972, Physiology of sperm in the female reproductive tract, in: *Biology of Mammalian Fertilization and Implantation* (K. S. Moghissi, and E. S. E. Hafez, eds.), Charles C. Thomas, Springfield, IL, pp. 203–212.

Harper, M. J. K., 1982, Sperm and egg transport, in: *Reproduction in Mammals*, 2nd ed., Vol. 1 (C. R. Austin and R. V. Short, eds.), Cambridge University Press, Cambridge, pp. 102–127.

Hartmann, J. F., 1983, Mammalian fertilization: Gamete surface interactions *in vitro*, in: *Mechanism and Control of Animal Fertilization* (J. F. Hartmann, ed.), Academic Press, New York, pp. 325–364.

Hartmann, J. F., and Gwatkin, R. B. L., 1971, Alteration of sites on the mammalian sperm surface following capacitation, *Nature* **234**:479–481.

Hartmann, J. F., Gwatkin, R. B. L., and Hutchison, C. F., 1972, Early contact interactions between mammalian gametes *in vitro*: Evidence that the vitellus influences adherence between sperm and the zona pellucida, *Proc. Natl. Acad. Sci. U.S.A.* **69**:2767–2769.

Hawk, H. W., 1987, Transport and fate of spermatozoa after insemination of cattle, *J. Dairy Sci.* **70**:1487–1503.

Hirao, Y., and Yanagimachi, R., 1978, Effects of various enzymes on the ability of hamster egg plasma kmembrane to fuse with spermatozoa, *Gamete Res.* **1**:3–12.

Hunter, R. H. F., 1975, Transport, migration and survival of spermatozoa in the female genital tract: Species with intra-uterine deposition of semen, in: *The Biology of Spermatozoa* (E. S. E. Hafez and C. G. Thibault, eds.), S. Karger, Basel, pp. 145–155.

Hunter, R. H. F., 1980, Transport and storage of spermatozoa in the female tract, in: *Proceedings 9th International Congress on Animal Reproduction*, Vol. 2, Padilla Publishing, Madrid, Spain, pp. 227–233.

Hunter, R H. F., 1987, Human fertilization *in vivo*, with special reference to progression, storage and release of competent spermatozoa, *Hum. Reprod.* **2:**329–332.

Ishijima, S., and Mohri, H., 1985, A quantitative description of flagellar movement in golden hamster spermatozoa, *J. Exp. Biol.* **114:**463–475.

Kan, F. W. K., St. Jacques, S., and Bleau, G., 1988, Immunoelectron microscopic localization of an oviductal antigen in hamster zona pellucida by use of a monoclonal antibody, *J. Histochem. Cytochem.* **36:** 1441–1447.

Katz, D. F., and Drobnis, E. Z., 1990, The forces generated by mammalian sperm, in: *Proceedings of the Serono Symposium on Fertilization in Mammals* (B. D. Bavister, J. Cummins, and E. R. S. Roldan, eds.), Serono Symposia, Norwell, Massachusetts, pp. 125–137.

Katz, D. F., and Yanagimachi, R., 1980, Movement characteristics of hamster spermatozoa within the oviduct, *Biol. Reprod.* **22:**759–764.

Katz, D. F., Yanagimachi, R., and Dresdner, R. D., 1978, Movement characteristics and power output of guinea-pig and hamster spermatozoa in relation to activation, *J. Reprod. Fertil.* **52:**167–172.

Katz, D. F., Cherr, G. N., and Lambert, H., 1986, The evolution of hamster sperm motility during capacitation and interaction with the ovum vestments *in vitro*, *Gamete Res.* **14:**333–346.

Katz, D. F., Drobnis, E. Z., Baltz, J., Cherr, G. N., Yudin, A. I., Cone, R. A., and Cheng, L. Y., 1987, The biophysics of sperm penetration of the cumulus and zona pellucida, in: *New Horizons in Sperm Cell Research* (H. Mohri, ed.), Japan Scientific Societies Press, Tokyo, pp. 275–285.

Katz, D. F., Drobnis, E. Z., and Overstreet, J. W., 1989, Factors regulating mammalian sperm migration through the female reproductive tract and oocyte vestments, *Gamete Res.* **22:**443–469.

Martin, G. G., and Talbot, P., 1981a, The role of follicular smooth muscle cells in hamster ovulation. *J. Exp. Zool.* **216:**469–482.

Martin, G. G., and Talbot, P., 1981b, Drugs that block smooth muscle contraction inhibit *in vivo* ovulation in hamsters, *J. Exp. Zool.* **216:**483–491.

Meizel, S., 1978, The mammalian sperm acrosome reaction. A biochemical approach, in: *Development in Mammals* (M. H. Johnson, ed.), North Holland, New York, pp. 1–64.

Meizel, S., 1984, The importance of hydrolytic enzymes to an exocytotic event, the mammalian sperm acrosome reaction, *Biol. Rev.* **59:**125–157.

Meizel, S., 1985, Molecules that initiate or help stimulate the acrosome reaction by their interaction with the mammalian sperm surface, *Am. J. Anat.* **174:**285–302.

Meizel, S., Lui, C. W., Working, P. K., and Mrsny, R. J., 1980, Taurine and hypotaurine: Their effects on motility, capacitation and the acrosome reaction of hamster sperm *in vitro* and their presence in sperm and reproductive tract fluids of several mammals, *Dev. Growth Diff.* **22:**483–494.

Miyazaki, S., 1988, Inositol 1,4,5-trisphosphate-induced calcium release and guanine nucleotide-binding protein-mediated periodic calcium rises in golden hamster eggs, *J. Cell Biol.* **106:**345–353.

Miyazaki, S., 1989, Signal transduction of sperm–egg interaction causing periodic calcium transients in hamster eggs, in: *Mechanisms of Egg Activation* (R. Nuccitelli, G. N. Cherr, and W. H. Clark, Jr., eds.), Plenum Press, New York, pp. 231–246.

Miyazaki, S., and Igusa, Y., 1981, Fertilization potential in golden hamster eggs consists of recurring hyperpolarizations, *Nature* **290:**702–704.

Miyazaki, S., and Igusa, Y., 1982, Ca-mediated activation of a K current at fertilization of golden hamster eggs, *Proc. Natl. Acad. Sci. U.S.A.* **79:**931–935.

Mohri, H., and Yano, Y., 1980, Analysis of mechanism of flagellar movement with golden hamster spermatozoa, *Biomed. Res.* **1:**552–555.

Moller, C. C., Bleil, J. D., Kinloch, R. A., and Wassarman, P. M., 1990, Structural and functional relationships between mouse and hamster zona pellucida glycoproteins, *Dev. Biol.* **137:**276–286.

Moore, H. D. M., and Bedford, J. M., 1983, The interaction of mammalian gametes in the female, in: *Mechanism and Control of Animal Fertilization* (J. R. Hartmann, ed.), Academic Press, New York, pp. 453–497.

Mortimer, D., 1983, Sperm transport in the human female reproductive tract, *Oxford Rev. Reprod. Biol.* **5:**30–61.

Morton, B., and Albagli, L., 1973, Modification of hamster sperm adenylcyclase by capacitation *in vitro*, *Biochem. Biophys. Res.Commun.* **50:**695–703.

Mrsny, R. J., and Meizel, S., 1980, Evidence suggesting a role for cyclic nucleotides in acrosome reactions of hamster sperm *in vitro, J. Exp. Zool.* **211:**153–158.

Mrsny, R. J., Siiteri, J. E., and Meizel, S., 1984, Hamster sperm Na^+, K^+ -adenosine triphosphatase: Increased activity during capacitation *in vitro* and its relationship to cyclic nucleotides, *Biol. Reprod.* **30:**573–584.

Murdoch, W. J., and Cavender, J. L., 1987, Mechanisms of ovulation. *Adv. Contracept. Deliv. Sys.* **3**(4):353–366.

Nicolson, G. L., Yanagimachi, R., and Yanagimachi, H., 1975, Ultrastructural localization of lectin-binding sites on the zonae pellucidae and plasma membranes of mammalian eggs, *J. Cell Biol.* **66:**263–274.

Oikawa, T., Yanagimachi, R., and Nicolson, G. L., 1973, Wheat germ agglutinin blocks mammalian fertilization, *Nature* **241:**256–259.

Oikawa, T., Nicolson, G. L., and Yanagimachi, R., 1974, Inhibition of hamster fertilization by phytoagglutinins, *Exp. Cell Res.* **83:**239–246.

Oikawa, T., Sendai, Y., Kurata, S., and Yanagimachi, R., 1988, Glycoprotein of oviductal origin alters biochemical properties of the zona pellucida of hamster egg, *Gamete Res.* **19:**113–122.

Orsini, M. W., 1961, The external vaginal phenomena characterizing the stages of the estrous cycle, pregnancy, pseudopregnancy, lactation, and the anestrous hamster, *Mesocricetus auratus, Waterhouse Proc. Anim. Care Panel* **11:**193–206.

Overstreet, J. W., 1983, Transport of gametes in the reproductive tract of the female mammal, in: *Mechanism and Control of Animal Fertilization* (J. F. Hartmann, ed.), Academic Press, New York, pp. 499–543.

Overstreet, J. W., and Katz, D. F., 1977, Sperm transport and selection in the female genital tract, in: *Development in Mammals*, Vol. 2, (M. Johnson, ed.), Elsevier/North-Holland Biomedical Press, Amsterdam, pp. 31–65.

Overstreet, J. W., Cooper, G. W., and Katz, D. F., 1978, Sperm transport in the reproductive tract of the female rabbit. II. The sustained phase of transport, *Biol. Reprod.* **19:**115–132.

Phillips, D. M., and Shalgi, R., 1980, Surface architecture of the mouse and hamster zona pellucida and oocyte, *J. Ultrastruct. Res.* **72:**1–12.

Schini, S. A., and Bavister, B. D., 1988a, Development of golden hamster embryos through the "two-cell block" in chemically defined medium, *J. Exp. Zool.* **245:**111–115.

Schini, S. A., and Bavister, B. D., 1988b, Two-cell block development of cultured hamster embryos is caused by phosphate and glucose, *Biol. Reprod.* **39:**1183–1192.

Schmell, E. D., Gulyas, B. J., and Hedrick, J. L., 1983, Egg surface changes during fertilization and the molecular mechanism of the block to polyspermy, in: *Mechanism and Control of Animal Fertilization* (J. F. Hartmann, ed.), Academic Press, New York, pp. 356–413.

Schroeder, P. C., and Talbot, P., 1982, Intrafollicular pressure decreases during contraction of hamster follicular smooth muscle cells *in vitro, J. Exp. Zool.* **224:**417–426.

Shalgi, R., and Phillips, D., 1980, Mechanics of sperm entry in cycling hamsters, *J. Ultrastruct. Res.* **71:**154–161.

Smith, T. T., Koyanage, F., and Yanagimachi, R., 1987, Distribution and number of spermatozoa in the oviduct of the golden hamster after natural mating and artificial insemination, *Biol. Reprod.* **37:**225–234.

Steinhardt, R. A. S., Epel, D., Carroll, E. J., and Yanagimachi, R., 1974, Is calcium ionophore a universal activator for unfertilized eggs? *Nature* **252:**41–43.

Strauss, F., 1956, The time and place of fertilization of the golden hamster egg., *J. Embryol. Exp. Morphol.* **4:**42–56.

Suarez, S. S., 1987, Sperm transport and motility in the mouse oviduct: Observations *in situ, Biol. Reprod.* **36:**203–210.

Suarez, S. S., 1988, Hamster sperm motility transformations during epididymal maturation and the development of hyperaction, *in vitro, Gamete Res.* **19:**51–65.

Suarez, S. S., Katz, D. F., and Meizel, S., 1984, Changes in motility that accompany the acrosome reaction in hyperactivated hamster spermatozoa, *Gamete Res.* **10:**253–265.

Sullivan, R., and Bleau, G., 1985, Interaction of isolated components from mammalian sperm and egg, *Gamete Res.* **12:**101–116.

Talbot, P., 1983, Videotape analysis of hamster ovulation *in vitro, J. Exp. Zool.* **225:**141–148.

Talbot, P., 1984, Hyaluronidase dissolves a component in the hamster zona pellucida, *J. Exp. Zool.* **229:**309–316.

Talbot, P., 1985, Sperm penetration through oocyte investments in mammals, *Am. J. Anat.* **174:**331–346.

Talbot, P., and Chacon, R., 1982, *In vitro* ovulation of hamster oocytes depends on contraction of follicular smooth muscle cells, *J. Exp. Zool.* **24:**409–415.

Uto, N., Yoshimatsu, N., Lopata, A., and Yanagimachi, R., 1988, Zona-induced acrosome reaction of hamster spermatozoa, *J. Exp. Zool.* **248:**113–120.

Ward, M. C., 1946, A study of the estrous cycle and the breeding of the golden hamster, *Cicretus auratus, Anat. Rec.* **94:**139–162.

Wassarman, P. M., 1987, Early events in mammalian fertilization, *Annu. Rev. Cell Biol.* **3:**109–142.

Wolf, D. P., 1981, The mammalian egg's block to polyspermy, in: *Fertilization and Embryonic Development In vitro* (L. Mastroianni and J. D. Biggers, eds.), Plenum Press, New York, pp. 183–197.

Woolley, D. M., 1977, Evidence for "twisted plane" undulations in golden hamster sperm tails, *J. Cell Biol.* **75:**851–865.

Yamanaka, H. S., and Soderwall, A. L., 1960, Transport of spermatozoa through the female genital tract of hamsters, *Fertil. Steril.* **11:**470–474.

Yanagimachi, R., 1969, *In vitro* acrosome reaction and capacitation of golden hamster spermatozoa by bovine follicular fluid and its fractions, *J. Exp. Zool.* **170:**269–280.

Yanagimachi, R., 1978, Sperm–egg association in mammals, *Curr. Top. Dev. Biol.* **12:**83–105.

Yanagimachi, R., 1981, Mechanisms of fertilization in mammals, in: *Fertilization and Embryonic Development in Vitro* (L. Mastroianni and J. D. Biggers, eds.), Plenum Press, New York, pp. 81–182.

Yanagimachi, R., 1982, Requirement of extracellular calcium ions for various stages of fertilization and fertilization related phenomena in the hamster, *Gamete Res.* **5:**323–344.

Yanagimachi, R., 1988a, Sperm–egg fusion, in: *Current Topics in Membranes and Transport* (F. Bonner and N. Duzgunes, eds.), Academic Press, Orlando, FL, pp. 3–43.

Yanagimachi, R., 1988b, Mammalian fertilization, in: *The Physiology of Reproduction* (E. Knobil and J. Neill, eds.), Raven Press, New York, pp. 135–185.

Yanagimachi, R., and Chang, M. C., 1963, Fertilization of hamster eggs *in vitro*, *Nature* **200:**281–282.

Yanagimachi, R., and Chang, M. C., 1964, *In vitro* fertilization of golden hamster ova, *J. Exp. Zool.* **156:**361–376.

Yanagimachi, R., and Phillips, D. M., 1984, The status of acrosomal caps of hamster spermatozoa immediately before fertilization *in vitro*, *Gamete Res.* **9:**1–19.

Yoshimatsu, N., and Yanagimachi, R., 1988, Effects of cations and other medium components on the zona-induced acrosome reaction of hamster spermatozoa, *Dev. Growth Differ.* **30**(6):651–659.

Yudin, A. I., Cherr, G. N., and Katz, D. F., 1988, Structure of the cumulus matrix and zona pellucida in the golden hamster: A new view of sperm interaction with oocyte-associated extracellular matrices, *Cell Tissue Res.* **251:**555–564.

Zao, P., and Talbot, P., 1986, Isolation of an insoluble glycoprotein component of golden hamster sperm acrosomes, *J. Cell Biol.* **103**(5, pt. 2):239a.

12

Fertilization in the Rat

Ruth Shalgi

1. INTRODUCTION

The rat has been used by reproductive biologists as an animal model for many years. Intensive studies have been carried out using rat oocytes (Austin, 1961) to study the resumption of meiosis (reviewed by Tsafriri, 1985), *in vitro* maturation of oocytes in culture or enclosed in a follicle, ovulation, and the hormonal control of this process (reviewed by Dekel, 1986). Studies on spermiogenesis and sperm maturation as well as endocrinology of the male have also employed the rat as an animal model (reviewed by Bedford, 1975; Bardin *et al.*, 1988; Eddy *et al.*, 1988). With the accumulated information already obtained for both the male and the female gamete, the rat could clearly serve as an ideal animal for studying the process of fertilization. This review attempts to summarize our current knowledge of fertilization in the rat *in vitro* as compared to fertilization *in vivo*.

2. *IN VIVO* STUDIES OF FERTILIZATION

2.1. Sperm Transport and Capacitation

In the rat, the ejaculate is deposited directly into the cornua of the uterus. Sperm gain access to the oviduct within a few minutes post-ejaculation as determined by oviductal flushings (reviewed in Blandau, 1975). Although spermatozoa can be seen accumulating near the utero-tubal junction, there is no delay at this point. The number of sperm in the oviduct increases rapidly until each oviduct contains hundreds of spermatozoa. Similar to sperm of other species, rat sperm require capacitation as described originally for other mammalian species by Austin (1952). In females that have recently ovulated, the fertilizing spermatozoon arrives at the ampulla a few hours after mating (Austin, 1952; Shalgi and Kraicer, 1978). Fertilization begins soon after spermatozoa arrive in the ampulla. Thus, the minimum time for capacitation *in vivo* is 3–4 hr. If

RUTH SHALGI • Department of Embryology and Teratology, Sackler School of Medicine, Tel-Aviv University, Ramat Aviv, Tel-Aviv 69978, Israel.

A Comparative Overview of Mammalian Fertilization, edited by Bonnie S. Dunbar and Michael G. O'Rand. Plenum Press, New York, 1991.

sperm arrive in the ampulla before ovulation has occurred, in the absence of oocytes they will not accumulate in the ampulla but will continue to the infundibular end of the oviduct (Blandau and Odor, 1949).

Although a large number of spermatozoa gain access to the oviduct, very few succeed in traversing its length. The number of spermatozoa in the ampulla at the time of fertilization has been quantified in several studies and found to be very low (Blandau and Odor, 1949; Moricard and Bossu, 1951; Braden and Austin, 1954; Shalgi and Kraicer, 1978; Shalgi and Phillips, 1988). It is possible that the low number of sperm reaching the ampulla is the major factor not only in the regulation of the timing of fertilization, but also in the prevention of polyspermy. When superovulated immature females were used as a model (Shalgi and Phillips, 1988), the number of spermatozoa in the ampulla was found to be larger, and fertilization of the oocytes began earlier and was terminated within a shorter time period (Table 1).

2.2. Cumulus Penetration

Rat oocytes can be fertilized *in vivo* in the presence of cumulus cells. Even when all of the oocytes are fertilized, the cumulus cell complex remains (Shalgi and Phillips, 1988). This is probably because of the low number of spermatozoa present in the ampulla at the time of fertilization. In the hyperstimulated model, however, the cumulus cells are more dispersed than those of cycling rats, and at 7 hr, when all of the oocytes have been fertilized, no cumulus cells remain. This may result from a higher concentration of acrosomal hydrolytic enzymes from the greater number of spermatozoa present in the ampulla (Table 1). Although the number of spermatozoa in the ampulla of the oviduct is small, the sperm concentration is relatively large. The volume of the ampullary region of the oviduct in the rat is 0.5–1.5 μl (Shalgi *et al.*, 1977). Thus, at the time of fertilization, the concentration of spermatozoa in the ampullae of mature females is 2×10^3/ml as compared to 2×10^4/ml in superovulated immature rats (Table 1). Sperm can be found in low numbers within the cumulus mass at the time that fertilization commences and some of them are still motile. The time required for sperm to traverse the cumulus in the rat has not yet been determined.

Table 1. The Number of Spermatozoa per Ampulla at Different Times after Mating[a]

Hours after mating	Mature cycling rats			Immature superovulated rats		
	Cum. present	Eggs fertilized (%)	Average[b] sperm n	Cum. present	Eggs fertilized (%)	Average[b] sperm n
4.5				+	3 (4/143)	3.1
5.0				+	4 (6/143)	4.3
5.5	+	0 (0/27)	0	±	28 (27/97)	13.4
6.0	+	23 (8/34)	0.4	±	83 (78/94)	36.3
6.5	+	52 (16/31)	2.1	±	56 (42/75)	26.1
7.0	+	49 (20/41)	5.8	−	40 (81/201)	41.3
7.5	+	26 (7/27)	0.8	−	77 (82/107)	47.7
8.0	+	55 (17/31)	3.5			
8.5	+	86 (32/37)	4.5			
Average		44%	2.4		37%	24.5

[a]Summarized from Shalgi and Phillips (1988).
[b]Includes all spermatozoa present: immotile sperm and motile sperm (hyperactivated and nonhyperactivated), both free and in cumulus.

2.3. Zona Interaction

Until about 10 years ago, it was generally accepted that the acrosome reaction of the mammalian spermatozoa occurred in the ampulla of the oviduct near or within the cumulus oophorus. Enzymes released from the acrosome were believed to induce lysis of extracellular matrix constituents, thereby facilitating penetration of the fertilizing spermatozoon through the cumulus cell mass. Considerable evidence supports this idea. Recently, however, a number of workers have concluded that mouse spermatozoa reach and attach to the zona pellucida before undergoing the acrosome reaction (Saling and Storey, 1979), and only then is the acrosome reaction induced (Florman and Storey, 1982).

In the rat, it is very rare to observe the fertilizing spermatozoon at the point of initial attachment with the zona, since the zona penetration begins shortly after it binds. In those few sperm that were bound to the zona pellucida, we were able to observe changes in their acrosomal caps (Shalgi *et al.*, 1989b).

2.4. Sperm–Vitellus Interaction

The events that occur following passage of the fertilizing sperm through the zona pellucida have been described in detail by Austin and Braden (1956) using light microscopy. The initial fusion of the egg plasma membrane and the fertilizing spermatozoon appears to occur at the convex surface of the sperm in the area anterior to the post-acrosomal region, which is presumably part of the equatorial segment (Piko, 1969; Phillips and Shalgi, 1982). At early stages of interaction, the tip of the sperm head is directed away from the oocyte surface (Fig. 1). The spermatozoon subsequently appears to rotate so that it comes to lie on its side. The egg membrane over the sperm head becomes free of microvilli, and a large incorporation cone is formed (Austin and Braden, 1956; Shalgi *et al.*, 1978; Phillips and Shalgi, 1982; Batagalia and Gaddum-Rosse, 1984).

3. *IN VITRO* STUDIES OF FERTILIZATION

Attempts to fertilize rat eggs *in vitro* (IVF) were made for many years until Toyoda and Chang (1968) described the incorporation of rat spermatozoon into the vitellus of the rat egg after dissolution of the zona pellucida. A few years later, Miyamoto and Chang (1973) reported that rat spermatozoa recovered from the uteri of mated females could be capacitated *in vitro* in a medium containing a high proportion of rat serum, and penetrate intact eggs. Further evaluation of the conditions for successful fertilization was done in later studies.

The *in vitro* fertilization system of the rat is extremely delicate. This is, in part, because of the great sensitivity of the sperm to environmental changes such as heat and to manipulations such as centrifugation. However, under controlled conditions the fertilization rates obtained can be consistently high and reproducible.

3.1. *In Vitro* Capacitation

The conditions for capacitation *in vitro* of ejaculated spermatozoa recovered from the uterus of mated females were reported by Miyamoto and Chang (1973). They achieved very poor rates of penetration when using epididymal sperm. The capacitation system was later defined for epididymal sperm as well (Toyoda and Chang, 1974a). It has been established that bovine serum albumin plays an important role in the rat IVF system, and its presence is required to allow capacitation *in vitro* (Miyamoto and Chang, 1973). In a series of studies published some years later, it was suggested that lipids are exchanged between plasma membrane of the rat sperm and

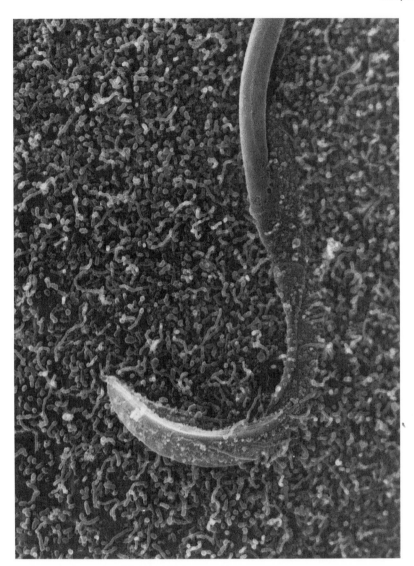

Figure 1.   Rat spermatozoa associated with the rat egg *in vivo*. The oocyte surface is covered with microvilli
(×7000) (D. M. Phillips and R. Shalgi, unpublished).

serum albumin during capacitation *in vitro* and that a decreased cholesterol/phospholipid ratio in
the sperm plasma membrane facilitates this transformation (Davis *et al.*, 1979). A new defined
medium (high K^+/Na^+ ratio, cAMP, and the unimportance of rat serum) was reported by Toyoda
and Chang (1974b), and in the same year Niwa and Chang (1974a) determined the necessity for
supplementary CO_2 and the possibility of cross-strain fertilization.

   Since then, a number of other laboratories have published studies using this system or
modifications to it, mainly with increased calcium concentration (Kaplan and Kraicer, 1978;
Davis, 1978; Shalgi *et al.*, 1981; Gaddum-Rosse *et al.*, 1984; Vanderhyden *et al.*, 1986). In our
laboratory, we capacitate spermatozoa in a modified RFM (RFMm). This medium is similar to

other media used for the IVF (reviewed by Yanagimachi, 1988) and is a modified Krebs–Ringer bicarbonate supplemented with energy sources, albumin, and the pH stabilized by HEPES (Table 2). Capacitation occurs during incubation in a CO_2 incubator in 200-μl drops under paraffin oil (Shalgi *et al.*, 1981). Preferable sperm concentration was reported to be between 0.3–1.4 × 10⁶/ ml (Niwa and Chang, 1973, 1974b; Shalgi *et al.*, 1981). This relatively high gamete ratio required for successful fertilization *in vitro* as compared to *in vivo* can be explained by the lower percentage of fertile sperm in the *in vitro* systems (Stewart-Savage and Bavister, 1988).

Because no clear morphological marker is yet available for capacitation, and only a fraction of the spermatozoa in the suspension show a clear hyperactivated motility, we use the ability of sperm to bind to the ZP as a biological marker. Immotile spermatozoa can be seen adhering to the zona pellucida, but motile noncapacitated sperm cannot bind to it. This ability is acquired in parallel to the period during which we observe penetration (unpublished observation). It has been reported that when rat spermatozoa are incubated with zona-free rat eggs, noncapacitated sperm can attach to the vitellus, but capacitation is required for penetration into the vitellus and enlargement of the sperm head, as judged by the time required for penetration of the zona-free eggs (Niwa and Chang, 1975). Sperm attachment to the zona pellucida and subsequent penetration of the zona pellucida can be observed in this system as early as 3.5–4 hr following sperm suspension (Shalgi *et al.*, 1983) or 5–5.5 hr according to Toyoda and Chang (1974a). The fertilizing life span of capacitated rat sperm is limited. In our hands, eggs exposed to capacitated sperm later than 8 hr following the beginning of incubation, can no longer be fertilized (Shalgi *et al.*, 1989a).

3.2. Cumulus Penetration

In vivo, the fertilizing spermatozoon penetrates through an intact cumulus oophorus, but it appears that *in vitro* the cumulus is not essential. It is unlikely that the cumulus cells facilitate or enable capacitation of the spermatozoa, as the rates of attachment to the ZP and subsequent penetration are similar in cultures where capacitation is allowed to occur in the presence or absence of cumulus and eggs. Cumulus cells surrounding ovulated oocytes disperse as soon as they are exposed *in vitro* to a sperm suspension, yet binding to the ZP cannot be observed before 3.5 hr post-suspension (Shalgi *et al.*, 1983).

Table 2. Composition of Rat Fertilization Media (mM)

Components	Toyoda and Chang (1974a)	Evans and Armstrong (1984)	RFMm
NaCl[a]	94.60	94.60[b]	70.32
KCl[a]	4.78	15.95[b]	4.08
CaCl$_2$[a]	1.71	7.38[b]	3.42
KH$_2$PO$_4$[a]	1.19	7.31[b]	1.18
MgSO$_4$[a]	1.19	8.10[b]	1.19
NaHCO$_3$[a]	25.07	27.07[b]	25.07
HEPES[a]	—		25.12
Na-lactate (60%)[c]	0.40 ml[b]	0.37 ml	0.37 ml
Na-pyruvate[c]	5.5 mg[b]	5.5 mg	5.5 mg
Glucose[c]	100 mg	100 mg	100 mg
BSA[c]	400 mg	400 mg	400 mg
Streptomycin[c]	5 mg	5000 units	5 mg
Penicillin[c]	7.5 mg	5 mg	10,000 units

[a]Expressed in millimoles per liter.
[b]Calculated from original data.
[c]Added to 100 ml final solution volume.

Lewin *et al.* (1982) measured the amount of hyaluronidase that was released from rat spermatozoa during incubation and found that after 4 hr in capacitating medium, only a small portion (3.5%) of the total hyaluronidase could be detected. They concluded that the cell-bound hyaluronidase was accessible to extracellular substrates. Our understanding is that *in vitro* capacitation is not required to enable rat spermatozoa to disperse the cumulus, and it is not yet clear whether it is required *in vivo*.

3.3. Sperm–Zona Interaction and the Acrosome Reaction

Capacitated spermatozoa can attach to the zona pellucida immediately after exposure to cumulus-free oocytes. Similar to results found in other species (reviewed by O'Rand, 1988), sperm–egg interaction in the rat involves carbohydrates, possibly fucose and mannose residues (Shalgi *et al.*, 1986). Maximum sperm binding under capacitating conditions is observed after 30 min of coincubation, and it is followed by a gradual decrease (Shalgi *et al.*, 1986). This might be the result of the interaction of the first fertilizing sperm with the egg plasma membrane. Sperm interaction with the vitellus probably leads to inactivation or modification of the zona as demonstrated in the mouse and the hamster (Kurasawa *et al.*, 1989; Moller and Wassarman 1989; Moller *et al.*, 1990) thus causing the detachment of sperm already bound. The block to polyspermy is dependent both on the vitelline block and, only later, on the zona reaction. However, polyspermy has been reported in several studies (Niwa and Chang, 1974a, 1975; Toyoda and Chang, 1974a; Shalgi *et al.*, 1981; Evans and Armstrong, 1984). No information is available on the biochemical modifications of the ZP but, similar to the mouse (Bleil and Wassarman, 1980, 1983), rat sperm are unable to bind and penetrate ZP of fertilized eggs.

Rat spermatozoa interact with the ZP by their acrosomes. The acrosomal cap of the rat spermatozoon is very narrow, a fact that makes it difficult to follow the acrosome reaction (Shalgi *et al.*, 1989b). With the scanning electron microscope, it is possible to observe both smooth and vesiculated acrosomes on the ZP surface. This is also consistent with the hypothesis that acrosomal changes occur on the ZP and that acrosome-intact sperm can penetrate the cumulus matrix and bind to the ZP (reviewed by Wassarman, 1987, 1990). Using monoclonal antibodies that recognize antigens in the acrosomal membranes, we could demonstrate that during this interaction, some acrosomal contents detach from the spermatozoon and remain on the ZP (Shalgi *et al.*, 1989b). ZP passage generally occurs within 30 min of binding (Shalgi *et al.*, 1986).

3.4. Sperm–Vitellus Interaction

The major portion of the egg surface is characterized by numerous microvilli, and it is this region of the vitellus with which the sperm head is associated. The mode of sperm interaction *in vitro* appears to be different than *in vivo*, at least morphologically. The head penetrates the zona pellucida with its long axis perpendicular to the zona, and this orientation is maintained during subsequent incorporation into the vitellus. It is the anteriormost tip of the sperm head that first associates with and then incorporates into the egg. Incorporation progresses caudally, so the postacrosomal region is the last portion of the acrosome to be incorporated into the ooplasm (Shalgi and Phillips, 1982). This tip-to-base incorporation contrasts with the situation observed when oocytes are penetrated *in vivo*, (Phillips and Shalgi, 1982) probably because *in vitro*, by the time the oocyte is penetrated, it is usually denuded of its cumulus cells, and the sperm approaches and penetrates the ZP at a different angle than *in vivo*. A few minutes after the initial contact with the vitellus, the motility of the flagellum is reduced. The head begins to decondense, and later, the entire flagellum is incorporated into the vitellus (Gaddum-Rosse *et al.*, 1984).

3.5. Source of the Egg for Fertilization

There are various reports on preferable sources of eggs. Miyamoto and Chang (1973) reported higher penetration rates into eggs recovered from naturally ovulated females (52–55%) than into those recovered from superovulated mature (22–32%) or immature females (26%). Niwa and Chang (1974b) found that fertilization rates were higher in the eggs from superovulated immature females (69–82%) than in the normally ovulated eggs from mature females (22%). These results are difficult to compare, as the source of sperm, sperm concentration, and the ages of the immature females were different. It appears as though eggs from immature females are more easily penetrated (Toyoda and Chang, 1974a). This observation may be correlated to different properties of the ZP of the two groups as the ZP of eggs from mature females are more resistant to proteolysis than zonae from immature females (Rufas and Shalgi, 1990).

In 1973, Miyamoto and Chang reported that even within eggs recovered from immature superovulated females, there are different types of eggs, as eggs recovered 1–3 hr after ovulation had better chances of being fertilized as those recovered 8–9 hr after ovulation (26–32% versus 8–10%). It appears as though the overall low rates of sperm penetration led to their conclusion, because in a later study it was shown that the dynamics of sperm penetration into freshly ovulated eggs compared to eggs recovered 7 hr after ovulation was similar (Shalgi et al., 1985). In both cases, care should be taken to compare the dose of hormones used to superovulate the females (Evans and Armstrong, 1984).

Maturation of the oocytes can be induced in vivo in intact females or even following hypophysectomy, and the hormonal induction can be either with hCG or a GnRH analogue (GnRHa) or in vitro in isolated follicles (Dekel and Shalgi, 1987). In all the above experiments, fertilization of eggs could result in normal embryos (Toyoda and Chang, 1974a; Shalgi, 1984; Vanderhyden et al., 1986; Shalgi and Dekel, 1990).

Although capacitated spermatozoa attach and probably also bind to immature oocytes before completion of their maturation division, penetration through the ZP is rarely seen. This is probably because of the sperm's inability to penetrate the ZP, as even cumulus-free oocytes failed to be penetrated (Fig. 2). Oocytes achieve their fertilization capacity gradually, concomitant with other maturational changes occurring in the cumulus–oocyte complex, as reviewed by Dekel (1986). Spermatozoa, even when capacitated, cannot penetrate the cumulus of oocytes that have not undergone maturation, probably because of their inability to dissolve the extracellular matrix, which has not undergone these maturational changes.

4. FROM PENETRATION TO FERTILIZATION

Rat IVF enables nearly all ovulated eggs to be penetrated in vitro by sperm. A high proportion can develop in vitro to the two-cell stage (Toyoda and Chang, 1974a; Evans and Armstrong, 1984; Shalgi et al., 1985). First, cleavage occurs in vitro at an average of 21 hr following the estimated time of sperm penetration into oocytes collected 7 hr after ovulation or 23 hr when freshly ovulated eggs are used (Shalgi et al., 1985), as compared to the calculated interval of 19.6 hr in vivo (Shalgi and Kraicer, 1978). First, cleavage can serve as a fairly good marker for development. Oocytes that were matured in vitro from various stages during their maturation process and then fertilized in vitro presented an improved rate of cleavage that was correlated to the maturational stage achieved in vivo (Niwa et al., 1976).

Almost no further development beyond the two-cell stage can be achieved in vitro when zygotes produced by IVF are used (Toyoda and Chang, 1974a) or when zygotes obtained by in vivo fertilization are cultured (R. Shalgi, unpublished observation). However, successful development to term can be achieved on transferring such embryos (Toyoda and Chang, 1974a), or

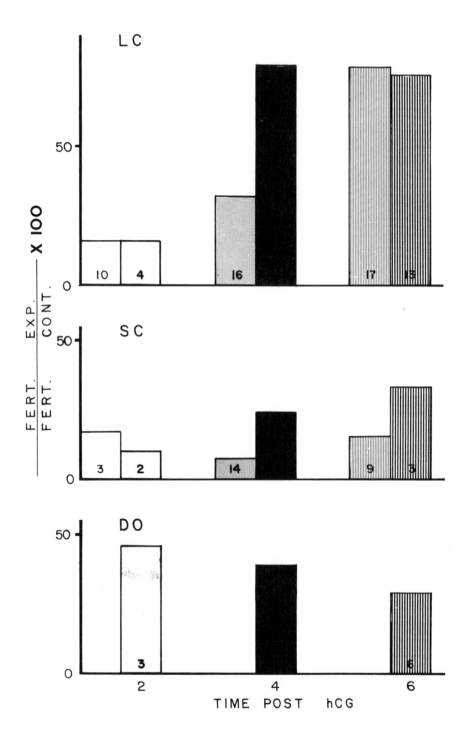

even embryos produced following induction by GnRHa (Shalgi and Dekel, 1990), at the two-cell stage to pseudopregnant females.

Two-cell-stage embryos produced by *in vivo* or *in vitro* fertilization are morphologically similar, yet they present a different developmental potential. The chances for embryos produced by *in vitro* fertilization to develop to term are less than for embryos produced by *in vivo* fertilization, even when oocytes from mature cycling females are used (Shalgi, 1984). The percentage of loss of the transferred embryos is high at early stages of pregnancy, probably because of their retardation during development to the blastocyst stage (Shalgi, 1984; Vanderhyden *et al.*, 1986; Vanderhyden and Armstrong, 1988).

Both the *in vivo* and *in vitro* systems of rat fertilization enable experiments to be conducted. Although the *in vitro* conditions are probably inferior, it is clear that the *in vitro* derived embryos can develop to term. Improvement of culture conditions, media, and handling can be expected to increase the number of viable embryos.

5. REFERENCES

Austin, C. R., 1952, The "capacitation" of mammalian sperm, *Nature* **170**:326.

Austin, C. R., 1961, *The Mammalian Egg*, Blackwell Scientific Publications, Oxford.

Austin, C. R., and Braden, A. W. H., 1956, Early reaction of the rodent egg to spermatozoon penetration, *J. Exp. Biol.* **33**:358–365.

Bardin, C. W., Yan Chang, C., Mustow, N. A., and Gunsalus, G. L., 1988, The sertoli cell, in: *Physiology of Reproduction* (E. Knobil and J. Neill, eds.), Raven Press, New York, pp. 933–974.

Battagalia, D. E., and Gaddum-Rosse, P., 1984, Rat eggs normally exhibit a variety of surface phenomenon during fertilization, *Gamete Res.* **19**:107–118.

Bedford, J. M., 1975, Maturation, transport, and fate of spermatozoa in the epididymis, in: *Handbook of Physiology*, Vol. 5 (D. W. Hamilton and R. O. Greep, eds.), American Physiological Society, Washington, pp. 303–317.

Blandau, J. R., 1975, Gamete transport in the female animal, in: *Handbook of Physiology*, Vol. 7, Part II, (R. O. Greep, ed.) American Physiological Society, Bethesda, pp. 153–163.

Blandau, J. R., and Odor, D. L., 1949, The total number of spermatozoa reaching various segments of the reproductive tract in the female albino rat at intervals after insemination, *Anat. Rec.* **103**:93–109.

Bleil, J. D., and Wassarman, P. M., 1980, Mammalian sperm–egg interaction: Identification of a glycoprotein in mouse egg zonae pellucidae possessing receptor activity for sperm, *Cell* **20**:873–882.

Bleil, J. D., and Wassarman, P. M., 1983, Sperm–egg interactions in the mouse: Sequence of events and induction of the acrosome reaction by zona pellucida glycoprotein, *Dev. Biol.* **95**:317–324.

Braden, A. W. H., and Austin, C. R., 1954, The number of sperm about the eggs in mammals and its significance for normal fertilization, *Aust. J. Biol. Sci.* **7**:552–565.

Davis, B. K., 1978, Effect of calcium on motility and fertilization by rat spermatozoa *in vitro*, *Proc. Soc. Exp. Biol. Med.* **157**:54–59.

Davis, B. K., Byrne, R., and Hungund, B., 1979, Studies on the mechanism of capacitation II. Evidence for lipid transfer between plasma membrane of rat sperm and serum albumin during capacitation *in vitro*, *Biochim, Biophys. Acta* **558**:257–266.

Figure 2. Fertilization results of oocytes isolated 2, 4, and 6 hr after hCG induction. Immature females were induced to superovulate by injection of pregnant mare serum gonadotropin followed 48 hr later by human chorionic gonadotropin (hCG). Oocytes were isolated from the follicles 2, 4, and 6 hr following hCG. Control ovulated eggs were isolated from the ampulla 18 hr after hCG. Oocytes isolated from the follicles were characterized morphologically as (1) surrounded by a large number of cumulus cells (LC), (2) surrounded by a small number of cumulus cells (SC), or (3) denuded of cumulus cells (DO). Oocytes from groups 1 and 2 were divided into two subgroups. To allow accessibility for sperm, eggs in the first subgroup were denuded of cumulus cells by exposure to hyaluronidase followed by pipetting through a narrow heat-drawn capillary. Fifteen to 20 eggs were placed in a drop of capacitated spermatozoa, and fertilization was assessed the following day. The right bar in each group of two bars represents the eggs denuded to their cumulus cells. Numbers inside bars represent experimental days.

Dekel, N., 1986, Hormonal control of ovulation, in: *Biochemical Action of Hormones*, Vol. 13 (G. Litwack, ed.), Academic Press, New York, pp. 57–90.

Dekel, N., and Shalgi, R., 1987, Fertilization *in vitro* of rat oocytes undergoing maturation in response to a GnRH analogue, *J. Reprod. Fertil.* **80:**531–535.

Eddy, E. M., 1988, The spermatozoon, in: *Physiology of Reproduction* (E. Knobil and J. Neill, eds.), Raven Press, New York, pp. 27–68.

Evans, G., and Armstrong, D. T., 1984, Reduction in fertilization rate *in vitro* of oocytes from immature rats induced to superovulate, *J. Reprod. Fertil.* **70:**131–135.

Florman, H. M., and Storey, B. T., 1982, Mouse gamete interactions: The zona pellucida is the site for acrosome reaction leading to fertilization *in vitro*, *Dev. Biol.* **91:**121–130.

Gaddum-Rosse, P., Blandau, R. J., Langley, L. B., and Battagalia, D. E., 1984, *In vitro* fertilization in the rat: Observation on living eggs, *Fertil. Steril.* **42:**285–292.

Kaplan, R., and Kraicer, P. F., 1978, Effect of elevated calcium concentration on fertilization of rat oocytes *in vitro*, *Gamete Res.* **1:**281–285.

Kurasawa, S., Schults, R. M., and Kopf, G. S., 1989, Egg-induced modifications of the zona pellucida of mouse eggs: Effect of microinjected inositol 1,4,5-triphosphate, *Dev. Biol.* **133:**295–304.

Lewin, L. M., Nevo, Z., Gabsu, A., and Weissenberg, R., 1982, The role of sperm bound hyaluronidase in the dispersal of the cumulus oophorus surrounding rat ova, *Int. J. Androl.* **5:**37–44.

Miyamoto, H., and Chang, M. C., 1973, Fertilization of rat eggs *in vitro*, *Biol. Reprod.* **9:**384–393.

Moller, C. C., and Wassarman, P. M., 1989, Characterization of a proteinase that cleaves zona pellucida glycoprotein ZP2 following activation of mouse eggs, *Dev. Biol.* **132:**103–112.

Moller, C. C., Bleil, J. D., Kinloch, R. A., and Wassarman, P. M., 1990, Structural and functional relationships between mouse and hamster zona pellucida glycoproteins, *Dev. Biol.* **137:**276–286.

Moricard, R., and Bossu, J., 1951, Arrival of fertilizing sperm at the follicular cell of the secondary oocyte: A study of the rat, *Fertil. Steril.* **2:**260–266.

Niwa, K., and Chang, M. C., 1973, Fertilization *in vitro* of rat eggs as affected by the maturity of the females and the sperm concentration, *J. Reprod. Fertil.* **35:**577–580.

Niwa, K., and Chang, M. C., 1974a, Various conditions for the fertilization of rat eggs *in vitro*, *Biol. Reprod.* **11:**463–469.

Niwa, K., and Chang, M. C., 1974b, Optimal sperm concentration and minimal number of spermatozoa for fertilization *in vitro* of rat eggs, *J. Reprod. Fertil.* **40:**471–474.

Niwa, K., and Chang, M. C., 1975, Requirement of capacitation for sperm penetration of zona-free rat eggs, *J. Reprod. Fertil.* **44:**305–308.

Niwa, K., Miyake, M., Iritani, A., and Nishikawa, Y., 1976, Fertilization of rat oocytes cultured *in vitro* from various stages of maturation, *Reprod. Fertil.* **47:**105–106.

O'Rand, M. G., 1988, Sperm-egg recognition and barries to interspecies fertilization, *Gamete Res.* **19:**315–328.

Phillips, D. M., and Shalgi, R., 1982, Sperm penetration into rat ova fertilized *in vivo*, *J. Exp. Zool.* **221:**373–378.

Piko, L., 1969, Gamete structure and sperm entry in mammals, in: *Fertilization*, Vol. 2 (C. B. Metz and A. Monroy, eds.), Academic Press, New York, pp. 325–403.

Rufas, O., and Shalgi, R., 1990, Maturation associated changes in the rat zona pellucida, *Mol. Reprod. Dev.* **26:**324–330.

Saling, P. M., and Storey, B. T., 1979, Mouse gamete interaction during fertilization *in vitro*: Chlorotetracycline as fluorescence probe for the mouse sperm acrosome reaction, *J. Cell Biol.* **83:**544–555.

Shalgi, R., 1984, Developmental capacity of rat embryos produced by *in vivo* or *in vitro* fertilization, *Gamete Res.* **10:**77–82.

Shalgi, R., and Dekel, N., 1990, Embryonic development of oocytes induced to mature by an analogue of gonadotrophin releasing hormone, *J. Reprod. Fertil.* **89:**681–687.

Shalgi, R., and Kraicer, P. F., 1978, Timing of sperm transport, sperm penetration and cleavage in the rat, *J. Exp. Zool.* **204:**353–360.

Shalgi, R., and Phillips, D. M., 1982, Sperm penetration into rat ova fertilized *in vitro*, *J. Androl.* **3:**382–387.

Shalgi, R., and Phillips, D. M., 1988, Motility of rat spermatozoa at the site of fertilization, *Biol. Reprod.* **39:**1207–1213.

Shalgi, R., Kaplan, R., and Kraicer, P. F., 1977, Proteins of follicular, bursal and ampullar fluids of rats, *Biol. Reprod.* **17:**333–338.

Shalgi, R., Phillips, D. M., and Kraicer, P. F., 1978, Observation on the incorporation cone in the rat, *Gamete Res.* **1:**27–37.

Shalgi, R., Kaplan, R., Nebel, L., and Kraicer, P. F., 1981, The male factor in fertilization of rat eggs *in vitro*, *J. Exp. Zool.* **217:**399–402.

Shalgi, R., Kaplan, R., Nebel, L., and Kraicer, P. F., 1983, The capacitation rate of rat sperm *in vitro*, in: *The Sperm Cell* (J. Andre, ed.), Martinus Nijhoff, The Hague, pp. 47–50.

Shalgi, R., Kaplan, R., and Kraicer, P. F., 1985, The influence of post-ovulatory age on the rate of cleavage of *in vitro* fertilized rat oocytes, *Gamete Res.* **11:**99–106.

Shalgi, R., Matityahu, A., and Nebel, L., 1986, The role of carbohydrates in sperm egg interaction in rats, *Biol. Reprod.* **34:**446–452.

Shalgi, R., Matityahu, A., and Rufas, O., 1989a, Sperm–egg interaction during oocyte maturation, in: *Follicular Development and the Ovulatory Response Serono Symposium Reviews*, Vol. 23 (A. Tzafriri and N. Dekel, eds.), Arcs-Serono Symposia, Italy, pp. 279–287.

Shalgi, R., Phillips, D. M., and Jones, R., 1989, Status of the rat acrosome during sperm-zona pellucida interactions, *Gamete Res.* **22:**1–13.

Stewart-Savage, J., and Bavister, B., 1988, Success of fertilization in golden hamsters is a function of the relative gamete ratio, *Gamete Res.* **21:**1–10.

Toyoda, Y., and Chang, M. C., 1968, Sperm penetration of rat eggs *in vitro* after dissolution of zona pellucida by chymotrypsin, *Nature* **220:**589–591.

Toyoda, Y., and Chang, M. C., 1974a, Fertilization of rat eggs *in vitro* by epididymal spermatozoa and the development of eggs following transfer, *J. Reprod. Fertil.* **36:**9–22.

Toyoda, Y., and Chang, M. C., 1974b, Capacitation of epididymal spermatozoa in a medium with high K/Na ratio and cyclic AMP for the fertilization of rat eggs *in vitro*, *J. Reprod. Fertil.* **36:**125–134.

Tsafriri, A., 1985, The control of meiotic maturation in mammals, in: *Biology of Fertilization*, Vol. I (C. Metz and A. Monroy, eds.), Academic Press, New York, pp. 221–252.

Vanderhyden, B. C., and Armstrong, D. T., 1988, Decreased embryonic survival of *in vitro* fertilized oocytes in rats is due to retardation of preimplantation development, *J. Reprod. Fertil.* **83:**851–857.

Vanderhyden, B. C., Rouleau, A., Walton, E. A., and Armstrong, D. T., 1986, Increased mortality during early embryonic development after *in vitro* fertilization of rat oocytes, *J. Reprod. Fertil.* **77:**401–409.

Wassarman, P. M., 1987, Early events in mammalian fertilization, *Annu. Rev. Cell Biol.* **3:**109–142.

Wassarman, P. M., 1990, Profile of mammalian sperm receptor, *Development* **108:**1–17.

Yanagimachi, R., 1988, Mammalian fertilization, in: *Physiology of Reproduction* (E. Knobil and J. Neill, eds.), Raven Press, New York, pp. 387–445.

13

Fertilization in the Guinea Pig

George L. Gerton

1. INTRODUCTION

Since the study of guinea pig fertilization and early development began over 137 years ago (see references cited by Hunter *et al.*, 1969), the guinea pig has been an important experimental animal for investigators trying to understand mechanisms of sperm–egg interaction that lead to fertilization. The purpose of this review is not to provide an exhaustive treatise on the subject. Rather, advantages and disadvantages of the guinea pig system are discussed with particular reference to specific areas of research that have received considerable attention. In addition, prospective areas for future investigations are highlighted.

2. EARLY STUDIES ON OVULATION AND COPULATION

In the late 19th and early 20th centuries, the guinea pig was used for many studies of mammalian ovulation. Classic reports from Loeb (1911, 1914, 1918) described the cyclical changes occurring in the sexual cycle of the ovary, including changes in the uterus and the development of the corpus luteum. Stockard and Papanicolaou (1917) further defined the estrous cycle of the guinea pig, providing criteria to use for the microscopic examination of the vaginal fluid as a means for interpreting the state of the uterine wall and ovarian follicles. Two years later, Stockard and Papanicolaou (1919) detailed guinea pig copulation and provided a description of features of the vaginal plug.

3. EPIDIDYMAL MATURATION OF GUINEA PIG SPERM

In the late 1910s, studies on spermatogenesis and sperm physiology in the guinea pig were also initiated by several investigators. One interesting question at the time concerned the function of the scrotum. It was recognized that the scrotum maintained the testis and epididymis at

GEORGE L. GERTON • Division of Reproductive Biology, Department of Obstetrics and Gynecology, University of Pennsylvania School of Medicine, Philadelphia, Pennsylvania 19104-6080.

A Comparative Overview of Mammalian Fertilization, edited by Bonnie S. Dunbar and Michael G. O'Rand. Plenum Press, New York, 1991.

temperatures lower than normal body temperatures, so experiments were designed to determine the effects of elevated scrotal temperatures on testicular and epididymal physiology. Stigler (1918) found that sperm in the epididymis were more resistant to heating than sperm in the testis. Furthermore, he discovered that sperm in the posterior end of the epididymis were more resistant to the effects of heat than sperm in the anterior end. Young (1927) also examined the effects of heating on epididymal sperm as well as the testis, also concluding that sperm from the epididymis were more resistant to heating than testicular sperm and spermatogenic cells.

The differential effects of heating on guinea pig sperm found in the epididymis stimulated Young to pursue further studies of the function of this tissue. Young (1929a) recognized that at least two possibilities existed to explain the effects of epididymal maturation on sperm motility and resistance to heating. One idea was that epididymal secretions acted on the sperm to modify the sperm during transit through this organ. The other concept was that the changes observed in sperm were inherent in these cells and occurred as a result of the aging or maturation of the sperm, i.e., the residence time within the epididymis. In contrast to Stigler (1918), Young reported that sperm from the distal portion of the epididymis were less resistant to heating than sperm from the proximal region. The same results were found for ultraviolet-irradiated sperm. Based on these results, Young concluded that the "ripening" of the sperm was initiated in the testis and continued in the epididymis without influence from the secretions found there. That is, the function of the epididymal fluid was simply to provide an appropriate medium to maintain the viability of sperm, not induce or promote their maturation.

In later studies, Young investigated the fertility of sperm in male guinea pigs whose epididymides had been surgically isolated from their testes. He found that these animals remained fertile for up to 25–30 days following the operative procedure yet contained motile sperm in the epididymides for at least 59 days, indicating that sperm became infertile before they became immotile (Young, 1929b). From these results, he modified his previous stance slightly, concluding that the epididymis is an organ that allows the maturation of sperm (that began in the testis) to continue but that there is no influence of the epididymis to keep sperm in an optimal fertilizing condition. Thus, sperm that stay within the the epididymis become "old" and infertile. Finally, he concluded that since "old" sperm were not known to exist in normal males, there must be a mechanism, in conjunction with the steady stream of spermatozoa through the epididymis, that serves to eliminate the "old" sperm.

In a third paper in his series on the function of the epididymis, Young (1931) investigated the functional changes in spermatozoa as they pass through the epididymis. In particular, he isolated spermatozoa from the proximal and distal regions of the epididymis and demonstrated that the sperm from the distal region were more capable of impregnating female guinea pigs than were the sperm from the proximal region. Epididymides were also ligated so that sperm were not able pass down the ducts. After 20 and 25 days, there was a realignment of the fertility of the sperm such that the sperm from the proximal epididymis became more fertile than those from the distal segment. Unfortunately, no attempt was made in these studies to inseminate females with equal numbers of sperm from each region. Also, Young did not assess the motility and viability of the sperm used in the inseminations. Thus, it is difficult to know whether the effect seen was caused by an increase in the fertility of corpus sperm, a decrease in the viability of the cauda sperm, or both. Nevertheless, these experiments helped to stimulate further experiments concerning the function of the epididymis in the maturation of mammalian sperm.

Following the development of the modern electron microscope, Fawcett and Hollenberg (1963) returned to the work of Young to determine whether there was any structural correlate of the functional differences that were described for sperm from the proximal and distal segments of the guinea pig epididymis. These investigators determined that there is a major change in the shape of the head of the guinea pig sperm as it travels down the epididymis. In particular, the acrosome of the sperm is modified from a relatively straight configuration in longitudinal sections of caput sperm to a curved, "pruning hook" shape with a prominent dorsal bulge in similar sections of cauda epididymal sperm. These changes were gradual, providing support for Young's view that sperm development is a continuous process beginning in the testis and continuing in the epididymis. Fawcett and Hollenberg (1963) proposed that the changes observed in sperm as they travel down the epididymis are intrinsic to the sperm, but they were careful to state they could not determine whether these changes were independent of specific stimuli or environmental factors provided by epididymal secretions.

Later studies investigated the hormonal control of the structural alteration of sperm maturing in the epididymis. The epididymis-associated alteration in morphology is under hormonal control since castration of guinea pigs decreases the percentage of cauda sperm that undergo the morphological maturation (Blaquier *et al.*, 1972). The conversion to a normal caudal sperm structure can be partially rescued in castrated animals by replacement therapy with testosterone propionate. Although this report does not preclude the possibility that testosterone may act on sperm to cause the observed ultrastructural changes, it is difficult to imagine circumstances whereby this hormone could directly cause these alterations. It is more likely that the hormone stimulates the epididymis to secrete components into the fluid that cause the sperm to change shape.

More recently, additional changes during epididymal transport have been detected. The cell-surface protein recognized by monoclonal antibody (MAb) PH-20 originally is distributed over the whole head of guinea pig sperm. After epididymal maturation, the surface PH-20 protein is found only in the posterior head region of guinea pig sperm (Phelps and Myles, 1987). Also, proacrosin, the zymogen precursor to the serine protease acrosin, starts with a molecular weight of 55,000 in round spermatids but ends up with a molecular weight of 47,000–50,000 in cauda epididymal sperm (Arboleda and Gerton, 1988). This change occurs in the corpus region of the epididymis and is caused by the modification of proacrosin oligosaccharides (Anakwe *et al.*, 1991).

4. FERTILIZATION

4.1. Fertilization and Early Development *in Vivo*

The study of Hunter *et al.* (1969) helped to usher in the modern era of the study of guinea pig reproduction. These investigators were primarily interested in describing guinea pig sperm penetration, pronuclear development, and early embryogenesis. Specifically, they identified the timetable of fertilization and early development. The information obtained from this study has helped to position the guinea pig as an important species for the study of mammalian fertilization.

Using phase-contrast microscopy and stained whole-mount preparations, Hunter *et al.* studied eggs recovered from guinea pigs that had been naturally mated. Animals were carefully monitored to determine the time of mating so that the timetable of early embryonic development could be carefully defined. The expected number of eggs was judged by postmortem examination of ovaries for corpora lutea. Fallopian tubes and uteri were flushed with Ringer's solution to collect eggs and embryos. The recovery rate was 89% of the expected values, and the authors surmised that some of those eggs that were not recovered were lost because they had been very recently ovulated and had not entered the oviducts. Ninety-seven percent of the recovered eggs were fertilized. Only three of the 174 eggs recovered were dispermic.

Embryonic development proceeded from the pronuclear stage to the two-cell stage during the first day following fertilization. By day 2, about half of the embryos were at the four-cell stage, with close synchronization (within one cell division) of the embryos recovered from a single animal. In most embryos of this stage (four cell), the tail of the fertilizing sperm remained visible.

During the first half of day 3, all of the embryos were found in the fallopian tubes. In the second half of this 24-hr period, most of the embryos were recovered from the uterus. During this period, 79% of the embryos were at the four- to eight-cell stage. As development proceeded, it became more difficult to define the embryonic stages. During day 4, most embryos were in the eight-cell to 32-cell stages, with hints of developing blastocoels in the more advanced embryos. On day 5, all of the embryos were classified as blastocysts. On the morning of day 6, all embryos were blastocysts, still enclosed by the zona pellucida. The blastocysts of one animal appeared to be more adhesive, as judged by handling difficulties. This effect was attributed to a presumed secretory product of the embryo.

4.2. Fertilization *in Vitro*

Yanagimachi (1972) was the first to demonstrate conclusively that guinea pig eggs could be fertilized *in vitro*. As cited by Yanagimachi, the first actual report of the *in vitro* fertilization of follicular eggs from the guinea pig had been published almost 100 years before Yanagimachi's report (Schenk, 1878). However, the previous conclusions were based solely on the observations of polar body extrusion and cleavage of the egg, two phenomena known to occur artifactually (Austin, 1961). Yanagimachi used a defined medium, oocytes flushed from oviducts, and cauda epididymal sperm for his *in vitro* fertilization experiments. In addition to observing polar body extrusion and egg cleavage, he carefully demonstrated sperm penetration of the egg and male pronuclear formation following *in vitro* fertilization. Two years later, the studies were extended. Following up on a report by Jagiello (1969) that guinea pig ovarian oocytes could be matured *in vitro*, Yanagimachi (1974) described a method for maturing guinea pig oocytes *in vitro* with the result that they could be subsequently fertilized *in vitro*.

5. SPERM CAPACITATION AND ACROSOME REACTION

5.1. Sperm Capacitation: Effect on Guinea Pig Sperm

With the ability to fertilize guinea pig eggs *in vitro*, the 1970s ushered in a new era in studies of guinea pig fertilization. Many laboratories began studies of sperm capacitation and the acrosome reaction. In fact, one set of the original IVF experiments carried out by Yanagimachi (1972) were designed to examine sperm capacitation. In these experiments, he found that if oocytes and freshly recovered cauda epididymal sperm were mixed together, fertilization took place at least 2 hr later. However, if sperm were preincubated 11–18 hr under the experimental conditions prior to mixing with the eggs, sperm penetration through the zona pellucida occurred within 10–15 min, with fertilization completed by 1 hr post-insemination. This result suggested that sperm from the epididymis were required to undergo additional changes before they were capable of fertilization.

Capacitation is a poorly defined physiological process. It is a conditioning phase that naturally takes place within the female reproductive tract but, as discussed above, can be mimicked *in vitro* by incubation of guinea pig sperm in a chemically defined medium. Capacitation is believed to be necessary for successful completion of the acrosome reaction and for sperm–egg fusion to occur (Yanagimachi, 1972). Currently, a complete understanding of the process is lacking at the molecular level principally because capacitation is such a difficult phenomenon to define.

Epididymal fluid is believed to block capacitation in several species, maintaining the sperm in a quiescent state. Thus, the idea arose that there were substances coating the sperm, sometimes called "decapacitation factors," that were removed from the cell surface during the capacitation process. Aonuma *et al.* (1973) incubated guinea pig sperm in Ca-free Krebs–Ringer phosphate buffer and recovered the substances released in the medium over periods of 1, 3, and 6 hr. This material was fractionated into two peaks by Sephadex G-200 chromatography. The first peak was used to immunize rabbits and yielded an antiserum recognizing four antigens by immunodiffusion and immunoelectrophoresis. This fraction, when added to rabbit sperm that were capacitated *in utero* in female rabbits, decreased their fertilization rate over a 24-hr period unless the sperm were "recapacitated" *in utero*. These results are provocative, but the molecules involved have not been identified in the guinea pig, although they may be similar to the "decapacitation" factor or acrosome-stabilizing factor described for rabbits (Wilson and Oliphant, 1987).

To examine surface changes in guinea pig sperm during capacitation, several laboratories assessed the binding of plant lectins to sperm. Differences in the binding of lectins between noncapacitated and capacitated sperm were used as indicators of the accessibility of carbohydrate ligands on the sperm surface. Talbot and Franklin (1978) quantitated the lectin-induced agglutination of live sperm before and after incubating the sperm in defined media known to capacitate guinea pig sperm. They found that with increasing times of incubation, the agglutinability of sperm by the lectin soybean agglutinin (N-acetylgalactosamine specific) and, to a lesser extent, concanavalin A (mannose and glucose specific) also increased. The increase in agglutination occurred in the tail region of the sperm, suggesting that new ligands for the lectins became available during capacitation or that the properties of the ligands on the sperm surface changed during this time, allowing the lectins to agglutinate the sperm more effectively. It is also interesting that *Ricinus communis* agglutinin, specific for N-acetylgalactosamine (like soybean agglutinin) and galactose, exhibited no changes in agglutinability during the incubation. Talbot and Franklin (1978) suggest that this dichotomy may result from the ligands for each of these lectins being on different molecules. The difference in soybean agglutinin binding was confirmed by Schwarz and Koehler (1979), who demonstrated with fluorescently labeled lectins that soybean agglutinin binding on the sperm tail increased during capacitation.

The role of cell-surface carbohydrates in sperm capacitation has also been examined by Srivastava *et al.* (1988). These investigators treated guinea pig spermatozoa with increasing concentrations of neuraminidase from *Arthrobacter ureafasciens* and found that high levels of this enzyme could promote capacitation as defined by the number of sperm that had undergone the acrosome reaction within the 1-hr incubation period. Unfortunately, although these authors also assessed sperm motility as a function of added neuraminidase, the percentage of sperm undergoing the acrosome reaction was reported for the total population. High concentrations of the enzyme preparation were lethal to the sperm, although this effect and the acrosome-inducing effect could be blocked by inhibitors of neuraminidase. These results are provocative and must be reexamined. For example, how pure is the *Arthrobacter* preparation? Why did neuraminidase from *Clostridium perfringens* not work in this assay? Is the difference in results seen with neuraminidase from the two sources related to linkage specificity? What are sialic acid linkages in the guinea pig sperm, and what glycoconjugates are effected? Nevertheless, these results are consistent with the lectin data, suggesting an increased number of N-acetylgalactosamine residues, since removal of terminal sialic acids from mammalian glycoproteins generally exposes penultimate galactose or N-acetylgalactosamine residues (Sharon, 1975).

5.2. Sperm Capacitation: Examination of Mechanism

The actual stimulus of the capacitation of guinea pig sperm has not been identified. Although surface changes appear to be related to capacitation, the mechanism(s) underlying these alterations are not known. Several studies have examined the effect of various treatments on capacitation with the result that certain agents have been reported to stimulate the process, whereas other treatments seem to retard or inhibit capacitation.

Fleming and Armstrong (1985) demonstrated that polycationic compounds such as poly-arginine and compound 48/80 appeared to stimulate spontaneous acrosome reactions that occurred when guinea pig sperm were transferred from a Ca^{2+}-free medium to a Ca^{2+}-containing medium. These results were concluded to be caused by an effect on capacitation, since the polycations were required during the preincubation (capacitation) period prior to exposure of the sperm to the calcium-containing medium. Curiously, the stimulatory effect of polyarginine could not be reversed by washing the sperm before transfer to the Ca^{2+}-containing medium, but polyarginine-treated sperm were able to fertilize intact guinea pig eggs. On the other hand, the stimulatory effect of compound 48/80 treatment could be reversed by washing. In addition, if the sperm or eggs were incubated in compound 48/80 and were not washed prior to *in vitro* fertilization, there was no sperm–egg fusion. This effect could be removed by washing the treated gametes prior to *in vitro* fertilization. Although these experiments provide some interesting observations on the possible role in the acrosome reaction and gamete fusion of anionic sites complementary to the polycationic compounds used in this study, the involvement of naturally occurring polyamines (e.g., spermidine and spermine) in capacitation is still unclear.

In other studies, Fleming and Kuehl (1985) examined the effects of preincubating sperm at different temperatures prior to incubating sperm at 37°C in a Ca^{2+}-containing medium. They found that at temperatures 25°C and below, few sperm acquired the ability to undergo the acrosome reaction. When the preincubation and incubation were carried out at temperatures of 37°C or 44°C, the sperm underwent the acrosome reaction in response to Ca^{2+}-containing buffer, and this event was greatly accelerated at the higher temperature. When the sperm were preincubated at 44°C and then incubated in Ca^{2+}-containing medium at 37°C, they underwent the acrosome reaction at about the same rate as sperm incubated throughout at 37°C. The authors discuss these results in terms of the effect of the different temperatures on sperm enzymatic rates or possible phase transitions in sperm lipids. It is very likely that the latter possibility can be investigated with membrane probes and methodology such as fluorescence recovery after photobleaching, which has been successfully used in other contexts on guinea pig sperm (Myles *et al.*, 1984; Phelps *et al.*, 1988).

In other studies, Shi and Friend (1985) examined the effect of gossypol acetate on guinea pig forward progressive motility, capacitation, and the acrosome reaction. They found that 10^{-4} M gossypol inhibited forward progressive motility of the sperm, whereas much lower levels (5×10^{-6} M) inhibited the acrosome reaction. By examining when it was necessary to have gossypol in the incubation medium, the authors concluded that this drug was affecting capacitation. If added after the sperm were incubated under capacitating conditions, gossypol had no effect. Again, the site of action was not identified, but, as for the temperature experiments above, these authors suggested that gossypol could either inhibit an enzyme involved in the acrosome reaction or affect membrane structure through hydrophobic interactions.

5.3. Guinea Pig Sperm Acrosome Reaction

Capacitation can be loosely defined as the process preparing the sperm for the acrosome reaction. Fortunately, much more is known about the acrosome reaction of guinea pig sperm. Guinea pig sperm have very large acrosomes, which makes it easy for investigators to score acrosome reactions. However, proper controls must always be run to verify that the "acrosome reactions" resulting from a particular experiment treatment did not result from the mechanical shearing of this structure. Indeed, if guinea pig sperm are centrifuged or pipetted too vigorously, the apical segments of acrosomes can be dislodged. In fact, such treatments have been used to isolate this structure (Olson *et al.*, 1987; Stojanoff *et al.*, 1987).

The natural course of the guinea pig acrosome reaction during fertilization *in vivo* has been the subject of debate. This has arisen from the observations, and general agreement, that the acrosome reaction of mouse sperm follows the binding of the mouse sperm to the zona pellucida (reviewed by Storey and Kopf, Chapter 10, this volume). Earlier observations in the guinea pig system suggested that it was only the acrosome-reacted sperm that were capable of binding to the zona pellucida (Huang *et al.*, 1981). Later it was shown that both acrosome-intact and acrosome-reacted guinea pig sperm were capable of binding to the zona pellucida during *in vitro* fertilization (Myles *et al.*, 1987).

Because the guinea pig sperm can be readily observed to undergo the acrosome reaction in response to various perturbants, this approach has been used to identify specific requirements for the guinea pig acrosome reaction. The calcium dependence of the guinea pig acrosome reaction has been established (Yanagimachi, 1974). The acrosome reaction can be stimulated by agents such as ionophores that elevate intracellular Ca^{2+} concentrations (Summers *et al.*, 1976; Green, 1978). At high pH values (pH 8.3–8.5), an antagonist of calmodulin, W-7, was reported to induce the acrosome reaction of guinea pig sperm, with the drug apparently affecting capacitation (Nagae and Srivastava, 1986).

The role of calcium during the capacitation and acrosome reaction steps has been studied in several laboratories. Capacitation apparently does not require calcium (Yanagimachi and Usui, 1974), but the acrosome reaction is strictly dependent on this divalent metal ion. Singh *et al.* (1978) examined the influx of calcium ions from a chemically defined medium into guinea pig sperm over a period of time and found that the net uptake of calcium occurred in two phases. Initially, Ca^{2+} loosely associated with the sperm in a manner that was unaffected by Mg^{2+} or inhibitors of mitochondrial function. Coincident with or occurring just prior to the acrosome reaction, there was a secondary uptake of calcium that could be accelerated with agents such as ionophores that stimulate the acrosome reaction. Conditions that delayed the acrosome reaction such as Mg^{2+} or low levels of Ca^{2+} also slowed the uptake of calcium from the medium. The regulation of calcium levels within the sperm during normal capacitation, acrosome reactions, and fertilization is still not completely understood, but recently, Coronel *et al.* (1988) identified caltrin-like proteins in the reproductive tract fluids and seminal vesicles of guinea pig sperm. These are small polypeptides with molecular weights ranging from about 5100 to 6200. Apparently they help to regulate calcium levels within guinea pig sperm by inhibiting calcium uptake.

The sperm acrosome reaction is an exocytotic event requiring the fusion of two membranes to cause release of acrosomal components. Like other membrane fusion events, calcium has been proposed to assist in the coalescence of the two membranes involved (Lawson *et al.*, 1978). The influx of calcium may also affect enzymatic processes by stimulating calcium-regulated activities. Specifically, calcium influx during the acrosome reaction of guinea pig spermatozoa has been shown to cause a rise in cAMP concentrations within the sperm (Hyne and Garbers, 1979). However, a cell-permeable analogue of cAMP was unable to stimulate the acrosome reaction in the absence of added calcium, suggesting that calcium did not stimulate the acrosome reaction via a cAMP-dependent pathway. Several other small molecules have been implicated as important regulators of the guinea pig acrosome reaction. These include prostaglandins (Joyce *et al.*, 1987), bicarbonate ions (Hyne, 1984; Bhattacharyya and Yanagimachi, 1988), fatty acid moieties of phospholipids (Fleming and Yanagimachi, 1984), and factors from serum (Hyne and Garbers, 1981).

As mentioned above, calcium may be having an effect at the level of the membrane. Friend *et al.* (1977) examined the freeze-fracture images of guinea pig sperm just prior to and just after the induction of the acrosome reaction. They found a variety of intracellular particle changes associated with the acrosome reaction. By omitting calcium during the capacitation step, they found that, prior to the fusion of the plasma membrane with the outer acrosomal membrane, fibrillar intramembranous particles were deleted from the E fracture faces of both membranes. In addition, globular particles were cleared from the P face of the plasma membrane. The clearing of these particles occurred near the equatorial segment, the region of membrane fusion that later becomes the boundary between the "old" plasma membrane and the "new" plasma membrane produced by the exposure of what was previously the inner acrosomal membrane. Following the addition of calcium to these sperm, the acrosome reaction was observed to occur with subsequent rearrangement of membrane particles. Additional membrane regions with cleared particles were observed in the acrosome-reacted sperm. These regions were thought to represent areas where sperm–egg interaction might occur. The reader is referred to this paper and a review by Friend (1982) for further information concerning the distribution of intramembranous particles in various regions of guinea pig sperm membranes.

More recently, the fusion of guinea pig membranes during the acrosome reaction has been studied using transmission electron microscopy (Flaherty and Olson, 1988). These authors observed elements that bridged the space between the plasma membrane and outer acrosomal membrane of the ventral surface of the apical segment of the guinea pig sperm acrosome. These bridging elements were absent in the corresponding dorsal region. In contrast, a parallel array of filaments was present on the side of the dorsal outer acrosomal membrane that faces the acrosomal contents. A complex pattern of membrane fusion resulted in the production of hybrid (plasma membrane–outer acrosomal membrane) tubules and vesicles. These authors also observed the presence of stable nonfusigenic regions of membrane. They believe that these domains result from the occurrence of membrane-associated assemblies that may be involved in the regulation of membrane fusion.

After the fusion of the outer acrosomal membrane with the plasma membrane overlying the acrosome, the fate of the acrosomal components must be considered. The contents of the acrosome do not exist in a simple fluid state. This has been shown by the ability of several laboratories to isolate the apical segment of the acrosome in a particulate state with or without attached membranes (Huang *et al.*, 1985; Olson *et al.*, 1987; Stojanoff *et al.*, 1987). Depending on the condition of isolation, the apical segments generally lack the plasma membrane that previously covered this portion of the acrosome; however, the outer acrosomal membrane usually remains attached. If the pH is maintained slightly acidic (pH 6.0), the apical segments remain particulate. Raising the pH to near neutrality causes the rapid dissolution of the matrix. If protease inhibitors are present, the matrix is stabilized at higher pH values, suggesting that proteases (e.g., acrosin) are responsible for the dissolution of the acrosomal matrix under these conditions (Perreault *et al.*, 1982). Additional experiments along these lines have shown that acrosin and hyaluronidase activities remain associated with the isolated apical segments (Stojanoff *et al.*, 1987). When apical segments were isolated at pH 6.0 and then adjusted to pH 7.5, a protein reactive with antitesticular proacrosin was converted from a molecular weight of 50,000 to 45,000, suggesting that this effect resulted from the proteolytic activation of the proacrosin zymogen; similar effects were seen for sperm undergoing the acrosome reaction in the presence of calcium and ionophore A23187 (Noland *et al.*, 1989).

6. THE SPERM SURFACE

6.1. Involvement of Sperm Surface Molecules in Guinea Pig Fertilization

Several laboratories have used the guinea pig system to begin characterizing the molecules present in different regions of the sperm membranes. For example, an added benefit of the apical segment isolation method of Olson *et al.*, (1987) is the ability to purify both the plasma membrane and the outer acrosomal membrane overlying the apical segment. These fractions can then been utilized for further biochemical studies. Earlier, Primakoff *et al.* (1980) stimulated guinea pig sperm to undergo the acrosome reaction in the presence of calcium with the ionophore A23187.

After removing reacted and unreacted sperm by differential centrifugation, the hybrid membrane vesicles (comprised of plasma membrane and outer acrosomal membrane) were isolated by sucrose density centrifugation. Membranes prepared in this manner were later used as immunogens for the preparation of monoclonal antibodies (Myles *et al.*, 1981). This approach demonstrated that certain antigens were restricted to specific regions, allowing Primakoff and Myles (1983) to map the guinea pig sperm surface with their panel of antibodies. Five distinct domains were defined: the anterior head, posterior head (postacrosomal region), whole head, whole tail, and posterior tail (principal piece and end piece).

There are several possible explanations for the maintenance of the domains in the sperm. One possibility is that specific proteins are anchored in place by some sort of an immobilizing mechanism such as a cytoskeletal network. Another possibility is that there are diffusion barriers within the membrane that trap specific molecules within their borders. Morphological support for this concept was provided by freeze-fracture studies (Friend, 1982; Friend *et al.*, 1977). Additional evidence for this mechanism was presented by Myles *et al.* (1984) for the posterior tail protein recognized by monoclonal antibody PT-1. Using fluorescence redistribution after photobleaching, these investigators demonstrated that the PT-1 antigen diffuses very rapidly within its domain, a result that rules out immobilization of the protein as a means for maintaining its regionalized localization. Curiously, the restricted localization of PT-1 protein changes as a function of the sperm state. If sperm are capacitated by incubation in 102.3 mM NaCl, 25.1 mM NaHCO$_3$, 0.25 mM sodium pyruvate, and 21.7 mM sodium lactate at 37°C for 1 hr, much of the PT-1 protein migrated into the midpiece of the sperm tail (Myles and Primakoff, 1984). Thus, capacitation may be accompanied by changes in the diffusion barriers present in the plasma membranes of sperm.

Other proteins of the guinea pig sperm also exhibit complex patterns of distribution and diffusion. One of these, the protein recognized by MAb PH-20, is a case in point. This protein has a dual localization in sperm, being found on the sperm surface and in the membrane of the acrosome (Phelps and Myles, 1987). Initially, surface PH-20 protein is distributed all over testicular sperm. As the cells mature in the epididymis, PH-20 protein becomes localized to the postacrosomal region of the sperm head. Following the acrosome reaction, surface PH-20 protein migrates from the postacrosomal region to the new plasma membrane (previously the inner acrosomal membrane) of the anterior head created by the loss of the former plasma membrane and outer acrosomal membranes (Myles and Primakoff, 1984). The migration of the PH-20 protein from the total surface of testicular sperm to the posterior head of acrosome-intact cauda epididymal sperm is accompanied by a tenfold increase in the diffusion constant. After the acrosome reaction has occurred, the diffusion rate is 250-fold higher than that observed for testicular sperm (Phelps *et al.*, 1988). The rate of diffusion of PH-20 protein on the surface of testicular sperm is more than a thousand times slower than lipid diffusion. This is significant because the PH-20 protein is anchored to the plasma membrane by the lipid phosphatidylinositol. The basis for this difference is not completely understood but apparently is not related to the density of the PH-20 protein molecules in the membrane, since the PH-20 protein is more freely diffusible when it is at its highest concentration within the membrane, i.e., after the acrosome reaction (Phelps *et al.*, 1988).

What, if any, is the significance of the observed rearrangements of the sperm surface? Based on comparisons with other rodent species, the corpus–proximal cauda region of the epididymis is probably the site where sperm gain the ability to fertilize eggs (Horan and Bedford, 1972; Dyson and Orgebin-Crist, 1973). The change in the distribution of the PH-20 protein from a total surface pattern to the posterior head pattern occurs here (Phelps and Myles, 1987). In addition, another antigen, recognized by MAb PH-30, appears at this point during epididymal maturation. What are the functions of these proteins?

6.2. Sperm Binding to Eggs

Two requisite processes for successful fertilization include sperm binding to the egg and sperm–egg fusion. Sperm–egg binding is best understood in the mouse, where it is accepted that sperm have intact acrosomes when they initially encounter the zona pellucida of the egg (comprised of three zona molecules ZP1, ZP2, and ZP3). Specifically, the ZP3 glycoprotein of the zona is known to contain the initial sperm-binding site and acrosome-reaction-inducing activity (reviewed by Wassarman, Chapter 9, this volume). Following the acrosome reaction, it is believed that the sperm interact with the ZP2 molecule while penetrating the zona. However, the actual sperm surface molecule(s) responsible for mouse sperm–zona binding have not been identified. In the guinea pig system, Primakoff *et al.* (1985) have proposed that the PH-20 protein may be involved in binding of sperm from this species to the zona. In this study, three monoclonal antibodies, MAb PH-20, MAb PH-21, and MAb PH-22, were used that recognized the PH-20 protein. At equal antibody concentrations (50 μg/ml), MAb PH-20 and MAb PH-21 inhibited sperm–zona binding, whereas MAb PH-22 had no effect. This inhibition was not caused simply by steric effects, since MAb PH-20 and MAb PH-22 could block each other's binding to sperm, but only MAb PH-20 could inhibit sperm–zona binding.

As mentioned earlier, acrosome-intact and acrosome-reacted guinea pig sperm are capable of binding to the zona, so the question arises whether the inhibition of sperm–zona binding by MAb PH-20 results from the blocking of acrosome-intact or acrosome-reacted sperm or both? To examine this question, Myles *et al.* (1987) used video microscopy and an *in vitro* sperm–egg binding assay to look at the effect of MAb PH-20 on sperm attachment to the zona. Because of the large size of the guinea pig sperm acrosome, it was quite easy to distinguish binding of acrosome-intact from acrosome-reacted sperm by phase-contrast microscopy. Surprisingly, these investigators found that MAb PH-20 inhibited the binding of acrosome-reacted but not acrosome-intact sperm to the zona. Since the PH-20 protein is present on the surface of both types of sperm, there are several possible interpretations of this result. The PH-20 protein may not be the actual sperm receptor for the zona. It may act in concert with other proteins that are exposed following the acrosome reaction. There might be a concentration effect of threshold level of PH-20 protein molecules that need to be exposed before MAb PH-20 is capable of blocking sperm–egg binding (keep in mind that additional PH-20 protein emanating from the inner acrosomal membrane is exposed following the acrosome reaction). Acrosome-intact and acrosome-reacted sperm might bind to guinea pig zona by independent mechanisms. This is apparently the case in the mouse, where sperm initially bind to the ZP3 molecule of the zona and subsequently bind to ZP2 following the acrosome reaction (Bleil and Wassarman, 1986).

What sort of recognition signal is required for the binding of guinea pig sperm to the zona pellucida? Huang *et al.* (1982) examined the effect of various carbohydrates and glycoconjugates on the binding of guinea pig sperm to the zona pellucida. L-Fucose, at a concentration of 50 mM, inhibited the binding by 70–80%, whereas other monosaccharides (D-glucose, D-galactose, N-acetylglucosamine, N-acetyl-D-galactosamine, and D-fucose) only blocked 30–40% of the binding at the same concentrations. Fucoidin, a polymer high in sulfated L-fucose residues, blocked 100% of the sperm attachment at a concentration of 100 μg/ml. Controls with other sulfated glycoconjugates (heparin and chondroitin sulfate A) were inactive at the same concentration. Other glycoconjugates such as horseradish peroxidase, yeast mannan, galactan, and bovine submaxillary mucin were also ineffective at concentrations as high as 2.0 mg/ml. No effects were seen on sperm motility. In a subsequent study, Huang and Yanagimachi (1984) demonstrated that the fucoidin binds to the inner acrosomal membrane and equatorial region of live, acrosome-reacted guinea pig sperm. In dead sperm, the fucoidin label bound to the postacrosomal region of the sperm. These results are interesting in light of the localization of the PH-20 protein in guinea pig sperm to the inner acrosomal membrane after the acrosome reaction.

6.3. Sperm Fusion with the Egg

After a sperm has successfully penetrated the zona and encountered the plasma membrane of the egg, interaction between the two gametes leads to the fusion of their respective membranes. Primakoff *et al.* (1987) have proposed that another posterior head molecule, the PH-30 antigen, may have a role in sperm–egg membrane fusion. With zona-free eggs, MAb PH-30 was found to block sperm–egg fusion. However, MAb PH-1, which recognizes a different epitope on the same molecule, failed to block membrane fusion. As mentioned above, during the development of sperm, the PH-30 antigen first appears in sperm from the proximal cauda epididymis. This is also the predicted point during epididymal maturation when the sperm would be expected to gain fertility. Obviously, additional studies need to be done to demonstrate the role of the PH-30 antigen in sperm–egg fusion. However, if this protein is involved, the question arises as to why it becomes detectable at this time. Is there a precursor for the antigen that becomes modified in the epididymis? Is the PH-30 antigen added to the sperm during epididymal transit? Is the PH-30 antigen very diffusely localized on sperm and then does it regionalize in the proximal cauda epididymis? Both MAb PH-30 and MAb PH-1 immunoprecipitate polypeptides with molecular weights of 44,000 and 60,000 that do not appear to be structurally related. Do these polypeptides represent a complex protein with two dissimilar subunits?

7. THE USE OF THE PH-20 PROTEIN AS AN IMMUNOCONTRACEPTIVE

The identification of the antigens recognized by MAb PH-20 and MAb PH-30 has allowed the Primakoff and Myles group to explore the possible functions of these proteins in fertilization. Since the PH-20 protein appears to be involved in the binding of sperm to the zona pellucida of guinea pig eggs, they have also been examining the possibility of using this surface protein as an immunocontraceptive (Primakoff *et al.*, 1988). Results to date are excellent, with these authors reporting a 100% effective rate of contraception of males and females immunized against the PH-20 protein. The antisera from immunized females had high titers, specifically recognized the PH-20 protein by Western immunoblotting of sperm extracts, and blocked sperm adhesion to the zona pellucida *in vitro*. The antifertility effect lasted for several months before the females began to become fertile again. These results are very exciting, since the antigen was very potent. Immunogen amounts of 5 μg for females and 2.5 μg for males were completely effective, causing one to wonder what the minimal dose for effect is. Once the cDNA for the PH-20 protein becomes available, it will be very interesting to determine whether a similar protein exists in the sperm of other species such as humans. If so, such a discovery could lead to the development of an immunocontraceptive based on a sperm protein that is effective at low doses, is long lasting, but eventually might be reversible.

8. CONCLUDING COMMENTS: FUTURE STUDIES

Much has yet to be learned about the process of fertilization in mammals. Certainly, over the last decade, the mouse has been the system of choice for dissecting many of the basic steps in the fertilization process, in particular the zona-pellucida-induced acrosome reaction. However, other systems such as the guinea pig offer specific advantages. For sperm physiologists, the guinea pig sperm has a beautiful acrosome that can be easily identified in the light microscope for studies of the acrosome reaction. The large acrosome also allows for the subfractionation of guinea pig sperm, yielding apical segments, plasma membranes from the apical segment, outer acrosomal membranes from this region, and the particulate component from the lumen of the acrosomal vesicle. Biochemical characterization of acrosomal components is possible because large numbers of sperm can be obtained from a single animal. The large surface area of the guinea pig sperm allows the examination of surface protein movement using procedures such as fluorescence recovery after photobleaching. There are, of course, disadvantages with the guinea pig system. Perhaps the largest problems are that the expense of the animals and the poor rate of superovulation of female guinea pigs make it difficult to obtain large numbers of eggs for characterization of oocyte-derived proteins such as the zona pellucida.

Future studies of fertilization using guinea pigs will undoubtedly provide new insights concerning this important process. Questions that need to be answered include: How is fertility increased during epididymal maturation? Is the increase in sperm fertility during epididymal transport related to the observed changes in sperm head structure and rearrangement of sperm surface molecules? What exactly is capacitation, and how does it come about? What is the natural stimulus for the acrosome reaction of guinea pig sperm? Is it the zona pellucida or some other accessory substance around or near the egg? Although acrosome-reacted sperm are capable of binding to the zona pellucida of eggs, can these sperm actually penetrate the zona and fertilize the egg? What role do the components of the acrosomal matrix play in the fertilization process? Do the enzymes of the acrosome enable the sperm to penetrate the zona or some other investment around the guinea pig egg? If the PH-20 protein is involved in binding the sperm to the zona, what component of this extracellular matrix does it interact with? What is the significance of the rearrangement of the PH-20 protein on the sperm surface following the acrosome reaction? Are homologues of the PH-20 protein found in the sperm of other species? Are these homologues also involved in sperm–egg binding? What is the role of the PH-30 antigen in sperm–egg fusion? What are the fates of guinea pig surface components following fertilization? Certainly, with the progress promised by current techniques of cell biology and molecular biology, the answers to many of these questions will be forthcoming in the next decade.

9. REFERENCES

Anakwe, O. O., Sharma, S., Hoff, H. B., Hardy, D. M., and Gerton, G. L., 1991, Maturation of guinea pig sperm in the epididymis involves the modification of proacrosin oligosaccharide side-chains, *Molec. Reprod. Dev.* **29**:294–301.

Aonuma, S., Mayumi, T., Suzuki, K., Noguchi, T., Iwai, M., and Okabe, M., 1973, Studies on sperm capacitation. I. The relationship between guinea pig sperm coating antigen and a sperm capacitation phenomenon, *J. Reprod. Fertil.* **35**:425–432.

Arboleda, C. E., and Gerton, G. L., 1988, Proacrosin/acrosin during guinea pig spermatogenesis, *Dev. Biol.* **125**:217–225.

Austin, C. R., 1961, Fertilization of eggs *in vitro*, *Int. Rev. Cytol.* **12**:337–359.

Bhattacharyya, A., and Yanagimachi, R., 1988, Synthetic organic pH buffers can support fertilization of guinea pig eggs, but not as efficiently as bicarbonate buffer, *Gamete Res.* **19**:123–129.

Blaquier, J. A., Cameo, M. S., and Burgos, M. H., 1972, The role of androgens in the maturation of epididymal spermatozoa in the guinea pig, *Endocrinology* **90**:839–842.

Bleil, J. D., and Wassarman, P. M., 1986, Autoradiographic visualization of the mouse egg's sperm receptor bound to sperm, *J. Cell Biol.* **102**:1363–1371.

Coronel, C. E., San Augustin, J., and Lardy, H. A., 1988, Identification and partial characterization of caltrin-like proteins in the reproductive tract of the guinea pig, *Biol. Reprod.* **38**:713–722.

Dyson, A. L. M. B., and Orgebin-Crist, M.-C., 1973, Effect of hypophysectomy, castration and androgen replacement upon the fertilizing ability of rat epididymal spermatozoa, *Endocrinology* **93**:391–402.

Fawcett, D. W., and Hollenberg, R. D., 1963, Changes in the acrosome of guinea pig spermatozoa during passage through the epididymis, *Z. Zellforsch.* **60**:276–292.

Flaherty, S. P., and Olson, G. E., 1988, Membrane domains in guinea pig sperm and their role in the membrane fusion events of the acrosome reaction, *Anat. Rec.* **220**:267–280.

Fleming, A. D., and Armstrong, D. T., 1985, Effects of polyamines upon capacitation and fertilization in the guinea pig, *J. Exp. Zool.* **233**:93–100.

Fleming, A. D., and Kuehl, T. J., 1985, Effects of temperature upon capacitation of guinea pig spermatozoa, *J. Exp. Zool.* **233**:405–411.

Fleming, A. D., and Yanagimachi, R., 1984, Evidence suggesting the importance of fatty acids and the fatty acid moieties of sperm membrane phospholipids in the acrosome reaction of guinea pig spermatozoa, *J. Exp. Zool.* **229**:485–489.

Friend, D. S., 1982, Plasma-membrane diversity in a highly polarized cell, *J. Cell Biol.* **93**:243–249.

Friend, D. S., Orci, L., Perrelet, A., and Yanagimachi, R., 1977, Membrane particle changes attending the acrosome reaction in guinea pig spermatozoa, *J. Cell Biol.* **74**:561–577.

Green, D. P. L., 1978, The induction of the acrosome reaction in guinea-pig sperm by the divalent metal cation ionophore A23187, *J. Cell Sci.* **32**:137–151.

Horan, A. H., and Bedford, J. M., 1972, Development of the fertilizing ability of spermatozoa in the epididymis of the Syrian hamster, *J. Reprod. Fertil.* **30**:417–423.

Huang, T. T. F., and Yanagimachi, Y., 1984, Fucoidin inhibits attachment of guinea pig spermatozoa to the zona pellucida through binding to the inner acrosomal membrane and equatorial domains, *Exp. Cell Res.* **153**:363–373.

Huang, T. T. F., Fleming, A. D., and Yanagimachi, Y., 1981, Only acrosome-reacted spermatozoa can bind and penetrate into zona pellucida: A study using guinea pig, *J. Exp. Zool.* **217**:286–290.

Huang, T. T. F., Ohzu, E., and Yanagimachi, Y., 1982, Evidence suggesting that L-fucose is part of a recognition signal for sperm–zona pellucida attachment in mammals, *Gamete Res.* **5**:355–361.

Huang, T. T. F., Hardy, D., Yanagimachi, H., Teuscher, C., Tung, K., Wild, G., and Yanagimachi, R., 1985, pH and protease control of acrosomal content stasis and release during the guinea pig sperm acrosome reaction, *Biol. Reprod.* **32**:451–462.

Hunter, R. H. F., Hunt, D. M., and Chang, M. C., 1969, Temporal and cytological aspects of fertilization and early development in the guinea pig, *Cavia porcellus*, *Anat. Rec.* **165**:411–430.

Hyne, R. V., 1984, Bicarbonate and calcium-dependent induction of rapid guinea pig sperm acrosome reactions by monovalent ionophores, *Biol. Reprod.* **31**:312–323.

Hyne, R. V., and Garbers, D. L., 1979, Calcium-dependent increase in adenosine $3',5'$-monophosphate and induction of the acrosome reaction in guinea pig spermatozoa, *Proc. Natl. Acad. Sci. USA* **76**:5699–5703.

Hyne, R. V., and Garbers, D. L., 1981, Requirement of serum factors for capacitation and the acrosome reaction of guinea pig spermatozoa in buffered medium below pH 7.8, *Biol. Reprod.* **24**:257–266.

Jagiello, G. M., 1969, Some cytologic aspects of meiosis in female guinea pig, *Chromosoma* **27**:95–101.

Joyce, C. L., Nuzzo, N. A., Wilson, L., Jr., and Zaneveld, L. J. D., 1987, Evidence for a role of cyclooxygenase (prostaglandin synthetase) and prostaglandins in the sperm acrosome reaction, *J. Androl.* **8**:74–82.

Lawson, D., Fewtrell, C., and Raff, M., 1978, Localized mast cell degranulation induced by concanavalin A–sepharose beads: Implications for the Ca^{2+} hypothesis of stimulus–secretion coupling, *J. Cell Biol.* **79**:394–400.

Loeb, L., 1911, The cyclic changes in the ovary of the guinea pig, *J. Morphol.* **22**:39–70.

Loeb, L., 1914, The correlation between the cyclic changes in the uterus and ovaries in the guinea pig, *Biol. Bull.* **27**:1–44.

Loeb, L., 1918, Corpus luteum and the periodicity in the sexual cycle, *Science* **48**:273–277.

Myles, D. G., and Primakoff, P., 1984, Localized surface antigens of guinea pig sperm migrate to new regions prior to fertilization, *J. Cell Biol.* **99**:1634–1641.

Myles, D. G., Primakoff, P., and Bellvé, A. R., 1981, Surface domains of the guinea pig sperm defined with monoclonal antibodies, *Cell* **23**:433–439.

Myles, D. G., Primakoff, P., and Koppel, D. E., 1984, A localized surface protein of guinea pig sperm exhibits free diffusion in its domain, *J. Cell Biol.* **98**:1905–1908.

Myles, D. G., Hyatt, H., and Primakoff, P., 1987, Binding of both acrosome-intact and acrosome-reacted guinea pig sperm to the zona pellucida during *in vitro* fertilization, *Dev. Biol.* **121**:559–567.

Nagae, T., and Srivastava, P. N., 1986, Induction of the acrosome reaction in guinea pig spermatozoa by calmodulin antagonist W-7, *Gamete Res.* **14**:197–208.

Noland, T. D., Davis, L. S., and Olson, G. E., 1989, Regulation of proacrosin conversion in isolated guinea pig sperm acrosomal apical segments, *J. Biol. Chem.* **264**:13586–13590.

Olson, G. E., Winfrey, V. P., Winer, M. A., and Davenport, G. R., 1987, Outer acrosomal membrane of guinea pig spermatozoa: Isolation and structural characterization, *Gamete Res.* **17**:77–94.

Perreault, S. D., Zirkin, B. R., and Rogers, B. J., 1982, Effect of trypsin inhibitors on acrosome of guinea pig spermatozoa, *Biol. Reprod.* **26**:343–351.

Phelps, B. M., and Myles, D. G., 1987, The guinea pig sperm plasma membrane protein, PH-20, reaches the surface via two transport pathways and becomes localized to a domain after an initial uniform distribution, *Dev. Biol.* **123**:63–72.

Phelps, B. M., Primakoff, P. M., Koppel, D. E., Low, M. G., and Myles, D. G., 1988, Restricted lateral diffusion of PH-20, a PI-anchored sperm membrane protein, *Science* **240**:1780–1782.

Primakoff, P., and Myles, D. G., 1983, A map of the guinea pig sperm surface constructed with monoclonal antibodies, *Dev. Biol.* **98**:417–428.

Primakoff, P., Myles, D. G., and Bellvé, A. R., 1980, Biochemical analysis of the released products of the mammalian acrosome reaction, *Dev. Biol.* **80**:324–331.

Primakoff, P., Hyatt, H., and Myles, D. G., 1985, A role for the migrating sperm surface antigen PH-20 in guinea pig sperm binding to the egg zona pellucida, *J. Cell Biol.* **101**:2239–2244.

Primakoff, P., Hyatt, H., and Tredick-Kline, J., 1987, Identification and purification of a sperm surface protein with a potential role in sperm–egg membrane fusion, *J. Cell Biol.* **104**:141–149.

Primakoff, P., Lathrop, W., Woolman, L., Cowan, A., and Myles, D., 1988, Fully effective contraception in male and female guinea pigs immunized with the sperm protein PH-20, *Nature* **335**:543–546.

Schenk, S. L., 1878, Das Sangethierei kunstlich befruchtet ausserhalb des Mutter thieves, *Mitt. Embryol. Inst. K.K. Univ. Wien* **1**:107–118.

Schwarz, M. A., and Koehler, J. K., 1979, Alterations in lectin binding to guinea pig spermatozoa accompanying *in vitro* capacitation and the acrosome reaction, *Biol. Reprod.* **21**:1295–1307.

Sharon, N., 1975, *Complex Carbohydrates*, Addison-Wesley, Reading, MA, pp. 155–166.

Shi, Q. X., and Friend, D. S., 1985, Effect of gossypol acetate on guinea pig epididymal spermatozoa *in vivo* and their susceptibility to capacitation *in vitro*, *J. Androl.* **6**:45–62.

Singh, J. P., Babcock, D. F., and Lardy, H. F., 1978, Increased calcium ion influx is a component of capacitation of spermatozoa, *Biochem. J.* **172**:549–556.

Srivastava, P. N., Kumar, V. M., and Arbtan, K. D., 1988, Neuraminidase induces capacitation and acrosome reaction in mammalian spermatozoa, *J. Exp. Zool.* **245**:106–110.

Stigler, R., 1918, Der Einfluss des Nebenhodens auf die Vitalität der Spermatozoen, *Pflügler's Arch.* **171**:273–282.

Stockard, C. R., and Papacinolaou, G., 1917, The existence of a typical oestrous cycle in the guinea pig—with a study of its histological and physiological changes, *Am. J. Anat.* **22**:225–265.

Stockard, C. R., and Papacinolaou, G., 1919, The vaginal closure membrane, copulation, and the vaginal plug in the guinea-pig, with further considerations of the oestrous rhythym, *Biol. Bull.* **37**:222–245.

Stojanoff, A., Bourne, H., Andrews, A. G., and Hyne, R. V., 1987, Isolation of a stable apical segment of the guinea pig sperm acrosome, *Gamete Res.* **17**:321–332.

Summers, R. G., Talbot, P., Keough, E. M., Hylander, B. L., and Franklin, L. E., 1976, Ionophore A23187 induces acrosome reactions in sea urchin and guinea pig spermatozoa, *J. Exp. Zool.* **196**:381–385.

Talbot, P., and Franklin, L. E., 1978, Surface modifications of guinea pig sperm during *in vitro* capacitation: An assessment using lectin-induced agglutination of living sperm, *J. Exp. Zool.* **203**:1–14.

Wilson, W. L., and Oliphant, G., 1987, Isolation and biochemical characterization of the subunits of the rabbit sperm acrosome stabilizing factor, *Biol. Reprod.* **37**:159–169.

Yanagimachi, R., 1972, Fertilization of guinea pig eggs *in vitro*, *Anat. Rec.* **174**:9–20.

Yanagimachi, R., 1974, Maturation and fertilization *in vitro* of guinea pig ovarian oocytes, *J. Reprod. Fertil.* **38**:485–488.

Yanagimachi, R., and Usui, N., 1974, Calcium dependence of the acrosome reaction and activation of guinea pig spermatozoa, *Exp. Cell Res.* **89**:161–174.

Young, W. C., 1927, The influence of high temperature on the guinea-pig testis: Histological changes and effects on reproduction, *J. Exp. Zool.* **49**:459–499.

Young, W. C., 1929a, A study of the function of the epididymis. I. Is the attainment of full spermatozoon maturity attributable to some specific action of the epididymal secretion? *J. Morphol. Physiol.* **47**:479–495.

Young, W. C., 1929b, A study of the function of the epididymis. II. The importance of an aging process in sperm for the length of the period during which fertilizing capacity is retained by sperm isolated in the epididymis of the guinea-pig, *J. Morphol. Physiol.* **48**:475–491.

Young, W. C., 1931, A study of the function of the epididymis. III. Functional changes undergone by spermatozoa during their passage through the epididymis and vas deferens in the guinea-pig, *J. Expo. Biol.* **8**:151–162.

14

Fertilization in the Rabbit

Michael G. O'Rand and Barbara S. Nikolajczyk

1. INTRODUCTION

Discovered by the Phoenicians in ancient times, the rabbit has been bred in captivity for hundreds of years. In the monasteries of the 16th century, domestication of the rabbit (*Oryctolagus*) and hare (*Lepus*) led to the continued breeding of rabbits because of their ability to reproduce easily in leporaria (Fox, 1974). Today's modern laboratory rabbit is a decendent of the European rabbit *Oryctolagus cuniculus* and now exists in dozens of different breeds, from the 15-lb Flemish giant to the 2½-lb. pound Polish white. Because of such a long history of domestication, evolutionary changes from the true wild type have undoubtedly occurred; nevertheless, researchers have continued to study the reproductive biology of the domestic rabbit for over 300 years. In fact, early observations on ovarian follicles, which later became known as "Graafian follicles," were made in the rabbit by De Graaf in the late 17th century. For the next 200 years numerous investigators reinvestigated ovulation and early development in the rabbit, using the rabbit as a model for mammalian development because it is easily bred in captivity and ovulation is induced by coitus. By 1839, Barry had clearly established the time of ovulation as 10 hr after coitus, and he made the significant observation that the interaction of the spermatozoon with the egg was responsible for the development of the embryo.

During the late 1800s, most work on fertilization utilized invertebrate material, and in 1891, Weismann emphasized that the male and female gametes each made equal genetic contributions to the zygote. However, it was not until the the first decade of the 20th century that the importance of fertilization as a process on which evolution depended was established. During the 1920s and 1930s studies on the rabbit revealed much about mammalian reproductive biology. In his book *Reproduction in the Rabbit*, Hammond (1925) brought together a number of observations on rabbit biology. He observed that in vasectomized bucks some of the seminiferous tubules were abnormal, after coitus the ovary went through the estrous cycle, and additionally if no fertilization occurred, pseudopregnancy resulted. He also made the important observation that for successful fertilization and development to occur after coitus, the sperm had to arrive in the

MICHAEL G. O'RAND AND BARBARA S. NIKOLAJCZYK • Department of Cell Biology and Anatomy, University of North Carolina at Chapel Hill, Chapel Hill, North Carolina 27599.

A Comparative Overview of Mammalian Fertilization, edited by Bonnie S. Dunbar and Michael G. O'Rand. Plenum Press, New York, 1991.

oviduct between 2 hr before ovulation and 2 hr after ovulation, ovulation occurring approximately 10 hr after coitus. The fertile life span of the egg was established to be between 2 and 4 hr, and normal pregnancy lasted 31 to 33 days.

Almost 40 years after Weismann's observations on the contributions from male and female gametes, P. W. Gregory (1930) described the early embryology of the rabbit. In living fertilized eggs, he observed two polar bodies, whereas in fixed material two unfused pronuclei were observed at 21.5 hr post-coitus, and fused pronuclei at 22.5 hr post-coitus. Over the next 20 years, rabbit fertilization came under rather intense experimental study through the work of Hammond (1934), Pincus (1939), and M. C. Chang. As a student of Arthur Walton and John Hammond, in 1939 Dr. Chang began his studies on fertilization in the rabbit (see Chang, 1984, for a review). Early studies by Chang (1951a, 1952) and by C. E. Adams (1956) centered around the timing of fertilization and the capacity of spermatozoa to fertilize eggs both *in vivo* and *in vitro*.

2. MATURATION OF THE GAMETES AND THE TIMING OF FERTILIZATION

The rabbit is a reflex ovulator whose oocytes mature to the first polar body stage and are ovulated as a result of an LH surge triggered by the stimulus of mating (see Ramirez and Beyer, 1988, for a review of the rabbit ovarian cycle). The fertilizability of rabbit ova has been reported to last up to 8 hr after ovulation (Chang, 1952), but after 3¼ to 4 hr post-ovulation the fertility level drops below 50% (Chang, 1952; Dukelow *et al.*, 1967).

Spermatozoa mature in the rabbit testis in 43.6 days (Romrell and O'Rand, 1978; Swierstra and Foote, 1963, 1965) and move through the epididymis in 4 to 7 days. During their transit through the distal corpus epididymis, spermatozoa become capable of undergoing capacitation in the female reproductive tract and consequently capable of undergoing the acrosome reaction and fertilizing ova. Epididymal maturation is a testosterone-dependent process in which the spermatozoa also gain progressive motility (Bedford, 1966; Orgebin-Crist, 1973; Orgebin-Crist and Jahad, 1979; see Robaire and Hermo, 1988, for a review of the epididymis, structure and function). There is considered to be no difference between mature epididymal spermatozoa and ejaculated spermatozoa with regard to their ability to reach the site of fertilization in the female tract, their ability to pentrate ova, or in the ability of their fertilized ova to develop normally (Overstreet and Bedford, 1974; Brackett *et al.*, 1978). Both epididymal and ejaculated spermatozoa require capacitation in the female reproductive tract.

Immediately after coitus spermatozoa are transported to the upper regions of the oviduct (infundibulum and ampulla) in what has been described as a rapid transit phase (Overstreet and Cooper, 1978a). This phase is a rapid but primarily passive transfer of spermatozoa that does not require sustained motility, and most of the spermatozoa undoubtedly end up in the peritoneal cavity. Spermatozoa in this population would not be responsible for fertilizing the ova. A second phase of transport then ensues, which brings the fertilizing spermatozoon to the ovum in the correct time and place (Harper, 1973; Overstreet and Cooper, 1978b, 1979). During the second, sustained phase, spermatozoa move from the initial vaginal pool into a cervical pool and ascend slowly into the uterus and oviduct. Uterine contractions and sperm motility propel the spermatozoa through the uterotubal junction and into the lower isthmus, where they remain until the time of ovulation (Overstreet and Cooper, 1978b). In an ovulation-dependent manner, spermatozoa in the isthmus ascend into the ampulla and are simultaneously "activated," showing the typical hyperactivation motility pattern of capacitated spermatozoa before they enter the cumulus (Overstreet and Cooper, 1979; Cooper *et al.*, 1979; Johnson *et al.*, 1981; Suarez *et al.*, 1983).

Consequently, in the rabbit, coitus initiates a highly orchestrated series of events in which mature ova are ovulated approximately 10 hr later and capacitated spermatozoa are waiting in the ampulla to fertilize them.

3. CAPACITATION AND THE ACROSOME REACTION

It is now well established that rabbit spermatozoa, similar to other eutherian spermatozoa, require capacitation in order to fertilize the ovum (Chang, 1951b; Austin, 1951). This physiological change in the spermatozoon requires 6 hr in the rabbit (Chang, 1951a) and is brought about by the synergistic action of the uterus and oviduct (Bedford, 1970). Although capacitation can occur in an ovariectomized rabbit, it is optimally achieved in an estrogen-dominated female. Attempts at *in vitro* capacitation have been controversial (Brackett and Oliphant, 1975; Akruk *et al.*, 1979; Hosoi *et al.*, 1981; Viriyapanich and Bedford, 1981), although Hosoi's modification of Brackett's method seems to work well with epididymal spermatozoa. More recently, combining a percoll gradient with Brackett's method has been shown to result in hyperactivated spermatozoa that are capable of penetrating ova (O'Rand and Fisher, 1987).

Although there is no known morphological difference between capacitated and noncapacitated spermatozoa, subtle changes in the number and distribution of intramembraneous particles may occur, as has been demonstrated in the guinea pig (Friend *et al.*, 1977). In rabbit spermatozoa capacitation has been shown to include removal of seminal plasma components from the sperm surface (Brackett and Oliphant, 1975). One such component is an 84,000 molecular weight (84-kDa) component that is present on ejaculated spermatozoa bound to the plasma membrane glycoprotein RSA (O'Rand and Romrell, 1981; O'Rand, 1982). Another component is the acrosome-stabilizing factor (ASF), which is synthesized in the distal corpus epididymis and binds to the acrosomal surface of spermatozoa (Thomas *et al.*, 1984). The ASF is a 360,000 mol. wt. dimer, each monomer being composed of a 92,000 and a 41,000–35,000 mol. wt. subunit as determined by SDS-PAGE (Reynolds and Oliphant, 1984). The removal of ASF and 84-kDa as well as possibly other components from the spermatozoa during capacitation could account for the change in surface charge and lectin-binding properties observed to be coincident with capacitation (O'Rand, 1982).

Sperm surface changes that occur during capacitation undoubtedly allow the spermatozoa to undergo the next phase of capacitation, namely, the rearrangement of intrinsic plasma membrane proteins and changes in cholesterol and phospholipid content. These changes in turn allow calcium influx and receptor-mediated induction of the acrosome reaction (see O'Rand 1979, 1982, 1988; Langlais and Roberts, 1985, for review). Thus, one can conclude that capacitation in the rabbit, as in other mammals, is the physiological change to the spermatozoon that allows the acrosome reaction to occur in the correct time and place for optimal fertilization.

As mentioned above, concomitant with the movement of spermatozoa to the site of fertilization in the ampulla is the change to hyperactivated motility characteristic of capacitation. In the oviduct, hyperactivated spermatozoa, characterized by rapid circular swimming, enter the cumulus mass of the ovulated rabbit egg with their acrosomes intact (Suarez *et al.*, 1983). Once within the cumulus, the flagellar activity of the spermatozoon is constrained by the viscoelastic properties of the cumulus matrix (Katz *et al.*, 1989), and the spermatozoa move "in a snaking, erratic manner around the granulosa cells" (Suarez *et al.*, 1983). During this movement through the cumulus matrix, the rabbit sperm enzyme hyaluronidase is thought to play a major role in penetration. This is based on the demonstration of hyaluronidase on the surface of spermatozoa (O'Rand and Metz, 1974) and in isolated membrane fractions (O'Rand and Metz, 1976) as well as on the inhibition of fertilization (cumulus penetration) *in vitro* by Fab antihyaluronidase antibodies (Dunbar *et al.*, 1976).

From numerous observations it is clear that acrosome-intact spermatozoa enter the cumulus mass, but it is not at all certain whether the fertilizing spermatozoon is acrosome intact when it reaches the zona pellucida surface. Most spermatozoa near the zona surface have begun the acrosome reaction, but the morphologically identifiable cap is probably still in place and subsequently left behind only on the zona surface (Bedford, 1970; Esaguy *et al.*, 1988).

Consequently, the acrosome reaction *in vivo* in a rabbit ovum with cumulus is most likely a slow but progressive change in the permeability of the plasma and acrosomal membranes over the apical acrosomal region that allows the acrosome to swell. Such changes may proceed in an anterior-to-posterior direction over the head surface and progressively activate and utilize the acrosomal enzymes for penetration.

Kuzan *et al.* (1984) have demonstrated that rabbit spermatozoa recovered from the perivitelline space and therefore acrosome reacted can penetrate fresh cumulus-intact ova. These results, if substantiated, must mean that acrosome-reacted spermatozoa contain the necessary enzymes and binding sites for both cumulus and zona penetration. With regard to zona pellucida binding sites, O'Rand and Fisher (1987) have demonstrated that spermatozoa lose approximately one-third of their zona binding sites after the acrosome reaction, as assayed by quantitative fluorescence microscopy.

4. ZONA PELLUCIDA BINDING AND PENETRATION

Having reached the surface of the zona pellucida, the spermatozoon will shed its acrosomal cap or "ghost" and proceed through the zona matrix. During the process of penetration of the zona the spermatozoon uses an array of zona-binding proteins (ZBP) (O'Rand *et al.*, 1985; O'Rand, 1988). Initially ZBP cover the head and middle piece of the spermatozoon. In the rabbit the predominant ZBP is the sperm lectin RSA, which preferentially binds sulfated carbohydrates such as chondroitin sulfate B, dextran sulfate, and fucoidin but not chondroitin sulfates A or C, cholesterol-3-sulfate, or dextran. The spatial orientation of the monosaccharides in an $\alpha 1 \rightarrow 3$ linkage within the carbohydrate chain and the position of the sulfate on the C4 carbon play an important role in the specificity of the lectin (O'Rand *et al.*, 1988). With a dissociation constant of 5.6×10^{-13} M, RSA has a very high affinity for the zona and would dominate over other molecules with lesser affinity during the initial sperm–zona binding. Once the acrosome reaction is under way, however, other ZBPs may play an increasingly important role.

The best-studied example of another ZBP is the sperm enzyme acrosin (see Section 5 below). Antibodies to rabbit sperm acrosin do not bind to the intact, live spermatozoon, indicating that the enzyme is not on the sperm surface. However, once the acrosome reaction has begun, acrosin becomes available both as a soluble enzyme and as a surface-bound enzyme (M. G. O'Rand, unpublished observations). Acrosin strongly binds zona and acts as both a lectin and an enzyme. Consequently, both acrosin and RSA act together to allow the spermatozoon to penetrate the zona. Penetration is achieved through a cyclic mechanism that has been previously described (O'Rand *et al.*, 1986). In this model, called the binding–release cycle, there are two requirements. First, the ZBP must have a high affinity for the homologous zona, and second, the bound ligand must be removed or degraded from the sperm surface so that when the motility of the spermatozoon pushes it forward, the binding site is again available. The RSA and acrosin molecules fulfill these requirements, including the fact that both are present on the sperm surface after the acrosome reaction occurs on the zona surface.

5. SPERM ENZYMES

The presence of enzymatic activity in rabbit testes and spermatozoa was first detected in the 1920s and 1930s through two independent series of experiments. In the first series, the "Reynals factor" was isolated from rabbit testis (Duran-Reynals, 1929; Hoffmann and Duran-Reynals, 1931; McClean, 1930). At the same time, Pincus and Enzmann (1932) recognized that rabbit spermatozoa could disperse the rabbit cumulus oophorus via a "dispersion factor." Over a decade later, both the Reynals factor and the cumulus dispersion factor were identified as hyaluronidase (Swyer, 1947; McClean and Rowlands, 1942).

Hyaluronidase is a rabbit sperm surface enzyme (O'Rand and Metz, 1974) with a well-defined function: dispersal of the oocyte cumulus cells (Pincus and Enzmann, 1932). Rabbit sperm treated with hyaluronidase inhibitors (Joyce and Zaneveld, 1985) or antihyaluronidase antibodies (Metz et al., 1972; Dunbar et al., 1976) are unable to disperse the cumulus or fertilize oocytes in vivo. A functionally related enzyme, arylsulfatase A, has also been well characterized as a soluble rabbit sperm enzyme (Yang and Srivastava, 1974, 1976). Similar to hyaluronidase, purified arylsulfatase can disperse the cumulus of rabbit oocytes without visibly affecting the corona radiata or zona pellucida. This finding led Farooqui and Srivastava (1979) to hypothesize that hyaluronidase and arylsulfatase act synergistically in cumulus dispersal.

Although hyaluronidase dominated early enzyme research, several classes of enzymes, including proteases, glycosidases, sulfatases, and phosphatases, have been detected in rabbit spermatozoan extracts. Many of these enzymes are identical to their somatic counterparts, but a testis-specific isozyme of lactate dehydrogenase has been identified.

Acrosin, an acrosomal protease, was first purified from rabbit spermatozoa lipoglycoprotein preparations as a proteolytic factor responsible for dispersion of corona radiata cells of rabbit oocytes (Srivastava et al., 1965). Later, removal of the zona pellucida was attributed to acrosin (Zaneveld et al., 1969; Stambaugh and Buckley, 1968), and it was suggested that acrosin is initially inhibited in ejaculated rabbit spermatozoa (Zaneveld et al., 1969). Proacrosin, an autoactivatable acrosin zymogen, was described soon thereafter (Polakoski et al., 1972; Joy Huang-Yang and Meizel, 1975, Mukerji and Meizel, 1979). Because acrosin is a trypsin-like enzyme capable of zona dissolution, inhibitors such as soybean trypsin inhibitor and lima bean trypsin inhibitor are able to block zona dissolution by rabbit spermatozoa (Stambaugh and Buckley, 1968). Inhibitor studies also suggest the importance of acrosin in rabbit fertilization: trypsin inhibitors are able to decrease fertility of rabbit spermatozoa (Zaneveld et al., 1971) and also block rabbit in vitro fertilization (Stambaugh et al., 1969). In more recent studies, acrosin inhibitors were identified as rabbit antifertility agents, and those that were the best inhibitors of acrosin were also the best contraceptive agents (Kaminski et al., 1985). The ability of anti-rabbit-acrosin antibodies to block rabbit fertilization further suggests that acrosin has a role in fertilization processes (Yang et al., 1976). However, because crude rabbit acrosomal extracts are better at digesting the zona pellucida than might be predicted by their acrosin content alone (Bedford and Cross, 1978), the precise site of acrosin action in rabbit fertilization is unclear.

Other well-characterized rabbit spermatozoan enzymes include neuraminidase, lactate dehydrogenase, and β-galactosidase. Neuraminidase has been localized on the inner acrosomal membrane and is optimally active at acid pH (Srivastava and Abou-Issa, 1977). Though the exact role of neuraminidase in fertilization is not clear, Soupart and Clewe (1965) demonstrated that pretreatment of rabbit oocytes with neuraminidase decreases the incidence of sperm penetration and dissolves the zona pellucida on prolonged exposure. This may, however, have been a result of protease activity in the neuraminidase preparation. In contrast, lactate dehydrogenase has metabolic functions (Storey and Kayne, 1978), but it also may serve a role in fertilization. Injection of female rabbits with sperm lactate dehydrogenase results in decreased embryo numbers, probably because of low sperm numbers in the oviduct (Goldberg, 1973, 1975; Killie and Goldberg, 1980), but a precise mechanism has not been elucidated. Finally, the glycosidase β-galactosidase has been purified from rabbit testis and spermatozoa. Its acidic pH optimum, concentration in nonmembrane sperm fractions, and inability to bind rabbit zona pellucida on a nitrocellulose blot suggest a role in alteration of the sperm acrosomal contents or hydrolysis of zona carbohydrates rather than direct participation in sperm–zona binding.

Other enzymes found in rabbit spermatozoa include acid phosphatase (Bernstein and Teichman, 1968; Gonzales and Meizel, 1973), catalase, carbonic anhydrase (Stambaugh and

Buckley, 1969), pyruvate kinase, and ATPase (Storey and Kayne, 1980). Though metabolic functions have been suggested for many of these enzymes, no direct involvement in rabbit sperm–egg interaction has been found.

6. SPERMATOZOON–OOCYTE FUSION

Having used its enzymes judiciously and freed itself of the last vestiges of zona, the spermatozoon enters the perivitelline space. In rabbit ova several spermatozoa usually penetrate the zona and enter the perivitelline space (Overstreet and Bedford, 1974). Although spermatozoa can penetrate the zonae of immature oocytes, they can not fuse with the immature oocyte plasma membrane (Overstreet and Bedford, 1974). It is only after the LH surge, the breakdown of the germinal vesicle, and the initiation of meiosis in the oocyte that fusion can occur. In spite of the fact that several spermatozoa are present in the perivitelline space, only one normally fuses with the ovum. The reasons for this are not entirely clear, but one would expect that there is a fast block to polyspermy operating at the level of the plasma membrane. The zona evidently plays some role in preventing polyspermy, since without it the ova are often polyspermic (McCulloh et al., 1987).

Following spermatozoon–oocyte fusion, the cortical reaction occurs, resulting in discharge of cortical granule contents into the perivitelline space and fusion of the cortical granule membrane with the ovum plasma membrane (Fraser et al., 1972). The sperm plasma membrane fuses with the ovum plasma membrane into a true mosaic as evidenced by the appearance of sperm-specific antigens in the zygote plasma membrane only after fusion (O'Rand, 1977). Sperm nuclear decondensation begins almost immediately, and the male pronucleus has been observed as early as 4 hr post-insemination in in vitro fertilized eggs (Brackett, 1970). With the fusion of the male and female pronuclei, the fertilization process is finished, and the first cleavage division occurs.

7. SUMMARY OF THE STEPS OF FERTILIZATION

Step 1 is the migration of the gametes to the site of fertilization.

Step 2 is the capacitation of spermatozoa in the female reproductive tract. It can be divided into two phases. During the first phase there is the loss of spermatozoan surface-coating proteins. During the second phase there are changes in the plasma membrane of the spermatozoa, including but not limited to loss of cholesterol, rearrangement of components of the lipid bilayer (e.g., glycoproteins), permeability changes, and influx of calcium.

Step 3 is the acrosome reaction, including internal pH changes, enzyme activation, and membrane fusion between plasma membrane and outer acrosomal membrane. The acrosome reaction is complete when the "ghost" is left behind on the zona pellucida surface.

Step 4 is penetration of the spermatozoon through the zona pellucida. At this time the surface of the spermatozoon has been reorganized and has aquired new surface properties to facilitate penetration.

Step 5 is spermatozoon–oocyte fusion and formation of the second polar body by the oocyte.

Step 6 is pronuclear formation and fusion.

ACKNOWLEDGMENTS. This study was supported by NIH grant HD-14232 to M. O'Rand and NSF graduate fellowship to B. Nikolajczyk.

8. REFERENCES

Adams, C. E., 1956, A study of fertilization in the rabbit: The effect of post-coital ligation of the fallopian tube or uterine horn, *J. Endocrinol.* **13**:296–308.

Akruk, S. R., Humphreys, W. J., and Williams, W. L., 1979, *In vitro* capacitation of ejaculated rabbit spermatozoa, *Differentiation* **13**:125–131.

Austin, C. R., 1951, Observations on the penetration of sperm into the mammalian egg, *Aust. J. Sci. Res. B* **4**:581–596.

Barry, M., 1839, Researches in embryology, Second series, *Phil. Trans. R. Soc. Lond.* **129**(Part II):307–380.

Bedford, J. M., 1966, Development of the fertilizing ability of spermatozoa in the epididymis of the rabbit, *J. Exp. Zool.* **163**:319–329.

Bedford, J. M., 1970, Sperm capacitation and fertilization in mammals, *Biol. Reprod. Suppl.* **2**:128–158.

Bedford, J. M., and Cross, N. L., 1978, Normal penetration of rabbit spermatozoa through a trypsin- and acrosin-resistant zona pellucida, *J. Reprod. Fertil.* **54**:385–392.

Bernstein, M. H., and Teichman, R. J., 1968, Localization of acid phosphatase in the rabbit spermatozoon, *J. Cell Biol.* **39**:14a.

Brackett, B. G., 1970, *In vitro* fertilization of rabbit ova: Time sequence of events, *Fertil. Steril.* **21**:169–176.

Brackett, B. G., and Oliphant, G., 1975, Capacitation of rabbit spermatozoa *in vitro*, *Biol. Reprod.* **12**:260–274.

Brackett, B. G., Hall, J. L., and Oh, Y. K., 1978, *In vitro* fertilizing ability of testicular, epididymal, and ejaculated rabbit spermatozoa, *Fertil. Steril.* **29**:571–582.

Chang, M. C., 1951a, Fertility and sterility as revealed in the study of fertilization and development of rabbit eggs, *Fertil. Steril.* **2**:205–222.

Chang, M. C., 1951b, Fertilizing capacity of spermatozoa deposited into the fallopian tubes, *Nature* **168**:697–698.

Chang, M. C., 1952, Fertilizability of rabbit ova and the effects of temperature *in vitro* on their subsequent fertilization *in vivo*, *J. Exp. Zool.* **121**:351–370.

Chang, M. C., 1984, Experimental studies of mammalian fertilization, *Zool. Sci.* **1**:349–364.

Cooper, G. W., Overstreet, J. W., and Katz, D. F., 1979, The motility of rabbit spermatozoa recovered from the female reproductive tract, *Gamete Res.* **2**:35–42.

Dukelow, W. R., Chernoff, H. N., and Williams, W. L., 1967, Fertilizable life of the rabbit ovum relative to sperm capacitation, *Am. J. Physiol.* **213**:1397–1400.

Dunbar, B. S., Munoz, M. G., Cordle, C. T., and Metz, C. B., 1976, Inhibition of fertilization *in vitro* by treatment of rabbit spermatozoa with univalent isoantibodies to rabbit sperm hyaluronidase, *J. Reprod. Fertil.* **47**:381–384.

Duran-Reynals, F., 1929, The effect of extracts of certain organs from normal and immunized animals on the infecting power of vaccine virus, *J. Exp. Med.* **50**:327–340.

Esaguy, N., Welch, J. E., and O'Rand, M. G., 1988, Ultrastructural mapping of a sperm plasma membrane autoantigen before and after the acrosome reaction, *Gamete Res.* **19**:387–399.

Farooqui, A. A., and Srivastava, P. N., 1979, Isolation, characterization and the role of rabbit testicular aryl-sulphatase A in fertilization, *Biochem. J.* **181**:331–337.

Fox, R. R., 1974, Taxonomy and genetics, in: *The Biology of the Laboratory Rabbit* (S. H. Weisbroth, R. E. Flatt, and A. L. Kraus, eds.), Academic Press, New York, pp. 1–22.

Fraser, L. R., Dandekar, P. V., and Gordon, M. K., 1972, Loss of cortical granules in rabbit eggs exposed to spermatozoa *in vitro*, *J. Reprod. Fertil.* **29**:295–297.

Friend, D. S., Orci, L., Perrelet, A., and Yanagimachi, R., 1977, Membrane particle changes attending the acrosome reaction in guinea pig spermatozoa, *J. Cell Biol.* **74**:561–577.

Goldberg, E., 1973, Infertility in female rabbits immunized with lactate dehydrogenase X, *Science* **181**:458.

Goldberg, E., 1975, Effects of immunization with LDH-X on fertility, *Acta Endocrinol [Suppl.]* **194**:202–222.

Gonzales, L. W. and Meizel, S., 1973, Acid phosphatases of rabbit spermatozoa I. Electrophoretic characterization of the multiple forms of acid phosphatase in rabbit spermatozoa and other semen constituents, *Biochim. Biophys. Acta* **320**:166–179.

Gregory, P. W., 1930, The early embryology of the rabbit, *Contrib. Embryol.* **21**:141–172.

Hammond, J., 1925, *Reproduction in the Rabbit*, Oliver and Reed, Edinburgh.

Hammond, J., 1934, The fertilization of rabbit ova in relation to time. A method of controlling the litter size, the duration of pregnancy and the weight of the young at birth, *J. Exp. Biol.* **11**:140–161.

Harper, M. J., 1973, Stimulation of sperm movement from the isthmus to the site of fertilization in the rabbit oviduct, *Biol. Reprod.* **8**:369–377.

Hoffmann, D. C., and Duran-Reynals, F., 1931, The influence of testicle extract on the intradermal spread of injected fluids and particles, *J. Exp. Med.* **53**:387–398.

Hosoi, K., Niwa, S., Hatanaka, S., and Iritani, A., 1981, Fertilization *in vitro* of rabbit eggs by epididymal spermatozoa capacitated in a chemically defined medium, *Biol. Reprod.* **24**:637–642.

Johnson, L. L., Katz, D. F., and Overstreet, J. W., 1981, The movement characteristics of rabbit spermatozoa before and after activation, *Gamete Res.* **4**:275–282.

Joyce, C. L. and Zaneveld, L. J. D., 1985, Vaginal contraceptive activity of hyaluronidase and cyclooxygenase (prostaglandin synthetase) inhibitors in the rabbit, *Fertil. Steril.* **44**:426–428.

Joy Huang-Yang, Y. H., and Meizel, S., 1975, Purification of rabbit testis proacrosin and studies of its active form, *Biol. Reprod.* **12**:232–238.

Kaminski, J. M., Nuzzo, N. A., Bauer, L., Waller, D. P., and Zaneveld, L. J. D., 1985, Vaginal contraceptive activity of aryl 4-guanidino-benzoates (acrosin inhibitors) in rabbits, *Contraception* **32**:183–189.

Katz, D. F., Drobnis, E. Z., and Overstreet, J. W., 1989, Factors regulating mammalian sperm migration through the female reproductive tract and oocyte vestments, *Gamete Res.* **22**:443–469.

Killie, J. W., and Goldberg, E., 1980, Inhibition of oviducal sperm transport in rabbits immunized against sperm-specific lactate dehydrogenase (LDH-C$_4$), *J. Reprod. Immunol.* **2**:15–21.

Kuzan, F. B., Fleming, A. D., and Seidel, G. E., 1984, Successful fertilization *in vitro* of fresh intact oocytes by perivitelline (acrosome reacted) spermatozoa of the rabbit, *Fertil. Steril.* **41**:766–770.

Langlais, J., and Roberts, K. D., 1985, A molecular membrane model of sperm capacitation and the acrosome reaction of mammalian spermatozoa, *Gamete Res.* **12**:183–224.

McClean, D., 1930, The influence of testicular extract on dermal permeability and the response to vaccine virus, *J. Pathol. Bacteriol.* **33**:1045–1070.

McClean, D., and Rowlands, I. W., 1942, Role of hyaluronidase in fertilization, *Nature* **150**:627–628.

McCulloh, D. H., Wall, R. J., and Levitan, H., 1987, Fertilization of rabbit ova and the role of ovum investments in the block to polyspermy, *Dev. Biol.* **120**:385–391.

Metz, C. B., Seiguer, A. C., and Castro, A. E., 1972, Inhibition of the cumulus dispersing and hyaluronidase activities of sperm by heterologous and isologous antisperm antibodies, *Proc. Soc. Exp. Biol. Med*, **140**:776–781.

Mukerji, S. K., and Meizel, S., 1979, Rabbit testis proacrosin: Purification, molecular weight estimation, and amino acid and carbohydrate composition of the molecule, *J. Biol. Chem.* **254**:11721–11728.

O'Rand, M. G., 1977, The presence of sperm-specific surface isoantigens on the egg following fertilization, *J. Exp. Zool.* **202**:267–273.

O'Rand, M. G., 1979, Changes in sperm surface properties correlated with capacitation, in: *The Spermatozoon: Maturation, Motility and Surface Properties* (D. Fawcett and M. J. Bedford, eds.), Urban and Schwarzenberg, Baltimore, pp. 412–430.

O'Rand, M. G., 1982, Modification of the sperm membrane during capacitation, *Ann. N.Y. Acad. Sci.* **383**:392–402.

O'Rand, M. G., 1988, Sperm–egg recognition and barriers to interspecies fertilization, *Gamete Res.* **19**:315–328.

O'Rand, M. G., and Fisher, S. J., 1987, Localization of the zona pellucida binding sites on rabbit spermatozoa and induction of the acrosome reaction by solubilized zonae, *Dev. Biol.* **119**:551–559.

O'Rand, M. G., and Metz, C. B., 1974, Tests for rabbit sperm surface iron-binding protein and hyaluronidase using the "exchange agglutination" reaction, *Biol. Reprod.* **11**:326–334.

O'Rand, M. G., and Metz, C. B., 1976, Isolation of an "immobilizing antigen" from rabbit sperm membranes, *Biol. Reprod.* **14**:586–598.

O'Rand, M. G., and Romrell, L. J., 1981, Localization of a single sperm membrane autoantigen (RSA-1) on spermatogenic cells and spermatozoa, *Dev. Biol.* **84**:322–331.

O'Rand, M. G., Matthews, J. E., Welch, J. E., and Fisher, S. J., 1985, Identification of zona binding proteins of rabbit, pig, human and mouse spermatozoa on nitrocellulose blots, *J. Exp. Zool.* **235**:423–428.

O'Rand, M. G., Welch, J. E., and Fisher, S. J., 1986, Sperm membrane and zona pellucida interactions during fertilization, in: *Molecular and Cellular Aspects of Reproduction* (D. S. Dhindsa and O. P. Bahl, eds.), Plenum Press, New York, pp. 131–144.

O'Rand, M. G., Widgren, E. E., and Fisher, S. J., 1988, Characterization of the rabbit sperm membrane autoantigen, RSA, as a lectin-like zona binding protein, *Dev. Biol.* **129**:231–240.

Orgebin-Crist, M. C., 1973, Maturation of spermatozoa in the rabbit epididymis: Effect of castration and testosterone replacement, *J. Exp. Zool.* **185**:301–310.

Orgebin-Crist, M. C., and Jahad, N., 1979, The maturation of rabbit epididymal spermatozoa in organ culture: Stimulation by epididymal cytoplasmic extracts, *Biol. Reprod.* **21**:511–515.

Overstreet, J. W., and Bedford, J. M., 1974, Comparison of the penetrability of the egg vestments in follicular oocytes. Unfertilized and fertilized ova of the rabbit, *Dev. Biol.* **41**:185–192.

Overstreet, J. W., and Cooper, G. W., 1978a, Sperm transport in the reproductive tract of the female rabbit: I. The rapid transit phase of transport, *Biol. Reprod.* **19**:101–114.

Overstreet, J. W., and Cooper, G. W., 1978b, Sperm transport in the reproductive tract of the female rabbit: II. The sustained phase of transport, *Biol. Reprod.* **19:**115–132.

Overstreet, J. W., and Cooper, G. W., 1979, The time and location of the acrosome reaction during sperm transport in the female rabbit, *J. Exp. Zool.* **209:**97–104.

Pincus, G., 1939, The comparative behavior of mammalian eggs *in vivo* and *in vitro*. IV. The development of fertilized and artificially activated rabbit eggs, *J. Exp. Zool.* **82:**85–129.

Pincus, G., and Enzmann, E. V., 1932, Fertilisation in the rabbit, *J. Exp. Biol.* **9:**403–408.

Polakoski, K. L., Zaneveld, L. J. D., and Williams, W. L., 1972, Purification of a proteolytic enzyme from rabbit acrosomes, *Biol. Reprod.* **6:**23–29.

Ramirez, V. D., and Beyer, C., 1988, The ovarian cycle in the rabbit: Its neuroendocrine control, in: *The Physiology of Reproduction*, Vol. 2 (E. Knobil and J. D. Neill, eds.), Raven Press, New York, pp. 1873–1892.

Reynolds, A. B., and Oliphant, G., 1984, Production and characterization of monoclonial antibodies to the sperm acrosome stabilizing factor (ASF): Utilization for purification and molecular analysis of ASF, *Biol. Reprod.* **30:**775–786.

Robaire, B., and Hermo, L., 1988, Efferent ducts, epididymis, and vas deferens: Structure, functions, and their regulation, in: *The Physiology of Reproduction*, Vol. 1 (E. Knobil and J D. Neill, eds.), Raven Press, New York, pp. 999–1080.

Romrell, L. J., and O'Rand, M. G., 1978, Capping and ultrastructural localization of sperm surface isoantigens during spermatogenesis, *Dev. Biol.* **63:**76–93.

Soupart, P., and Clewe, T. H., 1965, Sperm penetration of rabbit zona pellucida inhibited by treatment of ova with neuraminidase, *Fertil. Steril.* **16:**667–689.

Srivastava, P. N., and Abou-Issa, H., 1977, Purification and properties of rabbit spermatozoal acrosomal neuraminidase, *Biochem. J.* **161:**193–200.

Srivastava, P. N., Adams, C. E., and Hartree, E. F., 1965, Enzymatic action of acrosomal preparations on the rabbit ovum *in vitro*, *J. Reprod. Fertil.* 10:61–67.

Stambaugh, R., and Buckley, J., 1968, Zona pellucida dissolution enzymes of the rabbit sperm head, *Science* **161:**585–586.

Stambaugh, R., and Buckley, J., 1969, Identification and subcellular localization of the enzymes affecting penetration of the zona pellucida by rabbit spermatozoa, *J. Reprod. Fertil.* **19:**423–432.

Stambaugh, R., Brackett, B. G., and Mastroianni, L., 1969, Inhibition of *in vitro* fertilization of rabbit ova by trypsin inhibitors, *Biol. Reprod.* **1:**223–227.

Storey, B. T., and Kayne, F. J., 1978, Energy metabolism of spermatozoa VII. Interactions between lactate, pyruvate and malate as oxidative substrates for rabbit sperm mitochondria, *Biol. Reprod.* **18:**527–536.

Storey, B. T., and Kayne, F. J., 1980, Properties of pyruvate kinase and flagellar ATPase in rabbit spermatozoa: Relation to metabolic strategy of the sperm cell, *J. Exp. Zool.* **211:**361–367.

Suarez, S. S., Katz, D. F., and Overstreet, J. W., 1983, Movement characteristics and acrosomal status of rabbit spermatozoa recovered at the site and time of fertilization, *Biol. Reprod.* **29:**1277–1287.

Swierstra, E. E., and Foote, R. H., 1963, Cytology and kinetics of spermatogenesis in the rabbit, *J. Reprod. Fertil.* **5:**390–322.

Swierstra, E. E., and Foote, R. H., 1965, Duration of spermatogenesis and spermatozoan transport in the rabbit based on cytological changes, DNA synthesis and labeling with tritiated thymidine, *Am. J. Anat.* **116:**401–412.

Swyer, G. I. M., 1947, The release of hyaluronidase from spermatozoa, *Biochem. J.* **41:**413–417.

Thomas, T. S., Reynolds, A. B., and Oliphant, G., 1984, Evaluation of the site of synthesis of rabbit sperm acrosome stabilizing factor using immunocytochemical and metabolic labeling techniques, *Biol. Reprod.* **30:**693–705.

Viriyapanich, P., and Bedford, J. M., 1981, The fertilization performance *in vivo* of rabbit spermatozoa capacitated *in vitro*, *J. Exp. Zool.* **216:**169–174.

Weismann, A., 1891, Amphimixis, or the essential meaning of conjugation and sexual reproduction, *Essays Hered.* **2:**121.

Yang, C. H., and Srivastava, P. N., 1974, Purification and properties of aryl sulfatases from rabbit sperm acrosomes, *Soc. Exp. Biol. Med.* **145:**721–725.

Yang, C. H., and Srivastava, P. N., 1976, Purification and properties of arylsulphatase A from rabbit testis, *Biochem. J.* **159:**133–142.

Yang, S. L., Lourens, M. D., Zaneveld, L. J. D., and Schumacher, G. F. B., 1976, Effect of serum proteinase inhibitors on the fertilizing capacity of rabbit spermatozoa, *Fertil. Steril.* **27:**577–581.

Zaneveld, L. J. D., Srivastava, P. N., and Williams, W. L., 1969, Relationship of a trypsin-like enzyme in rabbit spermatozoa to capacitation, *J. Reprod. Fertil.* **20:**337–339.

Zaneveld, L. J. D., Robertson, R. T., Kessler, M., and Williams, W. L., 1971, Inhibition of fertilization *in vitro* by pancreatic and seminal plasma trypsin inhibitors, *J. Reprod. Fertil.* **25:**387–392.

15

Fertilization in Dogs

Cherrie A. Mahi-Brown

1. INTRODUCTION

Most knowledge of the basic reproductive biology of companion animals such as the dog has been derived from studies directed at developing contraceptive technologies to combat overpopulation of pets. The bitch has been used as a model in the study of some steroid hormones (El Etreby *et al.*, 1979; Frank *et al.*, 1979), and the male dog has been used extensively in studies of prostate function (Huggins, 1945). In addition, spermatozoa of the dog have been used in studies of sperm metabolism and motility (Wales and White, 1958; Tash *et al.*, 1986, 1988). However, there are numerous unique aspects of canine reproductive biology that make the dog worthy of study for its own sake even though these same aspects make the use of the dog as a model for human reproductive biology questionable. Furthermore, a more complete study of fertilization in the dog is indicated by the need for effective contraceptive technology for this species (Mahi-Brown, 1986).

In this review I describe these unique features of canine reproduction and discuss how they influence the process of fertilization. I relate what is known of the canine gametes and fertilization process, and I outline the questions I consider most intriguing for future studies. Finally, I discuss alternative technologies for fertilization in the dog (artificial insemination and embryo transfer) and fertilization in other canids.

2. BASIC FEATURES OF CANINE REPRODUCTIVE BIOLOGY

2.1. The Estrous Cycle of the Bitch

Although a detailed description of the estrous cycle of the bitch is not within the scope of this review, I feel it is necessary to provide a general outline of the cycle as the context in which fertilization takes place. The estrous cycle of the bitch has been studied in some detail (for review, see Concannon, 1986) and is unusual in several aspects. Although the bitch is monestrous, there is little or no seasonal association. The cycle length varies considerably (5–12 months), with the

CHERRIE A. MAHI-BROWN • California Primate Research Center, University of California, Davis, California 95616.

A Comparative Overview of Mammalian Fertilization, edited by Bonnie S. Dunbar and Michael G. O'Rand. Plenum Press, New York, 1991.

variation dependent on the length of anestrus, the quiescent phase of the cycle. Anestrus is followed by proestrus, the follicular phase, which lasts an average of 10 days. During proestrus, estrogen levels rise to a peak and then decline abruptly at the onset of estrus, approximately 24 hr before the LH surge. Ovulation occurs spontaneously (Bischoff, 1845) during a short interval approximately 30 hr after the LH surge. However, the onset of behavioral estrus is frequently not closely correlated with either of these endocrinological landmarks. What is most interesting from the viewpoint of fertilization in this species is that estrus lasts an average of 10 days and conception can be achieved by a single mating at any time during this period (Doak *et al.*, 1967, Concannon, 1986). Concannon (1986) reports that single matings as early as 3 days before and as late as 8 days or more after the LH peak can be fertile (this would be 4–5 days before and 6 days after ovulation, respectively). However, he states that peak fertility is associated with matings 0–5 days after the LH peak. Progesterone levels begin to rise in late proestrus and rise rapidly during estrus as the pre- and postovulatory follicles luteinize. The mural granulosa of the preovulatory follicle characteristically proliferates and folds into the follicle as it begins to luteinize prior to ovulation (Andersen and Simpson, 1973; Concannon, 1986). At the end of estrus, the corpus luteum is fully active, and the bitch enters the pregnant or pseudopregnant state called variously metestrus (Concannon, 1986) or diestrus (Holst and Phemister, 1974). The overlap of estrus with the luteal phase complicates this terminology (for discussion, see Shille and Stabenfeldt, 1980).

Because the corpus luteum is activated and maintained spontaneously regardless of mating, the pseudopregnant state is normal for the bitch and lasts slightly longer than pregnancy, which averages 64 days (Concannon, 1986). The corpus luteum is necessary for the entire pregnancy, so that ovariectomy at any time results in abortion (Sokolowski, 1971). Detection of pregnancy in bitches carrying one or two fetuses is often difficult because of this normal pseudopregnancy and because no chorionic gonadotropin has been identified in the bitch. However, recent studies have been directed at identifying other products of embryonic origin that may aid in early pregnancy detection (Shille *et al.*, 1988). At the end of pregnancy or pseudopregnancy, the endometrium is slowly repaired, and the bitch enters anestrus.

The reproductive organs and tissues of the bitch and their changes during development and the estrous cycle have been described in considerable detail (Evans and Cole, 1931; Andersen and Simpson, 1973; Sokolowski, *et al.*, 1973). For descriptions of the embryology of the dog, the reader is referred to Bischoff (1845), Holst and Phemister (1971), and Evans (1974, 1979).

2.2. The Male Dog

The male dog also has some unusual reproductive features. The dog is not a seasonal breeder, although there have been reports that sperm production declines during the summer (Kuroda and Hiroe, 1972). This is in contrast to most other canids, which are strongly seasonal (for example, the coyote: C. Hodges and M. Amos, personal communication). The dog completely lacks seminal vesicles, Cowper's glands, and coagulating glands (Christensen, 1979; Setchell and Brooks, 1988); therefore, semen consists almost entirely of prostatic fluid (Huggins, 1947; Huggins *et al.*, 1939). The semen is ejaculated in three distinct fractions (Hancock and Rowlands, 1949; Harrop, 1960), with spermatozoa contained in the second fraction. The volume of the first fraction ranges from 0.25 to 2.8 ml; the sperm-rich second fraction ranges from 0.4 to 4 ml; and the third fraction ranges from 1.1 to 25.0 ml (Boucher *et al.*, 1958; Seager and Fletcher, 1972). Although smaller dogs tend to have smaller ejaculates, there is considerable overlap with the volumes ejaculated by large dogs. During ejaculation of the second and third fractions, the dog is usually "tied" to the bitch as a result of enlargement of the bulbus glandis at the base of the penis within the vagina of the bitch. The dog dismounts, lifts one hind leg over the bitch, and turns to stand facing away from the bitch so that they are linked tail to tail throughout the 5–30 min it takes to ejaculate the third fraction and for detumescence of the penis. The very unusual anatomy of the canine penis that allows this behavior has been described by Grandage (1972) in a paper entitled "The erect canine penis: A paradox of flexible rigidity."

2.3. Sperm Storage in the Female Reproductive Tract

Because of the tie, semen is forced through the cervix into the uterine lumen. Evans (1933) studied this phenomenon by means of a uterine fistula. The cranial end of the uterine horn on one or both sides was pierced and anchored outside the body. A sperm suspension was ejected from the fistula within 25 to 50 seconds of the end of "active copulation," i.e., the end of the period when the male was actively thrusting and the first fraction was ejaculated. The female at this time began abdominal straining, which further forced the semen out the fistula, sometimes in small streams to a distance of 20–25 cm. Thus, the cervical barrier to sperm transport is not active in the bitch, at least during natural mating. Evans believed that the primary mechanism for rapid transport of semen into the uterus is the straining of the female, with the engorged penis in the vagina preventing the semen from going anywhere but through the cervix.

It has been found that once they reach the uterine horns, the spermatozoa survive in large numbers in the uterine glands (Fig. 1) throughout the duration of estrus (Doak *et al.*, 1967). This

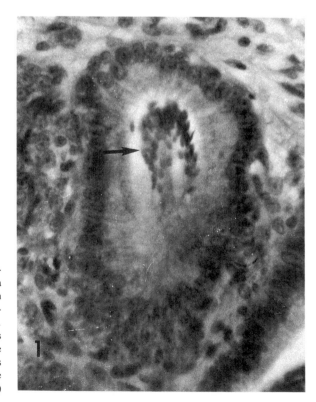

Figure 1. A cluster of canine spermatozoa (arrow) within the lumen of a uterine endometrial gland of a bitch mated 24 hr before the uterus was removed and fixed in Bouin's solution. Spermatozoa flushed from the uterus before it was fixed were motile. Note also the tall columnar epithelial cells lining the gland, which are indicative of estrus. (Hematoxylin and eosin stain.)

is a useful adaptation for a species in which ovulation and mating are not closely linked. In contrast, large numbers of spermatozoa are not observed in the oviducts at any time; in fact, Doak *et al.* (1967) found only occasional spermatozoa in the oviducts more than 40 hr after mating. These authors mated bitches on the first day of acceptance and then killed them at various times by injecting an overdose of sodium pentobarbitol. After the uterotubal junction was clamped, the oviducts and uterine horns were cut into segments, and the sperm numbers in each segment were determined by both flushing and examination of stained sections. It is possible that the method of killing the bitches caused the spermatozoa to be expelled from the oviducts or that the clamp obscured a site of storage in the isthmus, but at present we must conclude that the isthmus of the oviduct is not a major site of sperm storage in this species.

3. CANINE GAMETES

3.1. The Canine Oocyte

One of the most unusual features of canine reproductive biology is that the oocyte of the bitch is ovulated in the dictyate state; i.e., the germinal vesicle is intact, and the first polar body has not been shed (Van der Stricht, 1923; Evans and Cole, 1931; Anderson and Simpson, 1973; Tsutsui, 1975). The first meiotic division is not completed until at least 48 hr later (Holst and Phemister, 1971; Tsutsui, 1975). The usual situation among mammals is for the the oocyte to be arrested at metaphase of the second meiotic division at the time of ovulation. The consequences of this unique feature for the fertilization process are discussed later. The canine oocyte is approximately 120 μm in diameter at ovulation (Holst and Phemister, 1971) and contains a large quantity of lipid yolk bodies (Fig. 2), which render it opaque. Thus, it appears black with transmitted light and creamy yellow with incident illumination. The internal structure of the living oocyte cannot be discerned except that a less dense area in the compressed oocyte marks the location of the germinal vesicle (Fig. 3). In the mature ovulated oocyte, the germinal vesicle is elliptical in location, whereas in the immature ovarian oocyte it is located centrally.

The growth process (Andersen and Simpson, 1973; Tesoriero, 1981) and the process of lipid deposition within the cytoplasm of the canine ovarian oocyte (Tesoriero, 1982) have been described. As the oocyte enlarges in the ovary, a thick zona pellucida is secreted around it. By the time of ovulation the zona is approximately 18 μm thick. The zona pellucida is pierced by processes extending from the surrounding granulosa cells. These processes make gap junctional contact with the oocyte plasma membrane to provide communication between the granulosa cells and oocyte. In most mammals, these junctional contacts are terminated and the granulosa processes withdrawn following LH stimulation prior to ovulation (for review see Wassarman, 1988). However, the kinetics of this process have not been studied in the bitch. The ovulated oocyte has little or no expanded cumulus oophorus, but it retains a compact layer of corona radiata (Fig. 3; Andersen and Simpson, 1973), which is not shed for several days after ovulation. It would be interesting to determine by electron microscopic analysis whether the corona radiata cells retain junctional contact with each other after ovulation and, if so, when these contacts are terminated.

The fertilizable life of the canine oocyte after ovulation has been estimated as 5 days (Tsutsui and Shimizu, 1975). However, Concannon's (1986) report of rare fertile matings as late as 10 days after the LH surge (8 days after ovulation) suggests that the end of the fertile period may be related as much to the end of estrus as to the viability of the oocytes, since very few bitches remain in estrus 10 days after the LH surge. This long viability of the oocytes combined with the long viability of the spermatozoa provide an unusually wide range for fertile mating in a species with spontaneous ovulation.

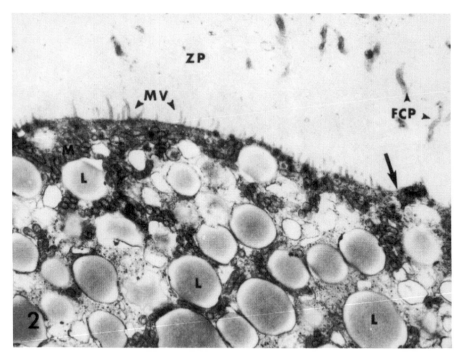

Figure 2. Cortex of canine oocyte in antral follicle. Note the large number of lipid yolk bodies in the cortex as well as the granulosa cell processes within the zona pellucida and in contact with the oolemma (arrow). FCP, follicle cell processes; L, lipid yolk bodies; M, mitochondria; MV, microvilli; ZP, zona pellucida. (Reproduced with permission from Tesoriero, 1981.)

3.2. The Canine Spermatozoon

Canine spermatozoa appear to share more morphological features with other mammalian spermatozoa. The head is shaped like a flattened, blunt ellipse without a hook or dish. The acrosome is relatively large so that its presence or absence on motile cells can be discerned by an experienced observer using a phase-contrast microscope (Fig. 4). The size of canine spermatozoa has been analyzed by Woodall and Johnstone (1988), who found the total length to be 61.4 ± 0.244 μm (mean ± S.E.), the head length 6.1 ± 0.037 μm, head width 3.8 ± 0.023 μm, midpiece 10.1 ± 0.074 μm, and principal piece 45.2 ± 0.214 μm. Spermatogenesis in the dog has been described by Foote *et al.* (1972), who reported that one cycle of the seminiferous epithelium lasts 13.6 days. Ghosal *et al.* (1983) found that spermatogenesis and spermiogenesis combined require 42.15 days. Both studies were conducted by injecting dogs with a single dose of tritiated thymidine and then using autoradiography to time the appearance of labeled cells in the seminiferous epithelium.

Spermatozoa are ejaculated in semen of low viscosity, which does not coagulate because of the absence of coagulating glands. Also, because there is no seminal vesicle fluid, the semen usually is very low in cellular debris and epithelial cells. It has been my experience that the presence of epithelial cells in an ejaculate is associated with poor performance of the spermatozoa in subsequent *in vitro* fertilization assays (C. A. Mahi-Brown, unpublished observations). Other characteristics of the ejaculate have been described in detail elsewhere (Harrop, 1960; Seager and Fletcher, 1972; Isaacs and Coffey, 1984).

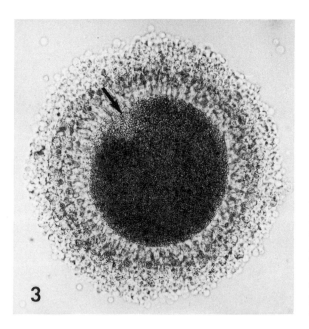

3

Figure 3. Living canine oocyte flushed from the oviduct of an estrous bitch at an undetermined time after ovulation. The oocyte has been strongly compressed to reveal the location of the germinal vesicle at the periphery of the oocyte (clear area, arrow). Note also the compact corona radiata and the absence of an expanded cumulus oophorus. The perivitelline space is not clearly discernible.

4. THE FERTILIZATION PROCESS *IN VITRO*

4.1. Sperm Capacitation and Acrosome Reaction

There has been very little work on the capacitation process in the dog, even though canine semen is easily collected by manual stimulation of the penis in conjunction with use of a simple artificial vagina. The artificial vagina consists of a 10-inch latex cone with a 12-to-15-ml centrifuge tube at the small end. It is only necessary to collect the first two fractions, since the third contains only prostatic fluid and its absence appears to have no effect on sperm performance. Spermatozoa are isolated from the ejaculate by centrifugation and resuspension in CCM. A sperm "swim-up" procedure to separate motile from immotile spermatozoa has not proven

a b

Figure 4. Phase-contrast micrographs of canine spermatozoa showing intact (A) and reacted (B) acrosomes. The spermatozoa were incubated for 7 hr in canine capacitation medium (CCM) before they were photographed. Acrosome reactions can be scored in motile spermatozoa by an experienced observer. (Reproduced with permission from Mahi and Yanagimachi, 1978.)

effective for canine semen, probably because of the low viscosity of the semen. If the viscosity of the semen were increased artificially, this procedure might prove effective, but this has not been attempted. Canine spermatozoa can be capacitated in a defined medium based on Brinster's medium (Brinster, 1965a,b,c,d) and called CCM for canine capacitation medium (Mahi and Yanagimachi, 1978). The composition of CCM is listed in Table 1. During incubation in CCM the initially highly motile spermatozoa become somewhat quiescent for several hours and form head-to-head associations. Motility of spermatozoa within clumps increases in intensity approximately 6 hr after the start of incubation. Free-swimming, highly motile spermatozoa, many of which (46.3 ± 6.3%) have lost their acrosomes, are seen at 7 hr (Mahi and Yanagimachi, 1978).

Although CCM has worked well in our studies, other types of media may also be effective. For instance, acrosome reactions have been induced in a modified Ham's F10 medium containing bovine serum albumin and the calcium ionophore A23187 (J. A. Metzler, personal communication). Studies have not been conducted to determine whether acrosome reaction is induced normally by proteins of the canine zona pellucida. Acrosome-intact spermatozoa do, however, adhere loosely to the zona surface and can be removed by pipetting (C. A. Mahi-Brown and R. Yanagimachi, unpublished observations). Critical studies will have to be done to determine whether the acrosome reaction normally occurs on the zona surface as in some other mammals (Meizel, 1985).

4.2. Zona Pellucida Penetration and Sperm–Egg Fusion

Canine ovarian oocytes for fertilization studies have been collected by puncturing antral follicles in ovaries obtained from a spay–neuter clinic immediately after surgery (Mahi and Yanagimachi, 1976, 1978). The corona radiata can be removed either by vigorous pipetting or by vortexing the oocytes vigorously in 75 mM sodium citrate buffer (pH 7.8). Spermatozoa will penetrate the oocytes regardless of whether the corona is removed, but removal facilitates scoring the oocytes for penetration. The oocytes can be exposed to spermatozoa immediately after isolation or first matured by culture *in vitro* for 24–72 hr in TC medium 199 (Hank's salts, Difco Labs, Detroit, MI) supplemented with 20% heat-inactivated fetal calf serum, sodium bicarbonate, and antibiotics (Mahi and Yanagimachi, 1976).

Table 1. Composition of Canine Capacitation Medium (CCM)[a]

Component	Quantity (g/liter)	Concentration (mM)
NaCl	4.880	83.49
KCl	0.356	4.78
$CaCl_2$	0.189	1.71
KH_2PO_4	0.162	1.19
$NaHCO_3$	3.159	37.61
Na pyruvate	0.028	0.25
Na lactate (60% syrup)	3.38 ml	21.55
Glucose	0.500	2.78
Bovine serum albumin (fraction V)	2.000	
Phenol red	0.020	
Gentamicin SO_4	0.050	
pH under 5% CO_2 in air		7.8
Osmolality		302 mOsm

[a]Modified from Mahi and Yanagimachi (1978). Gentamicin has been substituted for penicillin and streptomycin.

The kinetics of zona pellucida penetration and sperm–egg fusion have not been determined in the dog. Although they have been observed only at the light microscopic level (Mahi and Yanagimachi, 1976), the penetration and fusion processes in the dog do not appear to differ in any substantial manner from those of other mammals that have been studied in greater detail. Penetration of spermatozoa through the zona can be evaluated by examining the living oocytes with a phase-contrast microscope at 400×. It is essential to support the coverslip with wax to prevent crushing the oocyte and to roll the oocyte by sliding the coverslip to assure that the spermatozoa are actually passing through the zona and are not on the surface. The sperm path through the zona is angular (Fig. 5). Fusion of spermatozoa with these opaque oocytes can be ascertained only after the oocytes are fixed in acid alcohol and stained with aceto-carmine or aceto-lacmoid. Decondensed sperm nuclei are clearly visible after staining (Fig. 6). The sperm tail is often detached from the nucleus but can usually be identified nearby. We have observed cleavage of canine oocytes fertilized *in vitro* in rare cases (Fig. 7: C. A. Mahi-Brown and R. Yanagimachi, unpublished observations), but no attempt has been made to culture canine embryos.

4.3. Egg Activation

Fusion of the spermatozoon and egg appears to induce a zona reaction, since large, living fertilized oocytes have a greatly reduced number of spermatozoa bound to the zona surface compared with small or degenerating oocytes, which are penetrated by spermatozoa but do not fuse with them (Mahi and Yanagimachi, 1976). *In vitro* fertilization of naturally ovulated oocytes has not been attempted in the dog to the best of my knowledge, largely because of our previous inability to accurately predict or induce ovulation in the bitch. Such studies would greatly aid our understanding of the egg activation process and are now possible because of recent advances in ovulation induction in this species (Cain *et al.*, 1988).

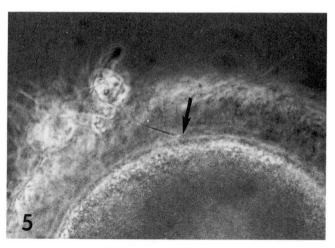

Figure 5.   Canine spermatozoon that has penetrated the zona pellucida of an oocyte *in vitro*. The arrow indicates the sperm head.

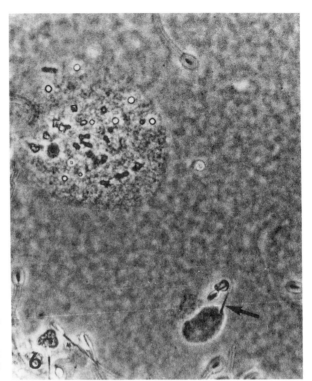

Figure 6. Greatly decondensed sperm head within the cytoplasm of a canine oocyte inseminated *in vitro*. In this plane of focus, the sperm tail (arrow) appears to remain attached to the head. The degree of decondensation can be seen by comparing this sperm head with several undecondensed spermatozoa that are visible at the egg surface (lower left-hand corner of figure). Note also that the germinal vesicle of this oocyte is beginning to break down, and the chromosomes are condensing as the egg begins to complete the first meiotic division.

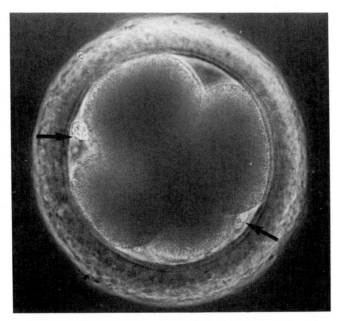

Figure 7. Four-cell canine embryo resulting from *in vitro* fertilization of an ovarian oocyte that had been cultured *in vitro* prior to insemination. Arrows indicate the two polar bodies.

4.4. Penetration of Zona-Free Hamster Eggs

The canine spermatozoon is unusual in that it does not fuse with zona-free hamster eggs (Yanagimachi, 1988a). This suggests that the receptor system may be unique in this species. Canine spermatozoa fuse readily with zona-free canine oocytes (C. A. Mahi-Brown and R. Yanagimachi, unpublished observations), so the failure is not caused by absence of a zona-induced acrosome reaction.

5. FERTILIZATION *IN VIVO*

Although the fertilization process in the dog appears to differ little from that of other well-studied mammals and offers no unique mysteries, the time at which spermatozoa penetrate canine oocytes *in vivo* is not certain because of the considerable asynchrony among ovulation, insemination, and egg maturation. Much of the current literature states that fertilization awaits completion of egg maturation to metaphase II and extrusion of the first polar body in the oviduct (Holst and Phemister, 1971; Phemister *et al.*, 1973; Concannon, 1986). However, no direct information has been presented to support this conclusion. It may be that this has been an assumption based on the fact that most mammalian eggs are fertilized at this stage. Most mammalian eggs, however, mature to metaphase II before ovulation and thus are not available in the oviduct for fertilization at earlier stages. In the usual case, the egg is arrested in metaphase II until it is activated by fusion with a spermatozoon. It then completes the second meiotic division, extrudes the second polar body, and forms a female pronucleus in parallel with formation of the male pronucleus from the sperm nucleus (for review, see Yanagimachi, 1988b). We found to our surprise (Mahi and Yanagimachi, 1976) that capacitated, acrosome-reacted canine spermatozoa would fuse *in vitro* with any large living canine oocyte regardless of the state of nuclear maturation. Figure 8 shows a highly decondensed sperm nucleus in the cytoplasm of an egg with an intact germinal vesicle.

What has been observed to take place *in vitro* may not be an accurate reflection of what happens *in vivo*. In our *in vitro* studies we always used ovarian oocytes that had been cultured *in vitro* for 0–48 hr (Mahi and Yanagimachi, 1976). Ovulated canine oocytes may be resistant to penetration by spermatozoa until they have matured, although there is no direct evidence for this. Another possibility is that sperm transport to the site of fertilization may be delayed until near the time of egg maturation. However, Evans (1933) found that spermatozoa are transported to the tubal end of the uterine horns within seconds, and Doak *et al.* (1967) recovered spermatozoa from the oviducts of bitches within minutes after ejaculation. The latter workers bred the bitches on the first day of behavioral estrus, which in most cases would be at least 1 day before ovulation. This gives the impression that spermatozoa must be in the oviduct at the time of ovulation. Interestingly, the latter investigators could identify spermatozoa in the oviducts later than 70 hr after copulation in only two of 26 bitches (135 and 168 hr after copulation, respectively). If ovulation occurs most frequently on the second or third day after the onset of estrus (although the time of ovulation is highly variable, the highest incidence is on day 2 or 3: Phemister *et al.*, 1973), and oocyte maturation requires another 48–72 hr. (Tsutsui, 1975), it appears that spermatozoa disappear from the oviducts before completion of egg maturation. Of course, it is possible that spermatozoa enter the oviduct in numbers too small for detection by the techniques of Doak *et al.* (1967). The few that enter may be trapped in the corona radiata and not detected. The largest number of spermatozoa is clearly present before ovulation, however, according to the observations of Doak *et al.* (1967). Studies of fertilization *in vitro* prove that canine spermatozoa can penetrate and fuse with immature oocytes (Mahi and Yanagimachi, 1976). Therefore, it is quite possible that canine spermatozoa penetrate ovulated oocytes as soon as they are available, but

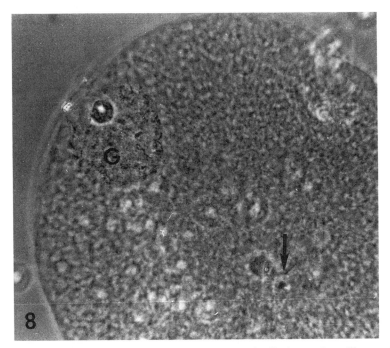

Figure 8. Dictyate primary oocyte with a highly decondensed sperm head in the cytoplasm. The arrow indicates the sperm tail. The germinal vesicle (G) is intact. The oocyte was exposed to spermatozoa overnight and then fixed overnight in 25% acetic acid in ethanol before it was stained with aceto-carmine. (Reproduced with permission from Mahi and Yanagimachi, 1976.)

male pronucleus formation and syngamy await completion of nuclear maturation by the oocytes. Indeed, Van der Stricht (1923) observed penetration of oocytes in naturally mated bitches at several stages of maturation including dictyate and concluded that spermatozoa may enter at any time but that there are no changes in the sperm nucleus until after extrusion of the second polar body.

My hypothesis for the events in the canine fertilization process *in vivo* is as follows. Viable spermatozoa stored in the uterus ascend the oviduct in small numbers throughout estrus. There they are capacitated, and if oocytes are present, they acrosome react, penetrate the zona pellucida, and probably fuse with the egg plasma membrane. After decondensation, the sperm nucleus arrests without forming a pronucleus. The oocyte undergoes a zona reaction to prevent entry of additional spermatozoa. Regardless of whether a spermatozoon has penetrated the oocyte, meiotic maturation of the oocyte proceeds (that is, maturation to metaphase II is not dependent on activation by a spermatozoon). If the oocyte has been penetrated before the completion of maturation, it continues to complete meiosis II with extrusion of the second polar body and formation of the female pronucleus. Simultaneously, the decondensed sperm nucleus is released from its arrest and completes formation of the male pronucleus. Events then proceed as in other mammals. If no spermatozoon has penetrated the oocyte before maturation, the oocyte arrests at metaphase II and resumes meiosis only on penetration, as in most mammals. Of course, further studies with timed ovulations and matings will be required to test each portion of this hypothesis. *In vitro* studies with ovulated oocytes are also indicated.

6. ARTIFICIAL INSEMINATION AND EMBRYO TRANSFER

Artificial insemination with fresh semen is widely practiced in canine husbandry as a method of overcoming behavioral and physical breeding problems (for review of methodologies, see Fielden, 1971). The first successful pregnancies with frozen canine semen were reported 20 years ago (Seager, 1969), and pregnancy rates comparable to natural breeding have since been reported (Seager *et al.*, 1975; Takeisi *et al.*, 1976). Whereas these investigators deposited the thawed semen in the anterior vagina or cervix with great success, other investigators have found that surgical deposition of semen into the uterine body is much more successful (Andersen, 1974; Smith and Graham, 1984). Most studies have used sperm motility and morphology as criteria for postthaw viability, but Garner *et al.* (1986) reported a method for flow cytometric analysis of previously frozen spermatozoa of several species, including the dog. The two stains used, propidium iodide and carboxyfluorescin diacetate, gave a measure of membrane integrity but did not necessarily predict the functional capacity of the spermatozoa. I know of only one study in which *in vitro* fertilization techniques have been utilized to evaluate frozen-thawed canine spermatozoa (C. A. Province and P. M. Riek, unpublished data, cited in Froman *et al.*, 1984). It was found that previously frozen spermatozoa incubated in CCM penetrated the zonae of only 5% of the canine oocytes in contrast to rates of 70% when fresh spermatozoa were tested. No data were given for the ability of the same frozen spermatozoa to fertilize *in vivo*, i.e., to produce pregnancies. However, the data do suggest that this zona penetration assay may not adequately predict the fertilizing ability of frozen-thawed dog spermatozoa *in vivo*. Further studies in this area are warranted.

Extenders have also been developed for storage of canine spermatozoa at 5°C (Province *et al.*, 1984). Such extenders preserve the motility of canine spermatozoa for several days, allowing transport of semen around the globe without its being frozen. They are not designed for long-term storage of semen.

Canine embryos have been successfully transferred from naturally mated donors to host bitches. Kinney *et al.* (1979) transferred 28 embryos into five bitches. Two of the bitches became pregnant, and three puppies were born for a success rate of 11%. This low rate may be caused by uterine asynchrony, since the recipients and donors entered estrus within as much as 4 days of each other, and the duration of estrus is highly variable in the bitch. Takeishi *et al.* (1980) induced estrus and superovulation in bitches using sequential administration of estrone, human chorionic gonadotropin (hCG), pregnant mare serum gonadotropin (PMSG), estradiol, hCG, and PMSG. Four of six bitches ovulated as assessed by the presence of a total of 69 corpora lutea, but only three fertilized eggs (at the morula stage) were harvested from a single mongrel bitch. These were transferred to an identically treated beagle bitch, which subsequently gave birth to a litter of one mongrel (the supposed donated embryo) and three beagle puppies. This outcome is difficult to interpret, since the beagle bitch was artificially inseminated with semen from a mongrel. How could there have been any beagle puppies? Apparent differences in the puppies could have reflected random variation in the sire's mongrel genotype. This question is not clearly resolved by the authors' statement that the breed of the puppies was confirmed by blood group testing. However, if the single "mongrel" puppy was indeed the product of a transferred embryo, the success rate of the transfer was 33%, three times that of Kinney *et al.* (1979), suggesting that artificial synchronization of cycles may increase the success rate of embryo transfer in the bitch.

Large numbers of corpora lutea (10–27) were counted in each bitch that ovulated in Takeishi's study (Takeishi *et al.*, 1980). It was not stated whether any of the bitches that were not used as transfer recipients whelped litters, but the recipient delivered four puppies. Since three puppies were presumed to be her own and she had ten corpora lutea, the implication is that seven ovulations did not result in viable offspring. It is possible that not all corpora lutea represented actual ovulations. The procedure for inducing superovulation may have caused the follicles to luteinize without ovulating. This is supported by the recovery of only three embryos although there were a total of 69 corpora lutea in the four bitches. Newer techniques utilizing GnRH pumps, although not increasing the number of follicles ovulating, may be more favorable for ovulation of viable oocytes and subsequent pregnancy (Cain *et al.*, 1988). The next step will be to transfer embryos resulting from *in vitro* fertilization of canine oocytes, and I predict that it is only a matter of time before these transfers are done.

7. FERTILIZATION IN OTHER CARNIVORA

7.1. Canidae

Estrous cycles have been examined in captive wolves (*Canis lupus*: Seal *et al.*, 1979) and coyotes (*Canis latrans*: Kennelly and Johns, 1976). Unlike the domestic dog, *Canis familiaris*, these species enter estrus only once yearly and are highly seasonal. Subordinant females do not breed. Studies under way on the reproductive physiology of the coyote seek to determine whether the failure of subordinant females to breed is caused by their failure to fully express estrus and to ovulate or by purely behavioral mechanisms (C. Hodges and M. Amos, personal communication). Fertilization itself has not been studied in either species. The coyote characteristically has a prolonged proestrus of 2–3 months, as evidenced by vulval swelling and sanguineous vaginal discharge (Kennelly and Johns, 1976; C. Hodges and M. Amos, personal communication), in contrast to the 9–10 days seen in dogs. As in the dog, however, the time of ovulation in estrus appears to be highly variable (Kennelly and Johns, 1976). These authors reported that the ovulated oocytes were immature, since no polar bodies were visible. A compact corona radiata was present for at least the first few days after ovulation. Motile spermatozoa could be flushed from the uterus at least 3 days after mated females were isolated from males, suggesting that spermatozoa are stored in the female tract in a manner similar to that in the dog (Doak *et al.*, 1967).

Ovulation and fertilization in the silver fox, *Vulpes fulva*, were studied by Pearson and Enders (1943). There are both similarities and differences between the dog and the fox. Estrus is shorter in the fox, usually lasting 3 days. Oocytes are ovulated spontaneously from follicles with greatly folded mural granulosa late on the first day or early on the second day of estrus. As in the dog, the oocytes are ovulated with intact germinal vesicles. The germinal vesicle begins to break down within 24 hr and the oocytes reach metaphase II within the next 24 hr. The authors found no evidence of fertilization before the oocytes had matured to metaphase II even though the females were mated on the first day of estrus (as determined by receptivity to the male).

The estrous cycle of the racoon dog (*Nyctereutes procyonoides*), a wild canid from Asia that has been farmed for fur in Europe to some extent, has been described by Valtonen *et al.* (1977). These animals differ from domestic dogs in failing to shed erythrocytes in the vaginal discharge during proestrus. Proestrus (as judged by vaginal swelling and mucopurulent discharge) averages 7.6 days, and estrus averages 3.9 days. During mating the dog and bitch tie as in other canids, but unlike other canids, they often lie down during the tie with their abdomens towards each other. Ovulation and fertilization were not studied.

To the extent that the canidae have been studied, it appears that there are many similarities among them. When they have been examined, it has been found that yolk-rich oocytes are ovulated in an immature state (dictyate) with a compact layer of corona radiata cells. The estrous cycles are similar, but with differences in the duration of various phases. Estrus is long compared to most mammals, varying from 3 days in the fox to 10 days in the dog. The tie is a characteristic of canid copulation. The domestic dog is the only canid studied that is not a seasonal breeder and that is not naturally monogomous.

7.2. Other Carnivores

The order Carnivora is represented by groups of mammals with diverse reproductive biology. Because my subject is fertilization in dogs, I do not provide extensive details of the reproductive biology of other carnivores. However, it is worth noting that conclusions derived from studies of one family of carnivores cannot be applied to members of other families. As an example of a familiar carnivore with reproductive biology extremely different from that of the dogs, I cite the cat. The domestic cat, *Felis catus*, and its close relatives are seasonally polyestrous (for review, see Shille and Stabenfeldt, 1980). Whereas canids ovulate spontaneously, cats are reflex ovulators, with ovulation dependent on appropriate vaginal stimulation. According to Van der Stricht (1923), feline oocytes are ovulated at metaphase II of meiosis. In absence of mating, ovulation does not normally occur, and the corpus luteum does not form; thus, there is no pseudopregnancy. Even when pseudopregnancy is induced by an infertile mating or other vaginal stimulation, its duration is shorter than the normal gestation period of 63 days. Thus, unlike that of the dog, the feline corpus luteum requires a signal from the uterus for its persistence. Furthermore, the ovaries are not necessary after day 45 in the cat, although they are necessary until the end of canine pregnancy. *In vitro* fertilization of ovulated feline oocytes has been achieved with spermatozoa capacitated *in utero* (Hamner *et al.*, 1970) or *in vitro* (Bowen, 1977). The requirements for *in vitro* fertilization in the cat do not appear to be demanding.

Another reproductive adaptation observed in carnivores is diapause or delayed implantation, which occurs in mink and other mustelids (Mead, 1981) but not in canids or felids. Obviously, the various families of carnivores have little in common in their approaches to reproduction.

8. CONCLUSIONS

Extended estrus combined with variable time of ovulation within estrus and ovulation of immature dictyate oocytes make the bitch an interesting model for studies of reproductive adaptation. One such adaptation may be the prolonged storage and viability of spermatozoa in the uterus. The most intriguing feature of canine fertilization is the potential for spermatozoa to penetrate and fuse with dictyate oocytes in the oviduct. Although *in vitro* studies have clearly demonstrated that spermatozoa can fuse with immature ovarian oocytes, it is still an open question as to whether such fusion occurs *in vivo* and, if not, what prevents it. Studies of the fertilization process in dogs have been hampered by the long estrous cycle of the bitch and the lack of consistently effective methods for induction of estrus and ovulation. In the past even the time of ovulation during the 10 days of estrus was a topic of controversy, and there was no quick way to determine that it had occurred. Thus, it is not surprising that there is only conjecture as to the fertilization process *in vivo*. *In vitro* studies, in contrast, are cheap and easy, since ovaries are readily available from spay-neuter clinics, and large volumes of ejaculated spermatozoa can be easily collected from a single donor male. However, the *in vitro* studies must be validated by appropriate *in vivo* studies. There are now methods for rapid detection of ovulation in the bitch as well as for induction of estrus. These techniques open the way for a more detailed examination of fertilization in dogs. They also will facilitate the technology of embryo transfer in this species.

ACKNOWLEDGMENTS. Some of the studies described in this review and performed in the laboratory of Dr. Ryuzo Yanagimachi were supported by grants from the Morris Animal Foundation and the National Institutes of Health (HD-03402). I wish to thank Dr. Yanagimachi for providing the facilities for these studies as well as the stimulating intellectual environment. I am also grateful to Dr. John Tesoriero for allowing reproduction of his figure (Fig. 2) and to Dr. Max Amos, Dr. Connie Hodges, and Ms. Jean Metzler for sharing their unpublished data. Finally, I wish to thank Dr. James W. Overstreet for his critical review of the manuscript.

9. REFERENCES

Andersen, A. C., and Simpson, M. E., 1973, *The Ovary and Reproductive Cycle of the Dog (Beagle)*, Geron-X, Inc., Los Altos.

Andersen, K., 1974, Intrauterine insemination with frozen semen in dogs, in: *Proceedings 12th Nordic Veterinary Congress, Reykjavik*, Iceland, pp. 153–154.

Bischoff, T. L. W., 1845, *Entwicklungsgeschichte des Hunde-Eies*, Friedrich Vieweg und Sohn, Braunschweig.

Boucher, J. H., Foote, R. H., and Kirk, R. W., 1958, The evaluation of semen quality in the dog and the effects of frequency of ejaculation upon semen quality, libido, and depletion of sperm reserves, *Cornell Vet.* **48:**67–86.

Bowen, R. A., 1977, Fertilization *in vitro* of feline ova by spermatozoa from the ductus deferens, *Biol. Reprod.* **17:**144–147.

Brinster, R. L., 1965a, Studies on the development of mouse embryos *in vitro*. I. The effect of osmolality and hydrogen ion concentration, *J. Exp. Zool.* **158:**49–58.

Brinster, R. L., 1965b, Studies on the development of mouse embryos *in vitro*. II. The effect of energy source, *J. Exp. Zool.* **158:**59–68.

Brinster, R. L., 1965c, Studies on the development of mouse embryos *in vitro*. III. The effect of fixed nitrogen source, *J. Exp. Zool.* **158:**69–78.

Brinster, R. L., 1965d, Studies on the development of mouse embryos *in vitro*. IV. Interaction of energy sources, *J. Reprod. Fertil.* **10:**227–240.

Cain, J. L., Cain, G. R., Feldman, E. C., Lasley, B. L., and Stabenfeldt, G. H., 1988, Use of pulsatile intravenous administration of gonadotropin-releasing hormone to induce fertile estrus in bitches, *Am. J. Vet. Res.* **49:**1993–1996.

Christensen, G. C., 1979, The urogenital system, in: *Miller's Anatomy of the Dog*, ed. 2 (H. E. Evans and G. C. Christensen, eds.), W. B. Saunders, Philadelphia, pp. 544–601.

Concannon, P. W., 1986, Canine physiology of reproduction, in: *Small Animal Reproduction and Infertility: A Clinical Approach to Diagnosis and Treatment* (T. J. Burke, ed.), Lea and Febiger, Philadelphia, pp. 23–77.

Doak, R. L., Hall, A., and Dale, H. E., 1967, Longevity of spermatozoa in the reproductive tract of the bitch, *J. Reprod. Fertil.* **13:**51–58.

El Etreby, M. F., Graft, K.-J., Beier, S., Elger, W., Gunzel, P., and Neuman, F., 1979, Suitability of the beagle dog as a test model for the tumorigenic potential of contraceptive steroids, short review, *Contraception* **20:**237–255.

Evans, E. I., 1933, The transport of spermatozoa in the dog, *Am. J. Physiol.* **105:**287–293.

Evans, H. E., 1974, Prenatal development of the dog, in: *Gaines Veterinary Symposium at Cornell University*, pp. 18–28.

Evans, H. E., 1979, Reproduction and prenatal development, in: *Miller's Anatomy of the Dog*, ed. 2 (H. E. Evans and G. C. Christensen, eds.), W. B. Saunders, Philadelphia, pp. 13–77.

Evans, H. M., and Cole, H. H., 1931, An introduction to the study of the oestrous cycle in the dog, *Mem. Univ. Calif.* **9:**65–119.

Fielden, E. D., 1971, Artificial insemination in the dog, *N. Z. Vet. J.* **19:**178–184.

Foote, R. H., Swierstra, E. E., and Hunt, W. L., 1972, Spermatogenesis in the dog, *Anat. Rec.* **173:**341–352.

Frank, D. W., Kirton, K. T., Murchison, T. E., Quinlan, W. J., Coleman, M. E., Gilbertson, T. J., Feenstra, E. S., and Kimball, F. A., 1979, Mammary tumors and serum hormones in the bitch treated with medroxy-progesterone acetate or progesterone for four years, *Fertil. Steril.* **31:**340–346.

Froman, D. P., Amann, R. P., Riek, P. M., and Olar, T. T., 1984, Acrosin activity of canine spermatozoa as an index of cellular damage, *J. Reprod. Fertil.* **70:**301–308.

Garner, D. L., Pinkel, D., Johnson, L. A., and Pace, M. M., 1986, Assessment of spermatozoal function using dual fluorescent staining and flow cytometric analyses, *Biol. Reprod.* **34:**127–138.

Ghosal, S. K., LaMarche, P. H., Bhanja, P., Joardar, S., Sengupta, A., and Midya, T., 1983, Duration of meiosis and spermiogenesis in the dog, *Can. J. Genet. Cytol.* **25:**678–681.

Grandage, J., 1972, The erect dog penis: A paradox of flexible rigidity, *Vet. Rec.* **91:**141–147.

Hamner, C. E., Jennings, L. L., and Sojka, N. J., 1970, Cat (*Felis catus* L.) spermatozoa require capacitation, *J. Reprod. Fertil.* **23:**477–480.

Hancock, J. L., and Rowlands, I. W., 1949, The physiology of reproduction in the dog, *Vet. Rec.* **47:** 771–776.

Harrop, A. E., 1960, *Reproduction in the Dog*. Bailliere, Tindal and Cox, London.

Holst, P. A., and Phemister, R. D., 1971, The prenatal development of the dog: Preimplantation events, *Biol. Reprod.* **5:**194–206.

Holst, P. A., and Phemister, R. D., 1974, Onset of diestrus in the beagle bitch: Definition and significance, *Am. J. Vet. Res.* **35:**401–406.

Huggins, C., 1945, The physiology of the prostate gland, *Physiol Rev.* **25:**281–295.

Huggins, C., 1947, The prostatic secretion, *Harvey Lect.* **42:**148–193.

Huggins, C., Masina, M. H., Eichelberger, L., and Wharton, J. D., 1939, Quantitative studies of prostatic secretion. I. Characteristics of the normal secretion; the influence of thyroid, suprarenal, and testis extirpation and androgen substitution on the prostatic output, *J. Exp. Med.* **70:**543–556.

Isaacs, W. B., and Coffey, D. S., 1984, The predominant protein of canine seminal plasma is an enzyme, *J. Biol. Chem.* **259:**11520–11526.

Kennelly, J. J., and Johns, B.E., 1976, The estrous cycle of coyotes, *J. Wildl. Manag.* **40:**272–277.

Kinney, G. M., Pennycook, J. W., Schriver, M. D., Templeton, J. W., and Kraemer, D. C., 1979, Surgical collection and transfer of canine embryos, *Biol. Reprod.* **20** (Suppl. 2):96A.

Kuroda, H., and Hiroe, K., 1972, Studies on the metabolism of dog spermatozoa. I. Seasonal variation on the semen quality and of the aerobic metabolism of spermatozoa, *Jpn. J. Anim. Reprod.* **17:**89–98.

Mahi, C. A., and Yanagimachi, R., 1976, Maturation and sperm penetration of canine ovarian oocytes *in vitro*, *J. Exp. Zool.* **196:**189–196.

Mahi, C. A., and Yanagimachi, R., 1978, Capacitation, acrosome reaction, and egg penetration by canine spermatozoa in a simple defined medium, *Gamete Res.* **1:**101–109.

Mahi-Brown, C. A., 1986, Prospects for control of fertility in the female dog by active immunization with porcine zona pellucida proteins, in: *Immunological Approaches to Contraception and Promotion of Fertility* (G. P. Talwar, ed.), Plenum Press, New York, pp. 301–309.

Mead, R. A., 1981, Delayed implantation in mustelids with special emphasis on the spotted skunk, *J. Reprod. Fertil. Suppl.* **29:**11–24.

Meizel, S., 1985, Molecules that initiate or help stimulate the acrosome reaction by their interaction with the mammalian sperm surface, *Am. J. Anat.* **174:**285–302.

Pearson, O. P., and Enders, R. K., 1943, Ovulation, maturation and fertilization in the fox, *Anat. Rec.* **85:**69–83.

Phemister, R. D., Holst, P. A., Spano, J. S., and Hopwood, M. L., 1973, Time of ovulation in the beagle bitch, *Biol. Reprod.* **8:**74–82.

Province, C. A., Amann, R. P., Pickett, B. W., and Squires, E. L., 1984, Extenders for preservation of canine and equine spermatozoa at 5°C, *Theriogenology* **22:**409–415.

Seager, S. W. J., 1969, Successful pregnancies utilizing frozen dog semen, *A.I. Digest* **17:**1–2.

Seager, S. W. J., and Fletcher, W. S., 1972, Collection, storage, and insemination of canine semen, *Lab. Anim. Sci.* **22:**177–182.

Seager, S. W., J., Platz, C. C., and Fletcher, W. S., 1975, Conception rates and related data using frozen dog semen, *J. Reprod. Fertil.* **45:**189–192.

Seal, U. S., Plotka, E. D., Packard, J. M., and Mech, L.D., 1979, Endocrine correlates of reproduction in the wolf. I. Serum progesterone, estradiol and LH during the estrous cycle, *Biol. Reprod.* **21:**1057–1066.

Setchell, B. P., and Brooks, D. E., 1988, Anatomy, vasculature, innervation, and fluids of the male reproductive tract, in: *The Physiology of Reproduction*, Vol. I (E. Knobil and J. D. Neill, eds.), Raven Press, New York, pp. 753–836.

Shille, V. M., and Stabenfeldt, G. H., 1980, Current concepts in reproduction of the dog and cat, in: *Advances in Veterinary Science and Comparative Medicine*, Vol. 24 (C. A. Brandly and C. E. Cornelius, eds.), Academic Press, New York, pp. 211–243.

Shille, V. M., Thatcher, M.-J., Buhi, W. C., Alvarez, I. M., and Lannon, A. P., 1988, Protein synthesis by endometrium and embryo during early pregnancy in the bitch, *Biol. Reprod.* **38** (Suppl. 1):132.

Smith, F. O., and Graham, E. F., 1984, Cryopreservation of canine semen: Technique and performance, in: *Proceedings of the 10th International Congress of Animal Reproduction and Artificial Insemination* Vol. 2, Urbana-Champaign, p. 216.

Sokolowski, J. H., 1971, The effects of ovariectomy on pregnancy maintenance in the bitch, *Lab. Anim. Sci.* **21:**696–699.

Sokolowski, J. N., Zimbelman, R. G., and Goyings, L.S., 1973, Canine reproduction: Reproductive organs and related structures of the nonparous, parous, and postpartum bitch, *Am. J. Vet. Res.* **34**:1001–1013.

Takeisi, M., Mikami, T., Kodama, Y., Tsunekane, T., and Iwaki, T., 1976, Studies on the reproduction in the dog. VII. Artificial insemination using the frozen semen, *Jpn. J. Anim. Reprod.* **22**:28–33.

Takeishi, M., Akai, R., Tsunekane, T., Iwaki, T., and Nakanowatari, K., 1980, Studies on the reproduction in dogs—A trial of ova transplantation in dogs, *Jpn. J. Anim. Reprod.* **26**:151–153.

Tash, J. S., Hidaka, H., and Means, A. R., 1986, Axokinin phosphorylation by cAMP-dependent protein kinase is sufficient for activation of sperm flagellar motility, *J. Cell Biol.* **103**:649–655.

Tash, J. S., Krinks, M., Patel, J., Means, R. L., Klee, C. B., and Means, A. R., 1988, Identification, characterization, and functional correlation of calmodulin-dependent protein phosphatase in sperm, *J. Cell. Biol.* **106**:1625–1633.

Tesoriero, J. V., 1981, Early ultrastructural changes of developing oocytes in the dog, *J. Morphol.* **168**:171–179.

Tesoriero, J. V., 1982, A morphologic, cytochemical, and chromatographic analysis of lipid yolk formation in the oocytes of the dog, *Gamete Res.* **6**:267–279.

Tsutsui, T., 1975, Studies on the reproduction in the dog. V. On cleavage and transport of fertilized ova in the oviduct, *Jpn. J. Anim. Reprod.* **21**:70–75.

Tsutsui, T., and Shimizu, T., 1975, Studies on the reproduction in the dog. IV. On the fertile period of ovum after ovulation, *Jpn. J. Anim. Reprod.* **21**:65–69.

Valtonen, M. H., Rajakoski, E. J.,and Makela, J. I., 1977, Reproductive features in the female racoon dog (*Nyctereutes procyonoides*), *J. Reprod. Fertil.* **51**:517–518.

Van der Stricht, O., 1923, Etude comparée des ovules des mammifères aux différentes périodes de l'ovogenèse, d'après les travaux du Laboratoire d'Histologie et d'Embryologie de l'Université de Gand, *Arch. Biol., Paris* **33**:229–300.

Wales, R. G., and White, I. G., 1958, The interaction of pH, tonicity and electrolyte concentration on the motility of dog spermatozoa, *J. Physiol. (Lond.)* **141**:273–280.

Wassarman, P. M., 1988, The mammalian ovum, in: *The Physiology of Reproduction*, Vol. 1 (E. Knobil and J. D. Neill, eds.), Raven Press, New York, pp. 69–102.

Woodall, P. F., and Johnstone, I. P., 1988, Dimensions and allometry of testes, epididymides and spermatozoa in the domestic dog (*Canis familiaris*), *J. Reprod. Fertil.* **82**:603–609.

Yanagimachi, R., 1988a, Sperm–egg fusion, in: *Current Topics in Membranes and Transport*, Vol. 32, *Membrane Fusion in Fertilization, Cellular Transport and Viral Infection* (N. Duzgunes and F. Bronner, eds.), Academic Press, New York, pp. 3–43.

Yanagimachi, R., 1988b, Mammalian fertilization, in: *The Physiology of Reproduction*, Vol. 1 (E. Knobil and J. D. Neill, eds.), Raven Press, New York, pp. 135–185.

16

Fertilization in Cats

David E. Wildt

1. INTRODUCTION

The domestic cat is a member of the order Carnivora, a taxon consisting of 231 species, most of which exhibit vastly different reproductive strategies. The data base on sperm and oocyte interaction is poor for almost all carnivores. However, the mechanisms of fertilization for the domestic cat are beginning to receive attention because these studies generate new information of both fundamental and applied benefit. Unlike most conventional laboratory species, which ovulate spontaneously during estrus, the cat is an "induced" ovulator. Ovulation occurs only after multiple copulations trigger pituitary release of sufficient amounts of luteinizing hormone (LH), which causes final maturation of follicular oocytes. Therefore, the cat is useful for studying the kinetics of LH effects on oocyte nuclear maturation. It also is possible that induced ovulatory species have different mechanisms for sustaining oocyte viability in the ovarian follicle. If for some reason copulation is delayed in the estrous cat, then, in theory, the intrafollicular oocyte must innately sustain its longevity, moreso than in most mammals, which ovulate spontaneously near the onset of sexual receptivity.

There is a need for a greater understanding of gamete interaction in the cat. This information would ensure the propagation of feline models useful for studying human diseases. The cat is used commonly as a research model for at least 36 human physiological abnormalities ranging from anatomic to oncologic dysfunctions (see review by Goodrowe *et al.*, 1989b). Certain natural genetic anomalies in cats (e.g., Klinefelter's syndrome) and similarities in chromosomal linkage homology with humans also make this species valuable for research. Particularly important studies are in progress on feline leukemia transmission (O'Brien, 1986), a condition caused by an immune-suppressing retrovirus similar to the T-cell trophic retrovirus, HTLV-III, causing acquired immunodeficiency syndrome (AIDS) in humans (Olsen, 1981; Hardy, 1984).

Whereas feral cats have a reputation for reproducing well in nature, the propagation of laboratory cats often is a cumbersome and inconsistent undertaking (Schmidt, 1986). It especially is difficult to propagate certain cat strains afflicted with human-type dysfunctions like muco-

DAVID E. WILDT • National Zoological Park, Smithsonian Institution, Washington, D.C. 20008.

A Comparative Overview of Mammalian Fertilization, edited by Bonnie S. Dunbar and Michael G. O'Rand. Plenum Press, New York, 1991.

polysaccharidosis and mannosidosis. The most persistent problems are physical inadequacies that prevent the copulatory act, but cat models as well as many random source cats also experience poor libido and sexually incompatibility. Artificial breeding approaches (e.g., artificial insemination, *in vitro* fertilization, embryo transfer) could be useful in circumventing these problems if we understood more about the mechanisms regulating sperm–oocyte interaction *in vivo* and *in vitro*. As discussed later in this chapter, individual cats produce different numbers of structurally normal spermatozoa, a finding that appears related directly to the incidence of zona pellucida penetration *in vitro*. Because men routinely produce high percentages of sperm pleiomorphisms, the cat may serve as an ethically attractive, scientifically sound model for studying the impact of teratospermia on the fertilization event. Finally, an improved data base on fertilization and artificial breeding in the domestic cat could have conservation potential if these techniques can be applied to other felid species. The Felidae family consists of 37 species, and all but the domestic cat are listed as threatened or endangered by extinction. For this reason, the domestic cat is becoming an important model for comparative studies of nondomestic felids including investigations into homologous and heterologous fertilization.

Available information on sperm–oocyte function, fertilization, and early embryogenesis in the cat is based almost solely on fewer than two dozen published papers and abstracts concerning gamete morphology and *in vitro* fertilization. This chapter summarizes the present state of knowledge by providing a synopsis of relevant facts on the general reproductive biology and endocrinology of the male and female cat followed by details of what is known about fertilization. For comparative purposes, information also is provided illustrating the similarities and dissimilarities in fertilization events between the domestic cat and several taxonomically related felid species.

2. GENERAL REPRODUCTIVE CHARACTERISTICS

2.1. The Male Cat

Mitotic activity occurs in the seminiferous tubules before puberty, usually at about 5 months of age (Scott and Scott, 1957). Three to 7 months later, puberty occurs as defined by the presence of sperm in the ejaculate. The effects of season on reproductive performance of the free-living male cat are unknown. However, laboratory males exposed daily to at least 12 hr of light and darkness are known to breed and sire offspring throughout the year (Beaver, 1977; Wildt *et al.*, 1978). Although unlimited promiscuous copulations sometimes decrease sperm concentration, most mature males are capable of mating repeatedly over a 4- to 5-day period without reducing conception rates (Sojka *et al.*, 1970).

Cat semen has been collected by surgically flushing the ductus deferens or cauda epididymis immediately after castration (Bowen, 1977; Niwa *et al.*, 1985), but the usual approach is by electroejaculation of anesthetized males (Platz and Seager, 1978; Wildt *et al.*, 1983). Cats also have been trained to ejaculate into an artificial vagina (AV) (Sojka *et al.*, 1970; Platz *et al.*, 1978). These ejaculates generally contain less volume (range 34 to 40 μl) and more total sperm per ejaculate (range, 57 to 61 × 10⁶) than semen obtained by electroejaculation (100 to 233 μl, 11 to 30 × 10⁶, respectively) (Sojka *et al.*, 1970; Platz *et al.*, 1978; Wildt *et al.*, 1983). Sperm motility usually ranges from 50% to 95% (Wildt *et al.*, 1983). In most ejaculates, approximately 20% to 30% of sperm are pleiomorphic or abnormally shaped (Platz and Seager, 1978; Wildt *et al.*, 1983). These proportions are relatively low compared to those measured in certain wild felid species including the cheetah (Wildt *et al.*, 1983, 1987b), leopard (Brown *et al.*, 1989), puma (Miller *et al.*, 1990b), and some populations of geographically isolated lions (Wildt *et al.*, 1987a). A recent study has identified individual domestic cats that consistently ejaculate more than 60%

pleiomorphic sperm/ejaculate (Howard *et al.*, 1990, 1991). A variety of abnormal forms are detected in these males, including defects in the sperm head, midpiece, and flagellum (Fig. 1). Clearly, the most prevalent anomalies, involve a bent midpiece with a laterally displaced cytoplasmic droplet (positioned within the bent angle), a bent flagellum (with or without a droplet), and a tightly coiled flagellum. Head defects (including macro- or microcephaly, bi- or tricephaly) constitute about 3.3% of all sperm pleiomorphisms. Acrosomal derangements, observed in sperm of some nondomestic cat species (Miller *et al.*, 1990), are uncommon in domestic cats.

The pH of cat semen is neutral to slightly alkaline (7.0 to 8.2) (Sojka *et al.*, 1970), and the osmolalities of seminal plasma, prostatic fluid, and bulbourethral gland secretions range from 323 to 331 mOsm/liter (Johnson *et al.*, 1988). The motility of undiluted cat sperm *in vitro* is short-lived at 37°C (usually less than 60 min). Diluting the semen in tissue culture medium and reducing the maintenance temperature to 23°C increase the duration of motility *in vitro* by ~35% (Goodrowe *et al.*, 1989b). More detailed laboratory processing (i.e., "swim-up processing," Section 3.3) enhances sperm viability even further *in vitro*. This approach recovers a high proportion of viable sperm that are capable of maintaining high rates of progressive motility for 6 (Howard *et al.*, 1990) to 30 hr *in vitro* (A. M. Donoghue, J. G. Howard, and D. E. Wildt, unpublished observations).

2.2. The Female Cat

The young cat usually demonstrates first estrus (puberty) at 6 to 9 months of age (Festing and Bleby, 1970; Jemmett and Evans, 1977). The free-living cat in North America is seasonally polyestrous; however, genetics and environment largely dictate seasonal reproductive activity. For example, most short-hair cat breeds are reproductively active throughout the year, whereas long-haired breeds experience periods of seasonal anestrus (Jemmett and Evans, 1977). In the northern hemisphere, anestrus coincides with the reduced photoperiod in the fall and early winter months; usually sexual activity resumes with increasing day length in January and February. Circannual fluctuations in cyclicity are eliminated under controlled laboratory conditions when a 12-hr light–dark cycle is provided and females are exposed daily to an adult male (Wildt *et al.*, 1978). Under these circumstances, the cat demonstrates a comparable number of estrous periods throughout the year, although the duration of each estrus is shorter in June, September, October, and November than in other months.

The unmated cat cycles every 14 to 21 days (Prescott, 1973; Paape *et al.*, 1975; Wildt *et al.*, 1978). The duration of estrus ranges from 5 to 7 days and does not appear affected by the mating event or the occurrence of ovulation (Wildt *et al.*, 1978). The copulation-induced reflex ovulatory response of the cat was first documented by Greulich (1934) and subsequently has been examined by a variety of neurophysiological, anatomic and endocrine approaches (see review by Goodrowe *et al.*, 1989b). Related studies have provided a basic understanding of folliculogenesis and the events leading to ovulation in the natural estrual cat. These findings are relevant because they can affect oocyte integrity, maturation, release, and fertilizability.

Several studies have integrated observations of sexual behavior with ovarian activity and endocrine patterns throughout the periestrual interval. Circulating estradiol-17β concentrations are elevated at the onset of estrus (Shille *et al.*, 1979; Wildt *et al.*, 1981; Schmidt *et al.*, 1983) and appear important for priming the hypothalamic–pituitary axis to the mating-induced LH peak (Banks and Stabenfeldt, 1982). Unlike the rabbit (Hilliard *et al.*, 1964), a single copulatory stimulus in the cat rarely causes an LH surge adequate to induce ovulation. However, multiple copulations transmit sufficient neural input via the medial basal hypothalamus (Robison and Sawyer, 1987) to cause the rapid and sustained release of LH. When multiple matings are allowed during less than a 2-hr interval, peak circulating LH values are achieved rapidly and decline to

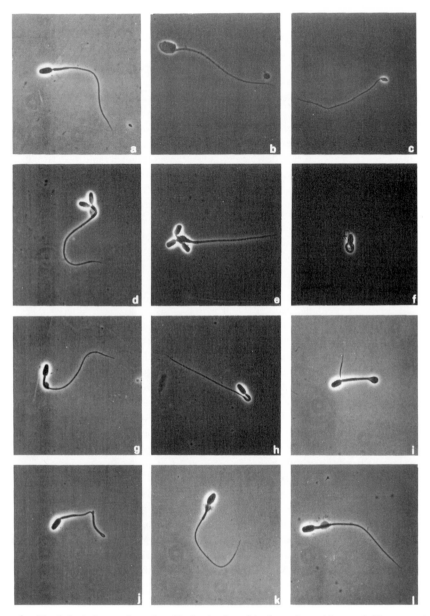

Figure 1. Morphological sperm forms found in the ejaculate of the domestic cat: (a) normal; (b) macrocephalic; (c) microcephalic with mitochondrial sheath aplasia; (d) bicephalic; (e) tricephalic; (f) tightly coiled flagellum; (g) bent midpiece with cytoplasmic droplet; (h) bent midpiece without cytoplasmic droplet; (i) bent flagellum with cytoplasmic droplet; (j) bent flagellum without cytoplasmic droplet; (k) proximal cytoplasmic droplet; (l) distal cytoplasmic droplet. (From Howard *et al.*, 1990).

baseline within 12 to 24 hr after mating (Concannon *et al.*, 1980, 1989; Johnson and Gay, 1981; Banks and Stabenfeldt, 1982; Shille *et al.*, 1983; Glover *et al.*, 1985). Spacing copulations at 3-hr intervals sustains elevated LH levels for as long as 38 hr (Wildt *et al.*, 1980, 1981; Schmidt *et al.*, 1983). Cats mated on consecutive days of estrus progressively produce lower-amplitude and shorter-duration LH surges until circulating LH concentrations remain basal by the third day of a mated estrus (Wildt *et al.*, 1981; Schmidt *et al.*, 1983). It now is evident that the mating interval, number of copulatory stimuli, and endocrine status affect ovulation onset post-mating (Goodrowe *et al.*, 1989b). Paape *et al.* (1975) mated cats *ad libitum* for 2 hr and estimated that ovulation (based on a sustained rise in plasma progesterone) occurred 46 hr after first coitus. Using an unspecified mating protocol, Verhage *et al.* (1976) detected no rise in systemic progesterone for the first 3 to 4 days following copulation. Wildt *et al.* (1981), using a three-times-daily mating protocol combined with laparoscopic and hormonal observations, estimated that ovulation occurred within 48 to 52 hr of the LH peak in five of 12 cats but was later than 52 hr after the LH peak in the other seven cats. Preovulatory luteinization of follicular granulosa cells does not occur in the cat. Steroidogenic changes from follicular estrogen to progesterone secretion occur coincident with ovulation; circulating progesterone concentrations simultaneously increase with the onset of ovulation, the latter occurring some 44 to 60 hr after the LH peak (Wildt *et al.*, 1981; Schmidt *et al.*, 1983).

3. GAMETE FEATURES

3.1. The Ovum during the Natural Periestrous Interval

Laparoscopic examinations of mated cats have provided a gross overview of ovarian and follicular activity during the periestrous interval. During the 2- to 3-day period preceding estrus, about three to seven primary follicles develop from ~1 mm to 2 mm in diameter. There are no descriptions of the actual ovulatory event in the cat; however, some early postovulatory follicles are hemorrhagic, whereas others are not (Wildt and Seager, 1980). Freshly collected ovulated oocytes have a substantial number of cumulus cells attached and a rather compact corona radiata.

The historical studies of Van der Stricht (1911), Hill and Tribe (1924), and Dawson and Friedgood (1940) provide a histological perspective of ovarian, follicular, and oocyte conditions during the periovulatory interval. In these studies, cats apparently were mated randomly different numbers of times, and the ovaries were then removed and examined. On occasion, the reproductive tracts were flushed, and the resulting oocytes and embryos were described. According to these workers, the antra of cat ovarian follicles immediately preceding estrus are relatively large. The mural granulosa of the proestrous follicle is composed of several cell layers. At the basement membrane (the most peripheral layer), the cells are columnar, containing uniformly basal nuclei. In the upper layers, the cells are polyhedral or almost spherical and have centrally positioned nuclei. Here, the ovum and its surrounding cumulus usually are found on the follicular wall contralateral to the ovarian surface. At this early stage of development, the cumulus cells are compacted, and the germinal vesicle is eccentric, lying adjacent to the membrane of the oocyte. The theca interna is not yet distinguishable.

These early-stage follicles increase in size coincident with behavioral estrus (to 3 to 4 mm in diameter). The mating act itself contributes to this increase in size and slight follicular protrusion above the ovarian surface. Even before mating, the basal mural granulosa cells become less columnar and the nuclei more centralized. At this time, the theca interna is distinguishable from the theca externa, and the cells become locally hypertrophied. Mating and the LH release are accompanied by substantial follicular fluid production. The outer layers of cumulus around the oocyte remain relatively uniform, and individual cells remain separated by a fine granular matrix, which occupies the intercellular spaces. The corona radiata immediately investing the ovum remains compact, and, as the surrounding cells loosen, the corona becomes more prominent. At ovulation, the entire cumulus mass separates from the mural granulosa and is extruded through a break in the follicular wall.

The ovulated cat oocyte is in metaphase II of meiosis, and we suspect that this stage of maturation occurs coincident with ovulation (Section 3.2). The mean diameter of a mature cat zona pellucida and oocyte combined is 162.2 μm (Goodrowe *et al.*, 1988b). The mean zona pellucida width is 17.4 μm, and the mean oocyte diameter without the zona pellucida is 127.5 μm. The cat oocyte is dark in appearance (Fig. 2) because of a high intracellular concentration of lipid (Guraya, 1965); this is one feature that appears common to all carnivore oocytes including the ferret, dog, mink, leopard cat, puma, and tiger (Chang, 1968; Bowen, 1977; Mahi and Yanagimachi, 1978; Goodrowe *et al.*, 1989a; Donoghue *et al.*, 1990; Miller *et al.*, 1990). The dark appearance of the cytoplasm makes it difficult to identify intracellular structures and assess meiotic status easily. However, high-speed centrifugation (15,000 $\times$ g, 3 min) readily displaces intracellular lipid (Fig. 3), allowing the germinal vesicle and/or pronuclei to be identified (Goodrowe *et al.*, 1988b). The germinal vesicle, recognized as a membrane-bound structure, comprises about 15–20% of the intracellular volume. Relative to the entire oocyte, the polar bodies are small ($\sim$ 10 μm in diameter) and, in our experience, can be viewed effectively only by using differential interference optics. Because polar body extrusion is so difficult to identify, cumulus cell mass expansion appears to be the best subjective predictor for estimating the extent of oocyte maturation (Fig. 4a,b).

3.2. Oocyte Maturation

Our current understanding of oocyte maturation in the cat, which is rather rudimentary, is based almost exclusively on recent *in vitro* studies. We have reported the time course and conditions necessary for oocyte maturation *in vitro* in the domestic cat (Johnston *et al.*, 1989).

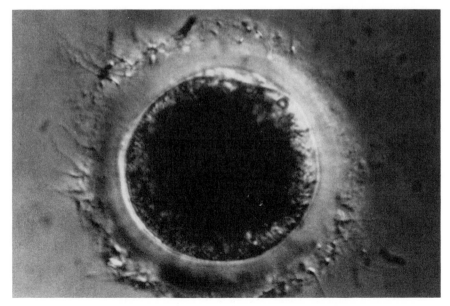

Figure 2. *In vivo*-matured domestic cat oocyte.

Figure 3.   A one-celled, *in vitro*-fertilized and centrifuged domestic cat embryo. (From Goodrowe *et al.*, 1988b.)

Immature oocytes surrounded by cumulus cells and a tight corona radiata were collected from ovaries removed at ovariohysterectomy. After culture ($38°C$, 5% CO_2, 95% air atmosphere, high humidity) in Eagle's essential medium (containing 0.23 mM pyruvate, 1% fetal calf serum, 3 mg/ml bovine serum albumin), oocytes were evaluated for nuclear maturation on the basis of meiotic chromosomal configuration. The kinetics of nuclear maturation *in vitro* are depicted in Fig. 5. At 12 hr, 75% of the oocytes were in the diplotene stage, 7% at the diakinesis/metaphase I, and the remainder (18%) were classified as degenerate. At 24 hr, 12% of the oocytes were in metaphase II, which increased threefold within 6 hr and eventually peaked (62.5%) at 48 hr. Diplotene oocytes were not detected by 30 hr of culture, a time when most oocytes were equally distributed between the diakinesis and metaphase II stages of maturation. Therefore, most domestic cat oocytes achieve metaphase II within 30 to 48 hr of being placed into culture. This interval is only slightly shorter than that described for the dog oocyte (48 to 72 hr) *in vitro* (Mahi and Yanagimachi, 1976). Because (1) the highest rate of *in vitro* oocyte maturation requires 48 hr of laboratory culture and (2) cats mated three times per day during estrus usually ovulate 48 hr or more later (Wildt *et al.*, 1981), we speculate that maturation of cat oocytes to metaphase II occurs coincident with ovulation, similar to observations in the mouse, sheep, pig, cow, monkey, and human (Edwards, 1965).

Our laboratory also has examined the influence of reproductive status of the donor and exogenous gonadotropin on the ability of cat oocytes to mature *in vitro*. The incidence of maturation is enhanced ($P < 0.05$) when oocytes are recovered from cats that either have no significant follicular activity (54%) or distinct ovarian follicles (56%) compared to those recovered from luteal-phase (29%) or pregnant (35%) animals. Differences in hormonal status among the various types of donors might well explain these variations. High, nonphysiological concentrations of steroids (including progesterone) inhibit maturation of mouse (Eppig and Koide, 1978)

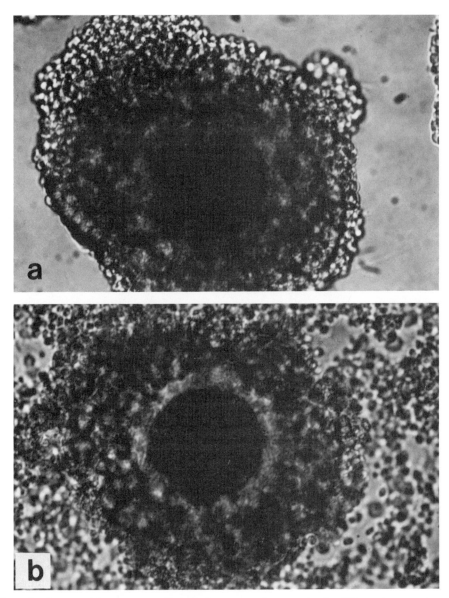

Figure 4. Immature (a) versus mature (b) domestic cat oocyte demonstrating the difference in compactness of the cumulus cells and corona radiata.

and pig (McGaughey, 1977) oocytes *in vitro*. Pregnant or luteal-phase cats consistently produce high concentrations of circulating progesterone (Wildt *et al.*, 1981; Schmidt *et al.*, 1983), which may be detrimental or inhibitory to the time course for subsequent oocyte maturation. Gonadotropins in culture medium also affect the cat oocyte's ability to achieve metaphase II. The proportion of oocytes successfully maturing *in vitro* in medium containing no hormone supplementation (37%) was less ($P < 0.05$) than counterparts cultured in follicle-stimulating hormone (FSH) only (48%) of FSH and LH (54%). In this context, the cat oocyte appears similar to that of the human

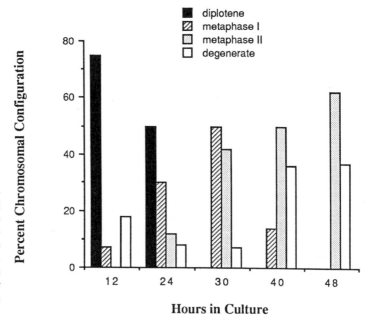

Figure 5. Time sequence of chromosomal configurations (percentage of identified chromosome configurations) during maturation of domestic cat oocytes in culture during a 48-hr interval. (From Johnston *et al.*, 1989.)

(Prins *et al.*, 1987) in that the addition of gonadotropins increases the incidence of nuclear maturation *in vitro*.

The influence of ovary and oocyte-processing procedures on *in vitro* maturation is especially important if these techniques are to be applied to "oocyte rescue" or the salvage of genetic material from rare domestic cats or endangered felids. For practical purposes, oocytes must remain viable within the follicle until recovered and placed in culture. Two factors have a major impact on the integrity and viability of these oocytes: (1) maintenance temperature of the ovaries and (2) time delay until oocytes are recovered. Johnston *et al.* (1989) recently studied the effect of holding/transport temperature on the ovaries and a delay in liberating oocytes from ovarian material on subsequent oocyte maturation. In the first trial, cat ovaries were immersed in phosphate-buffered saline (PBS) and maintained at 4°C or at room temperature (22°C) for 8 hr before oocyte recovery and culture. In trial 2, ovaries either were processed immediately (controls) or maintained at 4°C for 24 to 32 hr in PBS before oocyte recovery for *in vitro* maturation. Maintaining freshly collected ovaries at 22°C resulted in 43% of all oocytes reaching metaphase II compared to 54% for ovaries held at 4°C before processing ($P > 0.05$). Likewise, maturation of oocytes recovered from ovaries maintained for the delayed interval (54%) was the same ($P > 0.05$) as that observed for oocytes processed within 8 hr of ovariohysterectomy (54%). These observations suggest that cat antral oocytes do not degenerate rapidly *ex situ* but rather remain biologically viable when properly processed. This finding may have particular relevance in salvaging oocytes from rare individuals that die abruptly. If extrapolated to rare felid species, it may be possible to collect and ship ovarian material long distances to central laboratories designed to mature and fertilize such oocytes *in vitro*.

3.3. Sperm Capacitation

As with other mammals, the phenomenon of capacitation in the cat is thought to involve a series of sperm changes necessary for fertilization. The exact sequence of events occurring within the sperm plasma membrane that prepares the cell for the acrosome reaction and fertilization still is unknown. However, there is intriguing evidence that the cat differs from many species in the ease of sperm capacitation.

Most of our understanding of the requirements for cat sperm capacitation are based on four studies, all of which relied on *in vitro* fertilization of homologous oocytes. Hammer *et al.* (1970) first determined that cat sperm require capacitation. Fresh ejaculates, collected using an artificial vagina, were cocultured *in vitro* with ovulated oocytes flushed from oviducts. No evidence of fertilization was observed. However, using sperm previously incubated *in utero* for 30 min resulted in successful fertilization and cleavage in oocytes collected from one of three donors. Uterine incubation of sperm for 2 to 24 hr resulted in cleavage rates ranging from 53% to 90%. Later studies demonstrated that ductus deferens (Bowen, 1977) and epididymal (Niwa *et al.*, 1985) sperm were capable of fertilizing oocytes *in vitro* without any exposure to the female's reproductive tract. In the former study, ductus deferens tissue simply was minced and mixed with a tissue culture medium used to inseminate oviductal oocytes. In the latter, oviductal oocytes were cocultured with sperm recovered from the caudae epididymides of castrated males. Because oocyte penetration occurred as early as 30 min following insemination, it appeared that sperm capacitation probably required only 15 to 20 min.

Capacitation requirements are not completed so rapidly with ejaculated sperm. Goodrowe *et al.* (1988b) reported that freshly electroejaculated cat sperm could be readily induced to penetrate oocytes by swim-up processing. For this procedure, semen was diluted with modified Krebs Ringer bicarbonate (mKRB) solution (containing 4 mg/ml bovine serum albumin and 40 units heparin/ml of medium) and centrifuged for 8 min at $300 \times g$. The supernatant was aspirated and discarded; subsequently, 150 µl of mKRB was layered slowly onto the resulting pellet, and the sperm allowed a 1-hr swim-up at room temperature. The layered medium component was aspirated gently from the pellet surface and, after the appropriate dilution, used to inseminate follicular oocytes and achieve embryo cleavage. Whereas this study did not determine the time course of capacitation, Goodrowe *et al.* (1988a) explored the time required for freshly ejaculated cat sperm to initiate zona pellucida penetration. Immature oocytes were liberated from ovario-hysterectomy material. Homogeneously appearing oocytes with cumulus cell masses were transferred into mKRB medium and held in a 5% CO_2 in air humidified environment (37°C). The first of two experiments determined the approximate time necessary to initiate zona penetration. Intact cumulus cell–oocyte complexes were cocultured with the sperm cell suspension under oil for either 5 or 10 hr. Differential interference contrast optics ($400 \times$) demonstrated that there was no difference ($P > 0.05$) in the number of zonae pellucidae penetrated by ejaculated sperm after 5 (95%) or 10 (100%) hr. So a second experiment was designed to clarify the timing of the penetration event by testing shorter time periods. Based on the results of the first experiment, cumulus–oocyte complexes were incubated with sperm for 0 to 5 hr. Oocytes were removed at 30-min intervals for 5 hr, treated with hyaluronidase, and examined for sperm penetration of the zona pellucida as well as the number of sperm attached to the zona. The penetration rate was lowest ($P < 0.05$) in the 0.5- (36%) and 1.0-hr (34%) groups compared to all other times (range: 1.5 hr, 73% to 5.0 hr, 97%). Overall, the highest penetration rates ($\geq 90\%$) occurred when sperm and oocytes were cocultured for at least 3 hr. Whereas the number of sperm attached per oocyte did not differ ($P > 0.05$) among the 2.5- to 5.0-hr time periods (range 4.8 to 5.9 sperm), fewer ($P < 0.05$) sperm were detected bound to oocytes cultured for only 0.5 to 2 hr (mean range 2.3 to 3.4 sperm/oocyte).

The presence of seminal plasma appears to have little inhibitory effect on the ability of cat sperm to capacitate. Howard *et al.* (1991) recently compared the ovum-penetrating ability of cat sperm subjected to swim-up versus non-swim-up processing. Cat electroejaculates were collected, divided, and centrifuged at $300 \times g$ for 8 min. The supernatant (seminal plasma) from the swim-up aliquot was decanted, and sperm from the pellet allowed to migrate into fresh medium.

In contrast, the non-swim-up treatment consisted of simply reconstituting the pellet with the seminal plasma supernatant. When equal numbers of motile sperm from the swim-up and non-swim-up groups were cocultured with zona-pellucida-free hamster ova (Section 4.5), a similar ($P > 0.05$) number of ova were penetrated (12.6% and 13.1%, respectively, as demonstrated by a decondensed sperm head in the ovum cytoplasm). Likewise, 58.3% of zona-intact cat oocytes (Section 4.5) cultured with swim-up-processed sperm were penetrated (as demonstrated by sperm within the zona), which was similar ($P > 0.05$) to the rate observed using non-swim-up aliquots (56.7%).

Taken together, all evidence suggests that cat sperm undergo capacitation but that the requirements to achieve this condition are not very rigorous. This contrasts directly to observations made for many other species, which appear to have more specialized needs. For example, sperm of some animals require a high-ionic-strength medium (rabbit and bull: Brackett and Oliphant, 1975; Brackett, 1981; Brackett *et al.*, 1982), the addition of caffeine or cyclic adenosine $3',5'$-monophosphate (rhesus monkey: Boatman and Bavister, 1982), or simply slight variations in medium constituents (dog: Mahi and Yanagimachi, 1978). However, the high rates of *in vitro* ovum penetration reported by at least three different laboratories using four biologically different capacitation preparations indicate that cat sperm are relatively insensitive to variations in techniques used to induce capacitation.

4. FERTILIZATION *IN VITRO*

Almost all of our understanding of fertilization events in the cat are associated with *in vitro* studies. Recent reviews of *in vitro* fertilization (IVF) in the cat are available (Goodrowe, 1987; Goodrowe *et al.*, 1988b, 1989b). Hamner and associates (1970) were the first to fertilize cat ova successfully using an *in vitro* system. After injecting 150 IU of pregnant mares' serum gonadotropin (PMSG, s.c.) followed by 50 IU hCG (i.m.) 72 hr later, ova were flushed from the oviducts and placed in Brackett's medium (Brackett, 1970) under oil in a 5% CO_2 in air, humidified environment at 37°C. Ejaculated sperm, capacitated in the uterine horns of naturally estrous females for 0.5 to 24 hr, were retrieved by flushing and added to the ova at concentrations of 15 to 75×10^5 sperm/ml. After an 18-hr incubation, oocytes were removed and examined for cleavage. Spermatozoa incubated *in utero* for 2 to 24 hr resulted in cleavage rates of 53% to 90%.

Bowen (1977) compared the IVF rate using ductus deferens sperm suspended in a modified Biggers, Stern, and Whittingham (mBSW) medium or a modified Ham's F10. After treating anestrous females with PMSG (100 or 150 IU, i.m.) followed 72 hr later by commercially available exogenous LH (2.5 mg, i.m.), ova were flushed from the oviducts and cultured in one of the two sperm suspensions containing 5.5 to 23×10^5 sperm/ml. After 12 to 14 hr of culture (38°C), ova were washed free of sperm and assessed for fertilization. The presence of two or more blastomeres was indicative of fertilization. Each medium supported an equivalently ($P > 0.05$) high rate of fertilization (mBSW, 77.8%; mHam's, 80.1%). Although percentage culture rates were not provided, embryos allowed further cleavage developed to the blastocyst stage, whereas none of the 25 (noninseminated) ova underwent parthenogenetic cleavage.

Niwa *et al.* (1985) observed the early events of IVF of ovulated cat ova mixed with epididymal sperm. Following PMSG (150 IU, i.m.) and hCG (100 IU, 72 hr later) treatment, ova were collected from oviductal flushes 31 to 32 hr post-hCG. Diluted sperm (0.2 to 1.8×10^6/ml), collected from the caudae epididymides of castrated males, were placed with the ova in mKRB medium in a 37°C, 5% CO_2 in air, humidified environment and cultured from 0.25 to 5 hr. Following incubation, cumulus cells were removed with hyaluronidase (0.1%) treatment, and the ova were fixed, stained, and examined for sperm penetration. Penetration rates ranged from 0 to 100% with decondensation of sperm heads and male pronuclear formation observed 3 to 4 hr after insemination. Because none of the ova had supplemental spermatozoa in the perivitelline space, it was concluded that polyspermy in the cat is blocked at the zona pellucida rather than at the vitelline membrane.

Using these three reports for background information, our laboratory has conducted a series of studies to examine (1) the factors influencing successful *in vitro* gamete interaction and (2) the developmental competence of cat embryos resulting from IVF. The overall objective is to use follicular oocytes (collected by laparoscopic transabdominal aspiration) and electroejaculated sperm to develop a practical IVF system that consistently results in viable embryos capable of resulting in live young after embryo transfer.

4.1. Effect of hCG Dose and Interval from PMSG to hCG

The first study (Goodrowe *et al.*, 1988b) evaluated several variables that potentially could affect oocyte collection, maturation, quality, and *in vitro* fertilizability. Adult females with inactive ovaries were treated with a single injection of PMSG (150 IU, i.m.) followed by 100 or 200 IU (i.m.) of hCG 72 or 80 hr later. Follicular oocytes were aspirated laparoscopically 25 to 29 hr after hCG into mKRB medium. Mature oocytes were washed and placed in mKRB with swim-up processed sperm (2×10^4) in a 5% CO_2 in air, humidified environment (37°C). After 18 to 20 hr of coculture, the oocytes were washed in a hyaluronidase solution, cultured in fresh mKRB for 6 to 12 hr, and assessed for fertilization.

Overall and without regard to specific treatment, slightly more than half (52.7%) of the gonadotropin-injected cats exhibited behavioral estrus, and an average of 11.7 ± 0.7 ($\pm$S.E.M.) mature follicles were observed after hCG treatment. A high proportion (91.4%) of the oocytes from these follicles were recovered (overall mean, 10.7 ± 0.7 oocytes/cat). Neither the hCG dose nor the PMSG–hCG interval influenced the number of ovarian follicles available for aspiration, the number of oocytes collected, or oocyte recovery efficiency ($P > 0.05$). At 24 to 30 hr of culture, fewer ($P < 0.001$) degenerate oocytes were observed in cats receiving 100 IU hCG (8.2%) compared to those receiving 200 IU (20.6%). Likewise, the lower hCG dose resulted in a higher ($P < 0.025$) cleavage rate (100 IU, 38.2% versus 200 IU, 23.2%) (Fig. 6). The incidence of embryo cleavage was particularly sensitive to the PMSG-to-hCG interval, with the proportion increased ($P < 0.001$) from 19.6% for the 72-hr period to 41.3% for the 80-hr interval. Overall fertilization (48.1%) and cleavage (45.2%) rates were greatest following an 80-hr PMSG–hCG interval combined with the 100 IU hCG dose. We speculate that this effect is related to the extent of intrafollicular oocyte maturation (Goodrowe *et al.*, 1988b). The oocyte is unable to undergo independent metabolism, and, thus, maturation relies on communication with the surrounding granulosa cells (Gilula *et al*, 1978; Motlik *et al.*, 1986). The final stages of oocyte maturation are gonadotropin dependent; a follicle-stimulating-type signal initiates oocyte maturation, whereas an LH or hCG stimulus triggers nuclear maturation (Dekel *et al.*, 1981). Germinal vesicle breakdown and cytoplasmic metabolic maturation are independent events, and, therefore, it is possible for nuclear maturation to ensue within a metabolically immature oocyte (Moor *et al.*, 1981). Taken in the context of our observations in the cat, it is logical that an excessive ovulatory signal (provoked by hCG) or a poorly timed gonadotropin interval may cause premature granulosa cell uncoupling, thereby disrupting oocyte integrity or final oocyte maturation.

In most mammalian species, ovulation occurs within hours of the onset of sexual receptivity. However, the cat is an induced ovulator, and, thus, ova are released only after a series of matings, an event that may not occur until the later days of estrus. This reproductive strategy requires that the maturing oocyte be self-sustaining within the follicle for a rather extended (6 days or more) and unpredictable period after estrus onset. To test the sensitivity of the intrafollicular oocyte, cats with inactive ovaries were given 150 IU PMSG (i.m.) and then 100 IU hCG (i.m.) 80, 84, 88, 92, and 96 hr later (Miller *et al.*, 1989). Twenty-five to 27 hr after hCG, oocytes were aspirated

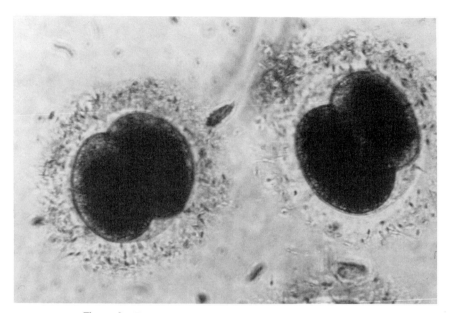

Figure 6. Two-cell stage, *in vitro*-fertilized domestic cat embryos.

laparoscopically, cocultured with electroejaculated, swim-up-processed sperm (under the conditions described by Goodrowe *et al.*, 1988b) for 30 hr, and examined for cleavage. The mean number of oocytes recovered from the 80- to 92-hr groups was similar (range 19.3 to 24.0, $P >$ 0.05); however, oocyte number was reduced ($P < 0.05$) by at least 50% in the 96-hr group (mean 10.3). More ($P < 0.05$) mature oocytes were collected when hCG was given 80 (95.9%), 84 (87.5%), or 88 (87.9%) hr compared to 92 (69.0%) and 96 (75.8%) hr after hCG. The IVF rate was greater ($P < 0.05$) for mature oocytes in the 84- (62.5%), 88- (58.2%), and 92-hr (55.7%) groups than for the 80- (48.3%) and 96-hr (40.7%) groups. Thus, under the conditions described here, cat oocytes stimulated to begin maturation with PMSG are incapable of a prolonged intrafollicular viability. On the contrary, there appears to be a time-related rise and fall in the incidence of fertilization *in vitro*. Likewise, as the interval between PMSG and hCG is lengthened, the total number of oocytes and mature oocytes recovered is reduced.

These observations suggest that the PMSG/hCG-triggered cat oocyte is not unlike follicular oocytes in spontaneously ovulating species. Once turned on, these oocytes are highly sensitive to the timing of the gonadotropic stimuli and experience critical maturational events up to the time of ovulation. If the interval between the FSH-like and LH-like hormone is too short, oocyte fertilizability is compromised, probably because maturation is incomplete. Likewise, extending the interval between the two stimuli reduces the number of mature oocytes recovered, probably because many of these oocytes rapidly degenerate within the follicle or fail to respond to hCG because of reduced sensitivity. A delay in the LH-like stimulus also reduces the fertilization rate of recovered oocytes, which probably is related to a number of simultaneous aging factors.

These results are not inconsistent with the theory that there still is an intrafollicular mechanism that maintains cat oocytes in an ovulation-ready state throughout estrus. No studies of oocyte integrity, maturation, and fertilizability have been conducted in natural estrous cats untreated with exogenous gonadotropins. Because circulating LH concentrations remain basal (Wildt *et al.*, 1980, 1981), the follicular oocytes of unmated, naturally estrous cats probably go unexposed to LH before copulation. The circumstances likely are very different in PMSG-challenged cats. Although consistently effective for eliciting follicular development in cats, this hormone also has substantial LH-like activity and a relatively long half-life. It is possible that giving PMSG to stimulate follicle growth also contributes partially to premature nuclear maturation. Therefore, some later-stage oocytes may be stimulated to complete maturation even before being exposed to hCG. This could be an alternative explanation for the reduced oocyte numbers and poor quality oocytes recovered from cats treated with hCG at later intervals. Studies are in progress to define the impact of interval between onset of natural estrus and hCG stimulus on the ability of follicular oocytes to mature and fertilize *in vitro*.

4.2. Effect of Culture Medium and Protein Supplementation

Johnston *et al.* (1991) recently examined the influence of culture medium and protein supplementation on fertilization of cat oocytes *in vitro*. The basic IVF system was that described by Goodrowe *et al* (1988b) using follicular oocytes and electroejaculated/swim-up-processed sperm. Study 1 explored the potential IVF/preimplantation embryo development differences among three media: (1) mKRB, a "simple" medium; (2) modified Tyrode's solution (TALP, another "simple" medium prepared without phosphate and glucose); and (3) Ham's F10, a "complex" medium. Each contained NaCl, KCl, $CaCl_2$, $MgSO_4$, $NaHCO_3$, and Na pyruvate. Modified KRB and Ham's F10 contained KH_2PO_4 and glucose, whereas mKRB and TALP contained NA lactate. Only Ham's F10 contained amino acids (glutamine) and vitamins. Each medium was supplemented with BSA (4 mg/ml) and equilibrated at 37° with 5% CO_2. Sperm and oocytes were cocultured in each test medium for 30 hr and assessed for fertilization on the basis of cleavage. Cleaved embryos were examined at 24 hr intervals for 7 days. The incidence of embryo cleavage was similar ($P > 0.05$) among mKRB (75%), TALP (70.6%), and Ham's F10 (80%), indicating that there was little effect of a "simple" versus "complex" medium on successful fertilization. In contrast, compared to TALP (78%), more ($P < 0.05$) embryos in Ham's F10 (95%) developed to the morula (Fig. 7) stage [the incidence of development for mKRB-cultured

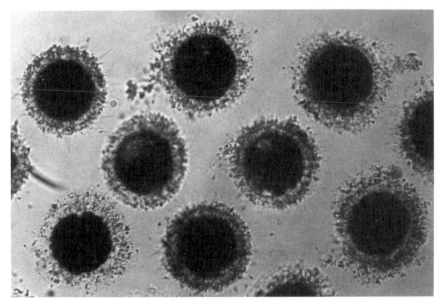

Figure 7. Morula stage, *in vitro*-fertilized domestic cat embryos.

embryos was intermediate (89%) and not different ($P > 0.05$)]. Based on these observations, Johnston *et al.* (1991) speculated that the cat embryo probably requires glucose and possibly glutamine to achieve later stages of preimplantation development.

In study 2, these investigators examined the effect of protein supplements on IVF and embryo culture success. Because Ham's F10 produced the highest embryo development rate to the morula stage in study 1, this solution was chosen as the test medium. Ham's F10 was fortified with (1) polyvinylalcohol (PVA), 2 mg/ml (a control, nonprotein supplement); (2) BSA, 4 mg/ml; (3) fetal calf serum (FCS), 5%; or (4) cat estrus serum (CES), 5%. The latter constituted heat-inactivated serum obtained from cats in behavioral estrus. Follicular oocytes and processed sperm were coincubated and monitored as described above. The IVF rate obtained using FCS (84%) and ECS (85.2%) was greater ($P < 0.05$) than that achieved with PVA (67.3%); oocytes exposed to BSA fertilized at a rate (76.1%) no different ($P > 0.05$) from the other treatments. Embryos exposed to each of the four treatments (PVA, 82.8%; BSA, 82.8%; FCS, 92.9%; or ECS, 97.8%) were equally capable ($P > 0.05$) of becoming morulae. However, more FCS- and ECS-supplemented morulae continued to the early blastocyst (Fig. 8) stage (30.8% and 22.2%, respectively, $P < 0.01$) than PVA- (10.3%) or BSA-exposed (13.8%) morulae. These observations, in the context of study 1, indicate that the type of medium and protein supplement used has a greater impact on embryo development *in vitro* than on fertilization. Presently, Ham's F10 containing FCS or ECS produces the highest incidence of fertilization (~85%), and about twofold more of these embryos develop in culture to blastocysts compared to embryos supplemented with PVA or BSA.

These results also provide evidence that domestic cat embryos experience a partial block in *in vitro* embryogenesis between the morula and blastocyst stage of development. In this context the cat appears similar to the rabbit but different from most other species which experience a block earlier in development (Johnston *et al.*, 1991). In the culture of bovine and ovine embryos, development becomes blocked *in vitro* at eight to 16 cells (Camous *et al.*, 1984; Gandolfi and Moor, 1987). Mouse embryos (Biggers *et al.*, 1962) arrest *in vitro* at the two-cell stage,

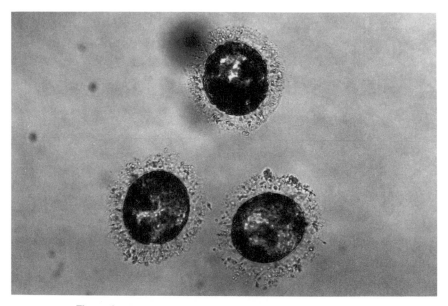

Figure 8. Blastocyst stage, *in vitro*-fertilized domestic cat embryos.

hamster embryos (Yanagimachi and Chang, 1964; Schini and Bavister, 1988) at both the two- and four-cell stage, and pig embryos (Davis and Day, 1978) at the four-cell stage. The ability of domestic cat embryos to overcome developmental arrest and progress to the early blastocyst stage appears partially dependent on protein source. This then suggests that there may be additional factors, some of which are in serum, that play an essential role in advancing *in vitro* growth beyond the morula stage. Overall, it appears that *in vitro* development requirements for the two- to 32-cell domestic cat embryo are relatively simple. However, the prerequisites for continued growth are more complex and are not yet established.

4.3. Features Associated with IVF of Mature Cat Oocytes

With the IVF system described in Section 4.1, a number of features are commonly observed in the cat, most of which are discussed elsewhere in more detail (Goodrowe, 1987; Goodrowe *et al.*, 1988b). Fewer than 10% (mean, 6%) of noninseminated, cultured cat oocytes undergo parthenogenetic cleavage, which is much less than the rate reported for the only other carnivore species studied (the ferret, 80%; Chang, 1957).

As discussed in Section 3.1 for the germinal vesicle and polar bodies, it is extremely tedious to identify pronuclei of fertilized cat oocytes. Distinguishing these organelles from intracellular vacuoles is difficult without DNA-specific fluorescent staining (e.g., Hoechst # 33342, H342; bisbenzamide; Sigma Chemical Co.). With fluorescence optics, these chromosomal structures are three-quarter "moon-shaped" and produce a "softer" brightness than somatic cell nuclei (Goodrowe *et al.*, 1988b). High-speed centrifugation (as described in Section 3.1) also allows displacing intracellular lipid while more readily exposing the presence of pronuclei (Fig. 3). In the one-cell cat embryo, the male pronucleus always is larger and situated more superficially than the female pronucleus. The first evidence of embryo cleavage in the cat occurs 20 to 28 hr after insemination.

Punctured and aspirated cat follicles undergo luteal development as demonstrated by the observation of normal-appearing corpora lutea (CL) and circulating progesterone concentrations (Goodrowe *et al.*, 1988b). The temporal progesterone profile in cats subjected to PMSG/hCG treatment and oocyte aspiration is no different from the pattern observed in natural estrous cats treated with hCG and allowed to ovulate. However, PMSG-treated cats have higher ($P < 0.05$) peripheral concentrations of progesterone than their natural estrous counterparts, probably because these animals produce more total follicles and, thus, more CL.

Embryos resulting from the IVF of laparoscopically recovered cat oocytes are biologically competent as demonstrated by the production of live-born offspring after embryo transfer (Fig. 9). Goodrowe *et al.* (1988b) placed two- to four-cell stage embryos into size 50 PE tubing in 2 μl of PB1 medium (Whittingham, 1971), which then was inserted into the fimbriated end of the oviducts at laparotomy. A total of six queens previously used as oocyte donors received 18, 13, eight, eight, seven, and six embryos, respectively, 42 to 52 hr after insemination. Five cats diagnosed pregnant by abdominal palpation at 35 days produced normal litters of four (four/18 embryos), one (one/13 embryos), three (three/eight embryos), one (one/eight embryos), and one (one/seven embryos) kitten, respectively. The birth of live young after embryo transfer unequivocally demonstrates the developmental competence of *in vitro* fertilized carnivore oocytes.

4.4. *In Vitro* Maturation and Fertilization

Section 3.2 discussed recent progress in the *in vitro* maturation of cat antral oocytes. Johnston *et al.* (1989) also has studied the ability of *in vitro* matured cat oocytes to fertilize *in vitro*. Oocytes from inactive donors were pooled, placed in standard culture conditions (Section 4.1), and inseminated 49, 52, or 55 hr later with electroejaculated sperm subjected to swim-up

Figure 9. Domestic cat kittens conceived by *in vitro* fertilization.

processing (see IVF procedures of Goodrowe *et al.*, 1988b). Swim-up aliquots containing sperm with an 80% motility rating were used for fertilization attempts (2×10^4 sperm/14 or fewer oocytes). After 12 hr of culture, oocytes were washed in a hyaluronidase solution to remove cumulus cells and loosely attached sperm. Oocytes were returned to culture in fresh medium for 12 to 18 hr and then assessed for fertilization.

The incidence of parthenogenetic cleavage for control, noninseminated oocytes was 4% (three/75 oocytes). Thirty-six percent and 33% of the oocytes inseminated 52 hr after culture were fertilized and produced two-cell embryos, respectively, which was greater ($P < 0.01$) than values observed at insemination times of 49 hr (11% and 5%, respectively) and 55 hr (26% and 8%, respectively). A total of 42 two-cell embryos were produced. Of the 38 two-cell embryos allowed to continue culture, four, 15, and 13 advanced to the four-cell, eight-cell, and 16-cell stage of development, respectively.

In vitro-matured cat oocytes are capable of fertilizing and developing *in vitro*, often to advanced stages of embryonic development. However, these oocytes appear highly sensitive to the timing of exposure to spermatozoa in culture, which likely is related to both nuclear and cytoplasmic maturation as well as aging processes within the oocyte. Although the majority of oocytes achieve metaphase II at 48 hr (Section 3.2), the fertilization rate *in vitro* is enhanced when insemination is delayed for an additional 4 hr. This probably is a result of incomplete cytoplasmic maturation. Although *in vitro*-matured oocytes are capable of fertilizing in culture, the highest IVF rate of 36% remains considerably less than the ~80% (Johnston *et al.*, 1991; Section 4.2) rates routinely observed for *in vivo*-matured oocytes subjected to similar fertilization conditions.

4.5. The "Male Component" to Fertilization *in Vitro*

As the rate of teratospermia increases, reproductive performance decreases in a variety of species including the human, sheep, horse, pig, and mouse (see review, Howard *et al.*, 1991). Compared to normal forms, structurally defective sperm probably are less capable of being transported through the reproductive tract (Nestor and Handel, 1984; Saacke *et al.*, 1988), although it is likely that many pleiomorphic cells reach the site of fertilization (oviduct). There is an indirect correlation between the incidence of abnormal spermatozoa in the inseminant and ovum penetration rate in human IVF trials (Mahadevan and Trounson, 1984). However, our actual understanding of the impact of abnormal sperm on fertilization and subsequent embryo viability is quite limited. Increased numbers of defective sperm may play a major role in infertility in a variety of species in which teratospermia is a common reproductive trait. The human and select members of the Felidae family, including the cheetah, leopard, puma, and certain populations of lions (Section 2.1), are examples of such species.

Gamete interaction studies in the cat are beginning to examine the significance of teratospermia on the fertilization event. Howard *et al.* (1990, 1991) have identified certain domestic cats that consistently produce more than 60% pleiomorphic sperm/ejaculate throughout the year. Compared to normospermic cats (producing less than 40% abnormally shaped sperm), these teratospermic males produce comparable serum FSH and LH concentrations but lower levels of testosterone. Two fertilization bioassays have been used to examine sperm function and specifically the influence of teratospermia on *in vitro* sperm–ovum interaction.

A heterologous zona-free hamster ova assay, based on the original test of Yanagimachi (1972), has been adapted for use with domestic cat sperm. Because this assay does not assess the ability of the spermatozoon to bind and penetrate the zona pellucida, Howard *et al.* (1991) also developed a complementary assay using homologous, zona-intact cat antral oocytes. For the hamster assay, oviductal ova were collected from PMSG/hCG-stimulated immature females. Recovered cumulus masses were treated in mKRB medium that included 0.1% bovine testicular hyaluronidase to remove the cumulus cells and then bovine pancreatic trypsin to dissolve the zona pellucida. For the zona-intact assay, domestic cat ovaries, obtained by ovariohysterectomy, were punctured repeatedly with a 22-gauge needle to release cumulus–oocyte complexes into surrounding mKRB medium. Only oocytes with homogeneously dark vitelli and tightly compacted corona radiata and cumulus cell masses (i.e., immature) were used. To allow *in vitro* maturation and loosening of the cumulus oophorus cells, oocytes were transferred into fresh mKRB and cultured for 16 hr. Cumulus cells then were removed by transferring oocytes to mKRB containing 0.2% hyaluronidase. For insemination, electroejaculates from both normospermic and teratospermic cats were subjected to swim-up processing (Section 3.3). Processed semen (2×10^6 motile sperm) was transferred to culture dishes containing either zona-free hamster ova or zona-intact cat oocytes and incubated for 3 hr in a 5% CO_2 in air, humidified environment (37°C). Ova and oocytes then were washed and microscopically examined for ovum penetration. For the hamster assay, this was defined by the presence of a swollen or decondensed sperm head with a corresponding flagellum in the vitellus. For the cat assay, this was defined by the presence of a spermatozoon penetrating the zona pellucida in the same focal plane as the vitellus.

Subjecting teratospermic ejaculates to swim-up processing increased the percentage of normal sperm forms from an average of 33.8 to 61.3%, a value no different ($P > 0.05$) from that measured in normospermic ejaculates (66.6%). Sperm from both cat populations were capable of binding with and penetrating zona-free hamster ova (Fig. 10) and zona-intact cat oocytes (Fig. 11).

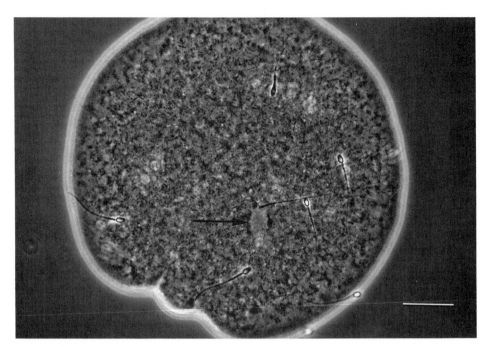

Figure 10.   Penetration of zona-free hamster ovum by domestic cat sperm. One swollen sperm head (arrow) with attached flagellum is visible in the vitellus of each ovum. (From Howard *et al.*, 1991.)

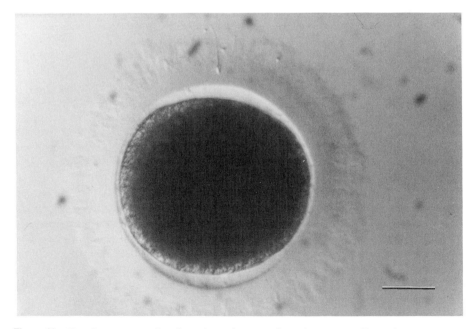

Figure 11.   Homologous penetration of zona-intact, immature domestic cat oocyte. Domestic cat spermatozoon is visible within the zona pellucida. (From Howard *et al.*, 1991.)

However, for the former assay, the penetration rate (i.e., the number of ova with decondensed sperm heads divided by the total number of ova × 100) was different between groups (normospermic, 12.6% versus teratospermic, 2.2%, $P < 0.05$). Likewise, in the latter assay, the penetration rate (i.e., the number of ova with sperm penetrating the zona pellucida divided by the total number of ova × 100) was different (normospermic, 58.3%, versus teratospermic, 15.0%, $P < 0.05$). For the zona-intact cat assay, there also was a fivefold increase ($P < 0.05$) in the number of bound sperm per oocyte.

These results are the first indicating that teratospermia in the cat adversely affects gamete interaction and fertilization *in vitro*. Clearly sperm from teratospermic cats are less capable of penetrating heterologous ova and binding and penetrating homologous oocytes. Particularly fascinating is the observation that sperm from ejaculates subjected to swim-up processing (to increase the number of normal sperm forms) are still five times less likely to penetrate zona-free hamster ova and almost four times less likely to penetrate zona-intact cat oocytes compared to similarly treated normospermic ejaculates. Because sperm concentration was maintained constant and motility ratings were similar between the two cat populations, there appears to be a fundamental functional deficit in spermatozoa from teratospermic cats. This compromised ability appears inherent to the teratospermic ejaculate, is independent of the absolute number of pleiomorphisms, and reduces the ability of sperm to bind and penetrate the zona pellucida and vitelline membrane. Howard *et al.* (1991) speculate that the inherent infertility factor may be related to ultrastructural or biochemical defects or epididymal dysfunction. Regardless, these findings suggest that the zona-free hamster ovum and zona-intact cat oocyte bioassays provide important tools for exploring many factors controlling fertilization as well as studying the impact and perhaps etiology of teratospermia.

5. FERTILIZATION IN NONDOMESTIC FELID SPECIES

Although wild felids play a valuable role in the ecology of our planet and in the emerging field of conservation biology (Wildt, 1990), we understand little about fertilization processes in these species. Even though gamete biology studies in nondomestic species present unique challenges (Wildt *et al.*, 1986b; Wildt, 1989, 1990), there have been a growing number of such reports in the scientific literature. Many of the original studies focused on semen and, specifically, sperm characteristics of a variety of felids including the cheetah (Wildt *et al.*, 1983, 1987b, 1988), clouded leopard (Wildt *et al.*, 1986a), lion (Wildt *et al.*, 1987a), North Chinese leopard (Wildt *et al.*, 1988), Sri Lankan leopard (Brown *et al.*, 1989), tiger (Wildt *et al.*, 1988; Byers *et al.*, 1989), and puma (Wildt *et al.*, 1988; Miller *et al.*, 1990). Of 28 felid species surveyed in one comparative study, 20 were found to ejaculate more than 40% pleiomorphic sperm forms (Howard *et al.*, 1984). Specific defects range from simple bending of the midpiece or flagellar regions to extensive derangements of the mitochondrial sheath and acrosome. For example, the sperm of one specific subspecies of puma, the endangered Florida panther (*Felis concolor coryi*), is afflicted with 50% abnormal acrosomes (Miller *et al.*, 1990). Although the etiology of this reproductive trait is not completely understood, there is evidence that the number of abnormal sperm in felid ejaculates is related to a lack of genetic variability (Wildt *et al.*, 1983, 1987a,b). Such observations are supported by molecular genetic (isozyme) data and certain morphometric features. For example, Florida panthers have fewer polymorphic loci than western pumas and experience a very high incidence of cryptorchidism (47%), a genetically inherited condition (Miller *et al.*, 1990).

In vitro fertilization studies are being used in wild felids for determining (1) the influence of teratospermia on fertilization, (2) the relative ease of inducing sperm capacitation, and (3) the utility of domestic cat IVF techniques as an approach for producing embryos and offspring from rare species.

5.1. Studies Using Heterologous Oocytes

The zona-free hamster assay has been used to determine capacitation requirements for tiger (Byers *et al.*, 1987, 1989) and leopard cat sperm (Howard and Wildt, 1990). The sperm of both species are capable of undergoing nuclear decondensation in zona-free hamster ova. Tiger sperm are able to penetrate hamster ova after a 2-hr preincubation at 37°C, and the presence of seminal plasma did not affect penetration ability. Leopard cat sperm exposed to hamster ova respond in a fashion similar to domestic cats: four males were tested on six occasions over a 6-month interval, and the average penetration rate (number of ova with at least one decondensed sperm head: Fig. 12) was 10.2%. Leopard cat sperm also have a high affinity for zona-intact domestic cat oocytes (using bioassay techniques described in Section 4.5). An average of 53.1% of domestic cat oocytes were penetrated by leopard cat sperm (Fig. 13), a rate approximating that of domestic cat sperm. A comparison of individual data revealed that these males did not differ ($P > 0.05$) in most ejaculate traits including sperm motility characteristics. However, one male consistently produced higher percentages of structurally abnormal sperm than his three counterparts. For this male, the penetration rate of zona-free hamster ova (mean 3%) and zona-intact cat oocytes (28%) was considerably less than that for the other three males (hamster bioassay range 6% to 12%; cat bioassay range 47% to 73%). This observation appears consistent with the notion (discussed earlier for the domestic cat, Section 4.5) that a major factor dictating nuclear decondensation within the hamster ovum and binding and penetration of zona-intact oocytes is the morphological

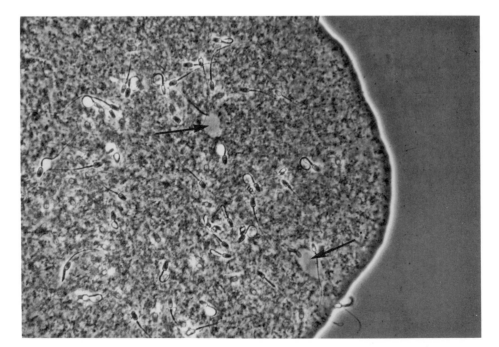

Figure 12.   Penetration of zona-free hamster ovum by leopard cat spermatozoa. Multiple swollen sperm heads (arrows) with attached flagellae are visible in the vitellus of the ovum. (From Howard and Wildt, 1990.)

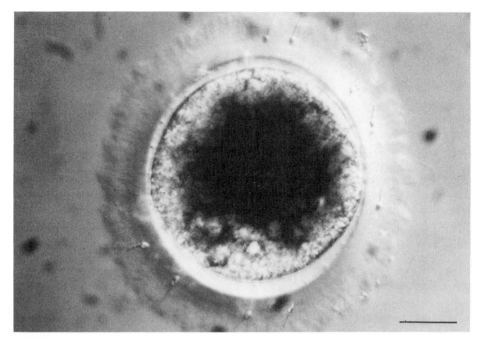

Figure 13. Penetration of zona-intact domestic cat ovum by leopard cat spermatozoa. Sperm are visible within the zona pellucida. (From Howard and Wildt, 1990.)

integrity of the sperm in the preprocessed ejaculate. These results also suggest that (1) these two IVF bioassays likely can be adapted across a range of Felidae species and (2) there may be no mechanism in this taxon for excluding penetration by a "foreign" felid spermatozoon. In accord with these assertions, we recently observed the penetration of domestic cat oocytes by puma or cheetah spermatozoa (J. G. Howard and D. E. Wildt unpublished observations).

These IVF bioassays have other potential uses including exploring how well specific mechanisms have been conserved among related species, especially those events that ensure gamete interaction and subsequent fertilization (Wildt, 1990). From a practical perspective, heterologous IVF assays will allow assessment of such critical events as sperm capacitation, binding, and general viability before commitments are made to costly artificial insemination (AI) or IVF attempts. This will be especially important for the captive breeding of endangered species because few females will be available as AI recipients or oocyte donors.

5.2. Studies Using Homologous Oocytes

There are four published reports indicating that conspecific sperm and follicular oocytes from nondomestic felid species will interact to form embryos *in vitro*. Our laboratory has been particularly interested in the question: Can the IVF system developed for the domestic cat be applied to other felid species? In this context, we have extended the basic IVF system described by Goodrowe *et al.* (1988b) to three nondomestic species, the leopard cat (Goodrowe *et al.*, 1989a), puma (Miller *et al.*, 1990), and tiger (Donoghue *et al.*, 1990). Although in the same genus (*Felis*) and the same size as domestic cats, leopard cats treated with the same PMSG/hCG regimen (Section 4.1) produce fewer ovarian follicles (range 2 to 11) and recovered oocytes (0 to 11) than domestic cats. However, as in the domestic cat, the proportion of mature leopard cat oocytes fertilized *in vitro* (Fig. 14) was increased by extending the PMSG-to-hCG interval from 80 (17.5%) to 84 (52.4%) hr. Regardless of treatment, a much higher percentage of leopard cat oocytes were degenerate (45.9%) at 30 hr post-insemination compared to the incidence observed in the domestic cat (8.5%: Goodrowe *et al.*, 1988b). Together, these findings demonstrate that even closely related species may vary substantially in response characteristics to a given IVF protocol.

In the puma IVF studies, three sperm donors (two western pumas and one Florida panther) and seven western pumas and one Florida panther oocyte donors were used. Three of the females and one of the males were free-living animals that were captured, used as gamete donors, and then released back into the wild. Females were treated with a single injection of PMSG (1000 or 2000 IU, i.m.) followed 84 hr later by hCG (800 IU, i.m.). Oocytes were aspirated 24 to 26 hr post-hCG from ovarian follicles via laparoscopy. Mature eggs were inseminated 4 to 6 hr later, whereas immature oocytes were cultured 24 hr and then inseminated. Seven of eight pumas responded with follicle development, and 140 oocytes were recovered from 145 follicles (96.6%; 77 mature, 43 immature, 20 degenerate; mean, 20.0 ± 5.9 oocytes/female). At 30 hr after insemination, 14.3% of the oocytes were classified as degenerate, and the overall fertilization rate was 43.5% (total oocytes fertilized = 40) despite the use of semen containing 82% to 99% pleiomorphic sperm. Of the 36 immature oocytes matured *in vitro* and inseminated with the Florida panther ejaculate, 12 were fertilized even though 50% of the sperm contained an acrosomal defect.

Female tigers responded to PMSG (2500 or 5000 IU, i.m.) and hCG (2000 IU, i.m., 84 hr later) by developing normal-appearing ovarian follicles. Oocytes were recovered in this largest of all felid species by transabdominal laparoscopic aspiration. Semen, collected by electroejaculation, contained sperm with a comparatively low incidence of structural pleiomorphisms (mean, 18.5%). Coculturing conditions were the optimal procedures recently described by Johnston *et al.* (1991; Section 4.2). A total of 456 oocytes were collected from 468 follicles (97.4% recovery). Of these, 378 (82.9%) qualified as mature, 48 (10.5%) as immature, and 30 (6.6%) as degenerate. Of the 358 mature oocytes inseminated, 227 (63.4%) were fertilized. Of the 195 fertilized

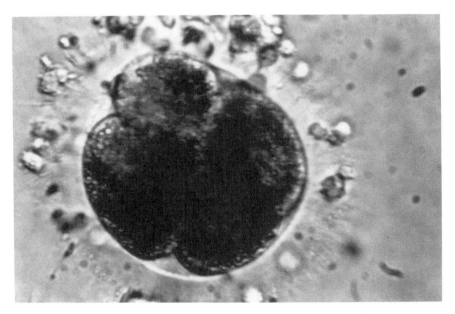

Figure 14. Four-cell leopard cat embryo resulting from *in vitro* fertilization.

oocytes cultured further, 187 (96.0%) cleaved to the two-cell stage (Fig. 15). Of the 56 cleaved embryos cultured *in vitro* in Ham's F10 for 72 hr (5% CO_2 in humidified air, 37°C), 14 (25.0%) were at the 16-cell stage, and 15 (26.8%) were morulae (Fig. 16). Of the 46 embryos cultured for 96 hr, 20 (43.5%) advanced to morulae, and 14 (30.4%) to early blastocyst. Eighty-six two- to four-cell embryos were transferred surgically into the oviducts of five of the original oocyte donors and one PMSG/hCG-stimulated recipient not subjected to oocyte aspiration. The latter tiger became pregnant and delivered three live-born cubs by cesarian section 107 days after embryo transfer (Fig. 17).

Together, these studies illustrate three important points. First, it appears that general IVF techniques developed in one felid species are largely adaptable to another species. However, certain species specificities are to be expected, particularly in the number and quality of oocytes recovered after exogenous gonadotropin treatment. However, it is rather remarkable that such a high percentage of cleaved embryos can be generated in a species (i.e., tiger) during early IVF attempts. Secondly, it is especially important that fertilized and cleaved embryos can be produced using semen containing high proportions of structurally abnormal sperm. In the case of the puma fertilization study, the Florida panther sperm donor was a rare and genetically valuable male that had failed to reproduce naturally in captivity for more than 4 years (Miller *et al.*, 1990). Our findings suggest that IVF may be one approach for propagating these rare animals. Third, it is important to emphasize that the puma and tiger fertilization studies described here were not performed at the investigator's "home" laboratory. Rather, the IVF laboratory (including incubators) was "mobilized" and moved to collaborating institutions where the animals were located. Most investigators are well aware of the sensitivity of gamete culture systems within a given laboratory setting. Frequently minor perturbations can disrupt existing fertilization and culture rates. Therefore, it was gratifying that the domestic cat IVF system could be mobilized effectively to collaborating institutions, thereby allowing the more efficient exploitation of existing expertise and often limited resources. Lastly, these early findings demonstrate the biological competence of nondomestic felid embryos generated by IVF and suggest that this technology might contribute significantly to the actual captive breeding and conservation of rare species.

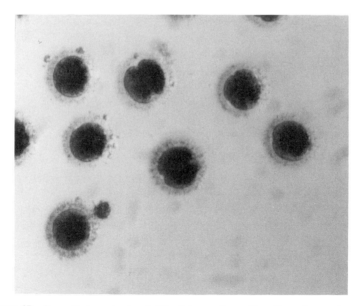

Figure 15. Two-cell stage, *in vitro*-fertilized tiger embryo. (From Donoghue *et al.*, 1990.)

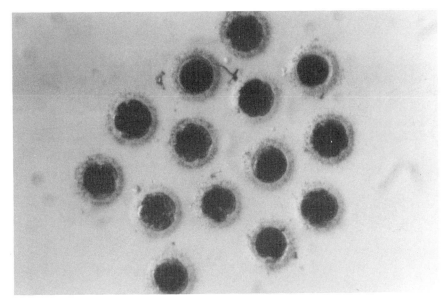

Figure 16.   Morula stage, *in vitro*-fertilized and cultured tiger embryos. (From Donoghue *et al.*, 1990.)

Figure 17.   Tiger cub produced by *in vitro* fertilization and embryo transfer.

In addition to our documented pregnancy in the tiger, Pope *et al.* (1989) recently reported the birth of an Indian desert cat kitten produced by IVF and transferred to a domestic cat surrogate mother. In this study, Indian desert cat oocyte donors were treated with exogenous FSH followed by hCG 48 hr after the last FSH injection. Oocytes, collected by laparoscopic aspiration, were cocultured with electroejaculated sperm in modified Ham's F10 medium. Eight and four embryos were surgically transferred to the uteri of two domestic cats, and one surrogate became pregnant and produced one live young. Therefore, all evidence suggests that IVF offers an exciting, potentially valuable strategy for producing offspring in nondomestic felids. There is little doubt that progress will be accelerated by more fundamental studies of basic gamete biology.

6. CONCLUSIONS

Although relatively few laboratories and investigators study carnivores, our understanding of general reproduction and gamete biology of the domestic cat has increased exponentially during the past decade. Data now are available on the basic characteristics of the cat spermatozoon and follicular oocyte as well as on the fundamental requirements for capacitation, oocyte maturation, fertilization *in vitro*, and preimplantation embryo development. Because most of these studies have been conducted at the level of the entire cell using *in vivo*-matured oocytes, more efforts need to be directed at the molecular and intracellular mechanisms regulating fertilization. Likewise, we still have little true comprehension about fertilization events in the "natural" environment (i.e., the cat oviduct), an area deserving much more attention. Further studies also are warranted because results could have important implications to our understanding of intrafollicular mechanisms for prolonging oocyte longevity in induced ovulators as well as the impact of teratospermia on sperm–ovum interaction. Fundamental information could have practical potential by increasing our ability to propagate important strains of laboratory cats used as experimental models for human disease. Sufficient evidence also is available supporting the assertion that parallel studies among the felids will be useful for comparing fertilization mechanisms, some of which may be unique. Such approaches may offer the first step toward using this strategy for preserving genetic diversity and increasing reproductive rates in rare populations or entire species.

ACKNOWLEDGMENTS. The author is grateful for the dedication of Drs. Karen Goodrowe, JoGayle Howard, Leslie Johnston, and Ann Donoghue, a talented group of former and current postdoctoral fellows/graduate students who were responsible for most of the recent research findings presented in the text. The author also wishes to thank them for their comments on the manuscript as well as the critique provided by Patricia Schmidt. Projects are supported by NIH Grants HD 23853 (to D.E.W.) and RR00045 (to J.G.H.) and grants from Friends of the National Zoo, the Ralston Purina Company, and NOAHS Center (to D.E.W.).

7. REFERENCES

Banks, D. R., and Stabenfeldt, G. H., 1982, Luteinizing hormone release in the cat in response to coitus on consecutive days of estrus, *Biol. Reprod.* **26:**603–611.

Beaver, B. V., 1977, Mating behavior in the cat, *Vet. Clin. North Am.* **16:**729–733.

Biggers, J. D., Gwatkin, R. B. L., and Brinster, R. L., 1962, Development of mouse embryos in organ culture of fallopian tubes on a chemically defined medium, *Nature* **194:**747–749.

Boatman, D. E., and Bavister, B. D., 1982, Cyclic nucleotide mediated stimulation of rhesus monkey sperm fertilizing ability, *J. Cell Biol.* **95:**153.

Bowen, R. A., 1977, Fertilization *in vitro* of feline ova by spermatozoa from the ductus deferens, *Biol. Reprod.* **17:**144–147.

Brackett, B. G., 1970, *In vitro* fertilization of mammalian ova, in: *Schering Symposium on Mechanisms Involved in Conception: Advances in the Biosciences*, Vol. 4 (G. Raspe, ed.), Pergamon Press-Vieweg, New York, p. 73.

Brackett, B. G., 1981, Applications of *in vitro* fertilization, in: *New Technologies in Animal Breeding* (G. E. Seidel, S. Seidel, and B. G. Brackett, eds.), Academic Press, New York, pp. 141–161.

Brackett, B. G., and Oliphant, G., 1975, Capacitation of rabbit spermatozoa *in vitro*, *Biol. Reprod.* **12**:260–274.

Brackett, B. G., Bousquet, D., Boice, M. L., Donawick, W. J., Evans, J. F., and Dressel, M. A., 1982, Normal development following *in vitro* fertilization in the cow, *Biol. Reprod.* **27**:147–158.

Brown, J. L., Wildt, D. E., Phillips, L. G., Seidensticker, J., Fernando, S. B. U., Miththapala, S., and Goodrowe, K. L., 1989, Ejaculate characteristics, and adrenal–pituitary–gonadal interrelationships in captive leopards (*Panthera pardus kotiya*) isolated on the island of Sri Lanka, *J. Reprod. Fertil.* **85**:605–613.

Byers, A. P., Hunter, A. G., Hensleigh, H. C., Kreeger, T. J., Binczik, G., Reindl, N. J., Seal, U. S., and Tilson, R. L., 1987, *In vitro* capacitation of Siberian tiger spermatozoa, *Zoo Biol.* **6**:297–304.

Byers, A. P., Hunter, A. G., Seal, U. S., Binczik, G. A., Graham, E. F., Reindl, N. J., and Tilson, R. L., 1989, *In vitro* induction of capacitation of fresh and frozen spermatozoa of the Siberian tiger (*Panthera tigris*), *J. Reprod. Fertil.* **86**:599–607.

Camous, S., Heyman, Y., Meziou, W., and Menezo, Y., 1984, Cleavage beyond the block stage and survival after transfer of early bovine embryos cultured with trophoblastic vesicles, *J. Reprod. Fertil.* **72**:479–485.

Chang, M. C., 1957, Natural occurrence and artificial induction of parthenogenetic cleavage of ferret ova, *Anat. Rec.* **128**:187–200.

Chang, M. C., 1968, Reciprocal insemination and egg transfer between ferrets and mink, *J. Exp. Zool.* **168**:49–60.

Concannon, P., Hodgson, B., and Lein, D., 1980, Reflex LH release in estrous cats following single and multiple copulations, *Biol. Reprod.* **23**:111–117.

Concannon, P. W., Lein, D. H., and Hodgson, B. G., 1989, Self-limiting reflex luteinizing hormone release and sexual behavior during extended periods of unrestricted copulatory activity in estrous domestic cats, *Biol. Reprod.* **40**:1179–1187.

Davis, D. L., and Day, B. N., 1978, Cleavage and blastocyst formation by pig eggs *in vitro*, *J. Anim. Sci.* **46**:1043–1053.

Dawson, A. B., and Friedgood, H. B., 1940, The time and sequence of preovulatory changes in the cat ovary after mating or mechanical stimulation of the cervix uteri, *Anat. Rec.* **76**:411–429.

Dekel, N., Lawrence, T. S., Gilula, N. B., and Beers, W. H., 1981, Modulation of cell-to-cell communication in the cumulus–oocyte complex and the regulation of oocyte maturation by LH, *Dev. Biol.* **86**:356–362.

Donoghue, A. M., Johnston, L. A., Seal, U. S., Armstrong, D. L., Tilson, R. L., Wolf, P., Petrini, K., Simmons, L. G., Gross, T., and Wildt, D. E., 1990, *In vitro* fertilization and embryo development *in vitro* and *in vivo* in the tiger (*Panthera tigris*), *Biol. Reprod.* **43**:733–747.

Edwards, R. G., 1965, Maturation *in vitro* of mouse, sheep, cow, pig, rhesus monkey and human ovarian oocytes, *Nature* **208**:349–351.

Eppig, J., and Koide, S., 1978, Effects of progesterone and oestradiol-17β on the spontaneous meiotic maturation of mouse oocytes, *J. Reprod. Fertil.* **53**:99–101.

Festing, M. F. W., and Bleby, J., 1970, Breeding performance and growth of SPF cats (*Felis catus*), *J. Small Anim. Pract.* **11**:533–538.

Gandolfi, F., and Moor, R. M., 1987, Stimulation of early development in the sheep by coculture with oviductal epithelial cells, *J. Reprod. Fertil.* **81**:23–28.

Gilula, N. B., Epstein, M. L., and Beers, W. N., 1978, Cell-to-cell communication and ovulation: A study of the cumulus–oocyte complex, *J. Cell Biol.* **78**:58–75.

Glover, T. E., Watson, P. F., and Bonney, R. C., 1985, Observations on variability in LH release and fertility during oestrus in the domestic cat (*Felis catus*), *J. Reprod. Fert.* **75**:145–152.

Goodrowe, K. L., 1987, Consequence of gonadotropin administration on fertilization and early embryonic development in the domestic cat and the *in vitro* fertilization of feline follicular oocytes, Ph.D. Dissertation, Uniformed Services University of the Health Sciences, Bethesda.

Goodrowe, K. L., Miller, A. M., and Wildt, D. E., 1988a, Capacitation of domestic cat spermatozoa as determined by homologous zona pellucida penetration, *Proc. XI Int. Cong. Anim. Reprod. Artif. Insem.* **3**:245–247.

Goodrowe, K. L., Wall, R. J., O'Brien, S. J., Schmidt, P. M., and Wildt, D. E., 1988b, Developmental competence of domestic cat follicular oocytes after fertilization *in vitro*, *Biol. Reprod.* **39**:355–372.

Goodrowe, K. L., Miller, A. M., and Wildt, D. E., 1989a, *In vitro* fertilization of gonadotropin-stimulated leopard cat (*Felis bengalensis*) follicular oocytes, *J. Exp. Zool.* **252**:89–95.

Goodrowe, K. L., Howard, J. G., Schmidt, P. M., and Wildt, D. E., 1989b, Reproductive biology of the domestic cat with special reference to endocrinology, sperm function and *in vitro* fertilization, *J. Reprod. Fertil. Suppl.* **39**:73–90.

Greulich, W. W., 1934, Artificially induced ovulation in the cat (*Felis domestica*), *Anat. Rec.* **58**:217–224.

Guraya, S. S., 1965. A histochemical analysis of lipid yolk deposition in the oocytes of cat and dog, *J. Exp. Zool.* **160**:123–136.

Hamner, C. E., Jennings, L. L., and Sojka, N. J., 1970, Cat (*Felis catus L.*) spermatozoa require capacitation, *J. Reprod. Fertil.* **23**:477–480.

Hardy, W. D., 1984, Feline leukemia virus as an animal retrovirus model for the human T-cell leukemia virus, in: *Human T-cell Leukemia/Lymphoma Virus*, (R. Gallo, M. Essex, and L. Gross, eds.), Cold Spring Harbor Laboratory, New York, pp. 35–43.

Hill, J. P., and Tribe, M., 1924, The early development of the cat (*Felis domestica*), *Q. J. Micr. Sci.* **68**:513–613.

Hilliard, J., Hayward, J. N., and Sawyer, C. H., 1964, Post-coital patterns of secretion of pituitary gonadotropin and ovarian progestin in the rabbit, *Endocrinology* **75**:957–961.

Howard, J. G., and Wildt, D. E., 1990, Ejaculate and hormonal characteristics in the leopard cat (*Felis bengalensis*) and sperm function as measured by *in vitro* penetration of zona-free hamster ova and zona-intact domestic cat oocytes, *Mol. Reprod. Dev.* **26**:163–174.

Howard, J. G., Bush, M., Hall, L. L., and Wildt, D. E., 1984, Morphological abnormalities in spermatozoa of 28 species of nondomestic felids, *Proc. X Int. Cong. Anim. Reprod. Artif. Insem.* **2**:57–59.

Howard, J. G., Brown, J. L., Bush, M., and Wildt, D. E., 1990, Teratospermic and normospermic domestic cats: Ejaculate traits, pituitary–gonadal hormones and improvement of spermatozoal viability and morphology after swim-up processing, *J. Androl.* **11**:204–215.

Howard, J. G., Bush, M., and Wildt, D. E., 1991, Teratospermic in domestic cats compromises penetration of zona-free hamster ova and cat zona pellucidae, *J. Androl.* **12**:36–45.

Jemmett, J. E., and Evans, J. M., 1977, A survey of sexual behavior and reproduction in female cats, *J. Small Anim. Pract.* **18**:31–37.

Johnson, L. M., and Gay, V. L., 1981, Luteinizing hormone in the cat. II. Mating-induced secretion, *Endocrinology* **109**:247–252.

Johnston, L. A., O'Brien, S. J., and Wildt, D. E., 1989, *In vitro* maturation and fertilization of domestic cat follicular oocytes, *Gamete Res.* **24**:343–356.

Johnston, L. A., Donoghue, A. M., O'Brien, S. J., and Wildt, D. E., 1991, Culture medium and protein supplementation influence *in vitro* fertilization and embryo development in the domestic cat, *J. Exp. Zool.* **257**:350–359.

Johnston, S. D., Osborn, C. A., and Lipowitz, A. J., 1988, Characterization of seminal plasma, prostatic fluid and bulbourethral gland secretions in the domestic cat, *Proc. 11th Int. Cong. Anim. Reprod. Artif. Insem.* **4**: 560–562.

Mahadevan, M. M., and Trounson, A. O., 1984. The influence of seminal characteristics on the success rate of human *in vitro* fertilization, *Fertil. Steril.* **42**:400–405.

Mahi, C. A., and Yanagimachi, R., 1976, Maturation and sperm penetration of canine ovarian oocytes *in vitro*, *J. Exp. Zool.* **196**:189–196.

Mahi, C. A., and Yanagimachi, R., 1978, Capacitation, acrosome reaction and egg penetration by canine spermatozoa in a simple defined medium, *Gamete Res.* **1**:101–109.

McGaughey, R. W., 1977, The culture of pig oocytes in minimal medium, and the influence of progesterone and estradiol-17β on meiotic maturation, *Endocrinology* **100**:39–45.

Miller, A. M., Johnston, L. A. Hurlbut, S. L., and Wildt, D. E., 1989, Intrafollicular domestic cat oocytes and later corpus luteum function are sensitive to the timing of gonadotropin stimulation, *Proc. Soc. Stud. Reprod. Biol. Reprod. Suppl.* **1**:40:96A.

Miller, A. M., Roelke, M. E., Goodrowe, K. L., Howard, J. G., and Wildt, D. E., 1990, Oocyte recovery, maturation and fertilization *in vitro* in the puma (*Felis concolor*), *J. Reprod. Fertil.* **88**:249–258.

Moor, R. M., Osborn, J. C., Cran, D. G., Walters, D. E., 1981, Selective effect of gonadotrophins on cell coupling, nuclear maturation and protein synthesis in mammalian oocytes, *J. Embryol. Exp. Morphol.* **61**:347–365.

Motlik, J., Fulka, J., and Flechon, J. E., 1986, Changes in intercellular coupling between pig oocytes and cumulus cells during maturation *in vivo* and *in vitro*, *J. Reprod. Fertil.* **76**:31–37.

Nestor, A., and Handel, M. A., 1984, The transport of morphologically abnormal sperm in the female reproductive tract of mice, *Gamete Res.* **10**:119–125.

Niwa, K., Ohara, K., Hosoi, Y., and Iritani, A., 1985, Early events of *in vitro* fertilization of cat eggs by epididymal spermatozoa, *J. Reprod. Fertil.* **74**:657–660.

O'Brien, S. J., 1986, Molecular genetics in the domestic cat and its relatives, *Trends Genet.* **2**:137–142.

Olsen, R. J., 1981, *Feline Leukemia* CRC Press, Boca Raton, FL.

Paape, S. R., Shille, V. M., Seto, H., and Stabenfeldt, G. H., 1975, Luteal activity in the pseudopregnant cat, *Biol. Reprod.* **13**:470–474.

Platz, C. C., and Seager, S. W. J., 1978, Semen collection by electroejaculation in the domestic cat, *J. Am. Vet. Med. Assoc.* **173**:1353–1355.

Platz, C. C., Wildt, D. E., and Seager, S. W. J., 1978, Pregnancy in the domestic cat after artificial insemination with previously frozen spermatozoa, *J. Reprod. Fertil.* **52**:279–282.

Pope, C. E., Glewicks, E. J., Wachs, K. B., Keller, G. L., Maruska, E. J., and Dresser, B. L., 1989, Successful interspecies transfer of embryos from the Indian desert cat (*Felis silvestris ornata*) to the domestic cat (*Felis catus*) following *in vitro* fertilization, *Proc. Soc. Stud. Reprod. Biol. Reprod. Suppl* **1:40**:61A.

Prescott, C. W., 1973, Reproduction patterns in the domestic cat, *Aust. Vet. J.* **49**:126–129.

Prins, G. S., Wagner, C., Weidel, L., Gianfortoni, J., Marut, E. L., and Scommegna, A., 1987, Gonadotropins augment maturation and fertilization of human immature oocytes cultured *in vitro*, *Fertil. Steril.* **47**:1035–1037.

Robison, B. L., and Sawyer, C. H., 1987, Hypothalamic control of ovulation and behavioral estrus in the cat, *Brain Res.* **418**:41–51.

Saacke, R. G., Bame, J. H., Karabinus, D. S., Mullins, K. J., and Whitman, S., 1988, Transport of abnormal spermatozoa in the artificially inseminated cow based upon accessory spermatozoa in the zona pellucida, *Proc. XI Int. Cong. Anim. Reprod. Artif. Insem.* **3**:292–294.

Schini, S. A., and Bavister, B. D., 1988, Two-cell block to development of cultured hamster embryos is caused by phosphate and glucose, *Biol. Reprod.* **40**:607–614.

Schmidt, P. M., 1986, Feline breeding management, *Vet. Clin. North Am.* **16**:435–452.

Schmidt, P. M., Chakraborty, P. K., and Wildt, D. E., 1983, Ovarian activity, circulating hormones and sexual behavior in the cat. II. Relationships during pregnancy, parturition, lactation and the postpartum estrus, *Biol. Reprod.* **28**:657–671.

Scott, M. A., and Scott, P. P., 1957, Post-natal development of the testis and epididymis in the cat, *J. Physiol. (Lond.)* **136**:40–41.

Shille, V. M., Lundstrom, K. E., and Stabenfeldt, G. H., 1979, Follicular function in the domestic cat as determined by estradiol-17β concentrations in plasma: Relation to estrous behavior and cornification of exfoliated vaginal epithelium, *Biol. Reprod.* **21**:953–963.

Shille, V. M., Munro, C., Farmer, S. W., Papkoff, H., and Stabenfeldt, G. H., 1983, Ovarian and endocrine responses in the cat after coitus, *J. Reprod. Fertil.* **68**:29–39.

Sojka, N. J., Jennings, L. L., and Hamner, C. E., 1970, Artificial insemination in the cat (*Felis catus L.*), *Lab. Anim. Care* **20**:198–204.

Van der Stricht, R., 1911, Vitellogenese dans l'ovule de chatte, *Arch. Biol. T.* **26**:365–481.

Verhage, H. G., Beamer, N. B., and Brenner, R. M., 1976, Plasma levels of estradiol and progesterone in the cat during polyestrus, pregnancy and pseudopregnancy, *Biol. Reprod.* **14**:579–585.

Whittingham, D. G., 1971, Survival of mouse embryos after freezing and thawing, *Nature* **233**:592–593.

Wildt, D. E., 1989, Reproductive research in conservation biology: Priorities and avenues for support, *J. Zoo Wildl. Med.* **20**:391–395.

Wildt, D. E., 1990, Potential applications of IVF technology for species conservation, in: *Fertilization in Mammals* (B. D. Bavister, J. Cummins, and R. S. Roldan, eds.), Serono Symposium, Norwell, pp. 349–364.

Wildt, D. E., and Seager, S. W. J., 1980, Laparoscopic determination of ovarian and uterine morphology during the reproductive cycle (of the cat), in: *Current Therapy in Theriogenology* (D. A. Morrow, ed.), W. B. Saunders, Philadelphia, pp. 828–832.

Wildt, D. E., Guthrie, S. C., and Seager, S. W. J., 1978, Ovarian and behavioral cyclicity of the laboratory maintained cat, *Horm. Behav.* **10**:251–257.

Wildt, D. E., Seager, S. W. J., and Chakraborty, P. K., 1980, Effect of copulatory stimuli on incidence of ovulation and on serum luteinizing hormone in the cat, *Endocrinology* **107**:1212–1217.

Wildt, D. E., Chan, S. Y. W., Seager, S. J., and Chakraborty, P. K., 1981, Ovarian activity, circulating hormones and sexual behavior in the cat. I. Relationships during the coitus-induced luteal phase and the estrous period without mating, *Biol. Reprod.* **25**:15–28.

Wildt, D. E., Bush, M., Howard, J. G., O'Brien, S. J., Meltzer, D., van Dyk, A., Ebedes, H., and Brand, D. J., 1983, Unique seminal quality in the South African cheetah and a comparative evaluation in the domestic cat, *Biol. Reprod.* **29**:1019–1025.

Wildt, D. E., Howard, J. G., Hall, L. L., and Bush, M., 1986a, The reproductive physiology of the clouded leopard. I. Electroejaculates contain high proportions of pleiomorphic spermatozoa throughout the year, *Biol. Reprod.* **34**:937–947.

Wildt, D. E., Schiewe, M. C., Schmidt, P. M., Goodrowe, K. L., Howard, J. G., Phillips, L. G., O'Brien, S. J., and Bush, M., 1986b, Developing animal model systems for embryo technologies in rare and endangered wildlife species, *Theriogenology* **25:**33–51.

Wildt, D. E., Bush, M., Goodrowe, K. L., Packer, C., Pusey, A. C., Brown, J. L., Joslin, P., and O'Brien, S. J., 1987a, Reproductive and genetic consequences of founding isolated lion populations, *Nature* **329:**328–331.

Wildt, D. E., O'Brien, S. J., Howard, J. G., Caro, T. M., Roelke, M. E., Brown, J. L., and Bush, M., 1987b, Similarity in ejaculate–endocrine characteristics in captive versus free-ranging cheetahs of two subspecies, *Biol. Reprod.* **36:**351–360.

Wildt, D. E., Phillips, L. G., Simmons, L. G., Chakraborty, P. K., Brown, J. L., Howard, J. G., Teare, A., and Bush, M., 1988. A comparative analysis of ejaculate and hormonal characteristics of the captive male cheetah, tiger, leopard and puma, *Biol. Reprod.* **38:**245–255.

Yanagimachi, R., 1972, Penetration of guinea-pig spermatozoa into hamster eggs *in vitro*, *J. Reprod. Fertil.* **28:**477–480.

Yanagimachi, R., and Chang, M. C., 1964, *In vitro* fertilization of hamster ova, *J. Exp. Zool.* **156:**361–376.

17

Fertilization in the Pig and Horse

R. H. F. Hunter

1. INTRODUCTION

This review attempts to give a balanced description of the overall process of fertilization in two species of placental mammal. Although a strictly limited view of fertilization might simply focus on the steps of sperm penetration and activation of the oocyte and restoration of the diploid condition, this essay also considers the preliminaries to a successful meeting of the gametes together with the consequences—in terms of successful embryonic development or otherwise. Because the two species under discussion, pig and horse, have a long preovulatory interval during which the female will stand for mating, it becomes essential to consider the events of sperm transport and storage—and likewise of ovulation and egg transport—before examining interactions between the gametes themselves. The prolonged preovulatory interval also enables analysis of gonadal control of these transport and storage events in the female genital tract and the consequent process of capacitation.

In the situation after activation of the secondary oocyte, understanding the first stages of embryonic development enables the process of fertilization to be placed in a broader perspective and similarly enables description of its anomalies. The extent to which errors of fertilization contribute to early embryonic loss cannot be predicted for individual animals, but it remains a fact that the overall incidence of prenatal loss in a population of placental mammals may reach 30–40% or more. Among the domestic farm animals, such a loss assumes considerable economic significance. Once again, therefore, an understanding of critical events before, during, and shortly after fertilization has a high priority.

Reviews of fertilization in mammals have been relatively frequent and extensive. The earlier classical ones include those of Thibault (1949), Chang and Pincus (1951), Austin and Walton (1960), Austin (1961), Hancock (1962), Bedford (1970), and Polge (1978). More recent endeavors include those by Yanagimachi (1981, 1988), Bedford (1982), Harper (1982), Hartmann (1983), Fraser (1984), Austin (1985), Bavister (1987), and Hunter (1988). Those focusing on events in pigs

R. H. F. HUNTER • Center for Research on Animal Reproduction, Faculty of Veterinary Medicine, University of Montreal, Saint-Hyacinthe, Quebec J2S 7C6, Canada. *Present address*: University of Edinburgh, Edinburgh EH16 5NT, Scotland, United Kingdom.

A Comparative Overview of Mammalian Fertilization, edited by Bonnie S. Dunbar and Michael G. O'Rand. Plenum Press, New York, 1991.

include the works of Thibault (1959), Hancock (1961), Alanko (1973), Baker and Polge (1976), Einarsson (1980), and Hunter (1982, 1990). In marked contrast, there is perhaps still insufficient detailed information on fertilization in horses to justify a full review. Individual papers on equine gametes are, of course, discussed below.

2. PIG

2.1. Preliminaries to Fertilization in Pigs

The duration of female receptivity to mating (estrus) is usually some 2–2½ days, exceptionally 3 days, in the improved Western breeds of domestic pig. During this period, animals will stand for repeated mating, although there is sound evidence that a single mating by a competent, mature boar at the onset of estrus will ensure fertilization in most animals (see below). Clearly, the time relationships between deposition of a sperm suspension and release of the eggs underlie such a statement. This review of pertinent processes commences with insemination, followed by details of ovulation and egg transport, and then considers the events of sperm release, capacitation, oocyte activation, and pronuclear development, together with subsequent processes.

2.2. Semen Deposition and Uterine Transit

The specific stimulus provoking ejaculation in the boar is the interlocking of the glans penis with the cervical ridges of an estrous gilt or sow. Because of the influence of ovarian steroid hormones, these muscular ridges become taut and edematous at estrus (Burger, 1952; Smith and Nalbandov, 1958), and their arrangement is such that the lumen of the cervix tapers towards the internal os, enabling the glans penis to be gripped by the female tissues. During thrusting of the penis through the anterior vagina and into the cervical canal, there is also a twisting component to the penile shaft, causing the spiral folds of the glans to interdigitate tightly with the cervical wall. The pressure derived from this "lock" leads to ejaculation, a situation that is mimicked during manual collection of boar semen.

As a result of these anatomic specializations, and because of the large volume of the boar ejaculate (150–500 ml), semen makes only passing contact with the innermost portion of the cervical canal before entering the body of the uterus. Deposition is therefore effectively intrauterine, and in the context of sperm transport to the site of fertilization, little further consideration needs to be given to the structure of the cervix. An initial distribution of semen in the uterus may be achieved as a result of the force of ejaculation and the volume of fluid involved, assuming that little or no leakage occurs into the vagina. However, contractions of the myometrium are vigorous during estrus (Corner, 1923; Keye, 1923), and these should assist transport of the male secretions and, in the preovulatory interval, redistribution of fluid between the two horns. Reflex release of oxytocin would enhance uterine contractions (see Knifton, 1962), and a stimulatory role of the semen itself should also be considered. Although very low concentrations of prostaglandin E_2 and $F_{2\alpha}$ were reported in boar semen (Hunter, 1973a), an increased concentration of prostaglandin $F_{2\alpha}$ was found in the uterine venous blood of a gilt sampled 15 min after mating, when the uterus was still full of semen (Hunter, 1982). This result was not endorsed in six other animals sampled 35–60 min after mating, but the notion that semen may activate a local release of smooth muscle stimulants should be examined further.

During the protracted period of coitus (e.g., commonly 5 to exceptionally 15 min), semen is not emitted as a homogeneous fluid. If the ejaculate is collected as a series of fractions, these will consist first of watery and gel presperm secretions, then a sperm-rich fraction, followed by a postsperm and gelatinous fraction (Rodolfo, 1934a; McKenzie *et al.*, 1938). Such a sequence may take 3–5 min and can be considered as one full wave of ejaculation. After renewed thrusting and reestablishment of the cervical lock, a second and indeed a third wave of ejaculation may follow. Whether the fractions have a special significance in the transport of spermatozoa is uncertain, but they may favor displacement of the sperm-rich portion to the region of the uterotubal junctions (McKenzie *et al.*, 1938; Du Mesnil du Buisson and Dauzier, 1955). In any event, under conditions of natural mating performed in the preovulatory interval, the uterine horns should be fully distended with semen by the time mating is finished, a situation that has been verified by laparotomy and at autopsy. The gelatinous material of the ejaculate acts as a cervical plug for a variable period after mating, thereby preventing immediate leakage of the uterine contents.

The total number of spermatozoa deposited may vary from 4 to 8 $\times$ 10^{10} cells, with the concentration of spermatozoa in the whole ejaculate being 1–3 $\times$ 10^8 cells/ml (Rodolfo, 1934b; McKenzie *et al.*, 1938; Polge, 1956). A concentration of this order, or even higher if there is preferential transport of the sperm-rich fraction, will bathe the uterotubal junction by the completion of mating with a mature boar; uterine transport of semen should not therefore be a physiological problem. Accordingly, discussion of the process of sperm transport after mating needs to focus on the role of the tract between the uterotubal junction and the site of fertilization.

2.3. Ovulation and Egg Transport

The duration of the estrous cycle in the domestic pig is approximately 21 days, of which 6 days represent the follicular phase. During this period, Graafian follicles increase in mean diameter from 4–5 mm on day 15 to the preovulatory diameter of 9–11 mm on day 21 (Corner, 1921; McKenzie, 1926; Burger, 1952). The actual number of follicles achieving this final diameter in any one cycle is reduced to an ovulation rate characteristic of the breed or strain, the number of mature follicles being well defined by the onset of estrus (commonly 12–20). A selection of follicles therefore occurs in this latter phase of the cycle, although evidence from experiments involving superovulation (Hunter, 1964) suggests that this may result solely from competition for circulating gonadotrophic hormones.

Ovulation occurs during the period of estrus and is precipitated by a specific surge of gonadotrophic hormone(s). The preovulatory peak of luteinizing hormone may reach serum concentrations of 4–5 ng/ml (Niswender *et al.*, 1970), and ovulation is noted 40–48 hr after this peak (Liptrap and Raeside, 1966) or 40–42 hr after injection of an LH-rich preparation given shortly before estrus to simulate the endogenous release of gonadotrophin (Dziuk *et al.*, 1964; Hunter, 1967a; Polge, 1967).

Primary oocytes within the population of preovulatory follicles also respond to the gonadotrophin surge and progress through a well-characterized series of nuclear changes. Resumption of meiosis has been examined under conditions of spontaneous estrus (Spalding *et al.*, 1955) or following injection of hCG (Hunter and Polge, 1966). The time scale of response is closely similar in both situations, with diakinesis at 17–18 hr, arrangement of chromosomes on the first metaphase spindle at 25–26 hr, and extrusion of the first polar body at 35–37 hr, that is, shortly before ovulation. Cytoplasmic changes in such oocytes have yet to be documented in full but may concern principally the spectrum and quantity of proteins.

Conspicuous modification of the granulosa cell investment of each oocyte occurs shortly before release into the reproductive tract. Modifications include expansion and mucification (Fléchon *et al.*, 1986) and doubtless facilitate (1) transport from the ovarian surface into the ostium of the oviduct and onward to the site of fertilization and (2) subsequent sperm-egg interactions leading to penetration. Although each ovary usually releases a group of eggs (each preovulatory follicle sheds only a single oocyte), individual oocytes become aggregated within a cumulus mass or egg plug (Hancock, 1961). Transport of this aggregate from the fimbria to the site of fertilization—a distance commonly of some 18–20 cm—is rapid: eggs arrive at the ampullary–isthmic junction of the oviduct within 30–45 min of ovulation, still invested with cells of the cumulus oophorus (Hunter, 1974). Transport is accomplished largely by segmental waves of peristaltic contraction together with a synchronized and direction-orientated beat of cilia. A positive role for fluid flow in these transport processes seems improbable, since the direction of bulk flow at this time is towards the peritoneal cavity.

2.4. Sperm Physiology in the Oviducts

Because a dense suspension of spermatozoa should already be bathing the uterotubal junction by the completion of mating, it is not surprising that some viable sperm cells will already have entered the oviducts within 15 min (Hunter and Hall, 1974a). In fact, in a majority of animals mated at the onset of estrus, sufficient spermatozoa have entered the oviducts within 1 hr subsequently to fertilize nearly all eggs in nearly all animals (Hunter, 1981). Passage from the uterus to the oviducts is achieved primarily by active sperm motility. However, once within the lumen of the oviducts, the preovulatory distribution of spermatozoa in animals mated at the onset of estrus is remarkable. Instead of progressing onward along the 8 cm or so towards the site of fertilization at the ampullary–isthmic junction, the population of fertilizing spermatozoa—the so-called functional population—is arrested in the caudal 1–2 cm of the oviduct isthmus and only activated and released shortly before ovulation (Hunter, 1981, 1984).

At this stage, a few comments are justified to put changes in sperm motility into a meaningful perspective. Spermatozoa in the male tract are essentially immotile until their admixture with the accessory secretions at the moment of ejaculation, whereas sperm sampled from the uterus soon after mating show excellent progressive motility. By contrast, sperm sampled from the caudal portion of the oviduct isthmus in the prolonged preovulatory interval (~36–38 hr in pigs) show a conspicuously reduced or negligible motility within the highly viscous mucous secretions of this portion of the tract. Thus, it is appropriate to use the term "arrested" in the preovulatory interval, a change in motility first described for rabbit spermatozoa (Overstreet and Cooper, 1975) and subsequently for ram spermatozoa (Cummins, 1982). The full nature of the process of arrest has yet to be clarified, but, apart from a physical influence of the viscous secretions, K^+ ions are thought to inhibit and pyruvate to stimulate sperm motility in the isthmus (Burkman et al., 1984). Temperature gradients within the oviduct may also play a role, for the caudal isthmus of the pig oviduct is cooler in the preovulatory interval than the ampulla, although the temperature gradient disappears close to the time of ovulation (Hunter and Nichol, 1986). To what extent small changes in temperature have an influence on sperm structural proteins is uncertain, but the potential conformation of the sperm flagellum could be affected directly.

These preovulatory changes in sperm motility bear on the topic of capacitation. Whereas this final maturational change in the sperm cell within the female tract has traditionally been considered to require a fixed number of hours, more recent interpretations suggest that completion of capacitation is closely related to the time of ovulation (Hunter, 1987, 1988, 1989a, 1990). Indeed, since capacitated spermatozoa are hyperactive, fragile, and short-lived, and because the preovulatory interval may be some 38–40 hr in pigs, it is essential that completion of capacitation is not achieved until release of the secondary oocytes into the oviducts is imminent. Specific observations supporting this proposition concern the acrosome reaction and hyperactivation, changes in the sperm cell considered to be consequent on capacitation. Both of these changes in the spermatozoa of domestic farm animals are seen to be periovulatory events (Hunter, 1987, 1988). A model indicating the manner whereby ovarian follicular secretions achieve control over sperm physiology has been presented elsewhere (Hunter, 1987, 1989a, 1990). In essence, the oviduct epithelium transduces endocrine information from the preovulatory Graafian follicles transmitted by a local (countercurrent) vascular route (Fig. 1).

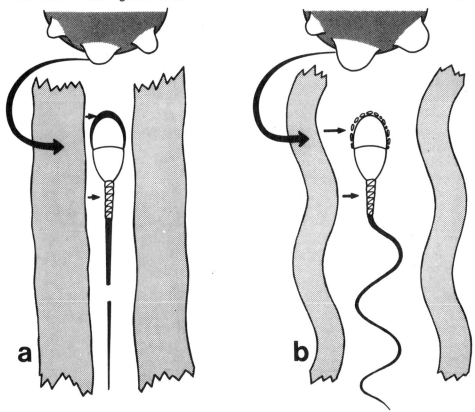

Figure 1.   Model to illustrate the manner whereby the endocrine activity of pre- or periovulatory Graafian follicles acts locally to program the membrane configuration and motility of spermatozoa in the lumen of the oviduct isthmus. Gonadal hormones from mature follicles act on the oviduct epithelium, the transudates and secretions of which, in turn, influence the composition of the luminal fluids. Expression of capacitation is reasoned to be a periovulatory event, at least in the large farm species such as pigs with a protracted interval between the gonadotrophin surge and ovulation (Hunter, 1987). (a) Intact, relatively quiescent spermatozoon under the overall influence of preovulatory follicle secretions. Membrane vesiculation on the anterior part of the sperm head is suppressed, as is the development of whiplash activity in the flagellum, presumably by local molecular control mechanisms. The lumen of the distal oviduct isthmus is extremely narrow and contains viscous secretions, and myosalpingeal contractions are reduced. (b) An acrosome-reacted, hyperactive spermatozoon under the influence of Graafian follicles on the point of ovulation. The patency of the oviduct isthmus has commenced to increase, enabling expression of the whiplash pattern of flagellar beat. Progression of such spermatozoa to the site of fertilization is also aided by enhanced contractile activity of the myosalpinx. Adapted from Hunter (1990).

2.5. Activation of Secondary Oocytes

As noted above, eggs pass relatively rapidly (30–45 min) to the site of fertilization at the ampullary–isthmic junction. Although they are aggregated within a cumulus plug, it is certainly worth questioning whether modifications to the egg surface occur during this phase of transport. As discussed elsewhere (Hunter, 1988), there are significant changes in the fluid milieu between the antrum of Graafian follicles and the lumen of the oviducts. There are also significant changes in the fluid composition of the oviduct lumen between the fimbriated end of the ampulla and the site of fertilization (unpublished observations).

In situations of preovulatory mating, the group of eggs in each oviduct is penetrated relatively synchronously, that is, within 2–3 hr (Hunter, 1972a). As yet, there is no specific evidence that chemotaxis between eggs and spermatozoa is involved. Key stages for spermatozoa preceding actual activation of the oocytes concern penetration through the cumulus investment, confrontation with the highly uneven surface of the zona pellucida, binding to the zona pellucida, and then penetration of the zona substance (see Dunbar, 1983). Despite controversy as to whether penetration is facilitated primarily by acrosomal enzymes or incisive cutting by the rigid head of a highly motile sperm, it is prudent to realize that both of these components of sperm activity are probably involved in this step. Once the spermatozoon is within the fluid-filled perivitelline space, a functional contact is made between the sperm head and egg plasmalemma. At the moment of such contact, the sperm flagellum is still traversing the zona pellucida, with the greater proportion of it remaining to the exterior (see Hunter and Dziuk, 1968). Progressive incorporation of the sperm head by the microvilli of the egg surface leads rapidly to loss of flagellar motion. Before this stage is reached, however, the still-active sperm appears to promote pulsating movements of the vitellus and also to cause gradual rotation of the vitellus within the zona, drawing the sperm tail into the perivitelline space. Detailed observations on these aspects were reported by Hunter and Dziuk (1968).

In eutherian mammals, the first functional contact of the sperm head with the egg plasma membrane is generally thought to be in the region of the postacrosomal cap (Barros et al., 1967). But, in a large series of phase-contrast observations on boar spermatozoa at this vital stage, incorporation was noted consistently to be initiated in the region of the equatorial segment (Fig. 2). Thereafter, contact of the egg microvilli with the surface of the sperm head proceeds in a posterior direction.

Three components of egg activation are known classically as the following: (1) resumption of the second meiotic division from the metaphase resting condition, culminating in extrusion of the second polar body; (2) exocytosis of the contents of the cortical granules into the perivitelline space, causing the establishment of a block to polyspermy in the innermost portion of the zona pellucida; and (3) contraction of the vitellus as a result of changes in the cytoskeleton of the egg associated with modification of structural proteins such as actin and tubulin. Such contraction leads to a conspicuous enlargement of the fluid-filled perivitelline space.

Precisely which component(s) of the boar sperm head or its membranes trigger the events of egg activation has yet to be discovered. It is nonetheless worth noting that a variety of artificial stimuli (cold shock, anesthetics, electrical fields) can initiate activation responses in mammalian eggs and indeed some degree of parthenogenetic development. Because there may be a dissociation of the various components of activation under such artificial conditions, there is a strong suggestion that activation by the fertilizing spermatozoon is a multimolecular event.

2.6. Block to Polyspermy

The general nature and function of the block to polyspermy have been known at least since the paper of Braden et al. (1954). Even so, because polyspermic penetration is a surprisingly frequent observation in experimental studies of pig eggs (Fig. 3), it is worth emphasizing aspects that are not yet understood. These include:

1. Measurements on the rate of establishment of a block to polyspermy, that is, from the time of a first functional contact between the activating sperm and completion of the block. Although there is strong circumstantial evidence that development of the block is progressive over the surface of the sphere from the point of sperm contact, the actual rate of its formation remains unknown. Evidence derived from studies on polyspermic penetration suggests a period measured in minutes rather than seconds.

2. The precise nature and role of the contents of cortical granules. Although there is some suggestion that the contents of the cortical granules act by incorporation into the substance of the

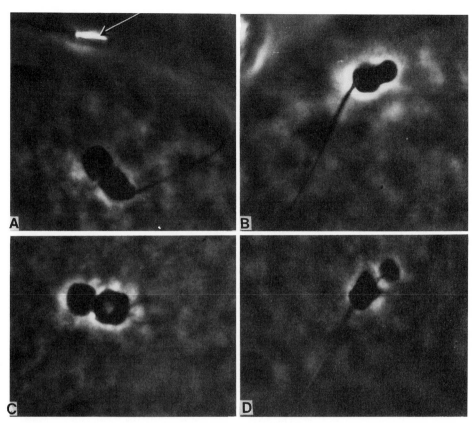

Figure 2. Examples of sperm–egg interactions in specimens recovered from the oviducts of pigs shortly after ovulation. The specimens had been prepared as whole-mounts, fixed, stained, and examined by phase-contrast microscopy. The head of each spermatozoon shows evidence of initial incorporation in the equatorial region by the microvilli of the egg plasmalemma (vitelline surface), a phase that is thought to be of short duration. Although the relative proportions of the anterior and posterior parts of the sperm head appear little changed, the overall dimensions have increased when compared with those of the spermatozoon (arrowed) that was in the zona pellucida. Adapted from Hunter (1976).

zona pellucida and there inhibit penetration by subsequent spermatozoa, the molecular events remain unclear for porcine gametes. Inhibition of acrosomal enzymes in the manner suggested needs to take account of the fact that accessory spermatozoa continue to penetrate the outer portions of the zona (Thibault, 1959), often in very large numbers (up to 400). Therefore, material from the cortical granules must be localized and/or initiate modifications in some quite specific way in the innermost portion of the zona. Structural changes in the zonae of activated pig eggs have yet to be demonstrated by electron microscopy. The use of sensitive staining techniques will probably be required to demonstrate concentric domains within the zona pellucida.

3. Whether there is only a zona block against polyspermic penetration. Recent observations suggest that there may also be a defense mechanism at the vitelline surface (Hunter and Nichol, 1988). However, such a secondary block to polyspermy is perhaps only revealed under conditions of excessive sperm numbers at the site of fertilization.

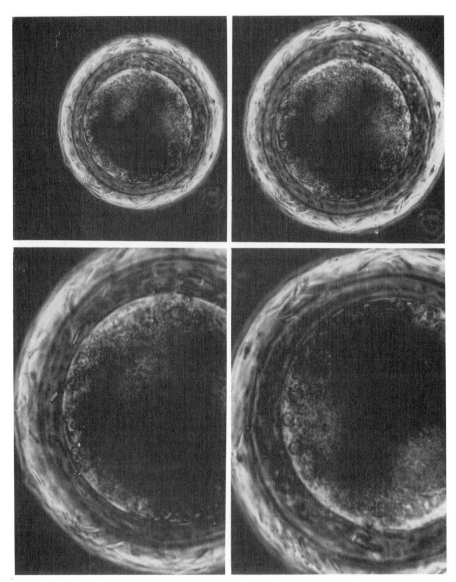

Figure 3. Examples of polyspermic penetration in freshly mounted pig eggs. Although the block to polyspermy is located principally in the zona pellucida of pig eggs, note the presence of several spermatozoa in the perivitelline space in each of these instances. On staining and examination by phase-contrast microscopy, each vitellus was also noted to have undergone polyspermic penetration.

Once fully established, and whatever its precise nature, there is sound evidence that the zona block is stable and long-lasting. The primary evidence is that spermatozoa continue to penetrate into the zona pellucida of zygotes in the oviduct for 40–48 hr, this reflecting in part a continual passage of viable spermatozoa from reserves in the caudal portion of the isthmus. There is also the experimental evidence that surgical insemination with a further suspension of spermatozoa in mated animals at a time when fertilization would be expected to be complete does not disturb the ploidy of the embryos or lead to an accumulation of accessory spermatozoa in the perivitelline space (unpublished observations).

2.7. Chronology of Fertilization and Early Development

The time scale of formation, migration, and apposition of the pronuclei has been reported in detail (Hunter, 1972a), as have events of the first and second mitotic divisions (Fig. 4; Hunter, 1974). Developmental steps are summarized in Table 1, and described in the ultrastructural study by Szollosi and Hunter (1973). The chronology is based on the normal physiological situation of a preovulatory mating. Observations from quite limited experiments on *in vitro* fertilization indicate that cleavage is frequently retarded under conditions of culture and that there may be development of a "block" at the four-cell stage. Such a "block" is, of course, a reflection of embryos being refractory (or extrasensitive) to conditions of culture *in vitro*. Although the four-cell stage of development is relatively long in pig embryos, there is no indication of specific developmental hurdles to be overcome at this stage or of an enhanced incidence of early embryonic loss specific to this stage. Nonetheless, the four-cell stage is critical in the sense that it is the stage when the embryonic genome is first becoming active.

2.8. Postovulatory Aging

Detailed information is available on the influence of postovulatory aging of pig eggs on subsequent sperm–egg interactions (Thibault, 1959; Hancock, 1961; Dziuk and Polge, 1962). Such information has been sought (1) because females of this species may remain receptive to

Table 1. Chronology of Development of the Pig Embryo during the First 6 Days[a]

Stage of embryonic development	Approximate interval from activation[b] until developmental stage (hours)	Comment
Appearance of pronuclei	1.5–2	Rapid formation of pronuclear membranes
Apposition of pronuclei	5–6	Phase of relatively long duration
First mitotic metaphase	12–13	Restoration of diploid condition
Two cells	14–16	Two-celled stage of short duration (6–8 hrs)
Three to four cells	20–24	Relatively long four celled stage (20–24 hrs)
Five to eight cells	46–52	Asynchronous cleavage becomes prominent
Nine to sixteen cells	64–[c]	Range in development found within animals
>16 cells (morulae)	70–[c]	Mitotic figures frequently visible
>32 cells (intact blastocysts)	92–[c]	Blastocoele and inner cell mass become conspicuous
Zona-free blastocysts	120–	>200 cells may have formed by hatching

[a]The figures represent the calculated time intervals from sperm penetration of the oocyte until appearance of a given stage (Hunter, 1974).
[b]Activation indicating morphological evidence of resumption of the second meiotic division.
[c]Difficult to determine because of range in cell numbers and the problem of counting more than 16 cells with accuracy.

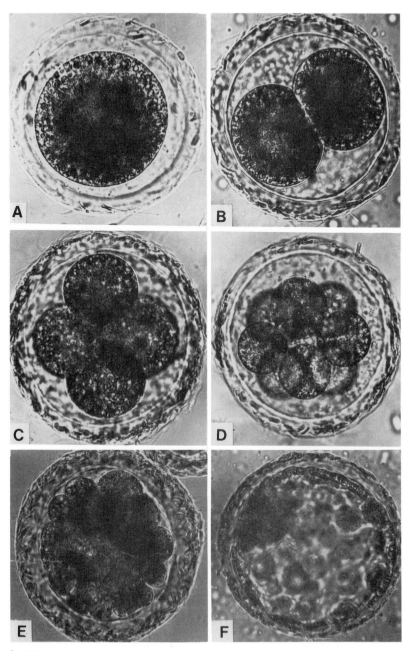

Figure 4.   Whole-mount preparations of freshly-recovered pig embryos to illustrate developmental stages from pronucleate single cell up to that of a blastocyst. Note the progressively reduced size of individual blastomeres from the two- to the eight-celled stage (B–C) and the increasing number of spermatozoa embedded in the zona pellucida of each embryo as development advances. Note also that the blastocyst shows a distinct inner cell mass and a relative thinning of the zona pellucida as a result of stretching. After Hunter (1974).

mating for 12–18 hr or more after ovulation and (2) because under conditions of artificial insemination, there is especially a risk of introducing the sperm suspension too late vis à vis ovulation.

Although the general problems of delayed mating or insemination were outlined by both Thibault (1959) and Hancock (1961), notably the reduced incidence of fertilization and the various abnormalities such as digyny (Thibault, 1959) and polyspermy (Hancock, 1959), precise control of ovulation time was required to enable anomalies of fertilization to be related to the postovulatory age of the egg. In a detailed study involving HCG-regulated ovulation and at least 150 eggs in each treatment group, aging of pig eggs was revealed within 6–8 hr of ovulation. Eggs penetrated by sperm after these intervals showed an increased risk of polyspermic penetration and subsequent fragmentation (Table 2; Hunter, 1967a). Nuclear and cytoplasmic disorders were extensive. At the electron microscopic level, disorganization of the arrangement of cortical granules and microtubules of the meiotic spindle would have been anticipated. Loss of potential sperm-binding sites on the zona pellucida would have occurred with a sufficient degree of aging, preventing the possibility of penetration.

2.9. Polyspermy in Pig Eggs

Polyspermic penetration of pig eggs has been observed under a variety of experimental conditions (Table 3). In addition to that of postovulatory aging just described, a much elevated incidence of polyspermy has been noted after (1) treatment of animals with progestagens for synchronization of estrous cycles (Dziuk and Polge, 1962), (2) ovulations induced during the luteal phase of the cycle coupled with artificial insemination (Hunter, 1967b), (3) injecting animals with a solution of progesterone in oil, either systemically (Day and Polge, 1968) or locally under the serosal layer at the uterotubal junction (Hunter, 1972b). The common factor in all these situations leading to polyspermy was the artificially elevated concentration of circulating progesterone or progestagens, raising the possibility of steroid-mediated changes in oviduct fluids influencing the nature of the egg membranes. However, further experiments showed that the principal cause of polyspermy in all these studies was an increased number of competent spermatozoa confronting the eggs at the time of fertilization.

A demonstrable role of increased concentrations of circulating progesterone—as in situations of postovulatory aging—is to increase the patency of the oviduct lumen by relaxing the smooth muscle coat and reducing the edematous condition of the mucosa (Hunter, 1972b, 1977). Such changes remove the steeply decreasing preovulatory gradient in sperm concentration along the oviduct isthmus and cause the eggs to be overwhelmed by competent spermatozoa at the time

Table 2. The Influence of Postovulatory Aging of Eggs in the Oviducts on Fertilization and on Embryonic Survival at 25 Days Post-Insemination[a]

Estimated age of eggs at fertilization (hours)	Eggs fertilized normally		Viable embryos at 25 days		
	Percentage	S.E.	Survival (%)	S.E.	Mean no.
0 (Control)	90.8	4.5	87.9	2.9	12.0
4	92.1	2.7	72.9	14.9	11.7
8	94.6	2.3	60.5	13.2	8.7
12	70.3	7.8	53.3	15.7	6.8
16	48.3	8.4	27.9	14.5	4.8
20	50.9	7.5	32.3	15.2	5.0

[a] Note particularly the inferred loss of embryos by day 25 of gestation in the group of animals in which eggs were fertilized some 8 hr after ovulation. Eighteen animals were inseminated in each group (Adapted from Hunter, 1967a).

Table 3. The Incidence of Polyspermic Fertilization Exhibited in Mature Pigs in
Diverse Experimental Situations After Mating or Insemination at the Time of Estrus
(after Hunter, 1979)

Treatment	Eggs examined (No.)	Polyspermic eggs		Reference
		No.	%	
Delayed mating	53	6	11.0[a]	Thibault (1959)
Delayed mating	41	12	29.2	Hancock (1959)
Delayed insemination	149	23	15.4	Hunter (1967a)
Tubal surgery	34	11	32.4	Hunter and Léglise (1971)
Progesterone micro-injections	198	64	32.3	Hunter (1972b)
Tubal insemination	77	26	33.8	Hunter (1973b)

[a]A further 21% of the eggs were considered digynic, giving a total of 32%.

of penetration. Increasing the population of spermatozoa in the isthmus by direct surgical insemination (Polge *et al.*, 1970; Hunter, 1973b) or increasing numbers of spermatozoa at the site of fertilization after surgical resection of the isthmus (Hunter and Léglise, 1971) gives the same result. Overall, therefore, polyspermy has been shown to be a consequence of too many spermatozoa at the site of fertilization at the time of egg activation. Studies of *in vitro* fertilization—when successful—suffer from the same problem (reviewed by Hunter, 1990).

An important point related to polyspermy is that the proportion of eggs showing the condition is frequently 30–40% in many experiments (Table 3). This finding raises the question of whether a consistent proportion of the eggs ovulated might be especially susceptible to polyspermic penetration and indeed might be those that, in normal conditions, would be destined to undergo embryonic loss. It remains a fact that the proportion of eggs released at the time of mating but not represented by viable foetuses is commonly some 30–40% in this species. Thus, polyspermy might be a means of identifying retrospectively a subnormal population of oocytes destined to be at risk from embryonic loss.

Polyspermy has been of interest in yet another experimental situation. Polyspermic penetration has been used to monitor the potential of the pig oviduct for capacitating different populations of spermatozoa. Although the uterus and oviduct appear to act synergistically to facilitate the process of capacitation (Adams and Chang, 1962; Bedford, 1968; Hunter and Hall, 1974b), there have been several tentative suggestions that the potential of the oviduct alone for promoting capacitation may be strictly limited (Bedford, 1968; Brown and Hamner, 1971); in other words, the number of sperm cells rendered functionally mature at any one time may be regulated with sensitivity. Evidence that this is indeed the case has now been obtained using the incidence and degree of polyspermy in recently ovulated oocytes as a means of revealing this subtle component of tubal physiology after surgical insemination (Hunter and Nichol, 1988).

2.10. Early Embryonic Loss

This discussion of events related to the process of fertilization in pigs cannot conclude without noting some recent views on embryonic loss. Although a longstanding question has been whether subtle errors of fertilization might underlie a proportion of the embryonic loss, there has been no convincing evidence that this is so, with the exception of some quite dramatic karyotyping results of McFeely (1967). (The consensus is that these results were not representative.) However, during the last 2–3 years, the studies of Pope and colleagues have promoted the idea that much of the embryonic loss may be an expression of a lack of synchrony between oocytes during preovulatory resumption of meiosis and/or during actual penetration of the newly ovulated eggs

(Pope, 1990). Such an asynchrony would be held to increase the chances of nutritional substrate requirements of the preattachment embryo not being met by the fluids of the oviduct or uterine lumen, thereby compromising development of the zygote. As already noted, there is a slight asynchrony in the stages of both resumption of meiosis (Spalding *et al.*, 1955; Hunter and Polge, 1966) and penetration and activation of the egg. The current experimental studies of Pope and colleagues (e.g., Pope, 1990) are therefore worth following with attention. Perhaps the role of asynchrony between the stages of development of the embryo and that of the reproductive tract leading to embryonic loss has been shown most clearly so far by means of transplantation studies in sheep (Wilmut and Sales, 1981; Wilmut *et al.*, 1986). Even so, the precise role of asynchrony in spontaneous embryonic loss, rather than that demonstrated by experimental manipulation, remains to be evaluated critically.

3. HORSE

3.1. Preliminaries to Fertilization in Equids

Under suitable environmental conditions, mares are polyestrous seasonal breeders; during the prolonged period of estrus, the preliminaries to fertilization in this species offer several unique aspects, and the same is true of the postovulatory situation. In comparison with domestic ruminants (cow, sheep, goat), the volume of the ejaculate is surprisingly large, and yet, paradoxically, these voluminous secretions are introduced more or less directly into the lumen of the uterus rather than simply into the vagina. This is achieved by quite specific anatomic adjustments.

In contrast to the tortuous cervix of ruminants, the equine cervix is short and droops in a flaccid manner during the period of estrus towards the floor of the vagina. Introduction of the penis changes this situation mechanically, for the engorged glans passes along the vagina and abuts against the external cervical os, thereby raising and aligning the cervical canal with the orifice of the male urethra. In this coitally reoriented state, the cervical canal is dilated, enabling the contractions at ejaculation, assisted by the pumping action of the penis, to transmit the male secretions almost directly into the body of the uterus. Despite this ready passage for male fluids, the process of mating may take several minutes or more, the principal limitation being the rate of liberation of semen and the interval between successive ejaculatory jets.

3.2. Semen Volume and Sperm Concentration

The volume of the ejaculate depends largely on the breed and maturity of the stallion and on the season of the year. It is also influenced by the degree of sexual stimulation, the frequency of ejaculation, and the method of semen collection. The volume commonly ranges from 15 to 200 ml, with an average of 50–70 ml. Different fractions may be distinguished in the ejaculate, representing the secretions of the various male accessory glands. The most conspicuous fraction is the highly viscous gelatinous material derived largely from the seminal vesicles (Mann *et al.*, 1956a), which is thought to function as a temporary cervical plug preventing immediate backflow and loss of semen from the uterine lumen. Because of the excitable state of the postcoital myometrium, together with the patency of the cervix, this is clearly an important adaptation. Even so, the gross presence of seminal fluid in the uterus seems only to be required until the establishment of a viable population of spermatozoa in the oviducts (see below).

Sperm concentration in the ejaculate may vary from 0.3 to 8×10^8 cells/ml. A single ejaculate may contain from 4 to 13×10^9 spermatozoa. Most of the spermatozoa are emitted in the first two or three jets of semen, but it remains uncertain whether the sequence of ejaculation leads to a preferential establishment of a more concentrated cell suspension in the region of the uterotubal junctions. In any event, the volume of semen is such that its initial distribution throughout the uterus should present little or no problem. As is the case in pigs, male secretions would be bathing the upper portion of the uterine horns shortly after the completion of mating. Over and above purely volumetric considerations, the concentration of prostaglandin E_2 in stallion semen (24.1 ng/ml: Hunter, 1973a) might facilitate initial seminal distribution by enhancing myometrial activity.

The uterotubal junction forms a substantial barrier during estrus, the edematous condition of its papillae preventing the free access of semen and/or uterine fluid. Spermatozoa are thought to gain access to the lumen of the oviducts from semen bathing the uterotubal junction principally through their own motility. Indeed, the act of swimming across the papillae of the junction may divest sperm cells of much of the surface-associated seminal plasma, an event of major importance in the preliminaries to capacitation. Quite apart from such removal of stabilizing influences from the male gametes, it is essential to exclude seminal plasma from the upper reaches of the oviducts close to the time of ovulation, since it is known to contain an ovicidal factor (Chang, 1950). Nonetheless, biochemical evidence for some of the components of stallion seminal plasma entering the lower portion of the oviducts cannot be overlooked (Mann et al., 1956b). In this regard, it is worth considering whether the process of slaughter may have altered the normal distribution of spermatozoa and seminal plasma.

3.3. Estrus, Sperm Storage, and Survival

Ovulation occurs towards the end of the 5- to 7-day period of estrus, commonly 24–48 hr before this time when the follicle has achieved a diameter of 33–55 mm (Stabenfeldt and Hughes, 1977). There is evidence that a single mating at the onset of estrus may frequently be adequate to ensure fertilization of the egg released 120 hr or so later. Indeed, Day (1942) recorded pregnancies in mares inseminated 5 days before ovulation, and, in one instance, conception followed a sperm survival interval of 6 days. These figures received support from the subsequent study of Burkhardt (1949), but, although the latter author makes the statement that ". . . in services given early in oestrus, the sperms are quite capable of reaching the fimbriae of the Fallopian tube and maintaining fertilizing capacity for 90 hours or more, . . ." no evidence was presented as to the site of sperm storage. Even so, such evidence for prolonged sperm storage in mares could strengthen the notion of a preovulatory sperm reservoir in the caudal portion of the oviduct isthmus. As described for pigs, sheep, and cows (Hunter, 1985, 1987, 1990), spermatozoa would be arrested in this preovulatory reservoir under the influence of follicular endocrine activity and only fully matured (i.e., capacitated), activated, and displaced towards the presumed site of fertilization as ovulation became imminent (Fig. 1). In other words, there would be ovarian programming of sperm storage and activation. Unfortunately, however, a preliminary scanning electron microscopic study of the horse oviduct by Boyle et al. (1987) did not provide clear evidence of a preovulatory sperm reservoir in the caudal portion of the isthmus. Irrespective of the precise site of sperm storage, prolonged maintenance (5–6 days) of viable spermatozoa in the female tract could have considerable implications for the use of stud stallions. It is still common practice to mate mares every second day throughout the period of estrus, a procedure that may turn out to be wasteful of a stallion's mating potential.

3.4. Fate of Seminal Plasma and Spermatozoa

Concerning the fate of the voluminous seminal plasma, only relatively small volumes (3–5 ml) can be recovered from the uterine horns when 30–50 min have elapsed since mating. Moreover, the concentration of spermatozoa in this fluid may have been reduced significantly, in part as a result of adhesion to epithelial surfaces. In time, backflow (i.e., leakage) through the cervical canal probably accounts for most of the seminal plasma, but a small volume may remain temporarily in the endometrial folds and crypts. On the other hand, storage of a fertilizing population of spermatozoa in the endometrial glands seems unlikely, for such cells would become vulnerable to the population of polymorphonuclear leukocytes on liberation into the uterine lumen. There is an important nuance here: the argument is not that sperm cells are not found in the endometrial glands but rather that those that are are unlikely to be directly concerned in the events of fertilization. The reservoir for viable, functional spermatozoa seems to be in the distal isthmus of the oviducts.

3.5. Ovulation and Egg Transport

Although many of the classical texts state that horse eggs are shed from mature Graafian follicles as primary oocytes, recent studies demonstrate that this is usually not the case. Eggs are released at ovulation as secondary oocytes with a distinct first polar body (Enders *et al.*, 1987a,b; King *et al.*, 1987). Occasional failure to complete the first meiotic division before ovulation might be related to the slow and protracted increase in circulating concentrations of LH, in contrast to the acute preovulatory gonadotrophin surge of other large farm animals. It has also been incorrectly claimed that the horse egg is released without an investment of cumulus cells. In fact, the oocyte is surrounded by a sticky gelatinous matrix formed by a highly expanded and mucified cumulus oophorus. Cumulus cells can be seen clearly within the gelatinous mass in the photographs presented by Ginther (1979) and McKinnon and Squires (1988), and they are also described in the ultrastructural study of Enders *et al.* (1987a,b). Stages in the depolymerization of the cumulus matrix would seem especially worth examining in equine oocytes, as would the role of hyaluronic acid and proteoglycans in general. The extent to which the gelatinous mass acts to facilitate sperm–egg interactions and not least the process of capacitation remains to be clarified.

Precise timing of the first phase of egg transport within the oviduct requires further study, as does the actual site of fertilization within this duct, which may have a length of 20–30 cm or more (Hunter, 1988). A relatively rapid adovarian displacement towards the ampullary–isthmic junction would parallel events in other domestic species. It is close to this region of narrowing of the tube that penetration and activation by the fertilizing spermatozoon would be anticipated. The principal manner whereby the fertilizing sperm traverses the gelatinous mass is unknown, although both incisive motility and a lytic influence of acrosomal enzymes are doubtless important. The gelatinous mass is reported to have separated from the egg within 2 days, suggesting a susceptibility to further lytic influences.

3.6. Fertilization

In contrast to studies on ram, bull, and boar spermatozoa (e.g., Brown and Cheng, 1985; Jones *et al.*, 1988), acrosomal enzymes of stallion spermatozoa and their involvement in the events of fertilization appear to have received little specific study. The same comment seems to be true of sperm interactions with the surface of the zona pellucida and, in particular, with sperm-binding sites, that is, zona receptors. Neither of these preliminaries to fertilization received coverage in the large 1987 volume reporting the Fourth International Symposium on Equine Reproduction.

Ultrastructural studies of fertilization have been the subject of recent reports using the approach of postovulatory mating to establish chronology (Enders *et al.*, 1987a,b). In this situation, five eggs collected between 7 and 9 hr post-coitum were not activated, whereas two eggs collected at 10 and 14 hr, respectively, were activated by a fertilizing sperm. Pronuclear stages were seen from 12 to 22 hr after mating, but no special features distinguishing Equids from other mammals were noted in this respect. The flagellum of the activating spermatozoon was fully incorporated into the vitellus. As to the timing of sperm transport and capacitation, it would seem wise not to pronounce on these processes using a postovulatory mating model and only observing such a small number of eggs, which themselves may have commenced to age. Even so, in an earlier phase-contrast study of whole mounts of horse eggs released in response to hCG injection, formation of the second polar body was suggested to be complete 10–12 hr after ovulation (Webel *et al.*, 1977).

3.7. Abnormal Fertilization

The pathological condition of polyspermic penetration of the vitellus has not received specific comment in reviews of fertilization in horses, although Betteridge *et al.* (1982) observed a large number of spermatozoa in the perivitelline space of a morula. Because of the intrauterine accumulation of the ejaculate, with relatively high concentrations of spermatozoa bathing the base of the oviducts by the completion of mating, polyspermic fertilization would perhaps have been expected at a higher incidence than reported in intravaginal semen depositors such as rabbits, ruminants, and primates—and may indeed have occurred. On the other hand, the oviduct isthmus has doubtless developed remarkable powers of imposing a steeply declining sperm gradient and of arresting spermatozoa during the 4 or 5 days of estrus preceding ovulation. Such a specialization would contribute to a seemingly low incidence of polyspermic fertilization, as might the specialized gelatinous "egg coat" (see above).

In this connection, the stability of the zona reaction is worth considering. If mares remain in estrus for 24–36 hr or more after ovulation, with the possibility of repeated mating by fertile stallions, then it is clear that the block to polyspermy needs to remain functional in the face of vigorous ascending spermatozoa if embryonic development is not to be compromised by further sperm penetration. Future studies might usefully look at the influence of surgical insemination directly into the oviducts at different intervals before and after ovulation. Deposition of enhanced sperm numbers at appropriate times relative to release of the egg might clarify the susceptibility of horse eggs to polyspermy, not least under conditions of postovulatory aging. Preliminary observations by B. Ball (personal communication) after insemination directly into the proximal ampulla noted only monospermic pronuclear eggs. This would endorse the suggestion that the gelatinous egg coat may contribute to the low incidence of polyspermy. Reports of other anomalies of fertilization, especially of digyny, have not been noted during preparation of this review.

3.8. Cleavage

As to observations on the rate of cleavage, Webel *et al.* (1977), using an hCG-induced ovulation model, concluded that the first mitotic division occurred about 20 hr after activation. This view was endorsed by the figure of 21 hr given by Enders *et al.* (1987a) as the earliest time at which first cleavage would be anticipated. By 48 hr after ovulation, embryos are commonly at the four- to eight-cell stage. A 32-cell morula is the latest stage to be expected in the oviduct.

Morphological details of cleavage are presented in many of the earlier studies of horse embryos, including those of Hamilton and Day (1945), Van Niekerk and Gerneke (1966), and Betteridge *et al.* (1982). The process seems not to differ significantly from mitotic division in embryos of other domestic species. Indeed, more attention seems to have been devoted to the loss of cumulus cells and development of the gelatinous clot, followed—in uterine embryos—by formation of the transparent acellular coat or capsule, than to events occurring within the zona pellucida.

3.9. Egg Progression in Equine Oviducts

It is now widely accepted that unfertilized eggs from successive cycles remain trapped in the oviducts for several months or more to undergo progressive degeneration, whereas fertilized eggs—embryos—pass along the oviducts and enter the uterus at a developmental stage of late morula or young blastocyst (Van Niekerk and Gerneke, 1966; Betteridge and Mitchell, 1974). Descriptions of this remarkable phenomenon usually infer that mares can actively discriminate between fertilized and unfertilized eggs. In fact, a small proportion of unfertilized eggs may enter the uterus at about the same time as passage of an embryo (Hamilton and Day, 1945; Steffenhagen *et al.*, 1972; Allen, 1979), but this occurrence remains unpredictable.

Explanations for the differential transport of the two types of eggs have included (1) a difference in the surface characteristics of the zona pellucida between fertilized and unfertilized eggs, enabling the mechanisms underlying tubal transport to distinguish between the two conditions (Rowson, 1971) [such differences have not so far been demonstrated (Betteridge *et al.*, 1976)] and (2) release of a humoral substance(s) or metabolite by the developing embryo that promotes its transport at the expense of unfertilized eggs (Hunter, 1977; Flood *et al.*, 1979). A systemic response of the tubal structures to a trophic substance of embryonic origin would be expected with the latter explanation, thereby leading to displacement of both fertilized and unfertilized eggs. Nonetheless, the recent discovery of early pregnancy factors, such as those produced by mouse zygotes within 6 hr of sperm penetration (O'Neill, 1985), has once more drawn attention to the involvement of a putative secretion by the very young horse embryo. Despite this possibility, facts already at hand appear to provide an adequate explanation for the differential transport of fertilized and unfertilized eggs within the oviducts of mares (Hunter, 1989b).

The first fact of relevance is the duration of egg transport through the oviducts. Among the large farm species, pig embryos require 2 days to pass to the uterus (Pomeroy, 1955), and embryos of sheep and cows require approximately 3 days (Betteridge, 1977; Hunter, 1980), whereas horse embryos enter the uterus 5–6 days after ovulation (Van Niekerk and Gerneke, 1966; Marrable and Flood, 1975). Thus, embryos remain in the oviducts of mares for significantly longer than in other domestic farm animals. The second fact concerns the rate and nature of degeneration of unfertilized eggs relative to their location in the tract. Although unfertilized eggs of pigs, sheep, and cows commence various forms of degeneration in the oviducts, these events are concerned principally with the cytoplasmic organelles. More conspicuous forms of degeneration, such as fragmentation of the vitellus itself, tend not to occur until eggs have entered the uterus (Dziuk, 1960). By contrast, unfertilized horse eggs exhibit degenerative features of both the zona pellucida and the cytoplasm while still in the oviducts (Van Niekerk and Gerneke, 1966). Indeed, the major contention of a recent paper is that it is the very softening and distortion of the zona pellucida and underlying vitellus during the 5- to 6-day sojourn that compromises transport of unfertilized eggs to the uterus (Hunter, 1989b).

Whatever the precise mechanism responsible for propelling eggs from the ampulla through the isthmus—such as myosalpingeal contractions, beat of cilia, flow of fluid, or a combination of all three (see Hunter, 1988)—these mechanisms will undoubtedly be rendered less effective when acting on an egg no longer spherical and resistant but instead deformed and soft and susceptible to further distortion by contractile forces. In essence, therefore, failure of egg transport may be explained primarily on mechanical (i.e., structural) grounds. Indeed, the deformed egg(s) may frequently lodge in epithelial folds away from the central lumen of the oviduct, thereby facilitating passage of a subsequently fertilized egg (embryo). Conversely, unfertilized eggs remaining more centrally in the duct and appropriately oriented (i.e., flattened in a transverse plane) could occasionally pass to the uterus by displacement immediately in front of an embryo. These proposals receive strong support from observations in various species of Chiroptera, in which the embryos also have a prolonged residence in the oviduct and which show differential transport of fertilized and unfertilized eggs.

Acknowledgments. I am grateful to a number of colleagues in Montreal for comments on a preliminary draft and also wish to thank Sylvie Lagacé for preparing the manuscript.

NOTE ADDED IN PROOF

After submission of this manuscript, an excellent review has appeared: Bézard, J., Magistrini, M., Duchamp, G., and Palmer, E., 1989, Chronology of equine fertilisation and embryonic development *in vivo* and *in vitro*, *Equine Veterinary Journal (Suppl.)* **8**:105–110.

4. REFERENCES

Adams, C. E., and Chang, M. C., 1962, Capacitation of rabbit spermatozoa in the Fallopian tube and in the uterus, *J. Exp. Zool.* **151**:159–166.

Alanko, M., 1973, Fertilisation and early development of ova in artificially inseminated gilts, Doctoral thesis, University of Helsinki.

Allen, W. R., 1979, Maternal recognition of pregnancy and immunological implications of trophoblast–endometrium interactions in equids, in: *Maternal Recognition of Pregnancy, Ciba Foundation Symposium No. 64*, Excerpta Medica, Amsterdam, pp. 323–352.

Austin, C. R., 1961, *The Mammalian Egg*, Blackwell Scientific Publications, Oxford.

Austin, C. R., 1985, Sperm maturation in the male and female genital tracts, in: *Biology of Fertilization*, Vol. 2 (C. B. Metz and A. Monroy, eds.), Academic Press, New York, pp. 121–155.

Austin, C. R., and Walton, A., 1960, Fertilisation, in: *Marshall's Physiology of Reproduction*, Vol. 1, Part 2 (A. S. Parkes, ed.), Longmans Green, London, pp. 310–416.

Baker, R. D., and Polge, C., 1976, Fertilisation in swine and cattle, *Can. J. Anim. Sci.* **56**:105–119.

Barros, C., Bedford, J. M., Franklin, L. E., and Austin, C. R., 1967, Membrane vesiculation as a feature of the mammalian acrosome reaction, *J. Cell Biol.* **34**:C1–C5.

Bavister, B. D. (ed.), 1987, *The Mammalian Preimplantation Embryo*, Plenum Press, New York.

Bedford, J. M., 1968, Importance of the Fallopian tube for capacitation in the rabbit, in: *Proceedings 6th International Congress of Animal Reproduction, Paris*, Institut National de la Recherche Agronomique (INRA), Paris, pp. 35–37.

Bedford, J. M., 1970, Sperm capacitation and fertilisation in mammals, *Biol. Reprod. [Suppl.]* **2**:128–158.

Bedford, J. M., 1982, Fertilisation, in: *Reproduction in Mammals*, Book 1 (C. R. Austin and R. V. Short, eds.), Cambridge University Press, Cambridge, pp. 128–163.

Betteridge, K. J., 1977, *Embryo Transfer in Farm Animals*, Agriculture Canada Monograph No. 16, Ottawa.

Betteridge, K. J., and Mitchell, D., 1974, Direct evidence of retention of unfertilised ova in the oviduct of the mare, *J. Reprod. Fertil.* **39**::145–148.

Betteridge, K. J., Flood, P. F., and Mitchell, D., 1976, Possible role of the embryo in the control of oviductal transport in mares, in: *Symposium on Ovum Transport and Fertility Regulation* (M. J. K. Harper and C. J. Pauerstein, eds.), Scriptor, Copenhagen, pp. 381–389.

Betteridge, K. J., Eaglesome, M. D., Mitchell, D., Flood, P. F., and Bériault, R., 1982, Development of horse embryos up to twenty-two days after ovulation: Observations on fresh specimens, *J. Anat.* **135**:191–209.

Boyle, M. S., Cran, D. G., Allen, W. R., and Hunter, R. H. F., 1987, Distribution of spermatozoa in the mare's oviduct, *J. Reprod. Fertil. [Suppl.]* **35**:79–86.

Braden, A. W. H., Austin, C. R., and David, H. A., 1954, The number of sperms about the eggs in mammals and its significance for normal fertilisation, *Aust. J. Biol. Sci.* **7**:543–551.

Brown, C. R., and Cheng, W. T. K., 1985, Limited proteolysis of the porcine zona pellucida by homologous sperm acrosin, *J. Reprod. Fertil.* **74**:257–260.

Brown, S. M., and Hamner, C. E., 1971, Capacitation of sperm in the female reproductive tract of the rabbit during estrus and pseudopregnancy, *Fertil. Steril.* **22**:92–97.

Burger, J. F., 1952, Sex physiology of pigs, *Onderstepoort J. Vet. Res. [Suppl.]* **2**:218.

Burkhardt, J., 1949, Sperm survival in the genital tract of the mare, *J. Agr. Sci.* **39**:201–203.

Burkman, L. J., Overstreet, J. W., and Katz, D. F., 1984, A possible role for potassium and pyruvate in the modulation of sperm motility in the rabbit oviducal isthmus, *J. Reprod. Fertil.* **71**:367–376.

Chang, M. C., 1950, Ovicidal effect of seminal plasma, *Proc. Natl. Acad. Sci. U.S.A.* **36**:188–191.

Chang, M. C., and Pincus, G., 1951, Physiology of fertilisation in mammals, *Physiol. Rev.* **31**:1–26.

Corner, G. W., 1921, Cyclic changes in the ovaries and uterus of the sow, and their relation to the mechanism of implantation, *Contrib. Embryol.* **13**:117–145.

Corner, G. W., 1923, Cyclic variation in uterine and tubal contraction waves, *Am. J. Anat.* **32**:345–351.

Cummins, J. M., 1982, Hyperactivated motility patterns of ram spermatozoa recovered from the oviducts of mated ewes. *Gamete Res.* **6**:53–63.

Day, B. N., and Polge, C., 1968, Effects of progesterone on fertilization and egg transport in the pig, *J. Reprod. Fertil.* **17**:227–230.

Day, F. T., 1942, Survival of spermatozoa in the genital tract of the mare, *J. Agri. Sci.* **32**:108–111.

Du Mesnil Du Buisson, F., and Dauzier, L., 1955, La remontée des spermatozoïdes du verrat dans le tractus de la truie en oestrus, *C. R. Seanc. Soc. Biol.* **149**:76–79.

Dunbar, B. S., 1983, Morphological, biochemical and immunochemical characterisation of the mammalian zona pellucida, in: *Mechanism and Control of Animal Fertilisation* (J. F. Hartmann, ed.), Academic Press, New York, pp. 139–175.

Dziuk, P. J., 1960, Frequency of spontaneous fragmentation of ova in unbred gilts, *Proc. Soc. Exp. Biol. Med.* **103**:91–92.

Dziuk, P. J., and Polge, C., 1962, Fertility in swine after induced ovulation, *J. Reprod. Fertil.* **4**:207–208.

Dziuk, P. J., Polge, C., and Rowson, L. E. A., 1964, Intra-uterine migration and mixing of embryos in swine following egg transfer, *J. Anim. Sci.* **23**:37–42.

Einarsson, S., 1980, Site, transport and fate of inseminated semen, in: *Proceedings 9th International Congress of Animal Reproduction and Artificial Insemination, Madrid*, Vol. 1, Gráficas Orbo, S. L., Madrid, pp. 147–158.

Enders, A. C., Liu, I. K. M., Bowers, J., Cabot, C., and Lantz, K. C., 1987a, Fertilisation and pronuclear stages in the horse, *J. Reprod. Fertil. [Suppl.]* **35**:691–692.

Enders, A. C., Liu, A. K. M., Bowers, J., Lantz, K. C., Schlafke, S., and Suarez, S., 1987b, The ovulated ovum of the horse: Cytology of nonfertilised ova to pronuclear stage, *Biol. Reprod.* **37**:453–466.

Fléchon, J. E., Motlik, J., Hunter, R. H. F., Fléchon, B., Pivko J., and Fulka, J., 1986, Cumulus oophorus mucification during resumption of meiosis in the pig. A scanning electron microscope study, *Reprod. Nutr. Dev.* **26**:989–998.

Flood, P. F., Jong, A., and Betteridge, K. J., 1979, The location of eggs retained in the oviducts of mares, *J. Reprod. Fertil.* **57**:291–294.

Fraser, L., 1984, Mechanisms controlling mammalian fertilisation, *Oxford Rev. Reprod. Biol.* **6**:174–225.

Ginther, O. J., 1979, *Reproductive Biology of the Mare—Basic and Applied Aspects*, McNaughton and Gunn, Ann Arbor.

Hamilton, W. J., and Day, F. T., 1945, Cleavage stages of the ova of the horse, with notes on ovulation, *J. Anat.* **79**:127–130.

Hancock, J. L., 1959, Polyspermy of pig ova, *Anim. Prod.* **1**:103–106.

Hancock, J. L., 1961, Fertilisation in the pig, *J. Reprod. Fertil.* **2**:307–331.

Hancock, J. L., 1962, Fertilisation in farm animals, *Anim. Breed. Abstr.* **30**:285–310.

Harper, M. J. K., 1982, Sperm and egg transport, in: *Reproduction in Mammals*, Book 1 (C. R. Austin and R. V. Short, eds.), Cambridge University Press, Cambridge, pp. 102–127.

Hartmann, J. F., 1983, Mammalian fertilisation: Gamete surface interactions *in vitro*, in: *Mechanism and Control of Animal Fertilisation*, Academic Press, New York, pp. 325–364.

Hunter, R. H. F., 1964, Superovulation and fertility in the pig, *Anim. Prod.* **6**:189–194.

Hunter, R. H. F., 1967a, The effects of delayed insemination on fertilisation and early cleavage in the pig, *J. Reprod. Fertil.* **13**:133–147.

Hunter, R. H. F., 1967b, Polyspermic fertilisation in pigs during the luteal phase of the estrous cycle, *J. Exp. Zool.* **165:**:451–460.

Hunter, R. H. F., 1972a, Fertilisation in the pig: Sequence of nuclear and cytoplasmic events, *J. Reprod. Fertil.* **29**:395–406.

Hunter, R. H. F., 1972b, Local action of progesterone leading to polyspermic fertilisation in pigs, *J. Reprod. Fertil.* **31**:433–444.

Hunter, R. H. F., 1973a, Transport, migration and survival of spermatozoa in the female genital tract: Species with intra-uterine deposition of semen, *I.N.S.E.R.M. Colloque on Sperm Transport, Survival and Fertilizing Ability* **26**:309–342.

Hunter, R. H. F.,1973b, Polyspermic fertilisation in pigs after tubal deposition of excessive numbers of spermatozoa, *J. Exp. Zool.* **183**:57–64.

Hunter, R. H. F., 1974, Chronological and cytological details of fertilisation and early embryonic development in the domestic pig, *Sus scrofa, Anat. Rec.* **178**:169–186.

Hunter, R. H. F., 1976, Sperm–egg interactions in the pig: Monospermy, extensive polyspermy, and the formation of chromatin aggregates, *J. Anat.* **122**:43–59.

Hunter, R. H. F., 1977, Function and malfunction of the Fallopian tubes in relation to gametes, embryos and hormones, *Eur. J. Obstet. Gynecol. Reprod. Biol.* **7**:267–283.

Hunter, R. H. F., 1979, Ovarian follicular responsiveness and oocyte quality after gonadotrophic stimulation of mature pigs, *Ann. Biol. Anim. Biochim. Biophys.* **19**:1511–1520.

Hunter, R. H. F., 1980, *Physiology and Technology of Reproduction in Female Domestic Animals*, Academic Press, London, New York.

Hunter, R. H. F., 1981, Sperm transport and reservoirs in the pig oviduct in relation to the time of ovulation, *J. Reprod. Fertil.* **63**:109–117.

Hunter, R. H. F., 1982, Interrelationships between spermatozoa, the female reproductive tract, and the egg investments, in: *Proceedings University of Nottingham 34th Easter School*, Butterworths, London, pp. 49–63.

Hunter, R. H. F., 1984, Pre-ovulatory arrest and peri-ovulatory redistribution of competent spermatozoa in the isthmus of the pig oviduct, *J. Reprod. Fertil.* **72**:203–211.

Hunter, R. H. F., 1985, Experimental studies of sperm transport in sheep, cows and pigs, *Vet. Rec.* **116**:188–189.

Hunter, R. H. F., 1987, Peri-ovulatory physiology of the oviduct, with special reference to progression, storage and capacitation of spermatozoa, in: *New Horizons in Sperm Cell Research* (H. Mohri, ed.), Japan Scientific Press, Tokyo, pp. 31–45.

Hunter, R. H. F., 1988, *The Fallopian Tubes: Their Role in Fertility and Infertility*, Springer-Verlag, Berlin, Heidelberg, New York.

Hunter, R. H. F., 1989a, Ovarian programming of gamete progression and maturation in the female genital tract, *Zool. J. Linn. Soc.* **95**:117–124.

Hunter, R. H. F., 1989b, Differential transport of fertilised and unfertilised eggs in equine Fallopian tubes: A straightforward explanation, *Vet. Rec.* **125**:304–305.

Hunter, R. H. F., 1990, Fertilisation of pig eggs *in vivo* and *in vitro*, *J. Reprod. Fertil. [Suppl.]* **40**:211–226.

Hunter, R. H. F., and Dziuk, P. J., 1968, Sperm penetration of pig eggs in relation to the timing of ovulation and insemination, *J. Reprod. Fertil.* **15**:199–208.

Hunter, R. H. F., and Hall, J. P., 1974a, Capacitation of boar spermatozoa: The influence of post-coital separation of the uterus and Fallopian tubes, *Anat. Rec.* **180**:597–604.

Hunter, R. H. F., and Hall, J. P., 1974b, Capacitation of boar spermatozoa: Synergism between uterine and tubal environments, *J. Exp. Zool.* **188**:203–214.

Hunter, R. H. F., and Léglise, P. C., 1971, Polyspermic fertilisation following tubal surgery in pigs, with particular reference to the role of the isthmus, *J. Reprod. Fertil.* **24**:233–246.

Hunter, R. H. F., and Nichol, R., 1986, A preovulatory temperature gradient between the isthmus and ampulla of pig oviducts during the phase of sperm storage, *J. Reprod. Fertil.* **77**:599–606.

Hunter, R. H. F., and Nichol, R., 1988, Capacitation potential of the Fallopian tube: A study involving surgical insemination and the subsequent incidence of polyspermy, *Gamete Res.* **21**:255–266.

Hunter, R. H. F., and Polge, C., 1966, Maturation of follicular oocytes in the pig after injection of human chorionic gonadotrophin, *J. Reprod. Fertil.* **12**:525–531.

Jones, R., Brown, C. R., and Lancaster, R. T., 1988, Carbohydrate-binding properties of boar sperm proacrosin and assessment of its role in sperm–egg recognition and adhesion during fertilisation, *Development* **102**:781–792.

Keye, J. D., 1923, Periodic variations in spontaneous contractions of uterine muscle, in relation to the oestrous cycle and early pregnancy, *Bull. Johns Hopkins Hosp.* **34**:60–63.

King, W. A., Bézard, J., Bousquet, D., Palmer, E., and Betteridge, K. J., 1987, The meiotic stage of preovulatory oocytes in mares, *Genome* **29**:679–682.

Knifton, A., 1962, The response of the pig uterus to oxytocin at different stages in the oestrous cycle, *J. Pharm. Pharmacol. [Suppl.]* **14**:42T-43T.

Liptrap, R. M., and Raeside, J. I., 1966, Luteinising hormone activity in blood and urinary oestrogen excretion by the sow at oestrus and ovulation, *J. Reprod. Fertil.* **11**:439–446.

Mann, T., Leone, E., and Polge, C., 1956a, The composition of the stallion's semen, *J. Endocrinol.* **13**:279–290.

Mann, T., Polge, C., and Rowson, L. E. A., 1956b, Participation of seminal plasma during the passage of spermatozoa in the female reproductive tract of the pig and horse, *J. Endocrinol.* **13**:133–140.

Marrable, A. W., and Flood, P. F., 1975, Embryological studies on the Dartmoor pony during the first third of gestation, *J. Reprod. Fertil. [Suppl.]* **23**:499–502.

McFeely, R. A., 1967, Chromosome abnormalities in early embryos of the pig, *J. Reprod. Fertil.* **13**:579–581.

McKenzie, F. F., 1926, The normal oestrous cycle in the sow, *Missouri Agr. Exp. Sta. Res. Bull.* **86**.

McKenzie, F. F., Miller, J. C., and Bauguess, L. C., 1938, The reproductive organs and semen of the boar, *Res. Bull. Missouri Exp. Sta.* **279**.

McKinnon, A. O., and Squires, E. L., 1988, Morphologic assessment of the equine embryo, *J. Am. Vet. Med. Assoc.* **192**:401–406.

Niswender, G. D., Reichert, L. E., and Zimmerman, D. R., 1970, Radioimmunoassay of serum levels of luteinising hormone throughout the estrous cycle in pigs, *Endocrinology* **87**:576–580.

O'Neill, C., 1985, Thrombocytopenia is an initial maternal response to fertilisation in mice, *J. Reprod. Fertil.* **73**:559–566.

Overstreet, J. W., and Cooper, G. W., 1975, Reduced sperm motility in the isthmus of the rabbit oviduct, *Nature* **258**:718–719.

Polge, C., 1956, Artificial insemination in pigs, *Vet. Rec.* **68**:62–76.

Polge, C., 1967, Control of ovulation in pigs, *Proc. R. Soc. Med.* **60**:654–655.

Polge, C., 1978, Fertilisation in the pig and horse, *J. Reprod. Fertil.* **54**:461–470.

Polge, C., Salamon, S., and Wilmut, I., 1970, Fertilizing capacity of frozen boar semen following surgical insemination, *Vet. Rec.* **87**:424–428.

Pomeroy, R. W., 1955, Ovulation and the passage of the ova through the Fallopian tubes in the pig, *J. Agr. Sci.* **45**:327–330.

Pope, W. F., 1990, Causes and consequences of early embryonic diversity in pigs, *J. Reprod. Fertil. [Suppl.]* **40**:251–260.

Rodolfo, A., 1934a, The physiology of reproduction in swine. I. The semen of boars under different intensivenesses of mating, *Philipp. J. Sci.* **53**:183–203.

Rodolfo, A., 1934b, The physiology of reproduction in swine. II. Some observations on mating, *Philipp. J. Sci.* **55**:13–18.

Rowson, L. E. A., 1971, The role of reproductive research in animal production, *J. Reprod. Fertil.* **26**:113–126.

Smith, J. C., and Nalbandov, A. V., 1958, The role of hormones in the relaxation of the uterine portion of the cervix in swine, *Am. J. Vet. Res.* **19**:15–18.

Spalding, J. F., Berry, R. O., and Moffit, J. G., 1955, The maturation process of the ovum of swine during normal and induced ovulations, *J. Anim. Sci.* **14**:609–620.

Stabenfeldt, G. H., and Hughes, J. P., 1977, Reproduction in horses, in: *Reproduction in Domestic Animals*, ed. 3 (H. H. Cole and P. T. Cupps, eds.), Academic Press, New York and London, pp. 401–431.

Steffenhagen, W. P., Pineda, M. H., and Ginther, O. J., 1972, Retention of unfertilized ova in the uterine tubes of mares, *Am. J. Vet. Res.* **33**:2391–2398.

Szollosi, D., and Hunter, R. H. F., 1973, Ultrastructural aspects of fertilisation in the domestic pig: Sperm penetration and pronucleus formation, *J. Anat.* **116**:181–206.

Thibault, C., 1949, L'oeuf des mammifères. Son développement parthénogénétique, *Ann. Sci. Nat. Zool.* **11**: 136–219.

Thibault, C., 1959, Analyse de la fécondation de l'oeuf de la truie après accouplement ou insémination artificielle, *Ann. Zootech. D. (Suppl.)* **8**:165–177.

Van Niekerk, C. H., and Gerneke, W. H., 1966, Persistence and parthenogenetic cleavage of tubal ova in the mare, *Onderstep. J. Vet. Res.* **33**:195–232.

Webel, S. K., Franklin, V., Harland, B., and Dziuk, P. J., 1977, Fertility, ovulation and maturation of eggs in mares injected with HCG, *J. Reprod. Fertil.* **51**:337–341.

Wilmut, I., and Sales, D. I., 1981, Effect of an asynchronous environment on embryonic development in sheep, *J. Reprod. Fertil.* **61**:179–184.

Wilmut, I., Sales, D. I., and Ashworth, C. J. 1986, Maternal and embryonic factors associated with prenatal loss in mammals, *J. Reprod. Fertil.* **76**:851–864.

Yanagimachi, R., 1981, Mechanisms of fertilisation in mammals, in: *Fertilisation and Embryonic Development In vitro* (L. Mastroianni and J. D. Biggers, eds.), Plenum Press, New York, pp. 81–182.

Yanagimachi, R., 1988, Mammalian fertilization, in: *The Physiology of Reproduction* (E. Knobil and J. Neill, eds.), Raven Press, New York, pp. 135–185.

18

Bovine *in Vitro* Fertilization

J. J. Parrish and N. L. First

1. INTRODUCTION

In vitro fertilization in cattle was first reported by Bracket *et al.* (1982) using surgically recovered oocytes that were matured *in vivo* and surgically transferred as zygotes to the oviducts of cows. The procedure was later simplified by laparoscopic removal of *in vivo*-matured oocytes and development of the embryos in oviducts of rabbits (Sirard and Lambert, 1985; Sirard *et al.*, 1985). Although successful, these experiments were not highly efficient in sperm capacitation and required *in vivo* oocyte maturation, surgery, and development of the embryos in oviducts. Recently systems for producing and culturing bovine embryos totally *in vitro* have been developed. As a result, bovine embryos are now produced in large numbers for research purposes or in some cases for commercial embryo transfer (Leibfried-Rutledge *et al.*, 1989; Gordon and Lu, 1990). The efficient production of embryos *in vitro* has required an understanding of the mechanisms of sperm capacitation, maturation of oocytes, and sperm-egg interaction in the fertilization process. Understanding these processes for bovine gametes is the focus of this review.

2. CAPACITATION

Mammalian sperm cannot fertilize oocytes immediately on ejaculation but must first undergo a series of changes referred to as capacitation. The culmination of capacitation is the ability of a spermatozoon to undergo the acrosome reaction in response to the physiological stimulator of that event. For the bovine, this stimulator of the acrosome reaction is likely the zona pellucida (Florman and First, 1988a). The exact mechanism of capacitation is poorly understood for any mammalian species.

Studies on *in vivo* capacitation in the bovine have been hampered by the cost of animals and difficulty in accessing sperm in the uterus and oviduct. Additionally, the bovine is a spontaneous

J. J. PARRISH AND N. L. FIRST · Department of Meat and Animal Science, University of Wisconsin, Madison, Wisconsin 53706.

A Comparative Overview of Mammalian Fertilization, edited by Bonnie S. Dunbar and Michael G. O'Rand. Plenum Press, New York, 1991.

ovulator, and the exact time of ovulation is not known for any particular female. Almost all our knowledge of capacitation in the bovine has been based on *in vitro* studies (First and Parrish, 1987). These studies do not take into account that capacitation *in vivo* and sperm transport in the female are likely two processes that are intimately linked (Parrish, 1989). Capacitation of bovine sperm *in vitro* was first demonstrated by Brackett *et al.* (1982) using a high-ionic-strength (HIS) medium. Later studies have confirmed that spermatozoa from some but not all bulls will capacitate after exposure to HIS media (Bousquet and Brackett, 1982; Sirard and Lambert, 1985; Sirard *et al.*, 1985). Previous studies in the rabbit suggest that HIS capacitates sperm by removing an acrosomal stabilizing or decapacitation factor from the sperm surface (Oliphant and Singhas, 1979). Normally *in vivo* such a factor could be removed from sperm as they move through cervical mucus or are exposed to secretions of the uterus or oviduct. Although sperm capacitated with HIS were responsible for producing the first calves with *in vitro* fertilization in the bovine (Bracket *et al.*, 1982; Sirard *et al.*, 1985), the variability among bulls in its success (Lambert *et al.*, 1985) has resulted in other capacitation procedures being utilized.

The first report of glycosaminoglycans (GAGs) affecting capacitation or the acrosome reaction was by Lenz *et al.* (1982), who noted that follicular fluid proteoglycans caused caudal epididymal bovine sperm to undergo the acrosome reaction during a 22-hr incubation. The active portion of the proteoglycan appeared to be a chondroitin sulfate glycosaminoglycan (GAG). Subsequent studies demonstrated similar effects of other GAGs on epididymal bovine sperm (Handrow *et al.*, 1982), suggesting that the more highly sulfated a GAG is, the more potent is its effect on sperm. The most potent GAG in terms of its ability to induce an acrosome reaction or capacitate sperm was heparin. At the time of these studies it was not known that the zona pellucida could induce the acrosome reaction in capacitated bovine sperm (Florman and First, 1988a). The important question was then whether the effect of GAGs on bovine sperm was on capacitation or the acrosome reaction. Heparin appears to affect capacitation, as a 4- to 6-hr incubation in media with heparin is required before sperm gain the ability to fertilize zona-intact oocytes (Parrish *et al.*, 1985, 1988). Surprisingly, the effects of chondroitin sulfate on ejaculated or epididymal sperm could not be verified (Parrish *et al.*, 1985), but heparin clearly increased the ability of sperm to fertilize oocytes *in vitro*. The effects of heparin on capacitation were further confirmed by its ability to prepare sperm for the acrosome reaction on exposure to the fusogenic agent lysophosphatidylcholine (LC; Parrish *et al.*, 1988). Both the time course of LC effects and heparin dose–response curves paralleled the ability of sperm to fertilize oocytes (Parrish *et al.*, 1988). The results implied that heparin caused changes in the membrane status of sperm and in particular the membranes overlying the acrosome. Heparin is the first substance shown to efficiently capacitate frozen-thawed bovine sperm (Parrish *et al.*, 1986; Leibfried-Rutledge *et al.*, 1989).

An important question is whether the mechanism by which heparin capacitates bovine sperm represents a physiological one and if heparin or a heparin-like GAG such as heparan sulfate is the endogenous capacitating agent in the bovine. It is unlikely that a follicular fluid component could capacitate sperm *in vivo* because oocytes are generally fertilized within a short time of ovulation. Numerous GAGs are known to be present in flushings of bovine reproductive tracts (Lee and Ax, 1984) and include large amounts of heparin-like GAG. These heparin-like GAGs are potential sources of capacitating activity. If results of sperm transport studies are considered, the fertilizing sperm appears to require 6–8 hr after deposition in the female to move up through the cervix and uterus to reach the lower isthmus of the oviduct (Hunter and Wilmut, 1984). Sperm remain in this region until ovulation, at which time they move to the junction of the isthmus and ampulla where they meet the oocyte.

The majority of the time (up to 22 hr) when sperm are undergoing capacitation *in vivo* occurs when sperm are in the oviduct. Oviduct fluid was therefore examined for its ability to capacitate sperm (Parrish *et al.*, 1989a). Oviduct fluid was collected from chronically cannulated heifers throughout the estrous cycle. Oviduct fluid contains capacitating activity, which peaks at the time of estrus. The capacitating activity is not sensitive to 60°C for 30 min. The capacitating activity is also not affected by extensive proteolysis, is precipitable by ethanol, is size-excluded by Sephadex G-25, and after those treatments in sequential order the capacitating activity is then destroyed by nitrous acid under conditions specific for heparin-like GAG destruction (Parrish *et al.*, 1989a). The results suggest that a heparin-like GAG could be an *in vivo* capacitating agent in the bovine. As such, the mechanism by which heparin capacitates sperm is important to understand.

Capacitation of bovine sperm by heparin was first suggested with epididymal sperm when long incubations of 22 hr resulted in sperm undergoing the acrosome reaction in the presence of heparin (Handrow *et al.*, 1982). Later, similar effects were noted when ejaculated sperm were capacitated after a 9-hr incubation with heparin (Lee *et al.*, 1985) and the capacitating effects were inhibited by seminal plasma. Whether the effect of heparin on bovine sperm was on capacitation or on the acrosome reaction was clarified when sperm were demonstrated to require 4–6 hr of incubation in heparin before developing an increased ability to fertilize oocytes (Parrish *et al.*, 1985, 1988). The increased fertilizing ability occurred before any changes in acrosomal status. As these changes in fertilizing ability occurred, changes in sperm membranes also occurred as evidenced by acquisition of the ability of sperm to acrosome react in response to LC (Parrish *et al.*, 1988). Similar doses of LC had no effect on noncapacitated sperm.

Several features of capacitation by heparin are known. First, capacitation requires extracellular Ca^{2+}, and an uptake of Ca^{2+} occurs during capacitation (Handrow *et al.*, 1989). The Ca^{2+} taken up by sperm either goes to an intracellular pool or is tightly bound to sperm, as both lanthanum and EGTA were unable to displace it. The uptake of Ca^{2+} and capacitation could be prevented by including glucose in the capacitation medium along with heparin (Handrow *et al.*, 1989; Parrish *et al.*, 1989b). The inhibition of capacitation by glucose was prevented by incubation with 8-bromo-cAMP, but 8-bromo-cAMP on its own did not capacitate bovine sperm (Handrow *et al.*, 1989). Interestingly, 8-bromo-cAMP can stimulate Ca^{2+} uptake into sperm above control levels but less than what is seen when sperm are incubated with heparin. The Ca^{2+} uptake when sperm are incubated with heparin, glucose, and 8-bromo-cAMP, by contrast, was four times that seen with heparin alone. The inhibition of capacitation by glucose can be reversed in a dose-dependent manner with 8-bromo-cAMP, IBMX, and caffeine (Susko-Parrish and Parrish, 1988). In addition, intracellular cAMP levels increased twofold in sperm incubated with heparin (J. L. Susko-Parrish, unpublished data). This suggests that increasing cAMP levels and not nonspecific effects of the cAMP modulators can restore the ability of heparin to capacitate sperm in the presence of glucose. Consistent with this, a number of investigators have manipulated cAMP levels in sperm incubated with heparin and glucose by the addition of caffeine (Niwa and Ohgoda, 1988; Niwa *et al.*, 1988; Ohgoda *et al.*, 1988; Park *et al.*, 1989). This appears to reduce the variation among bulls in ability of sperm to fertilize oocytes *in vitro*.

The effects of glucose on suppression of capacitation may at first be surprising, since only in the guinea pig has glucose ever been shown to not be needed for capacitation or induction of the acrosome reaction (see Fraser and Ahuja, 1988). The bovine is unique, however, in that oviduct fluid normally has only 54–92 μM glucose (Carlson *et al.*,1970). The K_m for the bovine sperm glucose transporter is 170 μM at 22°C (Hiipakka and Hammerstedt, 1978). Although it is difficult to know the concentration of sperm metabolizing glucose in the oviduct, the K_m for glucose metabolism in the oviduct suggests that flux through glycolysis is low during *in vivo* capacitation.

Although glucose may not be metabolized *in vivo* during capacitation, we have used its ability to inhibit capacitation by heparin to further understand the mechanism of capacitation. The glucose effect does not appear to be a direct effect of glucose itself on sperm, as 3-O-methylglucose fails to inhibit capacitation (Parrish *et al.*, 1989b) and glucose does not affect [3H]heparin binding to sperm (Susko-Parrish *et al.*, 1985). The effect of glucose on inhibiting capacitation by heparin appears to involve glucose metabolism through glycolysis, as other glycolyzable substrates such as fructose and mannose also inhibit capacitation, whereas sucrose and sorbitol, nonglycolyzable sugars, do not (Parrish *et al.*, 1989b). Further, 2-deoxyglucose and 3-O-methylglucose do not inhibit heparin-induced capacitation, but fluoride inhibits glucose metabolism by sperm in a dose-dependent manner and at the same time restores the ability of heparin to capacitate sperm (Parrish *et al.*, 1989b).

Glycolytic metabolism results in an acid load excreted by sperm, but the change in extracellular pH during metabolism of glucose is not responsible for the glucose effects on capacitation (Parrish *et al.*, 1989b). By contrast, there is an alkalinization of intracellular pH (pH$_i$) during capacitation of 0.4 pH units. Metabolism of glucose over 5 hr alone results in an acidification of pH$_i$ of 0.2 pH units. In the presence of heparin and glucose there is no change in pH$_i$, suggesting that a competition between heparin effects and glucose metabolism is occurring. Results suggested that sperm incubated with heparin and glucose should reach the same pH$_i$ as the heparin controls in 12–15 hr and should be capacitated if pH$_i$ was involved in capacitation. Indeed, this was the case when over a 12-hr incubation both glucose and external pH were kept constant. Glucose therefore delays but does not absolutely inhibit capacitation.

The change in pH$_i$ could be linked to the Ca^{2+} uptake by bovine sperm, as both appear to be linear over the capacitation interval. Rises in pH$_i$ or Ca^{2+} could also activate adenylate cyclase and/or inhibit phosphodiesterases and thereby increase cAMP during capacitation. Such a model would also explain the glucose effects on capacitation. In support of this model for capacitation, others have found that bovine sperm can be stimulated to fertilize oocytes *in vitro* by caffeine or caffeine plus the calcium–proton ionophore A23187 (Hanada, 1986; Aoyagi *et al.*, 1988; Fukuda *et al.*, 1990). The resulting changes in cAMP, pH$_i$, and intracellular calcium presumably affect capacitation. However, a direct effect on capacitation, the acrosome reaction, or some other step of sperm–egg interaction has not been conclusively demonstrated. In our work, 8-bromo-cAMP was not able to capacitate sperm in the absence of heparin, and we concluded that heparin must also have additional effects on sperm in addition to elevating cAMP.

A potential additional effect of heparin on sperm could be displacement of surface proteins such as decapacitation factors. Heparin is well known in other systems to displace proteins from cell surfaces (Casu, 1985). This could lead to surface alterations in addition to modulating changes in ionic pumps and channels. Heparin is known to bind with high affinity to bovine sperm (Handrow *et al.*, 1984), and the binding appears essential for heparin to capacitate sperm as the amount of bound heparin parallels capacitating activity, and blocking of heparin binding with protamine sulfate prevents capacitation (Parrish *et al.*, 1989b). In addition the chemical N-desulfation of heparin negates both its ability to bind and capacitate sperm (Miller and Ax, 1989). Recently, heparin-binding proteins have been described in bovine seminal plasma that bind to sperm at ejaculation and appear to be required for bovine sperm to be capacitated as judged by the ability of solubilized zona proteins to induce the acrosome reaction (Miller, 1989).

Calmodulin is a ubiquitous Ca^{2+}-binding protein that can potentially regulate a number of enzymes (Means *et al.*, 1982). Although it is located in bovine sperm (Olson *et al.*, 1985), there has previously not been a role established for modulation of calmodulin during capacitation. The one exception was an observation that trifluoperizine could inhibit the ability of follicular fluid GAG to induce the acrosome reaction (Lenz *et al.*, 1982). Capacitation of bovine sperm by heparin, however, has now been shown to modulate the ability of a 28- and 30-kDa protein to bind calmodulin (Leclerc *et al.*, 1989). The ability of these proteins to bind calmodulin decreases with capacitation time and appears to be both Ca^{2+} dependent and independent. These changes could reflect increases in intracellular pH or Ca^{2+} and activation of a cAMP-dependent protein kinase. Phosphorylation of the protein or Ca^{2+} modulation of calmodulin binding could alter activity of these two proteins. The identity and location of these two proteins are currently unknown.

The proposed events of capacitation in the bovine are summarized in Fig. 1. During

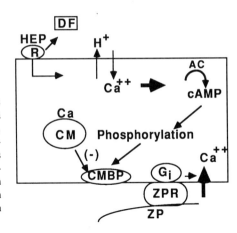

Figure 1. A proposed model for the sequence of events in capacitation of bovine spermatozoa. See the text for a description of events. Abbreviations: heparin (HEP), heparin receptor on sperm plasma membrane (R), decapacitation factor on sperm surface (DF), hydrogen ions (H^+), calcium (Ca^{++}), adenylate cyclase (AC), adenosine 3',5'-cyclic monophosphate (cAMP), calmodulin (CM), calmodulin-binding proteins (CMBP), G_i protein (G_i), zona pellucida receptor for capacitated sperm (ZPR), zona pellucida (ZP).

epididymal maturation and at ejaculation, decapacitation and heparin-binding proteins are added to the spermatozoon's surface. When sperm enter the female reproductive tract, they are removed from seminal plasma and exposed to heparin-like GAGs, probably in both the uterus and oviduct. The heparin-like GAGs bind to specific proteins, and the binding triggers the release of a decapacitation or acrosomal stabilizing factor. This triggers the activation of proton channels perhaps linked to Na^+–Ca^{2+} exchange. The cytoplasmic alkalinization ultimately results in an uptake of Ca^{2+}. Both the rise in Ca^{2+} and pH_i could lead to activation of adenylate cyclase and/or inhibition of phosphodiesterases, either of which would increase cAMP. The cAMP would activate protein kinases to modulate membrane or cellular proteins that are also regulated by calmodulin. The culmination of these events would be that the signal transduction pathway for the zona-pellucida-induced acrosome reaction would be coupled (Florman and First, 1988a; Florman *et al.*, 1989).

Strikingly absent from the model in Fig. 1 is any mention of cholesterol modulation during capacitation. Decreases in cholesterol : phospholipid ratios by removing cholesterol from sperm have been suggested to be the primary event occurring during capacitation (Davis, 1981; Langlais and Roberts, 1985). Molecules that could remove cholesterol from sperm are present in reproductive tract fluids (Langlais *et al.*, 1988; Ehrenwald *et al.*, 1990), and bovine oviduct fluid has recently been reported to be capable of removing cholesterol from bovine sperm. However, when cholesterol was removed from bovine sperm with liposomes (Ehrenwald *et al.*, 1988a,b), despite the ability of LC to now induce acrosome reactions, these sperm were not able to fertilize zona-intact bovine oocytes without LC addition, presumably because an LC-induced acrosome reaction was required. This suggests that although the sperm plasma membranes were modified by cholesterol depletion, the signal transduction system for a zona-pellucida-induced acrosome reaction was not coupled. Bovine sperm have also been shown to capacitate with heparin when polyvinylalcohol replaced albumin and no external receptor for sperm cholesterol was present (Parrish *et al.*, 1989b). Albumin was required for sperm to penetrate oocytes *in vitro*, suggesting a final cholesterol removal step in the zona-pellucida-induced acrosome reaction might occur. Although addition of cholesterol to sperm membranes can clearly modify the ability of sperm to capacitate (Go and Wolf, 1985; Fayrer-Hosken *et al.*, 1987; Ehrenwald *et al.*, 1990), the removal of cholesterol as a cause of capacitation is not clear and was therefore not included in Fig. 1.

3. OOCYTE MATURATION

An important prerequisite to fertilization is completion of the female gamete's growth and differentiation. This includes completion of maturation of the nucleus from the arrested germinal vesicle stage to arrest again at metaphase II, mucification and expansion of the cumulus cells surrounding the oocyte, and maturation of the cytoplasm such that it is competent to carry the oocyte, zygote, and embryo through fertilization and early embryo development (reviewed by First et al., 1988). As with other mammals, bovine oocytes complete growth and storage of messenger RNA while in antral follicles (Motlik and Fulka, 1986). Competence for resumption of meiosis is reached when the follicle is ≥ 1.6 mm (Motlik and Fulka, 1986), but competence for early embryo development may be reached at a later stage of follicular development (see review by First et al., 1988). The sequential development of competence by bovine oocytes for resumption of germinal vesicle breakdown (GVBD), completion of meiosis, fertilization, and embryo development has been described recently by Sato et al. (1990) from study of oocytes of various sizes and growth stages.

The mechanisms maintaining meiotic arrest in bovine oocytes appear to be similar to those in the mouse with a few exceptions. Whereas culture in or on the follicle (Foote and Thibault, 1969) maintains arrest for some bovine oocytes, the frequency of arrested oocytes is less than for rodents (reviewed by Tsafriri et al., 1987). The agents that prevent resumption of meiosis in the mouse, such as cAMP, follicular fluid, hypoxanthine, and adenosine, all cause a transient delay in meiotic resumption of bovine oocytes, but none prevents GVBD in the majority of oocytes for a prolonged period (Sirard et al., 1989). Prolonged maintenance of meiotic arrest in the germinal vesicle stage has been accomplished for bovine by combining IBMX and an invasive adenylate cyclase in the culture system (Aktas et al., 1990). However, the mechanisms controlling resumption of oocyte maturation in cattle are not well defined. Once removed from the follicle, oocytes of most species complete germinal vesicle breakdown (Tsafriri et al., 1987). A schedule of events and the timing of the events in resumption of bovine meiosis are shown in Table 1. There is much yet to be learned about the mechanisms and steps in this meiotic resumption and how it may be similar to or different from the comparable period of a mitotic cell cycle.

The bovine oocyte, like that of the mouse, becomes competent for resuming and completing meiosis I in progressive stages. The relationship of oocyte diameter to development has been described by Sato et al. (1990). As shown in Table 1, the time required for an oocyte to progress from the germinal vesicle stage to metaphase II of meiosis is ≥ 18 hr in culture. An irreversible commitment to resumption of meiosis occurs within the first hour after removal from the suppressive influence of the follicle (Sirard et al., 1988). Synthesis of stage-specific proteins is required for germinal vesicle breakdown, metaphase I, metaphase II, and arrest at metaphase II

Table 1. Schedule of Events in Resumption of Meiosis by Bovine Oocytes

Nucleus	In vitro[a]	In vivo[b]	
		Normal	Superovulation
Commitment to GVBD	<1		
Required protein synthesis	3		
GVBD	6–10	4–8	6–14
Metaphase I	12–18	20	19
Metaphase II	18–24		
First polar body lost	40–48		

[a]Hours after removal from follicle.
[b]Hours post-LH surge.

(Sirard *et al.*, 1989). Germinal vesicle breakdown requires RNA synthesis (Hunter and Moor, 1987) and likely the same cascade of Ca^{2+} signaling and sequential phosphorylation of kinases, including cyclin and maturation promoting factor (MPF), known in other organisms for regulating cell cycle transition from G_2 to M phase (Murray and Kirschner, 1989).

Unlike many other species, the polar body resulting from the first bovine meiotic division is short-lived and often not present in late metaphase oocytes (Table 1).

Expansion of cumulus cells is believed necessary in mammals to provide a large sticky mass for pickup and transport by the cells of the oviduct (Mahi-Brown and Yanagimachi, 1983). In the bovine cumulus mucification and expansion are caused primarily by follicle-stimulating hormone (Thibault *et al.*, 1975; Ball *et al.*, 1980), and the principal product of expansion is hyaluronic acid (Ball *et al.*, 1982). The extent of expansion is also influenced by components of body fluids such as blood serum (Eppig, 1980; Leibfried-Rutledge *et al.*, 1986). Recent evidence in the mouse indicates that the ability of a cumulus–oocyte complex to respond to FSH with secretion of a hyaluronic acid matrix and cumulus expansion depends on presence of the oocyte and a factor from the oocyte (Buccione *et al.*, 1990).

The most critical problem in bovine oocyte maturation is development of the competence to complete embryo development. Incompetence is displayed primarily by oocytes from small follicles that have not completed mRNA synthesis and storage, follicles less than 3 mm (Motlik and Fulka, 1986), and follicles not receiving or responding to a gonadotropin surge *in vivo*. The incompetence can be nearly completely corrected by coculture of the oocytes with $\geq 10^6$ hormonally treated granulosa or cumulus cells per milliliter (Leibfried-Rutledge *et al.*, 1989). Beneficial hormones are FSH, LH, and estradiol (Critser *et al.*, 1986; Sirard *et al.*, 1988; Saeki *et al.*, 1990). The variables influencing the developmental competence of bovine oocytes after fertilization include size of enclosing follicle, stage of estrous cycle at recovery, number of cocultured cells, number of oocytes per culture drop, presence of cumulus cells, hormone treatment and exclusion of morphologically abnormal cumulus–oocyte complexes (Leibfried-Rutledge *et al.*, 1989; Gordon and Lu, 1990).

At present the frequency of bovine blastocysts developing from *in vitro* fertilization of *in vivo*-matured oocytes is $\geq 45\%$; from immature oocytes cocultured with granulosa cells it is 23% to 63%; from culture with five to ten cumulus oocyte complexes per 50 μl medium, it is 20% to 30% (unpublished observations; Leibfried-Rutledge *et al.*, 1989).

4. INTERACTION OF SPERM AND EGG

In the mouse, sperm first attach loosely to the zona of the egg; then the O-linked oligosaccharide end of the zona protein, ZP3 interacts with a glycoprotein ligand of the sperm, suspected to be galactosyl transferase (Shur, 1989), resulting in tight binding of sperm to egg. Once bound, the peptide portion of the ZP3 protein promotes an acrosome reaction by the spermatozoa. Penetration of the sperm through the zona and sperm activation of the egg including zona hardening rapidly follow (Wassarman, 1987).

Fertilization in the bovine may be similar to that in the mouse. Bovine sperm loosely attached to the egg zona can be washed off. When capacitated bovine sperm are exposed to solubilized oocyte zonae pellucidae, they undergo acrosomal exocytosis (Florman and First, 1988a) with associated elevation of internal Ca^{2+} concentration and pH_i (Florman *et al.*, 1989). Sperm that have been pretreated with pertussis toxin have diminished response to the solubilized zonae (Florman *et al.*, 1989). These results indicate that G proteins are involved in coupling the effect of solubilized zona proteins to internal mediators of the acrosome reaction that control ion concentrations. It is interesting to note that epididymal sperm do not respond to solubilized zonae with acrosomal exocytosis (Florman and First, 1988a,b) and do not show an elevation in either internal Ca^{2+} concentration or pH_i (Florman *et al.*, 1988b, 1989). If seminal plasma is added to the epididymal sperm, they then become competent to respond to zona protein with an acrosome reaction (Florman and First, 1988a,b). It seems that seminal plasma contains one or more proteins mediating the binding or coupling of the signal transduction system to a mechanism that in part elevates Ca_i and pH_i.

When bovine oocytes are fertilized *in vitro*, the conditions required for fertilization that result in developmentally competent zygotes include incubation at bovine body temperature (39°C; Ball *et al.*, 1983), addition of capacitated sperm to the fertilization media, elimination of serum from the fertilization media, use of reduced sperm number to prevent polyspermy, and choice of bull, which effects fertilization and development rate (Leibfried-Rutledge *et al.*, 1989; Hillery *et al.*, 1990).

Historically the first calf from *in vitro* fertilization was born in 1982 (Brackett *et al.*, 1982). The first calves from *in vitro* fertilization of *in vitro*-matured oocytes were born in late 1985 (Hanada, 1986) and early 1986 (Critser *et al.*, 1986). By now the understanding of bovine fertilization discussed herein has resulted in commercial systems for *in vitro* production of embryos in use by two commercial companies and the production of large numbers of embryos for research on embryo development, embryo cloning, and gene transfer in several laboratories.

5. REFERENCES

Aktas, H., Leibfried-Rutledge, M. L., Wheeler, M. B., Rosenkrans, C. F., Jr., and First, N. L., 1990, Maintenance of meiotic arrest in bovine oocytes, *Biol. Reprod.* **42(Suppl. 1)**:91.

Aoyagi, Y., Fukii, K., Iwazumi, Y., Furudate, M., Fukui, Y., and Ono, H., 1988, Effects of two treatments on semen from different bulls on *in vitro* fertilization results of bovine oocytes, *Theriogenology* **30**:973–985.

Ball, G. D., Ax, R. L., and First, N. L., 1980, Mucopolysaccharide synthesis accompanies expression of bovine cumulus–oocyte complexes *in vitro*, in: *Functional Correlates of Hormone Receptors in Reproduction* (V. B. Mahesha, T. G. Muldoon, B. B. Saxena, and W. A. Sadler, eds.), Elsevier/North-Holland, New York, pp. 561–563.

Ball, G. D., Bellin, M. E., Ax, R. L., and First, N. L., 1982, Glycosaminoglycans in bovine cumulus oocyte complexes: Morphology and chemistry, *Mol. Cell. Endocrinol.* **28**:113–122.

Ball, G. D., Leibfried, M. L., Lenz, R. W., Ax, R. L., Bavister, B. D., and First, N. L., 1983, Factors affecting successful *in vitro* fertilization of bovine follicular oocytes, *Biol. Reprod.* **28**:717–725.

Bousquet, D., and Brackett, B. G., 1982, Penetration of zona-free hamster ova as a test to assess fertilizing ability of bull sperm after frozen storage, *Theriogenology* **17**:199–213.

Brackett, B. G., Bousquet, D., Boice, M. L., Donawich, W. J., Evans, J. F., and Dressel, M. A., 1982, Normal development following *in vitro* fertilization in the cow, *Biol. Reprod.* **27**:147–158.

Buccione, R. B. C., Caron, P. J., and Eppig, J. J., 1990, FSH-induced expansion of the mouse cumulus oophorus *in vitro* is dependent upon a specific factor(s) secreted by the oocyte, *Dev. Biol.* **138**:16–25.

Carlson, D., Black, D. L., and Howe, G. R., 1970, Oviduct secretion in the cow, *J. Reprod. Fertil.* **22**:549–552.

Casu, B., 1985, Structure and biological activity of heparin, *Adv. Carbohyd. Chem. Biochem.* **43**:51–134.

Crister, E. S., Leibfried-Rutledge, M. L., Eyestone, W. E., Northey, D. L., and First, N. L., 1986, Acquisition of developmental competence during maturation *in vitro*, *Theriogenology* **25**:150.

Davis, B. K., 1981. Timing of fertilization in mammals: Sperm cholesterol phospholipid ratio as a determinant of the capacitation interval, *Proc. Natl. Acad. Sci. U.S.A.* **78**:7560–7564.

Ehrenwald, E., Parks, J. E., and Foote, R. H., 1988a, Cholesterol efflux from bovine sperm. I. Induction of the acrosome reaction with lysophosphatidylcholine after reducing sperm cholesterol, *Gamete Res.* **20**:145–157.

Ehrenwald, E., Parks, J. E., and Foote, R. H., 1988b, Cholesterol efflux from bovine sperm: II. Effect of reducing sperm cholesterol on penetration of zona-free hamster and *in vitro* matured bovine ova, *Gamete Res.* **20**:413–420.

Ehrenwald, E., Foote, R. H., and Parks, J. E., 1990, Cholesterol efflux from bovine sperm: Bovine oviductal fluid components and their potential role in sperm cholesterol efflux, *Exp. Biol. Med.* **25**:195–204.

Eppig, J. J., 1980, Role of serum FSH stimulated cumulus expansion by mouse oocyte–cumulus cell complexes *in vitro*, *Biol. Reprod.* **23**:629–633.

Fayrer-Hosken, R. A., Brackett, B. G., and Brown, J., 1987, Reversible inhibition of rabbit sperm-fertilizing ability by cholesterol sulfate, *Biol. Reprod.* **36**:878–883.

First, N. L., and Parrish, J. J., 1987, *In vitro* fertilization of ruminants, *J. Reprod. Fertil.* **34**(Suppl.):151–165.

First, N. L., Leibfried-Rutledge, M. L., Sirard, M. A., 1988, Cytoplasmic control of oocyte maturation and species differences in the development of maturational competence, in: *Meiotic Inhibition: Molecular Control of Meiosis*, (F. P. Haseltine, N. L. First, P. Patinelli, and C. O. Liss, eds.), Alan R. Liss, New York, pp. 1–46.

Florman, H. M., and First, N. L., 1988a, The regulation of acrosomal exocytosis. I. Sperm capacitation is required for the induction of acrosome reactions by the bovine zona pellucida *in vitro*, *Dev. Biol.* **128**:453–463.

Florman, H. M., and First, N. L., 1988b, The regulation of acrosomal exocytosis. II. The zona pellucid-inducing acrosome reaction of bovine spermatozoa is controlled by extrinsic positive regulatory elements, *Dev. Biol.* **128**:464–473.

Florman, H. M., Tombes, R. M., First, N. L., and Babcock, D. F., 1989, An adhesion-associated agonist from the zona pellucida activates G protein-promoted elevations of internal Ca^{2+} and pH that mediate mammalian sperm acrosomal exocytosis, *Dev. Biol.* **135**:133–146.

Foote, W. D., and Thibault, C., 1969, Recherches experimentales sur la maturation *in vitro* des ovocytes de truie et de veau, *Ann. Biol. Anim. Biochem. Biophys.* **9**:327–349.

Fraser, L. R., and Ahuja, K. K., 1988, Metabolic and surface events in fertilization, *Gamete Res.* **20**:491–519.

Fukuda, A., Ichikawa, M., Naito, K., and Toyoda, Y., 1990, Birth of normal calves resulting from bovine oocytes matured, fertilized, and cultured with cumulus cells *in vitro* up to the blastocyst stage, *Biol. Reprod.* **42**:114–119.

Go, K. J., and Wolf, D. P., 1985, Albumin-mediated changes in sperm sterol content during capacitation, *Biol. Reprod.* **32**:145–153.

Gordon, I., and Lu, K. H., 1990, Production of embryos *in vitro* and its impact on livestock production, *Theriogenology* **33**:77–88.

Hanada, A., 1986, *In vitro* fertilization of bovine oocytes, *Consult. Anim. Sci.* **258**:10–15.

Handrow, R. R., Lenz, R. W., and Ax, R. L., 1982, Structural comparisons among glycosaminoglycans to promote an acrosome reaction in bovine spermatozoa, *Biochem. Biophys. Res. Commun.* **107**:1326–1332.

Handrow, R. R., Boehm, S. K., Lenz, R. W., Robinson, J. A., and Ax, R. L., 1984, Specific binding of the glycosaminoglycan ^{3}H-heparin to bovine, monkey and rabbit spermatozoa *in vitro*, *J. Androl.* **5**:51–63.

Handrow, R. R., First, N. L., and Parrish, J. J., 1989, Calcium requirement and increased association with bovine sperm during capacitation by heparin, *J. Exp. Zool.* **252**:174–182.

Hiipakka, R. A., and Hammerstedt, R. H., 1978, 2-Deoxyglucose transport and phosphorylation by bovine sperm, *Biol. Reprod.* **19**:368.

Hillery, F. L., Parrish, J. J., and First, N. L., 1990, Bull specific effect on fertilization and embryo development *in vitro*, *Theriogenology* **33**:249.

Hunter, A., and Moor, R. M., 1987, Stage dependent effects of inhibiting ribonucleic acids and protein synthesis on meiotic maturation of bovine oocytes *in vitro*, *J. Diary Sci.* **70**:1646–1652.

Hunter, R. H. F., and Wilmut, I., 1984, Sperm transport in the cow: Periovulatory redistribution of viable cells within the oviduct, *Reprod. Nutr. Dev.* **24**:597–608.

Lambert, R. D., Sirard, M. A., Bernard, C., Beland, R., Rioux, J. E., Leclerc, P., Menard, D. P., and Bedoya, M., 1985, *In vitro* fertilization of bovine oocytes matured *in vivo* and collected at laparoscopy, *Theriogenology* **25**:117–133.

Langlais, J., and Roberts, K. D., 1985, A molecular membrane model of sperm capacitation and the acrosome reaction of mammalian spermatozoa, *Gamete Res.* **12**:183–224.

Langlais, J., Kan, F. W. K., Granger, L., Raymond, L., Bleau, G., and Roberts, K. D., 1988, Identification of sterol acceptors that stimulate cholesterol efflux from human spermatozoa during *in vitro* capacitation, *Gamete Res.* **20**:185–201.

Leclerc, P., Langlais, J., Lambert, R. D., Sirard, M. A., and Chafouleas, J. G., 1989, Effect of heparin on the expression of calmodulin-binding proteins in bull spermatozoa, *J. Reprod. Fertil.* **85**:615–622.

Lee, C. N., and Ax, R. L., 1984, Concentrations and composition of glycosaminoglycans in the female bovine reproductive tract, *J. Diary Sci.* **67**:2006–2009.

Lee, C. N., Handrow, R. R., Lenz, R. W., and Ax, R. L., 1985, Interactions of seminal plasma and glycosaminoglycans on acrosome reactions in bovine spermatozoa, *Gamete Res.* **12**:345–355.

Leibfried-Rutledge, M. L., Critser, E. S., and First, N. L., 1986, Effects of fetal calf serum and bovine serum albumin on *in vitro* maturation and fertilization of bovine and hamster cumulus–oocyte complexes, *Biol. Reprod.* **35**:850–857.

Leibfried-Rutledge, M. L., Critser, E. S., Parrish, J. J., and First, N. L., 1989, *In vitro* maturation and fertilization of bovine oocytes, *Theriogenology* **31**:61–74.

Lenz, R. W., Ax, R. L., Grimek, H. J., and First, N. L., 1982, Proteoglycan from bovine follicular fluid enhances an acrosome reaction in bovine spermatozoa, *Biochem. Biophys. Res. Commun.* **106**:1092–1098.

Mahi-Brown, C. A., and Yanagimachi, R., 1983, Parameters influencing ovum pickup by oviductal fimbria in the golden hamster, *Gamete Res.* **8**:1–10.

Means, A. R., Tash, J. S., and Chafouleas, J. G., 1982, Physiological implications of the presence, distribution and regulation of calmodulin in eukaryotic cells, *Physiol. Rev.* **62**:1–38.

Miller, D. J., 1989, The interaction between glycosaminoglycans and sperm and its regulation by seminal plasma, Ph.D. thesis, University of Wisconsin, Madison.

Miller, D. J., and Ax, R. L., 1989, Chemical N-desulfation of heparin negates its ability to capacitate bovine spermatozoa, *Gamete Res.* **23**:451–465.

Motlik, J., and Fulka, J., 1986, Factors affecting meiotic competence in pig oocytes, *Theriogenology* **251**:87–96.

Murray, A. W., and Kirschner, M. W., 1989, Dominos and clocks: The union of two views of the cell cycle, *Science* **246**:614–621.

Niwa, K., and Ohgoda, O., 1988, Synergistic effect of caffeine and heparin on *in-vitro* fertilization of cattle oocytes matured in culture, *Theriogenology* **30**:733–741.

Niwa, K., Ohgoda, O., and Yuhara, M., 1988, Effects of caffeine in media for pretreatment of frozen-thawed sperm on *in vitro* penetration of cattle oocytes, in: *Proceedings 11th International Congress of Animals Reproduction and Artificial Insemination, Vol. 3, International Congress of Animal Reproduction and Artificial Insemination*, Dublin, pp. 346–348.

Ohgoda, O., Niew, K., Yuhara, M., Takahashi, S., and Kanoya, K., 1988, Variation in penetration rates *in vitro* of bovine follicular oocytes do not reflect conception rates after artificial insemination using frozen semen from different bulls, *Theriogenology* **29**:1375–1381.

Oliphant, G., and Singhas, C. A., 1979, Iodination of rabbit sperm plasma membrane: Relationship of specific surface proteins to epididymal function and sperm capacitation, *Biol. Reprod.* **21**:937–944.

Olson, G. E., Winfrey, V. P., Garbers, D. L., and Noland, T. D., 1985, Isolation and characterization of a macromolecular complex associated with the outer acrosomal membrane of bovine spermatozoa, *Biol. Reprod.* **33**:761–779.

Park, C.-K., Ohgoda, O., and Niwa, K., 1989, Penetration of bovine follicular oocytes by frozen-thawed spermatozoa in the presence of caffeine and heparin, *J. Reprod. Fertil.* **86**:577–582.

Parrish, J. J., 1989, Capacitation of spermatozoa *in vivo*: Physiochemical and molecular mechanisms, *J. Mol. Androl.* **1**:87–111.

Parrish, J. J., Susko-Parrish, J. L., and First, N. L., 1985, Effect of heparin and chondroitin sulfate on the acrosome reaction and fertility of bovine sperm *in vitro*, *Theriogenology* **24**:537–549.

Parrish, J. J., Susko-Parrish, J. L., Leibfried-Rutledge, M. L., Critser, E. S., Eyestone, W. H., and First, N. L., 1986, Bovine *in vitro* fertilization with frozen-thawed semen, *Theriogenology* **25**:591–600.

Parrish, J. J., Susko-Parrish, J. L., Winer, M. A., and First, N. L., 1988, Capacitation of bovine sperm by heparin, *Biol. Reprod.* **38**:1171–1180.

Parrish, J. J., Susko-Parrish, J. L., Handrow, R. H., Sims, M. M., and First, N. L., 1989a, Capacitation of bovine sperm by oviduct fluid, *Biol. Reprod.* **40**:1020–1025.

Parrish, J. J., Susko-Parrish, J. L., and First, N. L., 1989b, Capacitation of bovine sperm by heparin: Inhibitory effect of glucose and role of intracellular pH, *Biol. Reprod.* **41**:683–699.

Saeki, K., Leibfried-Rutledge, M. L., and First, N. L., 1990, Are fetal calf serum and hormones necessary during *in vitro* maturation of cattle oocytes for subsequent development? *Theriogenology* **33**:316.

Sato, E., Matsuo, M., and Miyamoto, H., 1990, Meiotic maturation of bovine oocytes *in vitro*: Improvement of meiotic competence by dibutyryl cyclic adenosine 3′,5′-monophosphate, *J. Anim. Sci.* **68**:1182–1187.

Shur, B. D., 1989, Galactosyltransferase as a recognition molecule during fertilization and development, in: *The Molecular Biology of Fertilization* (H. Schatten and G. Schatten, eds.), Academic Press, New York, pp. 37–71.

Sirard, M. A., and Lambert, R. D., 1985, *In vitro* fertilization of bovine follicular oocytes obtained by laparoscopy, *Biol. Reprod.* **33**:487–494.

Sirard, M. A., Lambert, R. D., Menard, D. P., and Bedoya, M., 1985, Pregnancies after *in-vitro* fertilization of cow follicular oocytes, their incubation in rabbit oviduct and their transfer to the cow uterus, *J. Reprod. Fertil.* **75**:551–556.

Sirard, M. A., Leibfried-Rutledge, M. L., Parrish, J. J., Ware, C. M., and First, N. L., 1988, The culture of bovine oocytes to obtain developmentally competent embryos, *Biol. Reprod.* **39**:546–552.

Sirard, M. A., Florman, H. M., Leibfried-Rutledge, M. L., Barnes, F. L., Sims, M. L., and First, N. L., 1989, Timing of nuclear progression and protein synthesis necessary for meiotic maturation of bovine oocytes, *Biol. Reprod.* **40**:1257–1263.

Susko-Parrish, J. L., and Parrish, J. J., 1988, Glucose and cAMP modulators effect capacitation of bovine sperm by heparin, *Biol. Reprod.* **38**(Suppl. 1):94.

Susko-Parrish, J. L., Parrish, J. J., and Handrow, R. R., 1985, Glucose inhibits the action of heparin by an indirect mechanism, *Biol. Reprod.* **32**(Suppl. 1):80.

Thibault, C., Gerard, M., and Menezo, Y., 1975, Preovulatory and ovulatory mechanisms in oocyte maturation, *J. Reprod. Fertil.* **45**:604–610.

Tsafriri, A., Reich, R., and Abisogun, A. O., 1987, The ovarian egg and ovulation, in: *Marshall's Physiology of Reproduction*, Vol. III (G. E. Lemming, ed.), Churchill Livingstone, London (in press).

Wassarman, P. M., 1987, Early events in mammalian fertilization, *Annu. Rev. Cell Biol.* **3**:109–142.

19

Fertilization in the Rhesus Monkey

Barry D. Bavister, Dorothy E. Boatman, Patricia M. Morgan, and Pradeep K. Warikoo

1. INTRODUCTION

The nonhuman primates occupy a unique niche among species suitable for experimental studies on fertilization and early development. Studies with nonhuman primates provide an important link between the huge literature derived from work on rodents and other common laboratory species and the relatively small amount of data on humans. Research with rodent species continues to provide the bulk of basic information on early development, in large part because of the ready availability of oocytes and embryos, but it is difficult to know to what extent information can be extrapolated to primate species, including humans. Information is increasingly available from the numerous human *in vitro* fertilization (IVF) programs now in operation. However, the amount and quality of basic scientific data on early development that can be obtained from this source are restricted, in part by ethical constraints (Austin, 1990) and in part by conflicting priorities between research and clinical needs for the supply of oocytes and embryos. For example, most "excess" human IVF embryos in the U.S.A. are destined for cryopreservation or for use in a donor program and are thus unsuitable for any research protocol that might compromise their viability. In contrast, there are no prohibitions on the use of fertilized eggs or embryos of nonhuman primates for research, while the reproductive physiology and embryology of the Old-World nonhuman primates are sufficiently similar to humans that direct extrapolation of concepts is possible.

At present, the major rate-limiting factor in research on fertilization and early development with nonhuman primates is the scarcity of oocytes (Boatman, 1987), which is largely a result of the inadequacy of ovarian superstimulation protocols. Additionally, a major commitment of effort and expense is required to work with many of these animals. Some investigators believe that the potential benefits of research with nonhuman primates outweigh the difficulties, and substantial

BARRY D. BAVISTER, DOROTHY E. BOATMAN, PATRICIA M. MORGAN, AND PRADEEP K. WARIKOO • Wisconsin Regional Primate Research Center, University of Wisconsin, Madison, Wisconsin 53715. *Present address of P.M.M.*: Department of Biochemistry, University College, Galway, Ireland. *Present address of P.K.W.*: Department of IVF, Saginaw General Hospital, Saginaw, Michigan 48602.

A Comparative Overview of Mammalian Fertilization, edited by Bonnie S. Dunbar and Michael G. O'Rand. Plenum Press, New York, 1991.

progress has been made in establishing *in vitro* protocols for oocyte maturation, fertilization, and early embryonic development, such that it is now feasible to make serious, in-depth studies on these topics.

Information on mechanisms regulating sperm fertilizing ability, oocyte maturation, the fertilization process, and early cleavage events is most readily obtained from IVF experiments. The value of IVF in nonhuman primates has been recognized at least since the pioneering days of Lewis and Hartman and their colleagues 50 years ago (Hartman and Corner, 1941; Lewis and Hartman, 1941), but proven successful techniques for IVF in these animals have not been available until recently. The first demonstrations of IVF (sperm penetration) in a nonhuman primate species were by Cline, Gould, and their colleagues in the early 1970s (Cline *et al.*, 1972; Gould *et al.*, 1973) using the squirrel monkey (*Saimiri sciureus*). In a large series of experiments with the squirrel monkey, IVF was achieved quite routinely, but cleavage of the embryos was limited, and no pregnancies were obtained after transfer of IVF embryos (Kuehl and Dukelow, 1979). A significant advance by Hodgen and his co-workers in 1982 was the demonstration of IVF in 22 eggs from cynomolgus monkeys (*Macaca fascicularis*); some of the embryos reached the early morula stage (Kreitman *et al.*, 1982).

Our laboratory began to study IVF in the rhesus monkey in 1981. The rationale for this research was to provide a reliable primate model for investigating very early events in embryonic development, specifically the regulation of sperm fertilizing ability, oocyte maturation, the process of fertilization, embryo cleavage, and differentiation. In this chapter, we describe the development of procedures for supporting these events *in vitro* using the rhesus monkey, and we outline some of the basic information that has been gathered to date. Data are presented on the maturation of rhesus monkey oocytes, on sperm capacitation and egg activation *in vitro*, and on the development of IVF eggs as a criterion for their viability and normality. Much remains to be done, and it is hoped that this account may stimulate renewed interest in research on fertilization and associated events in nonhuman primates, because of the unique contribution that such studies can bring to our total knowledge of early development.

2. CAPACITATION OF RHESUS SPERM *IN VITRO*

Spermatozoa of several species, e.g., mouse, hamster, rat, human, rabbit, cattle, can be capacitated (acquire the ability to undergo a physiological acrosome reaction: Austin, 1951; Chang, 1951; Yanagimachi, 1981; Bavister, 1986) by incubation for several hours in a simple culture solution containing protein, usually serum albumin (Bavister, 1986). We first tested this approach for capacitation of rhesus sperm. Rhesus monkey semen was collected by electroejaculation (Mastroianni and Manson, 1963) from proven breeders. This is a simple, nontraumatic procedure that has been used for several years to obtain sperm for artificial insemination of the breeding colony at the Wisconsin Regional Primate Center. After the coagulum had liquefied at 37°C, the sperm were washed twice in a HEPES-buffered version of our standard TALP culture medium (Boatman, 1987; Bavister *et al.*, 1983a,b; Bavister, 1989). Sperm capacitation was evaluated as (1) expression of hyperactivated motility (Yanagimachi, 1970, 1981; Boatman and Bavister, 1984), (2) ability to fuse with hamster zona-free eggs (Yanagimachi *et al.*, 1976; Boatman and Bavister, 1984), and (3) ability to penetrate the intact zonae pellucidae of rhesus oocytes (Boatman and Bavister, 1984). Hyperactivated motility, which had not previously been described in sperm of Old-World primates, is known to be closely linked to acquisition of fertilizing ability in other species (Yanagimachi, 1970, 1981; Boatman and Bavister, 1984).

When washed rhesus sperm were incubated in TALP medium under 5% CO_2 in air at 37°C, a high proportion showed vigorous motility for 36 hr or longer. However, the sperm trajectories were linear, and there was no hyperactivated motility (Boatman and Bavister, 1984). The ability of rhesus sperm to penetrate hamster zona-free eggs was also assessed as an indicator of capacitation (see Bavister, 1990). Although rhesus sperm showed a tenacious binding affinity for the plasma membrane of hamster zona-free eggs, very few eggs contained unequivocally decondensing sperm heads (Fig. 1). We concluded that rhesus monkey sperm were not capacitated (or failed to undergo acrosome reactions) during incubation in simple protein-containing culture medium and that the zona-free penetration assay was unsuitable for evaluating the capacitation status of rhesus sperm.

Next, we tested the hamster reproductive tract as a xenogenous "incubator" for rhesus sperm. Washed rhesus sperm were instilled into the hamster uterus *in situ*, and after 4 hr of incubation, some of the recovered sperm showed large-amplitude flagellar bending, similar to hyperactivated motility (B. D. Bavister and D. E. Boatman, unpublished data). Since these changes had not been found in sperm even during prolonged incubation (up to 48 hr) in simple culture medium containing serum albumin, it seemed that hyperactivated motility might be indicative of capacitation in rhesus sperm. Attention was next turned to stimulation of hyperactivated sperm motility by cyclic nucleotide mediators.

Changes in sperm cAMP levels have been implicated in capacitation in guinea pig sperm (Hyne and Garbers, 1979). Furthermore, in the golden hamster, epinephrine is known to be capable of stimulating capacitation and the acrosome reaction (Leibfried and Bavister, 1982; Meizel *et al.*, 1984), which might imply a regulatory role for cAMP in this species also. The hypothesis was tested that cyclic nucleotide mediators could induce capacitation/acrosome reactions in rhesus monkey sperm (Boatman and Bavister, 1984). Sperm were incubated at 5 to 10 $\times 10^6$/ml for up to 24 hr with various combinations of cyclic nucleotide mediators and sampled at regular intervals for evaluation of motility. Sperm motility was assessed qualitatively at 37°C for changes in progressive motility and for signs of hyperactivation. In addition, changes in the proportions of agglutinated sperm were noted.

The combination of 1.0 mM caffeine plus 1.0 mM dibutyryl cAMP (dbcAMP) produced substantial levels of hyperactivation (about 30% to 60% of the freely motile sperm) (Table 1). The maximal hyperactivation response was obtained about 1.5 hr from the start of treatment. Agglutination of sperm began soon after addition of caffeine/dbcAMP and increased progressively. When dbcGMP was substituted for dbcAMP in the presence of caffeine, it was ineffective in stimulating hyperactivation. Also, dbcAMP, used alone, was ineffective, as was caffeine alone at 1.0 mM. The hyperactivation observed in rhesus sperm was discontinuous; that is, a spermatozoon would typically show a high-amplitude flagellar waveform in the classic "figure of eight" pattern (Yanagimachi, 1970, 1981) for a second or two, then dart off with a straight trajectory for a few more seconds before repeating the sequence. The overall path of hyperactivated sperm was characterized by frequent changes of direction. This was clearly demonstrated using videomicrographic recordings of sperm motility patterns (Boatman and Bavister, 1984). Although the trajectories of stimulated (hyperactivated) and control sperm were strikingly different, the actual distance traveled per unit time (velocity) was not significantly different between the two categories (e.g., 155 and 144 μm/sec, respectively: Boatman and Bavister, 1984).

Sperm treated with 1 mM cAMP and 1 mM caffeine were able to penetrate the zonae of nonmature eggs and to fertilize mature eggs obtained from cycling rhesus monkeys (Table 2). Some of the fertilized eggs cleaved once or twice. In contrast, sperm that were incubated without caffeine/dbcAMP were completely unable to penetrate the zonae, and, with the inverted microscope, these sperm could be seen frequently colliding with the zonae but failing to remain attached.

These experiments showed that rhesus monkey sperm are not readily capacitated by incubation in simple culture media that have proven successful for sperm of many other species (Rogers, 1978) but, like other sperm, do respond to agents mediating the cyclic nucleotide pathway (Tash and Means, 1983). This result, which has been confirmed by other laboratories

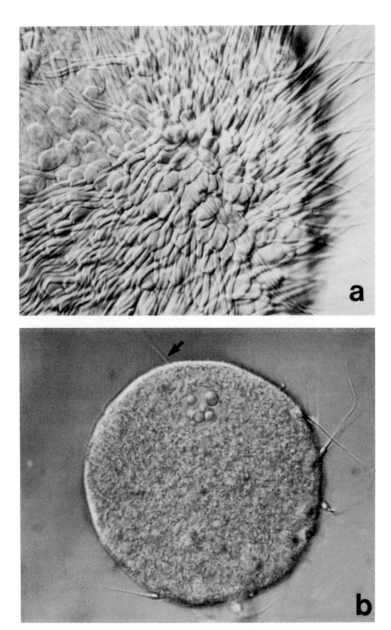

Figure 1. Interaction of rhesus monkey sperm with hamster zona-pellucida-free eggs. Rhesus sperm show very strong affinity for the hamster egg plasma membrane and can bind in large numbers (a). Most sperm cannot be removed from the egg plasma membrane by pipetting. However, only rarely were decondensation and male pronucleus formation seen (b) following incubation of sperm in medium TALP. Arrow points to sperm tail belonging to pronucleus. DIC optics.

Table 1. Effects of Cyclic Nucleotide Mediators on
the Motility of Rhesus Monkey Spermatozoa *in Vitro*[a]

Additions to medium	Change in progressive velocity	Increase in sperm agglutination	Acquistition of hyperactivated motility
Dibutyryl cGMP, 1–5 mM	0	0	0
Dibutyryl cAMP, 1–5 mM	0	0	0
Caffeine or theophylline: 10 μm–1 mM	0 to + 3	+ 1 to + 2	0
10 mM	− 1 to + 1	+ 5	+ 1
Caffeine and:			
dbcGMP 1mM	0 to + 3	+ 1 to + 2	0
dbcAMP 1mM	− 1 to + 1	+ 2	+ 3
Control	0	0	0

[a]Sperm were incubated in medium TALP containing 3 mg/ml bovine serum albumin (Fraction V, Sigma Chemical Co., St. Louis, MO). Mediators were added at 0.5 hr, and observations were made from 1.5 to 6.0 hr of incubation, using a minimum of six sperm samples from different males. Values are subjective assessments of change in characteristic measured relative to control value (velocity and agglutination) or of degree of hyperactivation. Scale represents no change or no response (0) up to maximum (5). Modified from Boatman and Bavister (1984).

Table 2. Stimulatory Effect of Caffeine and Dibutyryl cAMP on
Rhesus Monkey Sperm Penetration *in Vitro* of Rhesus Oocytes[a]

Treatment of sperm	No. of replicates	Number of oocytes penetrated/no. inseminated			No. of oocytes cleaved/ no. inseminated
		Nonmature oocytes	Mature oocytes	Total oocytes	
Caffeine + dbcAMP	4	$^{13}/_{16}$	$^{18}/_{18}$	$^{31}/_{54}a$	$^{16}/_{18}a$
Control	4	$^{0}/_{34}$	$^{0}/_{10}$	$^{0}/_{44}b$	$^{0}/_{10}b$

[a]Oocytes were randomly selected for incubation with control or with treated sperm. Caffeine and dbcAMP were at 1.0 mM each during the last 1.5 hr of sperm preincubation, then reduced to 0.1 mM each (by dilution) at insemination of IVF media drops. Nonmature (nonviable) oocytes were counted as penetrated if sperm heads were observed deeply embedded in the zona pellucida, within the perivitelline space, or lying on the vitellus. Totals with different superscripts are significantly different ($p < 0.001$). Modified from Boatman and Bavister (1984).

using macaque sperm (Balmaceda *et al.*, 1984; Binor *et al.*, 1988; Cranfield *et al.*, 1989; Wolf *et al.*, 1989), raises several interesting questions. First, why do rhesus sperm behave differently from human sperm in regard to capacitation in simple culture media? The answer to this question must await further knowledge about the biochemical pathways involved in the regulation of fertilizing ability in primate sperm. Secondly, can stimulating factors present in the rhesus oviduct or in the cumulus oophorus perform a similar function to the cyclic nucleotide mediators used in these experiments? There is evidence that factors in the hamster cumulus oophorus are capable of stimulating acrosome reactions in fertilizing sperm, leading to increased sperm penetration into eggs (Bavister, 1982; Boatman, 1990). To date, we have not been able to demonstrate penetration of cumulus-intact, *in vivo*-matured rhesus eggs by sperm that were not treated with cyclic nucleotide mediators.

A third basic question is: Is the zona pellucida necessary for induction of the acrosome reaction in rhesus sperm? In the mouse, a zona protein (ZP3) is involved in the induction of sperm acrosome reactions (Wassarman and Bleil, 1982). Although there is some indirect evidence supporting the concept of zona-induced acrosome reactions in human sperm (Cross *et al.*, 1988), definitive data are lacking. Despite the lack of direct evidence on this mechanism in primates, it is striking that rhesus sperm stimulated with cyclic nucleotide mediators can penetrate rhesus zonae (Table 2) but are substantially less reactive in fusing with zona-free hamster eggs (Binor *et al.*, 1988). It is assumed that the hamster zona-free egg penetration assay assesses the incidence of acrosome-reacted sperm (Bavister, 1990). If this is true, then if rhesus monkey sperm were acrosome-reacted in our experiments, they should have been capable of fusing with zona-free eggs, as in most other species (Yanagimachi *et al.*, 1976; Yanagimachi, 1981). One possibility is that rhesus sperm may be capable of fusing with zona-free eggs following the acrosome reaction but fail to decondense, which would make detection of sperm–egg fusion very difficult under the light microscope. Nuclei of sperm from different species show wide variations in decondensation ability in hamster egg cytoplasm (Naish *et al.*, 1988). Spontaneous activation of hamster eggs in culture (Bavister, 1989, 1990) may preclude detection of fusion (as evidenced by sperm decondensation) in sperm that were very slow in undergoing decondensation. Alternatively, more recent studies show that the hamster zona-free assay may be sensitive to factors, additional to occurrence of the acrosome reaction, that are related to fusability of sperm (Moreno and Barros, 1990). Additional data are required to determine why rhesus sperm treated with caffeine and dbcAMP together are more reactive with zona-intact homologous eggs than with zona-free hamster eggs.

3. OVARIAN STIMULATION AND OOCYTE COLLECTION

Having shown that cyclic nucleotide mediators can stimulate capacitation in rhesus monkey sperm, we next turned attention to the collection of viable oocytes. Oocytes were collected initially from gonadotropin-stimulated, regularly cycling rhesus monkeys. Pregnant mare serum gonadotropin (PMSG) was mostly used (approx. 2100 IU over days 4 through 14 to 16 of the cycle) followed by 4000 IU human chorionic gonadotropin (hCG) in a single dose 30 hr before laparoscopic aspiration of follicles (Bavister *et al.*, 1983a). As many as 40 oocytes have been obtained from one treatment cycle, although the mean number per animal was 21 ± 14.8 (S.D.), 80% of which were classified as mature (Boatman *et al.*, 1986). Unfortunately, PMSG is very immunogenic in the rhesus monkey, and only one treatment cycle per animal is possible (Bavister *et al.*, 1986). Use of clomiphene and Pergonal (human menopausal gonadotropin, hMG) was not very satisfactory in our experience; the numbers of mature oocytes recovered were very low (mean of four per cycle), and the animals' responses have been variable (Boatman *et al.*, 1986). Treatment regimens using Pergonal and Metrodin (urofollitropin) plus Pergonal in twice-daily injections have been more successful (VanderVoort *et al.*, 1989). However, the ideal stimulation regimen would use rhesus monkey FSH; we have recently begun to extract FSH activity from the urine of ovariectomized rhesus monkeys (Matteri, *et al.*, 1991). This preparation has been tested in radioreceptor and Sertoli-cell assays, and it is also potent in stimulating follicle development in golden hamsters (Matteri *et al.*, 1991). Trials with rhesus monkeys are planned.

In order to monitor the response to gonadotropin regimens, ovaries can be directly examined by laparoscopy, and serum estradiol concentrations are usually measured daily or twice daily. Laparoscopy is minimally invasive but requires general anesthesia, and rapid RIAs for estradiol are somewhat labor intensive. Consequently, we examined ultrasonography as a tool for monitoring the growth of follicles in both stimulated and normally cycling rhesus monkeys (Morgan *et al.*, 1987a). It was unnecessary to anesthetize animals for the procedure. Follicles with diameters < 2 mm were visible by ultrasonography; follicles of this size were not detectable by laparoscopy, possibly because they are located deep within the ovary. In unstimulated monkeys, the preovulatory follicle became invisible by ultrasonography 1 to 2 days prior to ovulation, a phenomenon also observed in some naturally cycling women (Renaud et al., 1980; Polan et al., 1982). In stimulated monkeys, the ovarian responses could easily be examined daily. We concluded (Morgan et al., 1987a) that ultrasonography is a useful, noninvasive procedure for monitoring ovarian responses that can be repeated frequently on the same animal. This technique is especially helpful for evaluation of new gonadotropin stimulation protocols.

Oocytes were collected from gonadotropin-stimulated monkeys using a vacuum aspiration system (Bavister et al., 1983a) or from unstimulated animals using a 1-ml hypodermic syringe and needle (Morgan et al., 1987b). For oocyte maturation experiments and for obtaining a supply of intact zonae pellucidae (see below), nonmature (GV) oocytes were collected from excised ovaries (Bavister, 1987). In all cases, care was taken to minimize trauma to the oocytes before they were placed in culture (Moor and Cran, 1980). Oocytes were collected initially in warm (35°C) HEPES-buffered TALP medium, pH 7.4. Oocytes were recovered from the follicular aspirates as quickly as possible and placed into drops of TALP medium for IVF (see below) or for continuation of maturation, as appropriate. The drops of culture medium were covered with silicone oil (Bavister, 1989) to prevent evaporation and to reduce pH changes during initial examination and manipulations (e.g., insemination).

Our original procedures and criteria for assessing oocyte maturity (Bavister et al., 1983a) have since been modified (Boatman, 1987). A rapid initial assessment is performed with a dissecting microscope (fitted with a warm stage) in order to process all of the oocytes as quickly as possible. This is followed by a more detailed examination with an inverted microscope equipped with Nomarski optics and an environment-control chamber that maintains the entire stage area at 37°C. A simple device enclosing the culture dish (Bavister, 1988) allows a constant flow of humidified 5% CO_2 in air to be passed over the culture medium so that pH is maintained during the examination period. If necessary, the cumulus and corona layers are stretched out with sterile hypodermic needles and attached to the bottom of the culture dish so that the oocyte becomes more visible. Criteria of oocyte normality and maturity include appearance of the cumulus and corona cell layers, visibility of the zona pellucida, presence or absence of a perivitelline space, the shape of the oocyte (spherical or otherwise), color of the oocyte cytoplasm (optimally light brown), and the presence of a polar body. On the basis of these criteria, oocyte–cumulus complexes are classified as immature, nearly mature, mature, or atretic/degenerate. Oocytes that appear to be capable of undergoing maturation in vitro are examined every 6 to 8 hr and reclassified.

As a general rule, oocytes remain in culture for a minimum of 6 to 8 hr before they are inseminated. If prolonged culture is necessary to achieve maturation, the medium is usually changed before insemination is carried out. The maturation status of oocytes collected from PMSG-stimulated rhesus monkeys has a marked effect on their development subsequent to IVF (Table 3) (Boatman et al., 1986). Females in which the overall development of mature oocytes was high (≥2 or ≥50% of maximum score; see footnotes to Table 3) had higher levels of plasma estradiol over the treatment cycle and yielded a lower percentage of dismature oocytes than females in which development of mature oocytes was low (<2 or <50% of maximum) (see also Section 6).

4. *IN VITRO* FERTILIZATION

Current procedures for rhesus IVF have undergone little change from our original published protocols (Bavister et al., 1983a). The IVF medium is a modified Tyrode's solution originally devised for hamster IVF (Bavister and Yanagimachi, 1977) that is supplemented with four amino acids (Bavister et al., 1983b; Bavister, 1989). Washed ejaculated sperm are incubated for 4 to >36

Table 3. Parameters Related to Mean Development of Matured
Rhesus Monkey Oocytes Resulting from PMSG Stimulation

Mean development[a] (n)	Plasma estradiol levels[b]	Oocytes matured[c] (% of total)	Oocytes initially mature (% of matured)	Mean no. matured oocytes/ experiment	Mean development score of matured oocytes/experiment[d] (% maximum)
Low	50–239%[e]	46.2[e]	34.8	5.8	(0.63 ± 0.44)
(4)		(± 23.7)	(± 29.9)	(± 6.2)	(15.8 ± 11.0)
High	266–1172%[e]	88.8[e]	69.4	20.8	2.38 ± 0.49
(4)		(± 8.9)	(± 36.8)	(± 11.7)	(59.5 ± 12.3)

[a]Experiments were divided into two groups: those with mean development scores ≥2 (high) and those <2 (low).
[b]Daily estradiol secretion was graphed and integrated over the stimulation cycle, then expressed as a percentage of maximal levels for an unstimulated cycle (daily mean estradiol + 1 S.D. integrated over days 1–12 of normal menstrual cycle).
[c]Includes oocytes that were initially mature [first polar body (PB1) expressed in <8 hr following follicular aspiration] and oocytes matured *in vitro* (PB1 in >8 hr).
[d]To each matured oocyte, a score was assigned: 0 = unfertilized; 1 = fertilized; 2 = cleavage (two cells to <20 cells); 3 = morula (compaction); 4 = blastocyst (early).
[e]$p < 0.05$ (Mann–Whitney test). Modified from Boatman *et al.* (1986).

hr in TALP and then with caffeine and dbcAMP for 90 min, as described in Section 2. Single oocytes are cultured in 90-μl drops of IVF medium for 6 hr or longer and then inseminated with (usually) 10 μl of sperm suspension to give a final sperm concentration of 2×10^6/ml (less if oocytes from unstimulated cycles are used) (Bavister *et al.*, 1983a). This dilutes the caffeine and dbcAMP in the fertilization drop to 0.1 mM each, which is compatible with normal fertilization (Bavister *et al.*, 1984; Boatman *et al.*, 1986).

Sperm and oocytes are coincubated for 12 to 15 hr at 37°C under 5% CO_2 in air. Then excess sperm and corona cells adhering to the zonae pellucidae are removed by pipetting, and eggs are examined for evidence of fertilization. The continuous-flow CO_2 device (Bavister, 1988) permits thorough examination of the eggs without jeopardizing their viability because of pH changes. At this time, the maturity of the egg is confirmed (or reconfirmed) by the presence of the first polar body (Boatman, 1987). The majority of the eggs should have two polar bodies and two well-developed pronuclei, which are easily seen using Nomarski optics (Boatman, 1987). Pronuclei may no longer be visible a few hours later because syngamy commences about 18 hr after insemination. Unlike the situation in rodent IVF eggs (Bavister, 1989), the fertilizing sperm tail cannot be seen in living primate eggs (human and monkey) with the light microscope because the sperm mitochondria and outer filaments are rapidly disassembled in the egg cytoplasm (Soupart and Strong, 1974; Bavister *et al.*, 1983a). This makes it difficult to guard against the possibility of parthenogenetic activation of eggs in primates, which can mimic normal fertilization.

One way to virtually ensure that sperm penetration is responsible for egg activation is to inseminate good-quality oocytes with killed sperm that cannot penetrate. If the control oocytes fail to activate or to cleave but comparable oocytes inseminated with live sperm are able to do so, then it is highly probable that the latter oocytes are undergoing or have undergone fertilization. In a careful control study of this type, there were no signs of activation (pronuclei or second polar body), cleavage, or fragmentation in randomly selected oocytes 48 to 73 hr after insemination with killed sperm (Bavister *et al.*, 1983a). In contrast, 79% of oocytes inseminated with live sperm showed two pronuclei and two polar bodies, and most of these reached the 8- to 16-cell stage within the same time period. This difference between results with killed and with live sperm was highly significant ($P < 0.001$). Even more rigorous evidence of normal fertilization is provided by birth of normal offspring after transfer of IVF embryos, which has been accomplished on several occasions in our laboratory (see Section 5.2).

It is important to check for multiple sperm penetration in IVF eggs because several aspects of the IVF procedures can predispose toward defects in the block to polyspermy, e.g., an inadequate cortical response to sperm penetration (because of oocyte maturation defects) and/or excessive numbers of competent sperm in the IVF drop. Polypronucleate primate embryos can develop at least as far as the morula or early blastocyst stages (Van Blerkom et al., 1984; Boatman, 1987) so that observation of several cleavage divisions in an embryo following IVF does not necessarily indicate that normal fertilization has taken place unless a careful examination of the pronuclear number was carried out at the presyngamy stage. It is also not uncommon to find that primate oocytes collected after gonadotropin stimulation show anucleate fragments or "blebs" pinched off from the main embryonic mass during development (Boatman, 1987). The presence of these blebs is not incompatible with normal fertilization and embryonic development; for example, in our own experience, one such embryo gave rise to a normal male offspring following a singleton embryo transfer (Boatman, 1987). However, these abnormalities are undesirable from the viewpoint of investigating developmental events in vitro. It is our hope that newer stimulation protocols will produce oocytes devoid of these blebs or other irregularities.

5. DEVELOPMENT OF IVF EMBRYOS

5.1. Development of IVF Embryos in Vitro

If useful information about development is to be obtained with IVF procedures, it is necessary to establish that embryos produced in this way are normal. Several approaches can be used for this purpose. By growing embryos in culture, information obtained about cleavage timing and/or metabolism (Boatman, 1987) can be compared with normal values. Fertilized eggs or embryos can be fixed and examined for evidence of defects (e.g., Chakraborty, 1981; Hegele-Hartung et al., 1988; Enders et al., 1989). However, the only incontrovertible evidence of normality is obtained by transferring IVF embryos to recipient females and obtaining normal offspring. We have used all of these approaches with IVF rhesus embryos.

After evaluation of rhesus eggs for IVF, all embryos (eggs with two pronuclei, etc.) are transferred to fresh drops of culture medium. Each embryo is aspirated through a fine Pasteur pipette to remove as much of the adhering corona cell layer as possible so that details of embryo development are not obscured. Embryos are examined thoroughly about once every 24 hr, at which time the culture medium is replaced. The "spent" medium may be stored for later studies on embryo secretion products (Pope et al., 1982a; Fishel et al., 1984; Simon and Hodgen, 1989).

Little is known about the culture requirements of primate embryos. The cleavage stages are of special interest because transfer of IVF embryos is usually performed at or before the eight-cell stage. Development of normal-looking IVF rhesus embryos has been obtained up to the eight-cell stage in TALP supplemented with four amino acids, but development in vitro beyond this stage was very limited (Bavister et al., 1983a). TALP with serum alone, in the absence of additional amino acids or other supplements, may be sufficient for the culture of IVF rhesus embryos (Wolf et al., 1989). Because the metabolic needs of embryos probably change at the eight-cell (early uterine) stage (Bavister et al., 1983b; Boatman, 1987), we substituted medium CMRL-1066, which was used by Hsu (1979) to culture mouse blastocysts through peri- and postimplantation stages and by Pope et al. (1982b) to culture baboon morulae and blastocysts. For culturing IVF rhesus embryos, we modified CMRL by the addition of sodium pyruvate (0.5 mM), sodium lactate (10 mM), and 20% serum (Boatman, 1987). Midcycle rhesus monkey serum, human fetal cord serum, and commercially available sera (horse, calf, and bovine: HyClone Laboratories, Logan, UT) have all been used successfully to culture IVF rhesus embryos to the hatched blastocyst stage (Boatman, 1987; Morgan et al., 1984; Goodeaux et al., 1989). However, despite the usefulness of this medium, it appears that uterine-stage rhesus embryos may depend on specific uterine factors to stimulate normal growth and development, even before hatching and attachment to the uterine epithelium occur (Goodeaux *et al.*, 1989).

In one early series of experiments, we found no difference in the percentages of IVF embryos that developed to the six- to eight-cell stage in TALP versus CMRL medium (Table 4). Moreover, since a normal offspring was born following transfer of a pair of cleavage-stage embryos grown to the four- to six-cell stage in serum-supplemented TALP, we concluded (Bavister *et al.*, 1984) that this medium is suitable for supporting early embryo development in the rhesus monkey (see also Wolf *et al.*, 1989). However, only 13% of fertilized ova in the TALP group developed to the morula stage or beyond, compared to 68% of fertilized ova grown in modified CMRL (Table 4).

The cleavage timing was the same for IVF embryos grown in TALP or CMRL media and was comparable to the timing of embryos grown *in vivo* (Fig. 2); the mean cell-number-doubling times were 19 hr and 18.4 hr for embryos grown *in vivo* and *in vitro*, respectively. The importance of establishing the cleavage timing for embryos grown *in vitro* ("timely cleavage": Morgan *et al.*, 1986, 1990; Boatman, 1987) is evident from our studies on maturation of rhesus oocytes *in vitro* (see Section 6).

After the eight-cell stage, the growth of embryos slowed appreciably (mean cell-number-doubling time 34 hr; Fig. 2). Rhesus (and baboon) embryos *in vivo* also experience a slowing of their growth rates during the uterine phase of preimplantation development, with a mean cell-doubling time of 27.4 hr (from published cell counts), but this is still substantially faster than the embryos grown *in vitro* (Fig. 2). By the time of zona shedding ("hatching"), IVF embryos grown *in vitro* were on average 2 to 3 days behind schedule compared to the normal timing of embryonic development in the rhesus monkey.

Because of the scarcity of information about the milieu of embryos developing *in vivo*, the design of culture media for supporting growth of primate embryos *in vitro* is empirical. As a result, the rate of attrition of embryos *in vitro* is high: about 9% of the remaining embryos ceased development each 24 hr in our experiments (Boatman, 1987). Overall, about 20% of IVF ova reached the expanded zonal and hatched blastocyst stages *in vitro*. These data should be considered in light of the high incidence of pre- and periimplantation embryonic losses *in vivo* in primates. Embryonic losses prior to establishment of clinical pregnancy are estimated to be between 25% and 50% (Hertig *et al.*, 1956; Hendrickx and Binkerd, 1980; Hurst *et al.*, 1980; Enders *et al.*, 1982). A substantial proportion of preimplantation primate embryos are abnormal, as shown by studies in which embryos were flushed from the reproductive tracts of mated rhesus

Table 4. Comparison of Development of IVF Rhesus Embryos in Two Culture Media

Culture medium[a]	Two cells		Six to eight cells		≥ Morula	
	No. (%)[b]	Time[c]	No. (%)[d]	Time[c]	No. (%)[d]	Time [c]
TALP	23/29 (79)	26 ± 5	20/23 (87)	54 ± 14	3/23 (13)	88 ± 11
CMRL	75/110 (68)	23 ± 5.5	56/72 (78)	59 ± 19	46/68 (68)	111 ± 16

[a]TALP is a modified Tyrode's solution supplemented with pyruvate, lactate, and 3 mg/ml bovine serum albumin (Bavister, 1989). CMRL-1066 is modified by addition of pyruvate and lactate and usually contains 20% fetal (human or bovine) or adult (rhesus) or horse) serum. Oocytes that were inseminated in TALP were transferred to CMRL at 12 to 18 hr post-insemination. Data from Boatman (1987).
[b]Percentage of inseminated oocytes.
[c]Time (hr) post-insemination (mean ± S.D.).
[d]Percentage of fertilized ova.

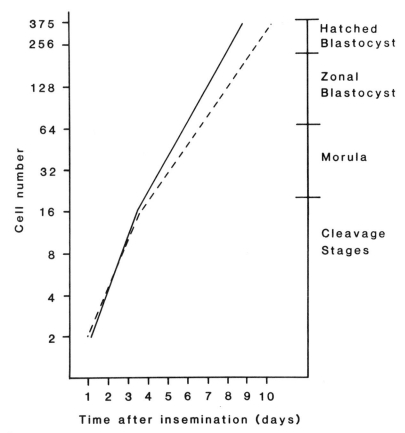

Figure 2. Timing of cleavage in rhesus monkey embryos. Solid lines are *in vivo* data, and dashed lines are from *in vitro* culture experiments. Data for two- to 16-cell stages from midpoints of cleavage timing for embryos grown *in vivo* (Lewis and Hartman, 1941) or *in vitro* (100 embryos: Boatman, 1987). *In vivo* data for 16-cell to hatched blastocyst from estimated fertilization age or midpoint of stage in rhesus or baboon embryos (Lewis and Hartman, 1941; Heuser and Streeter, 1941; Hendrickx and Kraemer, 1968; Kraemer and Hendrickx, 1971); corresponding *in vitro* data obtained by assigning the same mean cell number to *in vitro* hatched blastocysts (Boatman, 1987) as reported for comparable embryos grown *in vivo*. Data from Boatman (1987).

monkeys. Defective embryos with abnormal morphology and/or retarded cleavage timing have been found (Hurst *et al.*, 1980; Enders *et al.*, 1982). To what extent these losses in primates are related to defects in oocyte maturation and/or in fertilization is a topic that should be given a high priority for research. In view of the practice of the "rhythm" or "safe-day" contraceptive method and rather poorly developed mechanisms for synchronizing coitus with ovulation, a high incidence of "delayed fertilization" in humans seems inevitable. The detrimental consequences of delayed fertilization on embryonic development have been documented in nonprimate species (Blandau and Young, 1939; Hunter, 1980; Juetten and Bavister, 1983) and have been considered as a possible contributing factor in spontaneous abortion and developmental defects in humans (Guerrero and Lanctot, 1970; Abramson, 1973; Edwards, 1980), but little "hard" data are available for primates. This is an area that could benefit from carefully timed studies using IVF of nonhuman primate eggs.

5.2. Development of IVF Embryos following Transfer

The endpoints selected for evaluating embryo development are obviously of critical importance for establishing the normality of IVF. As mentioned earlier, observation of polar bodies and pronuclei in the oocyte constitutes good evidence of fertilization, especially if appropriate controls (such as insemination with killed sperm: Bavister *et al.*, 1983a) are used. However, this does not provide information on the normality of embryos following IVF. The ultimate test of viability is, of course, the production of normal offspring subsequent to embryo transfer. Since the main emphasis of our rhesus IVF program has been to culture embryos in order to obtain information about embryo development, relatively few of our IVF embryos have been transferred to recipients to assess viability.

The PMSG stimulation regimen that we have frequently used to produce numerous oocytes disturbs the luteal phase of the cycle by elevating estradiol and progesterone to abnormally high levels. Therefore, there is little point in transferring IVF embryos back to the donor animal. Instead, we have relied on surrogate recipients for embryo transfer. A problem with this approach is that there is no satisfactory way to synchronize menstrual cycles in Old-World monkeys, so several rhesus females are selected during the early luteal phase as being likely embryo transfer candidates, predictions being based on their previous cycle data. The cycles of these animals are monitored by rapid radioimmunoassay for estradiol (Morgan *et al.*, 1987a). As the anticipated day of embryo transfer approaches, fewer of the selected animals are found to be closely synchronized with the developing IVF embryo, so that only one or two good candidates remain by the time that the transfer procedure is performed. Although the recipient and embryo do not need to be as closely synchronous as in rodents, the success of embryo transfer still depends to a large degree on the recipient's day of ovulation being within 1 day of the time of oocyte recovery for IVF (Boatman, 1987; Wolf *et al.*, 1989). It seems to be preferable for the recipient to be as much as 1 day in arrears rather than 1 day ahead of the donor (embryo). With recent advances in development of freeze-storage protocols for human and monkey embryos (Mohr *et al.*, 1985; Pope *et al.*, 1986; Balmaceda *et al.*, 1986; Boatman, 1987; Summers *et al.*, 1987; Kuzan and Quinn, 1988), this dependency on chance synchrony between donor and recipient for success of embryo transfer should be reduced (Wolf *et al.*, 1989). In our experiments, most of our transfers were performed before a rapid RIA for estradiol became available for monitoring recipients' cycles, so we relied on the "sex skin breakdown" as the main criterion of synchrony (Czaja *et al.*, 1977). Retrospective estradiol measurements showed that we were frequently one to several days off in our estimates of synchrony (Boatman, 1987).

Based on about 30 embryo transfer procedures, with usually one to three IVF embryos per transfer, we have obtained three normal rhesus offspring. This represents a birth rate of about 6% per transferred embryo (Boatman, 1987). When more than one embryo was transferred to a single recipient, the pregnancy rate was 8% for a single IVF rhesus embryo and 15% for two embryos (Boatman, 1987). Two offspring were produced after nonsurgical (uterine) transfers of IVF embryos (two embryos in one case and a single embryo in the other). The parentage of one of these animals ("Petri") was confirmed by blood typing, which showed that he was unequivocally the genetic product of the sperm and egg donors (Bavister *et al.*, 1984). This animal (Fig. 3) is the first genetically confirmed IVF nonhuman primate. A third offspring resulted from a laparoscopic transfer of two IVF embryos into one oviduct of a synchronous recipient (unpublished data). Using laparoscopy to transfer embryos into the oviductal ampulla is preferable to laparotomy, since it avoids surgery on the recipient animals. This transfer method seems particularly promising for rare or endangered primates in which minimizing the invasiveness of procedures is desirable (see Section 7). Oviductal embryo transfer (e.g., Balmaceda *et al.*, 1988) is likely to be less damaging to embryos because there is no cervical mucus to interfere with deposition of

Figure 3.   The first IVF macaque monkey, "Petri," at about 4 years of age.

embryos in the reproductive tract, unlike with uterine embryo transfers. Moreover, since the position of the transfer catheter can easily be detected within the oviductal ampulla, there is no danger of crushing the embryos against the wall of the tract.

Each of the IVF monkeys (Petri, Orwell, and Murphy) has reached the age of sexual maturity and fathered normal offspring. They have become the first reported IVF primate parents, and their eight living offspring should help alleviate any anxiety that "test tube" primates, including humans, may have latent reproductive or genetic problems related to IVF conception.

6. USE OF IVF FOR STUDIES ON OOCYTE MATURATION

We have recently begun to examine the maturation of rhesus monkey oocytes *in vitro*. Studies on oocyte maturation in primates are urgently needed for several reasons. First, the basic mechanisms regulating oocyte maturation in primates are virtually unknown. This stands in contrast to the insights gained into the control of oocyte maturation in mice and sheep (Staigmiller and Moor, 1984; Schultz, 1985; Eppig and Schroeder, 1986). Second, following gonadotropin stimulation of nonhuman primates and women, immature oocytes may be recovered at follicular aspiration. Since we do not presently know how to achieve normal maturation of primate oocytes *in vitro*, this represents a loss of efficiency in the treatment cycle. This can have significant consequences, both for research and for the treatment of infertility. Third, if primate oocytes could be held in the dictyate condition [so-called "germinal vesicle" (GV) stage] by manipulation of the culture conditions, as in mice (Downs *et al.*, 1985), then IVF could be delayed until other features of the IVF protocol (e.g., semen collection, recipient availability) were optimal without resorting to cryopreservation of oocytes.

Much of the early data on oocyte maturation in mammals are not helpful toward understanding the underlying control mechanisms because a distinction between nuclear and cytoplasmic maturation was not often made. The combined applications of *in vitro* culture for maturation of oocytes, followed by IVF and embryo culture/transfer, have revealed the dual nature of the oocyte maturation process. The nucleus can escape meiotic arrest and undergo GV breakdown (=nuclear maturation), with attainment of the metaphase II chromosome configuration and extrusion of the first polar body (PB1), within the cytoplasm of an oocyte that has not yet completed its programming of synthetic machinery (cytoplasmic maturation). A substantial body of evidence demonstrates that such oocytes, after penetration by sperm, would be severely compromised in their ability to undergo normal activation and embryonic development. Defects in cytoplasmic maturation include (1) inability of the oocyte cytoplasm to support nuclear decondensation in the penetrating sperm, resulting in failure to complete fertilization (Thibault, 1977; Leibfried and Bavister, 1983), (2) errors in the distribution of cortical granules and/or in the timing of their release, resulting in either polyspermy (which causes embryonic death) or premature zona block (which leads to failure of sperm penetration) (Sathananthan and Trounson, 1982; Van Blerkom *et al.*, 1984), and (3) diminished viability and/or retarded development of embryos following apparently normal fertilization, resulting in death of embryos or in asynchrony between the embryos and the mother, thereby compromising pregnancy (Shalgi, 1984; Fleming *et al.*, 1985). As an example of such defects, a study in our laboratory showed that only 2% of hamster follicular oocytes "matured" *in vitro* were capable of decondensing the penetrating sperm nucleus and forming a male pronucleus, compared with 98% of *in vivo*-matured oocytes (Leibfried and Bavister, 1983). Mouse (Schroeder and Eppig, 1984) and sheep (Staigmiller and Moor, 1984) oocytes have been successfully matured *in vitro*, as evidenced by the production of live young following *in vitro* or *in vivo* fertilization. Using these studies as a guide, we began to study the maturation of rhesus monkey oocytes *in vitro*.

The majority of oocytes were obtained from gonadotropin-stimulated animals, using PMSG or hMG to recruit follicles and hCG to induce oocyte maturation, as described previously (Bavister *et al.*, 1983a; Boatman *et al.*, 1986). Oocytes were collected by aspiration of follicles about 30 hr after hCG, estimating the time of ovulation as 36 hr post-hCG. In one study, there was no significant difference in the fertilizability of mature oocytes (defined as extruding PB1 within 8 hr of collection) and oocytes matured *in vitro* (requiring >8 hr to extrude PB1) (66% and 71%, respectively: Table 5). However, the incidence of embryonic development to the six- to eight-cell stage *in vitro* was 95% versus 60%, respectively ($P < 0.001$). This result showed that (1) oocytes that underwent a substantial part of the maturation process *in vitro* were defective, and (2) embryo development is a more rigorous indicator of oocyte maturity than IVF. In a second study using an even more stringent criterion of embryo development, i.e., *timely* development to the six- to eight-cell stage (by 50 hr post-insemination: Morgan *et al.*, 1986), there was again no difference in fertilization between oocytes matured *in vivo* and those that completed maturation *in vitro* (72 versus 76%). However, only 23% of the timely embryos were derived from initially immature oocytes, whereas 61% were from oocytes defined as mature at the time of collection ($P < 0.01$ Morgan *et al.*, 1990). Clearly, timely embryonic development is a critical indicator for evaluation of oocyte maturity.

Table 5. Comparison of Fertilization and Development of IVF Rhesus Monkey Ova that Were Initially Mature or Matured *in Vitro*

Oocyte category	Number of ova fertilized/ no. inseminated (%)	Number of six- to eight-cell embryos developing (% of ova fertilized)
Mature[a]	40/61 (65.6)	35/37 (94.6)[c]
Maturation completed *in vitro*[b]	35/49 (71.4)	21/35 (60.0)[c]

[a]Incubated <8 hr before insemination; PB1 confirmed after insemination and cumulus dispersal.
[b]Incubated >8 hr before insemination; PB1 confirmed subsequently.
[c]Significant difference by χ^2, p <0.001. Data from Boatman (1987).

The conditions used for oocyte maturation in the preceding study were minimal, consisting of our IVF medium, TALP, supplemented with a few amino acids and serum (Boatman, 1987). Subsequently, conditions needed to support complete oocyte maturation *in vitro* were examined. In a pilot study, rhesus oocytes were collected from excised ovaries obtained at various times during the menstrual cycles of unstimulated monkeys. Unstimulated females were used to avoid the possibility that exposure to exogenous gonadotropin *in vivo* might affect the capacity of oocytes for subsequent embryonic development, as recently verified by Schroeder and Eppig (1989). Oocytes were cultured in TALP or in modified CMRL-1066 medium (Bavister, 1987; Morgan *et al.*, 1991). Supplements were 20% calf serum plus or minus 10 μg/ml FSH-P (an animal pituitary preparation) plus or minus 5 IU/ml hCG after 24 hr culture. Groups of five oocytes were cultured using a 2 × 3 factorial design to examine treatment effects. Oocytes were examined starting at 24 hr culture for evidence of PB1 or a distinctive perivitelline space indicative of PB1 (Boatman, 1987). When more than three oocytes per group showed one of these signs, all oocytes in the group were inseminated. The maximum culture period before insemination was 50 hr. Oocytes were coincubated with sperm for 16–20 hr, then examined for evidence of fertilization. If oocytes were considered to be fertilized, culture was continued until they stopped development. All oocytes and embryos were fixed and stained at the end of culture.

For all experimental treatments, the development of cultured oocytes can be summarized as follows: 58% of oocytes underwent germinal vesicle breakdown; 37% extruded a first polar body; 17% were fertilized; and 12% cleaved to at least the two-cell stage *in vitro* (Morgan *et al.*, 1991). There were no treatment differences in oocyte nuclear maturation, based on breakdown of the germinal vesicle nucleus or extrusion of the first polar body. Additionally, there were no differences in maturation or development attributable to the medium used during oocyte maturation. However, the presence of gonadotropins during oocyte maturation profoundly affected cytoplasmic maturation: fertilization (20.7% versus 6.7%) and embryo development (15.7% versus 2.2%) were significantly higher in treatments with, than in those without, gonadotropin supplementation. There were no differences between FSH alone or FSH + hCG. Despite the apparently low incidence of fertilization, 40% of the fertilized eggs developed beyond the four-cell stage, comparable to IVF results for *in vivo* matured rhesus monkey eggs (Bavister *et al.*, 1983a; Boatman *et al.*, 1986; Boatman, 1987; Morgan *et al.*, 1990). Thus, rhesus monkey oocytes obtained from nonstimulated follicles can undergo both nuclear and cytoplasmic maturation *in vitro* even though they have not been exposed *in vitro* to human FSH (which was not available to us at the time of this study). This preliminary study paved the way for more critical experiments to investigate oocyte maturation and to define protocols for obtaining normal maturation of primate oocytes *in vitro*.

In additional studies, evidence was obtained that the mechanisms for maintaining oocytes in meiotic arrest within the follicle may be similar in rodents and primates (Downs *et al.*, 1985;

Eppig and Downs, 1987; Warikoo and Bavister, 1989). Up to 66% of cultured rhesus oocytes could be maintained in meiotic arrest for up to 80 hr in the presence of hypoxanthine plus dbcAMP (Warikoo and Bavister, 1989). Removal of hypoxanthine plus dbcAMP at 40 hr permitted the majority of oocytes (75%) to undergo GVB. Except for the greatly extended periods of culture and prolonged inhibition by hypoxanthine plus dbcAMP for the monkey oocytes, these results (Warikoo and Bavister, 1989) are similar to those reported in the mouse (Downs et al., 1985).

The finding that embryo development is retarded following fertilization of oocytes matured under suboptimal conditions may have relevance to the etiology of primate embryo losses occurring in vivo. Anomalies of cytoplasmic oocyte maturation may account for a sizable proportion of defective embryos recovered from the reproductive tract (Hurst et al., 1980; Enders et al., 1982). Our preliminary data (Morgan et al., 1986, 1987b, 1990, 1991; Bavister, 1987; Boatman, 1987) show that the relationship between oocyte maturity and embryo viability can be examined critically using IVF. The use of more refined culture conditions for achieving oocyte maturation will permit the underlying regulatory mechanisms to be examined. At the same time, improved in vitro conditions for oocyte maturation, in combination with existing IVF technology, will help to provide a reliable supply of material for in-depth investigation of fertilization and early embryonic development.

7. USE OF IVF FOR CONSERVATION OF ENDANGERED SPECIES

The demonstrated success of our protocols for IVF, embryo culture, and transfer raises the possibility of using these experimental techniques to help conserve endangered primates. Since Balmaceda et al. (1984) have shown the feasibility of obtaining term pregnancy following cross-species transfer of an IVF (cynomolgus) embryo into a rhesus mother, it is reasonable to propose to transfer IVF embryos of endangered macaque species into recipient rhesus monkeys. Such a program was established between our laboratory and the Medical Division of the Baltimore Zoo under the supervision of Michael Cranfield, D.V.M. Preliminary work used pigtail (Macaca nemestrina) monkeys (Cranfield et al., 1989) as (1) a model species for testing the application of IVF-ET techniques to zoo facilities, and (2) surrogate mothers for transferred embryos of the endangered lion-tailed monkey (Macaca silenus). During the past 18 months, three clinical IVF pregnancies have been achieved at the Baltimore Zoo. One twin pregnancy aborted mid-gestation and one infant was lost during birth; however, a healthy infant was born in April 1990, after transfer of frozen-thawed embryos derived from M. nemestrina eggs and M. silenus sperm. The birth of this hybrid infant and the appropriate maternal behavior of the M. nemestrina embryo transfer recipient demonstrate the suitability of M. nemestrina as surrogate mothers for gestation and postnatal care of M. silenus IVF babies.

8. SUMMARY AND CONCLUSIONS

We have devised culture conditions and treatments for obtaining capacitation of rhesus monkey sperm in vitro, leading to IVF and embryo development to the hatched blastocyst stage and beyond. Some of the embryos produced by IVF were normal, as evidenced by the birth of offspring after embryo transfer. Other laboratories have confirmed the reproducibility of our basic protocols. These procedures will allow detailed studies to be undertaken of the regulatory events involved in primate fertilization and embryogenesis. One area that needs investigation is the control of sperm fertilizing ability by components of the egg investments or by factors present in the female reproductive tract (Boatman, 1990). Studies on the control of oocyte maturation in primates are now clearly feasible, although emphasis needs to be placed first on obtaining good quality ova by *in vitro* maturation before major efforts are made to analyze the biochemistry and molecular biology of oocyte maturation, as in the mouse (Downs *et al.*, 1985; Schultz, 1985; Eppig and Schroeder, 1986). Information on the quality of oocytes can be provided by IVF, so that the significance of constituents of the follicular milieu from which oocytes were derived can be more critically evaluated (Morgan *et al.*, 1986, 1990, 1991; Bavister, 1987). The nutritional requirements and metabolism of primate preimplantation embryos are poorly understood. For studies on early-cleavage-stage embryos, IVF is a useful procedure for obtaining a supply of material of precisely known developmental age. In order to examine later-stage embryos (morulae and blastocysts), it is more practical to recover embryos from the uteri of mated animals rather than to grow IVF embryos for several days in culture media that are presently suboptimal.

The use of nonhuman primates such as the rhesus monkey is not without problems. Apart from the expense of maintaining the animals, relatively few oocytes or embryos can be obtained in a given treatment cycle, which could limit the scope of research. These challenges must be met with new experimental approaches. Even single embryos can yield valuable data if utilized effectively. For example, an immature oocyte could be matured *in vitro*, providing information on the chemical control of maturation; the mature ovum can then be inseminated to examine the role of the cumulus oophorus in stimulating sperm fertilizing ability; the resulting embryo can be cultured to provide data on the regulation of embryo growth; and finally, the embryo can be fixed and examined by electron microscopy to provide data on subcellular organization (Enders *et al.*, 1989). Moreover, the development of exquisitely sensitive methods for measuring metabolism allows even single preimplantation embryos to provide useful data (Leese *et al.*, 1986; Hardy *et al.*, 1989). Another refinement is to split an embryo at an early stage of development, such as dissociating the blastomeres of a two-cell embryo. The two matched halves could then be subjected to control and experimental conditions to give statistically analyzable data on the treatment effects. We have studied this approach; although we have been unsuccessful to date in obtaining development of split cleavage-stage embryos, one bisected rhesus morula produced two normal-looking blastocysts (Warikoo, 1990).

Nonhuman primates have been underutilized during the past 40 years for the study of fertilization and preimplantation development. Now that reproducible procedures for IVF and embryo development *in vitro* are available for nonhuman primates, it seems time to devote substantially more attention to the potential of these procedures for providing data that are directly relevant to fertility problems and to embryological questions in humans.

ACKNOWLEDGMENTS. We are grateful to Kevin Collins, Emily Kraus, Michael Erwin, and Robert Robbins for skilled technical assistance during the course of our studies on rhesus monkey IVF, embryo culture, and embryo transfer. We also thank the Wisconsin Regional Primate Research Center (Base Operating Grant RR00167) and NICHD (grant HD14765) for supporting our research. This is publication no. 31-014 of the WRPRC.

9. REFERENCES

Abramson, F. D., 1973, Spontaneous fetal death in man, *Soc. Biol.* **20:**375–403.

Austin, C. R., 1951, Observations on the penetration of the sperm into the mammalian egg, *Aust. J. Sci. Res.* **4:** 581–596.

Austin, C. R., 1990, Dilemmas in human IVF practice, in: *Fertilization in Mammals* (B. D. Bavister, J. M. Cummins, and E. R. S. Roldan, eds.), Serono Symposia U.S.A., Norwell, Massachusetts, pp. 373–379.

Balmaceda, J. P., Pool, T. B., Arana, J. B., Heitman, T. S., and Asch, R. H., 1984, Successful *in vitro* fertilization and embryo transfer in cynomolgus monkeys, *Fertil. Steril.* **42:**791–795.

Balmaceda, J. P., Heitman, T. O., Garcia, M. R., Pauerstein, C. J., and Pool, T. B., 1986, Embryo cryopreservation in cynomolgus monkeys, *Fertil. Steril.* **45:**403–406.

Balmaceda, J. P., Gastaldi, C., Ord, T., Borrero, C., and Asch, R. H., 1988, Tubal embryo transfer in cynomolgus monkeys: Effects of hyperstimulation and synchrony, *Hum. Reprod.* **3:**441–443.

Bavister, B. D., 1982, Evidence for a role of post-ovulatory cumulus components in supporting fertilizing ability of hamster spermatozoa, *J. Androl.* **3:**365–372.

Bavister, B. D., 1986, Animal *in vitro* fertilization and embryo development, in: *Manipulation of Mammalian Development*, Vol. 4, *Developmental Biology: A Comprehensive Synthesis* (R. B. L. Gwatkin, ed.), Plenum Press, New York, pp. 81–148.

Bavister, B. D., 1987, Oocyte maturation and *in vitro* fertilization in the rhesus monkey, in: *The Primate Ovary: Proceedings of the Serono Symposium* (R. L. Stouffer, ed.), Plenum Press, New York, pp. 119–138.

Bavister, B. D., 1988, A mini-chamber device for maintaining a constant carbon dioxide in air atmosphere during prolonged culture of cells on the stage of an inverted microscope, *In Vitro Cell Dev. Biol.* **24:**759–763.

Bavister, B. D., 1989, A consistently successful procedure for *in vitro* fertilization of golden hamster eggs, *Gamete Res.,* **23:**139–158.

Bavister, B. D., 1990, Tests of sperm fertilizing ability, in: *Gamete Physiology* (R. H. Asch, J. P. Balmaceda, and I. Johnston, eds.), Serono Symposia U.S.A., Norwell, Massachusetts, pp. 77–105.

Bavister, B. D., and Yanagimachi, R., 1977, The effects of sperm extracts and energy sources on the motility and acrosome reaction of hamster spermatozoa *in vitro*, *Biol. Reprod.* **16:**228–237.

Bavister, B. D., Boatman, D. E., Leibfried, M. L., Loose, M., and Vernon, M. W., 1983a, Fertilization and cleavage of rhesus monkey oocytes *in vitro*, *Biol. Reprod.* **28:**983–999.

Bavister, B. D., Leibfried, M. L., and Lieberman, G., 1983b, Development of preimplantation embryos of the golden hamster in a defined culture medium, *Biol. Reprod.* **28:**235–247.

Bavister, B. D., Boatman, D. E., Collins, K., Dierschke, D. J., and Eisele, S. G., 1984, Birth of rhesus monkey infant following *in vitro* fertilization and non-surgical embryo transfer, *Proc. Natl. Acad. Sci. U.S.A.* **81:** 2218–2222.

Bavister, B. D., Dees, H. C., and Schultz, R. D., 1986, Refractoriness of rhesus monkeys to repeated gonadotropin stimulation is due to formation of nonprecipitating antibodies, *Am. J. Reprod. Immunol. Microbiol.* **11:**11–16.

Binor, Z., Rawlins, R. G., der Ven, H. V., and Dmowski, W. P., 1988, Rhesus monkey sperm penetration into zona-free hamster ova: Comparison of preparation and culture conditions, *Gamete Res.* **19:**91–100.

Blandau, R. J., and Young, W. C., 1939, The effects of delayed fertilization on the development of the guinea pig ovum, *Am. J. Anat.* **64:**303–329.

Boatman, D. E., 1987, *In vitro* growth of non-human primate pre- and periimplantation embryos, in: *The Mammalian Preimplantation Embryo: Regulation of Growth and Differentiation* (B. D. Bavister, ed.), Plenum Press, New York, pp. 273–308.

Boatman, D. E., 1990, Oviductal modulators of sperm fertilizing ability, in: *Fertilization in Mammals* (B. D. Bavister, J. M. Cummins, and E. R. S. Roldan, eds.), Serono Symposia U.S.A., Norwell, Massachusetts, pp. 223–238.

Boatman, D. E., and Bavister, B. D., 1984, Stimulation of rhesus monkey sperm capacitation by cyclic nucleotide mediators, *J. Reprod. Fertil.* **71:**357–366.

Boatman, D. E., Morgan, P. M., and Bavister, B. D., 1986, Variables affecting the yield and developmental potential of embryos following superstimulation and *in vitro* fertilization in rhesus monkeys, *Gamete Res.* **13:**327–338.

Chakraborty, J., 1981, Fine structural abnormalities in the developing mouse embryo, *Gamete Res.* **4:**535–545.

Chang, M. C., 1951, Fertilizing capacity of spermatozoa deposited into the fallopian tubes, *Nature* **168:** 697–698.

Cline, E. M., Gould, K. G., and Foley, C. W., 1972, Regulation of ovulation, recovery of mature ova and fertilization *in vitro* of mature ova of the squirrel monkey (*Saimiri sciureus*), *Fed. Proc.* **31:**360.

Cranfield, M. R., Schaffer, N., Bavister, B. D., Berger N., Boatman, D. E., Kempske, S., Miner, N., Panos, M., Adams, J., and Morgan, P. M., 1989, Assessment of oocytes retrieved from stimulated and unstimulated ovaries of pig-tailed macaques (*Macaca nemestrina*) as a model to enhance the genetic diversity of captive lion-tailed macaques (*Macaca silenus*), *Zoo Biol. [Suppl.]***1:**33–46.

Cross, N. L., Morales, P., Overstreet, J. W., and Hanson, F. W., 1988, Induction of acrosome reactions by the human zona pellucida, *Biol. Reprod.* **38:**235–244.

Czaja, J. A., Robinson, J. A., Eisele, S. G., Scheffler, G., and Goy, R. W., 1977, Relationship between sexual skin colour of female rhesus monkeys and midcycle plasma levels of oestradiol and progesterone, *J. Reprod. Fertil.* **49:**147–150.

Downs, S. M., Coleman, D. L., Ward-Bailey, P. F. and Eppig, J. J., 1985, Hypoxanthine is the principal inhibitor of murine oocyte maturation in a low molecular weight fraction of porcine follicular fluid, *Proc. Natl. Acad. Sci. U.S.A.* **82**:454–458.

Edwards, R. G., 1980, *Conception in the Human Female*, Academic Press, London, pp. 561, 631–634.

Enders, A. C., Hendrickx, A. G., and Binkerd, P. E., 1982, Abnormal development of blastocysts and blastomeres in the rhesus monkey, *Biol. Reprod.* **26**:353–366.

Enders, A. C., Schlafke, S., Boatman, D. E., Morgan, P. M., and Bavister, B. D., 1989, Differentiation of blastocysts derived from *in vitro* fertilized rhesus monkey oocytes, *Biol. Reprod.* **41**:715–727.

Eppig, J. J., and Downs, S. M., 1987, The effect of hypoxanthine on mouse oocyte growth and development *in vitro*: Maintenance of meiotic arrest and gonadotropin-induced oocyte maturation, *Dev. Biol.* **119**:313–321.

Eppig, J. J., and Schroeder, A. C., 1986, Culture systems for mammalian oocyte development: Progress and prospects, *Theriogenology* **25**:97–106.

Fishel, S. B., Edwards, R. G., and Evans, C. J., 1984, HCG secreted by preimplantation embryos cultured *in vitro* [human], *Science* **223**:816–818.

Fleming, A. D., Evans, G., Walton, E. A., and Armstrong, D. T., 1985, Developmental capability of rat oocytes matured *in vitro* in defined medium, *Gamete Res.* **12**:255–263.

Goodeaux, L. L., Voelkel, S. A., Anzalone, C. A., Menezo, Y., and Graves, K. H., 1989, The effect of rhesus uterine epithelial cell monolayers on *in vitro* growth of rhesus embryos, *Theriogenology* **31**:197.

Gould, K. G., Cline, E. M., and Williams, W. L., 1973, Observations on the induction of ovulation and fertilization *in vitro* in the squirrel monkey (*Saimiri sciureus*), *Fertil. Steril.* **24**:260–268.

Guerrero, R., and Lanctot, C. A., 1970, Aging of fertilizing gametes and spontaneous abortion, *Am. J. Obstet. Gynecol.* **107**:263–267.

Hardy, K., Hooper, M. A. K., Handyside, A. H., Rutherford, A. J., Winston, R. M. L., and Leese, H. J., 1989, Non-invasive measurement of glucose and pyruvate uptake by individual human oocytes and preimplantation embryos, *Hum. Reprod.* **4**:188–191.

Hartman, C. G., and Corner, G. W., 1941, The first maturation division of the macaque ovum, *Contrib. Embryol.* **29**:1–6.

Hegele-Hartung, C., Fisher, B., and Beier, H. M., 1988, Development of preimplantation rabbit embryos after *in-vitro* culture and embryo transfer: An electron microscopic study, *Anat. Rec.* **220**:31–42.

Hendrickx, A. G., and Binkerd, P. E., 1980, Fetal deaths in nonhuman primates, in: *Human Embryonic and Fetal Death* (I. H. Porter and E. B. Hook, eds.), Academic Press, New York, pp. 45–69.

Hendrickx, A. G., and Kraemer D. C., 1968, Preimplantation stages of baboon embryos (*Papio* sp.), *Anat. Rec.* **162**:111–120.

Hertig, A. L., Rock, J., and Adams, E. C., 1956, A description of 34 human ova within the first 17 days of development, *Am. J. Anat.* **98**:435–493.

Heuser, C. H., and Streeter, G. L., 1941, Development of the macaque embryo, in: *Embryology of the Rhesus Monkey (Macaca mulatta)*, Carnegie Institution, Washington, DC, pp. 17–65.

Hsu, Y.-C., 1979, *In vitro* development of individually cultured whole mouse embryos from blastocyst to early somite stage, *Dev. Biol.* **68**:453–461.

Hunter, R. H. F., 1980, *Physiology and Technology of Reproduction in Female Domestic Animals*, Academic Press, London, pp. 163–169.

Hurst, P. R., Wheeler, A. G., and Eckstein, P., 1980, A study of uterine embryos recovered from rhesus monkeys fitted with intrauterine devices, *Fertil. Steril.* **33**:69–76.

Hyne, R. V., and Garbers, D. L., 1979, Calcium-dependent increase in adenosine 3′,5′-monophosphate and induction of the acrosome reaction in guinea pig spermatozoa, *Proc. Natl. Acad. Sci. U.S.A.* **76**:5699–5703.

Juetten, J., and Bavister, B. D., 1983, Effects of egg aging on *in vitro* fertilization and first cleavage in the hamster, *Gamete Res.* **8**:219–230.

Kraemer, D. C., and Hendrickx, A. G., 1971, Descriptions of stages I, II and III, in: *Embryology of the Baboon* (A. G. Hendrickx, ed.), University of Chicago Press, Chicago, pp. 45–52.

Kreitman, O., Lynch, A., Nixon, W. E., and Hodgen, D. G., 1982, Ovum collection, induced luteal dysfunction, *in vitro* fertilization, embryo development and low tubal ovum transfer in primates, in: *In Vitro Fertilization and Embryo Transfer* (E. S. E. Hafez and K. Semm, eds.), MTP Press, Lancaster, England, pp. 303–324.

Kuehl, T. J., and Dukelow, W. R., 1979, Maturation and *in vitro* fertilization of follicular oocytes of the squirrel monkey (*Saimiri sciureus*), *Biol. Reprod.* **21**:545–556.

Kuzan, F. B., and Quinn, P., 1988, Cryopreservation of mammalian embryos, in: *In Vitro Fertilization and Embryo Transfer: A Manual of Basic Techniques* (D. P. Wolf, B. D. Bavister, M. Gerrity, and G. S. Kopf, eds.), Plenum Press, New York, pp. 301–347.

Leese, H. J., Hooper, M. A. K., Edwards, R. G., and Ashwood-Smith, M. J., 1986, Uptake of pyruvate by early human embryos determined by a non-invasive technique, *Hum. Reprod.* **1**:181–182.

Leibfried, M. L., and Bavister, B. D., 1982, Effects of epinephrine and hypotaurine on *in vitro* fertilization in the golden hamster, *J. Reprod. Fertil.* **66**:87–93.

Leibfried, M. L., and Bavister, B. D., 1983, Fertilizability of *in vitro* matured oocytes from golden hamsters, *J. Exp. Zool.* **226**:481–485.

Lewis, W. H., and Hartman, C. G., 1941, Tubal ova of the rhesus monkey, in: *Embryology of the Rhesus Monkey (Macaca mulatta)*, Carnegie Institution, Washington DC, pp. 9–15.

Mastroianni, L., and Manson, W. A., 1963, Collection of monkey semen by electroejaculation, *Proc. Soc. Exp. Biol. Med.* **112**:1025–1027.

Matteri, R. L., Warikoo, P. K., and Bavister, B. D., 1991, Biochemical and biological properties of urinary follicle-stimulating hormone (FSH) from the rhesus monkey (*Macaca mulatta*), *Am. J. Primatol.* (in press).

Meizel, S., Turner, K. O., and Thomas, P., 1984, The stimulation of hamster sperm capacitation and acrosome reactions by biogenic amines, in: *Catecholamines as Hormone Regulators* (N. Ben-Jonathon, J. M. Bahr, and R. I. Weiner, eds.), Raven Press, New York, pp. 329–343.

Mohr, L. R., Trounson, A., and Freemann, L., 1985, Deep-freezing and transfer of human embryos, *J. in Vitro Fertil. Embryo Transfer* **2**:1–10.

Moor, R. M., and Cran, D. G., 1980, Intercellular coupling in mammalian oocytes, in: *Development in Mammals*, Vol. 4 (M. H. Johnson, ed), Elsevier/North-Holland Biomedical Press, New York, pp. 3–37.

Moreno, R., and Barros, C., 1990, Effect of test-yolk upon the induction of the acrosome reaction and on the ability of human spermatozoa to fuse with zona-free hamster oocytes, in: *Fertilization in Mammals* (B. D. Bavister, J. M. Cummins, and E. R. S. Roldan, eds.), Serono Symposia U.S.A., Norwell, Massachusetts, p. 424.

Morgan, P. M., Boatman, D. E., Collins, K., and Bavister, B. D., 1984, Complete preimplantation development in culture of *in vitro* fertilized rhesus monkey oocytes, *Biol. Reprod. [Suppl.]* **1**:96a.

Morgan, P. M., Boatman, D. E., and Kraus, E. M., 1986, Relationship between follicular fluid steroid hormone concentrations and *in vitro* development of rhesus monkey embryos, *Biol. Reprod.* **34**(Suppl. 1):94a.

Morgan, P. M., Hutz, R. J., Kraus, E. M., Cormie, J. A., Dierschke, D. J., and Bavister, B. D., 1987a, Evaluation of ultrasonography for monitoring follicular growth in rhesus monkeys, *Theriogenology* **27**:769–781.

Morgan, P. M., Warikoo, P. K., Erwin, M. J., and Kraus, E. M., 1987b, Recovery of oocytes from spontaneously cycling rhesus monkeys: Timing, follicular steroids and *in vitro* fertilization, *Biol. Reprod.* **36**(Suppl. 1):130.

Morgan, P. M., Boatman, D. E., and Bavister, B. D., 1990, Relationships between follicular fluid steroid hormone concentrations, oocyte maturation, *in vitro* fertilization and embryo development in the rhesus monkey, *Molec. Reprod. Dev.* **27**:145–151.

Morgan, P. M., Warikoo, P. K., and Bavister, B. D., 1991, *In vitro* maturation of ovarian oocytes from unstimulated rhesus monkeys: assessment of cytoplasmic maturity by embryonic development after *in vitro* fertilization, *Biol. Reprod.* **45**:89–93.

Naish, S. J., Perreault, S. D., and Zirkin, B. R., 1988, DNA synthesis following microinjection of heterologous sperm and somatic cell nuclei into hamster oocytes, *Gamete Res.* **18**:109–120.

Polan, M. L., Totora, M., Caldwell, B. V., DeCherney, A. H., Haseltine, F. P., and Kase, N., 1982, Abnormal ovarian cycles as diagnosed by ultrasound and serum estradiol levels, *Fertil. Steril.* **37**:342–347.

Pope, V. Z., Pope, C. E., and Beck, L. R., 1982a, Gonadotropin production by baboon embryos *in vitro*, in: *In Vitro Fertilization and Embryo Transfer* (E. S. E. Hafez and K. Semm, eds.), MTP Press, Lancaster, England, pp. 129–134.

Pope, C. E., Pope, V. Z., and Beck, L. R., 1982b, Development of baboon preimplantation embryos to post-implantation stages *in vitro*, *Biol. Reprod.* **27**:915–923.

Pope, C. E., Pope, V. Z., and Beck, L. R., 1986, Cryopreservation and transfer of baboon embryos, *J. in Vitro Fertil. Embryo Transfer* **3**:33–39.

Renaud, R. L., Macler, J., Dervain, I., Ehret, M. C., Aron, C., Plas-Roser S., Spira, A., and Pollack, H., 1980, Echographic study of follicular maturation and ovulation during the normal menstrual cycle, *Fertil. Steril.* **33**:272–276.

Rogers, B. J., 1978, Mammalian sperm capacitation and fertilization *in vitro*: A critique of methodology, *Gamete Res.* **1**:165–223.

Sathananthan, A. H., and Trounson, A. O., 1982, Ultrastructure of cortical granule release and zona interaction in monospermic and polyspermic human ova fertilized *in vitro*, *Gamete Res.* **6**:225–234.

Schroeder, A. C., and Eppig, J. J., 1984, The developmental capacity of mouse oocytes that matured spontaneously *in vitro* is normal, *Dev. Biol.* **102**:493–497.

Schroeder, A. C., and Eppig, J. J., 1989, Developmental capacity of mouse oocytes that undergo maturation *in vitro*: Effect of the hormonal state of the oocyte donor, *Gamete Res.* **24**:81–92.

Schultz, R. M., 1985, Roles of cell-to-cell communication in development, *Biol. Reprod.* **32**:27–42.

Shalgi, R., 1984, Developmental capacity of rat embryos produced by *in vivo* or *in vitro* fertilization, *Gamete Res.* **10**:77–82.

Simon, J. A., and Hodgen, G. D., 1989, Macaque embryos produce stage specific angiogenic factors associated with viability, *Biol. Reprod.* **40** (Suppl 1):63.

Soupart, P., and Strong, P. A., 1974, Ultrastructural observations on human oocytes fertilized *in vitro*, *Fertil. Steril.* **25**:11–44.

Staigmiller, R. B., and Moor, R. M., 1984, Effect of follicle cells on the maturation and developmental competence of ovine oocytes matured outside the follicle, *Gamete Res.* **9**:221–229.

Summers, P. M., Shephard, A. M., Taylor, C. T., and Hearn, J. P., 1987, The effects of cryopreservation and transfer on embryonic development in the common marmoset monkey, *Callithrix jacchus*, *J. Reprod. Fertil.* **79**:241–250.

Tash, J. S., and Means, A. R., 1983, Cyclic adenosine 3′,5′monophosphate, calcium and protein phosphorylation in flagellar motility, *Biol. Reprod.* **28**:75–104.

Thibault, C., 1977, Are follicular maturation and oocyte maturation independent processes? *J. Reprod. Fertil.* **51**:1–15.

Van Blerkom, J., Henry, G., and Porreco, R., 1984, Preimplantation human embryonic development from polypronuclear eggs after *in vitro* fertilization, *Fertil. Steril.* **41**:686–696.

VanderVoort, C. A., Baughman, W. L., and Stouffer, R. L., 1989, Comparison of different regimens of human gonadotropins for superovulation of rhesus monkeys: Ovulatory response and subsequent luteal function, *J. in Vitro Fertil. Embryo Transfer* **6**:85–91.

Warikoo, P. K., 1990, Development of blastocysts following bisection of an *in vitro* fertilized rhesus monkey embryo, *Biol. Reprod.* **42(Suppl 1)**:54 (abstract no. 42).

Warikoo, P. K., and Bavister, B. D., 1989, Hypoxanthine and cAMP maintain meiotic arrest of rhesus monkey oocytes *in vitro*, *Fertil. Steril.* **51**:886–889.

Wassarman, P. M., and Bleil, J. D., 1982, The role of zona pellucida glycoproteins as regulators of sperm–egg interaction in the mouse, in: *Cellular Recognition* (W. Fraser, L. Glaser, and D. Gottlieb, eds.), Alan R. Liss, New York, pp. 845–863.

Wolf, D. P., VanderVoort, C. A., Meyer-Haas, G. R., Zelinski-Wooten, M. B., Hess, D. L., Baughman, W. L., and Stouffer, R. L., 1989, *In vitro* fertilization and embryo transfer in the rhesus monkey, *Biol. Reprod.* **41**:335–346.

Yanagimachi, R., 1970, The movement of golden hamster spermatozoa before and after capacitation, *J. Reprod. Fertil.* **23**:193–196.

Yanagimachi, R., 1981, Mechanisms of fertilization in mammals, in: *Fertilization and Embryonic Development In Vitro* (L. Mastroianni and J. D. Biggers, eds.), Plenum Press, New York, pp. 81–182.

Yanagimachi, R., 1984, Zona-free hamster eggs: Their use in assessing fertilizing capacity and examining chromosomes of human spermatozoa, *Gamete Res.* **10**:178–232.

Yanagimachi, R., Yanagimachi, H., and Rogers, B. J., 1976, The use of zona-free animal ova as a test-system for the assessment of the fertilizing capacity of human spermatozoa, *Biol. Reprod.* **15**:471–476.

20

Fertilization in Man

Don P. Wolf and Susan E. Lanzendorf

1. INTRODUCTION

Clinical IVF-ET was initiated with the birth of baby Louise Brown in 1978 following several years of pioneering effort by Patrick Steptoe and Robert Edwards. In the ensuing 12 years, human IVF-ET has evolved into standard medical practice for the treatment of many types of infertility, and the service is now offered by some 700 clinics in 48 countries. The application, confined originally to the hopelessly infertile patient with irreparably damaged or absent fallopian tubes, has expanded to include idiopathic, male, endometriosis, cervical, and immunologic etiologies. About 15% of egg retrieval cycles end in pregnancy and live birth resulting from the transfer of, on average, three embryos. Over 15,000 babies have been born by IVF-ET, which means that the experience base in man exceeds that in all other mammalian species with the possible exception of the mouse. A summary of the IVF experience in the United States for 1987 has appeared (Medical Research International *et al.*, 1989). Several peer-reviewed journals now devote major attention to the clinical and applied aspects of IVF—*Fertility and Sterility, Human Reproduction*, and *Journal of In Vitro Fertilization and Embryo Transfer*.

Simultaneous with the expansion of IVF has been the development of new approaches, which constitute the assisted reproductive technologies. These include GIFT (gamete intrafallopian transfer), ZIFT (zygote intrafallopian transfer), PROST (pronuclear stage transfer), TET (tubal embryo transfer), SOURCE (superovulation uterine replacement capacitation-enhanced sperm), the direct aspiration of epididymal sperm for IVF, the freezing of embryos, and the use of micromanipulation to assist fertilization. Despite the impressive growth and expansion of the assisted reproductive technologies, it is important to recognize that our ability to generate new insights into the physiology and cell biology associated with human fertilization is limited by

DON P. WOLF • Division of Reproductive Biology and Behavior, Oregon Regional Primate Research Center, Beaverton, Oregon 97006, and Department of Obstetrics/Gynecology, Oregon Health Sciences University, Portland, Oregon 97201.   SUSAN E. LANZENDORF • Division of Reproductive Biology and Behavior, Oregon Regional Primate Research Center, Beaverton, Oregon 97006. *Present address of S.E.L.*: The Jones Institute for Reproductive Medicine, Department of Obstetrics/Gynecology, Eastern Virginia Medical School, Norfolk, Virginia 23510.

A Comparative Overview of Mammalian Fertilization, edited by Bonnie S. Dunbar and Michael G. O'Rand. Plenum Press, New York, 1991.

ethical and practical concerns. Thus, controlled studies that jeopardize the viability of human eggs or embryos are largely precluded. This observation might suggest that the development of these clinical technologies is based on model mammalian studies; however, such is not always the case.

Frontiers for the assisted reproductive technologies in 1989 include the effective treatment of male infertility using micromanipulation and assisted fertilization; maturation of ovarian oocytes and their cryopreservation; and the genetic diagnosis of preimplantation-stage embryos.

2. OVARIAN HYPERSTIMULATION

Although limited success has been achieved in the recovery and fertilization of eggs from natural cycles (there is currently interest in returning to natural cycles), the majority of programs employ controlled ovarian hyperstimulation (COH) to increase the potential number of oocytes available and to allow a predictable intervention time for surgical pickup of eggs. The success of COH is predicated on monitoring the location, growth, and functional maturation of follicles using ultrasound imaging along with rapid, quantitative measurements of estradiol. Two regimens for increasing gonadotropic hormone levels have been employed in recruiting and/or supporting the growth of multiple ovarian follicles: stimulating endogenous hormone release and administering exogenous hormone (Diedrich et al., 1988). Clomiphene citrate represents the former approach. This drug is employed extensively in the clinical management of menstrual regularity and was used extensively during early IVF trials. Advantages that accrue to its use include relatively low cost and self-administration. This drug competes with estradiol for the latter's receptor and exerts an estrogenic effect on the pituitary. Clomiphene citrate administration during assisted reproductive technologies is associated with moderate OH and the recovery of three to five oocytes. This medication has also been used in conjunction with exogenous gonadotropins.

In the early 1980s, Howard and Georgeanna Jones at the Eastern Virginia Medical School in Norfolk, Virginia, initiated the first IVF program in the United States and pioneered the use of exogenous gonadotropins for COH. Their approach involved using Pergonal® (Serono), a combination of LH and FSH recovered from the urine of postmenopausal women. Metrodin® (Serono), a purified preparation of FSH, has also been employed more recently singly or in combination with Pergonal. Typically five to eight oocytes are recovered following COH with Pergonal.

All of these hyperstimulation protocols are initiated early in the follicular phase of the cycle and usually extend for 8 to 10 days. When follicles reach mature size (15–20 mm mean diameter, depending upon the COH protocol and the means of detection) and the circulating or urinary levels of estradiol peak or begin to plateau, a bolus injection of human chorionic gonadotropin (hCG) is administered to induce preovulatory maturation of the follicle and its enclosed egg. A major limitation to the COH protocols described here has been a high dropout rate associated with either an inadequate response or the occurrence of an endogenous LH surge—successful egg collection can follow the occurrence of an endogenous LH surge if frequent monitoring allows a definitive identification of surge initiation. GnRH agonists have been administered to down-regulate the pituitary, that is, to induce a transient medical hypophysectomy, thereby preventing an endogenous surge; COH and ovulation induction then become entirely dependent on administration of exogenous hormone. Down-regulation with GnRH agonists can involve prolonged exposure to drug, i.e., complete pituitary desensitization before the administration of exogenous gonadotropins, or a short "burst" protocol in which agonists and exogenous gonadotropins are administered simultaneously starting on day 3 of the treatment cycle. GnRH agonist administration is continued until hyperstimulation is completed and exogenous hCG is administered. This approach has led not only to a significant increase in the percentage of initiated cycles carried through to egg pickup but also to an increase in the implantation rate per embryo transfer and in pregnancy rate. A disadvantage of GnRH agonist use is the cost associated with the increased requirements (nearly twofold) for exogenous gonadotropins.

3. EGG COLLECTION

The collection of mature oocytes is obviously crucial to the success of IVF-ET, and egg pickup techniques have evolved substantially over the last 10 years. Oocytes are usually recovered by aspirating the contents of the preovulatory follicle with a needle introduced into the pelvic cavity 34 to 36 hr after the administration of exogenous hCG. The first successful attempts at egg pickup involved laparoscopic follicular aspiration (Beauchamp, 1984), a minor surgical procedure requiring major anesthesia. For egg pickup, three puncture sites are normally involved—one for the laparoscope, one for the aspirating needle, and one for a grasping forceps. Advantages associated with a laparoscopic approach include direct visualization of the ovary, allowing the aspiration of follicles sequentially by size and the reintroduction of the aspirating needle for retrograde flushing. Disadvantages include exposure of the ovary and its follicular contents to the pure CO_2 commonly used for insufflation, patient exposure to a general anesthetic, and an inability to proceed when the ovary is inaccessible or otherwise obscured. The GIFT procedure involves laparoscopy.

During the mid-1980s, ultrasound-guided egg pickup techniques were introduced (Wikland *et al.*, 1988). Sequential real-time ultrasound, which utilizes sound waves to create visual images, was of course being used during IVF to monitor follicular growth and development. Initially an abdominal transducer and a transvesicle approach was employed for egg pickup. However, the technology rapidly evolved through free-handed transvaginal or transurethral approaches to the development of a vaginal transducer with needle guide. The advantages associated with the widely used transvaginal technique include the avoidance of general anesthesia (usually only local anesthesia and analgesia are required), decreased cost, and savings in time. Disadvantages derive from the possibility of damage from energy produced by the transducer, infection associated with vaginal puncture, or trauma from the inadvertent puncturing of bowel or blood vessels.

The efficiency of egg collection from mature follicles ($\geq$15 mm mean diameter) by either laparoscopy or ultrasound-guided pickup is high (85–95%). If an egg is not recovered in the initial aspiration, retrograde flushing and reaspiration almost always result in egg recovery. The contents of the ovarian follicle are usually aspirated directly into a collecting tube and passed off to the embryologist, who identifies an egg if present. Follicular contents are commonly diluted immediately on collection with a pH-stable buffer such as phosphate-buffered saline to assist in pH control during the egg's transient exposure to room air.

4. EGG QUALITY

Egg quality or maturity is usually based on considerations of follicle size from which the egg originated and observations on the cumulus–corona cell matrix (Wolf, 1988b; Fig. 1A). Although recovered eggs in cumulus may occasionally be immature [germinal vesicle-intact (GV)], they usually have resumed meiosis from prophase I to metaphase II and are most often metaphase II (Sundström and Nilsson, 1988). The ability to mature GV-intact oocytes *in vitro* with subsequent fertilization, development, and pregnancy initiation on transfer is limited. Although sperm penetration of oocytes at varied nuclear states may occur, fertilization and the activation of development are dependent on nuclear as well as cytoplasmic maturation; thus, sperm penetration

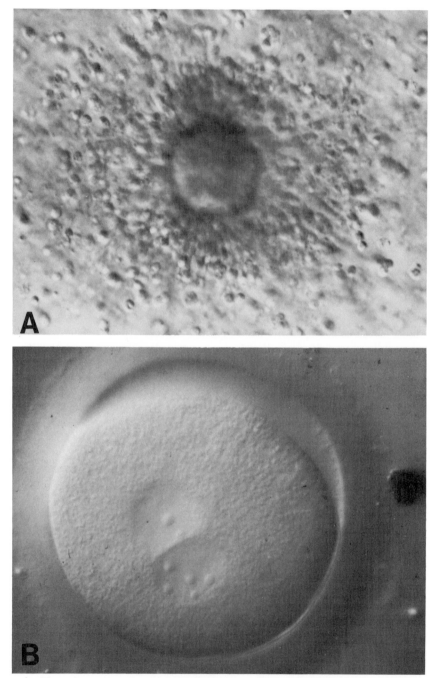

Figure 1. Photomicrographs of human eggs and embryos. A: Mature oocyte surrounded by radiating layer of cells (corona radiata) and the cumulus; optical magnification 100×. B: Pronuclear egg from the collection of Dr. Pierre Soupart. C: Two-cell embryo—note the presence of perivitelline space debris. D: Four-cell embryo; optical magnification for C and D is 200×.

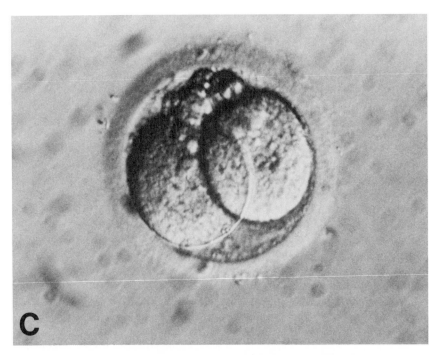

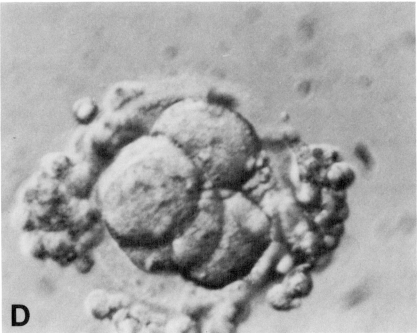

Figure 1. *(continued)*

into cytoplasmically immature oocytes may result in sperm nucleus differentiation but an improper oocyte response. Early indications from human IVF attempts suggested that cytoplasmic maturation was incomplete in eggs recovered from COH cycles, since maximal fertilization rates were achieved only after eggs were cultured for 4 to 6 hr before insemination (Trounson *et al.*, 1982). However, Harrision *et al.* (1988) recently reported no significant differences in fertilization rates when eggs were inseminated between 1 and 26 hr after pickup; pregnancy rates were identical for eggs inseminated within a 3- to 16-hr time window. Moreover, it is recognized that the GIFT procedure is conducted without egg preincubation *in vitro*.

The ultrastructural features of human oocytes have been described extensively (Baca and Zamboni, 1967; Zamboni *et al.*, 1972; Suzuki *et al.*, 1981; Sathananthan and Trounson, 1982a,b). Cytogenetic evaluations of oocytes recovered following COH have been conducted typically on mature oocytes that failed to fertilize. Early estimates suggested very high levels (>50%) of chromosomal abnormalities. However, the combined reports by Plachot *et al.* (1988) and Bongso *et al.* (1988) place this level at 25%—426 of 567 examined eggs were normal haploid, 21–24% aneuploid, and 2% diploid. Van Blerkom and Henry (1988) characterized 92% ($^{124}/_{135}$) of oocytes as normal haploid with only 8% showing chromosomal abnormalities. An association between increasing maternal age and aneuploidy in recovered oocytes has been noted (Plachot *et al.*, 1988).

The ability to cryopreserve unfertilized eggs has met with limited success. Although live births have been reported (Chen, 1986; van Uem *et al.*, 1987) following cryopreservation and IVF, a considerably higher level of success has been associated with embryo banking. This disappointing outcome with oocytes may reflect not only differences in the cryobiology between eggs and embryos but also the occurrence of cortical granule exocytosis induced by exposure to cryoprotectants such as dimethylsulfoxide or propanediol (Schalkoff *et al.*, 1989).

5. EPIDIDYMAL SPERM

A role for the epididymis in sperm maturation, although well established in rodents, is inferred in man from studies associating the acquisition of motility, fertility potential as measured with the sperm penetration assay, and *in vivo* fertility, with epididymal exposure. Androgen-induced proteins have been described that coat sperm during epididymal transit (Tezón *et al.*, 1985), and fertility potential increases in sperm recovered more distally from successive epididymal segments, i.e., from the caput to the cauda (Hinrichsen and Blaquier, 1980). A significant role for the epididymis in sperm maturation and storage, however, remains arguable. Schoysman and Bedford (1986) concluded from studies on epididymovasostomy patients that sperm exposure to only the caput epididymis was sufficient for maturation. More recently, Johnson and Varner (1988) estimated epididymal transit time to be short at 3.8–4.3 days, suggesting that maturation if required is rapid. The sperm storage capacity of this organ is probably low (200×10^6 cells). Finally, Silber (1988) and Silber *et al.* (1988) have successfully aspirated fertile sperm from the proximal caput epididymis in patients with congenital absence of the vas deferens.

6. EJACULATED SPERM

One of the hallmarks of a normal human ejaculate is the relatively wide range in sperm motility and morphology. In contrast with other mammals, the normal human ejaculate is "dirty." In addition to its cellular component, ejaculates contain complex secretions from the accessory glands—the Cowper's gland, prostate, and seminal vesicles—which, in turn, contain a variety of energy substrates, hormones, nonenzymatic and enzymatic proteins, and various ions.

Within the first few minutes following ejaculation, semen liquefaction occurs as a result of the enzymatic activities emanating from prostatic secretions (Mann and Lutwak-Mann, 1981). Prolonged sperm exposure to seminal plasma *in vitro* is associated with detrimental effects on motility and viability. *In vivo*, sperm are effectively separated from seminal plasma within minutes of coitus, on colonization of the cervix and its mucus.

Conventional semen analysis is a mainstay of a clinical workup for male infertility. The analysis includes a description of the ejaculate volume, viscosity, and color as well as a description of the cellular fraction. The World Health Organization has established normal values for sperm concentration, motility, progression, and morphology (World Health Organization, 1987; Kopf, 1988b). A role for electron microscopy in the evaluation of semen quality has also been advocated (Zamboni, 1987). Recent advances in technology support the quantitation of motility by computer-assisted video image analysis. Unfortunately, results from conventional semen analysis correlate poorly with fertility except in extreme cases such as azoospermia and/or asthenospermia. This lack of correlation may reflect the fact that the endpoint is crude when one considers the number of physiological and biochemical changes that occur during sperm transit in the female reproductive tract coupled with the requirement that only a few (one) capacitated sperm present at the right time and the right place are adequate to result in fertility. With regard to *in vitro* fertilization outcome, it is immediately obvious that the total number of morphologically normal, motile sperm required to achieve maximum fertilization levels is greatly reduced over that required *in vivo* (see Section 9).

The techniques employed for semen handling are, of course, dependent on the intended use. For cervical inseminations, whole semen is usually employed. For cryopreservation, liquefied semen can be directly diluted with cryoprotectant or with cryoprotectant plus extender(s) in preparation for freezing. For the *in vitro* isolation of sperm from seminal plasma, a number of protocols are currently in use ranging from a simple swim-up from whole semen or a washed sperm pellet to the use of buoyant density gradient centrifugation approaches involving commercial products such as Percoll® or Sperm Select®.

Recent success in the amplification and analysis of DNA sequences in single sperm may pave the way for genetic analysis of the male contribution to fertilization and development (Li *et al.*, 1988).

7. SPERM CAPACITATION

Freshly recovered mammalian sperm require a period of maturation, called capacitation, before acquiring the ability to bind to and penetrate the zona pellucida and fuse with the egg (Yanagimachi, 1981; Boldt and Wolf, 1984). This requirement in man is largely undefined. The evaluation of capacitation using human IVF is limited by practical and ethical considerations. Since no overt morphological correlations can be used to identify capacitation, its measurement usually involves monitoring acrosomal loss (see Section 7.3) or two rather cumbersome sperm–egg interaction systems: penetration of zona-enclosed nonviable human oocytes (Overstreet and Hembree, 1976) or of zona-free hamster eggs (Yanagimachi *et al.*, 1976). A disadvantage to these bioassays is that the performance of only a limited number of sperm is monitored, requiring substantial extrapolation to assess the capacitation status of the entire population. Perhaps the most important component in initiating capacitation of seminal sperm is seminal plasma removal, since this process presumably results in altered membrane structure and function. The persistent presence of even dilute seminal plasma prevents attachment and penetration of zona-free hamster and salt-stored human eggs (Kanwar *et al.*, 1979). A rather wide variety of media support the capacitation of human sperm, as evidenced by clinical IVF use. These include both simple and complex bicarbonate-buffered solutions, usually augmented with an albumin source. Albumin-mediated sperm capacitation may involve sterol efflux (Moubasher and Wolf, 1986); however, it has become clear that capacitation is not absolutely dependent on albumin, since it can be conducted in protein-free medium or in gelatin (Mack *et al.*, 1989b).

7.1. Hyperactivation

Over the past several years, with the availability of automated, quantitative methods for motility measurement (Katz *et al.*, 1985; Mack *et al.*, 1988), the concept that hyperactivation is a component of capacitation has been extended to man. Hyperactivation occurs when seminal or washed sperm, which typically swim with a linear trajectory, change trajectory to a complex, nonlinear form. Hyperactivation may represent the first overt sign of capacitation, followed by the ability of the capacitated cell to bind to the zona pellucida and undergo an acrosome reaction. Initially, human sperm in capacitating medium were not thought to show hyperactivation (Yanagimachi, 1981), although changes in the type of motilities displayed in synthetic media were recognized (Mortimer *et al.*, 1983, 1984). Subsequently, it was demonstrated that approximately 20% of washed cells developed hyperactivation (Burkman, 1984). The validation and use of a computer-assisted digital image analysis system for the rapid measurement of a range of motility parameters in individual sperm (Mack *et al.*, 1988) led to the development of an automated technique for the quantitation of hyperactivation (Robertson *et al.*, 1988). This transition in trajectory occurs in some sperm immediately on removal from seminal plasma and probably in all cases prior to spontaneous acrosomal loss. The conditions for optimally measuring hyperactivation in capacitating populations have been defined, and it appears that the extent of hyperactivation is a donor characteristic (Mack *et al.*, 1989a).

7.2. Sperm–Zona Binding

Capacitated, acrosome-intact sperm in many species are uniquely capable of binding to the zona pellucida. Binding is an important prerequisite step for zona penetration because it initiates events that culminate in induction of the acrosome reaction. A specific zona glycoprotein acts as a receptor and as a potent inducer of the acrosome reaction in the mouse (see Chapter 10, this volume). The human zona pellucida also contains three acidic glycoproteins at least grossly comparable to those in the mouse—ZP1, ZP2, and ZP3 (Shabanowtiz and O'Rand, 1988). Moreover, a fertilization-dependent modification of ZP1 has been described (see Section 10.2).

The kinetics of capacitated sperm binding to the zona have been studied using nonfertilizable ovarian oocytes (Overstreet, 1983; Singer *et al.*, 1985). Sperm acquired the ability to bind to and penetrate the zona rather quickly (within several hours) but surprisingly lost this ability after 20 hr of "capacitation." Both acrosome-intact and acrosome-reacted sperm were observed on the zona surface by transmission electron microscopic examination. Zonae discarded from IVF attempts have also been used to measure fertility potential (see Section 8).

7.3. Acrosome Reaction

A physiological acrosome reaction occurs uniquely in a capacitated cell; hence, the ability of free-swimming sperm to undergo acrosome reactions has been equated with capacitation or fertility potential. The development of indirect immunofluorescence techniques at the light microscope level has made it possible to quantitate acrosomal status in large populations of sperm rapidly and accurately. The use of this technology has been reviewed recently (Wolf, 1989), and the following general conclusions are warranted. Unlike their rodent counterparts, human sperm incubated under capacitating conditions do not readily undergo spontaneous loss of acrosomes *in vitro*; the reacting population may be restricted to only 10% of the motile cells. Moreover, sperm do not undergo degenerative acrosomal loss as an obligatory component of senescence. Acrosomal loss can be induced by biological agents such as follicular fluid, cumulus cells, or zonae pellucidae or by physiocochemical agents such as calcium ionophores, lysophosphatidylcholine, and electropermeabilization. The ability of capacitated cells to respond to ionophore induction of acrosomal loss may represent a convenient assay for capacitation (Byrd and Wolf, 1986). Washed motile sperm incubated under capacitating conditions can certainly be subdivided (based on their state of capacitation?) into inducible and noninducible populations. By using ionophore-induced acrosomal loss as a measure, the kinetics of capacitation can be inferred. The responsive population increases in size to a maximum value of 35% of the motile cells within 6 to 8 hr of incubation. These conclusions are derived largely from observations on free-swimming sperm populations. In the fertilizing or zona-penetrating sperm, the reaction presumably occurs after zona binding, as induced by ZP3, in analogy to the mouse. Evidence to support this sequence is available in man based on an assessment of acrosomal status of sperm associated with the zona (Cross et al., 1988). When sperm and eggs were coincubated for very short time periods, almost all attached sperm were acrosome-intact. In contrast, both acrosome-intact and acrosome-reacted sperm were found bound to the zona when a 60-min coincubation period was employed. In this study, the acrosome reaction could also be induced by acid-disaggregated zonae.

At the molecular level, capacitation and the acrosome reaction are dependent on the presence of exogenous calcium, implying that increased intracellular levels of this cation trigger the membrane fusion events associated with the reaction. Obviously, the ability to induce acrosomal loss with calcium ionophores implicates calcium as an intracellular messenger. Direct evidence is now available (Thomas and Meizel, 1988) documenting an increase in intracellular calcium as a concomitant of induced acrosomal loss.

8. FERTILITY POTENTIAL MEASUREMENT

Although the zona pellucida is the major species-specific barrier to fertilization, in many cases the zona-free egg also retains specificity. The hamster represents a major exception to this rule, since capacitated sperm from many mammals can fuse with the zona-free hamster egg. This observation, reported by Yanagimachi in the early 1970s, led to the development of a sperm penetration assay (SPA) using hamster eggs in the diagnostic evaluation of fertility potential in man (Yanagimachi, 1984; Rogers, 1985; Aitken, 1986). Clinical use of the SPA is predicated on differences in test outcome observed with sperm from males of proven fertility as opposed to infertility patients in whom there is no evidence of infertility in the female partner. Additionally, over the last several years, correlations have been established between test scores and the in vitro fertilization of wives' eggs.

Despite extensive application, the usefulness of the SPA in directing clinical treatment is still questioned (Mao and Grimes, 1988). Although the test may evaluate sperm capacitation and the acrosome reaction, it is clearly neither the optimum nor the ideal bioassay, for it is difficult to evaluate the relevance of sperm penetration of heterologous eggs to homologous fertilization, and sperm fusion with the zona-free egg is not always predictive of the ability to penetrate the zona-intact egg. Over and above these theoretical arguments, the methodology for conducting the SPA has varied widely. As originally described and applied, sperm were capacitated by exposure to albumin-containing physiological saline solutions for various time periods at 37°C. This approach resulted in penetration indices (mean number of penetrated sperm/inseminated egg) of less than 1. In order to increase the sensitivity of the assay, creative approaches to sperm capacitation have been reported, including exposure to the calcium ionophore A23187 (Aitken et al., 1984), to strontium (Mortimer et al., 1986), or to TES–Tris egg yolk buffer at 4° for prolonged time periods (Rogers, 1985). The latter approach has been associated with high penetration indices (greater than 6 and as high as 50) and has increased assay reliability. Since the test is dependent on the presence of acrosome-reacted sperm, these creative capacitation approaches presumably result in increased concentrations of viable acrosome-reacted sperm. This expectation has been realized; however, relatively small changes in acrosomal status stand opposed to large changes in mean penetration indices (D. P. Wolf, unpublished results).

Capacitation kinetics, based on SPA outcome, are donor-specific (Wolf and Sokoloski, 1982; Perreault and Rogers, 1982). In some individuals, functional competence is acquired within 3 to 6 hr of sperm removal from seminal plasma, whereas in others, more than 10 hr is required.

Sperm binding to human zonae pellucidae has been used as a means of testing the fertilizing potential of individuals with suspected male factor infertility (Overstreet, 1983). A recent variant on this theme, the hemizona assay (HZA) (Burkman et al., 1988), uses discarded or surplus nonliving, nonfertilizable oocytes obtained during IVF. With micromanipulative techniques, zonae pellucidae are bisected into equal halves, one half to be inseminated with donor sperm as an internal control for zona-to-zona variability and one half to be inseminated with test sperm. Sperm from men with IVF failure exhibited a significantly lower binding capacity to hemizonae when compared to the sperm of fertile men. It has been suggested that poor binding of sperm could result from morphological abnormalities or from interfering activity of substances in seminal plasma such as microorganisms or antibodies (Burkman et al., 1988). Using salt-stored zona pellucida from human oocytes that failed to fertilize in vitro, Liu et al. (1988) also demonstrated that individuals with poor zona-binding ability frequently demonstrated poor or failed IVF. In this study, fluorochome markers were used to differentiate test sperm from those already bound as a result of the IVF attempt.

9. FERTILIZATION

Human IVF has been successfully performed using a wide variety of culture media for sperm preparation and capacitation (see Sections 6 and 7), oocyte maturation, insemination, and embryo development (Edwards, 1980; Veeck and Maloney, 1986; Kopf, 1988a). These media range from simple balanced salt solutions to complex media containing amino acids and vitamins. Protein supplementation is common, and sources include maternal or human fetal cord serum and human or bovine serum albumin. Human IVF has also been performed successfully in media containing no protein (Menezo et al., 1984), in 100% heat-inactivated serum (Kemeter and Feichtinger, 1984), in amniotic fluid (Gianaroli et al., 1986), and in defined media (Feichtinger et al., 1986). The culture media in use also contain energy sources such as pyruvate and lactate for sperm, oocytes, and pronuclear embryos and pyruvate and glucose for cleaving embryos. Wales et al. (1987) concluded, based on glucose turnover experiments, that before day 3 (eight cells) of development, human embryos do not effectively utilize glucose as an energy source because of a blockage in glycolysis. These findings provide a basis for the inclusion of lactate and/or pyruvate in media used for culturing human embryos during the first cleavage divisions. Typical culture requirements also include incubation of gametes and embryos in a 5% CO_2 in air atmosphere at 37°C with a pH ranging from 7.2 to 7.5 and osmolality of 280 to 305 mOsm/kg.

Fertilization of human oocytes has been achieved with sperm concentrations ranging from 1.0×10^4 to 50×10^5 motile sperm per milliliter, with 2.5 to 5.0×10^4/ml as an optimal range for normal fertilization (Craft et al., 1981; Wolf et al., 1984; McDowell, 1986; Trounson, 1986). Where fertilization is attempted with sperm from a "subfertile male" as evidenced by abnormal semen parameters, the sperm concentration is often increased to 50×10^4 motile sperm/ml.

The incidence of polyspermic fertilization has been associated with sperm concentration (Wolf et al., 1984), oocyte maturational status at insemination, the age of the inseminated egg, and the stimulation protocol utilized (Rudak et al., 1984; Trounson and Webb, 1984; McDowell,

1986; Ben-Rafael *et al.*, 1987). Polyspermy rates typically vary from 2% to 6%.

As indicated in Section 6, human semen routinely contains many morphologically abnormal sperm such as those with large or small heads, double heads, tail defects, midpiece defects, cytoplasmic droplets, and immature forms (Hargreave and Nillson, 1983). The percentage of sperm with "normal" morphology has been correlated with fertilization outcome. Using strict criteria for evaluation, Kruger *et al.* (1988) demonstrated that fertilization rates were significantly higher for patients whose sperm samples contained greater than 14% normal morphology. A normal acrosomal morphology has also been shown to be important for successful human IVF (Jeulin *et al.*, 1986; Liu and Baker, 1988). The importance of the acrosome to successful sperm-egg interaction is inferred from the failure of acrosomeless sperm—round headed sperm syndrome—to penetrate hamster eggs (Weissenberg *et al.*, 1982; Syms *et al.*, 1984).

Approximately 12 to 18 hr post-insemination, fertilized eggs will display two readily visible pronuclei (Fig. 1B). Two polar bodies will also be present, which may be obscured by cumulus cells or fragmented, and, therefore, should not be used as a major criterion for normal fertilization (Veeck and Maloney, 1986). Failure of eggs to fertilize *in vitro* may result from immaturity (cytoplasmic and/or nuclear), postmaturity at the time of insemination, or a lack of sperm fertilizing ability (Veeck and Maloney, 1986; Kopf, 1988b).

9.1. Reinsemination

The successful reinsemination of unfertilized oocytes has been reported (Trounson and Webb, 1984; Ben-Rafael *et al.*, 1986; Boldt *et al.*, 1987), but the efficiency of the procedure is unknown, since it is uncertain if such embryos result in pregnancy following transfer. Visual assessment of oocyte nuclear maturation to determine the appropriate time for insemination may decrease the need for reinsemination.

In an effort to successfully reinseminate eggs or to treat male factor infertility, Malter and Cohen (1989) have developed a micromanipulative technique called partial zona dissection. With this technique, a microneedle is passed through the zona pellucida to create an opening through which sperm can pass and come in contact with the oocyte plasma membrane. Increased fertilization rates following reinsemination and live/born offspring have resulted following partial zona dissection.

9.2. Embryo Banking

Because the transfer of multiple embryos increases the risk of multiple pregnancy, many IVF programs cryopreserve "surplus" embryos for transfer in subsequent, unstimulated menstrual cycles. Three different cryoprotectants have been used: glycerol (Fehilly *et al.*, 1985; Cohen *et al.*, 1986), dimethylsulfoxide (Freeman *et al.*, 1986; Trounson and Sjöblom, 1988; Camus *et al.*, 1989), and 1,2-propanediol (Lassalle *et al.*, 1985). In addition, freezing media, the freezing vessel utilized, the speed of the freezing and thawing procedure, and the cell stage at which the embryo is frozen vary between IVF programs (see Kuzan and Quinn, 1988, for review). Cryosurvival is typically around 70%, with approximately 20% of embryo transfers producing a pregnancy.

10. EARLY DEVELOPMENT

10.1. Kinetics and Morphology

Fertilization, as demonstrated by the presence of male and female pronuclei, can be evaluated between 12 and 18 hr following insemination of human oocytes *in vitro* (Trounson,

1986; Veeck and Maloney, 1986; Wolf, 1988a). If allowed to continue in culture, such embryos are capable of cleavage and development to the blastocyst stage *in vitro* with differentiation of the trophectoderm, inner cell mass, and primitive ectoderm as well as initial production of hCG (Mohr and Trounson, 1982; Fishel *et al.*, 1984). Based on pregnancy as the ultimate outcome, Cummins *et al.* (1986) reported "ideal" growth (cleavage) rates for human embryos fertilized *in vitro* as: the two-cell stage at 33.6 hr; the four-cell stage at 45.5 hr; and the eight-cell stage at 56.4 hr following insemination (Fig. 1C,D). In general, however, the more rapid the cleavage rate, the more likely pregnancy will occur following transfer (Mohr, 1984; Cummins *et al.*, 1986). Additional criteria for the evaluation of embryo quality include the size and shape of blastomeres, the number and location of nuclei within the blastomeres, the presence of anucleate fragments, and the appearance of the cytoplasm (Trounson, 1986; Veeck and Maloney, 1986; Wolf, 1988a). The incidence of chromosomal abnormalities in embryos analyzed after IVF is in the range of 20–30% (for review, see Plachot *et al.*, 1988).

10.2. Cortical Activation and the Block to Polyspermy

Sperm penetration of the mammalian oocyte initiates the cortical reaction, an exocytotic fusion event between the oocyte plasma membrane and the cortical granule membrane. Containing hydrolytic enzymes, the released contents of the cortical granules diffuse into the zona pellucida and modify its structure. This process, referred to as the zona reaction, functions in the prevention of polyspermy, the lethal condition of penetration by more than one sperm (Lopata *et al.*, 1980; Bedford, 1982). Cortical and zona reactions in human oocytes and embryos have been examined by high-resolution analysis (Lopata *et al.*, 1980; Sathananthan and Trounson, 1982a,b, 1985; Schalkoff *et al.*, 1989). In unfertilized oocytes, cortical granules line the cortex in a uniform manner. Following insemination, fertilized oocytes and embryos lose cortical granules, and cortical granule material appears in the perivitelline space. Cortical granule contents have been associated with the inner surface of the zona pellucida as demonstrated by the presence of striae of dense material at the site of interaction (Sathananthan and Trounson, 1982b). In oocytes showing normal, monospermic fertilization, sperm are not usually found penetrating the inner zona pellucida or within the perivitelline space. In polyspermic oocytes, however, supernumerary sperm are visible penetrating the inner zona pellucida and in the perivitelline space, perhaps reflecting delayed cortical granule release (Sathananthan and Trounson, 1982b, 1985).

10.3. Maternal and Zygotic Control of Development

Early embryonic development is associated with quantitative and qualitative changes in protein synthetic activity originating, in general, from the differential activation of stored mRNAs. Genomic participation, as evidenced by qualitative changes in protein synthesis and developmental sensitivity to the transcription inhibitor α-amanitin, has been associated with the four- to eight-cell stage transition in the human (Braude *et al.*, 1988). This information is relevant to recent research advances in the area of preimplantation diagnosis and in the treatment of genetic disorders in animal models (Monk *et al.*, 1987; Wilton *et al.*, 1989) and in primates.

ACKNOWLEDGMENTS. The work described in this chapter, publication No. 1676 of the Oregon Regional Primate Research Center, was supported in part by NIH grant RR00163. The authors recognize the secretarial and editorial skills of Patsy Kimzey.

11. REFERENCES

Aitken, R. J. (ed.), 1986, The zona-free hamster oocyte penetration test and the diagnosis of male fertility, *Int. J. Androl.* **[Suppl.]6**:1–199.

Aitken, R. J., Ross, A., Hargreave, T., Richardson, D., and Best, F., 1984, Analysis of human sperm function following exposure to the ionophore A23187, *J. Androl.* **5**:321–329.

Baca, M., and Zamboni, L., 1967, The fine structure of human follicular oocytes, *J. Ultrastruct. Res.* **19**:354–381.

Beauchamp, P. J., 1984, Laparoscopic follicular aspiration, in: *Human In Vitro Fertilization and Embryo Transfer* (D. P. Wolf and M. M. Quigley, eds.), Plenum Press, New York, pp. 149–169.

Bedford, J. M., 1982, Fertilization, in: *Reproduction in Mammals*, Book 1, *Germ Cells and Fertilization* (C. R. Austin and R. V. Short, eds.), Cambridge University Press, New York, pp. 128–163.

Ben-Rafael, Z., Kopf, G. S., Blasco, L., Tureck, R. W., and Mastroianni, L., Jr., 1986, Fertilization and cleavage after reinsemination of human oocytes *in vitro, Fertil. Steril.* **45**:58–62.

Ben-Rafael, Z., Meloni, F., Strauss, J. F. III, Blasco, L., Mastroianni, L., Jr., and Flickinger, G. L., 1987, Relationships between polypronuclear fertilization and follicular fluid hormones in gonadotropin-treated women, *Fertil. Steril.* **47**:284–288.

Boldt, J., and Wolf, D. P., 1984, Sperm capacitation, in: *Human In Vitro Fertilization and Embryo Transfer* (D. P. Wolf and M. M. Quigley, eds.), Plenum Press, New York, pp. 171–211.

Boldt, J., Howe, A. M., Butler, W. J., McDonough, P. G., and Padilla, S. L., 1987, The value of oocyte reinsemination in human *in vitro* fertilization, *Fertil. Steril.* **48**:617–623.

Bongso, A., Chye, N. S., Ratnam, S., Sathananthan, H., and Wong, P. C., 1988, Chromosome anomalies in human oocytes failing to fertilize after insemination *in vitro, Hum. Reprod.* **3**:645–649.

Braude, P., Bolton, V., and Moore, S., 1988, Human gene expression first occurs between the four- and eight-cell stages of preimplantation development, *Nature* **332**:459–461.

Burkman, L. J., 1984, Characterization of hyperactivated motility by human spermatozoa during capacitation: Comparison of fertile and oligozoospermic sperm populations, *Arch. Androl.* **13**:153–165.

Burkman, L. J., Coddington, C. C., Franken, D. R., Kruger, T. F., Rosenwaks, Z., and Hodgen, G. D., 1988, The hemizona assay (HZA): Development of a diagnostic test for the binding of human spermatozoa to the human hemizona pellucida to predict fertilization potential, *Fertil. Steril.* **49**:688–697.

Byrd, W., and Wolf, D. P., 1986, Acrosomal status in fresh and capacitated human ejaculated sperm, *Biol. Reprod.* **34**:859–869.

Camus, M., Van den Abbeel, E., Van Waesberghe, L., Wisanto, A., Devroey, P., and Van Steirteghem, A. C., 1989, Human embryo viability after freezing with dimethylsulfoxide as a cryoprotectant, *Fertil. Steril.* **51**:460–465.

Chen, C., 1986, Pregnancy after human oocyte cryopreservation, *Lancet* **1**:884–886.

Cohen, J., Simons, R. S., Fehilly, C. B., and Edwards, R. G., 1986, Factors affecting survival and implantation of cryopreserved human embryos, *J. In Vitro Fert. Embryo Transf.* **3**:46–52.

Craft, I., McLeod, F., Bernard, A., Green, S., and Twigg, H., 1981, Sperm numbers and in-vitro fertilisation, *Lancet* **2**:1165–1166.

Cross, N. L., Morales, P., Overstreet, J. W., and Hanson, F. W., 1988, Induction of acrosome reactions by the human zona pellucida, *Biol. Reprod.* **38**:235–244.

Cummins, J. M., Breen, T. M., Harrison, K. L., Shaw, J. M., Wilson, L. M., and Hennessey, J. F., 1986, A formula for scoring human embryo growth rates in *in vitro* fertilization: Its value in predicting pregnancy and in comparison with visual estimates of embryo quality, *J. In Vitro Fert. Embryo Transf.* **3**:284–295.

Diedrich, K., van der Ven, H., Al-Hasani, S., and Krebs, D., 1988, Ovarian stimulation for in-vitro fertilization, *Hum. Reprod.* **3**:39–44.

Edwards, R. G., 1980, *Conception in the Human Female*, Academic Press, New York, pp. 668–776.

Fehilly, C. B., Cohen, J., Simons, R. F., Fishel, S. B., and Edwards, R. G., 1985, Cryopreservation of cleaving embryos and expanded blastocysts in the human: A comparative study, *Fertil. Steril.* **44**:638–644.

Feichtinger, W., Kemeter, P., and Menezo, Y., 1986, The use of synthetic culture medium and patient serum for human *in vitro* fertilization and embryo replacement, *J. In Vitro Fert. Embryo Transf.* **3**:87–92.

Fishel, S. B., Edwards, R. G., and Evans, C. J., 1984, Human chorionic gonadotropin secreted by preimplantation embryos cultured *in vitro, Science* **223**:816–818.

Freemann, L., Trounson, A., and Kirby, C., 1986, Cryopreservation of human embryos: Progress on the clinical use of the technique in human *in vitro* fertilization, *J. In Vitro Fert. Embryo Transf.* **3**:53–61.

Gianaroli, L., Seracchioli, R., Ferraretti, A. P., Trounson, A., Flamigni, C., and Bovicelli, L., 1986, The successful use of human amniotic fluid for mouse embryo culture and human *in vitro* fertilization, embryo culture, and transfer, *Fertil. Steril.* **46:**907–913.

Hargreave, T. B., and Nillson, S., 1983, Seminology, in: *Male Infertility* (T. B. Hargreave, ed.), Springer-Verlag, New York, pp. 56–74.

Harrison, K. L., Wilson, L. M., Breen, T. M., Pope, A. K., Cummins, J. M., and Hennessey, J. F., 1988, Fertilization of human oocytes in relation to varying delay before insemination, *Fertil. Steril.* **50:**294–297.

Hinrichsen, M. J., and Blaquier, J. A., 1980, Evidence supporting the existence of sperm maturation in the human epididymis, *J. Reprod. Fertil.* **60:**291–294.

Jeulin, C., Feneux, D., Serres, C., Jouannet, P., Guillet-Rosso, F., Belaisch-Allart, J., Frydman, R., and Testart, J., 1986, Sperm factors related to failure of human *in-vitro* fertilization, *J. Reprod. Fertil.* **76:**735–744.

Johnson, L., and Varner, D. D., 1988, Effect of daily spermatozoan production but not age on transit time of spermatozoa through the human epididymis, *Biol. Reprod.* **39:**812–817.

Kanwar, K. C., Yanagimachi, R., and Lopata, A., 1979, Effects of human seminal plasma on fertilizing capacity of human spermatozoa, *Fertil. Steril.* **31:**321–327.

Katz, D. F., Davis, R. O., Delandmeter, B. A., and Overstreet, J. W., 1985, Real-time analysis of sperm motion using automatic video image digitization, *Comput. Programs Biomed.* **21:**173–182.

Kemeter, P., and Feichtinger, W., 1984, Pregnancy following *in vitro* fertilization and embryo transfer using pure human serum as culture and transfer medium, *Fertil. Steril.* **41:**936–937.

Kopf, G. S., 1988a, Choice, preparation and use of culture medium, in: *In Vitro Fertilization and Embryo Transfer: A Manual of Basic Techniques* (D. P. Wolf, ed.; B. D. Bavister, M. Gerrity, and G. S. Kopf, assoc. eds.), Plenum Press, New York, pp. 47–56.

Kopf, G. S., 1988b, Preparation and analysis of semen samples, in: *In Vitro Fertilization and Embryo Transfer: A Manual of Basic Techniques* (D. P. Wolf, ed.; B. D. Bavister, M. Gerrity, and G. S. Kopf, assoc. eds.), Plenum Press, New York, pp. 77–102.

Kruger, T. F., Acosta, A. A., Simmons, K. F., Swanson, R. J., Matta, J. F., and Oehninger, S., 1988, Predictive value of abnormal sperm morphology in *in vitro* fertilization, *Fertil. Steril.* **49:**112–117.

Kuzan, F. B., and Quinn, P., 1988, Cryopreservation of mammalian embryos, in: *In Vitro Fertilization and Embryo Transfer: A Manual of Basic Techniques* (D. P. Wolf, ed.; B. D. Bavister, M. Gerrity, and G. S. Kopf, assoc. eds.), Plenum Press, New York, pp. 301–347.

Lassalle, B., Testart, J., and Renard, J.-P., 1985, Human embryo features that influence the success of cryopreservation with the use of 1,2 propanediol, *Fertil. Steril.* **44:**645–651.

Li, H., Gyllensten, U. B., Cui, X., Saiki, R. K., Erlich, H. A., and Arnheim, N., 1988, Amplification and analysis of DNA sequences in single human sperm and diploid cells, *Nature* **335:**414–417.

Liu, D. Y., and Baker, H. W. G., 1988, The proportion of human sperm with poor morphology but normal intact acrosomes detected with *Pisum sativum* agglutinin correlates with fertilization *in vitro*, *Fertil. Steril.* **50:**288–293.

Liu, D. Y., Lopata, A., Johnston, W. I. H., and Baker, H. W. G., 1988, A human sperm–zona pellucida binding test using oocytes that failed to fertilize *in vitro*, *Fertil. Steril,* **50:**782–788.

Lopata, A., Sathananthan, A. H., McBain, J. C., Johnston, W. I. H., and Speirs, A. L., 1980, The ultrastructure of the preovulatory human egg fertilized *in vitro*, *Fertil. Steril.* **33:**12–20.

Mack, S. O., Wolf, D. P., and Tash, J. S., 1988, Quantitation of specific parameters of motility in large numbers of human sperm by digital image processing, *Biol. Reprod.* **38:**270–281.

Mack, S. O., Tash, J. S., and Wolf, D. P., 1989a, Effect of measurement conditions on quantification of hyperactivated human sperm subpopulations by digital image analysis, *Biol. Reprod.* **40:**1162–1169.

Mack, S. O., Tash, J. S., and Wolf, D. P., 1989b, Human sperm capacitation in gelatin fortified medium, *Fertil. Steril.* **52:**1074–1076.

Malter, H. E., and Cohen, J., 1989, Partial zona dissection of the human oocyte: A nontraumatic method using micromanipulation to assist zona pellucida penetration, *Fertil. Steril.* **51:**139–148.

Mann, T., and Lutwak-Mann, C., 1981, *Male Reproductive Function and Semen*, Springer-Verlag, New York.

Mao, C., and Grimes, D. A., 1988, The sperm penetration assay: Can it discriminate between fertile and infertile men? *Am. J. Obstet. Gynecol.* **159:**279–286.

McDowell, J. S., 1986, Preparation of spermatozoa for insemination *in vitro*, in: *In Vitro Fertilization—Norfolk* (H. W. Jones, Jr., G. S. Jones, G. D. Hodgen, and Z. Rosenwaks, eds.), Williams & Wilkins, Baltimore, pp. 162–167.

Medical Research International, Society of Assisted Reproductive Technology, and The American Fertility Society, 1989, *In vitro* fertilization/embryo transfer in the United States: 1987 results from the National IVF-ET Registry, *Fertil. Steril.* **51:**13–19.

Menezo, Y., Testart, J., and Perrone, D., 1984, Serum is not necessary in human *in vitro* fertilization, early embryo culture, and transfer, *Fertil. Steril.* **42:**750–755.

Mohr, L. R., 1984, Assessment of human embryos, in: *In Vitro Fertilization and Embryo Transfer* (A. Trounson and C. Wood, eds.), Churchill Livingstone, Edinburgh, pp. 159–171.

Mohr, L. R., and Trounson, A. O., 1982, Comparative ultrastructure of hatched human, mouse and bovine blastocysts, *J. Reprod. Fertil.* **66:**499–504.

Monk, M., Handyside, A., Hardy, K., and Whittingham, D., 1987, Preimplantation diagnosis of deficiency of hypoxanthine phosphoribosyl transferase in a mouse model for Lesch–Nyhan syndrome, *Lancet* **2:**423–425.

Mortimer, D., Courtot, A. M., Giovangrandi, Y., and Jeulin, C., 1983, Do capacitated human spermatozoa shown an "activated" pattern of motility? in: *The Sperm Cell* (A. J. Martinus, ed.), Martinus Nijhoff, Hingham, MA, pp. 349–352.

Mortimer, D., Courtot, A. M., Giovangrandi, Y., Jeulin, C., and David, G., 1984, Human sperm motility after migration into, and incubation in, synthetic media, *Gamete Res.* **9:**131–144.

Mortimer, D., Curtis, E. F., and Dravland, J. E., 1986, The use of strontium-substituted media for capacitating human spermatozoa: An improved sperm preparation method for the zona-free hamster egg penetration test, *Fertil. Steril.* **46:**97–103.

Moubasher, A. E. D., and Wolf, D. P., 1986, The effect of exogenous cholesterol on human sperm function *in vitro*, *J. Androl.* **7:**22-P.

Overstreet, J. W., 1983, The use of the human zona pellucida in diagnostic tests of sperm fertilizing capacity, in: *In Vitro Fertilization and Embryo Transfer* (P. G. Crosignani and B. L. Rubin, eds.), Academic Press, London, pp. 145–166.

Overstreet, J. W., and Hembree, W. C., 1976, Penetration of the zona pellucida of nonliving human oocytes by human spermatozoa *in vitro*, *Fertil. Steril.* **27:**815–831.

Perreault, S. D., and Rogers, B. J., 1982, Capacitation pattern of human spermatozoa, *Fertil. Steril.* **38:**258–260.

Plachot, M., Veiga, A., Montagut, J., de Grouchy, J., Calderon, G., Lepretre, S., Junca, A.-M., Santalo, J., Carles, E., Mandelbaum, J., Barri, P., Degoy, J., Cohen, J., Egozcue, J., Sabatier, J. C., and Salat-Baroux, J., 1988, Are clinical and biological IVF parameters correlated with chromosomal disorders in early life: A multicentric study, *Hum. Reprod.* **3:**627–635.

Robertson, L., Wolf, D. P., and Tash, J. S., 1988, Temporal changes in motility parameters related to acrosomal status: Identification and characterization of populations of hyperactivated human sperm, *Biol. Reprod.* **39:**797–805.

Rogers, B. J., 1985, The sperm penetration assay: Its usefulness reevaluated, *Fertil. Steril.* **43:**821–840.

Rudak, E., Dor, J., Mashiach, S., Nebel, L., and Goldman, B., 1984, Chromosome analysis of multipronuclear human oocytes fertilized *in vitro*, *Fertil. Steril.* **41:**538–545.

Sathananthan, A. H., and Trounson, A. O., 1982a, Ultrastructural observations on cortical granules in human follicular oocytes cultured *in vitro*, *Gamete Res.* **5:**191–198.

Sathananthan, A. H., and Trounson, A. O., 1982b, Ultrastructure of cortical granule release and zona interaction in monospermic and polyspermic human ova fertilized *in vitro*, *Gamete Res.* **6:**225–234.

Sathananthan, A. H., and Trounson, A. O., 1985, The human pronuclear ovum: Fine structure of monospermic and polyspermic fertilization *in vitro*, *Gamete Res.* **12:**385–398.

Schalkoff, M. E., Oskowitz, S. P., and Powers, R. D., 1989, Ultrastructural observations of human and mouse oocytes treated with cryopreservatives, *Biol. Reprod.* **40:**379–393.

Schoysman, R. J., and Bedford, J. M., 1986, The role of the human epididymis in sperm maturation and sperm storage as reflected in the consequences of epididymovasostomy, *Fertil. Steril.* **46:**293–299.

Shabanowitz, R. B., and O'Rand, M. G., 1988, Characterization of the human zona pellucida from fertilized and unfertilized eggs, *J. Reprod. Fertil.* **82:**151–161.

Silber, S. J., 1988, Pregnancy caused by sperm from vasa efferentia, *Fertil. Steril.* **49:**373–375.

Silber, S. J., Balmaceda, J., Borrero, C., Ord, T., and Asch, R., 1988, Pregnancy with sperm aspiration from the proximal head of the epididymis: A new treatment for congenital absence of the vas deferens, *Fertil. Steril.* **50:**525–528.

Singer, S. L., Lambert, H., Overstreet, J. W., Hanson, F. W., and Yanagimachi, R., 1985, The kinetics of human sperm binding to the human zona pellucida and zona-free hamster oocyte *in vitro*, *Gamete Res.* **12:**29–39.

Sundström, P., and Nilsson, B. O., 1988, Meiotic and cytoplasmic maturation of oocytes collected in stimulated cycles is asynchronous, *Hum. Reprod.* **3:**613–619.

Suzuki, S., Kitai, H., Tojo, R., Seki, K., Oba, M., Fujiwara, T., and Iizuka, R., 1981, Ultrastructure and some biologic properties of human oocytes and granulosa cells cultured *in vitro*, *Fertil. Steril.* **35:**142–148.

Syms, A. J., Johnson, A. R., Lipshultz, L. I., and Smith, R. G., 1984, Studies on human spermatozoa with round head syndrome, *Fertil. Steril.* **42:**431–435.

Tezón, J. G., Vazquez, M. H., Piñeiro, L., de Larminat, M. A., and Blaquier, J. A., 1985, Identification of androgen-induced proteins in human epididymis, *Biol. Reprod.* **32:**584–590.

Thomas, P., and Meizel, S., 1988, An influx of extracellular calcium is required for initiation of the human sperm acrosome reaction induced by human follicular fluid, *Gamete Res.* **20:**397–411.

Trounson, A., 1986, Recent progress in human *in vitro* fertilization and embryo transfer, in: *Developmental Biology: A Comprehensive Synthesis*, Vol. 4, *Manipulation of Mammalian Development* (R. B. L. Gwatkin, ed.), Plenum Press, New York, pp. 149–194.

Trounson, A., and Sjöblom, P., 1988, Cleavage and development of human embryos *in vitro* after ultrarapid freezing and thawing, *Fertil. Steril.* **50:**373–376.

Trounson, A., and Webb, J., 1984, Fertilization of human oocytes following reinsemination *in vitro*, *Fertil. Steril.* **41:**816–819.

Trounson, A. O., Mohr, L. R., Wood, C., and Leeton, J. F., 1982, Effect of delayed insemination on *in-vitro* fertilization, culture and transfer of human embryos, *J. Reprod. Fertil.* **64:**285–294.

Van Blerkom, J., and Henry, G., 1988, Cytogenetic analysis of living human oocytes: Cellular basis and developmental consequences of perturbations in chromosomal organization and complement, *Hum. Reprod.* **3:**777–790.

van Uem, J. F. H. M., Siebzehnrübl, E. R., Schuh, B., Koch, R., Trotnow, S., and Lang, N., 1987, Birth after cryopreservation of unfertilised oocytes, *Lancet* **1:**752–753.

Veeck, L. L., 1986, Morphological estimation of mature oocytes and their preparation for insemination, in: *In Vitro Fertilization—Norfolk* (H. W. Jones, Jr., G. S. Jones, G. D. Hodgen, and Z. Rosenwaks, eds.), Williams & Wilkins, Baltimore, pp. 81–93.

Veeck, L. L., and Maloney, M., 1986, Insemination and fertilization, in: *In Vitro Fertilization—Norfolk* (H. W. Jones, Jr., G. S. Jones, G. D. Hodgen, and Z. Rosenwaks, eds.), Williams & Wilkins, Baltimore, pp. 168–200.

Wales, R. G., Whittingham, D. G., Hardy, K., and Craft, I. L., 1987, Metabolism of glucose by human embryos, *J. Reprod. Fertil.* **79:**289–297.

Weissenberg, R., Eshkol, A., Rudak, E., and Lunenfeld, B., 1982, Inability of round acrosomeless human spermatozoa to penetrate zona-free hamster ova, *Arch. Androl.* **11:**167–169.

Wikland, M., Hamberger, L., Enk, L., and Nilsson, L., 1988, Sonographic techniques in human *in-vitro* fertilization programmes, *Hum. Reprod.* **3:**65–68.

Wilton, L. J., Shaw, J. M., and Trounson, A. O., 1989, Successful single-cell biopsy and cryopreservation of preimplantation mouse embryos, *Fertil. Steril.* **51:**513–517.

Wolf, D. P., 1988a, Analysis of embryonic development, in: *In Vitro Fertilization and Embryo Transfer: A Manual of Basic Techniques* (D. P. Wolf, ed.; B. D. Bavister, M. Gerrity, and G. S. Kopf, assoc. eds.), Plenum Press, New York, pp. 137–145.

Wolf, D. P., 1988b, Analysis of oocyte quality and fertilization, in: *In Vitro Fertilization and Embryo Transfer: A Manual of Basic Techniques* (D. P. Wolf, ed.; B. D. Bavister, M. Gerrity, and G. S. Kopf, assoc. eds.), Plenum Press, New York, pp. 125–136.

Wolf, D. P., 1989, Acrosomal status quantitation in human sperm, *Am. J. Reprod. Immunol.* **20:**106–113.

Wolf, D. P., and Sokoloski, J. E., 1982, Characterization of the sperm penetration bioassay, *J. Androl.* **3:**445–451.

Wolf, D. P., Byrd, W., Dandekar, P., and Quigley, M. M., 1984, Sperm concentration and the fertilization of human eggs *in vitro*, *Biol. Reprod.* **31:**837–848.

World Health Organization, 1987, *WHO Laboratory Manual for the Examination of Human Semen and Semen–Cervical Mucus Interaction* 2, Cambridge University Press, Cambridge.

Yanagimachi, R., 1981, Mechanisms of fertilization in mammals, in: *Fertilization and Embryonic Development In Vitro* (L. Mastroianni, Jr., and J. D. Biggers, eds.), Plenum Press, New York, pp. 81–182.

Yanagimachi, R., 1984, Zona-free hamster eggs: Their use in assessing fertilizing capacity and examining chromosomes of human spermatozoa, *Gamete Res.* **10:**187–232.

Yanagimachi, R., Yanagimachi, H., and Rogers, B. J., 1976, The use of zona-free animal ova as a test-system for the assessment of the fertilizing capacity of human spermatozoa, *Biol. Reprod.* **15:**471–476.

Zamboni, L., 1987, The ultrastructural pathology of the spermatozoon as a cause of infertility: The role of electron microscopy in the evaluation of semen quality, *Fertil. Steril.* **48:**711–734.

Zamboni, L., Thompson, R. S., and Smith, D. M., 1972, Fine morphology of human oocyte maturation *in vitro*, *Biol. Reprod.* **7:**425–457.

Part III

NEW APPROACHES FOR STUDYING MAMMALIAN GAMETES

21

Micromanipulation of Mammalian Gametes

Kristen A. Ivani and George E. Seidel, Jr.

1. SPERM-CENTERED APPROACHES

1.1. Placing Gametes in Close Apposition

Many sperm are deposited in the female reproductive tract, but a very low percentage survive for more than a few hours in a motile condition after copulation; most never come near an egg. In the mouse the average ejaculate contains approximately 58 million sperm (Hogan *et al.*, 1986), but only 17 were found in the ampulla of the oviduct at the time of fertilization (Braden and Austin, 1954). Similar phenomena occur in other mammalian species. The proportion of sperm capable of fertilizing ova and supporting normal development is unknown. Despite development of sophisticated techniques and information on mammalian *in vitro* fertilization, we still know very little about the manner in which a single sperm is selected to fertilize each oocyte. Much of what we know results from *in vitro* studies in which thousands of sperm are added to a few oocytes; unfortunately, this approach gives us little information about the behavior of the fertilizing spermatozoon. In order to refine *in vitro* fertilization systems to more closely mimic fertilization *in vivo*, one must deal with "dilution effects." At very low sperm concentrations, sperm frequently lose their motility and, hence, their subsequent fertilizing ability, probably because of dilution of important components of reproductive tract fluids.

Bavister (1979) reported the first *in vitro* fertilization of intact mammalian (hamster) oocytes with physiological sperm numbers. Epinephrine and adrenal extracts were added to the sperm culture drops to improve capacitation and maintain motility, respectively. Sperm : egg ratios ranged from 4 : 1 to 1 : 1. Fertilization was "in progress" in 13 of 67 of the ova based on morphological criteria. These results indicate that many sperm are capable of fertilizing oocytes.

Wolf (1978) demonstrated that the speed and percentage of sperm penetration of zona-free ova in the mouse were dependent on sperm concentration. At the lowest sperm concentrations

Kristen A. Ivani and George E. Seidel, Jr. • Animal Reproduction and Biotechnology Laboratory, Colorado State University, Fort Collins, Colorado 80523.

A Comparative Overview of Mammalian Fertilization, edited by Bonnie S. Dunbar and Michael G. O'Rand. Plenum Press, New York, 1991.

examined, sperm to egg ratios were approximately 1 : 1, and most ova were penetrated (75%). An alternate method of exposing oocytes to low numbers of sperm is with the aid of a micromanipulator. Thadani (1981) inseminated drops containing one to three zona-free mouse ova with five to ten capacitated mouse sperm. The overall fertilization rate was 50%. Most of the resulting embryos developed to blastocysts *in vitro*; some were transferred to recipient mice, and a few gave rise to normal progeny. Under similar conditions, zona-intact ova were not fertilized. Ivani and Seidel (1988) selected individual capacitated, motile mouse spermatozoa with a micropipet (i.d. 40 μm) and placed each near a homologous zona-free oocyte. Fertilization resulted approximately 40% of the time; 75% of these embryos developed to the blastocyst stage *in vitro*.

From these experiments it appears that most sperm are genetically and physiologically competent to fertilize oocytes, providing they can gain access to the oocytes. These results should be interpreted cautiously, however, because randomly selected spermatozoa that are competent to fertilize zona-free ova *in vitro* may not have been able to penetrate the zona pellucida *in vivo*. Additional experiments in which randomly selected single sperm are evaluated for their ability to fertilize cumulus-intact and zona-intact oocytes are necessary to clear up this issue.

1.2. Manipulation of the Zona Pellucida for Assisted Fertilization

Fertilization of oocytes *in vitro* with low sperm numbers can be facilitated by manipulation of the zona pellucida. Currently there are two approaches other than zona removal: "drilling" of the zona pellucida with low-pH solution and mechanically opening the zona. Each of these manipulations is then followed by incubation with sperm.

Zona drilling is accomplished by pressing a microneedle loaded with acid Tyrode's solution against the zona in a tangential position until a thinning, swelling, and rupture of the zona is observed (Fig. 1). The microneedle is then withdrawn, and the drilled oocyte is mixed with sperm. Gordon and Talansky (1986) reported that the procedure decreased the number of sperm required to achieve fertilization in mice by a factor of nearly 100. Zona-drilled mouse ova had a similar incidence of polyspermy to undrilled control ova. Conover and Gwatkin (1988) used a monoclonal antibody against the zona sperm receptor ZP3 to block sperm entry through the zona pellucida of mouse oocytes. Fertilization was completely blocked by the monoclonal antibody in undrilled oocytes but proceeded normally in predrilled ova, confirming that ova are being fertilized by sperm getting through the acid-drilled aperture. When zona drilling was used for human oocytes, however (Gordon *et al.*, 1988), the rate of polyspermy was 50% (5/10). Diploid fertilization was achieved in five cases; three of these were transferred after cleavage, but none resulted in pregnancy. The high incidence of polyspermy observed in this clinical trial suggests that the zona pellucida has a major role in preventing polyspermy in human ova.

Potential applications of assisted fertilization by manipulating the zona pellucida led several investigators to develop alternate methods of making an opening in the zona pellucida without damaging the vitelline membrane (Fig. 1). "Zona cutting," developed by Tsunoda *et al.* (1986) for nuclear transplantation procedures, was recently adopted by Depypere *et al.* (1988) for assisted fertilization in the mouse. The oocyte is held in place by suction, and a fine glass needle (0.5 mm long with O.D. of 5–7 μm) is inserted into the perivitelline space. Suction is released, and the needle is withdrawn, rubbing the oocyte against the wall of the holding pipet to tear the zona pellucida. The procedure required about 30 min to cut the zonae of 100 mouse oocytes; no apparent damage occurred to the vitellus. Depypere *et al.* (1988) compared zona cutting to zona drilling of mouse oocytes and observed improved fertilization rates with suboptimal sperm numbers (<200,000/ml) with both zona drilling and zona cutting (4–90%) compared to control rates (0–45%). In addition, manipulation of the zona resulted in neither increased oocyte loss nor parthenogenetic activation of oocytes. Rates of polyspermy were not significantly different from

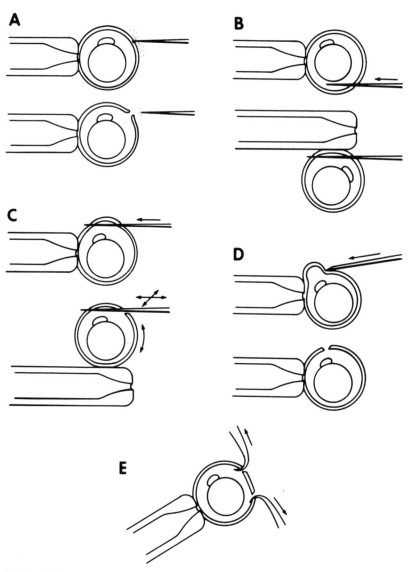

Figure 1. Manipulations of the zona pellucida. (A) Zona drilling. Acidified Tyrode's medium is released from a microneedle against the zona pellucida surface, resulting in a small thinned and eventually ruptured area of the zona pellucida. (B) Zona cutting. After penetration of the zona pellucida with a microneedle, the oocyte is moved to the underside of the holding pipet and rubbed against it to tear the zona pellucida. (C) Partial zona dissection. After penetration with the microneedle, the oocyte is brought on top of the holding pipet and rocked back and forth to slit the zona pellucida. (D) Zona chiseling. A small fold of zona pellucida is caught between the microchisel and holding pipet, resulting in a cut through the zona pellucida. (E) Opening the zona with microhooks. Two microhooks are inserted through the zona pellucida and pulled in opposite directions to tear the zona pellucida.

controls. An interesting finding was that embryos resulting from zona-drilled oocytes hatched about 12 hr earlier than zona-cut or control embryos.

A similar procedure to zona cutting has been referred to as partial zona dissection (Malter and Cohen, 1989a). Human oocytes are shrunken by incubation in 0.5 M sucrose prior to manipulation. As the oocyte is held in place by suction with the largest area of perivitelline space in the "12 o'clock" position, a sharp microneedle is used to pierce the zona between the 1 o'clock and 2 o'clock positions (Fig. 1). The needle is pushed completely through the perivitelline space (avoiding contact between the needle and the holding pipet) and out of the zona between the 10 o'clock and 11 o'clock positions. As suction is released, the oocyte is rolled on top of the holding pipet and gently rocked, causing a slit less than 20 μm in length in the zona. Sucrose is removed from oocytes by repeated washing in fresh medium. Three pregnancies with six fetal sacs resulted from transfer of 13 zona-dissected and six untreated human embryos. One of these patients had only zona-dissected embryos transferred, documenting the first pregnancy from assisted fertilization in the human. Twin pregnancies also have been confirmed after transfer of two zona-dissected and one control embryo to each of two patients. When Malter and Cohen (1989a) compared zona dissecting with zona drilling, they obtained fertilization rates of 50% and 30%, respectively. Furthermore, significantly fewer drilled oocytes than zona-dissected oocytes cleaved, and drilled oocytes never developed to the blastocyst stage. The authors suggested that this may result from the effect of acid on the vitelline membrane; they also noted that many of the drilled zygotes looked granular and misshapen.

A third method of mechanically opening the zona was developed by Odawara and Lopata (1989). Oocytes are treated with sucrose prior to manipulation. As the oocyte is held in place with a suction pipet, two "microhooks" are introduced from the right and left sides of the oocyte. The tips of the microhooks are pushed through the zona pellucida into the perivitelline space and then gently pulled to each side, resulting in a localized tear in the zona pellucida. After removal of the microhooks, the zona recoils to its original shape. The procedure takes about 1 min per oocyte. Fertilization rates with opened zonae were significantly higher than with intact zonae at 10^4 sperm/ml but not at 10^5 or 10^3 sperm/ml. As with the previous methods, rates of oocyte loss, polyspermy, and subsequent embryonic development were unaffected by zona opening.

Malter and Cohen (1989b) used still another technique to open the zona pellucida (zona chiseling) to study blastocyst formation and hatching *in vitro* in mouse and human embryos. The two-cell embryo is held in place by suction with the largest volume of perivitelline space at the "12 o'clock" position (embryos were treated previously with sucrose), and a microchisel is pressed against the zona pellucida between the 12 o'clock and 1 o'clock positions. A small fold in the zona is formed between the holding pipet and the chisel; as the microchisel is forced against the holding pipet, the zona pellucida is cut. Although the authors observed essentially the same incidence of hatching for manipulated and control mouse embryos, the percentage of "partial hatching" was greater in manipulated (20–25%) than control (9%) embryos. They also observed that hatching occurred a day earlier (on day 5) in manipulated (98%) than in control (27%) embryos.

Although there is a need for additional comparative studies, both zona drilling and zona cutting improve fertilization rates with low sperm numbers, result in comparable embryonic development rates, and are relatively easy to perform. These experiments provide encouraging evidence that manipulation of the zona pellucida may be an effective approach to correcting infertility caused by oligospermia.

1.3. Sperm Microinjection

One important issue in the study of fertilization has been whether the sperm needs to fertilize the oocyte by active penetration and membrane fusion for normal development to begin. Sperm injection techniques that enable sperm or sperm nuclei to be placed in the perivitelline space or directly into the ooplasm provide an opportunity to address this issue. Microinjection of sperm can be used to explore timing of fertilization, produce offspring from selected, treated, or abnormal sperm, and circumvent barriers to heterospecific fertilization.

1.3.1. INJECTION OF SPERM INTO THE OOPLASM

A typical protocol for sperm microinjection follows. Oocytes are collected from oviducts of superovulated females, and cumulus cells are removed by treatment with hyaluronidase. These oocytes are washed and kept in a drop of medium under paraffin oil. If whole sperm are desired, they are washed to remove seminal or epididymal plasma and may be incubated for capacitation or induced to undergo the acrosome reaction. Sperm nuclei can be isolated according to Perreault and Zirkin (1982). Briefly, cauda epididymal sperm are sonicated in Tris buffer to remove acrosomes and tails, centrifuged through a two-step sucrose gradient to isolate nuclei from the suspension, washed, and resuspended. Polyvinylpyrrolidone (PVP) (Thadani, 1980; Perreault and Zirkin, 1982) or methyl cellulose (Mann, 1988) may be added to the sperm suspension to retard motility and/or prevent sperm from sticking to the injection pipet. Micropipets for holding oocytes and injecting sperm are described by Thadani (1980). The tip of the injection pipet may be beveled on a grinder (Thadani 1980) or thinned with hydrofluoric acid and then broken against the holding pipet to create a sharp point (Perreault and Zirkin, 1982). After individual sperm (or sperm nuclei) are aspirated into the tip of the injection pipet, the pipet is moved to the drop containing ova. The zona pellucida and vitelline membrane are penetrated, a small amount of ooplasm is sucked into the injection pipet, and the sperm is expelled into the cytoplasm. Death of oocytes, which may be more than 50%, is characterized by filling of the perivitelline space with ooplasm. Ova surviving the injection procedure may be cultured and monitored for development or transferred to pseudopregnant recipient females.

Early efforts with sperm microinjection revealed that oocyte activation was necessary for normal development to continue and that oocytes of some species were activated by microinjection (Hiramoto, 1962; Brun, 1974; Uehara and Yanagimachi, 1976). Gametes from a number of different species have been combined via sperm microinjection (Table 1). Clearly, the ability of the sperm to undergo decondensation and pronuclear formation is not species specific; however, the timing of decondensation and formation of the pronucleus is dependent on sperm nuclear disulfide bond content (Naish et al., 1987; Perreault et al., 1988).

Historically, hamster oocytes have been used for microinjection studies because of their resilience to micromanipulation and their ability to direct decondensation and pronuclear formation of heterologous as well as homologous sperm. Unfortunately, hamster embryos do not develop well in culture, so study of embryonic development of injected ova was limited. Bavister (1989) has recently outlined culture conditions that may alleviate this problem. To date the rabbit and cow are the only species from which live young have been produced as a result of cytoplasmic sperm injection (Hosoi et al., 1988; Goto et al., 1990). Although rabbit oocytes are not as resilient to microinjection as hamster ova, more than 50% survive the procedure, and many undergo normal-appearing cleavage after 22 hr of culture (Keefer, 1989).

1.3.2. INJECTION OF SPERM INTO THE PERIVITELLINE SPACE

An alternative to cytoplasmic injection is injection of sperm under the zona pellucida into the perivitelline space. Oocytes may be shrunken by treatment with sucrose (Yang et al., 1988) or high-ionic-strength medium (Smith and Seidel, 1986) prior to microinjection to enlarge the perivitelline space. In the case of perivitelline injection, the zona pellucida is penetrated tangentially to the ooplasm rather than "head on."

Table 1. Comparison of Cytoplasmic Sperm Injection Studies

Species		Component[a]	Pronucleus formation	Cleavage	Reference
Sperm	Egg				
Urchin	Urchin	W	N	N	Hiramoto (1962)
Xenopus	Xenopus	W	Y	Y[b]	Brun (1974)
Frog	Toad	W	Y	—	Skoblina (1974)
Human, hamster	Hamster	N	Y	—	Uehara and Yanagimachi (1976)
Mouse	Rat	H	Y	Y	Thadani (1980)
Hamster	Hamster	N	Y	—	Perreault and Zirkin (1982); Perreault et al. (1984, 1987)
Human, Xenopus	Xenopus	W	Y	—	Ohsumi et al. (1986)
Bull, hamster	Hamster	N	Y	—	Keefer et al. (1986)
Ram	Sheep	W,N	Y,Y	Y,Y	Clarke et al. (1988)
Ram, bull, chinch, boar, hamster, vole	Hamster	N	—	—	Clarke and Johnson (1988a,b)
Rabbit	Rabbit	W	Y	Y[b]	Hosoi et al. (1988)
Human	Hamster	W	Y	—	Lanzendorf et al. (1988a)
Human	Human	W	Y	—	Lanzendorf et al. (1988b)
Human, mouse, hamster	Hamster	N	Y	—	Perreault et al. (1988)
Rabbit	Rabbit, hamster	W	Y	Y	Keefer (1989)
Bull	Cow	W	Y	Y	Keefer et al. (1990)
Bull	Cow	H	Y	Y	Goto et al. (1990)

[a]W, whole sperm; H, head; N, nucleus.
[b]Birth of live young.

Placement of the sperm under the zona pellucida allows the spermatozoon to fuse naturally with the oocyte membrane. This approach may be used to dissect the importance of capacitation, the acrosome reaction, and sperm motility to gamete interaction. A summary of recent studies of perivitelline sperm injection is presented in Table 2.

The ability of microinjected sperm to fertilize oocytes may be increased by allowing sperm to capacitate prior to injection. Human sperm capacitated by exposure to calcium-depleted medium containing strontium chloride fertilized more human oocytes (5/7) than uncapacitated sperm (0/6) (Laws-King et al., 1987). Mann (1988) reported 25% fertilization with capacitated mouse sperm microinjected into the perivitelline space of homologous oocytes. Fifty-four percent of these developed to normal fetuses or to term; these are the only live young reported to date resulting from perivitelline sperm injection.

It is widely accepted that spermatozoa must be acrosome-reacted to fuse with and penetrate the oolemma (reviewed by Yanagimachi, 1988). However, Barg et al. (1986) reported that immotile acrosome-reacted mouse sperm microinjected into the perivitelline space of homologous oocytes were unable to fertilize oocytes even when fusion of gametes was induced by exposure to Sendai virus. They offered two explanations for the inability of sperm to fertilize oocytes. The first was that sperm motility may be essential to the fertilization process. The second was that the treatments used to induce the acrosome reaction and immobilize sperm also rendered the sperm incapable of binding to the vitelline membrane. The latter conclusion seems more likely, since sperm from men with Kartagener's syndrome (immotile but alive and acrosome-reacted) are still capable of forming pronuclei in oocytes provided they can gain access to the vitelline membrane (Aitken et al., 1983). Lacham et al. (1989) microinjected capacitated and acrosome-reacted mouse sperm (Talansky et al., 1987) into the perivitelline space of mouse oocytes. The optimal treatment, which included cGMP, resulted in 43% fertilization. There was, however, no correlation between fertilization rates and the proportion of acrosome-reacted sperm identified after sperm treatment. They attributed this lack of correlation to biased selection of acrosome-reacted sperm for microinjection.

Microinjection of sperm into oocytes may provide a means of treating infertility in men. Fertilization and cleavage have already been achieved by perivitelline (Lassalle et al., 1987; Laws-King et al., 1987) and cytoplasmic injection (Lanzendorf et al., 1988b). In addition, Lanzendorf et al. (1988a) reported nuclear decondensation and pronuclear formation with acrosome-defective sperm microinjected into the cytoplasm of hamster oocytes, suggesting that once the inability to penetrate the oocyte is bypassed, sperm from otherwise infertile men may be able to participate in fertilization.

Table 2. Comparison of Perivitelline Sperm Injection Studies

Species		Sperm decondensation	Pronucleus formation	Cleavage	Reference
Sperm	Egg				
Human	Hamster	Y	?	—	Lassalle et al. (1985)
Mouse	Mouse	N	N	N	Barg et al. (1986)
Human	Human	Y	Y	Y	Laws-King et al. (1987)
Human	Hamster	Y	?	—	Lassalle and Testart (1988)
Mouse	Mouse	Y	Y	Y[a]	Mann (1988)
Rabbit	Rabbit	Y	Y	Y	Yang et al. (1988)

[a]Denotes birth of live young.

One important issue vis-à-vis curing human male infertility is whether or not microinjection will result in normal fertilization and development most of the time. Cohen (1983) suggested that the majority of sperm within a population may be abnormal. If this is the case, selection of sperm for microinjection may lead to lower numbers of oocytes fertilized or higher numbers of abnormal embryos. Martin *et al.* (1988) reported a higher frequency of chromosomal abnormalities in human sperm after microinjection than after *in vitro* fertilization of hamster ova. They observed sperm with simple chromosome breaks as well as spermatozoa with multiple breaks and rearrangements after treatment by brief sonication or TES–Tris–yolk buffer. Although the numbers are small, chromosomal normality of microinjected ova should be rigorously examined before the technique is used clinically in human *in vitro* fertilization programs.

1.4. Transgenic Animals from DNA Adsorbed to Sperm

A recent paper by Lavitrano *et al.* (1989) has stirred up excitement and controversy in the scientific community. These workers demonstrated that transgenic mice could be produced simply by incubating spermatozoa in the DNA vector of interest followed by *in vitro* fertilization. Earlier, Brackett *et al.* (1971) had demonstrated that incubation of rabbit sperm with SV-40 DNA prior to *in vitro* fertilization resulted in SV-40-positive embryos. This technique would greatly simplify the process of making transgenic animals, particularly in large domestic farm animal species and poultry.

At this writing, these techniques have not been confirmed by other workers. The work of Brackett *et al.* (1971) would be more convincing if the site of DNA incorporation were determined by modern molecular biology techniques. The work of Lavitrano *et al.* (1989) appears to be very convincing; the paper includes several independent confirmatory procedures. However, there is a very disturbing, consistent rearrangement of the DNA construct used in their independently generated transgenic lines. If confirmed, this propensity to mutate DNA in a predictable fashion is a very important finding. Of course, if similar rearrangements occur with all DNA constructs, this technique has limited application. Many laboratories have tried to repeat these procedures with other DNA constructs; results to date have been negative, suggesting that the success reported is construct-dependent, difficult to duplicate, or that some key aspect of the procedures was not given. Laboratory-to-laboratory differences, such as source and purity of water, may contribute to the disparate results. It will be interesting to see how this controversy is resolved.

It is likely that several safeguards have evolved to prevent such insertion of DNA (e.g., from viruses) into the germ line during natural reproduction. For example, seminal plasma contains nucleases that likely would destroy foreign DNA adsorbed or taken up by sperm. Fluids in the female reproductive tract may contain similar enzymes. Therefore, simple artificial insemination may not work as well as *in vitro* fertilization for these and related techniques.

2. OOCYTE-CENTERED APPROACHES

2.1. Physical Activation of Oocytes with Micropipets

The fertilization process can be divided into two parts: physical entry of the sperm into the oocyte and activation of the oocyte. The activation process changes the oocyte from a lethargic, slowly dying cell into a very active cell that increases energy utilization dramatically, releases contents of thousands of cortical granules into the perivitelline space to block polyspermy, completes the second meiotic division including extrusion of the second polar body, processes the sperm including decondensation of the sperm chromatin, and initiates the first cell cycle, which leads to DNA synthesis and the mitotic cell divisions that eventually form a new organism. Normally, activation is initiated by the sperm; however, a myriad of other stimuli can start the process, including bathing in dilute ethanol for several minutes, heat or cold shock, and mechanical perturbation of the oocyte. Such activation has been studied in the context of parthenogenesis by numerous investigators (e.g., Kaufman, 1983). Likely a common final pathway involves increasing concentrations of free intracellular Ca^{2+}.

Proper activation of the oocyte is required for many manipulations in basic research as well as applications to practical use. Examples include cloning by nuclear transplantation, fertilization by sperm injection, and studies in gynogenesis, parthenogenesis, and androgenesis. Without appropriate activation at the correct time, subsequent steps will not occur properly.

It turns out that mature oocytes of many mammalian species are activated by sucking out ooplasm or piercing them with a sufficiently large pipet. This was first demonstrated in mammals by Uehara and Yanagimachi (1976) and subsequently confirmed by Thadani (1980). Current methods of nuclear transplantation pioneered by Willadsen (1986) rely on inducing activation by removing on the order of 20% of the ooplasm when removing the second meiotic spindle. The mechanism of activation of these mechanical procedures is unknown, although increased intracellular free Ca^{2+} probably occurs. Recently, Keefer *et al.* (1990) demonstrated that cleavage rates of bovine oocytes with sperm injected into the cytoplasm were higher if oocytes were activated by incubation in A23187.

2.2. Removal of Chromosomes from Oocytes

For many manipulations (e.g., cloning by nuclear transplantation), the genetic material in oocytes must be removed or destroyed. The chromosomes are in the germinal vesicle (nucleus) until the last 12–24 hr of the maturation process, when the germinal vesicle breaks down. Although the germinal vesicle can be removed by aspiration with a micropipet, this is not a common approach, because a functional germinal vesicle is required for some aspects of oocyte maturation. Pretreatment with cytochalasin B or D appears to reduce cytoplasmic damage during such procedures.

An elegant experimental approach is to bisect oocytes into nucleate and anucleate halves. This can be done very effectively with a fine glass needle while keeping the oocytes at 5°C (Tarkowski, 1977). It also is possible to destroy the germinal vesicle with a laser, although this has been done more often with the larger amphibian oocyte (McKinnell, 1981). A novel approach was recently described by Tsunoda *et al.* (1988) in which pronuclei were stained with Hoechst 33342 and then exposed to ultraviolet light, effectively destroying the chromosomes. This same technique likely would work in oocytes and later-stage embryos.

After the germinal vesicle breaks down, it is difficult to locate the chromosomes precisely. However, they generally are near the site of extrusion of the first polar body. Removal of 20–25% of the ooplasm in this region seems to remove sufficient genetic material that the chromosomes do not take part in embryonic development (Willadsen, 1986).

2.3. Oocyte–Oocyte Fusion

The union of sperm and oocyte at fertilization is a widely recognized example of fusion between two cells. Yet, in spite of the wealth of information available (excellent reviews by Longo, 1973; Yanagimachi, 1981, 1988), the mechanism behind such spontaneous cell fusion remains poorly understood. Fusion of oocytes has been used to study oocyte maturation, sperm and oocyte fusion, gynogenesis, parthenogenesis, and the introduction of new genetic material into cells (Gulyas, 1986).

2.3.1. METHODOLOGY

Cell fusion is attained with one of the following fusogenic methods: polyethylene glycol (PEG) (Klebe and Mancuso, 1981), Sendai virus (Harris and Watkins, 1965; Tsunoda and Shioda, 1988); or electrofusion (Zimmermann and Vienken, 1982). Fusion occurs in three general stages regardless of the fusing method employed (Knutton and Pasternak, 1979): (1) agglutination of plasma membranes; (2) fusion of small localized areas of the cell membranes; and (3) increased size of the fused sites until fusion is complete.

2.3.2. FUSION OF IMMATURE OOCYTES

Oocyte fusion has been used to study maturation-promoting factor (MPF), which is responsible for germinal vesicle breakdown (GVBD), chromosome condensation, and subsequent oocyte maturation. This factor can induce similar changes in nuclei of interphase blastomeres (Czolowska et al., 1986; Balakier and Masui, 1986a,b) or somatic cells such as thymocytes (Szollosi et al., 1986) and brain cells (Ziegler and Masui, 1973).

Fulka (1983) demonstrated that immature porcine oocytes fused to each other could not be induced to undergo GVBD; however, when one of the oocytes was matured in vitro prior to fusion, GVBD occurred in the immature oocyte within 3–6 hr post-fusion. Balakier (1978) fused immature mouse oocytes with metaphase II mouse oocytes, which induced GVBD in the immature ova immediately after fusion. This effect was not species-specific, as immature mouse oocytes fused to metaphase II rabbit oocytes could be induced to undergo GVBD. These experiments suggest that growing oocytes do not produce MPF, but are able to respond when exposed to it (Tarkowski, 1982).

Recent experiments demonstrate that MPF's ability to induce GVBD and chromosome condensation can be diluted by multiple fusions (Fulka et al., 1986, 1988). Sperm decondensation properties of oocytes can be diluted similarly (Clarke and Masui, 1987). Fulka et al. (1986) showed that whereas fusion of one metaphase I oocyte to one growing oocyte (porcine or murine) resulted in rapid GVBD or chromosome condensation, fusion with more than one growing oocyte resulted in neither GVBD nor chromosome condensation. Clarke and Masui (1987) reported a quantitative relationship between oocyte volume and the number of sperm nuclei that could be transformed into metaphase II chromosomes. Oocytes that had their cell volume doubled by cell fusion could support metaphase chromosome formation of up to five sperm; tripling cell volume resulted in metaphase chromosome formation by up to eight sperm. The converse was also true: when cytoplasmic volume was halved by oocyte bisection, a maximum of two sperm could be transformed into metaphase chromosomes. Fulka et al. (1988) demonstrated that in the presence of cycloheximide, a protein synthesis inhibitor, fusion of one mature to one immature oocyte resulted in GVBD, whereas fusion of one mature oocyte with three immature oocytes did not. The workers concluded that MPF cannot "autocatalytically" amplify but requires active protein synthesis for its production.

2.3.3. FUSION OF METAPHASE II OOCYTES

Gulyas et al. (1984) and Gulyas (1986) used oocyte fusion to mimic fertilization, i.e., fertilization of an oocyte by an oocyte. Artificially activated (Cuthbertson et al., 1981; Cuthbertson, 1983) zona-free metaphase II mouse oocytes were fused with PEG. Nineteen percent of fused products developed to the blastocyst stage in vitro compared to 86% in untreated control embryos. Electron microscopy revealed that these fusion-derived gynogenetic blastocysts were nearly identical ultrastructurally to fertilization-derived blastocysts (Gulyas, 1986). Although one of seven gynogenetic embryos developed to the 14-somite stage after embryo transfer, no viable young was produced. Numerous explanations for the lack of development to term of oocyte fusion products and other parthenogenetically derived embryos have been offered over the years; recent observations suggest that maternal and paternal genomic contributions to embryogenesis are imprinted, and both are required for development to term (Surani *et al.*, 1987).

2.4. Fusion of Karyoplasts with Oocytes (Nuclear Transplantation)

Cloning has been described as the production of two or more genetically identical individuals from a single sexually derived zygote (Pappaioannou and Ebert, 1986). Methods for cloning mammalian embryos could be used for (1) rapid production of inbred or genetically superior lines; (2) removal of genetic variation in experiments, thus reducing the number of animals required for statistical validity (Biggers, 1986), and (3) examination of nuclear–cytoplasmic communication. Nuclear transplantation, the introduction of foreign nuclei into enucleated oocytes or zygotes, theoretically is the technique of choice to produce genetically identical individuals. In practice, many more genetically identical sets of animals have been produced by blastomere separation (Willadsen, 1979) or simple bisection of postimplantation embryos (Williams *et al.*, 1984; Baker and Shea, 1985).

Briggs and King (1952) reported successful transplantation of *Rana pipiens* nuclei into enucleated oocytes and demonstrated that nuclei transplanted from a single donor embryo could result in multiple identical blastula-stage embryos. These and other early experiments (reviewed by Gurdon, 1964) have revealed important information on the developmental pluripotency of transplanted nuclei during embryogenesis. The capacity for the donor nucleus to guide development after transplantation likely is related to its differentiated state, as will be discussed.

2.4.1. DEVELOPMENT OF CURRENT NUCLEAR TRANSPLANTATION TECHNIQUES

The technique of enucleation and nuclear transfer was described by McGrath and Solter (1983) for the mouse, and variations have been used in the sheep (Willadsen, 1986), cow (Prather *et al.*, 1987; Robl *et al.*, 1987), and rabbit (Stice and Robl, 1988). Usually the same oocyte is not used as donor and recipient of nuclei. Although procedures are very similar, oocytes to be used for recipients should be enucleated more carefully so as not to remove too much cytoplasm or cause extensive damage to the vitelline membrane, both of which could result in lower embryonic development. A typical protocol for enucleation of mouse embryos follows. The one-cell-stage embryo is held in place by a suction pipet, oriented with the second polar body opposite to the approach of the enucleation pipet (15–20 μm I.D.). The enucleation pipet is pushed into the embryo, indenting the zona pellucida and the plasma membrane until the zona pellucida is penetrated. The pipet tip is moved to a point adjacent to a pronucleus, and, with slight aspiration, the pronucleus and a small amount of ovum plasma membrane and surrounding cytoplasm are aspirated into the pipet. The pipet is then relocated near the second pronucleus, which is aspirated similarly. The enucleation pipet is withdrawn from the ovum, leaving a thin cytoplasmic bridge between the pronuclei in the pipet and the embryo, which eventually pinches off. Embryos are preincubated with cytoskeletal inhibitors such as cytochalasin D prior to micromanipulation to minimize damage to the ovum and to facilitate the enucleation process.

McGrath and Solter (1983) employed inactivated Sendai virus (2000–3000 hemagglutination units per milliliter) to induce fusion of the reintroduced karyoplast. Following enucleation, the pipet containing the pronuclei was moved to a second drop containing virus. A volume of virus approximately equal to the volume of the karyoplast was aspirated, and the pipet was moved to a third drop containing an enucleated embryo. The pipet was inserted through the zona pellucida at the previous enucleation site, and contents were injected into the perivitelline space.

Injected embryos were incubated at 37°C; fusion was usually achieved during the first hour of incubation. The overall efficiency of this nuclear transplantation procedure was 91%. This method allows use of large-bore enucleation pipets (necessary to accommodate pronuclei) without penetration of the plasma membrane of the recipient ovum. A caveat is that huge differences occur in fusion efficiency among substrains of Sendai virus. Note that this and related procedures can be used to transplant one or more pronuclei or one or more nuclei per ovum.

An alternative to inactivated Sendai virus for inducing fusion between the injected karyoplast and the recipient plasma membrane is electrofusion (Zimmermann and Vienken, 1982). Embryonic cells and/or karyoplasts are placed on a slide with two parallel electrodes 1 mm apart, aligned, and then exposed to a direct-current pulse to induce fusion. Electrofusion has been used successfully to fuse blastomeres of mouse (Kubiak and Tarkowski, 1985) and rabbit two-cell embryos with other blastomeres (Ozil and Modlinski, 1986) or oocytes (Stice and Robl, 1988), mouse karyoplasts with enucleated egg cytoplasm (Tsunoda *et al.*, 1987), sheep blastomeres with enucleated oocytes (Willadsen, 1986), and bovine karyoplasts with recipient oocytes (Prather *et al.*, 1987) or enucleated pronuclear embryos (Robl *et al.*, 1987). Tsunoda *et al.* (1987) found both fusion methods successful for mouse oocytes and suggested that electrofusion might be useful in laboratories where the infectious activity makes viruses inappropriate.

2.4.2. TRANSFER OF NUCLEI INTO ENUCLEATED EMBRYOS

Illmensee and Hoppe (1981) reported the first live young born as a result of nuclear transfer in mammals. Isolated trophectoderm (TE) or inner cell mass (ICM) cells from mouse blastocyst-stage embryos were microinjected into enucleated zygotes. Only 10% of zygotes injected with TE developed to the morula stage, and none developed to term, compared to the birth of live young as a result of injection of ICM cells. Although the authors used coat color and glucose phosphate isomerase variants to verify that the mice born were of the nuclear-donor genotype, numerous attempts to reproduce these results have been unsuccessful (McGrath and Solter, 1984; Robl *et al.*, 1986).

McGrath and Solter (1984) transplanted pronuclei from one-cell embryos and nuclei from subsequent developmental stages into one-cell mouse embryos. Nearly all zygotes fused with pronuclei from enucleated zygotes developed to the blastocyst stage *in vitro*; however, development rates decreased with transfer of late-stage nuclei. In order to determine if their karyoplast fusion method was somehow affecting *in vitro* development of transplanted nuclei, they microinjected ICM cell nuclei into zygotes according to Illmensee and Hoppe (1981). One-third of the embryos survived the injection, and although 40% of the embryos divided once, no embryo cleaved beyond the four-cell stage. Robl *et al.* (1986) transplanted nuclei from eight-cell embryos into enucleated mouse zygotes and observed only one or two cleavage divisions. When eight-cell nuclei were transplanted into enucleated two-cell embryos, however, 51% (45/89) developed to blastocysts *in vitro*, and 42% (25/60) initiated implantation. Robl *et al.* (1987) also found that bovine embryos from which pronuclei were removed and others reinserted developed well during *in vitro* and *in vivo* culture. Two such embryos transferred to recipient cows developed to live, normal calves. Pronuclear embryos that received nuclei from two-, four-, or eight-cell embryos failed to develop beyond one or two cleavage divisions.

The work of McGrath and Solter and Robl *et al.* supports the idea of transitional periods in early embryogenesis. Apparently, the more out of synchrony with respect to stage of embryonic development the donor nuclei and recipient cytoplasm are, the more restricted the development is after transplantation to the enucleated embryo. It is interesting that blastocyst nuclei transferred to enucleated *Rana pipiens* zygotes resulted in normal embryogenesis (Briggs and King, 1952). In addition, endodermal nuclei supported development of fertile adults when transplanted into *Xenopus* eggs (Gurdon, 1962). These results indicate that developmental potential of early embryonic nuclei is less restricted in amphibians than in mice. Many of these experiments involve transfer of pronuclei from a sexually derived zygote, but because they result in only one genetically identical animal, these procedures are not useful for cloning animals.

2.4.3. TRANSFER OF NUCLEI INTO OOCYTES

Recent nuclear transplantation experiments have focused on the mature oocyte as a nucleus recipient rather than the fertilized zygote or two-cell embryo (Wiladsen, 1986; Prather *et al.*, 1987; Stice and Robl, 1988). Willadsen (1986) fused blastomeres of eight- or 16-cell sheep embryos with enucleate or nucleated halves of unfertilized sheep oocytes. Oocytes were bisected by sucking the polar body and adjacent ooplasm into a pipet (30 μm I.D.), resulting in two cytoplasts of approximately equal size. Approximately 75% of the halves containing the polar body also contained the metaphase II chromosomes; these were termed "nucleated egg halves." The other halves from the same oocyte contained no nuclear structures and were termed "enucleated egg halves." After fusion of blastomeres to egg halves, embryos were transferred to ligated sheep oviducts, recovered 4½ to 5½ days later and examined for development. A larger number of blastocyst-stage embryos developed from enucleated (diploid) than nucleated (triploid) egg halves regardless of whether the half received a blastomere from an eight- cell or 16-cell embryo, possible because of polyploidy. Three of four blastocysts resulting from fusion of eight-cell blastomeres with enucleated halves developed to term when transferred to recipient ewes (Willadsen, 1986). Embryos resulting from fusion of 16-cell blastomeres to enucleated egg halves were transferred to recipient ewes, and three of three were found to be carrying normal fetuses when killed at day 60. Smith and Wilmut (1989) confirmed and expanded on these results. They transferred 16-cell ovine blastomeres or ICM cells into unfertilized, enucleated ovine oocytes by electrofusion and obtained four lambs with genotypes and phenotypes of the nuclear donor breed. Therefore ovine nuclei are totipotent as late as the ICM stage when transferred to a recipient oocyte. Interestingly, enucleated ovine oocytes appear capable of reprogramming transplanted nuclei from 16-cell and later stage embryos, whereas pronuclear embryos are not.

Prather *et al.* (1987) examined *in vivo* and *in vitro* development of enucleated bovine oocytes fused with blastomeres from two- to 32-cell bovine embryos. Rate of *in vitro* development to the four- to six-cell stage ranged from 33% for transfer of six-cell blastomeres to 71% for transfer of four-cell blastomeres. Nuclear transfer embryos cultured *in vivo* (ligated sheep oviducts) developed to morula- or blastocyst-stage embryos at rates of 12% for two- to eight-cell, 16% for nine- to 16-cell, and 8% for 17- to 32-cell blastomeres. Nineteen such embryos were transferred into 13 heifers, resulting in seven pregnancies and the birth of two live calves. In addition to the demonstration that cleavage-stage bovine embryo blastomeres may be totipotent, higher rate of development occurred when the oocytes were from a recent ovulation (36 versus 48 hr after onset of estrus) and were matured *in vivo* rather than *in vitro*.

Stice and Robl (1988) placed individual eight-cell blastomeres into the perivitelline space of rabbit oocytes and induced fusion with an electrical pulse, which also activated the oocytes. Six of the 164 micromanipulated embryos developed to live offspring, resulting in "genetically verified nuclear transplant rabbits." At present, the number of live young reported in the literature as a result of nuclear transplantation is discouragingly low. This may be caused more by technical difficulties than by any inherent problems with the reconstructed embryos. Recent studies have resulted in further refinement of the enucleation and transfer techniques as well as the elucidation of the importance of fusion technique, oocyte activation, and culture conditions. Also, there are unpublished reports of production of hundreds of cloned calves by private companies. Additional research should improve prospects of cloning mammalian embryos and improve understanding of communication between transplanted nuclei and recipient cytoplasm. Perhaps this information eventually can be used to clone adult mammals.

2.5. Injection of Mitochondria and Chromosomes

2.5.1. MITOCHONDRIA

The genetic information for mammalian mitochondrial formation and function resides primarily in the DNA of chromosomes in the nucleus. However, about 20% of this information resides in double-stranded circular DNA within the mitochondrion itself. The mitochondrial DNA specifies several proteins along with transfer RNAs and ribosomal RNA to synthesize the proteins. Mitochondrial DNA is inherited maternally from generation to generation via the ovum. Although sperm contain mitochondria, they do not transmit them genetically.

Differences in mitochondrial DNA between maternal lines are not large. Although most differences are silent mutations, several genetic differences have profound phenotypic effects. For example, Brown *et al.* (1989) recently demonstrated that one maternal line of Holstein cattle has a lowered butterfat percentage in milk; this appears to be associated with a mitochondrial genetic polymorphism. Another example, the *Mta* antigen in mice, is especially interesting because the mitochondrial genome influences expression of an antigen on the outside surface of the cell (Smith *et al.*, 1983). Many mitochondrial genetic differences may interact significantly with genetic differences in chromosomal DNA. Little effort has been made to look for such interactions.

Mitochondria can be moved from cell line to cell line by cytoplast fusion or microinjection. This was illustrated clearly by King and Attardi (1988) when mitochondria with genetic resistance to chloramphenicol were injected into cells cultured in chloramphenicol. This led to rapid replacement of endogenous mitochondria with the injected ones. Recently, Ebert *et al.* (1989) injected genetically distinguishable mitochondria into mouse zygotes. These were lost during embryonic development. In many respects, it is simpler to study interaction effects of mitochondrial and nuclear genomes by transplanting nuclei than mitochondria. Recently it has been clearly demonstrated that extranuclear DNA associated with centrioles specifies genetic characteristics of these organelles in Chlamydomonas (Hall *et al.*, 1989). This situation is complex because this DNA may shuttle back and forth to the nucleus. Centriole/basal body transplantation studies similar to those just described for mitochondria may prove to be very interesting in mammals if centriolar DNA is present.

2.5.2. CHROMOSOME TRANSPLANTATION

Injecting individual chromosomes might be thought of as partial fertilization. In most situations, this would lead to aneuploidy, although even aneuploidy would not be fatal in many cases, particularly with sex chromosomes. One might even correct conditions such as Turner's syndrome (XO).

A route for doing similar studies, without the advantage of having the change in the germ line, has been accomplished by so-called somatic cell genetic techniques such as cell fusion. In a different approach, human chromosomal fragments were injected in pronuclei of mouse zygotes, and in some cases viable transgenic lines of mice were produced (Richa and Lo, 1989). The mechanics of this procedure are very similar to the standard procedures of producing transgenic animals.

A particularly intriguing approach to genetic alteration of animals is to place the genes on artificial chromosomes (Murray and Szostak, 1983), a feat already accomplished in yeast. One theoretically would simply increase chromosomal numbers by one pair. One can think of chromosomes as having three parts—centromeres, telomeres, and other DNA—which contain the various genes, their regulatory sequences, and large stretches of nonfunctional DNA. Centromeric and telomeric DNA appear to be composed primarily of repeating subunits, and it may not be too difficult to synthesize such sequences in the near future. On the other hand, it may be difficult to maintain genetic stability from generation to generation with the artificial chromosome approach. For example, the correct sequences for proper function during meiosis could be a problem.

3. CLOSING COMMENTS

Micromanipulation procedures for mammalian gametes clearly are very useful for basic research. These methods also are being used for medical and agricultural purposes. Often, the objective is to facilitate or circumvent a naturally occurring process such as fertilization. However, in some cases a fundamentally new process is developed such as cloning mammals by nuclear transplantation.

The pace of new developments in this field is rapid. Some of the procedures are very easy to learn and apply, whereas others are more akin to art than technology and are very difficult to learn. Often these methods are sensationalized by both scientists and nonscientists. However, as we use these procedures to unlock nature's secrets, it is humbling to learn how little we understand. Although the cells and organelles that constitute the raw material described in this chapter are extremely fragile, in other senses they are amazingly resilient. This contributes to the wonder each time that normal offspring result from gestation that begins with micromanipulation of gametes *in vitro*.

4. REFERENCES

Aitken, R. J., Ross, A., and Lees, M., 1983, Analysis of sperm function in Kartagener's syndrome, *Fertil. Steril.* **40:**696–698.

Baker, R. D., and Shea, B. F., 1985, Commercial splitting of bovine embryos, *Theriogenology* **23:**3–16.

Balakier, H., 1978, Induction of maturation in small oocytes from sexually immature mice by fusion with meiotic or mitotic cells, *Exp. Cell. Res.* **112:**137–141.

Balakier, H., and Masui, Y., 1986a, Chromosome condensation activity in the cytoplasm of anucleate and nucleate fragments of mouse oocytes, *Dev. Biol.* **113:**155–159.

Balakier, H., and Masui, Y., 1986b, Interactions between metaphase and interphase factors in heterokaryons produced by fusion of mouse oocytes and zygotes, *Dev. Biol.* **117:**102–108.

Barg, P. E., Wahrman, M. Z., Talansky, B. E., and Gordon, J. W., 1986, Capacitated, acrosome reacted but immotile sperm, when microinjected under the mouse zona pellucida, will not fertilize the oocyte, *J. Exp. Zool.* **237:**365–374.

Bavister, B. D., 1979, Fertilization of hamster eggs *in vitro* at sperm : egg ratios close to unity, *J. Exp. Zool.* **210:**259–274.

Bavister, B. D., 1989, A consistently successful procedure for *in vitro* fertilization of golden hamster eggs, *Gamete Res.* **23:**139–158.

Biggers, J. D., 1986, The potential use of artificially produced monozygotic twins for comparative experiments, *Theriogenology* **26:**1–25.

Brackett, B. G., Baranska, W., Sawicki, W., and Kaprowski, H., 1971, Uptake of heterologous genome by mammalian spermatozoa and its transfer to ova through fertilization, *Proc. Natl. Acad. Sci. U.S.A.* **68:** 353–357.

Braden, A. W. H., and Austin, C. R., 1954, The number of sperms about the eggs in mammals and its significance for normal fertilization, *Aust. J. Biol. Sci.* **7:**543–551.

Briggs, R., and King, T. J., 1952, Transplantation of living nuclei from blastula cells into enucleated frogs' eggs, *Zoology* **38:**455–463.

Brown, D. R., Koehler, C. M., Lindberg, G. L., Freeman, A. E., Mayfield, J. E., Meyers, A. M., Schultz, M. M., and Beitz, D. C., 1989, Molecular analysis of cytoplasmic genetic variation in Holstein cows, *J. Anim. Sci.* **67:**1926–1932.

Brun, R. B., 1974, Studies on fertilization in *Xenopus laevis*, *Biol. Reprod.* **11:**513–518.

Clarke, H. J., and Masui, Y., 1987, Dose-dependent relationship between oocyte cytoplasmic volume and transformation of sperm nuclei to metaphase chromosomes, *J. Cell Biol.* **104:**831–840.

Clarke, R. N., and Johnson, L. A., 1988a, Factors related to successful sperm microinjection of hamster eggs: The effect of sperm species, technical expertise, needle dimensions, and incubation medium on egg viability and sperm decondensation following microinjection, *Theriogenology* **30:**447–460.

Clarke, R. N., and Johnson, L. A., 1988b, Sperm microinjection for the study of fertilization and early development in mammalian eggs, in: *11th International Congress on Animal Reproduction and Artificial Insemination*, Vol. 4, International Congress on Animal Reproduction and Artificial Insemination, Dublin, pp. 469–471.

Clarke, R. N., Rexroad, C. E., Powell, A. M., and Johnson, L. A., 1988, Microinjection of ram spermatozoa into homologous and heterologous oocytes, *Biol. Reprod.* **38:**75.

Cohen, J., 1983, Selection among spermatozoa, in: *The Sperm Cell, Fertilizing Power, Surface Properties, Motility, Nucleus and Acrosome, Evolutionary Aspects* (J. Andre, ed.), Martinus Nijhoff, The Hague, pp. 33–37.

Conover, J. C., and Gwatkin, R. B. L., 1988, Fertilization of zona-drilled mouse oocytes treated with a monoclonal antibody to the zona glycoprotein, ZP3, *J. Exp. Zool.* **247:**113–118.

Cuthbertson, K. S. R., 1983, Parthenogenetic activation of mouse oocytes *in vitro* with ethanol and benzyl alcohol, *J. Exp. Zool.* **226:**311–314.

Cuthbertson, K. S. R., Whittingham, D. G., and Cobbold, P. H., 1981, Free Ca^{2+} increases in exponential phases during mouse oocyte activation, *Nature* **294:**754–757.

Czolowska, R., Waksmundzka, M., Kubiak, J. Z., and Tarkowski, A. K., 1986, Chromosome condensation activity in ovulated metaphase II mouse oocytes assayed by fusion with interphase blastomeres, *J. Cell Sci.* **84:**129–138.

Depypere H. T., McLaughlin, K. J., Seamark, R. F., Warnes, G. M., and Matthews, C. D., 1988, Comparison of zona cutting and zona drilling as techniques for assisted fertilization in the mouse, *J. Reprod. Fertil.* **84:** 205–211.

Ebert, K. M., Alcivar, A., Liem, H., Goggins, R., and Hecht, N. B., 1989, Mouse zygotes injected with mitochondria develop normally but the exogenous mitochondria are not detectable in the progeny, *Molec. Reprod. Dev.* **1:**156–163.

Fulka, J., Jr., 1983, Nuclear maturation in pig and rabbit oocytes after interspecific fusion, *Exp. Cell. Res.* **146:** 212–218.

Fulka, J., Jr., Motlik, J., Fulka, J., and Crozet, N., 1986, Activity of maturation promoting factor in mammalian oocytes after its dilution by single and multiple fusions, *Dev. Biol.* **118:**176–181.

Fulka, J., Jr., Flechon, J. E., Motlik, J., and Fulka, J., 1988, Does autocatalytic amplification of maturation-promoting factor (MPF) exist in mammalian oocytes? *Gamete Res.* **21:**185–192.

Gordon, J. W., and Talansky, B. E., 1986, Assisted fertilization by zona drilling: A mouse model for correction of oligospermia, *J. Exp. Zool.* **239:**347–354.

Gordon, J. W., Grunfeld, L., Garrisi, G. J., Talansky, B. E., Richards, C., and Laufer, N., 1988, Fertilization of human oocytes from infertile males after zona pellucida drilling, *Fertil. Steril.* **50:**68–73.

Goto, K., Kinoshita, A., Takuma, Y., and Ogawa, K., 1990, Fertilization of bovine oocytes by the injection of immobilized, killed spermatozoa, *Vet. Rec.* **127:**517–520.

Gulyas, B. J., 1986, Oocyte fusion, in: *Developmental Biology*, Vol. 4: *Manipulation of Mammalian Development* (R. B. L. Gwatkin, ed.), Plenum Press, New York, pp. 57–80.

Gulyas, B. J., Wood, M., and Whittingham, D. G., 1984, Fusion of oocytes and development of oocyte fusion products in the mouse, *Dev. Biol.* **101:**246–250.

Gurdon, J. B., 1962, Adult frogs derived from the nuclei of single somatic cells, *Dev. Biol.* **4:**256–273.

Gurdon, J. B., 1964, The transplantation of living cell nuclei, *Adv. Morphol.* **4:**1–41.

Hall, J. L., Ramanis, Z., and Luck, D. J. L., 1989, Basal body/centriolar DNA: Molecular genetic studies in Chlamydomonas, *Cell* **59:**129–132.

Harris, H., and Watkins, J. F., 1965, Hybrid cells derived from mouse and man: Artificial heterokaryons of mammalian cells from different species, *Nature* **205:**640–646.

Hiramoto, Y., 1962, Microinjection of the live spermatozoa into sea urchin eggs, *Exp. Cell Res.* **27:**416–426.

Hogan, B., Costantini, F., and Lacy, E., 1986, *Manipulating the Mouse Embryo*, Cold Spring Harbor Laboratory, New York.

Hosoi, Y., Miyake, M., Utsumi, K., and Iritani, A., 1988, Development of rabbit oocytes after microinjection of spermatozoon, in: *11th International Congress on Animal Reproduction and Artificial Insemination*, Vol. 3, International Congress on Animal Reproduction and Artificial Insemination, Dublin, pp. 331–333.

Illmensee, K., and Hoppe, P. C., 1981, Nuclear transplantation in *Mus musculus*: Developmental potential of nuclei from preimplantation embryos, *Cell* **23**:9–18.

Ivani, K. A., and Seidel, G. E., Jr., 1988, One sperm per oocyte *in vitro* fertilization in mice, *Biol. Reprod.* **38**:74.

Kaufman, M. H., 1983, *Early Mammalian Development: Parthenogenetic Studies*, Cambridge University Press, Cambridge.

Keefer, C. L., 1989, Fertilization by sperm injection in the rabbit, *Gamete Res.* **22**:59–69.

Keefer, C. L., Brackett, B. G., and Perreault, S. D., 1986, Behavior of bull sperm nuclei following microinjection into hamster oocytes, *Biol. Reprod.* **34**:54.

Keefer, C. L., Younis, A. I., and Brackett, B. G., 1990, Cleavage development of bovine oocytes fertilized by sperm injection, *Molec. Reprod. Dev.* **25**:281–285.

King, M. P., and Attardi, G., 1988, Injection of mitochondria into human cells leads to a rapid replacement of the endogenous mitochondrial DNA, *Cell* **52**:811–819.

Klebe, R. J., and Mancuso, M. G., 1981, Chemicals which promote cell hybridization, *Somat. Cell Genet.* **7**: 473–488.

Knutton, S., and Pasternak, C. A., 1979, The mechanism of cell–cell fusion, *Trends Biochem. Sci.* **4**:220–223.

Kubiak, J. Z., and Tarkowski, A. J., 1985, Electrofusion of mouse blastomeres, *Exp. Cell Res.* **157**:561–566.

Lacham, O., Trounson, A., Holden, C., Mann, J., and Sathananthan, H., 1989, Fertilization and development of mouse eggs injected under the zona pellucida with single spermatozoa treated to induce the acrosome reaction, *Gamete Res.* **23**:233–243.

Lanzendorf, S., Maloney, M., Ackerman, S., Acosta, A., and Hodgen, G., 1988a, Fertilizing potential of acrosome-defective sperm following microsurgical injection into eggs, *Gamete Res.* **19**:329–337.

Lanzendorf, S. E., Maloney, M. K., Veeck, L. L., Slusser, J., Hodgen, G. D., and Rosenwaks, Z., 1988b, A preclinical evaluation of pronuclear formation by microinjection of human spermatozoa into human oocytes, *Fertil. Steril.* **49**:835–842.

Lassalle, B., and Testart, J., 1988, Human sperm injection into the perivitelline space (SI-PVS) of hamster oocytes: Effect of sperm pretreatment by calcium ionophore A23187 and freeze-thawing on the penetration rate and polyspermy, *Gamete Res.* **20**:301–311.

Lassale B., Courtot, A. M., and Testart, J., 1985, Fertilization of hamster oocytes by microinjection of human sperm in the perivitelline space, in: *Human In Vitro Fertilization—Actual Problems and Prospects* (J. Testart and R. Frydman, eds.), INSERM-Elsevier, Amsterdam, pp. 209–212.

Lassale, B., Courtot, A. M., and Testart, J., 1987, *In vitro* fertilization of hamster and human oocytes by microinjection of human sperm, *Gamete Res.* **16**:69–78.

Lavitrano, M., Camaioni, A., Fazio, V. M., Dolci, S., Farace, M. G., and Spadafora, C., 1989, Sperm cells as vectors for introducing foreign DNA into eggs: Genetic transformation of mice, *Cell* **57**:717–723.

Laws-King, A., Trounson, A., Sathananthan, H., and Kola, I., 1987, Fertilization of human oocytes by microinjection of a single sperm under the zona pellucida, *Fertil. Steril.* **48**:637–642.

Longo, F. J., 1973, Fertilization: A comparative ultrastructural review, *Biol. Reprod.* **9**:149–215.

Malter, H. E., and Cohen, J., 1989a, Partial zona dissection of the human oocyte: A nontraumatic method using micromanipulation to assist zona pellucida penetration, *Fertil. Steril.* **51**:139–148.

Malter, H. E., and Cohen, J., 1989b, Blastocyst formation and hatching *in vitro* following zona drilling of mouse and human embryos, *Gamete Res.* **24**:67–80.

Mann, J. R., 1988, Full term development of mouse eggs fertilized by a spermatozoon microinjected under the zona pellucida, *Biol. Reprod.* **38**:1077–1083.

Martin, R. H., Ko, E., and Rademaker, A., 1988, Human sperm chromosome complements after microinjection of hamster eggs, *J. Reprod. Fertil.* **84**:179–186.

McGrath, J., and Solter, D., 1983, Nuclear transplantation in the mouse embryo by microsurgery and cell fusion, *Science* **220**:1300–1302.

McGrath, J., and Solter, D., 1984, Inability of mouse blastomere nuclei transferred to enucleated zygotes to support development *in vitro*, *Science* **226**:1317–1319.

McKinnell, R. G., 1981, Amphibian nuclear transplantation: state of the art, in: *New Technologies In Animal Breeding* (B. G. Brackett, G. E. Seidel, Jr., and S. M. Seidel, eds.), Academic Press, New York, pp. 163–180.

Murray, A. W., and Szostak, J. W., 1983, Construction of artificial chromosomes in yeast, *Nature* **305**:189–193.

Naish, S. J., Perreault, S. D., and Zirkin, B. R., 1987, DNA synthesis following microinjection of heterologous sperm and somatic cell nuclei into hamster oocytes, *Gamete Res.* **18**:109–120.

Odawara, Y., and Lopata, A., 1989, A zona opening procedure for improving *in vitro* fertilization at low sperm concentrations: A mouse model, *Fertil. Steril.* **51**:699–704.

Ohsumi, K., Katagiri, C., and Yanagimachi, R., 1986, Development of pronuclei from human spermatozoa injected microsurgically into frog (*Xenopus*) eggs, *J. Exp. Zool.* **237**:319–325.

Ozil, J., and Modlinski, J. A., 1986, Effects of electric field on fusion rate and survival of 2-cell rabbit embryos, *J. Embryol. Exp. Morphol.* **96**:211–228.

Pappaioannou, V. E., and Ebert, K. M., 1986, Comparative aspects of embryo manipulation in mammals, in: *Experimental Approaches to Mammalian Embryonic Development* (J. Rossant and R. A. Pedersen, eds.), Cambridge University Press, Cambridge, pp. 67–96.

Perreault, S. D., and Zirkin, B. R., 1982, Sperm nuclear decondensation in mammals: Role of sperm-associated proteinase *in vitro*, *J. Exp. Zool.* **224**:253–257.

Perreault, S. D., Wolff, R. A., and Zirkin, B. R., 1984, The role of disulfide bond reduction during mammalian sperm nuclear decondensation *in vivo*, *Dev. Biol.* **101**:160–167.

Perreault, S. D., Naish, S. J., and Zirkin, B. R., 1987, The timing of hamster sperm nuclear decondensation and male pronucleus formation is related to sperm nuclear disulfide bond content, *Biol. Reprod.* **36**:239–244.

Perreault, S. D., Barbee, R. R., Elstein, K. H., Zucker, R. M., and Keefer, C. L., 1988, Interspecies differences in the stability of mammalian sperm nuclei assessed *in vitro* by sperm microinjection and *in vitro* by flow cytometry, *Biol. Reprod.* **39**:157–167.

Prather, R. S., Barnes, F. L., Sims, M. M., Robl, J. M., Eyestone, W. H., and First, N. L., 1987, Nuclear transplantation in the bovine embryo: Assessment of donor nuclei and recipient oocyte, *Biol. Reprod.* **37**:859–866.

Richa, J., and Lo, C. W., 1989, Introduction of human DNA into mouse eggs by injection of dissected chromosome fragments, *Science* **245**:175–177.

Robl, J. M., Gilligan, B., Critser, E. S., and First, N. L., 1986, Nuclear transplantation in mouse embryos: assessment of recipient cell stage, *Biol. Reprod.* **34**:733–739.

Robl, J. M., Prather, R., Barnes, F., Eyestone, W., Northey, D., Gilligan, B., and First, N. L., 1987, Nuclear transplantation in bovine embryos, *J. Anim. Sci.* **64**:642–647.

Skoblina, M. N., 1974, Behaviour of sperm nuclei injected into intact ripening and ripe toad oocytes and into oocytes ripening after removal of the germinal vesicle, *Sov. J. Dev. Biol.* (English translation of *Ontogenez*) **5**:294–299.

Smith, A. L., and Seidel, G. E., Jr., 1986, Factors affecting sperm penetration through fresh murine zonae pellucidae after inhibition of the zona block to polyspermy with high ionic strength phosphate buffered saline, *Biol. Reprod.* **34**:241.

Smith, L. C., and Wilmut, I., 1989, Influence of nuclear and cytoplasmic activity on the development *in vivo* of sheep embryos after nuclear transplantation, *Biol. Reprod.* **40**:1027–1035.

Smith, R. III, Huston, M. M., Jenkins, R. N., Huston, D. P. and Rich, R. R., 1983, Mitochondria control expression of a murine cell surface antigen, *Nature* **306**:599–601.

Stice, S. L., and Robl, J. M., 1988, Nuclear reprogramming in nuclear transplant rabbit embryos, *Biol. Reprod.* **39**:657–664.

Surani, M. A. H., Barton, S. C., and Norris, M. L., 1987, Experimental reconstruction of mouse eggs and embryos: An analysis of mammalian development, *Biol. Reprod.* **36**:1–16.

Szollosi, D., Czolowska, R., Soltynska, M. S., and Tarkowski, A. K., 1986, Remodelling of thymocyte nuclei in activated mouse oocytes: An ultrastructural study, *Eur. J. Cell Biol.* **42**:140–151.

Talansky, B. E., Barg, P. E., and Gordon, J. W., 1987, Ion pump ATPase inhibitors block the fertilization of zona-free mouse oocytes by acrosome-reacted spermatozoa, *J. Reprod. Fertil.* **79**:447–455.

Tarkowski, A. K., 1977, *In vitro* development of haploid mouse embryos produced by bisection of 1-cell fertilized eggs, *J. Embryol. Exp. Morphol.* **38**:187–202.

Tarkowski, A. K., 1982, Nucleo-cytoplasmic interaction in oogenesis and early embryogenesis in the mouse, in: *Progress in Clinical and Biological Research* (M. M. Burger and R. Weber, eds.), Alan R. Liss, New York, pp. 407–416.

Thadani, V. M., 1980, A study of hetero-specific sperm–egg interactions in the rat, mouse and deer mouse using *in vitro* fertilization and sperm injection, *J. Exp. Zool.* **212**:435–453.

Thadani, V. M., 1981, A Study of Oocyte Maturation and Sperm–Egg Interactions Using *in Vitro* Fertilization and Sperm Injection, Ph.D. dissertation, Yale University, New Haven, CT.

Tsunoda, Y., and Shioda, Y., 1988, Development of enucleated parthenogenones that received pronuclei or nuclei from fertilized mouse eggs, *Gamete Res.* **21**:151–155.

Tsunoda, Y., Yasui, T., Nakamura, K., Uchida, T., and Sugie, T., 1986, Effect of cutting the zona pellucida on the pronuclear transplantation in the mouse, *J. Exp. Zool.* **240**:119–125.

Tsunoda, Y., Kat, Y., and Shioda, Y., 1987, Electrofusion for the pronuclear transplantation of mouse eggs, *Gamete Res.* **17**:15–20.

Tsunoda, Y., Shioda, Y., Onodera, M., Nakamura, K., and Uchida, T., 1988, Differential sensitivity of mouse pronuclei and zygote cytoplasm to Hoechst staining and ultraviolet irradiation, *J. Reprod. Fertil.* **82**:173–178.

Uehara, T., and Yanagimachi, R., 1976, Microsurgical injection of spermatozoa into hamster eggs with subsequent transformation of sperm nuclei into male pronuclei, *Biol. Reprod.* **15**:467–470.

Willadsen, S. M., 1979, A method for culture of micromanipulated sheep embryos and its use to produce monozygotic twins, *Nature* **277**:298–300.

Willadsen, S. M., 1986, Nuclear transplantation in sheep embryos, *Nature* **320**:63–65.

Williams, T. J., Elsden, R. P., and Seidel, G. E., Jr., 1984, Pregnancy rates with bisected bovine embryos, *Theriogenology* **22**:521–531.

Wolf, D. P., 1978, The block to sperm penetration in zona-free mouse eggs, *Dev. Biol.* **64**:1–14.

Yanagimachi, R., 1981, Mechanisms of fertilization in mammals, in: *Fertilization and Embryonic Development in Vitro* (L. Mastroianni and J. D. Biggers, eds.), Plenum Press, New York, pp. 81–182.

Yanagimachi, R., 1988, Mammalian fertilization, in: *The Physiology of Reproduction* (E. Knobil and J. Neill, eds.), Raven Press, New York, pp. 135–186.

Yang, X., Chen, J., and Foote, R. H., 1988, Blastocyst development from rabbit ova fertilized by injected sperm, *J. Reprod. Fertil.* [Abstr. Ser.] **1**:13.

Zielgler, D. H., and Masui, Y., 1973, Control of chromosome behavior in amphibian oocytes. I. The activity of maturing oocytes inducing chromosome condensation in transplanted brain nuclei, *Dev. Biol.* **35**:283–292.

Zimmermann, U., and Vienken, J., 1982, Electric field-induced cell-to-cell fusion, *J. Memb. Biol.* **67**:165–182.

22

The Use of Molecular Biology to Study Sperm Function

Erwin Goldberg

1. INTRODUCTION

Current molecular biology technology provides an increased level of sensitivity for the analysis of sperm structure and function. Excellent laboratory manuals are available that provide familiarity with the few basic concepts required for mastery of these techniques (Maniatis *et al.*, 1982; Ausubel *et al.*, 1987; Berger and Kimmel, 1987). Therefore, this chapter does not detail such laboratory procedures but rather attempts to review the recent literature in which sperm function has been studied by using the tools of molecular biology. Basically, I focus on the investigations that involve DNA and RNA isolation, cloning, and analysis to shed light on the mechanisms whereby the spermatozoan acquires the unique properties necessary to achieve success in accomplishing its primary function, fertilization of the egg. This overview is intended to be instructive rather than exhaustive, and I hope it will provide a resource with which to initiate further study of fertilization.

2. SPERM PEPTIDES INVOLVED IN FERTILIZATION

Much of our basic knowledge on fertilization comes from pioneering studies with Echinoderm spermatozoa (cf. Dan, 1967). In the sea urchin, for example, fertilization involves chemoattraction, acrosome reaction, and sperm–egg fusion. Analyses of the peptides that mediate these processes could not be accomplished by conventional protein extraction and purification methods. Recombinant DNA technology was used by Dangott *et al.* (1989) to characterize the sperm membrane protein that cross-links to speract, the peptide from the egg jelly of the sea urchins *Strongylocentrotus purpuratus* and *Hemicentrotus pulcherrimus* (Dangott and Garbers, 1987). Bentley and Garbers (1986) reported the first studies to identify a receptor for

---

ERWIN GOLDBERG • Department of Biochemistry, Molecular Biology, and Cell Biology, Northwestern University, Evanston, Illinois 60208.

A Comparative Overview of Mammalian Fertilization, edited by Bonnie S. Dunbar and Michael G. O'Rand. Plenum Press, New York, 1991.

speract in isolated spermatozoan membranes. Speract and resact from the *Arbacia punctulata* egg do not cross-react with spermatozoa of the opposite species. These peptides may mediate a species-specific chemoattractant response involving substantial morphological and behavioral changes in the sperm. A synthetic analogue of speract was cross-linked to the putative receptor protein in sperm membrane vesicle preparations (Bentley and Garbers, 1986). The apparent receptor appeared to represent a glycoprotein of M_r 77,000. The cross-linked protein has now been purified to yield limited amino acid sequence information (Dangott *et al.*, 1989). Mixed antisense oligonucleotide probes made to the peptide sequence detected a cDNA clone in an *S. purpuratus* testis cDNA library packaged in λgtl0. From the nucleotide sequence of the full-length clone, an amino acid sequence could be deduced.

The apparent receptor can be divided into various tentative domains based on the deduced amino acid sequence, which suggests that the protein contains a 26-residue amino-terminal signal peptide, a large extracellular domain relatively rich in cysteine that includes a fourfold repeat of about 115 amino acids, a single membrane-spanning region, and only 12 amino acid residues extending into the cytoplasm (Dangott *et al.*, 1989). Northern blot analysis of total RNA prepared from testes of *S. purpuratus* and the closely related *Lytechinus pictus* demonstrated a 2.5-kb mRNA. Whereas speract activates sperm from both of these species, *A. punctulata* testes contain a cross-hybridizing mRNA that responds to resact rather than speract. This elegant study (Dangott *et al.*, 1989) clearly exemplifies the use of molecular biology to study sperm function. A more detailed review of the sperm-activating peptides that play a role in fertilization can be found in Dangott and Garbers (1987) and Garbers (1989).

Resact, the chemotactic peptide released by the sea urchin *Arbacia punctulata* egg, is specifically cross-linked to *A. punctulata* guanylate cyclase (Shimomura *et al.*, 1986), thereby implicating this enzyme as a cell surface receptor in spermatozoa. Singh *et al.* (1988) screened a sea urchin testis cDNA library with synthetic oligonucleotide probes and isolated clones that included the entire guanylate cyclase coding region and 5' and 3' untranslated sequences. An open reading frame of 986 codons within the 3444-nucleotide sequence of cloned guanylate cyclase was used to deduce the amino acid sequence of the translation product. These analyses predicted an intrinsic membrane protein of 986 amino acids with an amino-terminal signal sequence. From the deduced primary structure the authors suggested that this protein was separated by a single transmembrane domain into putative extracellular and cytoplasmic–catalytic domains. Furthermore, sequence comparisons revealed distinct similarities between sea urchin spermatozoan guanylate cyclase and highly conserved regions of protein kinases. Additional studies should elucidate the significance of this finding in relation to the presumed catalytic function of the protein.

A major protein of the sea urchin acrosome granule, bindin, mediates the species-specific adhesion and binding of sperm to egg (Vacquier, 1980). Bindin has been purified and characterized chemically (cf. Vacquier, 1980). Gao *et al.* (1986) reported the isolation of bindin cDNA clones and the complete sequence of the derived protein. These investigators constructed a λgtl0 cDNA library from testis poly(a)+ RNA extracted from *S. purpuratus* males when spermatogenesis was maximal. This library was screened with a synthetic 17-nt probe mixture including all possible sequences predicted by a known region of the bindin protein, and a clone was selected that included an 1873-nt sequence of the bindin cDNA insert. Characterization of this cDNA revealed that bindin mRNA had a 5' leader sequence. Apparently the mature 24-kDa bindin protein is produced by processing a 51-kDa precursor, an observation consistent with the acrosomal localization of bindin. The acrosomal granule is initially derived from the Golgi complex, where processing of larger precursors to mature proteins is often accomplished in the cell. Apparently, species specificity of sperm–egg interaction resides in sperm bindin–egg bindin receptor recognition of protein primary structure (Gao *et al.*, 1986). Also, the bindin gene is productively expressed only in the testis, with no detectable mRNA for this protein in female gonads or eggs (Gao et al., 1986).

Rabbit anti-sea-urchin-bindin antiserum recognizes a peptide sequence from the 14-kDa member of the rabbit sperm membrane autoantigen RSA (O'Rand and Widgren, 1989). The RSA functions as a sperm lectin and binds the spermatozoan to the zona pellucida with high affinity (O'Rand et al., 1988). O'Rand and Widgren (1990) sequenced a peptide from a kallikrein digest of 14-kDa RSA. The sequence appears to have a similar motif to several segments of sea urchin bindin. From this sequence a ten-amino-acid peptide, P10G, was synthesized, conjugated to thyroglobulin, and used for immunization. Affinity-purified anti-P10G antibodies reacted with RSA on ELISA, with the 14-kDa RSA band on Western blots, and showed strong equatorial localization by immunofluorescence on acrosome-intact ejaculated rabbit spermatozoa. Several clones have been isolated from a rabbit testis cDNA library screened with the affinity-purified anti-P10G antiserum and plaque purified for sequencing.

The role of the sperm acrosome during mammalian sperm–zona pellucida interactions has been the subject of numerous investigations (cf. Urch, 1986, for review). The primary structure of human proacrosin has been deduced from its cDNA sequence (Baba et al., 1989). Nucleotide sequencing of a cDNA clone isolated from a human testis cDNA library has predicted that the human proacrosin is synthesized with a 19-amino-acid signal peptide at the N terminus. The C-terminal portion is very rich in proline residues. Boar preproacrosin cDNA contains at the 3′ end a sequence that codes for an amino acid sequence rich in proline that is suggested to be involved in recognition and binding of the sperm to the zona pellucida of the egg (Adham et al., 1989). These authors have also determined by in situ hybridization and Northern analysis that the mRNA for preproacrosin is synthesized as an approximately 1.6-kb-long molecule only in the postmeiotic stages of boar and bull spermatogenesis. A human sperm acrosomal protein, SP-10, has been described by Herr et al. (1990) as a testes-specific intraacrosomal differentiation antigen that first localizes within the nascent acrosomal vesicle of Golgi phase spermatids. Wright et al. (1990) used a monoclonal antibody to SP-10 to probe a human testes λgt11 expression library (Millan et al., 1987). Complementary DNAs coding for SP-10 have been isolated and sequenced, and the deduced SP-10 amino acid sequence has been analyzed. It appears that SP-10 is an 18-kDa peptide containing a signal peptide sequence for membrane transport, two potential N-linked and possible O-linked glycosylation sites, and a central peptide core containing several hydrophilic domains comprised almost entirely of three repeating peptide motifs (Wright et al., 1990).

3. TESTIS-SPECIFIC GENE EXPRESSION

Unique proteins and enzymes have been identified in spermatozoa and are products of a finely tuned developmental program in the testes. There are examples of testis-specific proteins expressed during spermatogenesis both before and after meiosis. The mRNAs for some proteins are transcribed and then stored for translation at a later stage of differentiation for translation. For some sperm products immediate message utilization takes place. There are recent reviews on gene expression during spermatogenesis (Hecht, 1987) and on postmeiotic gene expression (Hecht, 1988).

Sperm histones are uniquely expressed in the testis and presumably function in sperm DNA condensation. Decondensation and expression of the paternal genome following fertilization occur in the two-cell-stage embryo (Maxson and Egrie, 1980) and probably involve replacement in chromatin by cleavage-stage histones (Newrock et al., 1977; Poccia et al., 1981). Lieber et al. (1986) analyzed the histone mRNA population found in several adult tissues of S. purpuratus and L. pictus. They found unique species of H1 and H2b mRNAs encoding the sperm-specific histones exclusively in testis RNA. In addition, these investigators prepared cDNA clones encoding the testis H2b and H2a proteins from two different sea urchin species. Northern blot analyses confirmed tissue-specific expression of these genes. Busslinger and Barberis (1985) cloned and characterized several cDNAs encoding sperm and late histone mRNA from *Psammechinus miliaris*. The two sperm H2b mRNAs were found in testes but not in ovaries and embryos of this sea urchin.

A sea urchin egg injection procedure was used to study developmental regulation of homologous histone genes (Vitelli *et al.*, 1988). Cloned DNA was injected into unfertilized sea urchin eggs. They were then fertilized and cultured to different developmental stages. Embryo RNA was isolated and analyzed to identify transcripts of the injected genes. Transcripts of the histone H2a gene, which is normally expressed in early embryonic stages, accumulated during cleavage and decayed in late embryos. The sperm H2b-1 gene, normally inactive in development, was very inefficiently expressed compared to H2a in transformed blastulae. A fusion gene with the early H2a promoter linked to the structural H2b was efficiently transcribed, suggesting that the 5' region of the early H2a gene contains the regulatory elements necessary for proper expression (Vitelli *et al.*, 1988).

A comparison of histone H1 from sea urchin sperm and calf thymus reveals that the former induces a stronger aggregation of DNA, presumably because of differences in amino acid sequences (Osipova *et al.*, 1985). These structural differences presumably account for the special chromatin-condensing ability of the sperm histone.

During mammalian spermatogenesis, somatic histones are replaced with testis-specific or enriched histone variants. Application of recombinant DNA techniques and *in situ* hybridization of a rat Sertoli–spermatogenic cell coculture system revealed that the somatic H2B gene is expressed predominantly in spermatogonia/preleptotene spermatocytes and that testis-specific TH2B gene is active in pachytene spermatocytes (Kim *et al.*, 1987). These studies are an important first step toward understanding the regulation of histone genes during spermatogenesis as well as establishing the biological significance of the replacement of somatic histones by testis-specific histones.

Hecht (1987) has reviewed recent data on nuclear and cytoplasmic structural proteins essential for sperm formation and function. The protamines constitute a major class of sperm nuclear proteins whose primary function may be compaction but not condensation of the sperm nuclear DNA (Hecht, 1987). The two different protamine variants in the mouse are the products of two distinct gene loci that appear to be differentially expressed during spermatogenesis. Possible upstream regulatory sequences common to temporally expressed testicular genes have been described by this laboratory for the protamine genes.

4. LACTATE DEHYDROGENASE C_4 TRANSCRIPTION AND TRANSLATION

One useful example of a temporally expressed testis-specific protein is the C_4 isozyme of lactate dehydrogenase (LDH), investigated in considerable detail in my laboratory. From a molecular standpoint it is clear that the *Ldh-c* gene is turned on during prophase of the first meiotic division. From some early work it appeared that in the mouse, LDH-C_4 was first detected in midpachytene spermatocytes (Hintz and Goldberg, 1977). Synthesis of the LDH-C subunit (Meistrich *et al.*, 1977) and functional LDH-C mRNA (Wieben, 1981) also have been measured in isolated pachytene spermatocytes and spermatids. A recent reinvestigation of the developmental expression of the *Ldh-c* gene (Li *et al.*, 1989) in germ cells purified from prepubertal and adult mice confirms the original finding of spermatocyte expression. However, these investigators found LDH-C_4 activity and synthesis earlier in meiosis in highly enriched populations of preleptotene and leptotene/zygotene spermatocytes. Whatever the exact timing, it is clear that once the type A spermatogonium differentiates into a type B spermatogonium and becomes committed to the spermatogenic pathway, expression of the *Ldh-c* is activated.

This isozyme of lactate dehydrogenase is predominantly a cytosolic enzyme; however, a small proportion appears to be associated with the structurally modified mitochondria of the germinal epithelium (Blanco *et al.*, 1976) as well as the surface of the spermatozoan (Erickson *et al.*, 1975; Storey and Kayne, 1977; Alvarez and Storey, 1984). Presumably LDH-C$_4$ functions metabolically in testes and/or spermatozoa (Goldberg, 1977; Wheat and Goldberg, 1983). However, one of the key questions about LDH-C$_4$ that can be approached by molecular technology concerns regulation of the developmental expression of the gene encoding this protein.

Millan *et al.* (1987) probed a human testes cDNA expression library in λgt11 with both monoclonal and polyclonal antisera specific to murine LDH-C$_4$. Positive plaques (Fig. 1A) and macroplaques (Fig. 1B) were subcloned, and the DNA isolated and purified. Restriction enzyme analyses (Fig. 2) revealed three clones that hybridized with each other on Southern blots. From one of these clones, a 1.6-kb fragment of DNA was isolated and sequenced and found to contain the entire 1.0-kb open reading frame coding for the human LDH-C subunit. Northern analysis was used to demonstrate testis specificity (Fig. 3) of the 1.6-kb DNA fragment.

This specific cDNA was used as a probe to map the *Ldh-c* gene to the short arm of human chromosome 11 (Edwards *et al.*, 1987). *In situ* hybridization data and analysis of mouse/human somatic cell hybrids carrying deletions of human chromosome 11 indicate that the gene is localized at p.15.3–p.15.4 close to the *Ldh-a* gene (Edwards *et al.*, 1989). These findings suggest that *Ldh-c*, which incidentally is restricted to mammals and the pigeon, may have evolved as a tandem duplication of *Ldh-a* rather than of *Ldh-b* as originally proposed (Holmes, 1972). Molecular cloning and nucleotide sequence of the cDNA for the mouse *Ldh-c* has been reported also (Sakai *et al.*, 1987).

Recently, recombinant DNA probes have been used to examine the expression of specific LDH genes during spermatogenesis. RNA:cDNA hybridization *in situ* in the testis confirmed previous findings that LDH-C mRNA first appears in the primary spermatocyte and continues to accumulate until the late spermatid stage (Fujimoto *et al.*, 1988; K. Salehi-Ashtiani and E. Goldberg, unpublished observations). Preliminary data from my laboratory revealed that silver grains above background levels accumulated in pachytene primary spermatocytes, increased in early round spermatids, and decreased in elongated late spermatids. Fujimoto *et al.* (1988) found, by Northern analysis, that meiotic and postmeiotic cells contained 1.2- and 1.3-kb size classes of hybridizing mRNA and that the difference reflected the poly(A) tail length of the *Ldh-c* mRNA. This study demonstrated that spermatids contained the *Ldh-c* mRNA with the longer poly(A) tract, leading the authors to suggest this as a mechanism whereby stable accumulation of *Ldh-c* mRNA is permitted for use during spermiogenesis.

In a developmental study by Northern analyses of mRNA in mouse testes extracts and isolated germinal cell populations (J. Trasler, unpublished observations), a specific message encoding the LDH-C subunit was detected in primary spermatocytes. The specificity of gene expression was confirmed by our inability to detect *Ldh-c* mRNA in somatic tissues. The mRNA level increased during testis differentiation, and it was far more abundant in round spermatids than in pachytene spermatocytes. There may be species-specific differences suggested by preliminary results of K. Salehi-Ashtiani and E. Goldberg (unpublished observations) that there is almost twice as much *Ldh-c* mRNA in primary spermatocytes than spermatids of the rat. In contrast to the findings of Fujimoto *et al.* (1988) in the mouse, there appears to be a single size class of mRNA for *Ldh-c* from rat testes, an observation consistent with the decrease in specific message during rat spermatogenesis. These observations on mRNA size classes, stability, and concentrations during germ cell differentiation require further study, especially of translation rates. For example, a scenario for the mouse would have a single size class of message to represent the original transcript. Rapid translation would reduce the size of the poly(A) tail. In spermatids

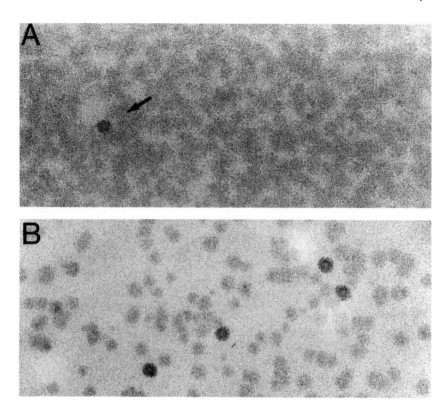

Figure 1. Primary (A) and secondary (B) screening of a λgt11 human testis cDNA library (Millan *et al.*, 1987) with rabbit antibodies against mouse LDH-C$_4$. β-Galactosidase-labeled goat antirabbit IgG was used to detect positive plaques.

in which stable product has accumulated and stable message exists, the larger message class would predominate. Although it has not been established that regulation of *Ldh-c* gene expression affects sperm function, the report by Gavella and Cvitkovic (1985) that human sperm lacking LDH-C$_4$ are abnormal and nonfunctional may be instructive in this respect.

5. TRANSCRIPTIONAL REGULATION OF TESTIS-SPECIFIC GENES

Other examples of testis-specific enzymes include phosphoglycerate kinase-2 (PGK-2) and testicular cytochrome c$_t$. A testis-specific cytochrome c first described in the mouse (Hennig, 1975) has also been demonstrated in rat, rabbit, and bull (Kim and Nolla, 1986). Goldberg *et al.* (1977) used immunofluorescent staining of sections of mouse seminiferous epithelium to detect cytochrome c$_t$ in midpachytene primary spermatocytes and subsequent stages of spermatogenesis. Cytochrome c$_t$ is clearly associated with the uniquely modified sperm mitochondria and presumably retains its function as an essential component of the respiratory chain. Just as with LDH-C$_4$, noted above, tissue specificity and temporal regulation of this testis-specific gene represent compelling questions. Virbasius and Scarpulla (1988) isolated cDNA clones encoding cytochrome c$_t$ from both rat and mouse as well as the corresponding genomic clone from rat. DNA sequence data reveal several notable differences between the somatic and testicular cytochromes. The testis gene is a single copy about 7 kb along with three introns totaling nearly

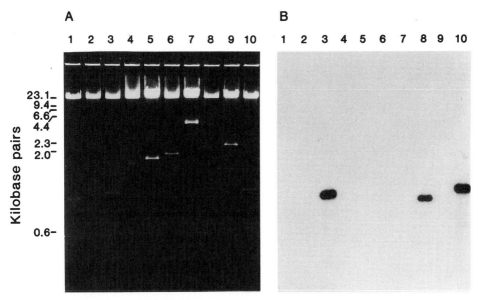

Figure 2. (A) Agarose gel electrophoresis of 10 λgt11 recombinant DNAs after *EcoRI* digestion. (B) After Southern transfer, the filter was probed with the nick-translated 0.9-kb isolated insert from lane 3 (from Millan *et al.*, 1987).

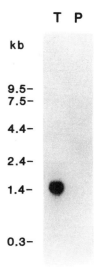

Figure 3. A Northern blot of human testis (T) and human placental (P) poly(A)+ RNA (10 μg) probed with the nick-translated 1.0-kb insert from the human *Ldh-c* cDNA clone.

6.5 kb with no detectable pseudogenes. In contrast, the somatic cytochrome c gene has given rise to a multipseudogene family, and the 2.1-kb gene contains two introns occupying only 0.9 kb (Virbasius and Scarpulla, 1988). Differences in presumed regulatory sequences were also detected by these investigators and are likely to account for the tissue-specific pattern of expression. It has recently been established that the somatic cytochrome c gene is not expressed in spermatogenic cells but is shut down in cells committed to meiosis (Hake *et al.*, 1990).

Phosphoglycerate kinase-2 (PGK-2), a key enzyme in glycolysis, is an example of a testis-specific autosomal gene that apparently is expressed to replace the X-chromosome inactivated PGK-1 isozyme in round spermatids (Gold *et al.*, 1983; VandeBerg *et al.*, 1981). McCarrey and Thomas (1987) reported that the human *Pgk-2* gene is an intronless retroposon that apparently evolved via gene processing from the intron-containing *Pgk-1* gene. These authors demonstrated that this *Pgk-2* retroposon was expressed during spermatogenesis, whereas pseudogenes formed from processed genes are usually not expressed. McCarrey (1987) compared the nucleotide sequences of the promoter regions of *Pgk-1* and *Pgk-2* and of *Pgk-2* with that of two other genes expressed uniquely in the postmeiotic stages of spermatogenesis, the mouse protamine-1 (MP1) and protamine-2 (MP2) genes. Consensus sequences present in MP1, MP2, and *Pgk-2* were not detected in the *Pgk-1* promoter. The *Pgk-1*/*Pgk-2* sequence comparison suggests that the promoter sequence of *Pgk-2* arose with the coding sequence by reverse-transcriptase-mediated processing of an aberrant transcript of *Pgk-1*, including its promoter elements (McCarrey, 1987). Sequence divergence presumably led to tissue- and developmentally specific regulation of expression of *Pgk-2*.

A testis-specific transcript for rat phosphoribosyl pyrophosphate synthetase (PRPS) has been reported by Taira *et al.* (1989). There are two genes designated PRPS1 and PRPS2 that are differentially regulated in different tissues. Recombinant DNA probes for Northern blot analysis revealed a unique 1.4-kb mRNA in testis of rat, mouse, and human. Further analyses indicated that the 1.4-kb mRNA was generated from another gene homologous to but distinct from the PRPS1 coding sequence. Taira *et al.* (1989) suggest, on the basis of preliminary human gene mapping that assigns PRPS1 and PRPS2 to chromosome X and the existence of two other PRPS1-related sequences on autosomes, that X-chromosome inactivation during meiosis is compensated by expression of the putative autosomal gene.

Testis-specific expression of the *Drosophila* β_2-tubulin gene was analyzed by Michiels *et al.* (1989). Germ-line transformation experiments with the upstream region of the *D. melanogaster* β_2-tubulin gene fused to the *Escherichia coli lac2* gene resulted in the correct tissue expression of the reporter gene. Upstream sequences of *D. lydei* were able to drive expression of *lac2* in testis of *D. melanogaster*. Deletion analysis showed that 53 bp of upstream and 23 bp (*D. melanogaster*) or 29 bp (*d. lydei*) of leader sequences are sufficient to confer tissue specificity. A short promoter region contains a 14-bp motif, acts as a position-dependent promoter, and is the only sequence element necessary for the testis-specific transcription of the β_2-tubulin gene in *Drosophila* (Michiels *et al.*, 1989).

An unusual strategy for testis specificity is used by a human α-tubulin gene, Hα44, which is similar to other mammalian α-tubulin genes. A testis-specific promoter is generated by recruiting a new 5' exon and translation-initiation site (Dobner *et al.*, 1987). The somatic promoter sequence, then, is incorporated into the 5' leader region of the message.

Hecht *et al.* (1984) have described a testis-specific α-tubulin (α-TT-1) transcript that encodes a variant α-tubulin. The nucleotide sequence was used to deduce the primary structure of a portion of the testicular α-tubulin. A synthetic peptide containing a portion of the amino acid sequence not found in other α-tubulins was used to generate specific antibodies that by immuno-fluorescence localized α-TT-1 to the manchette (Hecht *et al.*, 1988), a unique and transient structure formed during spermiogenesis.

Sperm of the nematode *Caenorhabditis elegans* contain a family of closely related, small basic proteins (Ward *et al.*, 1988) designated major sperm proteins (MSPs). These are encoded in the genome by a multigene family containing more than 50 genes that are not in a tandem array (Burke and Ward, 1983; Klass *et al.*, 1984) and have maintained strict regulation of expression in late primary spermatocytes. Ward *et al.* (1988) sequenced independent MSP cDNA clones to identify different MSP genes. Probes were then prepared from unique portions of the sequence and used to isolate each identified MSP gene. The organization in the genome of these and other MSP genes was determined by physical mapping and *in situ* hybridization. The organization of 40 MSP genes was described, with 37 of them found in clusters on only two chromosomes (Ward *et al.*, 1988). In addition, the level of expression of individual MSP genes was determined, and it was shown that transcribed genes have highly conserved 5' sequences that might contain *cis*-acting control signals contributing to the coordinate regulation of these genes (Klass *et al.*, 1988). These studies are of value in determining the role of MSPs in normal sperm production (Ward *et al.*, 1988).

6. TRANSLATIONAL CONTROL DURING SPERMATOGENESIS

Translational control is an important mechanism of regulation during spermatogenesis in many organisms. In the rainbow trout, protamine mRNA is stored for 15–30 days before it is translated (Iatrou and Dixon, 1977), in *Drosophila* there is no postmeiotic transcription (Olivieri and Olivieri, 1965), and in the mouse many proteins required for spermatozoan assembly are not synthesized until well after transcription ceases (O'Brien and Bellve, 1980; Kuhn *et al.*, 1988).

Evidence presently available on specific transcripts reveals that mRNAs or translation products for LDH-C$_4$ (Hintz and Goldberg, 1977; Meistrich *et al.*, 1977; Wieben, 1981; Li *et al.*, 1989), cytochrome c$_t$ (Goldberg *et al.*, 1977; Hake *et al.*, 1990), histones (Bhatnagar *et al.*, 1985), and actins (Hecht *et al.*, 1984) can be detected during meiotic prophase. This may be compared with the PGK-2 transcript that is synthesized in pachytene spermatocytes (Gold *et al.*, 1983) but translated only in early- to midstage spermatids (Kramer, 1981; VandeBerg *et al.*, 1981).

The exquisite sensitivity of translational control during spermiogenesis has been demonstrated by analysis of a transgene product that is expressed exclusively in postmeiotic germ cells (Braun *et al.*, 1989a). Fusions were constructed between the mouse protamine 1 gene (mP1) and the human growth hormone (hGH) structural gene. Their expression was analyzed in transgenic mice. Braun *et al.* (1989a) show that mP1 sequences 5' to the start of transcription confer spermatid-specific expression on the hGH gene and that a portion of the mP1 3' untranslated sequence is sufficient to confer mP1-like translational regulation on the hGH mRNA. The time during spermiogenesis at which different transgenic mRNAs are translated affects the subcellular localization of the product of the transgene within the developing spermatid. Synthesis of hGH in early, round spermatids resulted in localization in the acrosome, whereas synthesis in late, elongating spermatids resulted in intracellular but not acrosomal localization. Braun *et al.* (1989b) used the mP1–hGH construct to demonstrate in hemizygous transgenic mice that genetically distinct spermatids share the product of the transgene and can be phenotypically equivalent.

7. DEVELOPMENTALLY REGULATED GENE TRANSCRIPTS

A major gap in understanding the molecular aspects of gamete biology concerns the relationship of gene transcripts produced in spermatids to potential sperm fertilizing ability. A series of DNA-cloning studies have identified a number of novel cDNAs solely expressed in postmeiotic germ cells of the male (reviewed by Hecht, 1988; Olds-Clarke, 1988). Transcripts of the protooncogenes *c-ab1* (Ponzetto and Wolgemuth, 1985), *c-mos* (Propst *et al.*, 1988; Goldman *et al.*, 1987; Mutter and Wolgemuth, 1987), and *int-1* (Shackleford and Varmus, 1987) and a gene including a "homeo" box (Wolgemuth *et al.*, 1986) have been described. Translation products of these RNAs have not yet been detected. Propst *et al.* (1988) studied the pattern of *Mos* protooncogene RNA expression in gonads of several sterile mouse mutants and found it to be identical to the pattern of expression of the testis-specific transcripts for *Abl*, actin, and the mouse homeobox *Hox-1.4* gene. However, during normal spermatogenesis, differences in the time of appearance of the four transcripts were detected during haploid spermatid maturation. These protooncogenes may be involved in normal development and differentiation. Work on mutations that cause sterility is described in the excellent review by Olds-Clarke (1988), who emphasizes the utility of genetic analyses to relate specific transcripts and sperm function.

Recently Krawczyk *et al.* (1987) and Zakeri and Wolgemuth (1987) reported that testes of the adult rat and mouse contain abundant levels of the transcript related to the *hsp70* gene that encodes the heat shock protein of mol. wt. 70,000. The transcript, called hst70RNA, differs from other *hsp70*-related heat shock transcripts, and its level is not affected by hyperthermia (Krawczyk *et al.*, 1987; Zakeri and Wolgemuth, 1987). Its expression is developmentally regulated since it is absent in testes of prepubertal animals and appears when spermiogenesis is initiated (Krawczyk *et al.*, 1987; Zakeri and Wolgemuth, 1987). Krawczyk *et al.* (1988) investigated changes in the level of hst70RNA and its cellular localization during the cycle of rat seminiferous epithelium. *In situ* hybridization revealed that the *hst70* gene was activated in late pachytene primary spermatocytes during stage XII of the cycle and that mRNA was then present in cells during differentiation through diakinesis, meiotic divisions, and early spermiogenesis (steps 1 through early 7). The authors (Krawczyk *et al.*, 1988) suggest that activation of the gene encoding hst70RNA shortly before meiotic divisions indicates a need for the gene product either during differentiation of late spermatocytes into spermatids or later during spermiogenesis and that the mRNA may be stored in early spermatids.

8. STAGE-SPECIFIC mRNA LEVELS

Stage-specific levels of some mRNAs were determined in rat seminiferous tubules synchronized to a few related stages of the cycle of the seminiferous epithelium (Morales *et al.*, 1989). The germinal cell content of the synchronized testes was examined with Northern blots probed with nick-translated protamine 1 and transition protein 1 cDNAs. The tubules synchronized to stages VIII to XIV of the cycle had both high levels and the predicted nonpolysomal and polysomal sizes of these two mRNAs. Synchronization was achieved by a treatment using vitamin A depletion followed by repletion and has provided a valuable model to probe the molecular nature of the cycle of the seminiferous epithelium (Morales *et al.*, 1989).

Messenger RNAs encoding calmodulin, α- and β-tubulins, and actins were analyzed in nucleic acid isolated from purified rat testis cell populations (Slaughter *et al.*, 1989). The tubulin and actin, encoded by multigene families differentially regulated during development, are cytoskeletal proteins important in sperm cell modeling during spermiogenesis. Calmodulin may be involved in meiotic division. Slaughter *et al.* (1989) define the phases of testis cell differentiation in which alterations in specific RNA levels occur and propose that these changes may reflect activation or repression of transcription of coordinately regulated genes during spermatogenesis. Furthermore, they point out the importance of developing specific probes to distinguish members of gene families that may respond differentially to developmental signals.

A novel cytoskeletal role for actins in the testis has been suggested from identification of two size classes of actin mRNA differently distributed between spermatids and pachytene spermatocytes (Hecht, 1987). Kim *et al.* (1989) identified the testicular postmeiotic actin encoded by the 1.5-kb mRNA as a smooth-muscle λ actin (SMGA) and presented its cDNA sequence. An isoform-specific probe was used to demonstrate that testicular SMGA mRNA was first detectable during spermiogenesis and that it increased substantially in amount. Kim *et al.* (1989) discuss possible functional activities of actin in shaping the acrosome, sequestering the cytoplasm as residual bodies are pinched off from the elongating spermatids, or, in concert with calmodulin, playing a role in capacitation and in the acrosome reaction in mammalian spermatozoa.

9. CONCLUSION AND SUMMARY

The spermatozoan is a unique cell type with a highly specific function. Spermatogenesis is a complex process of cell differentiation. The studies described in this chapter exemplify such specialization with the synthesis of sperm-specific proteins that result from processes regulated at the level of both transcription and translation. There are examples of genes expressed both pre- and postmeiotically, the storage and utilization of stable messages, coordinate regulation of some events of transcription and of molecular events whose regulation must involve a cascade of mechanisms that has to be precisely coordinated temporally and spatially.

The application of the techniques and principles of molecular biology to problems of spermatogenesis and fertilization should enable investigators to understand more completely the complex cellular interactions involved and more precisely the paracrine–endocrine–cellular relationships that are necessary for successful reproduction. This knowledge may be applied to resolve problems of infertility as well as to perturb the process as an innovative method of fertility control. Furthermore, studies of specific gene expression during spermatogenesis promote the understanding of differential gene expression in the broader context of developmental biology.

ACKNOWLEDGMENTS. I thank Drs. Norman Hecht, Michael O'Rand, and John Herr for providing preprints of their recent publications and my students and research associates for their critical comments on the manuscripts. I am very grateful to Martine Benoit for superb editorial and secretarial assistance.

10. REFERENCES

Adham, I. M., Klemm, U., Maier, W.-M., Hoyer-Fender, S., Tsaousidou, S., and Engel, W., 1989, Molecular cloning of preproacrosin and analysis of its expression pattern in spermatogenesis, *Eur. J. Biochem.* **182:** 563–568.

Alvarez, J. G., and Storey, B. T., 1984, Assessment of cell damage caused by spontaneous lipid peroxidation in rabbit sperm, *Biol. Reprod.* **30:**323–331.

Ausubel, F. M., Brent, R., Kingston, R. E., Moore, D. D., Smith, J. A., Seidman, J. G., and Struhl, K., eds., 1987, *Current Protocols in Molecular Biology*, John Wiley & Sons, New York.

Baba, T., Watanabe, K., Kashiwabara, S.-I., and Arai, Y., 1989, Primary structure of human proacrosin deduced from its cDNA sequence, *FEBS Lett.* **244:**296–300.

Bentley, J. K., and Garbers, D. L., 1986, Retention of the speract receptor by isolated plasma membranes of sea urchin spermatozoa, *Biol. Reprod.* **34:**413–421.

Berger, S. L., and Kimmel, A. R., eds., 1987, *Methods in Enzymology*, Vol. 152, *Guide to Molecular Cloning Techniques*, Academic Press, Orlando, FL.

Bhatnagar, Y. M., Romrell, L. J., and Bellve, A. R., 1985, Biosynthesis of specific histones during meiotic prophase of mouse spermatogenesis, *Biol. Reprod.* **32:**599–609.

Blanco, A., Burgos, C., Gerez de Burgos, N. M., and Montamat, E. E., 1976, Properties of the testicular lactate dehydrogenase isoenzyme, *Biochem. J.* **153:**165–172.

Braun, R. E., Peschon, J. J., Behringer, R. R., Brinster, R. L., and Palmiter, R. D., 1989a, Protamine 3'-untranslated sequences regulate temporal translational control and subcellular localization of growth hormone in spermatids of transgenic mice. *Genes Dev.* **3:**793–802.

Braun, R. E., Behringer, R. R., Peschon, J. J., Brinster, R. L., and Palmiter, R. D., 1989b, Genetically haploid spermatids are phenotypically diploid, *Nature* **337:**373–376.

Burke, D. J., and Ward, S., 1983, Identification of a large multigene family encoding the major sperm protein of *Caenorhabditis elegans*, *J. Mol. Biol.* **171:**1–29.

Busslinger, M., and Barberis, A., 1985, Synthesis of sperm and late histone cDNAs of the sea urchin with a primer complementary to the conserved 3′ terminal palindrome: Evidence for tissue-specific and more general histone gene variants, *Proc. Natl. Acad. Sci. U.S.A.* **82:**5676–5680.

Dan, J. C., 1967, Acrosome reaction and lysins, in: *Fertilization*, Vol. 1 (C. B. Metz, and A. Monroy, eds.), Academic Press, New York, pp. 237–293.

Dangott, L. J., and Garbers, D. L., 1987, Further characterization of a speract receptor on sea urchin spermatozoa, in: *Cell Biology of the Testis and Epididymis*, Volume 513 (M.-C. Orgebin-Crist and B. J. Danzo, eds.), The New York Academy of Sciences, New York, pp. 274–285.

Dangott, L. J., Jordan, J. E., Bellet, R. A., and Garbers, D. L., 1989, Cloning of the mRNA for the protein that crosslinks to the egg peptide speract, *Proc. Natl. Acad. Sci. U.S.A.* **86:**2128–2132.

Dobner, P. R., Kislauskis, E., Wentworth, B. M., and Villa-Komaroff, L., 1987, Alternative 5′ exons either provide or deny an initiator methionine codon to the same α-tubulin coding region, *Nucleic Acids Res.* **15(1):**199–218.

Edwards, Y. H., Povey, S., LeVan, K. M., Driscoll, C. E., Millan, J. L., and Goldberg, E., 1987, Locus determining the human sperm- specific lactate dehydrogenase, *LDHC*, is syntenic with *LDHA*, *Dev. Biol.* **8:**219–232.

Edwards, Y., West, L., Van Heyningen, V., Cowell, J., and Goldberg, E., 1989, Regional localization of the sperm-specific lactate dehydrogenase, *LDHC*, gene on human chromosome 11, *Ann. Hum. Genet.* **53:**215–219.

Erickson, R. P., Friend, D. S., and Tennenbaum, D., 1975, Localization of lactate dehydrogenase-X on the surfaces of mouse spermatozoa, *Exp. Cell Res.* **91:**1–5.

Fujimoto, H., Erickson, R. P., and Toné, S., 1988, Changes in polyadenylation of lactate dehydrogenase-X mRNA during spermatogenesis in mice, *Mol. Reprod. Dev.* **1:**27–34.

Gao, B., Klein, L. E., Britten, R. J., and Davidson, E. H., 1986, Sequence of mRNA coding for bindin, a species-specific sea urchin sperm protein required for fertilization, *Proc. Natl. Acad. Sci. U.S.A.* **83:**8634–8638.

Garbers, D. L., 1989, Molecular basis of fertilization, *Annu. Rev. Biochem.* **58:**719–742.

Gavella, M., and Cvitkovic, P., 1985, Semen LDH-X deficiency and male infertility, *Arch. Androl.* **15:**173–176.

Gold, B., Fujimoto, H., Kramer, J. M., Erickson, R. P., and Hecht, N. B., 1983, Haploid accumulation and translational control of phosphoglycerate kinase-2 messenger RNA during mouse spermatogenesis, *Dev. Biol.* **98:**392–399.

Goldberg, E., 1977, Isozymes in testes and spermatozoa, in: *Isozymes: Current Topics in Biological and Medical Research*, Vol. 1 (M. C. Rattazzi, J. G. Scandalios, and G. S. Whitt, eds.), Alan R. Liss, New York, pp. 79–124.

Goldberg, E., Sberna, D., Wheat, T. E., Urbanski, G. J., and Margoliash, E., 1977, Cytochrome c: Immunofluorescent localization of the testis-specific form, *Science* **196:**1010–1012.

Goldman, D. S., Kiessling, A. A., Millette, C. F., and Cooper, G. M., 1987, Expression of *c-mos* RNA in germ cells of male and female mice, *Proc. Natl. Acad. Sci. U.S.A.* **84:**4509–4513.

Hake, L. E., Alcivar, A. A., and Hecht, N. B., 1990, Changes in mRNA length accompany translational regulation of the somatic and testis-specific cytochrome c genes during spermatogenesis in the mouse, *Development* **110:**249–257.

Hecht, N. B., 1987, Gene expression during spermatogenesis, in: *Cell Biology of the Testis and Epididymis*, Vol. 513 (M.-C. Orgebin-Crist and B. J. Danzo, eds.), The New York Academy of Sciences, New York, pp. 90–101.

Hecht, N. B., 1988, Post-meiotic gene expression during spermatogenesis, in: *Meiotic Inhibition: Molecular Control of Meiosis* (F. P. Haseltine, ed.), Alan R. Liss, New York, pp. 291–313.

Hecht, N. B., Kleene, K. C., Distel, R. J., and Silver, L. M., 1984, The differential expression of the actins and tubulins during spermatogenesis in the mouse, *Exp. Cell Res.* **153:**275–279.

Hecht, N. B., Distel, R. J., Yelick, P. C., Tanhauser, S. M., Driscoll, C. E., Goldberg, E., and Tung, K. S. K., 1988, Localization of a highly divergent mammalian testicular α tubulin that is not detectable in brain, *Mol. Cell. Biol.* **8(2):**996–1000.

Hennig, B., 1975, Change of cytochrome c structure during development of the mouse, *Eur. J. Biochem.* **55:**167–183.

Herr, J. C., Wright, R. M., John, E., Foster, J., and Flickinger, C. J., 1990, Identification of human acrosomal antigen SP-10 in primates and pigs, *Biol. Reprod.* **42:**377–382.

Hintz, M., and Goldberg, E., 1977, Immunohistochemical localization of LDH-X during spermatogenesis in mouse testes, *Dev. Biol.* **57:**375–384.

Holmes, R. S., 1972, Evolution of lactate dehydrogenase genes, *FEBS Lett.* **28:**51–55.

Iatrou, K., and Dixon, G. H., 1977, Messenger RNA sequences in the developing trout testis, *Cell* **10:**433–441.

Kim, E., Waters, S. H., Hake, L. E., and Hecht, N. B., 1989, Identification and developmental expression of a smooth-muscle γ-actin in postmeiotic male germ cells of mice, *Mol. Cell. Biol.* **9(5):**1875–1881.

Kim, I. C., and Nolla, H., 1986, Antigenic analysis of testicular cytochromes c using monoclonal antibodies, *Biochem. Cell Biol.* **64:**1211–1217.

Kim, Y.-J., Hwang, I., Tres, L. L., Kierszenbaum, A. L., and Chae, C.-B., 1987, Molecular cloning and differential expression of somatic and testis-specific H$_2$B histone genes during rat spermatogenesis, *Dev. Biol.* **124:**23–34.

Klass, M. R., Kinsley, S., and Lopez, L. C., 1984, Isolation and characterization of a sperm-specific gene family in the nematode *Caenorhabditis elegans*, *Mol. Cell. Biol.* **4(3):**529–537.

Klass, M., Ammons, D., and Ward, S., 1988, Conservation in the 5′ flanking sequences of transcribed members of the *Caenorhabditis elegans* major sperm protein gene family, *J. Mol. Biol.* **199:**15–22.

Kramer, J. M., 1981, Immunofluorescent localization of PGK-1 and PGK-2 isozymes within specific cells of the mouse testis, *Dev. Biol.* **87:**30–36.

Krawczyk, Z., Wisniewski, J., and Beisiada, E., 1987, A *hsp70*-related gene is constitutively highly expressed in testis of rat and mouse, *Mol. Biol. Rep.* **12:**27–34.

Krawczyk, Z., Mali, P., and Parvinen, M., 1988, Expression of a testis-specific *hsp70* gene-related RNA in defined stages of rat seminiferous epithelium, *J. Cell Biol.* **107:**1317–1323.

Kuhn, R., Schafer, U., and Schafer, M. 1988, *Cis*-acting regions sufficient for spermatocyte-specific transcriptional and spermatid-specific translational control of the *Drosophila melanogaster* gene *mst(3)gl-9*, *EMBO J.* **7:**447–454.

Li, S. S.-L., O'Brien, D. A., Hou, E. W., Versola, J., Rockett, D. L., and Eddy, E. M., 1989, Differential activity and synthesis of lactate dehydrogenase isozymes A (muscle), B (heart), and C (testis) in mouse spermatogenic cells, *Biol. Reprod.* **40:**173–180.

Lieber, T., Weisser, K., and Childs, G., 1986, Analysis of histone gene expression in adult tissues of the sea urchins *Strongylocentrotus purpuratus* and *Lytechinus pictus*: Tissue-specific expression of sperm histone genes, *Mol. Cell. Biol.* **6(7):**2602–2612.

Maniatis, T., Fritsch, E. F., and Sambrook, J., 1982, *Molecular Cloning, A Laboratory Manual*, Cold Spring Harbor Laboratory, Cold Spring Harbor, NY.

Maxson, R. E., and Egrie J. C., 1980, Expression of maternal and paternal histone genes during early cleavage stages of the echinoderm hybrid *Strongylocentrotus purpuratus* × *Lytechinus pictus*, *Dev. Biol.* **74:** 335–342.

McCarrey, J. R., 1987, Nucleotide sequence of the promoter region of a tissue-specific human retroposon: Comparison with its housekeeping progenitor, *Gene* **61:**291–298.

McCarrey, J. R., and Thomas, K., 1987, Human testis-specific PGK gene lacks introns and possesses characteristics of a processed gene, *Nature* **326:**501–505.

Meistrich, M. L., Trostle, P. K., Frapart, M., and Erickson, R. P., 1977, Biosynthesis and localization of lactate dehydrogenase X in pachytene spermatocytes and spermatids of mouse testis, *Dev. Biol.* **60:**428–441.

Michiels, F., Gasch, A., Kaltschmidt, B., and Renkawitz-Pohl, R., 1989, A 14 bp promoter element directs the testis specificity of the *Drosophila* β$_2$ tubulin gene, *EMBO J.* **8:**1559–1565.

Millan, J. L., Driscoll, C. E., LeVan, K. M., and Goldberg, E., 1987, Epitopes of human testis-specific lactate dehydrogenase deduced from a cDNA sequence, *Proc. Natl. Acad. Sci. U.S.A.* **84:**5311–5315.

Morales, C. R., Alcivar, A. A., Hecht, N. B., and Griswold, M. D., 1989, Specific mRNAs in Sertoli and germinal cells of testes from stage synchronized rats, *Mol. Endocrinol.* **3:**725–733.

Mutter, G. L., and Wolgemuth, D. J., 1987, Distinct developmental patterns of *c-mos* protooncogene expression in female and male mouse germ cells, *Proc. Natl. Acad. Sci. U.S.A.* **84:**5301–5305.

Newrock, K. J., Alfgame, C. R., Nardi, R. V., and Cohen, L. H., 1977, Histone changes during chromatin remodeling in embryogenesis, *Cold Spring Harbor Symp. Quant. Biol.* **42:**421–431.

O'Brien, D. A., and Bellve, A. R., 1980, Protein constituents of the mouse spermatozoon, *Dev. Biol.* **75:**405–418.

Olds-Clarke, P., 1988, Genetic analysis of sperm function in fertilization, *Gamete Res.* **20:**241–264.

Olivieri, G., and Olivieri, A., 1965, Autoradiographic study of nucleic acid synthesis during spermatogenesis in *Drosophila melanogaster*, *Mutat. Res.* **2:**366–380.

O'Rand, M. G., and Widgen, E. E., 1990, Molecular biology of a sperm antigen: Identification of the sequence of an autoantigenic epitope, in: *Reproductive Immunology 1989* (L. Mettler and W. D. Billington, eds.), Elsevier Scientific Publishers, Amsterdam, pp. 61–67.

O'Rand, M. G., Widgren, E. E., and Fisher, S. J., 1988, Characterization of the rabbit sperm membrane autoantigen, RSA, as a lectin-like zona binding protein, *Dev. Biol.* **129:**231–240.

Osipova, T. N., Triebel, H., Bär, H., Zalenskaya, I. A., and Hartmann, M., 1985, Interaction of histone H1 from sea urchin sperm with superhelical and relaxed DNA, *Mol. Biol. Rep.* **10:**153–158.

Poccia, D., Salik, J., and Krystal, G., 1981 Transitions in histone variants of the male pronucleus following fertilization and evidence for a maternal store of cleavage-stage histones in the sea urchin egg, *Dev. Biol.* **82:**287–296.

Ponzetto, C., and Wolgemuth, D. J., 1985, Haploid expression of a unique *c-abl* transcript in the mouse male germ line, *Mol. Cell Biol.* **5:**1791–1794.

Propst, F., Rosenberg, M. P., Oskarsson, M. K., Russell, L. B., Nguyen-Huu, M. C., Nadeau, J., Jenkins, N. A., Copeland, N. G., and Vande Woude, G. F., 1988, Genetic analysis and developmental regulation of testis-specific RNA expression of *Mos*, *Abl*, actin and *Hox-1.4*, *Oncogene* **2:**227–233.

Sakai, I., Sharief, F. S., and Li, S. S.-L., 1987, Molecular cloning and nucleotide sequence of the cDNA for sperm-specific lactate dehydrogenase-C from mouse, *Biochem. J.* **242:**619–622.

Shackleford, G. M., and Varmus, H. E., 1987, Expression of the proto-oncogene *int-1* is restricted to postmeiotic male germ cells and the neural tube of mid-gestational embryos, *Cell* **50:**89–95.

Shimomura, H., Dangott, L. J., and Garbers, D. L., 1986, Covalent coupling of a resact analogue to guanylate cyclase, *J. Biol. Chem.* **261:**15778–15782.

Singh, S., Lowe, D. G., Thorpe, D. S., Rodriguez, H., Kuang, W.-J., Dangott, L. J., Chinkers, M., Goeddel, D. V., and Garbers, L. D., 1988, Membrane guanylate cyclase is a cell-surface receptor with homology to protein kinases, *Nature* **334:**708–712.

Slaughter, G. R., Meistrich, M. L., and Means, A. R., 1989, Expression of RNAs for calmodulin, actins, and tubulins in rat testis cells, *Biol. Reprod.* **40:**395–405.

Storey, B. T., and Kayne, F. J., 1977, Energy metabolism of spermatozoa. VI. Direct intramitochondrial lactate oxidation by rabbit sperm mitochondria, *Biol. Reprod.* **16:**549–556.

Taira, M., Iizasa, T., Yamada, K., Shimada, H., and Tatibana, M., 1989, Tissue-differential expression of two distinct genes for phosphoribosyl pyrophosphate synthetase and existence of the testis-specific transcript, *Biochim. Biophys. Acta* **1007:**203–208.

Urch, U. A., 1986, The action of acrosin on the zona pellucida, in: *The Molecular and Cellular Biology of Fertilization*, Vol. 207 (J. L. Hedrick, ed.), Plenum Press, New York, pp. 113–132.

Vacquier, V. D., 1980, The adhesion of sperm to sea urchin eggs, in: *The Cell Surface, Mediator of Developmental Processes* (N. K. Wessells, ed.), Academic Press, New York, pp. 151–168.

VandeBerg, J. L., Lee, C.-Y., and Goldberg, E., 1981, Immunohistochemical localization of phosphoglycerate kinase isozymes in mouse testes, *J. Exp. Zool.* **217:**435–441.

Virbasius, J. V., and Scarpulla, R. C., 1988, Structure and expression of rodent genes encoding the testis-specific cytochrome c, *J. Biol. Chem.* **263:**6791–6796.

Vitelli, L., Kemler, I., Lauber, B., Birnstiel, M., and Busslinger, M., 1988, Developmental regulation of micro-injected histone genes in sea urchin embryos, *Dev. Biol.* **127:**54–63.

Ward, S., Burke, D. J., Sulston, J. E., Coulson, A. R., Albertson, D. G., Ammons, D., Klass, M., and Hogan, E., 1988, Genomic organization of major sperm protein genes and pseudogenes in the nematode *Caenorhabditis elegans*, *J. Mol. Biol.* **199:**1–13.

Wheat, T. E., and Goldberg, E., 1983, Sperm-specific lactate dehydrogenase C_4: Antigenic structure and immunosuppression of fertility, in: *Isozymes: Current Topics in Biological and Medical Research*, Vol. 7 (M. C. Rattazzi, J. G. Scandalios, and G. S. Whitt, eds.), Alan R. Liss, New York, pp. 113–130.

Wieben, E. D., 1981, Regulation of the synthesis of lactate dehydrogenase-X during spermatogenesis in the mouse, *J. Cell. Biol.* **88:**492–498.

Wolgemuth, D. J., Engelmyer, E., Duggal, R. N., Gizang-Ginsberg, E., Mutter, G. L., Ponzetto, C., Viviano, C., and Zakeri, Z. F., 1986, Isolation of a mouse cDNA coding for a developmentally regulated, testis-specific transcript containing homeo box homology, *EMBO J.* **5:**1229–1235.

Wright, R. M., John E., Klotz, K., Flickinger, C. J., and Herr, J. C., 1990, Cloning and sequencing of cDNAs coding for the human intra-acrosomal antigen SP-10, *Biol. Reprod.* **42:**693–701.

Zakeri, Z. F., and Wolgemuth, D. J., 1987, Developmental-stage-specific expression of the *hsp70* gene family during differentiation of the mammalian male germ line, *Mol. Cell. Biol.* **7:**1791–1796.

23

Use of Molecular Biology to Study Development and Function of Mammalian Oocytes

Debra J. Wolgemuth

1. INTRODUCTION

Elucidating the genetic program that controls the differentiation of mammalian gametes has been complicated by various factors, including the small size, small numbers, and relative inaccessibility of the cells. In this chapter, I review recent advances in molecular biological approaches and methods that now make this task feasible. Although particular emphasis is placed on analysis at the level of the genes and their mRNAs, analysis of proteins is equally important. Rather than providing a catalogue of the temporal expression of genes during oogenesis, this review illustrates how different approaches have been used to elucidate the pattern of expression of a few of the better-characterized genes.

2. GENERAL STRATEGIES TO IDENTIFY GENES EXPRESSED DURING MAMMALIAN OOGENESIS

Several strategies can be used to identify and characterize specific genes involved with oogenesis. Specific examples are cited that are relevant to oocyte development in particular, and the following approaches have been applied to many developmental systems. (1) Examine the

DEBRA J. WOLGEMUTH • Department of Genetics and Development, and Center for Reproductive Sciences, Columbia University College of Physicians and Surgeons, New York, New York 10032.

A Comparative Overview of Mammalian Fertilization, edited by Bonnie S. Dunbar and Michael G. O'Rand. Plenum Press, New York, 1991.

expression of genes that are transcribed in other cells and tissues but that exhibit unique patterns of expression in developing germ cells. (2) Identify genes expressed only in germ cells, using two general approaches: (a) subtractive hybridization of cDNA libraries made from RNA of purified populations of cells at specific stages of differentiation and (b) enhancer-trap or gene-trap cloning approaches in which the expression of reporter genes integrated into the genome is monitored in the gonads of transgenic mice. (3) Utilize conserved sequences from genes with known development- or differentiation-regulating function in other systems to identify genes with potential regulatory functions during oogenesis.

Each of the above approaches has certain advantages and inherent limitations. The first strategy usually permits an immediate assessment of the pattern of expression as well as the formulation of hypotheses as to potential function, since the genes in question have already been identified. Thus, the tools necessary for descriptive studies and experimental manipulation of gene expression are in place. Genes within this category relating to growth and differentiation of oocytes would include those genes with potential function as protooncogenes, genes that function as growth factors and their receptors, and genes that are involved with regulating the cell cycle. For example, the protooncogene *c-mos* has been shown to be expressed in a developmentally regulated pattern in ovarian oocytes (Mutter and Wolgemuth, 1987; Goldman *et al.*, 1987). Transcripts from several growth factors, including members of the transforming growth factor β (TGFB) family, have been detected in ovarian and ovulated oocytes and early embryos (Rappolee *et al.*, 1988). Studies on expression of cell cycle genes, such as the cyclins, in mammalian oocytes have not been reported to date. However, cyclin gene expression and function have been studied in oocytes from clams (Swenson *et al.*, 1986), starfish (Labbe *et al.*, 1988), frogs (Gautier *et al.*, 1988), sea urchins (Meijer *et al.*, 1989), and flies (Whitfield *et al.*, 1989), and mammalian cyclin genes have been identified (Pines and Hunter, 1989).

The second approach provides the opportunity to identify new genes important in oocyte differentiation. The production of the requisite cDNA libraries from mammalian oocyte mRNA for this strategy is complicated by the difficulty in obtaining sufficient quantities of starting material. Technological advances to overcome this limitation are discussed below. Various applications of the subtractive hybridization strategies currently are being used to identify genes expressed uniquely in specific spermatogenic cell types (e.g., Kleene *et al.*, 1985; Thomas *et al.*, 1989). Genes thus identified, such as the spermatid-specific gene protamine 1, can then be used to manipulate the expression of heterologous genes in male germ cells of transgenic mice (Peschon *et al.*, 1987). Oogenic stage-specific genes might prove similarly useful for manipulating expression of specific genes during oogenesis in transgenic systems.

An alternative approach for identifying new genes, that of enhancer or gene trap cloning, is based on a technique developed in the *Drosophila* system (O'Kane and Gehring, 1987). Essentially, a construct is produced in which a weak promoter (enhancer trap) or no promoter at all (gene trap) is linked to a reporter gene such as the bacterial gene *lacZ*, which encodes β-galactosidase (β-gal). These constructs are then introduced into the genome of flies by P-element transformation. If the construct integrates in the vicinity of an enhancer or promoter, it will be expressed in the progeny that carry the construct. Such progeny are screened for developmentally regulated patterns of expression (particularly within the germ line) of β-gal, which can be detected by histochemical staining with X-gal. Gehring and colleagues (Grossniklaus *et al.*, 1989) have used this approach to screen specifically for genes important in *Drosophila* oogenesis, underscoring the usefulness of this technique even in systems more amenable to classical genetic screening approaches. Although screening is more rapid in a system such as *Drosophila* (Bellen *et al.*, 1989), the approach has been used to identify developmentally regulated genes in mice as well (Gossler *et al.*, 1989). Unlike subtractive hybridization, this technique may be useful in identifying genes expressed at low levels that may have critical regulatory functions. The genes thus identified can then be cloned because they are now "tagged" by the presence of the *lacZ* sequences.

The third strategy also owes its inception to initial studies in *Drosophila*. This approach utilizes genes with known development-regulating function in simpler systems to screen for related genes in more complex organisms. Classical genetic analysis had identified a class of *Drosophila* genes that exhibited major regulatory effects on development, as assessed by the occurrence of homeotic mutations (Lewis, 1978). Molecular analysis of these homeotic genes revealed that many of them contained a region of DNA with similar sequences (McGinnis *et al.*, 1984a; Scott and Weiner, 1984). This domain, termed the homeobox, subsequently was shown to be conserved in the genomes of a variety of higher eukaryotes, including mammals (McGinnis *et al.*, 1984b). This raised the exciting possibility that genes containing this domain might have development regulating functions during mammalian development and differentiation as well. About 30 such genes have been identified in the mouse genome, and their pattern of expression has been studied extensively during embryonic and adult development (reviewed by Holland and Hogan, 1988). The exquisite specificity of expression of several of these genes in particular cell lineages (e.g., Wolgemuth *et al.*, 1986, 1987) and in spatially and temporally unique patterns during embryogenesis suggests that they might play an important role during development and differentiation (reviewed by Holland Hogan, 1988). Although no homeobox gene has been shown to exhibit specific patterns of expression in mammalian oocytes, a homeobox-containing frog gene appears to give rise to maternal mRNAs (Muller *et al.*, 1984), as does the *Drosophila* gene *caudal* (Mlodzik *et al.*, 1985; Mlodzik and Gehring, 1987).

Other development-regulating genes, exemplified by the *Drosophila* gene *kruppel*, code for proteins with zinc finger domains (Rosenberg *et al.*, 1986; reviewed by Evans and Hollenberg, 1988). The *kruppel* gene was used to isolate mouse genes with zinc finger domains (e.g., Chowdhury *et al.*, 1987; Chavrier *et al.*, 1988), which also exhibit temporally and spatially restricted patterns of expression during embryogenesis. One of these genes, originally named *mkr5*, may be of interest with respect to ovarian and possibly oocyte differentiation, since it is expressed in the adult ovary (Chowdhury *et al.*, 1988).

Zinc finger-containing genes have received considerable attention with respect to gonad (testis) differentiation since it was observed that Y-linked genes putatively involved with sex determination in man (*ZFY*) and in mouse (*Zfy*) contained zinc fingers (Page *et al.*, 1987). Although recent evidence suggests that *ZFY* (*Zfy*) is not likely the sex-determining gene (Palmer *et al.*, 1989; Koopman *et al.*, 1989), the pattern of expression of this and other structurally related genes suggests a role in male germ cell differentiation (Nagamine *et al.*, 1989; Koopman *et al.*, 1989).

Both homeobox and zinc finger domains are believed to be involved in DNA binding, suggesting that genes containing these motifs may function as transcription factors. Thus, these genes may regulate development as well as being developmentally regulated. Other putative DNA-binding motifs, such as the "leucine zipper" (Landschultz *et al.*, 1988), the basic "helix–loop–helix" motif (Davis *et al.*, 1987), or the "POU-domain" (Herr *et al.*, 1988), also appear to be characteristic of development-regulating genes. The POU domain, which consists of homeobox-related and POU-specific subdomains, has been identified in genes clearly implicated in transcriptional control as well as in genes important in cell lineage and cell differentiation (Herr *et al.*, 1988; He *et al.*, 1989). The POU domain is found in genes belonging to the OCT class of DNA-binding proteins. These proteins interact with defined octameric DNA sequences upstream of promoter start sites to convey tissue and developmental specificity of expression. The helix–loop–helix motif may also play a role in development, as exemplified by its presence in *MyoD1* and myogenin, genes apparently involved in muscle differentiation (Davis *et al.*, 1987; Schafer *et al.*, 1990).

3. PARTICULAR PROBLEMS OF MOLECULAR ANALYSIS OF MAMMALIAN OOCYTE GENE EXPRESSION AND RECENT ADVANCES TO CIRCUMVENT THESE DIFFICULTIES

3.1. The Biology of Mammalian Oogenesis Complicates Studying Gene Expression

Studying gene expression throughout oogenesis is made difficult by several aspects of this developmental program. The oocytes of virtually all mammals enter meiosis during embryonic development. This results in the lack of a renewing stem cell population and thus a continually diminishing pool of cells in the adult animal. The midgestation mouse embryo is not particularly amenable to experimental manipulation, making access to fetal ovaries, which contain oocytes in meiotic prophase, difficult. In addition, the absolute number of oocytes that can be obtained (at any stage) is limited. In the embryonic stages, there are hundreds of thousands of oocytes, but it is difficult to purify these cells away from the surrounding somatic cells. As development ensues, waves of atresia result in the loss of cells to the point that the female mouse is born with at most ~12,000 oocytes (Peters, 1969).

At about the time of pubertal development, a small number of cells enter into the growth phase, during which the cells increase in diameter and mass, thereby increasing the amount of RNA available for analysis several hundredfold. However, only a very small number of cells (<50) enter this stage at any one time, and again, the cells in this stage are difficult to isolate away from the surrounding follicular cells. Such manipulation, at least in the mouse, is best undertaken by hand and represents a tedious task. Use of superovulation can increase slightly the number of cells that enter the preovulatory phase. Ovulated eggs can be isolated in relatively pure (free of somatic granulosa cells) form, but the cells are harvested individually. The same limitations apply to the early, preimplantation embryo. The embryos can be obtained, but the amount of RNA available for analysis frequently is in the nanogram to microgram range and represents the pooled samples of hundreds of female mice. Thus, the mass of starting material is small, and the possibility for variation among developmental stages of different embryos from different females is high.

These limitations in obtaining female mammalian gametes are in marked contrast to the situation in working with male gametes. Sufficient numbers of meiotic and postmeiotic cells at various stages of spermatogenesis can be obtained in enriched populations (Bellve *et al.*, 1977; Wolgemuth *et al.*, 1985) such that RNA and proteins from these cells can be analyzed at the molecular level. The most obvious limitation in the male system is the inability to manipulate various meiotic stages *in vitro*. Furthermore, the small size of the postmeiotic male germ cell (<10 μm in diameter) in contrast with that of the oocyte (>70 μm in diameter) makes microinjection and other forms of *in vitro* manipulation beyond the technical expertise of all but the most highly skilled.

So is it impossible to study gene expression in female mammalian germ cells? Clearly not, as the following discussion illustrates.

3.2. Recent Technological Advances that Facilitate Molecular Analysis

3.2.1. *IN SITU* HYBRIDIZATION ANALYSIS

One of the most powerful approaches has involved the use of *in situ* hybridization. *In situ* hybridization permits the localization of specific transcripts to precise cell types and, as has been shown for frog oocytes to a precise subcellular distribution (Melton, 1987). However, the technique is only relatively quantitative (see discussion by Mutter *et al.*, 1988), and the sensitivity is limited to about 50 copies of a given mRNA per cell (D. Wolgemuth and R. Palmiter, unpublished observations). As an example of how *in situ* hybridization has been used to study oocyte gene expression, we describe results from our laboratory as well as studies from other investigators on the expression of the *c-mos* protooncogene.

The *c-mos* protooncogene encodes a cytoplasmic serine/threonine protein kinase that, in the mouse, is expressed most abundantly in adult gonads and at very low levels in adult epididymis and other somatic tissues (Mutter and Wolgemuth, 1987; Goldman *et al.*, 1987; Propst *et al.*, 1987). Very low levels of *c-mos* are also expressed in day-19 embryos (Propst *et al.*, 1987). In testis, *c-mos* is expressed as a 1.7-kb transcript almost exclusively in germ cells and most abundantly in early spermatids. The ovarian *c-mos* transcript is 1.4 kb in size. The tissue-specific variations in the size of *c-mos* transcripts appear to arise from the use of different promoters (Propst *et al.*, 1987). Such size, and thus presumably structural, differences in transcripts cannot be detected by *in situ* hybridization analysis.

Murine ovarian *c-mos* mRNA levels exceed testicular levels severalfold (Goldman *et al.*, 1987; Mutter and Wolgemuth, 1987). Since the adult ovary is comprised primarily of somatic cells, whereas the adult testis is primarily germ cells, we reasoned that the most abundant cellular compartment of the adult ovary would be the source of the abundant *c-mos* mRNAs.

To test this hypothesis, we hybridized [^{35}S]-labeled RNA probes synthesized in antisense (experimental) and sense (control) orientations to histological sections of mouse ovaries. Much to our surprise, we observed that *c-mos* transcripts were most abundant in fully grown oocytes and were not present, within the limits of sensitivity of detection, in the somatic cells (Mutter and Wolgemuth, 1987). The *c-mos* transcripts began to accumulate in the cytoplasm of growing oocytes, increasing 36- to 90-fold as the oocyte increased in volume (Mutter and Wolgemuth, 1987; Goldman *et al.*, 1987). Northern blot and *in situ* hybridization analyses indicated an absence of *c-mos* RNA in primordial or resting oocytes (Mutter and Wolgemuth, 1987; Keshet *et al.*, 1987).

In murine oocytes, germinal vesicle breakdown (GVBD) occurs within 4 hr of the resumption of meiosis, and meiosis I is complete by 6–8 hr. The mature oocyte will arrest again at metaphase II until fertilization, at which time miosis II is completed. The levels of *c-mos* transcripts appear to be stable as the oocytes resume meiosis but then drop precipitously at the metaphase II stage (Mutter *et al.*, 1988). No *c-mos* transcripts were detected in two-cell or early preimplantation embryos by *in situ* hybridization analysis (Mutter *et al.*, 1988).

Recently, mouse oocytes undergoing meiotic maturation have been shown to contain the *c-mos* protein product, pp39*mos*, whereas growing oocytes did not (Paules *et al.*, 1989). Thus, although *c-mos* transcripts accumulate in the growing oocyte, they are not translated until later, at maturation. This observation calls attention to the fact that simply observing the presence of mRNAs at a specific stage of differentiation does not necessarily mean that the gene product is functioning at this same stage. Indeed, translational regulation is believed to be a major regulatory mechanism in controlling gene expression during both gamete development and early embryogenesis.

3.2.2. APPLICATION OF PCR TECHNOLOGY

The polymerase chain reaction (PCR) was recently selected as the major scientific development of 1989, and the DNA polymerase molecule that drives this reaction was voted "the molecule of the year" (Guyer and Koshland, 1989). It is doubtful that the editors of *Science* magazine were focused on the application of this technology to the question of detecting the presence of specific mRNAs in mammalian oocytes, but it is exactly this technology that makes this task feasible. The reader is referred to several recent papers on this subject for detailed discussions and further applications (Erlich *et al.*, 1988; Higuchi *et al.*, 1988; Rappolee *et al.*, 1989; Williams, 1989).

Briefly, PCR involves the sequential amplification of DNA molecules by a cycle of priming with specific oligonucleotides and elongation of the strand with DNA polymerase, followed by melting of the resulting duplex and repriming and elongation on both the parent and daughter strands. This permits a millionfold amplification of the starting or "target" sequence in a matter of hours. Practical application of this procedure depended on the use of a heat-resistant form of the DNA polymerase (isolated from bacteria that live in hot springs!), such that repeated cycles of hybridization, elongation, and melting could be performed in the same reaction vial. The application of this procedure to analyzing gene expression in extremely small samples, even in single cells, first requires the production of a DNA copy of the mRNA with reverse transcriptase, followed by sequential amplification of the DNAs. This method has been used to measure the expression of several growth factor genes and protooncogenes in oocytes and early embryos (Rappolee *et al.*, 1988).

A limitation in the application of PCR to studying gene expression lies in its quantifiability. It has been very difficult to draw quantitative conclusions about the levels of specific transcripts because it is difficult to control precisely the yield of the amplification reaction, which for most applications is not critical. One can readily observe the presence or absence of a template, but the ability to quantify the amount of starting template is less precise (Rappolee *et al.*, 1989). Clearly, changes in the levels of mRNAs can greatly influence the potential function of the resulting gene product. Significant changes in the levels of relatively abundant transcripts such as actin, tubulin, and histones have been carefully documented in mouse oocytes and early embryos (reviewed by Bachvarova, 1985). Although the relevance of these changes to specific biological function has not been demonstrated directly, it is likely that quantitative as well as qualitative variation will be important.

This PCR technology may also be used to produce cDNA libraries from extremely small samples, thus providing the reagents for subtractive hybridization studies to identify new genes involved with particular stages of oocyte differentiation. Several strategies have been developed for production of cDNA libraries from small samples or that would enable the detection of low-abundance cDNAs, including "anchored" PCR cloning (Loh *et al.*, 1989) and "rapid amplification of cDNA ends" (RACE) (Frohman *et al.*, 1988), to name but two. Both methods again depend on the use of reverse transcriptase in the first step of the procedure, which may be rate limiting. Several laboratories are currently generating cDNA libraries from oocytes, ova, and various stages of early embryos, although the use of these libraries has not yet been reported.

4. APPROACHES TO IDENTIFY FUNCTION

4.1. Introduction to Genetic Strategies

Thus far we have been discussing strategies to identify genes potentially important in oocyte development. How does one make the next step—to demonstrating function? Genetic approaches have proved very powerful in elucidating the control of development in many systems. Several of these strategies can be used in studying gene function during oogenesis. One approach that could be used in the mammal involves altering the expression of genes during germ cell differentiation in a negative or positive manner in transgenic mice. Negative interference, accomplished by mutating the endogenous gene via homologous recombination or by overproducing an antisense orientation of the gene, would produce a loss-of-function mutation. Positive interference could be accomplished by ectopic or elevated levels of expression of a fully functional gene under the direction of the endogenous or a heterologous promoter, yielding a gain of function phenotype.

4.2. Negative Interference

4.2.1. HOMOLOGOUS RECOMBINATION IN MAMMALIAN CELLS

The classical approach of generating mutations involves disrupting the endogenous gene and examining the resulting phenotype in the heterozygous or homozygous state. In mammalian cells, exogenous DNA usually integrates at random. The frequency of integration of specific sequences (such as a gene) at the homologous region in the genome, which could result in disruption of the endogenous gene, was estimated to be too low for screening for such insertions in transgenic mice. Although studies on the specific properties of the inserted sequences that increase the probability of integration at the homologous site have been useful, it has been the development of targeted recombination in embryonic stem (ES) cells (Robertson et al., 1986; Thomas and Capecchi, 1987; discussed by Frohman and Martin, 1989) that has made gene disruption in mice a realistic experimental approach.

Embryonic stem cells can be established as pluripotent cell lines in vitro. Exogenous DNA containing a selectable marker such as the NEO gene can then be introduced in these cells by methods such as electroporation or microinjection. Cells that have integrated the foreign DNA are selected by their ability to express neomycin phosphotransferase. These cells are then screened, usually in pools, for the site of integration by PCR amplification of fragments primed with oligonucleotides designed such that specific fragments are produced only when the exogenous DNA has integrated properly (into the homologous endogenous gene). The clone of cells yielding this pattern is isolated from the pool of cells and used to generate chimeric embryos. Chimeras are produced by introducing ES cells into host blastocysts and producing liveborn animals. If some of the ES cells have contributed to the germ line, the modified gene can subsequently be transmitted, the animals bred to homozygosity, and the consequence of gene disruption can be assessed at the functional level.

Since many of the genes potentially involved in oocyte development might also be expressed in various embryonic and adult tissues, completely disrupting the function of these genes might result in embryonic or neonatal lethality. Recently, Schwartzberg et al. (1989) reported the disruption of the c-abl protooncogene and the generation of mice carrying the disrupted gene in the germ line. Given the previous studies on the expression of novel forms of the c-abl protooncogene in the male germ line (Ponzetto and Wolgemuth, 1985; Ponzetto et al., 1989), it will be most interesting to examine the resulting phenotype in these animals, especially with respect to reproductive functions. The fact that the heterozygous animals are fertile implies at the very least that mice need only one functional c-abl gene in the germ line.

4.2.2. ANTISENSE OLIGONUCLEOTIDES

It is possible to disrupt the function of a gene in a dominant fashion by direct introduction of sequences that will interfere with the processing or translation of the mRNA (Weintraub et al., 1985). These sequences may be short, in the range of synthetically obtained deoxyoligo-nucleotides (oligos). This approach has been particularly useful in in vitro systems but may have applications in vivo as well. For example, the function of c-mos in murine oocytes has been assessed by injection of c-mos antisense oligos (O'Keefe et al., 1989; Paules et al., 1989). Oocytes injected with c-mos oligos failed to arrest at metaphase II and often cleaved into two cells (O'Keefe et al., 1989). It was concluded that loss of c-mos function leads to premature reformation of a nucleus, essentially permitting ensuing DNA synthesis and cell cleavage. Although a second set of studies revealed a slightly different result, namely, arrest at the metaphase I stage (Paules *et al.*, 1989), both results implied that the *c-mos* product is needed for oocyte maturation.

4.2.3. ANTISENSE RNA

Antisense RNA itself can also be used to interfere with expression and hence to assess function. This approach was first used in developing vertebrate cells to examine the expression of exogenous genes in frog oocytes (Melton, 1985). Injected globin RNA in an antisense orientation formed a hybrid with globin mRNA and blocked its translation. Interference with mRNA translation was also observed in experiments with mouse oocytes and antisense RNAs (Strickland *et al.*, 1988). Growing mouse oocytes accumulate mRNA for tissue plasminogen activator (t-PA). Injection of maturing oocytes with different antisense RNAs complementary to both coding and noncoding portions of t-PA mRNA blocked t-PA protein synthesis. Injection of antisense RNA complementary to as little as 103 nt at the 3' end of the mRNA was sufficient to block translation, consistent with the observations using oligos described above. Although antisense approaches (using oligos or RNA) appear to be successful in oocytes, their effectiveness in eggs and early embryos, at least in the frog, may be complicated by the presence of an unwinding activity that modifies and destabilizes double-stranded RNA substrates (Bass and Weintraub, 1988).

4.2.4. ANTIBODIES

It is possible to interfere directly with the function of the protein product of a particular gene by introducing specific antibodies. For example, antibodies against the frog homeobox protein XIHbox 1 were injected into early embryos and caused the deletion of certain neural crest derivatives, including the dorsal fin and dorsal root ganglia (Wright *et al.*, 1989). Technological improvements in the microinjection system subsequently permitted introducing antibodies into the one-cell frog embryo, which improved the distribution of the antibodies into all the embryonic compartments. Depending on the homeobox protein being inhibited, specific abnormalities were observed in various regions of the spinal cord (Wright *et al.*, 1989). There are several problems with this approach, including the fact that antibodies cannot be repeatedly microinjected, and thus the antibody must be present at specific developmental windows in order to elicit an effect. Introduction of the antibody also requires *in vitro* manipulation, which may limit assessment of effects in subsequent developmental stages.

4.2.5. PRODUCTION OF MUTANT PROTEINS

The production of proteins in specific stages or in specific cell types is relatively straightforward to achieve (at least conceptually). As discussed previously, exogenous gene products can be targeted to particular cells by using tissue- or cell-type-specific promoters. The mutant protein approach would involve introducing an altered gene instead of the normal gene. Cells expressing the transgene would thus produce abnormal proteins. This approach would be most likely to generate an abnormal phenotype if the protein is in limited supply or if it is part of a larger complex of proteins.

4.3. Positive or Gain-of-Function Mutations

4.3.1. *IN VITRO* MANIPULATIONS

Gain-of-function mutations could result from the expression of a gene in an inappropriate cell type or at elevated levels in the correct cell type or stage of development. For example, microinjection of cytoplasm from mature oocytes into blastomeres of two-cell embryos induces cleavage arrest at metaphase, apparently because of a cytostatic factor (CSF) (Masui and Markert, 1971). When 5'-capped *c-mos* RNA (which directs the synthesis of pp39^{mos}) was injected into one of the blastomeres of the two-cell embryo, all of the blastomeres ceased to cleave within 30 to 90 min after injection (Sagata *et al.*, 1989a,b; Watanabe *et al.*, 1989). Injection of *c-mos* RNA preannealed with antisense oligos did not induce cleavage arrest, nor did *c-mos* RNA lacking the ATP binding domain. It has been suggested that the *c-mos* product may be the CSF responsible for arrest of metaphase II oocytes. Fertilization may generate a signal to degrade the *c-mos* product, thereby removing the block to division of the zygote.

Another example is illustrated in the studies on the function of the *Xenopus* homeobox gene *Xhox3* (Ruiz i Altaba and Melton, 1989). *Xhox3* shows a graded expression in the axial mesoderm of frog embryos, with highest expression in the posterior end of the embryo at the gastrula and neurula stages. The function of *Xhox3* was assessed by injecting capped synthetic *Xhox3* mRNA into different regions of developing embryos. Excess *Xhox3* mRNA in anterior regions of the embryo, which usually lack *Xhox3* products, produced a series of axial defects. The embryos gastrulated normally but failed to form head structures.

4.3.2. *IN VIVO* EXPERIMENTS

The usefulness of producing gain-of-function mutations is demonstrated by recent experiments in which we manipulated the expression of the homeobox gene *Hox-1.4* in transgenic mice (Wolgemuth *et al.*, 1989). Elevated levels of *Hox-1.4* in the mesodermal germ layer of midgestation embryos correlates with the developmental of congenital megacolon in neonatal and adult animals, apparently as a result of abnormal differentiation of the enteric nervous system. Whether the abnormality arises in the neural crest cells, which differentiate into the enteric ganglia, or the microenvironment of the gut into which the cells migrate remains to be determined.

The experiment nonetheless underscores the usefulness of manipulating the expression of particular genes in a specific tissue or cell type. For example, there are high levels of uniquely sized *c-mos* transcripts in growing oocytes. It therefore would be of interest to use an oocyte-specific promoter to elevate levels of the *c-mos* protein product at the appropriate stages or to produce such proteins at a developmental stage in which they are not expressed normally. The former experiment might be accomplished by using the promoter for a zona pellucida gene to drive expression (Chamberlin and Dean, 1989), whereas the latter experiment awaits the identification of genes expressed earlier in oogenesis but uniquely in the female germ line.

Such experiments depend on identifying the genes involved in gametogenesis, doubtless by using the molecular approaches that were outlined at the beginning of this chapter.

ACKNOWLEDGMENTS. I would like to thank Dr. Martin A. Winer for his critical reading of this manuscript. This work was supported in part by grants from the NIH (HD 05077, HD18122), the American Cancer Society, and NASA (NAGW-2-385).

5. REFERENCES

Bachvarova, R., 1985, Gene expression during oogenesis and oocyte development in mammals, in: *Developmental Biology*, Vol. 1 (L. W. Browder, ed.), Plenum Press, New York, pp. 453–524.

Bass, B. L., and Weintraub, H., 1988, An unwinding activity that covalently modifies its double-stranded RNA substrate, *Cell* **55:**1089–1098.

Bellen, H. J., O'Kane, C. J., Wilson, C., Grossniklaus, U., Pearson, R. K., and Gehring, W. J., 1989, P-element mediated enhancer detection: A versatile method to study development in *Drosophila*, *Genes Dev.* **3:** 1288–1300.

Bellve, A. R., Cavicchia, J. C., Millette, C. F., O'Brien, D. A., Bhatnagar, Y. M., and Dym, M., 1977, Spermatogenic cells of the prepuberal mouse. Isolation and morphological characterization, *J. Cell Biol.* **74:**68–85.

Chamberlin, M. E., and Dean, J., 1989, Genomic organization of a sex specific gene: The primary sperm receptor of the mouse zona pellucida, *Dev. Biol.* **131:**207–214.

Chavrier, P., Lemaire, P., Relevant, O., Bravo, R., and Charnay, P., 1988, Characterization of a mouse multigene family that encodes zinc finger structures, *Mol. Cell. Biol.* **8:**1319–1326.

Chowdhury, K., Deutsch, U., and Gruss, P., 1987, A multigene family encoding several "finger" structures is present and differentially active in mammalian genomes, *Cell* **48:**771–778.

Chowdhury, K., Rohdewohld, H., Gruss, P., 1988, Specific and ubiquitous expression of different zinc finger protein genes in mouse, *Nuc. Acids Res.* **16:**9995–10011.

Davis, R. L., Weintraub, H. A., and Lassar, A. B., 1987, Expression of a single transfected cDNA converts fibroblasts to myoblasts, *Cell* **51:**987–1000.

Erlich, H. A., Gelfand, D. H., and Saiki, R. K., 1988, Specific DNA amplification, *Science* **331:**461–462.

Evans, R., and Hollenberg, S. M., 1988, Zinc fingers: Gilt by association, *Cell* **52:**1–3.

Frohman, M. A., and Martin, G. R., 1989, Cut, paste, and save: New approaches to altering specific genes in mice, *Cell* **56:**145–147.

Frohman, M. A., Dush, M. K., and Martin, G. R., 1988, Rapid production of full-length cDNAs from rare transcripts: Amplification using a single gene-specific oligonucleotide primer, *Proc. Natl. Acad. Sci. U.S.A.* **85:**8998–9002.

Gautier, J., Norbury, C., Lohka, M., Nurse, P., and Maller, J., 1988, Purified maturation-promoting factor contains the product of a *Xenopus* homologue of the fission yeast cell cycle control gene *cdc2+*, *Cell* **54:**433–439.

Goldman, D. S., Kisseling, A. A., Millette, C. F., and Cooper, G. M., 1987, Expression of *c-mos* RNA in germ cells of male and female mice, *Proc. Natl. Acad. Sci. U.S.A.* **84:**4509–4513.

Gossler, A., Joyner, A. L., Rossant, J., and Skarnes, W. C., 1989, Mouse embryonic stem cells and reporter constructs to detect developmentally regulated genes, *Science* **244:**463–465.

Grossniklaus, U., Bellen, H. J., Wilson, C., and Gehring, W. J., 1989, P-element-mediated enhancer detection applied to the study of oogenesis in *Drosophila*, *Development* **107:**189–200.

Guyer, R. L., and Koshland, D. E., 1989, The molecule of the year, *Science* **246:**1543–1546.

He, X., Treacy, M. N., Simmons, D. M., Ingraham, H. A., Swanson, L. W., and Rosenfeld, M. G., 1989, Expression of a large family of POU-domain regulatory genes in mammalian brain development, *Nature* **340:**35–42.

Herr, W., Sturm, R. A., Clerc, R. G., Corcoran, L. M., Baltimore, D., Sharp, P. A., Ingraham, H. A., Rosenfeld, M. G., Finney, M., Ruvkun, G., and Horvitz, H. R., 1988, The POU domain: A large conserved region in the mammalian *pit-1*, *oct-1*, *oct-2*, and *Caenorrhabditis elegans unc-86* gene products, *Genes Dev.* **2:**1513–1516.

Higuchi, R., von Beroldingen, C. H., Sensabaugh, G. F., and Erlich, H. A., 1988, DNA typing from single hairs, *Nature* **332:**543–546.

Holland, P., and Hogan, B., 1988, Expression of homeo box genes during development, *Genes Dev.* **2:**773–782.

Keshet, E., Rosenberg, M. P., Mercer, J. A., Propst, F., Vande Woude, G. F., Jenkins, N. A., and Copeland, N. G., 1987, Developmental regulation of ovarian-specific *mos* expression, *Oncogene* **2:**235–240.

Kleene, K., Distel, R. J., and Hecht, N. B., 1985, Nucleotide sequence of a cDNA clone encoding mouse protamine 1, *Biochemistry* **24:**719–722.

Koopman, P., Gubbay, J., Collignon, J., and Lovell-Badge, R., 1989, *Zfy* gene expression patterns are not compatible with a primary role in mouse sex determination, *Nature* **342:**940–942.

Labbe, J. C., Lee, M. G., Nurse, P., Picard, A., and Doree, M., 1988, Activation at M-phase of a protein kinase encoded by a starfish homologue of the cell cycle control gene *cdc2+*, *Nature* **335:**251–254.

Landschulz, W. H., Johnson, P. F., and McKnight, S. L., 1988, The leucine zipper: A hypothetical structure common to a new class of DNA binding proteins, *Science* **240:**1759–1764.

Lewis, E. B., 1978, A gene complex controlling segmentation in *Drosophila*, *Nature* **276**:565–570.

Loh, E. Y., Elliott, J. F., Cwirla, S., Lanier, L. L., and Davis, M. M., 1989, Polymerase chain reaction with single-sided specificity: Analysis of T cell receptor delta chain, *Science* **243**:217–220.

Masui, Y., and Markert, C. L., 1971, Cytoplasmic control of nuclear behavior during meiotic maturation of frog oocytes, *J. Exp. Zool.* **177**:129–146.

McGinnis, W., Levine, M., Hafen, E., Kuroiwa, A., and Gehring, W. J., 1984a, A conserved DNA sequence in homeotic genes of the *Drosophila Antennapedia* and *Bithorax* complexes, *Nature* **308**:428–433.

McGinnis, W., Hart, C. P., Gehring, W. J., and Ruddle, F. H., 1984b, Molecular cloning and chromosome mapping of a mouse DNA sequence homologous to homeotic genes of *Drosophila*, *Cell* **38**:675–680.

Meijer, L., Arion, D., Golsteyn, R., Brizuela, L., Hunt, T., and Beach, D., 1989, Cyclin is a component of the sea urchin egg M-phase specific histone H_1 kinase, *EMBO J.* **8**:2275–2282.

Melton, D. A., 1985, Injected anti-sense RNAs specifically block messenger RNA translation *in vivo*, *Proc. Natl. Acad. Sci. U.S.A.* **82**:144–148.

Melton, D. A., 1987, Translocation of maternal mRNA to the vegetal pole of *Xenopus* oocytes, *Nature* **328**:80–82.

Mlodzik, M., and Gehring, W. J., 1987, Expression of the caudal gene in the germ line of *Drosophila*: Formation of an RNA and protein gradient during early embryogenesis, *Cell* **48**:465–478.

Mlodzik, M., Fjose, A. and Gehring, W. J., 1985, Isolation of *caudal*, a *Drosophila* homeobox-containing gene with maternal expression, whose transcripts form a concentration gradient at the pre-blastoderm stage, *EMBO J.* **4**:2961–2969.

Muller, M. M., Carrasco, A. E., and DeRobertis, E. M., 1984, A homeobox-containing gene expressed during oogenesis in *Xenopus*, *Cell* **39**:157–162.

Mutter, G. L., and Wolgemuth, D. J., 1987, Distinct developmental patterns of *c-mos* protooncogene expression in female and male germ cells, *Proc. Natl. Acad. Sci. U.S.A.* **84**:5301–5305.

Mutter, G. L., Grills, G. S., and Wolgemuth, D. J., 1988, Evidence for the involvement of the proto-oncogene *c-mos* in mammalian meiotic maturation and possibly very early embryogenesis, *EMBO J.* **7**:683–389.

Nagamine, C. M., Chan, K., Hake, L. E., and Lau, Y.-F. C., 1990, The two candidate testis-determining Y genes (*Zfy-1*, *Zfy-2*) are differentially expressed in fetal and adult mouse tissues, *Genes Dev.* **4**:63–74.

O'Kane, C. J., and Gehring, W. J., 1984, Detection *in situ* of genomic regulatory elements in *Drosophila*, *Proc. Natl. Acad. Sci. U.S.A.* **84**:9123–9127.

O'Keefe, S. J., Wolfes, H., Kiessling, A. A., and Cooper, G. M., 1989, Microinjection of antisense *c-mos* oligonucleotides prevents meiosis II in the maturing mouse egg, *Proc. Natl. Acad. Sci. U.S.A.* **86**:7038–7042.

Page, D. C., Mosher, R., Simpson, E. M., Fisher, E. M. C., Mardon, G., Pollack, J., McGillivray, B., de la Chapelle, A., and Brown, L. G., 1987, The sex-determining region of the human Y chromosome encodes a finger protein, *Cell* **51**:1091–1104.

Palmer, M. S., Sinclair, A. H., Berta, P., Ellis, N. A., Goodfellow, P. N., Abbas, N. E., and Fellous, M., 1989, Genetic evidence that ZFY is not the testis-determining factor, *Nature* **342**:937–939.

Paules, R. S., Buccione, R., Moschel, R. C., Vande Woude, G. F., and Eppig, J., 1989, Mouse *mos* proto-oncogene product is present and functions during oogenesis, *Proc. Natl. Acad. Sci. U.S.A.* **86**:5395–5399.

Peschon, J. J., Behringer, R. R., Brinster, R. L., and Palmiter, R. D., 1987, Spermatid-specific expression of protamine 1 in transgenic mice, *Proc. Natl. Acad. Sci. U.S.A.* **84**:5316–5319.

Peters, H., 1969, The development of the mouse ovary from birth to maturity, *Acta Endocrinol.* **62**:98–116.

Pines, J., and Hunter, T., 1989, Isolation of a human cyclin cDNA: Evidence for cyclin mRNA and protein regulation in the cell cycle and for interaction with $p34^{cdc2}$, *Cell* **58**:833–846.

Ponzetto, C., and Wolgemuth, D. J., 1985, Haploid expression of a unique *c-abl* transcript in the mouse male germ line, *Mol. Cell. Biol.* **5**:1791–1794.

Ponzetto, C., Wadewitz, A. G., Pendergast, A. M., Witte, O. N., and Wolgemuth, D. J., 1989, $P150^{c-abl}$ is detected in mouse male germ cells by an *in vitro* kinase assay and is associated with stage-specific phosphoproteins in haploid cells, *Oncogene* **4**:685–690.

Propst, F., Rosenberg, M. P., Iyer, A., Kaul, K., and Vande Woude, G. F., 1987, *C-mos* proto-oncogene RNA transcripts in mouse tissues: Structural features, developmental regulation, and localization in specific cell types, *Mol. Cell. Biol.* **7**:1629–1637.

Rappolee, D. A., Brenner, C. A., Schultz, R., Mark, D., and Werb, Z., 1988, Developmental expression of PDGF, TGF-alpha, and TGF-beta genes in preimplantation mouse embryos, *Science* **241**:1823–1825.

Rappolee, D. A., Wang, A., Mark, D., and Werb, Z., 1989, Novel method for studying mRNA phenotypes in single or small numbers of cells, *J. Cell. Biochem.* **39**:1–11.

Robertson, E. J., Bradley, A., Kuehn, M., and Evans, M., 1986, Germ-line transmission of genes introduced into cultured pluripotential cells by retroviral vectors, *Nature* **323**:445–447.

Rosenberg, U. B., Schroder, C., Preiss, A., Kienlin, A., Cote, S., Riede, I., and Jackle, H., 1986, Structural homology of the product of the *Drosophila kruppel* gene with *Xenopus* transcription factor IIIA, *Nature* **319**:336–339.

Ruiz i Altaba, A., and Melton, D. A., 1989, Involvement of the *Xenopus* homeobox gene *Xhox3* in pattern formation along the anterior–posterior axis, *Cell* **57**:317–326.

Sagata, N., Daar, I., Oskarsson, M., Showalter, S. D., and Vande Woude, G. F., 1989a, The produce of the *mos* protooncogene as a candidate "initiator" for oocyte maturation, *Science* **245**:643–645.

Sagata, N., Watanabe, N., Vande Woude, G. F., and Ikawa, Y., 1989b, The *c-mos* proto-oncogene product is a cytostatic factor responsible for meiotic arrest in vertebrate eggs, *Nature* **342**:512–518.

Schafer, B. W., Blakely, B. T., Darlington, G. J., and Blau, H. M., 1990, Effect of cell history on response to helix–loop–helix family of myogenic regulators, *Nature* **344**:454–456.

Schwartzberg, P., Goff, S. P., and Robertson, E. J., 1989, Germ-line transmission of a *c-abl* mutation produced by targeted gene disruption in ES cells *Science* **246**:799–803.

Scott, M. P., and Weiner, A. J., 1984, Structural relationships among genes that control development: Sequence homology between the *Antennapedia*, *Ultrbithorax*, and *Fushi tarazu* loci of *Drosophila*, *Proc. Natl. Acad. Sci. U.S.A.* **81**:4115–4119.

Strickland, S., Huarte, J., Belin, D., Vassalli, A., Rickles, R. J., and Vassalli, J.-D., 1988, Antisense RNA directed against the 3′ noncoding region prevents dormant mRNA activation in mouse oocytes, *Science* **241**:680–684.

Swenson, K. I., Farrell, K. M., and Ruderman, J. V., 1986, The clam embryo protein cyclin A induces entry into M phase and the resumption of meiosis in *Xenopus oocytes*, *Cell* **47**:861–870.

Thomas, K. H., Wilkie, T. M., Tomashefsky, P., Bellve, A. R., and Simon, M. I., 1989, Differential gene expression during mouse spermatogenesis, *Biol. Reprod.* **41**:729–739.

Thomas, K. R., and Capecchi, M. R., 1987, Site-directed mutagenesis by gene targeting in mouse embryo-derived stem cells, *Cell* **51**:503–512.

Watanabe, N., Vande Woude, G. F., Ikawa, Y., and Sagata, N., 1989, Specific proteolysis of the *c-mos* proto-oncogene product by calpain on fertilization of *Xenopus* eggs, *Nature* **342**:505–511.

Weintraub, H., Izant, J. C., and Harland, R. M., 1985, Anti-sense RNA as a molecular tool for genetic analysis, *Trends Genet.* **1**:22–25.

Whitfield, W. G., Gonzalez, C., Sanchez-Herrero, E., and Glover, D. M., 1989, Transcripts of one of two *Drosophila* cyclin genes become localized in pole cells during embryogenesis, *Nature* **338**:337–340.

Williams, J. F., 1989, Optimization strategies for the polymerase chain reaction, *BioTechniques* **7**:762–767.

Wolgemuth, D. J., Gizang-Ginsberg, E., Engelmyer, E., Gavin, B. J., and Ponzetto, C., 1985, Separation of mouse testis cells on a Celsep™ apparatus and their usefulness as a source of high molecular weight DNA or RNA, *Gamete Res.* **12**:1–10.

Wolgemuth. D. J., Engelmeyer, E., Duggal, R. N., Gizang–Ginsberg, E. E., Mutter, G. L., Ponzetto, C., Vivano, C., and Zakeri, Z. F., 1986, Isolation of a mouse cDNA coding for a developmentally regulated, testis-specific transcript containing homeo box homology, *EMBO J.* **5**:1229–1235.

Wolgemuth, D. J., Viviano, C. M., Gizang-Ginsberg, E., Frohman, M. A., Joyner, A. L., and Martin, G. R., 1987, Differential expression of the mouse homeobox-containing gene *Hox-1.4* during male germ cell differentiation and embryonic development, *Proc. Natl. Acad. Sci. U.S.A.* **84**:5813–5817.

Wolgemuth, D. J., Behringer, R. R., Mostoller, M. P., Brinster, R. L., and Palmiter, R. D., 1989, Transgenic mice overexpressing the mouse homeobox-containing gene *Hox-1.4* exhibit abnormal gut development, *Nature* **337**:464–467.

Wright, C. V. E., Cho, K. W. Y., Hardwicke, J., Collins, R. H., and DeRobertis, E. M., 1989, Interference with function of a homeobox gene in *Xenopus* embryos produces malformations of the anterior spinal cord, *Cell* **59**:81–93.

Index